PHYSICAL CHEMISTRY

PHYSICAL CHEMISTRY

Volume 2:
Quantum Chemistry, Spectroscopy, and Statistical Thermodynamics

Peter Atkins
Professor of Chemistry,
University of Oxford,
and Fellow of Lincoln College, Oxford

Julio de Paula
Professor and Dean of the College of Arts and Sciences
Lewis and Clark College,
Portland, Oregon

W. H. Freeman and Company
New York

Library of Congress Control Number: 2005936591

Physical Chemistry, Volume 2: Quantum Chemistry, Spectroscopy, and
Statistical Thermodynamics

ISBN-13: 978-0-7167-8569-9
ISBN-10: 0-7167-8569-2

Published in Great Britain by Oxford University Press
This edition has been authorized by Oxford University Press for sale in the
United States and Canada only and not for export therefrom.

First printing

W. H. Freeman and Company
41 Madison Avenue
New York, NY 10010
www.whfreeman.com

Preface

We have taken the opportunity to refresh both the content and presentation of this text while—as for all its editions—keeping it flexible to use, accessible to students, broad in scope, and authoritative. The bulk of textbooks is a perennial concern: we have sought to tighten the presentation in this edition. However, it should always be borne in mind that much of the bulk arises from the numerous pedagogical features that we include (such as *Worked examples* and the *Data section*), not necessarily from density of information.

The most striking change in presentation is the use of colour. We have made every effort to use colour systematically and pedagogically, not gratuitously, seeing as a medium for making the text more attractive but using it to convey concepts and data more clearly. A second striking change is the division of this version of the text into two parts, one focused on thermodynamics and kinetics and the other on quantum theory and spectroscopy. Certain chapters that are less widely used have been omitted from the two-volume version but may be found in the single-volume version.

Volume 1 collects the more 'classical' aspects of physical chemistry. We have taken the opportunity to streamline the presentation of thermodynamics by combining several chapters from the previous edition, bearing in mind that some of the material will already have been covered in earlier courses. We no longer, for instance, make a distinction between 'concepts' and 'machinery', and as a result have provided a more compact presentation of thermodynamics with less artificial divisions between the approaches. Similarly, equilibrium electrochemistry now finds a home within the chapter on chemical equilibrium, where space has been made by reducing the discussion of acids and bases. This volume also contains the chapters on kinetics that form Part 3 of the previous version, with a number of significant changes. For instance, the description of electron transfer reactions is now a part of this sequence.

Volume 2 collects together the parts of chemistry based on quantum mechanics, which means largely the foundations of quantum mechanics, atomic and molecular structure, and spectroscopy. In preparing this edition we have sought to bring into the discussion contemporary techniques of spectroscopy and approaches to computational chemistry. In recognition of the major role that physical chemistry plays in materials science, we have a short sequence of chapters on which deal respectively with hard and soft matter. Moreover, we have introduced concepts of nanoscience throughout much of this volume.

We have discarded the Boxes of earlier editions. They have been replaced by more fully integrated and extensive *Impact* sections, which show how physical chemistry is applied to biology, materials, and the environment. By liberating these topics from their boxes, we believe they are more likely to be used and read; there are end-of-chapter problems on most of the material in these sections.

In the preface to the 7th edition we wrote that there was vigorous discussion in the physical chemistry community about the choice of a 'quantum first' or a 'thermodynamics first' approach. That discussion continues. In response, we have paid particular attention to making the organization of the full text flexible. We have also created these two volumes of the text: Volume 1: Thermodynamics and Kinetics, and Volume 2: Quantum Chemistry, Spectroscopy, and Statistical Thermodynamics. The strategic aim of this revision is to make it possible to work through the text, using one volume or two, in a variety of orders according to the preferences of the instructor.

We are, of course, alert to the developments in electronic resources and have made a special effort in this edition to encourage the use of the resources on our Web site (at www.whfreeman.com/pchem8) where you can also access the eBook. In particular, we think it important to encourage students to use the *Living graphs* and their considerable extension as *Explorations in Physical Chemistry*. To do so, wherever we call out a *Living graph* (by an icon attached to a graph in the text), we include an *Exploration* in the figure legend, suggesting how to explore the consequences of changing parameters.

Overall, we have taken this opportunity to refresh the text thoroughly, to integrate applications, to encourage the use of electronic resources, and to make the text even more flexible and up to date.

Oxford P.W.A.
Portland J.de P.

About the book

There are numerous features in this edition that are designed to make learning physical chemistry more effective and more enjoyable. One of the problems that make the subject daunting is the sheer amount of information: we have introduced several devices for organizing the material: see ***Organizing the information***. We appreciate that mathematics is often troublesome, and therefore have taken care to give help with this enormously important aspect of physical chemistry: see ***Mathematics and Physics support***. Problem solving—especially, 'where do I start?'—is often a challenge, and we have done our best to help overcome this first hurdle: see ***Problem solving***. Finally, the web is an extraordinary resource, but it is necessary to know where to start, or where to go for a particular piece of information; we have tried to indicate the right direction: see ***About the Web site***. The following paragraphs explain the features in more detail.

Organizing the information

Checklist of key ideas

☐ 1. A gas is a form of matter that fills any container it occupies.

☐ 2. An equation of state interrelates pressure, volume, temperature, and amount of substance: $p = f(T,V,n)$.

☐ 3. The pressure is the force divided by the area to which the force is applied. The standard pressure is $p^{\ominus} = 1$ bar (10^5 Pa).

☐ 4. Mechanical equilibrium is the condition of equality of pressure on either side of a movable wall.

☐ 5. Temperature is the property that indicates the direction of the flow of energy through a thermally conducting, rigid wall.

☐ 6. A diathermic boundary is a boundary that permits the passage of energy as heat. An adiabatic boundary is a boundary that prevents the passage of energy as heat.

☐ 7. Thermal equilibrium is a condition in which no change of state occurs when two objects A and B are in contact through a diathermic boundary.

☐ 8. The Zeroth Law of thermodynamics states that, if A is in thermal equilibrium with B, and B is in thermal equilibrium with C, then C is also in thermal equilibrium with A.

☐ 9. The Celsius and thermodynamic temperature scales are related by $T/\text{K} = \theta/°\text{C} + 273.15$.

☐ 10. A perfect gas obeys the perfect gas equation, $pV = nRT$, exactly

☐ 12. The partial pressure of any gas i $x_J = n_J/n$ is its mole fraction in a pressure.

☐ 13. In real gases, molecular interac state; the true equation of state coefficients $B, C, \ldots : pV_m = R$

☐ 14. The vapour pressure is the pres with its condensed phase.

☐ 15. The critical point is the point a end of the horizontal part of th a single point. The critical cons pressure, molar volume, and te critical point.

☐ 16. A supercritical fluid is a dense f temperature and pressure.

☐ 17. The van der Waals equation of the true equation of state in wh by a parameter a and repulsion parameter b: $p = nRT/(V - nb)$

☐ 18. A reduced variable is the actual corresponding critical constant

Checklist of key ideas

Here we collect together the major concepts introduced in the chapter. We suggest checking off the box that precedes each entry when you feel confident about the topic.

IMPACT ON NANOSCIENCE

I20.2 Nanowires

We have already remarked (*Impacts* I9.1, I9.2, and I19.3) that research on nanometre-sized materials is motivated by the possibility that they will form the basis for cheaper and smaller electronic devices. The synthesis of *nanowires*, nanometre-sized atomic assemblies that conduct electricity, is a major step in the fabrication of nanodevices. An important type of nanowire is based on carbon nanotubes, which, like graphite, can conduct electrons through delocalized π molecular orbitals that form from unhybridized $2p$ orbitals on carbon. Recent studies have shown a correlation between structure and conductivity in single-walled nanotubes (SWNTs) that does not occur in graphite. The SWNT in Fig. 20.45 is a semiconductor. If the hexagons are rotated by 60° about their sixfold axis, the resulting SWNT is a metallic conductor.

Carbon nanotubes are promising building blocks not only because they have useful electrical properties but also because they have unusual mechanical properties. For example, an SWNT has a Young's modulus that is approximately five times larger and a tensile strength that is approximately 375 times larger than that of steel.

Silicon nanowires can be made by focusing a pulsed laser beam on to a solid target composed of silicon and iron. The laser ejects Fe and Si atoms from the surface of the

Impact sections

Where appropriate, we have separated the principles from their applications: the principles are constant and straightforward; the applications come and go as the subject progresses. The *Impact* sections show how the principles developed in the chapter are currently being applied in a variety of modern contexts.

A note on good practice We write $T = 0$, not $T = 0$ K for the zero temperature on the thermodynamic temperature scale. This scale is absolute, and the lowest temperature is 0 regardless of the size of the divisions on the scale (just as we write $p = 0$ for zero pressure, regardless of the size of the units we adopt, such as bar or pascal). However, we write 0°C because the Celsius scale is not absolute.

Notes on good practice

Science is a precise activity and its language should be used accurately. We have used this feature to help encourage the use of the language and procedures of science in conformity to international practice and to help avoid common mistakes.

5.8 The activities of regular solutions

The material on regular solutions presented in Section 5.4 gives further insight into the origin of deviations from Raoult's law and its relation to activity coefficients. The starting point is the expression for the Gibbs energy of mixing for a regular solution (eqn 5.31). We show in the following *Justification* that eqn 5.31 implies that the activity coefficients are given by expressions of the form

$$\ln \gamma_A = \beta x_B^2 \qquad \ln \gamma_B = \beta x_A^2 \tag{5.57}$$

These relations are called the **Margules equations.**

Justification 5.4 *The Margules equations*

The Gibbs energy of mixing to form a nonideal solution is

$$\Delta_{mix}G = nRT\{x_A \ln a_A + x_B \ln a_B\}$$

This relation follows from the derivation of eqn 5.31 with activities in place of mole fractions. If each activity is replaced by γx, this expression becomes

$$\Delta_{mix}G = nRT\{x_A \ln x_A + x_B \ln x_B + x_A \ln \gamma_A + x_B \ln \gamma_B\}$$

Now we introduce the two expressions in eqn 5.57, and use $x_A + x_B = 1$, which gives

$$\begin{aligned}\Delta_{mix}G &= nRT\{x_A \ln x_A + x_B \ln x_B + \beta x_A x_B^2 + \beta x_B x_A^2\} \\ &= nRT\{x_A \ln x_A + x_B \ln x_B + \beta x_A x_B(x_A + x_B)\} \\ &= nRT\{x_A \ln x_A + x_B \ln x_B + \beta x_A x_B\}\end{aligned}$$

as required by eqn 5.31. Note, moreover, that the activity coefficients behave correctly for dilute solutions: $\gamma_A \to 1$ as $x_B \to 0$ and $\gamma_B \to 1$ as $x_A \to 0$.

Justifications

On first reading it might be sufficient to appreciate the 'bottom line' rather than work through detailed development of a mathematical expression. However, mathematical development is an intrinsic part of physical chemistry, and it is important to see how a particular expression is obtained. The *Justifications* let you adjust the level of detail that you require to your current needs, and make it easier to review material.

Molecular interpretation 5.2 *The lowering of vapour pressure of a solvent in a mixture*

The molecular origin of the lowering of the chemical potential is not the energy of interaction of the solute and solvent particles, because the lowering occurs even in an ideal solution (for which the enthalpy of mixing is zero). If it is not an enthalpy effect, it must be an entropy effect.

The pure liquid solvent has an entropy that reflects the number of microstates available to its molecules. Its vapour pressure reflects the tendency of the solution towards greater entropy, which can be achieved if the liquid vaporizes to form a gas. When a solute is present, there is an additional contribution to the entropy of the liquid, even in an ideal solution. Because the entropy of the liquid is already higher than that of the pure liquid, there is a weaker tendency to form the gas (Fig. 5.22). The effect of the solute appears as a lowered vapour pressure, and hence a higher boiling point.

Similarly, the enhanced molecular randomness of the solution opposes the tendency to freeze. Consequently, a lower temperature must be reached before equilibrium between solid and solution is achieved. Hence, the freezing point is lowered.

Molecular interpretation sections

Historically, much of the material in the first part of the text was developed before the emergence of detailed models of atoms, molecules, and molecular assemblies. The *Molecular interpretation* sections enhance and enrich coverage of that material by explaining how it can be understood in terms of the behaviour of atoms and molecules.

Further information

In some cases, we have judged that a derivation is too long, too detailed, or too different in level for it to be included in the text. In these cases, the derivations will be found less obtrusively at the end of the chapter.

Further information

Further information 5.1 *The Debye–Hückel theory of ionic solutions*

Imagine a solution in which all the ions have their actual positions, but in which their Coulombic interactions have been turned off. The difference in molar Gibbs energy between the ideal and real solutions is equal to w_e, the electrical work of charging the system in this arrangement. For a salt M_pX_q, we write

$$w_e = \overbrace{(p\mu_+ + q\mu_-)}^{G_m} - \overbrace{(p\mu_+^{ideal} + q\mu_-^{ideal})}^{G_m^{ideal}}$$
$$= p(\mu_+ - \mu_+^{ideal}) + q(\mu_- - \mu_-^{ideal})$$

From eqn 5.64 we write

$$\mu_+ - \mu_+^{ideal} = \mu_- - \mu_-^{ideal} = RT \ln \gamma_\pm$$

So it follows that

$$\ln \gamma_\pm = \frac{w_e}{sRT} \qquad s = p + q \qquad (5.73)$$

This equation tells us that we must first find the final distribution of the ions and then the work of charging them in that distribution.

The Coulomb potential at a distance r from an isolated ion of charge $z_i e$ in a medium of permittivity ε is

$$\phi_i = \frac{Z_i}{r} \qquad Z_i = \frac{z_i e}{4\pi\varepsilon} \qquad (5.74)$$

The ionic atmosphere causes the potential to decay with distance more sharply than this expression implies. Such shielding is a familiar problem in electrostatics, and its effect is taken into account by replacing the Coulomb potential by the **shielded Coulomb potential**, an expression of the form

$$\phi_i = \frac{Z_i}{r} e^{-r/r_D} \qquad (5.75)$$

where r_D is called the **Debye length**. W potential is virtually the same as the un small, the shielded potential is much sı potential, even for short distances (Fig.

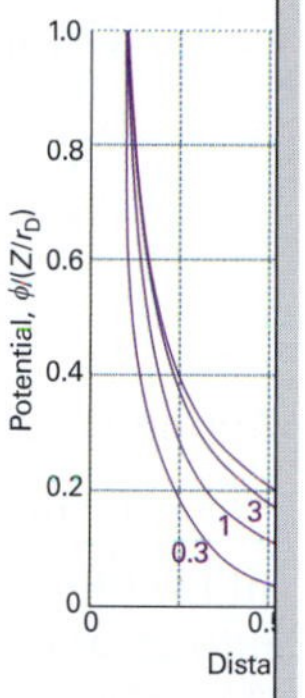

Fig. 5.36 The variation of the shielded C distance for different values of the Deb Debye length, the more sharply the pot case, a is an arbitrary unit of length.

Exploration Write an expression unshielded and shielded Coulon Then plot this expression against r_D an interpretation for the shape of the plot.

Appendices

Physical chemistry draws on a lot of background material, especially in mathematics and physics. We have included a set of *Appendices* to provide a quick survey of some of the information relating to units, physics, and mathematics that we draw on in the text.

966 Appendix 2 MATHEMATICAL TECHNIQUES

A2.6 Partial derivatives

A **partial derivative** of a function of more than one variabl of the function with respect to one of the variables, all th constant (see Fig. 2.*). Although a partial derivative sho when one variable changes, it may be used to determine when more than one variable changes by an infinitesimal tion of x and y, then when x and y change by dx and dy, re

$$df = \left(\frac{\partial f}{\partial x}\right)_y dx + \left(\frac{\partial f}{\partial y}\right)_x dy$$

where the symbol ∂ is used (instead of d) to denote a part df is also called the **differential** of f. For example, if $f = ax^3$

$$\left(\frac{\partial f}{\partial x}\right)_y = 3ax^2y \qquad \left(\frac{\partial f}{\partial y}\right)_x = ax^3 + 2by$$

Synoptic tables and the Data section

Long tables of data are helpful for assembling and solving exercises and problems, but can break up the flow of the text. We provide a lot of data in the *Data section* at the end of the text and short extracts in the *Synoptic tables* in the text itself to give an idea of the typical values of the physical quantities we are introducing.

1000 DATA SECTION

Table 2.8 Expansion coefficients, α, and isothermal compressibilities, κ_T

	$\alpha/(10^{-4}\,K^{-1})$	$\kappa_T/(10^{-6}\,atm^{-1})$
Liquids		
Benzene	12.4	92.1
Carbon tetrachloride	12.4	90.5
Ethanol	11.2	76.8
Mercury	1.82	38.7
Water	2.1	49.6
Solids		
Copper	0.501	0.735
Diamond	0.030	0.187
Iron	0.354	0.589
Lead	0.861	2.21

The values refer to 20°C.
Data: AIP(α), KL(κ_T).

Table 2.9 Inversion temperatures, n points, and Joule–Thomson coefficien

	T_I/K	T_f/K
Air	603	
Argon	723	83.8
Carbon dioxide	1500	194.7s
Helium	40	
Hydrogen	202	14.0
Krypton	1090	116.6
Methane	968	90.6
Neon	231	24.5
Nitrogen	621	63.3
Oxygen	764	54.8

s: sublimes.
Data: AIP, JL, and M.W. Zemansky, *Heat and* New York (1957).

Mathematics and Physics support

Comment 1.2
A hyperbola is a curve obtained by plotting y against x with xy = constant.

Comment 2.5
The partial-differential operation $(\partial z/\partial x)_y$ consists of taking the first derivative of $z(x,y)$ with respect to x, treating y as a constant. For example, if $z(x,y) = x^2y$, then

$$\left(\frac{\partial z}{\partial x}\right)_y = \left(\frac{\partial [x^2y]}{\partial x}\right)_y = y\frac{dx^2}{dx} = 2yx$$

Partial derivatives are reviewed in *Appendix 2*.

Comments

A topic often needs to draw on a mathematical procedure or a concept of physics; a *Comment* is a quick reminder of the procedure or concept.

978 Appendix 3 ESSENTIAL CONCEPTS OF PHYSICS

Classical mechanics

Classical mechanics describes the behaviour of objects in
expresses the fact that the total energy is constant in the a
other expresses the response of particles to the forces acti

A3.3 The trajectory in terms of the energy

The **velocity**, $\boldsymbol{v}$, of a particle is the rate of change of its po

$$\boldsymbol{v} = \frac{d\boldsymbol{r}}{dt}$$

The velocity is a vector, with both direction and magni
velocity is the **speed**, v. The **linear momentum**, $\boldsymbol{p}$, of a pa
its velocity, $\boldsymbol{v}$, by

$$\boldsymbol{p} = m\boldsymbol{v}$$

Like the velocity vector, the linear momentum vector po
of the particle (Fig. A3.1). In terms of the linear moment
ticle is

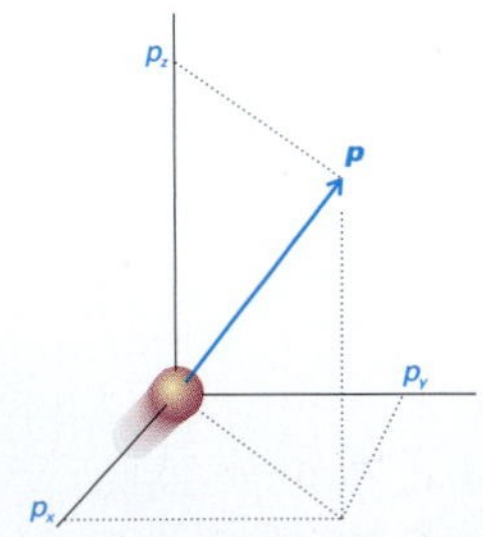

A3.1 The linear momentum of a particle is a vector property and points in the direction of motion.

Appendices

There is further information on mathematics and physics in Appendices 2 and 3, respectively. These appendices do not go into great detail, but should be enough to act as reminders of topics learned in other courses.

Problem solving

Illustration 5.2 *Using Henry's law*

To estimate the molar solubility of oxygen in water at 25°C and a partial pressure of 21 kPa, its partial pressure in the atmosphere at sea level, we write

$$b_{O_2} = \frac{p_{O_2}}{K_{O_2}} = \frac{21\ \text{kPa}}{7.9\times 10^4\ \text{kPa kg mol}^{-1}} = 2.9\times 10^{-4}\ \text{mol kg}^{-1}$$

The molality of the saturated solution is therefore 0.29 mmol kg^{-1}. To convert this quantity to a molar concentration, we assume that the mass density of this dilute solution is essentially that of pure water at 25°C, or $\rho_{H_2O} = 0.99709$ kg dm^{-3}. It follows that the molar concentration of oxygen is

$$[O_2] = b_{O_2} \times \rho_{H_2O} = 0.29\ \text{mmol kg}^{-1} \times 0.99709\ \text{kg dm}^{-3} = 0.29\ \text{mmol dm}^{-3}$$

A note on good practice The number of significant figures in the result of a calculation should not exceed the number in the data (only two in this case).

Self-test 5.5 Calculate the molar solubility of nitrogen in water exposed to air at 25°C; partial pressures were calculated in *Example 1.3*. [0.51 mmol dm^{-3}]

Illustrations

An *Illustration* (don't confuse this with a diagram!) is a short example of how to use an equation that has just been introduced in the text. In particular, we show how to use data and how to manipulate units correctly.

Worked examples

A *Worked example* is a much more structured form of *Illustration*, often involving a more elaborate procedure. Every *Worked example* has a Method section to suggest how to set up the problem (another way might seem more natural: setting up problems is a highly personal business). Then there is the worked-out Answer.

Example 8.1 *Calculating the number of photons*

Calculate the number of photons emitted by a 100 W yellow lamp in 1.0 s. Take the wavelength of yellow light as 560 nm and assume 100 per cent efficiency.

Method Each photon has an energy $h\nu$, so the total number of photons needed to produce an energy E is $E/h\nu$. To use this equation, we need to know the frequency of the radiation (from $\nu = c/\lambda$) and the total energy emitted by the lamp. The latter is given by the product of the power (P, in watts) and the time interval for which the lamp is turned on ($E = P\Delta t$).

Answer The number of photons is

$$N = \frac{E}{h\nu} = \frac{P\Delta t}{h(c/\lambda)} = \frac{\lambda P\Delta t}{hc}$$

Substitution of the data gives

$$N = \frac{(5.60\times10^{-7}\ \mathrm{m})\times(100\ \mathrm{J\ s^{-1}})\times(1.0\ \mathrm{s})}{(6.626\times10^{-34}\ \mathrm{J\ s})\times(2.998\times10^{8}\ \mathrm{m\ s^{-1}})} = 2.8\times10^{20}$$

Note that it would take nearly 40 min to produce 1 mol of these photons.

A note on good practice To avoid rounding and other numerical errors, it is best to carry out algebraic mainpulations first, and to substitute numerical values into a single, final formula. Moreover, an analytical result may be used for other data without having to repeat the entire calculation.

Self-test 8.1 How many photons does a monochromatic (single frequency) infrared rangefinder of power 1 mW and wavelength 1000 nm emit in 0.1 s? [5×10^{14}]

Self-tests

Each *Worked example*, and many of the *Illustrations*, has a *Self-test*, with the answer provided as a check that the procedure has been mastered. There are also free-standing *Self-tests* where we thought it a good idea to provide a question to check understanding. Think of *Self-tests* as in-chapter *Exercises* designed to help monitor your progress.

Self-test 3.12 Calculate the change in G_m for ice at -10°C, with density 917 kg m^{-3}, when the pressure is increased from 1.0 bar to 2.0 bar. [$+2.0$ J mol^{-1}]

Discussion questions

The end-of-chapter material starts with a short set of questions that are intended to encourage reflection on the material and to view it in a broader context than is obtained by solving numerical problems.

Discussion questions

1.1 Explain how the perfect gas equation of state arises by combination of Boyle's law, Charles's law, and Avogadro's principle.

1.2 Explain the term 'partial pressure' and explain why Dalton's law is a limiting law.

1.3 Explain how the compression factor varies with pressure and temperature and describe how it reveals information about intermolecular interactions in real gases.

1.4 What is the significance of the critical c

1.5 Describe the formulation of the van de rationale for one other equation of state in

1.6 Explain how the van der Waals equatio behaviour.

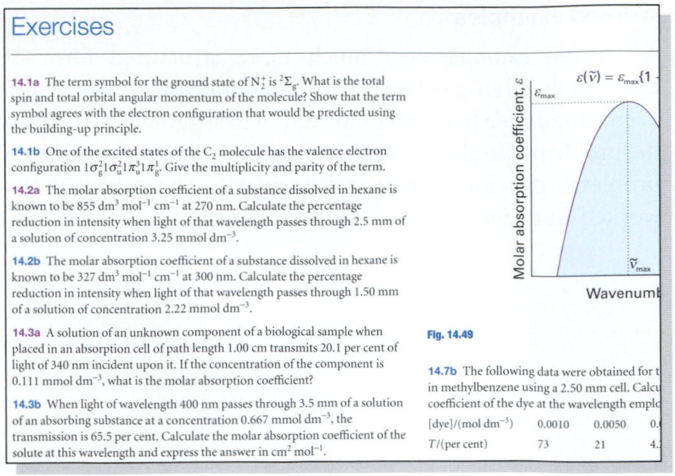

Exercises

14.1a The term symbol for the ground state of N_2^+ is $^2\Sigma_g$. What is the total spin and total orbital angular momentum of the molecule? Show that the term symbol agrees with the electron configuration that would be predicted using the building-up principle.

14.1b One of the excited states of the C_2 molecule has the valence electron configuration $1\sigma_g^2 1\sigma_u^2 1\pi_u^3 1\pi_g^1$. Give the multiplicity and parity of the term.

14.2a The molar absorption coefficient of a substance dissolved in hexane is known to be 855 $dm^3\ mol^{-1}\ cm^{-1}$ at 270 nm. Calculate the percentage reduction in intensity when light of that wavelength passes through 2.5 mm of a solution of concentration 3.25 mmol dm^{-3}.

14.2b The molar absorption coefficient of a substance dissolved in hexane is known to be 327 $dm^3\ mol^{-1}\ cm^{-1}$ at 300 nm. Calculate the percentage reduction in intensity when light of that wavelength passes through 1.50 mm of a solution of concentration 2.22 mmol dm^{-3}.

14.3a A solution of an unknown component of a biological sample when placed in an absorption cell of path length 1.00 cm transmits 20.1 per cent of light of 340 nm incident upon it. If the concentration of the component is 0.111 mmol dm^{-3}, what is the molar absorption coefficient?

14.3b When light of wavelength 400 nm passes through 3.5 mm of a solution of an absorbing substance at a concentration 0.667 mmol dm^{-3}, the transmission is 65.5 per cent. Calculate the molar absorption coefficient of the solute at this wavelength and express the answer in $cm^2\ mol^{-1}$.

Fig. 14.49

14.7b The following data were obtained for t in methylbenzene using a 2.50 mm cell. Calcu coefficient of the dye at the wavelength emplo

[dye]/(mol dm^{-3})	0.0010	0.0050	0.
T/(per cent)	73	21	4.

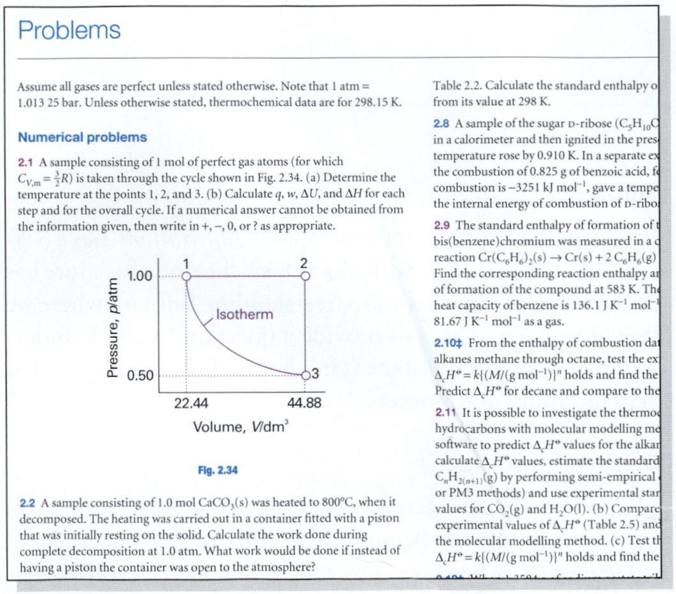

Problems

Assume all gases are perfect unless stated otherwise. Note that 1 atm = 1.013 25 bar. Unless otherwise stated, thermochemical data are for 298.15 K.

Numerical problems

2.1 A sample consisting of 1 mol of perfect gas atoms (for which $C_{V,m} = \frac{3}{2}R$) is taken through the cycle shown in Fig. 2.34. (a) Determine the temperature at the points 1, 2, and 3. (b) Calculate q, w, ΔU, and ΔH for each step and for the overall cycle. If a numerical answer cannot be obtained from the information given, then write in +, −, 0, or ? as appropriate.

Fig. 2.34

2.2 A sample consisting of 1.0 mol $CaCO_3$(s) was heated to 800°C, when it decomposed. The heating was carried out in a container fitted with a piston that was initially resting on the solid. Calculate the work done during complete decomposition at 1.0 atm. What work would be done if instead of having a piston the container was open to the atmosphere?

Table 2.2. Calculate the standard enthalpy o from its value at 298 K.

2.8 A sample of the sugar D-ribose (C_5H_{10}C in a calorimeter and then ignited in the pres temperature rose by 0.910 K. In a separate ex the combustion of 0.825 g of benzoic acid, fo combustion is −3251 kJ mol^{-1}, gave a tempe the internal energy of combustion of D-ribo

2.9 The standard enthalpy of formation of t bis(benzene)chromium was measured in a c reaction $Cr(C_6H_6)_2(s) \rightarrow Cr(s) + 2\,C_6H_6(g)$ Find the corresponding reaction enthalpy ar of formation of the compound at 583 K. Th heat capacity of benzene is 136.1 J K^{-1} mol^{-1} 81.67 J K^{-1} mol^{-1} as a gas.

2.10‡ From the enthalpy of combustion dat alkanes methane through octane, test the ex $\Delta_c H^\circ = k\{(M/(g\ mol^{-1}))\}^n$ holds and find the Predict $\Delta_c H^\circ$ for decane and compare to the

2.11 It is possible to investigate the thermo hydrocarbons with molecular modelling me software to predict $\Delta_c H^\circ$ values for the alkan calculate $\Delta_c H^\circ$ values, estimate the standard $C_nH_{2(n+1)}$(g) by performing semi-empirical or PM3 methods) and use experimental star values for CO_2(g) and H_2O(l). (b) Compare experimental values of $\Delta_c H^\circ$ (Table 2.5) and the molecular modelling method. (c) Test th $\Delta_c H^\circ = k\{(M/(g\ mol^{-1}))\}^n$ holds and find the

Exercises and Problems

The real core of testing understanding is the collection of end-of-chapter *Exercises* and *Problems*. The *Exercises* are straightforward numerical tests that give practice with manipulating numerical data. The *Problems* are more searching. They are divided into 'numerical', where the emphasis is on the manipulation of data, and 'theoretical', where the emphasis is on the manipulation of equations before (in some cases) using numerical data. At the end of the *Problems* are collections of problems that focus on practical applications of various kinds, including the material covered in the *Impact* sections.

Each end-of-chapter exercise and problem for which a solution is available in the Student Solution Manual has a corresponding bracketed number referring to the Student Solutions Manual number for this problem, for example [SSM 2.5]. Each end-of-chapter exercise and problem for which a solution is available in the Instructors Solution Manual has a corresponding bracketed number referring to the Instructor Solutions Manual number for this problem, for example [ISM 2.4].

About the Web site

The Web site to accompany *Physical Chemistry* is available at:

www.whfreeman.com/pchem8

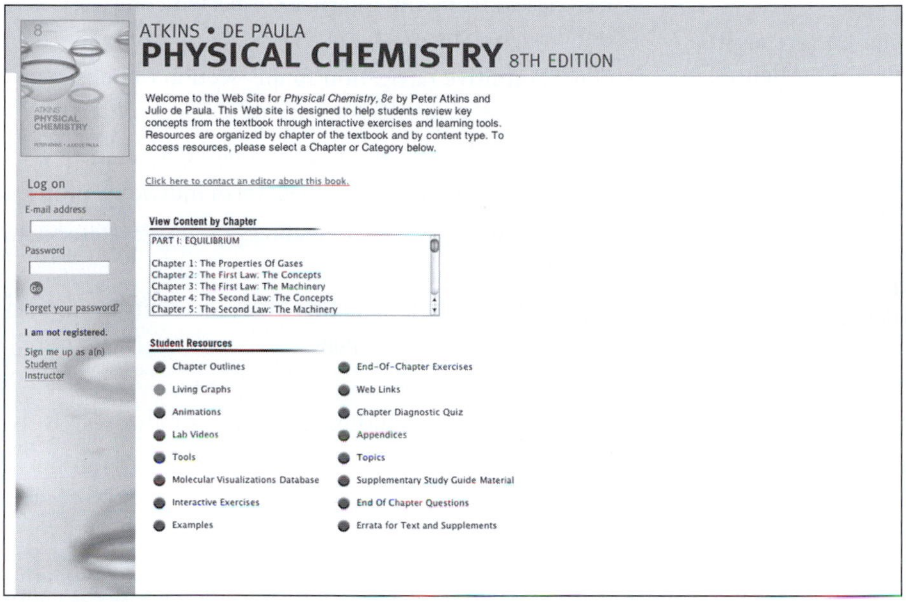

It includes the following features:

Living graphs

A *Living graph* is indicated in the text by the icon attached to a graph. This feature can be used to explore how a property changes as a variety of parameters are changed. To encourage the use of this resource (and the more extensive *Explorations in Physical Chemistry*) we have added a question to each figure where a *Living graph* is called out.

10.16 The boundary surfaces of *d* orbitals. Two nodal planes in each orbital intersect at the nucleus and separate the lobes of each orbital. The dark and light areas denote regions of opposite sign of the wavefunction.

Exploration To gain insight into the shapes of the *f* orbitals, use mathematical software to plot the boundary surfaces of the spherical harmonics $Y_{3,m_l}(\theta,\varphi)$.

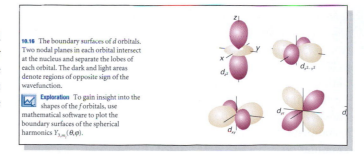

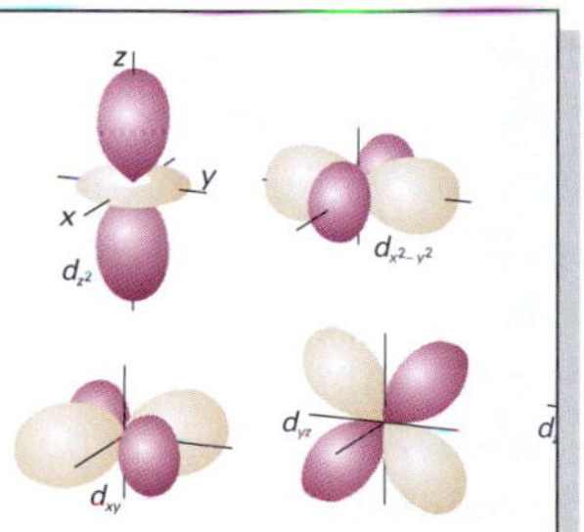

Artwork

An instructor may wish to use the illustrations from this text in a lecture. Almost all the illustrations are available and can be used for lectures without charge (but not for commercial purposes without specific permission). This edition is in full colour: we have aimed to use colour systematically and helpfully, not just to make the page prettier.

Tables of data

All the tables of data that appear in the chapter text are available and may be used under the same conditions as the figures.

Web links

There is a huge network of information available about physical chemistry, and it can be bewildering to find your way to it. Also, a piece of information may be needed that we have not included in the text. The web site might suggest where to find the specific data or indicate where additional data can be found.

Tools

Interactive calculators, plotters and a periodic table for the study of chemistry.

Group theory tables

Comprehensive group theory tables are available for downloading.

Explorations in Physical Chemistry

Now from W.H. Freeman & Company, the new edition of the popular *Explorations in Physical Chemistry* is available on-line at www.whfreeman.com/explorations, using the activation code card included with *Physical Chemistry* 8e. The new edition consists of interactive Mathcad® worksheets and, for the first time, interactive Excel® workbooks. They motivate students to simulate physical, chemical, and biochemical phenomena with their personal computers. Harnessing the computational power of Mathcad® by Mathsoft, Inc. and Excel® by Microsoft Corporation, students can manipulate over 75 graphics, alter simulation parameters, and solve equations to gain deeper insight into physical chemistry. Complete with thought-stimulating exercises, *Explorations in Physical Chemistry* is a perfect addition to any physical chemistry course, using any physical chemistry text book.

The *Physical Chemistry*, Eighth Edition eBook

A complete online version of the textbook. The eBook offers students substantial savings and provides a rich learning experience by taking full advantage of the electronic medium integrating all student media resources and adds features unique to the eBook. The eBook also offers instructors unparalleled flexibility and customization options not previously possible with any printed textbook. Access to the eBook is included with purchase of the special package of the text (0-7167-8586-2), through use of an activation code card. Individual eBook copies can be purchased on-line at www.whfreeman.com.

Key features of the eBook include:

- Easy access from any Internet-connected computer via a standard Web browser.
- Quick, intuitive navigation to any section or subsection, as well as any printed book page number.
- Integration of all Living Graph animations.
- Text **highlighting**, down to the level of individual phrases.
- A **book marking** feature that allows for quick reference to any page.
- A powerful **Notes** feature that allows students or instructors to add notes to any page.
- A full **index**.
- **Full-text search**, including an option to also search the glossary and index.
- Automatic saving of all notes, highlighting, and bookmarks.

Additional features for lecturers:

- Custom chapter selection: Lecturers can choose the chapters that correspond with their syllabus, and students will get a custom version of the eBook with the selected chapters only.
- Instructor notes: Lecturers can choose to create an annotated version of the eBook with their notes on any page. When students in their course log in, they will see the lecturer's version.
- Custom content: Lecturer notes can include text, web links, and even images, allowing lecturers to place any content they choose exactly where they want it.

Physical Chemistry is now available in multiple formats!

For maximum flexibility in your physical chemistry course, this text is now offered as a traditional, full text or in two volumes. The chapters from *Physical Chemistry*, 8e that appear in each volume are as follows:

Physical Chemistry 8th Edition (0-7167-8759-8)

PART 1 Equilibrium
1 The properties of gases
2 The First Law

Volume 1: Thermodynamics and Kinetics (0-7167-8567-6)

Volume 2: Quantum Chemistry, Spectroscopy, and Statistical Thermodynamics (0-7167-8569-2)

Solutions manuals

As with previous editions Charles Trapp, Carmen Giunta, and Marshall Cady have produced the solutions manuals to accompany this book. A *Student's Solutions Manual* (0-7167-6206-4) provides full solutions to the 'b' exercises and the odd-numbered problems. An *Instructor's Solutions Manual* (0-7167-2566-5) provides full solutions to the 'a' exercises and the even-numbered problems.

In the split volumes each end-of-chapter discussion question, exercise and problem for which a solution is available in the Student Solution Manual has a corresponding bracketed number referring to the Student Solutions Manual number for this problem, for example [SSM 2.5]. Each end-of-chapter exercise and problem for which a solution is available in the Instructors Solution Manual has a corresponding bracketed number referring to the Instructor Solutions Manual number for this problem, for example [ISM 2.4].

About the authors

Peter Atkins is Professor of Chemistry at Oxford University, a fellow of Lincoln College, and the author of more than fifty books for students and a general audience. His texts are market leaders around the globe. A frequent lecturer in the United States and throughout the world, he has held visiting prefessorships in France, Israel, Japan, China, and New Zealand. He was the founding chairman of the Committee on Chemistry Education of the International Union of Pure and Applied Chemistry and a member of IUPAC's Physical and Biophysical Chemistry Division.

Julio de Paula is Professor of Chemistry and Dean of the College of Arts & Sciences at Lewis & Clark College. A native of Brazil, Professor de Paula received a B.A. degree in chemistry from Rutgers, The State University of New Jersey, and a Ph.D. in biophysical chemistry from Yale University. His research activities encompass the areas of molecular spectroscopy, biophysical chemistry, and nanoscience. He has taught courses in general chemistry, physical chemistry, biophysical chemistry, instrumental analysis, and writing.

Acknowledgements

A book as extensive as this could not have been written without significant input from many individuals. We would like to reiterate our thanks to the hundreds of people who contributed to the first seven editions. Our warm thanks go Charles Trapp, Carmen Giunta, and Marshall Cady who have produced the *Solutions manuals* that accompany this book.

Many people gave their advice based on the seventh edition, and others reviewed the draft chapters for the eighth edition as they emerged. We therefore wish to thank the following colleagues most warmly:

Joe Addison, Governors State University
Joseph Alia, University of Minnesota Morris
David Andrews, University of East Anglia
Mike Ashfold, University of Bristol
Daniel E. Autrey, Fayetteville State University
Jeffrey Bartz, Kalamazoo College
Martin Bates, University of Southampton
Roger Bickley, University of Bradford
E.M. Blokhuis, Leiden University
Jim Bowers, University of Exeter
Mark S. Braiman, Syracuse University
Alex Brown, University of Alberta
David E. Budil, Northeastern University
Dave Cook, University of Sheffield
Ian Cooper, University of Newcastle-upon-Tyne
T. Michael Duncan, Cornell University
Christer Elvingson, Uppsala University
Cherice M. Evans, Queens College—CUNY
Stephen Fletcher, Loughborough University
Alyx S. Frantzen, Stephen F. Austin State University
David Gardner, Lander University
Roberto A. Garza-López, Pomona College
Robert J. Gordon, University of Illinois at Chicago
Pete Griffiths, Cardiff University
Robert Haines, University of Prince Edward Island
Ron Haines, University of New South Wales
Arthur M. Halpern, Indiana State University
Tom Halstead, University of York
Todd M. Hamilton, Adrian College
Gerard S. Harbison, University Nebraska at Lincoln
Ulf Henriksson, Royal Institute of Technology, Sweden
Mike Hey, University of Nottingham
Paul Hodgkinson, University of Durham
Robert E. Howard, University of Tulsa
Mike Jezercak, University of Central Oklahoma
Clarence Josefson, Millikin University
Pramesh N. Kapoor, University of Delhi
Peter Karadakov, University of York
Miklos Kertesz, Georgetown University
Neil R. Kestner, Louisiana State University
Sanjay Kumar, Indian Institute of Technology
Jeffry D. Madura, Duquesne University
Andrew Masters, University of Manchester
Paul May, University of Bristol
Mitchell D. Menzmer, Southwestern Adventist University
David A. Micha, University of Florida
Sergey Mikhalovsky, University of Brighton
Jonathan Mitschele, Saint Joseph's College
Vicki D. Moravec, Tri-State University
Gareth Morris, University of Manchester
Tony Morton-Blake, Trinity College, Dublin
Andy Mount, University of Edinburgh
Maureen Kendrick Murphy, Huntingdon College
John Parker, Heriot Watt University
Jozef Peeters, University of Leuven
Michael J. Perona, CSU Stanislaus
Nils-Ola Persson, Linköping University
Richard Pethrick, University of Strathclyde
John A. Pojman, The University of Southern Mississippi
Durga M. Prasad, University of Hyderabad
Steve Price, University College London
S. Rajagopal, Madurai Kamaraj University
R. Ramaraj, Madurai Kamaraj University
David Ritter, Southeast Missouri State University
Bent Ronsholdt, Aalborg University
Stephen Roser, University of Bath
Kathryn Rowberg, Purdue University Calumet
S.A. Safron, Florida State University
Kari Salmi, Espoo-Vantaa Institute of Technology
Stephan Sauer, University of Copenhagen
Nicholas Schlotter, Hamline University
Roseanne J. Sension, University of Michigan
A.J. Shaka, University of California
Joe Shapter, Flinders University of South Australia
Paul D. Siders, University of Minnesota, Duluth
Harjinder Singh, Panjab University
Steen Skaarup, Technical University of Denmark
David Smith, University of Exeter
Patricia A. Snyder, Florida Atlantic University
Olle Söderman, Lund University
Peter Stilbs, Royal Institute of Technology, Sweden
Svein Stølen, University of Oslo
Fu-Ming Tao, California State University, Fullerton
Eimer Tuite, University of Newcastle
Eric Waclawik, Queensland University of Technology
Yan Waguespack, University of Maryland Eastern Shore
Terence E. Warner, University of Southern Denmark

Richard Wells, University of Aberdeen
Ben Whitaker, University of Leeds
Christopher Whitehead, University of Manchester
Mark Wilson, University College London
Kazushige Yokoyama, State University of New York at Geneseo
Nigel Young, University of Hull
Sidney H. Young, University of South Alabama

We also thank Fabienne Meyers (of the IUPAC Secretariat) for helping us to bring colour to most of the illustrations and doing so on a very short timescale. We would also like to thank our two publishers, Oxford University Press and W.H. Freeman & Co., for their constant encouragement, advice, and assistance, and in particular our editors Jonathan Crowe, Jessica Fiorillo, and Ruth Hughes. Authors could not wish for a more congenial publishing environment.

Summary of contents

Contents

List of impact sections

Volume 1

Volume 2

Quantum theory: introduction and principles

1

This chapter introduces some of the basic principles of quantum mechanics. First, it reviews the experimental results that overthrew the concepts of classical physics. These experiments led to the conclusion that particles may not have an arbitrary energy and that the classical concepts of 'particle' and 'wave' blend together. The overthrow of classical mechanics inspired the formulation of a new set of concepts and led to the formulation of quantum mechanics. In quantum mechanics, all the properties of a system are expressed in terms of a wavefunction that is obtained by solving the Schrödinger equation. We see how to interpret wavefunctions. Finally, we introduce some of the techniques of quantum mechanics in terms of operators, and see that they lead to the uncertainty principle, one of the most profound departures from classical mechanics.

It was once thought that the motion of atoms and subatomic particles could be expressed using **classical mechanics**, the laws of motion introduced in the seventeenth century by Isaac Newton, for these laws were very successful at explaining the motion of everyday objects and planets. However, towards the end of the nineteenth century, experimental evidence accumulated showing that classical mechanics failed when it was applied to particles as small as electrons, and it took until the 1920s to discover the appropriate concepts and equations for describing them. We describe the concepts of this new mechanics, which is called **quantum mechanics**, in this chapter, and apply them throughout the remainder of the text.

The origins of quantum mechanics

The basic principles of classical mechanics are reviewed in *Appendix* 2. In brief, they show that classical physics (1) predicts a precise trajectory for particles, with precisely specified locations and momenta at each instant, and (2) allows the translational, rotational, and vibrational modes of motion to be excited to any energy simply by controlling the forces that are applied. These conclusions agree with everyday experience. Everyday experience, however, does not extend to individual atoms, and careful experiments of the type described below have shown that classical mechanics fails when applied to the transfers of very small energies and to objects of very small mass.

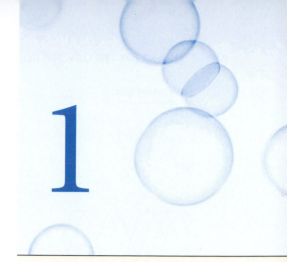

We shall also investigate the properties of light. In classical physics, light is described as electromagnetic radiation, which is understood in terms of the **electromagnetic field**, an oscillating electric and magnetic disturbance that spreads as a harmonic wave through empty space, the vacuum. Such waves are generated by the acceleration of electric charge, as in the oscillating motion of electrons in the antenna of a radio transmitter. The wave travels at a constant speed called the *speed of light*, c, which is about 3×10^8 m s^{-1}. As its name suggests, an electromagnetic field has two components, an **electric field** that acts on charged particles (whether stationary or moving) and a **magnetic field** that acts only on moving charged particles. The electromagnetic field is characterized by a **wavelength,** λ (lambda), the distance between the neighbouring peaks of the wave, and its **frequency**, ν (nu), the number of times per second at which its displacement at a fixed point returns to its original value (Fig. 1.1). The frequency is measured in *hertz*, where 1 Hz = 1 s^{-1}. The wavelength and frequency of an electromagnetic wave are related by

$$\lambda\nu = c \tag{1.1}$$

Therefore, the shorter the wavelength, the higher the frequency. The characteristics of the wave are also reported by giving the **wavenumber**, $\tilde{\nu}$ (nu tilde), of the radiation, where

$$\tilde{\nu} = \frac{\nu}{c} = \frac{1}{\lambda} \qquad [1.2]$$

Wavenumbers are normally reported in reciprocal centimetres (cm^{-1}).

Figure 1.2 summarizes the **electromagnetic spectrum**, the description and classification of the electromagnetic field according to its frequency and wavelength. White light is a mixture of electromagnetic radiation with wavelengths ranging from about 380 nm to about 700 nm (1 nm = 10^{-9} m). Our eyes perceive different wavelengths of radiation in this range as different colours, so it can be said that white light is a mixture of light of all different colours.

The wave model falls short of describing all the properties of radiation. So, just as our view of particles (and in particular small particles) needs to be adjusted, a new view of light also has to be developed.

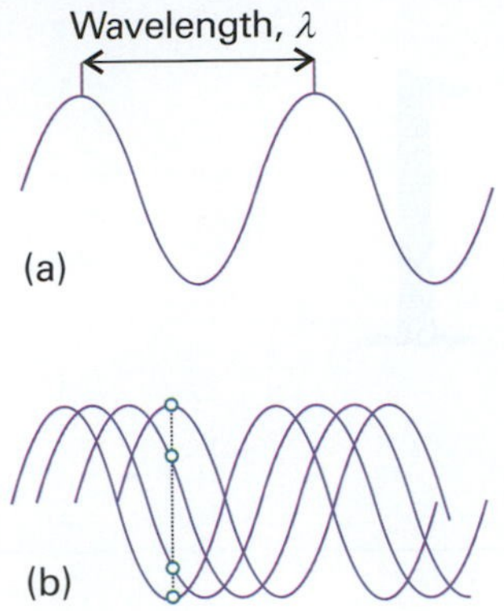

Fig. 1.1 The wavelength, λ, of a wave is the peak-to-peak distance. (b) The wave is shown travelling to the right at a speed c. At a given location, the instantaneous amplitude of the wave changes through a complete cycle (the four dots show half a cycle). The frequency, ν, is the number of cycles per second that occur at a given point.

Comment 1.1

Harmonic waves are waves with displacements that can be expressed as sine or cosine functions. The physics of waves is reviewed in *Appendix 3*.

1.1 The failures of classical physics

In this section we review some of the experimental evidence that showed that several concepts of classical mechanics are untenable. In particular, we shall see that observations of the radiation emitted by hot bodies, heat capacities, and the spectra of atoms and molecules indicate that systems can take up energy only in discrete amounts.

(a) Black-body radiation

A hot object emits electromagnetic radiation. At high temperatures, an appreciable proportion of the radiation is in the visible region of the spectrum, and a higher

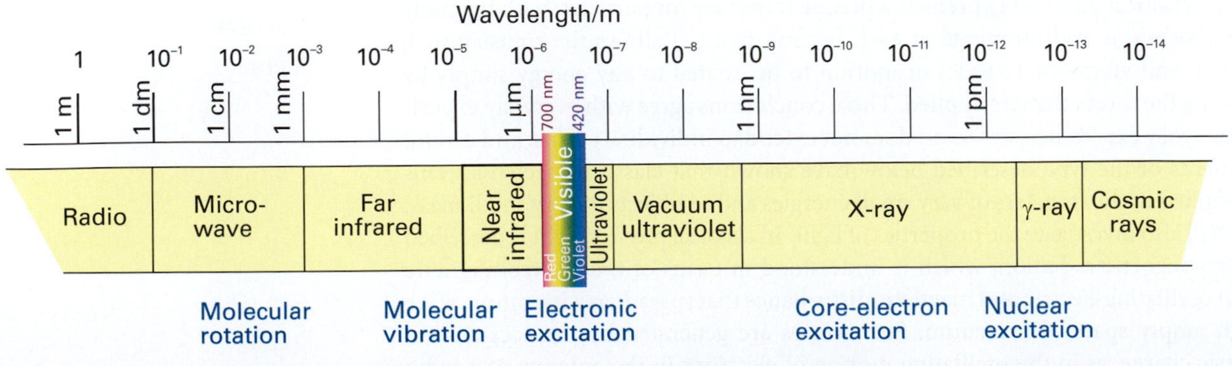

Fig. 1.2 The electromagnetic spectrum and the classification of the spectral regions.

proportion of short-wavelength blue light is generated as the temperature is raised. This behaviour is seen when a heated iron bar glowing red hot becomes white hot when heated further. The dependence is illustrated in Fig. 1.3, which shows how the energy output varies with wavelength at several temperatures. The curves are those of an ideal emitter called a **black body**, which is an object capable of emitting and absorbing all frequencies of radiation uniformly. A good approximation to a black body is a pinhole in an empty container maintained at a constant temperature, because any radiation leaking out of the hole has been absorbed and re-emitted inside so many times that it has come to thermal equilibrium with the walls (Fig. 1.4).

The explanation of black-body radiation was a major challenge for nineteenth-century scientists, and in due course it was found to be beyond the capabilities of classical physics. The physicist Lord Rayleigh studied it theoretically from a classical viewpoint, and thought of the electromagnetic field as a collection of oscillators of all possible frequencies. He regarded the presence of radiation of frequency ν (and therefore of wavelength $\lambda = c/\nu$) as signifying that the electromagnetic oscillator of that frequency had been excited (Fig. 1.5). Rayleigh used the equipartition principle (Volume 1, Section 2.2) to calculate the average energy of each oscillator as kT. Then, with minor help from James Jeans, he arrived at the **Rayleigh–Jeans law** (see *Further reading* for its justification):

$$d\mathcal{E} = \rho d\lambda \qquad \rho = \frac{8\pi kT}{\lambda^4} \tag{1.3}$$

where ρ (rho), the **density of states**, is the proportionality constant between $d\lambda$ and the energy density, $d\mathcal{E}$, in the range of wavelengths between λ and $\lambda + d\lambda$, k is Boltzmann's constant ($k = 1.381 \times 10^{-23}$ J K^{-1}). The units of ρ are typically joules per metre^4 (J m^{-4}), to give an energy density $d\mathcal{E}$ in joules per cubic metre (J m^{-3}) when multiplied by a wavelength range $d\lambda$ in metres. A high density of states at the wavelength λ simply means that there is a lot of energy associated with wavelengths lying between λ and $\lambda + d\lambda$. The total energy density (in joules per cubic metre) in a region is obtained by integrating eqn 1.3 over all wavelengths between zero and infinity, and the total energy (in joules) within the region is obtained by multiplying that total energy density by the volume of the region.

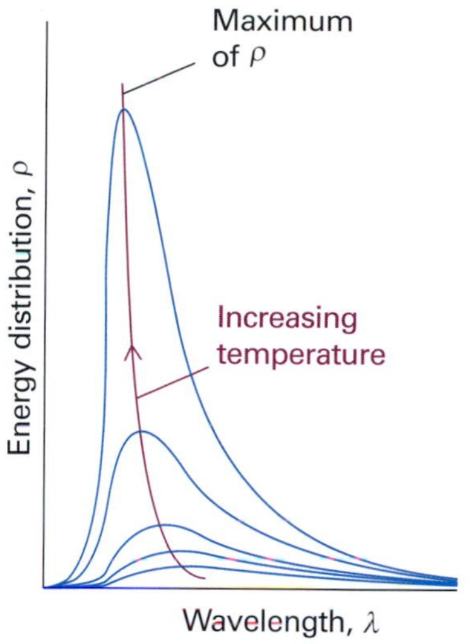

Fig. 1.3 The energy distribution in a black-body cavity at several temperatures. Note how the energy density increases in the region of shorter wavelengths as the temperature is raised, and how the peak shifts to shorter wavelengths. The total energy density (the area under the curve) increases as the temperature is increased (as T^4).

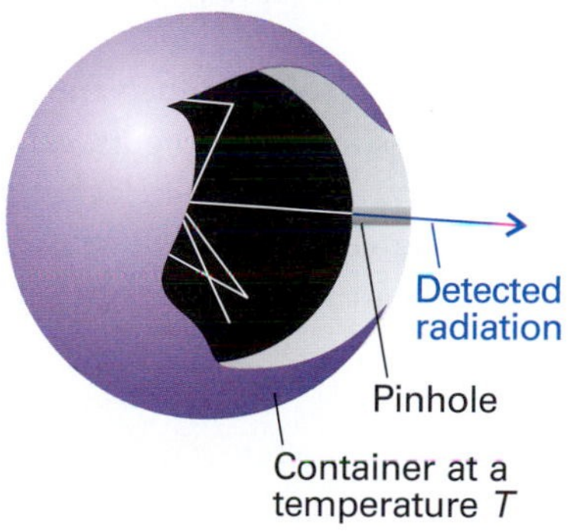

Fig. 1.4 An experimental representation of a black-body is a pinhole in an otherwise closed container. The radiation is reflected many times within the container and comes to thermal equilibrium with the walls at a temperature T. Radiation leaking out through the pinhole is characteristic of the radiation within the container.

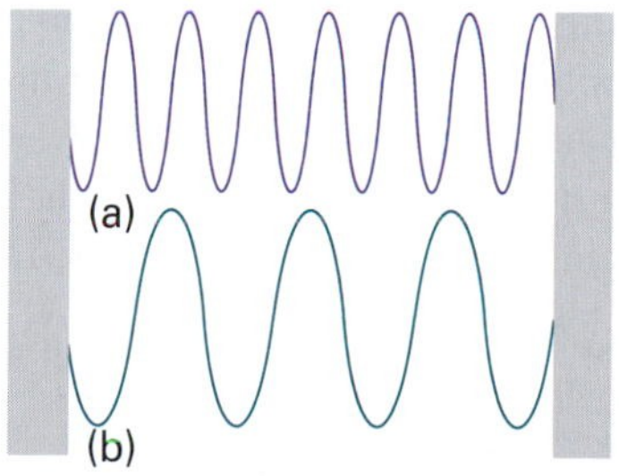

Fig. 1.5 The electromagnetic vacuum can be regarded as able to support oscillations of the electromagnetic field. When a high frequency, short wavelength oscillator (a) is excited, that frequency of radiation is present. The presence of low frequency, long wavelength radiation (b) signifies that an oscillator of the corresponding frequency has been excited.

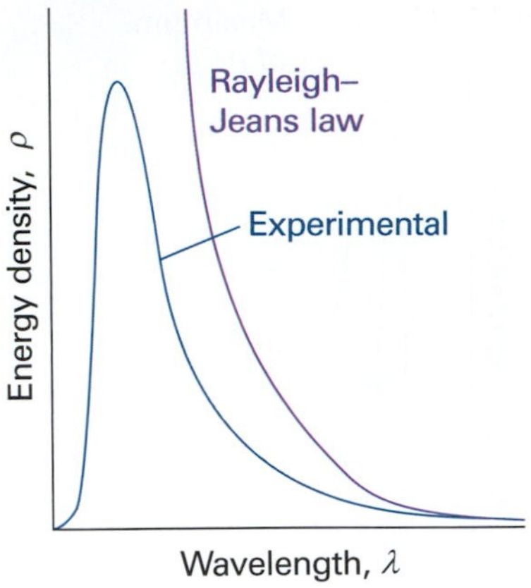

Fig. 1.6 The Rayleigh–Jeans law (eqn 1.3) predicts an infinite energy density at short wavelengths. This approach to infinity is called the *ultraviolet catastrophe*.

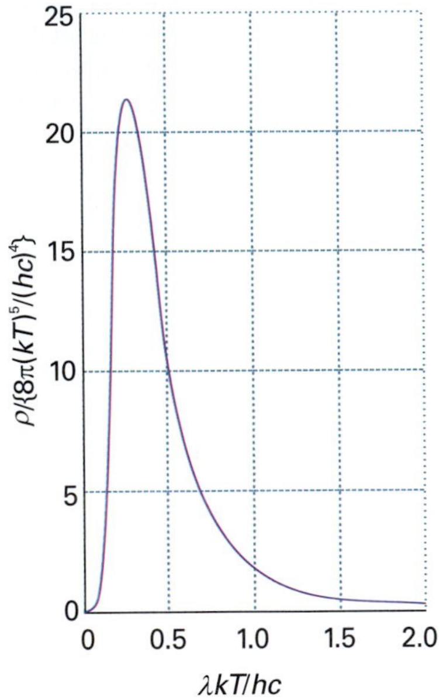

Fig. 1.7 The Planck distribution (eqn 1.5) accounts very well for the experimentally determined distribution of black-body radiation. Planck's quantization hypothesis essentially quenches the contributions of high frequency, short wavelength oscillators. The distribution coincides with the Rayleigh–Jeans distribution at long wavelengths.

Exploration Plot the Planck distribution at several temperatures and confirm that eqn 1.5 predicts the behaviour summarized by Fig. 1.2.

Unfortunately (for Rayleigh, Jeans, and classical physics), although the Rayleigh–Jeans law is quite successful at long wavelengths (low frequencies), it fails badly at short wavelengths (high frequencies). Thus, as λ decreases, ρ increases without going through a maximum (Fig. 1.6). The equation therefore predicts that oscillators of very short wavelength (corresponding to ultraviolet radiation, X-rays, and even γ-rays) are strongly excited even at room temperature. This absurd result, which implies that a large amount of energy is radiated in the high-frequency region of the electromagnetic spectrum, is called the **ultraviolet catastrophe**. According to classical physics, even cool objects should radiate in the visible and ultraviolet regions, so objects should glow in the dark; there should in fact be no darkness.

(b) The Planck distribution

The German physicist Max Planck studied black-body radiation from the viewpoint of thermodynamics. In 1900 he found that he could account for the experimental observations by proposing that the energy of each electromagnetic oscillator is limited to discrete values and cannot be varied arbitrarily. This proposal is quite contrary to the viewpoint of classical physics (on which the equipartition principle used by Rayleigh is based), in which all possible energies are allowed. The limitation of energies to discrete values is called the **quantization of energy**. In particular, Planck found that he could account for the observed distribution of energy if he supposed that the permitted energies of an electromagnetic oscillator of frequency ν are integer multiples of $h\nu$:

$$E = nh\nu \qquad n = 0, 1, 2, \ldots \tag{1.4}$$

where h is a fundamental constant now known as **Planck's constant**. On the basis of this assumption, Planck was able to derive the **Planck distribution**:

$$\mathrm{d}\mathcal{E} = \rho\mathrm{d}\lambda \qquad \rho = \frac{8\pi hc}{\lambda^5(\mathrm{e}^{hc/\lambda kT} - 1)} \tag{1.5}$$

(For references to the derivation of this expression, see *Further reading*.) This expression fits the experimental curve very well at all wavelengths (Fig. 1.7), and the value of h, which is an undetermined parameter in the theory, may be obtained by varying its value until a best fit is obtained. The currently accepted value for h is 6.626×10^{-34} J s.

The Planck distribution resembles the Rayleigh–Jeans law (eqn 1.3) apart from the all-important exponential factor in the denominator. For short wavelengths, $hc/\lambda kT \gg 1$ and $\mathrm{e}^{hc/\lambda kT} \to \infty$ faster than $\lambda^5 \to 0$; therefore $\rho \to 0$ as $\lambda \to 0$ or $\nu \to \infty$. Hence, the energy density approaches zero at high frequencies, in agreement with observation. For long wavelengths, $hc/\lambda kT \ll 1$, and the denominator in the Planck distribution can be replaced by

$$\mathrm{e}^{hc/\lambda kT} - 1 = \left(1 + \frac{hc}{\lambda kT} + \cdots\right) - 1 \approx \frac{hc}{\lambda kT}$$

When this approximation is substituted into eqn 1.5, we find that the Planck distribution reduces to the Rayleigh–Jeans law.

It is quite easy to see why Planck's approach was successful while Rayleigh's was not. The thermal motion of the atoms in the walls of the black body excites the oscillators of the electromagnetic field. According to classical mechanics, all the oscillators of the field share equally in the energy supplied by the walls, so even the highest frequencies are excited. The excitation of very high frequency oscillators results in the ultraviolet catastrophe. According to Planck's hypothesis, however, oscillators are excited only if they can acquire an energy of at least $h\nu$. This energy is too large for the walls to supply in the case of the very high frequency oscillators, so the latter remain unexcited. The effect of quantization is to reduce the contribution from the high frequency oscillators, for they cannot be significantly excited with the energy available.

(c) Heat capacities

In the early nineteenth century, the French scientists Pierre-Louis Dulong and Alexis-Thérèse Petit determined the heat capacities of a number of monatomic solids. On the basis of some somewhat slender experimental evidence, they proposed that the molar heat capacities of all monatomic solids are the same and (in modern units) close to $25\ \text{J K}^{-1}\ \text{mol}^{-1}$.

Dulong and Petit's law is easy to justify in terms of classical physics. If classical physics were valid, the equipartition principle could be used to calculate the heat capacity of a solid. According to this principle, the mean energy of an atom as it oscillates about its mean position in a solid is kT for each direction of displacement. As each atom can oscillate in three dimensions, the average energy of each atom is $3kT$; for N atoms the total energy is $3NkT$. The contribution of this motion to the molar internal energy is therefore

$$U_\text{m} = 3N_\text{A}kT = 3RT$$

because $N_\text{A}k = R$, the gas constant. The molar constant volume heat capacity (see *Comment* 1.3) is then predicted to be

$$C_{V,\text{m}} = \left(\frac{\partial U_\text{m}}{\partial T}\right)_V = 3R \tag{1.6}$$

This result, with $3R = 24.9\ \text{J K}^{-1}\ \text{mol}^{-1}$, is in striking accord with Dulong and Petit's value.

Unfortunately (this time, for Dulong and Petit), significant deviations from their law were observed when advances in refrigeration techniques made it possible to measure heat capacities at low temperatures. It was found that the molar heat capacities of all monatomic solids are lower than $3R$ at low temperatures, and that the values approach zero as $T \rightarrow 0$. To account for these observations, Einstein (in 1905) assumed that each atom oscillated about its equilibrium position with a single frequency ν. He then invoked Planck's hypothesis to assert that the energy of oscillation is confined to discrete values, and specifically to $nh\nu$, where n is an integer. Einstein first calculated the contribution of the oscillations of the atoms to the total molar energy of the metal (by a method described in Section 9.4) and obtained

$$U_\text{m} = \frac{3N_\text{A}h\nu}{e^{h\nu/kT} - 1}$$

in place of the classical expression $3RT$. Then he found the molar heat capacity by differentiating U_m with respect to T. The resulting expression is now known as the **Einstein formula**:

$$C_{V,\text{m}} = 3Rf \qquad f = \left(\frac{\theta_\text{E}}{T}\right)^2 \left(\frac{e^{\theta_\text{E}/2T}}{e^{\theta_\text{E}/T} - 1}\right)^2 \tag{1.7}$$

The **Einstein temperature**, $\theta_\text{E} = h\nu/k$, is a way of expressing the frequency of oscillation of the atoms as a temperature: a high frequency corresponds to a high Einstein temperature.

Comment 1.2

The series expansion of an exponential function is $e^x = 1 + x + \frac{1}{2}x^2 + \cdots$. If $x \ll 1$, a good approximation is $e^x \approx 1 + x$. For example, $e^{0.01} = 1.010\,050\ldots \approx 1 + 0.01$.

Comment 1.3

The internal energy, U, a concept from thermodynamics (Volume 1, Chapter 2), can be regarded as the total energy of the particles making up a sample of matter. The constant-volume heat capacity is defined as $C_V = (\partial U/\partial T)_V$. A small heat capacity indicates that a large rise in temperature results from a given transfer of energy.

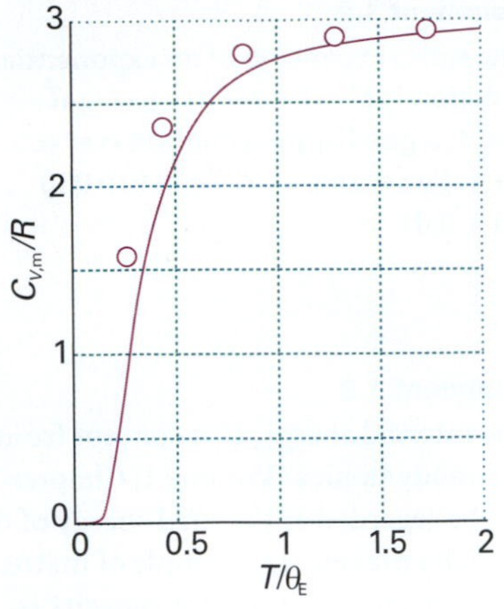

Fig. 1.8 Experimental low-temperature molar heat capacities and the temperature dependence predicted on the basis of Einstein's theory. His equation (eqn 1.7) accounts for the dependence fairly well, but is everywhere too low.

Exploration Using eqn 1.7, plot $C_{V,m}$ against T for several values of the Einstein temperature θ_E. At low temperature, does an increase in θ_E result in an increase or decrease of $C_{V,m}$? Estimate the temperature at which the value of $C_{V,m}$ reaches the classical value given by eqn 1.6.

At high temperatures (when $T \gg \theta_E$) the exponentials in f can be expanded as $1 + \theta_E/T + \cdots$ and higher terms ignored (see *Comment 1.2*). The result is

$$f = \left(\frac{\theta_E}{T}\right)^2 \left\{\frac{1 + \theta_E/2T + \cdots}{(1 + \theta_E/T + \cdots) - 1}\right\}^2 \approx 1 \tag{1.8a}$$

Consequently, the classical result ($C_{V,m} = 3R$) is obtained at high temperatures. At low temperatures, when $T \ll \theta_E$,

$$f \approx \left(\frac{\theta_E}{T}\right)^2 \left(\frac{e^{\theta_E/2T}}{e^{\theta_E/T}}\right)^2 = \left(\frac{\theta_E}{T}\right)^2 e^{-\theta_E/T} \tag{1.8b}$$

The strongly decaying exponential function goes to zero more rapidly than $1/T$ goes to infinity; so $f \to 0$ as $T \to 0$, and the heat capacity therefore approaches zero too. We see that Einstein's formula accounts for the decrease of heat capacity at low temperatures. The physical reason for this success is that at low temperatures only a few oscillators possess enough energy to oscillate significantly. At higher temperatures, there is enough energy available for all the oscillators to become active: all $3N$ oscillators contribute, and the heat capacity approaches its classical value.

Figure 1.8 shows the temperature dependence of the heat capacity predicted by the Einstein formula. The general shape of the curve is satisfactory, but the numerical agreement is in fact quite poor. The poor fit arises from Einstein's assumption that all the atoms oscillate with the same frequency, whereas in fact they oscillate over a range of frequencies from zero up to a maximum value, ν_D. This complication is taken into account by averaging over all the frequencies present, the final result being the **Debye formula**:

$$C_{V,m} = 3Rf \qquad f = 3\left(\frac{T}{\theta_D}\right)^3 \int_0^{\theta_D/T} \frac{x^4 e^x}{(e^x - 1)^2} dx \tag{1.9}$$

where $\theta_D = h\nu_D/k$ is the **Debye temperature** (for a derivation, see *Further reading*). The integral in eqn 1.9 has to be evaluated numerically, but that is simple with mathematical software. The details of this modification, which, as Fig. 1.9 shows, gives improved agreement with experiment, need not distract us at this stage from the main conclusion, which is that quantization must be introduced in order to explain the thermal properties of solids.

Illustration 1.1 *Assessing the heat capacity*

The Debye temperature for lead is 105 K, corresponding to a vibrational frequency of 2.2×10^{12} Hz, whereas that for diamond and its much lighter, more rigidly bonded atoms, is 2230 K, corresponding to 4.6×10^{13} Hz. As we see from Fig. 1.9, $f \approx 1$ for $T > \theta_D$ and the heat capacity is almost classical. For lead at 25°C, corresponding to $T/\theta_D = 2.8$, $f = 0.99$ and the heat capacity has almost its classical value. For diamond at the same temperature, $T/\theta_D = 0.13$, corresponding to $f = 0.15$, and the heat capacity is only 15 per cent of its classical value.

(d) Atomic and molecular spectra

The most compelling evidence for the quantization of energy comes from **spectroscopy**, the detection and analysis of the electromagnetic radiation absorbed, emitted, or scattered by a substance. The record of the intensity of light intensity transmitted

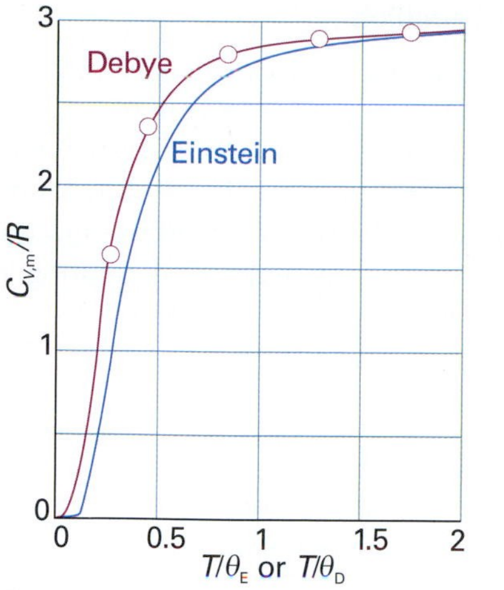

Fig. 1.9 Debye's modification of Einstein's calculation (eqn 1.9) gives very good agreement with experiment. For copper, $T/\theta_D = 2$ corresponds to about 170 K, so the detection of deviations from Dulong and Petit's law had to await advances in low-temperature physics.

Exploration Starting with the Debye formula (eqn 1.9), plot $dC_{V,m}/dT$, the temperature coefficient of $C_{V,m}$, against T for $\theta_D = 400$ K. At what temperature is $C_{V,m}$ most sensitive to temperature?

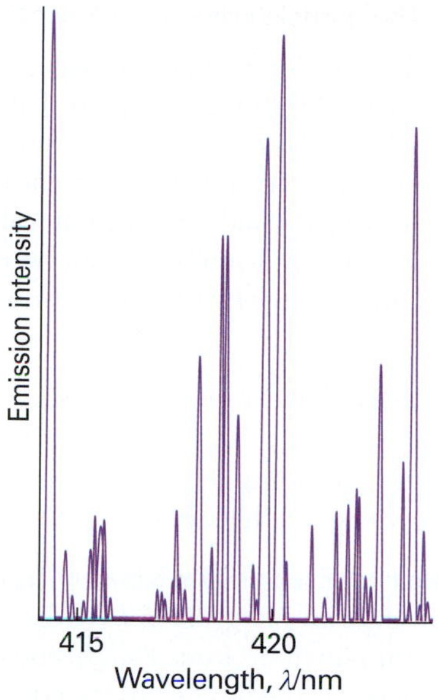

Fig. 1.10 A region of the spectrum of radiation emitted by excited iron atoms consists of radiation at a series of discrete wavelengths (or frequencies).

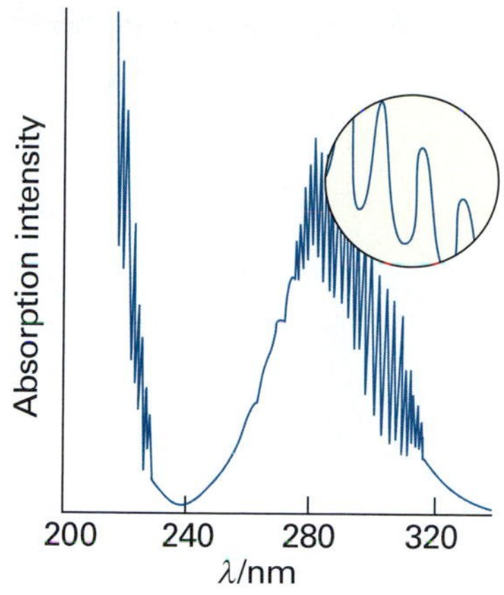

Fig. 1.11 When a molecule changes its state, it does so by absorbing radiation at definite frequencies. This spectrum is part of that due to the electronic, vibrational, and rotational excitation of sulfur dioxide (SO_2) molecules. This observation suggests that molecules can possess only discrete energies, not an arbitrary energy.

or scattered by a molecule as a function of frequency (ν), wavelength (λ), or wavenumber ($\tilde{\nu} = \nu/c$) is called its **spectrum** (from the Latin word for appearance).

A typical atomic spectrum is shown in Fig. 1.10, and a typical molecular spectrum is shown in Fig. 1.11. The obvious feature of both is that radiation is emitted or absorbed at a series of discrete frequencies. This observation can be understood if the energy of the atoms or molecules is also confined to discrete values, for then energy can be discarded or absorbed only in discrete amounts (Fig. 1.12). Then, if the energy of an atom decreases by ΔE, the energy is carried away as radiation of frequency ν, and an emission 'line', a sharply defined peak, appears in the spectrum. We say that a molecule undergoes a **spectroscopic transition**, a change of state, when the **Bohr frequency condition**

$$\Delta E = h\nu \qquad (1.10)$$

is fulfilled. We develop the principles and applications of atomic spectroscopy in Chapter 3 and of molecular spectroscopy in Chapters 6–8.

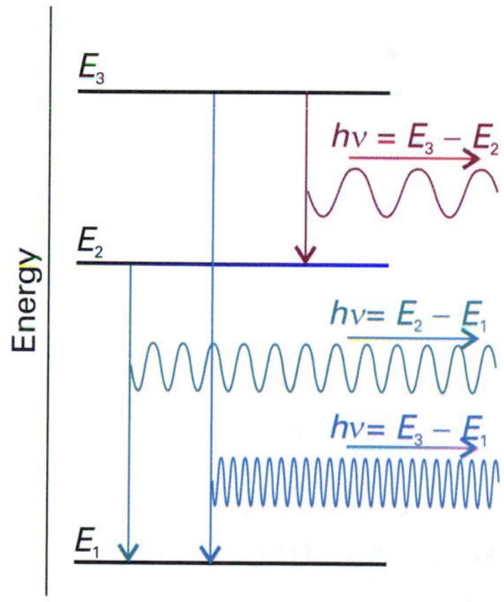

Fig. 1.12 Spectroscopic transitions, such as those shown above, can be accounted for if we assume that a molecule emits a photon as it changes between discrete energy levels. Note that high-frequency radiation is emitted when the energy change is large.

1.2 Wave–particle duality

At this stage we have established that the energies of the electromagnetic field and of oscillating atoms are quantized. In this section we shall see the experimental evidence that led to the revision of two other basic concepts concerning natural phenomena. One experiment shows that electromagnetic radiation—which classical physics treats as wave-like—actually also displays the characteristics of particles. Another experiment shows that electrons—which classical physics treats as particles—also display the characteristics of waves.

(a) The particle character of electromagnetic radiation

The observation that electromagnetic radiation of frequency ν can possess only the energies 0, $h\nu$, $2h\nu$, . . . suggests that it can be thought of as consisting of 0, 1, 2, . . . particles, each particle having an energy $h\nu$. Then, if one of these particles is present, the energy is $h\nu$, if two are present the energy is $2h\nu$, and so on. These particles of electromagnetic radiation are now called **photons**. The observation of discrete spectra from atoms and molecules can be pictured as the atom or molecule generating a photon of energy $h\nu$ when it discards an energy of magnitude ΔE, with $\Delta E = h\nu$.

Example 1.1 *Calculating the number of photons*

Calculate the number of photons emitted by a 100 W yellow lamp in 1.0 s. Take the wavelength of yellow light as 560 nm and assume 100 per cent efficiency.

Method Each photon has an energy $h\nu$, so the total number of photons needed to produce an energy E is $E/h\nu$. To use this equation, we need to know the frequency of the radiation (from $\nu = c/\lambda$) and the total energy emitted by the lamp. The latter is given by the product of the power (P, in watts) and the time interval for which the lamp is turned on ($E = P\Delta t$).

Answer The number of photons is

$$N = \frac{E}{h\nu} = \frac{P\Delta t}{h(c/\lambda)} = \frac{\lambda P\Delta t}{hc}$$

Substitution of the data gives

$$N = \frac{(5.60 \times 10^{-7}\ \text{m}) \times (100\ \text{J s}^{-1}) \times (1.0\ \text{s})}{(6.626 \times 10^{-34}\ \text{J s}) \times (2.998 \times 10^{8}\ \text{m s}^{-1})} = 2.8 \times 10^{20}$$

Note that it would take nearly 40 min to produce 1 mol of these photons.

A note on good practice To avoid rounding and other numerical errors, it is best to carry out algebraic manipulations first, and to substitute numerical values into a single, final formula. Moreover, an analytical result may be used for other data without having to repeat the entire calculation.

Self-test 1.1 How many photons does a monochromatic (single frequency) infrared rangefinder of power 1 mW and wavelength 1000 nm emit in 0.1 s? [5×10^{14}]

Further evidence for the particle-like character of radiation comes from the measurement of the energies of electrons produced in the **photoelectric effect**. This effect is the ejection of electrons from metals when they are exposed to ultraviolet radiation. The experimental characteristics of the photoelectric effect are as follows:

1 No electrons are ejected, regardless of the intensity of the radiation, unless its frequency exceeds a threshold value characteristic of the metal.

2 The kinetic energy of the ejected electrons increases linearly with the frequency of the incident radiation but is independent of the intensity of the radiation.

3 Even at low light intensities, electrons are ejected immediately if the frequency is above the threshold.

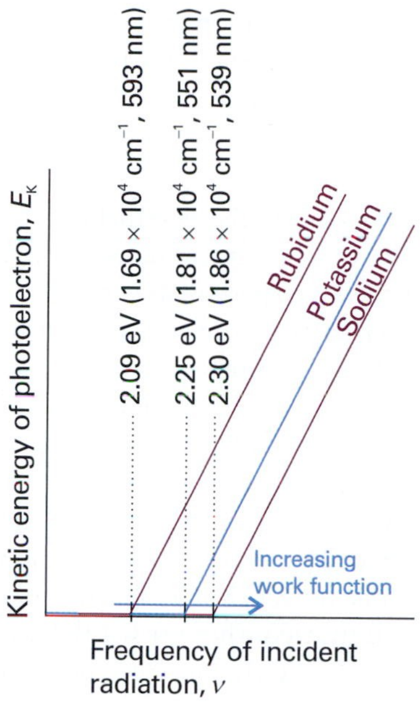

Fig. 1.13 In the photoelectric effect, it is found that no electrons are ejected when the incident radiation has a frequency below a value characteristic of the metal and, above that value, the kinetic energy of the photoelectrons varies linearly with the frequency of the incident radiation.

Exploration Calculate the value of Planck's constant given that the following kinetic energies were observed for photoejected electrons irradiated by radiation of the wavelengths noted.

λ_i/nm	320	330	345	360	385
E_K/eV	1.17	1.05	0.885	0.735	0.511

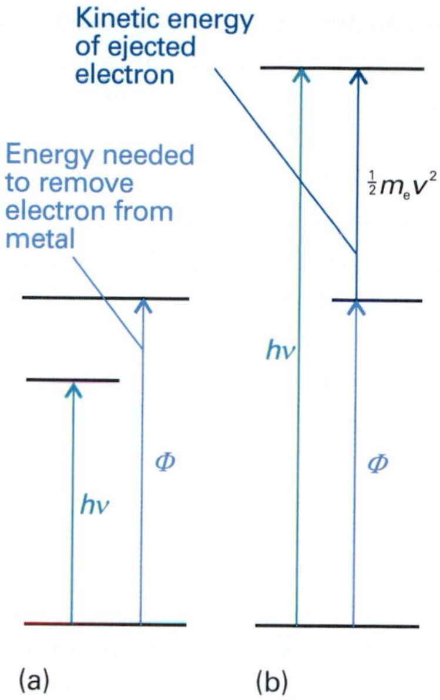

Fig. 1.14 The photoelectric effect can be explained if it is supposed that the incident radiation is composed of photons that have energy proportional to the frequency of the radiation. (a) The energy of the photon is insufficient to drive an electron out of the metal. (b) The energy of the photon is more than enough to eject an electron, and the excess energy is carried away as the kinetic energy of the photoelectron (the ejected electron).

Figure 1.13 illustrates the first and second characteristics.

These observations strongly suggest that the photoelectric effect depends on the ejection of an electron when it is involved in a collision with a particle-like projectile that carries enough energy to eject the electron from the metal. If we suppose that the projectile is a photon of energy $h\nu$, where ν is the frequency of the radiation, then the conservation of energy requires that the kinetic energy of the ejected electron should obey

$$\tfrac{1}{2}m_e v^2 = h\nu - \Phi \qquad (1.11)$$

In this expression Φ is a characteristic of the metal called its **work function**, the energy required to remove an electron from the metal to infinity (Fig. 1.14), the analogue of the ionization energy of an individual atom or molecule. Photoejection cannot occur if $h\nu < \Phi$ because the photon brings insufficient energy: this conclusion accounts for observation (1). Equation 1.11 predicts that the kinetic energy of an ejected electron should increase linearly with frequency, in agreement with observation (2). When a photon collides with an electron, it gives up all its energy, so we should expect electrons to appear as soon as the collisions begin, provided the photons have sufficient energy;

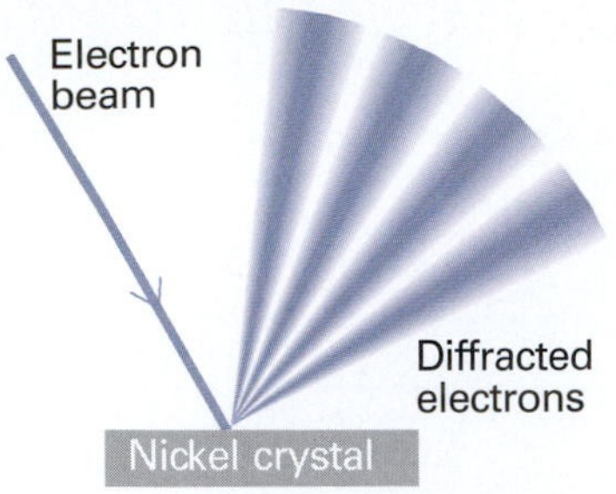

Fig. 1.15 The Davisson–Germer experiment. The scattering of an electron beam from a nickel crystal shows a variation of intensity characteristic of a diffraction experiment in which waves interfere constructively and destructively in different directions.

Comment 1.4

A characteristic property of waves is that they interfere with one another, giving a greater displacement where peaks or troughs coincide, leading to constructive interference, and a smaller displacement where peaks coincide with troughs, leading to destructive interference (see the illustration: (a) constructive, (b) destructive).

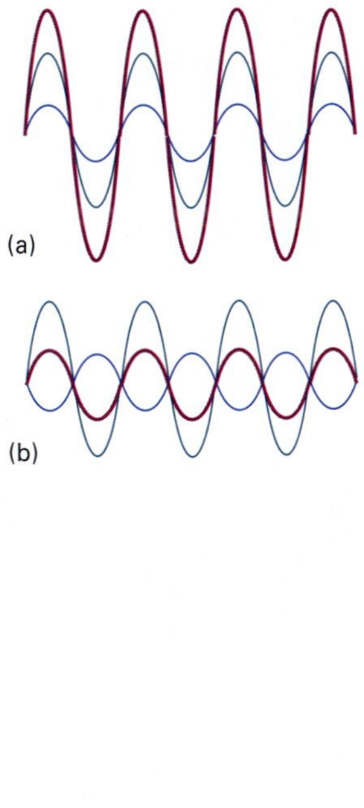

this conclusion agrees with observation (3). A practical application of eqn 1.11 is that it provides a technique for the determination of Planck's constant, for the slopes of the lines in Fig. 1.13 are all equal to h.

(b) The wave character of particles

Although contrary to the long-established wave theory of light, the view that light consists of particles had been held before, but discarded. No significant scientist, however, had taken the view that matter is wave-like. Nevertheless, experiments carried out in 1925 forced people to consider that possibility. The crucial experiment was performed by the American physicists Clinton Davisson and Lester Germer, who observed the diffraction of electrons by a crystal (Fig. 1.15). Diffraction is the interference caused by an object in the path of waves. Depending on whether the interference is constructive or destructive, the result is a region of enhanced or diminished intensity of the wave. Davisson and Germer's success was a lucky accident, because a chance rise of temperature caused their polycrystalline sample to anneal, and the ordered planes of atoms then acted as a diffraction grating. At almost the same time, G.P. Thomson, working in Scotland, showed that a beam of electrons was diffracted when passed through a thin gold foil. Electron diffraction is the basis for special techniques in microscopy used by biologists and materials scientists (*Impact I1.1*).

The Davisson–Germer experiment, which has since been repeated with other particles (including α particles and molecular hydrogen), shows clearly that particles have wave-like properties, and the diffraction of neutrons is a well-established technique for investigating the structures and dynamics of condensed phases. We have also seen that waves of electromagnetic radiation have particle-like properties. Thus we are brought to the heart of modern physics. When examined on an atomic scale, the classical concepts of particle and wave melt together, particles taking on the characteristics of waves, and waves the characteristics of particles.

Some progress towards coordinating these properties had already been made by the French physicist Louis de Broglie when, in 1924, he suggested that any particle, not only photons, travelling with a linear momentum p should have (in some sense) a wavelength given by the **de Broglie relation**:

$$\lambda = \frac{h}{p} \tag{1.12}$$

That is, a particle with a high linear momentum has a short wavelength (Fig. 1.16). Macroscopic bodies have such high momenta (because their mass is so great), even when they are moving slowly, that their wavelengths are undetectably small, and the wave-like properties cannot be observed.

Example 1.2 *Estimating the de Broglie wavelength*

Estimate the wavelength of electrons that have been accelerated from rest through a potential difference of 40 kV.

Method To use the de Broglie relation, we need to know the linear momentum, p, of the electrons. To calculate the linear momentum, we note that the energy acquired by an electron accelerated through a potential difference V is eV, where e is the magnitude of its charge. At the end of the period of acceleration, all the acquired energy is in the form of kinetic energy, $E_K = p^2/2m_e$, so we can determine p by setting $p^2/2m_e$ equal to eV. As before, carry through the calculation algebraically before substituting the data.

Answer The expression $p^2/2m_e = eV$ solves to $p = (2m_eeV)^{1/2}$; then, from the de Broglie relation $\lambda = h/p$,

$$\lambda = \frac{h}{(2m_e eV)^{1/2}}$$

Substitution of the data and the fundamental constants (from inside the front cover) gives

$$\lambda = \frac{6.626 \times 10^{-34}\ \text{J s}}{\{2 \times (9.109 \times 10^{-31}\ \text{kg}) \times (1.609 \times 10^{-19}\ \text{C}) \times (4.0 \times 10^{4}\ \text{V})\}^{1/2}}$$

$$= 6.1 \times 10^{-12}\ \text{m}$$

where we have used 1 V C = 1 J and 1 J = 1 kg m^2 s^{-2}. The wavelength of 6.1 pm is shorter than typical bond lengths in molecules (about 100 pm). Electrons accelerated in this way are used in the technique of electron diffraction for the determination of molecular structure.

Self-test 1.2 Calculate (a) the wavelength of a neutron with a translational kinetic energy equal to kT at 300 K, (b) a tennis ball of mass 57 g travelling at 80 km/h.

[(a) 178 pm, (b) 5.2×10^{-34} m]

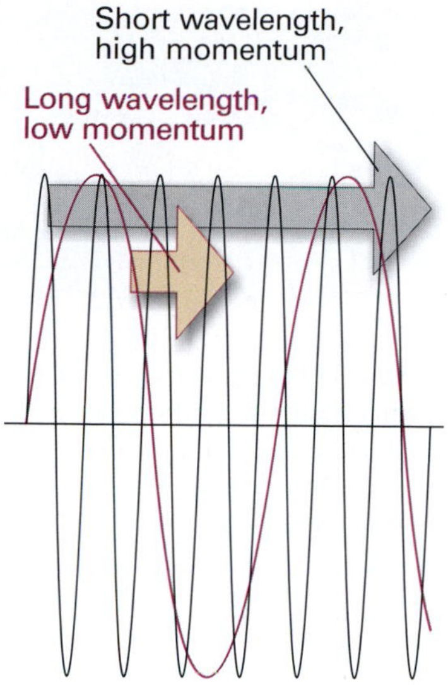

Fig. 1.16 An illustration of the de Broglie relation between momentum and wavelength. The wave is associated with a particle (shortly this wave will be seen to be the wavefunction of the particle). A particle with high momentum has a wavefunction with a short wavelength, and vice versa.

We now have to conclude that, not only has electromagnetic radiation the character classically ascribed to particles, but electrons (and all other particles) have the characteristics classically ascribed to waves. This joint particle and wave character of matter and radiation is called **wave–particle duality**. Duality strikes at the heart of classical physics, where particles and waves are treated as entirely distinct entities. We have also seen that the energies of electromagnetic radiation and of matter cannot be varied continuously, and that for small objects the discreteness of energy is highly significant. In classical mechanics, in contrast, energies could be varied continuously. Such total failure of classical physics for small objects implied that its basic concepts were false. A new mechanics had to be devised to take its place.

IMPACT ON BIOLOGY
I1.1 Electron microscopy

The basic approach of illuminating a small area of a sample and collecting light with a microscope has been used for many years to image small specimens. However, the *resolution* of a microscope, the minimum distance between two objects that leads to two distinct images, is on the order of the wavelength of light used as a probe (see *Impact I6.1*). Therefore, conventional microscopes employing visible light have resolutions in the micrometre range and are blind to features on a scale of nanometres.

There is great interest in the development of new experimental probes of very small specimens that cannot be studied by traditional light microscopy. For example, our understanding of biochemical processes, such as enzymatic catalysis, protein folding, and the insertion of DNA into the cell's nucleus, will be enhanced if it becomes possible to image individual biopolymers—with dimensions much smaller than visible wavelengths—at work. One technique that is often used to image nanometre-sized objects is *electron microscopy*, in which a beam of electrons with a well defined de Broglie wavelength replaces the lamp found in traditional light microscopes. Instead of glass or quartz lenses, magnetic fields are used to focus the beam. In *transmission electron microscopy* (TEM), the electron beam passes through the specimen and the

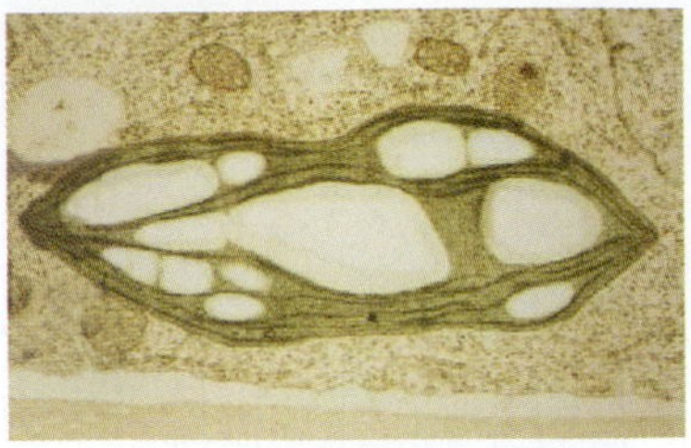

Fig. 1.17 A TEM image of a cross-section of a plant cell showing chloroplasts, organelles responsible for the reactions of photosynthesis. Chloroplasts are typically 5 μm long. (Image supplied by Brian Bowes.)

image is collected on a screen. In *scanning electron microscopy* (SEM), electrons scattered back from a small irradiated area of the sample are detected and the electrical signal is sent to a video screen. An image of the surface is then obtained by scanning the electron beam across the sample.

As in traditional light microscopy, the wavelength of and the ability to focus the incident beam—in this case a beam of electrons—govern the resolution. Electron wavelengths in typical electron microscopes can be as short as 10 pm, but it is not possible to focus electrons well with magnetic lenses so, in the end, typical resolutions of TEM and SEM instruments are about 2 nm and 50 nm, respectively. It follows that electron microscopes cannot resolve individual atoms (which have diameters of about 0.2 nm). Furthermore, only certain samples can be observed under certain conditions. The measurements must be conducted under high vacuum. For TEM observations, the samples must be very thin cross-sections of a specimen and SEM observations must be made on dry samples. A consequence of these requirements is that neither technique can be used to study living cells. In spite of these limitations, electron microscopy is very useful in studies of the internal structure of cells (Fig. 1.17).

The dynamics of microscopic systems

Quantum mechanics acknowledges the wave–particle duality of matter by supposing that, rather than travelling along a definite path, a particle is distributed through space like a wave. This remark may seem mysterious: it will be interpreted more fully shortly. The mathematical representation of the wave that in quantum mechanics replaces the classical concept of trajectory is called a **wavefunction**, ψ (psi).

1.3 The Schrödinger equation

In 1926, the Austrian physicist Erwin Schrödinger proposed an equation for finding the wavefunction of any system. The **time-independent Schrödinger equation** for a particle of mass m moving in one dimension with energy E is

$$-\frac{\hbar^2}{2m}\frac{d^2\psi}{dx^2}+V(x)\psi=E\psi \qquad (1.13)$$

The factor $V(x)$ is the potential energy of the particle at the point x; because the total energy E is the sum of potential and kinetic energies, the first term must be related (in a manner we explore later) to the kinetic energy of the particle; $\hbar$ (which is read h-cross or h-bar) is a convenient modification of Planck's constant:

$$\hbar=\frac{h}{2\pi}=1.054\,57\times10^{-34}\text{ J s} \qquad (1.14)$$

For a partial justification of the form of the Schrödinger equation, see the *Justification* below. The discussions later in the chapter will help to overcome the apparent arbitrariness of this complicated expression. For the present, treat the equation as a quantum-mechanical postulate. Various ways of expressing the Schrödinger equation, of incorporating the time-dependence of the wavefunction, and of extending it to more dimensions, are collected in Table 1.1. In Chapter 2 we shall solve the equation for a number of important cases; in this chapter we are mainly concerned with its significance, the interpretation of its solutions, and seeing how it implies that energy is quantized.

Table 1.1 The Schrödinger equation

For one-dimensional systems:

$$-\frac{\hbar^2}{2m}\frac{d^2\psi}{dx^2}+V(x)\psi=E\psi$$

Where $V(x)$ is the potential energy of the particle and E is its total energy. For three-dimensional systems

$$-\frac{\hbar^2}{2m}\nabla^2\psi+V\psi=E\psi$$

where V may depend on position and ∇^2 ('del squared') is

$$\nabla^2=\frac{\partial^2}{\partial x^2}+\frac{\partial^2}{\partial y^2}+\frac{\partial^2}{\partial z^2}$$

In systems with spherical symmetry three equivalent forms are

$$\nabla^2=\frac{1}{r}\frac{\partial^2}{\partial r^2}r+\frac{1}{r^2}\Lambda^2$$

$$=\frac{1}{r^2}\frac{\partial}{\partial r}r^2\frac{\partial}{\partial r}+\frac{1}{r^2}\Lambda^2$$

$$=\frac{\partial^2}{\partial r^2}+\frac{2}{r}\frac{\partial}{\partial r}+\frac{1}{r^2}\Lambda^2$$

where

$$\Lambda^2=\frac{1}{\sin^2\theta}\frac{\partial^2}{\partial\phi^2}+\frac{1}{\sin\theta}\frac{\partial}{\partial\theta}\sin\theta\frac{\partial}{\partial\theta}$$

In the general case the Schrodinger equation is written

$$\hat{H}\psi=E\psi$$

where $\hat{H}$ is the hamiltonian operator for the system:

$$\hat{H}=-\frac{\hbar^2}{2m}\nabla^2+V$$

For the evolution of a system with time, it is necessary to solve the time-dependent Schrödinger equation:

$$\hat{H}\Psi=i\hbar\frac{\partial\Psi}{\partial t}$$

Justification 1.1 *Using the Schrödinger equation to develop the de Broglie relation*

Although the Schrödinger equation should be regarded as a postulate, like Newton's equations of motion, it can be seen to be plausible by noting that it implies the de Broglie relation for a freely moving particle in a region with constant potential energy V. After making the substitution $V(x) = V$, we can rearrange eqn 1.13 into

$$\frac{d^2\psi}{dx^2}=-\frac{2m}{\hbar^2}(E-V)\psi$$

General strategies for solving differential equations of this and other types that occur frequently in physical chemistry are treated in *Appendix 2*. In the case at hand, we note that a solution is

$$\psi=e^{ikx}\qquad k=\left\{\frac{2m(E-V)}{\hbar^2}\right\}^{1/2}$$

Comment 1.5

Complex numbers and functions are discussed in *Appendix 2*. Complex numbers have the form $z = x + iy$, where $i = (-1)^{1/2}$ and the real numbers x and y are the real and imaginary parts of z, denoted $Re(z)$ and $Im(z)$, respectively. Similarly, a complex function of the form $f = g + ih$, where g and h are functions of real arguments, has a real part $Re(f) = g$ and an imaginary part $Im(f) = h$.

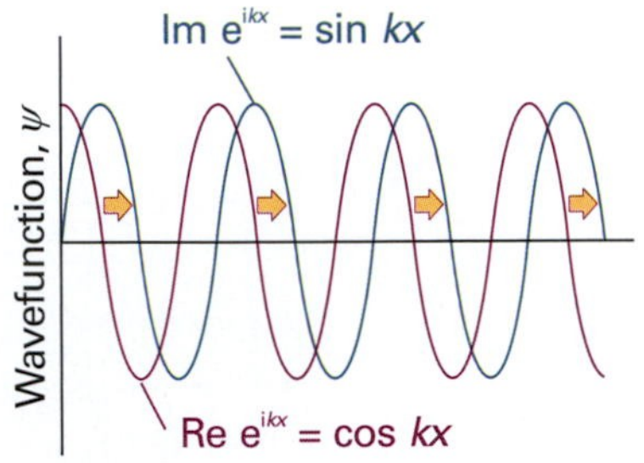

Fig. 1.18 The real (purple) and imaginary (blue) parts of a free particle wavefunction corresponding to motion towards positive x (as shown by the arrow).

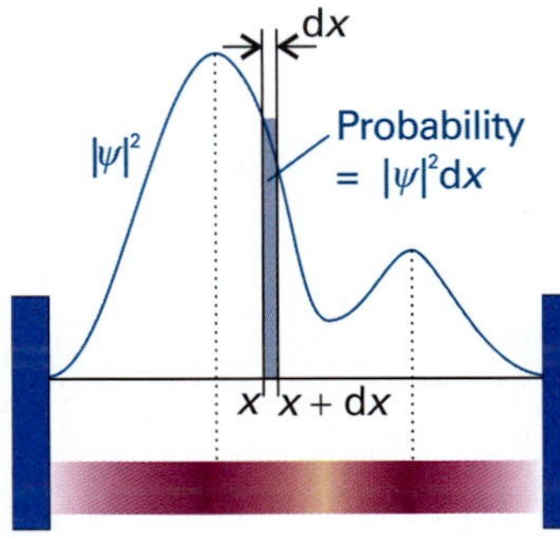

Fig. 1.19 The wavefunction ψ is a probability amplitude in the sense that its square modulus ($\psi^*\psi$ or $|\psi|^2$) is a probability density. The probability of finding a particle in the region dx located at x is proportional to $|\psi|^2$dx. We represent the probability density by the density of shading in the superimposed band.

Comment 1.6

To form the complex conjugate, ψ^*, of a complex function, replace i wherever it occurs by −i. For instance, the complex conjugate of e^{ikx} is e^{-ikx}. If the wavefunction is real, $|\psi|^2 = \psi^2$.

In quantum mechanics, a wavefunction that describes the spatial distribution of a particle (a 'spatial wavefunction') is complex if the particle it describes has a net motion. In the present case, we can use the relation $e^{i\theta} = \cos\theta + i\sin\theta$ to write

$$\psi = \cos kx + i\sin kx$$

The real and imaginary parts of ψ are drawn in Fig. 1.18, and we see that the imaginary component $Im(\psi) = \sin kx$ is shifted in the direction of the particle's motion. That is, both the real and imaginary parts of the wavefunction are 'real', in the sense of being present, and we express ψ as a complex function simply to help with the visualization of the motion of the particle the wavefunction desribes.

Now we recognize that $\cos kx$ (or $\sin kx$) is a wave of wavelength $\lambda = 2\pi/k$, as can be seen by comparing $\cos kx$ with the standard form of a harmonic wave, $\cos(2\pi x/\lambda)$. The quantity $E - V$ is equal to the kinetic energy of the particle, E_K, so $k = (2mE_K/\hbar^2)^{1/2}$, which implies that $E_K = k^2\hbar^2/2m$. Because $E_K = p^2/2m$, it follows that

$$p = k\hbar$$

Therefore, the linear momentum is related to the wavelength of the wavefunction by

$$p = \frac{2\pi}{\lambda} \times \frac{h}{2\pi} = \frac{h}{\lambda}$$

which is the de Broglie relation.

1.4 The Born interpretation of the wavefunction

A principal tenet of quantum mechanics is that *the wavefunction contains all the dynamical information about the system it describes.* Here we concentrate on the information it carries about the location of the particle.

The interpretation of the wavefunction in terms of the location of the particle is based on a suggestion made by Max Born. He made use of an analogy with the wave theory of light, in which the square of the amplitude of an electromagnetic wave in a region is interpreted as its intensity and therefore (in quantum terms) as a measure of the probability of finding a photon present in the region. The **Born interpretation** of the wavefunction focuses on the square of the wavefunction (or the square modulus, $|\psi|^2 = \psi^*\psi$, if ψ is complex). It states that the value of $|\psi|^2$ at a point is proportional to the probability of finding the particle in a region around that point. Specifically, for a one-dimensional system (Fig. 1.19):

> If the wavefunction of a particle has the value ψ at some point x, then the probability of finding the particle between x and x + dx is proportional to $|\psi|^2$dx.

Thus, $|\psi|^2$ is the **probability density**, and to obtain the probability it must be multiplied by the length of the infinitesimal region dx. The wavefunction ψ itself is called the **probability amplitude**. For a particle free to move in three dimensions (for example, an electron near a nucleus in an atom), the wavefunction depends on the point d$\boldsymbol{r}$ with coordinates x, y, and z, and the interpretation of $\psi(\boldsymbol{r})$ is as follows (Fig. 1.20):

> If the wavefunction of a particle has the value ψ at some point $\boldsymbol{r}$, then the probability of finding the particle in an infinitesimal volume $d\tau = dxdydz$ at that point is proportional to $|\psi|^2 d\tau$.

The Born interpretation does away with any worry about the significance of a negative (and, in general, complex) value of ψ because $|\psi|^2$ is real and never negative. There is no *direct* significance in the negative (or complex) value of a wavefunction: only the square modulus, a positive quantity, is directly physically significant, and both negative and positive regions of a wavefunction may correspond to a high probability of finding a particle in a region (Fig. 1.21). However, later we shall see that the presence of positive and negative regions of a wavefunction is of great *indirect* significance, because it gives rise to the possibility of constructive and destructive interference between different wavefunctions.

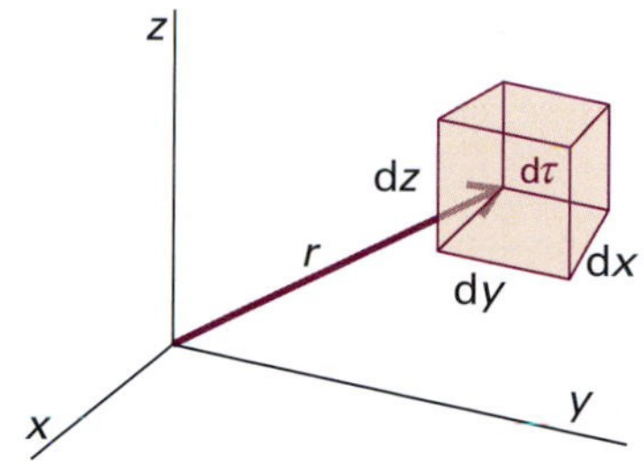

Fig. 1.20 The Born interpretation of the wavefunction in three-dimensional space implies that the probability of finding the particle in the volume element $d\tau = dxdydz$ at some location $\boldsymbol{r}$ is proportional to the product of $d\tau$ and the value of $|\psi|^2$ at that location.

Example 1.3 *Interpreting a wavefunction*

We shall see in Chapter 5 that the wavefunction of an electron in the lowest energy state of a hydrogen atom is proportional to e^{-r/a_0}, with a_0 a constant and r the distance from the nucleus. (Notice that this wavefunction depends only on this distance, not the angular position relative to the nucleus.) Calculate the relative probabilities of finding the electron inside a region of volume 1.0 pm^3, which is small even on the scale of the atom, located at (a) the nucleus, (b) a distance a_0 from the nucleus.

Method The region of interest is so small on the scale of the atom that we can ignore the variation of ψ within it and write the probability, P, as proportional to the probability density (ψ^2; note that ψ is real) evaluated at the point of interest multiplied by the volume of interest, δV. That is, $P \propto \psi^2 \delta V$, with $\psi^2 \propto e^{-2r/a_0}$.

Answer In each case $\delta V = 1.0$ pm^3. (a) At the nucleus, $r = 0$, so

$$P \propto e^0 \times (1.0\ \text{pm}^3) = (1.0) \times (1.0\ \text{pm}^3)$$

(b) At a distance $r = a_0$ in an arbitrary direction,

$$P \propto e^{-2} \times (1.0\ \text{pm}^3) = (0.14) \times (1.0\ \text{pm}^3)$$

Therefore, the ratio of probabilities is $1.0/0.14 = 7.1$. Note that it is more probable (by a factor of 7) that the electron will be found at the nucleus than in a volume element of the same size located at a distance a_0 from the nucleus. The negatively charged electron is attracted to the positively charged nucleus, and is likely to be found close to it.

A note on good practice The square of a wavefunction is not a probability: it is a probability density, and (in three dimensions) has the dimensions of 1/length^3. It becomes a (unitless) probability when multiplied by a volume. In general, we have to take into account the variation of the amplitude of the wavefunction over the volume of interest, but here we are supposing that the volume is so small that the variation of ψ in the region can be ignored.

Self-test 1.3 The wavefunction for the electron in its lowest energy state in the ion He^+ is proportional to e^{-2r/a_0}. Repeat the calculation for this ion. Any comment?

[55; more compact wavefunction]

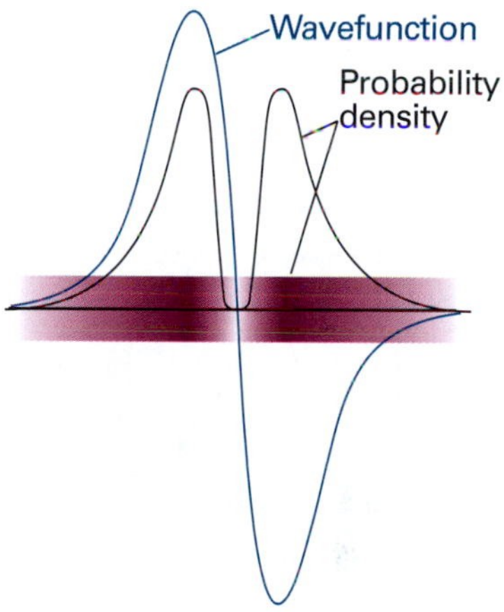

Fig. 1.21 The sign of a wavefunction has no direct physical significance: the positive and negative regions of this wavefunction both correspond to the same probability distribution (as given by the square modulus of ψ and depicted by the density of shading).

(a) Normalization

A mathematical feature of the Schrödinger equation is that, if ψ is a solution, then so is $N\psi$, where N is any constant. This feature is confirmed by noting that ψ occurs in every term in eqn 1.13, so any constant factor can be cancelled. This freedom to vary the wavefunction by a constant factor means that it is always possible to find a **normalization constant**, N, such that the proportionality of the Born interpretation becomes an equality.

We find the normalization constant by noting that, for a normalized wavefunction $N\psi$, the probability that a particle is in the region dx is equal to $(N\psi^*)(N\psi)\mathrm{d}x$ (we are taking N to be real). Furthermore, the sum over all space of these individual probabilities must be 1 (the probability of the particle being somewhere is 1). Expressed mathematically, the latter requirement is

$$N^2\int_{-\infty}^{\infty}\psi^*\psi\mathrm{d}x=1 \tag{1.15}$$

Almost all wavefunctions go to zero at sufficiently great distances so there is rarely any difficulty with the evaluation of this integral, and wavefunctions for which the integral in eqn 1.15 exists (in the sense of having a finite value) are said to be 'square-integrable'. It follows that

$$N=\frac{1}{\left(\int_{-\infty}^{\infty}\psi^*\psi\mathrm{d}x\right)^{1/2}} \tag{1.16}$$

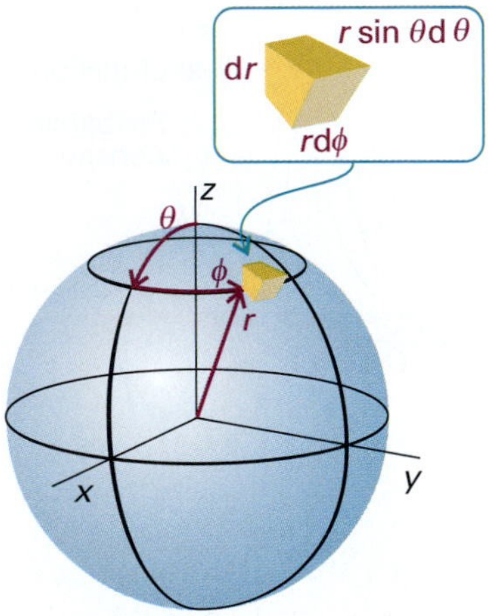

Fig. 1.22 The spherical polar coordinates used for discussing systems with spherical symmetry.

Therefore, by evaluating the integral, we can find the value of N and hence 'normalize' the wavefunction. From now on, unless we state otherwise, we always use wavefunctions that have been normalized to 1; that is, from now on we assume that ψ already includes a factor that ensures that (in one dimension)

$$\int_{-\infty}^{\infty}\psi^*\psi\,\mathrm{d}x=1 \tag{1.17a}$$

In three dimensions, the wavefunction is normalized if

$$\int_{-\infty}^{\infty}\int_{-\infty}^{\infty}\int_{-\infty}^{\infty}\psi^*\psi\,\mathrm{d}x\mathrm{d}y\mathrm{d}z=1 \tag{1.17b}$$

or, more succinctly, if

$$\int\psi^*\psi\,\mathrm{d}\tau=1 \tag{1.17c}$$

where $\mathrm{d}\tau=\mathrm{d}x\mathrm{d}y\mathrm{d}z$ and the limits of this definite integral are not written explicitly: in all such integrals, the integration is over all the space accessible to the particle. For systems with spherical symmetry it is best to work in **spherical polar coordinates** r, θ, and ϕ (Fig. 1.22): $x=r\sin\theta\cos\phi$, $y=r\sin\theta\sin\phi$, $z=r\cos\theta$. The volume element in spherical polar coordinates is $\mathrm{d}\tau=r^2\sin\theta\,\mathrm{d}r\mathrm{d}\theta\mathrm{d}\phi$. To cover all space, the radius r ranges from 0 to ∞, the colatitude, θ, ranges from 0 to π, and the azimuth, ϕ, ranges from 0 to 2π (Fig. 1.23), so the explicit form of eqn 1.17c is

$$\int_0^{\infty}\int_0^{\pi}\int_0^{2\pi}\psi^*\psi r^2\sin\theta\mathrm{d}r\mathrm{d}\theta\mathrm{d}\phi=1 \tag{1.17d}$$

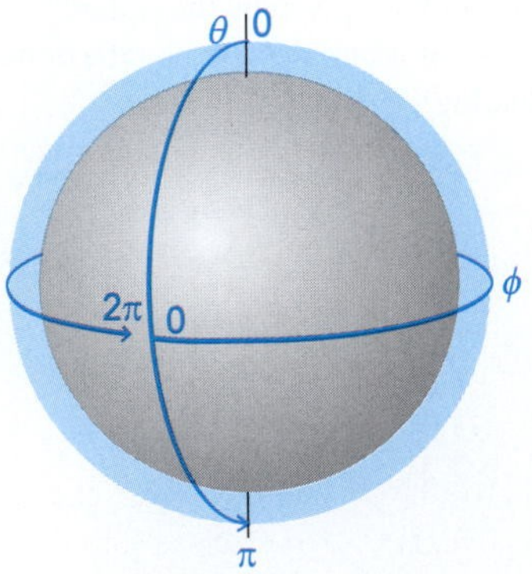

Fig. 1.23 The surface of a sphere is covered by allowing θ to range from 0 to π, and then sweeping that arc around a complete circle by allowing ϕ to range from 0 to 2π.

Example 1.4 *Normalizing a wavefunction*

Normalize the wavefunction used for the hydrogen atom in Example 1.3.

Method We need to find the factor N that guarantees that the integral in eqn 1.17c is equal to 1. Because the system is spherical, it is most convenient to use spherical coordinates and to carry out the integrations specified in eqn 1.17d. Note that the limits on the first integral sign refer to r, those on the second to θ, and those on the third to ϕ. A useful integral for calculations on atomic wavefunctions is

$$\int_0^\infty x^n e^{-ax} dx = \frac{n!}{a^{n+1}}$$

where $n!$ denotes a factorial: $n! = n(n-1)(n-2)\ldots 1$.

Answer The integration required is the product of three factors:

$$\int \psi^* \psi d\tau = N^2 \overbrace{\int_0^\infty r^2 e^{-2r/a_0} dr}^{\frac{1}{4}a_0^3} \overbrace{\int_0^\pi \sin\theta \, d\theta}^{2} \overbrace{\int_0^{2\pi} d\phi}^{2\pi} = \pi a_0^3 N^2$$

Therefore, for this integral to equal 1, we must set

$$N = \left(\frac{1}{\pi a_0^3}\right)^{1/2}$$

and the normalized wavefunction is

$$\psi = \left(\frac{1}{\pi a_0^3}\right)^{1/2} e^{-r/a_0}$$

Note that, because a_0 is a length, the dimensions of ψ are 1/length$^{3/2}$ and therefore those of ψ^2 are 1/length^3 (for instance, 1/m^3) as is appropriate for a probability density.

If Example 1.3 is now repeated, we can obtain the actual probabilities of finding the electron in the volume element at each location, not just their relative values. Given (from Section 3.1) that $a_0 = 52.9$ pm, the results are (a) 2.2×10^{-6}, corresponding to 1 chance in about 500 000 inspections of finding the electron in the test volume, and (b) 2.9×10^{-7}, corresponding to 1 chance in 3.4 million.

Self-test 1.4 Normalize the wavefunction given in Self-test 1.3.

$[N = (8/\pi a_0^3)^{1/2}]$

(b) Quantization

The Born interpretation puts severe restrictions on the acceptability of wavefunctions. The principal constraint is that ψ must not be infinite anywhere. If it were, the integral in eqn 1.17 would be infinite (in other words, ψ would not be square-integrable) and the normalization constant would be zero. The normalized function would then be zero everywhere, except where it is infinite, which would be unacceptable. The requirement that ψ is finite everywhere rules out many possible solutions of the Schrödinger equation, because many mathematically acceptable solutions rise to infinity and are therefore physically unacceptable. We shall meet several examples shortly.

The requirement that ψ is finite everywhere is not the only restriction implied by the Born interpretation. We could imagine (and in Section 2.6a will meet) a solution of the Schrödinger equation that gives rise to more than one value of $|\psi|^2$ at a single point. The Born interpretation implies that such solutions are unacceptable, because it would be absurd to have more than one probability that a particle is at the same point. This restriction is expressed by saying that the wavefunction must be *single-valued*; that is, have only one value at each point of space.

Comment 1.7
Infinitely sharp spikes are acceptable provided they have zero width, so it is more appropriate to state that the wavefunction must not be infinite over any finite region. In elementary quantum mechanics the simpler restriction, to finite ψ, is sufficient.

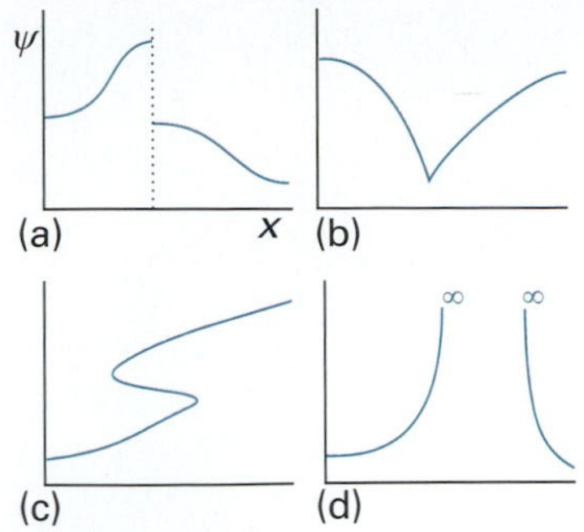

Fig. 1.24 The wavefunction must satisfy stringent conditions for it to be acceptable. (a) Unacceptable because it is not continuous; (b) unacceptable because its slope is discontinuous; (c) unacceptable because it is not single-valued; (d) unacceptable because it is infinite over a finite region.

Comment 1.8

There are cases, and we shall meet them, where acceptable wavefunctions have kinks. These cases arise when the potential energy has peculiar properties, such as rising abruptly to infinity. When the potential energy is smoothly well-behaved and finite, the slope of the wavefunction must be continuous; if the potential energy becomes infinite, then the slope of the wavefunction need not be continuous. There are only two cases of this behaviour in elementary quantum mechanics, and the peculiarity will be mentioned when we meet them.

The Schrödinger equation itself also implies some mathematical restrictions on the type of functions that will occur. Because it is a second-order differential equation, the second derivative of ψ must be well-defined if the equation is to be applicable everywhere. We can take the second derivative of a function only if it is continuous (so there are no sharp steps in it, Fig. 1.24) and if its first derivative, its slope, is continuous (so there are no kinks).

At this stage we see that ψ must be continuous, have a continuous slope, be single-valued, and be square-integrable. An acceptable wavefunction cannot be zero everywhere, because the particle it describes must be somewhere. These are such severe restrictions that acceptable solutions of the Schrödinger equation do not in general exist for arbitrary values of the energy E. In other words, a particle may possess only certain energies, for otherwise its wavefunction would be physically unacceptable. That is, *the energy of a particle is quantized.* We can find the acceptable energies by solving the Schrödinger equation for motion of various kinds, and selecting the solutions that conform to the restrictions listed above. That is the task of the next chapter.

Quantum mechanical principles

We have claimed that a wavefunction contains all the information it is possible to obtain about the dynamical properties of the particle (for example, its location and momentum). We have seen that the Born interpretation tells us as much as we can know about location, but how do we find any additional information?

1.5 The information in a wavefunction

The Schrödinger equation for a particle of mass m free to move parallel to the x-axis with zero potential energy is obtained from eqn 1.13 by setting $V=0$, and is

$$-\frac{\hbar^2}{2m}\frac{\mathrm{d}^2\psi}{\mathrm{d}x^2}=E\psi \qquad (1.18)$$

The solutions of this equation have the form

$$\psi=A\mathrm{e}^{\mathrm{i}kx}+B\mathrm{e}^{-\mathrm{i}kx} \qquad E=\frac{k^2\hbar^2}{2m} \qquad (1.19)$$

where A and B are constants. To verify that ψ is a solution of eqn 1.18, we simply substitute it into the left-hand side of the equation and confirm that we obtain $E\psi$:

$$\begin{aligned}-\frac{\hbar^2}{2m}\frac{\mathrm{d}^2\psi}{\mathrm{d}x^2}&=-\frac{\hbar^2}{2m}\frac{\mathrm{d}^2}{\mathrm{d}x^2}(A\mathrm{e}^{\mathrm{i}kx}+B\mathrm{e}^{-kx})\\&=-\frac{\hbar^2}{2m}\{A(\mathrm{i}k)^2\mathrm{e}^{\mathrm{i}kx}+B(-\mathrm{i}k)^2\mathrm{e}^{-\mathrm{i}kx}\}\\&=\frac{\hbar^2k^2}{2m}(A\mathrm{e}^{\mathrm{i}kx}+B\mathrm{e}^{-\mathrm{i}kx})=E\psi\end{aligned}$$

(a) The probability density

We shall see later what determines the values of A and B; for the time being we can treat them as arbitrary constants. Suppose that $B=0$ in eqn 1.19, then the wavefunction is simply

$$\psi=A\mathrm{e}^{\mathrm{i}kx} \qquad (1.20)$$

Where is the particle? To find out, we calculate the probability density:

$$|\psi|^2 = (Ae^{ikx})^*(Ae^{ikx}) = (A^*e^{-ikx})(Ae^{ikx}) = |A|^2 \quad (1.21)$$

This probability density is independent of x; so, wherever we look along the x-axis, there is an equal probability of finding the particle (Fig. 1.25a). In other words, if the wavefunction of the particle is given by eqn 1.20, then we cannot predict where we will find the particle. The same would be true if the wavefunction in eqn 1.19 had $A = 0$; then the probability density would be $|B|^2$, a constant.

Now suppose that in the wavefunction $A = B$. Then eqn 1.19 becomes

$$\psi = A(e^{ikx} + e^{-ikx}) = 2A\cos kx \quad (1.22)$$

The probability density now has the form

$$|\psi|^2 = (2A\cos kx)^*(2A\cos kx) = 4|A|^2\cos^2 kx \quad (1.23)$$

This function is illustrated in Fig. 1.25b. As we see, the probability density periodically varies between 0 and $4|A|^2$. The locations where the probability density is zero correspond to *nodes* in the wavefunction: particles will never be found at the nodes. Specifically, a **node** is a point where a wavefunction passes *through* zero. The location where a wavefunction approaches zero without actually passing through zero is not a node. Nodes are defined in terms of the probability amplitude, the wavefunction itself. The probability density, of course, never passes through zero because it cannot be negative.

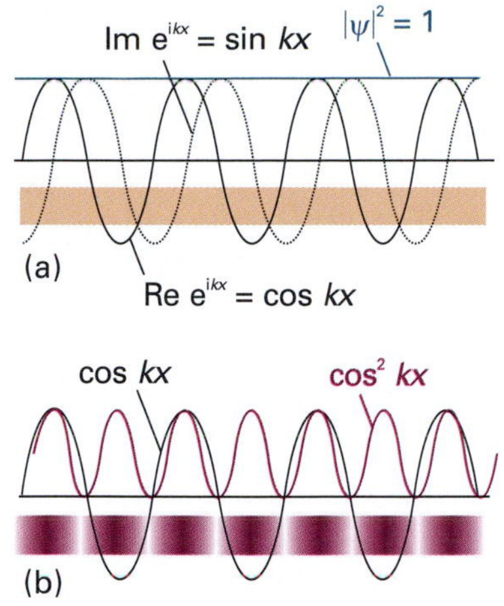

Fig. 1.25 (a) The square modulus of a wavefunction corresponding to a definite state of linear momentum is a constant; so it corresponds to a uniform probability of finding the particle anywhere. (b) The probability distribution corresponding to the superposition of states of equal magnitude of linear momentum but opposite direction of travel.

Comment 1.9

If the probability density of a particle is a constant, then it follows that, with x ranging from $-\infty$ to $+\infty$, the normalization constants, A or B, are 0. To avoid this embarrassing problem, x is allowed to range from $-L$ to $+L$, and L is allowed to go to infinity at the end of all calculations. We ignore this complication here.

(b) Operators, eigenvalues, and eigenfunctions

To formulate a systematic way of extracting information from the wavefunction, we first note that any Schrödinger equation (such as those in eqns 1.13 and 1.18) may be written in the succinct form

$$\hat{H}\psi = E\psi \quad (1.24a)$$

with (in one dimension)

$$\hat{H} = -\frac{\hbar^2}{2m}\frac{d^2}{dx^2} + V(x) \quad (1.24b)$$

The quantity $\hat{H}$ is an **operator**, something that carries out a mathematical operation on the function ψ. In this case, the operation is to take the second derivative of ψ and (after multiplication by $-\hbar^2/2m$) to add the result to the outcome of multiplying ψ by V. The operator $\hat{H}$ plays a special role in quantum mechanics, and is called the **hamiltonian operator** after the nineteenth century mathematician William Hamilton, who developed a form of classical mechanics that, it subsequently turned out, is well suited to the formulation of quantum mechanics. The hamiltonian operator is the operator corresponding to the total energy of the system, the sum of the kinetic and potential energies. Consequently, we can infer—as we anticipated in Section 1.3—that the first term in eqn 1.24b (the term proportional to the second derivative) must be the operator for the kinetic energy. When the Schrödinger equation is written as in eqn 1.24a, it is seen to be an **eigenvalue equation**, an equation of the form

$$(\text{Operator})(\text{function}) = (\text{constant factor}) \times (\text{same function}) \quad (1.25a)$$

If we denote a general operator by $\hat{\Omega}$ (where Ω is uppercase omega) and a constant factor by ω (lowercase omega), then an eigenvalue equation has the form

$$\hat{\Omega}\psi = \omega\psi \quad (1.25b)$$

The factor ω is called the **eigenvalue** of the operator $\hat{\Omega}$. The eigenvalue in eqn 1.24a is the energy. The function ψ in an equation of this kind is called an **eigenfunction** of the operator $\hat{\Omega}$ and is different for each eigenvalue. The eigenfunction in eqn 1.24a is the wavefunction corresponding to the energy E. It follows that another way of saying 'solve the Schrödinger equation' is to say 'find the eigenvalues and eigenfunctions of the hamiltonian operator for the system'. The wavefunctions are the eigenfunctions of the hamiltonian operator, and the corresponding eigenvalues are the allowed energies.

Example 1.5 *Identifying an eigenfunction*

Show that e^{ax} is an eigenfunction of the operator d/dx, and find the corresponding eigenvalue. Show that e^{ax^2} is not an eigenfunction of d/dx.

Method We need to operate on the function with the operator and check whether the result is a constant factor times the original function.

Answer For $\hat{\Omega} = \mathrm{d}/\mathrm{d}x$ and $\psi = e^{ax}$:

$$\hat{\Omega}\psi = \frac{\mathrm{d}}{\mathrm{d}x}e^{ax} = ae^{ax} = a\psi$$

Therefore e^{ax} is indeed an eigenfunction of d/dx, and its eigenvalue is a. For $\psi = e^{ax^2}$,

$$\hat{\Omega}\psi = \frac{\mathrm{d}}{\mathrm{d}x}e^{ax^2} = 2axe^{ax^2} = 2ax \times \psi$$

which is not an eigenvalue equation even though the same function ψ occurs on the right, because ψ is now multiplied by a variable factor ($2ax$), not a constant factor. Alternatively, if the right-hand side is written $2a(xe^{ax^2})$, we see that it is a constant ($2a$) times a *different* function.

Self-test 1.5 Is the function cos ax an eigenfunction of (a) d/dx, (b) $\mathrm{d}^2/\mathrm{d}x^2$?

[(a) No, (b) yes]

The importance of eigenvalue equations is that the pattern

(Energy operator) ψ = (energy) × ψ

exemplified by the Schrödinger equation is repeated for other **observables**, or measurable properties of a system, such as the momentum or the electric dipole moment. Thus, it is often the case that we can write

(Operator corresponding to an observable) ψ = (value of observable) × ψ

The symbol $\hat{\Omega}$ in eqn 1.25b is then interpreted as an operator (for example, the hamiltonian, $\hat{H}$) corresponding to an observable (for example, the energy), and the eigenvalue ω is the value of that observable (for example, the value of the energy, E). Therefore, if we know both the wavefunction ψ and the operator $\hat{\Omega}$ corresponding to the observable Ω of interest, and the wavefunction is an eigenfunction of the operator $\hat{\Omega}$, then we can predict the outcome of an observation of the property Ω (for example, an atom's energy) by picking out the factor ω in the eigenvalue equation, eqn 1.25b.

A basic postulate of quantum mechanics tells us how to set up the operator corresponding to a given observable:

Observables, Ω, are represented by operators, $\hat{\Omega}$, built from the following position and momentum operators:

$$\hat{x} = x\times \qquad \hat{p}_x = \frac{\hbar}{\mathrm{i}}\frac{\mathrm{d}}{\mathrm{d}x} \qquad [1.26]$$

Comment 1.10
The rules summarized by eqn 1.26 apply to observables that depend on spatial variables; intrinsic properties, such as spin (see Section 2.8) are treated differently.

That is, the operator for location along the x-axis is multiplication (of the wavefunction) by x and the operator for linear momentum parallel to the x-axis is proportional to taking the derivative (of the wavefunction) with respect to x.

Example 1.6 *Determining the value of an observable*

What is the linear momentum of a particle described by the wavefunction in eqn 1.19 with (a) $B = 0$, (b) $A = 0$?

Method We operate on ψ with the operator corresponding to linear momentum (eqn 1.26), and inspect the result. If the outcome is the original wavefunction multiplied by a constant (that is, we generate an eigenvalue equation), then the constant is identified with the value of the observable.

Answer (a) With the wavefunction given in eqn 1.19 with $B = 0$,

$$\hat{p}_x\psi = \frac{\hbar}{\mathrm{i}}\frac{\mathrm{d}\psi}{\mathrm{d}x} = \frac{\hbar}{\mathrm{i}}A\frac{\mathrm{d}e^{\mathrm{i}kx}}{\mathrm{d}x} = \frac{\hbar}{\mathrm{i}}B\times \mathrm{i}ke^{\mathrm{i}kx} = k\hbar Be^{-\mathrm{i}kx} = k\hbar\psi$$

This is an eigenvalue equation, and by comparing it with eqn 1.25b we find that $p_x = +k\hbar$. (b) For the wavefunction with $A = 0$,

$$\hat{p}_x\psi = \frac{\hbar}{\mathrm{i}}\frac{\mathrm{d}\psi}{\mathrm{d}x} = \frac{\hbar}{\mathrm{i}}B\frac{\mathrm{d}e^{-\mathrm{i}kx}}{\mathrm{d}x} = \frac{\hbar}{\mathrm{i}}B\times(-\mathrm{i}k)e^{-\mathrm{i}kx} = -k\hbar Be^{-\mathrm{i}kx} = -k\hbar\psi$$

The magnitude of the linear momentum is the same in each case ($k\hbar$), but the signs are different: In (a) the particle is travelling to the right (positive x) but in (b) it is travelling to the left (negative x).

Self-test 1.6 The operator for the angular momentum of a particle travelling in a circle in the xy-plane is $\hat{l}_z = (\hbar/\mathrm{i})\mathrm{d}/\mathrm{d}\phi$, where ϕ is its angular position. What is the angular momentum of a particle described by the wavefunction $e^{-2\mathrm{i}\phi}$?

[$l_z = -2\hbar$]

We use the definitions in eqn 1.26 to construct operators for other spatial observables. For example, suppose we wanted the operator for a potential energy of the form $V = \frac{1}{2}kx^2$, with k a constant (later, we shall see that this potential energy describes the vibrations of atoms in molecules). Then it follows from eqn 1.26 that the operator corresponding to V is multiplication by x^2:

$$\hat{V} = \tfrac{1}{2}kx^2\times \qquad (1.27)$$

In normal practice, the multiplication sign is omitted. To construct the operator for kinetic energy, we make use of the classical relation between kinetic energy and linear momentum, which in one dimension is $E_K = p_x^2/2m$. Then, by using the operator for p_x in eqn 1.26 we find:

$$\hat{E}_K = \frac{1}{2m}\left(\frac{\hbar}{i}\frac{d}{dx}\right)\left(\frac{\hbar}{i}\frac{d}{dx}\right) = -\frac{\hbar^2}{2m}\frac{d^2}{dx^2} \quad (1.28)$$

It follows that the operator for the total energy, the hamiltonian operator, is

$$\hat{H} = \hat{E}_K + \hat{V} = -\frac{\hbar^2}{2m}\frac{d^2}{dx^2} + \hat{V} \quad (1.29)$$

with $\hat{V}$ the multiplicative operator in eqn 1.27 (or some other relevant potential energy).

The expression for the kinetic energy operator, eqn 1.28, enables us to develop the point made earlier concerning the interpretation of the Schrödinger equation. In mathematics, the second derivative of a function is a measure of its curvature: a large second derivative indicates a sharply curved function (Fig. 1.26). It follows that a sharply curved wavefunction is associated with a high kinetic energy, and one with a low curvature is associated with a low kinetic energy. This interpretation is consistent with the de Broglie relation, which predicts a short wavelength (a sharply curved wavefunction) when the linear momentum (and hence the kinetic energy) is high. However, it extends the interpretation to wavefunctions that do not spread through space and resemble those shown in Fig. 1.26. The curvature of a wavefunction in general varies from place to place. Wherever a wavefunction is sharply curved, its contribution to the total kinetic energy is large (Fig. 1.27). Wherever the wavefunction is not sharply curved, its contribution to the overall kinetic energy is low. As we shall shortly see, the observed kinetic energy of the particle is an integral of all the contributions of the kinetic energy from each region. Hence, we can expect a particle to have a high kinetic energy if the average curvature of its wavefunction is high. Locally there can be both positive and negative contributions to the kinetic energy (because the curvature can be either positive, ⌣, or negative, ⌢), but the average is always positive (see Problem 1.22).

The association of high curvature with high kinetic energy will turn out to be a valuable guide to the interpretation of wavefunctions and the prediction of their shapes. For example, suppose we need to know the wavefunction of a particle with a given total energy and a potential energy that decreases with increasing x (Fig. 1.28). Because the difference $E - V = E_K$ increases from left to right, the wavefunction must become more sharply curved as x increases: its wavelength decreases as the local contributions to its kinetic energy increase. We can therefore guess that the wavefunction will look like the function sketched in the illustration, and more detailed calculation confirms this to be so.

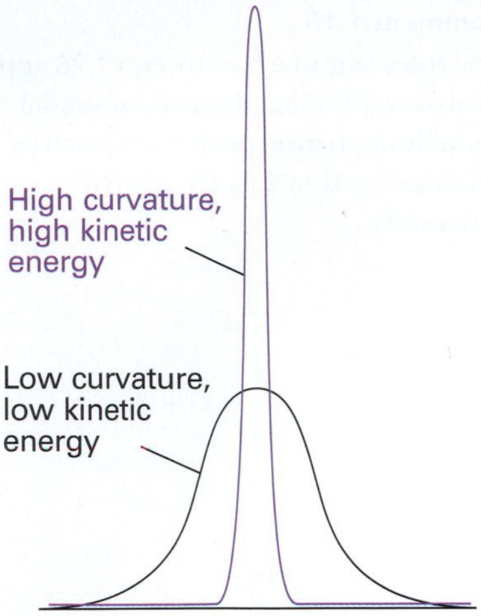

Fig. 1.26 Even if a wavefunction does not have the form of a periodic wave, it is still possible to infer from it the average kinetic energy of a particle by noting its average curvature. This illustration shows two wavefunctions: the sharply curved function corresponds to a higher kinetic energy than the less sharply curved function.

Comment 1.11

We are using the term 'curvature' informally: the precise technical definition of the curvature of a function f is $(d^2f/dx^2)/\{1 + (df/dx)^2\}^{3/2}$.

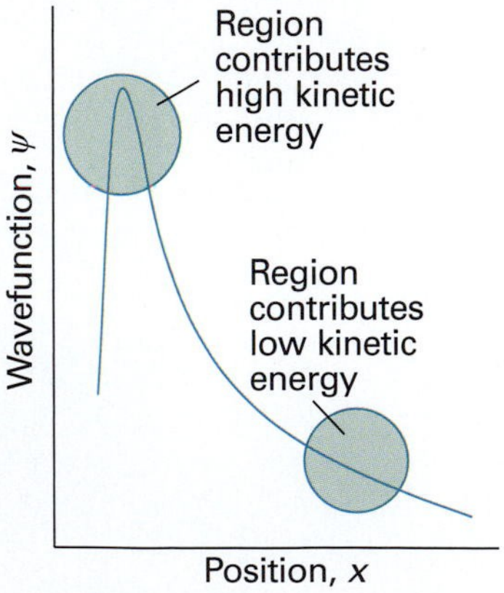

Fig. 1.27 The observed kinetic energy of a particle is an average of contributions from the entire space covered by the wavefunction. Sharply curved regions contribute a high kinetic energy to the average; slightly curved regions contribute only a small kinetic energy.

(c) Hermitian operators

All the quantum mechanical operators that correspond to observables have a very special mathematical property: they are 'hermitian'. An **hermitian operator** is one for which the following relation is true:

$$\text{Hermiticity:} \quad \int \psi_i^* \hat{\Omega} \psi_j \, dx = \left\{ \int \psi_j^* \hat{\Omega} \psi_i \, dx \right\}^* \quad [1.30]$$

It is easy to confirm that the position operator ($x\times$) is hermitian because we are free to change the order of the factors in the integrand:

$$\int_{-\infty}^{\infty} \psi_i^* x \psi_j \, dx = \int_{-\infty}^{\infty} \psi_j x \psi_i^* \, dx = \left\{ \int_{-\infty}^{\infty} \psi_j^* x \psi_i \, dx \right\}^*$$

The demonstration that the linear momentum operator is hermitian is more involved because we cannot just alter the order of functions we differentiate, but it is hermitian, as we show in the following *Justification*.

Justification 1.2 *The hermiticity of the linear momentum operator*

Our task is to show that

$$\int_{-\infty}^{\infty} \psi_i^* \hat{p}_x \psi_j \mathrm{d}x = \left\{ \int_{-\infty}^{\infty} \psi_j^* \hat{p}_x \psi_i \mathrm{d}x \right\}^*$$

with $\hat{p}_x$ given in eqn 1.26. To do so, we use 'integration by parts', the relation

$$\int f \frac{\mathrm{d}g}{\mathrm{d}x} \mathrm{d}x = fg - \int g \frac{\mathrm{d}f}{\mathrm{d}x} \mathrm{d}x$$

In the present case we write

$$\int_{-\infty}^{\infty} \psi_i^* \hat{p}_x \psi_j \mathrm{d}x = \frac{\hbar}{\mathrm{i}} \int_{-\infty}^{\infty} \psi_i^* \frac{\mathrm{d}\psi_j}{\mathrm{d}x} \mathrm{d}x$$

$$= \frac{\hbar}{\mathrm{i}} \psi_i^* \psi_j \Big|_{-\infty}^{\infty} - \frac{\hbar}{\mathrm{i}} \int_{-\infty}^{\infty} \psi_j \frac{\mathrm{d}\psi_i^*}{\mathrm{d}x} \mathrm{d}x$$

The first term on the right is zero, because all wavefunctions are zero at infinity in either direction, so we are left with

$$\int_{-\infty}^{\infty} \psi_i^* \hat{p}_x \psi_j \mathrm{d}x = -\frac{\hbar}{\mathrm{i}} \int_{-\infty}^{\infty} \psi_j \frac{\mathrm{d}\psi_i^*}{\mathrm{d}x} \mathrm{d}x = \left\{ \frac{\hbar}{\mathrm{i}} \int_{-\infty}^{\infty} \psi_j^* \frac{\mathrm{d}\psi_i}{\mathrm{d}x} \mathrm{d}x \right\}^*$$

$$= \left\{ \int_{-\infty}^{\infty} \psi_j^* \hat{p}_x \psi_i \mathrm{d}x \right\}^*$$

as we set out to prove.

Self-test 1.7 Confirm that the operator $\mathrm{d}^2/\mathrm{d}x^2$ is hermitian.

Hermitian operators are enormously important by virtue of two properties: their eigenvalues are real (as we prove in the *Justification* below), and their eigenfunctions are 'orthogonal'. All observables have real values (in the mathematical sense, such as $x = 2$ m and $E = 10$ J), so all observables are represented by hermitian operators. To say that two different functions ψ_i and ψ_j are **orthogonal** means that the integral (over all space) of their product is zero:

$$\text{Orthogonality:} \int \psi_i^* \psi_j \, \mathrm{d}\tau = 0 \qquad [1.31]$$

For example, the hamiltonian operator is hermitian (it corresponds to an observable, the energy). Therefore, if ψ_1 corresponds to one energy, and ψ_2 corresponds to a different energy, then we know at once that the two functions are orthogonal and that the integral of their product is zero.

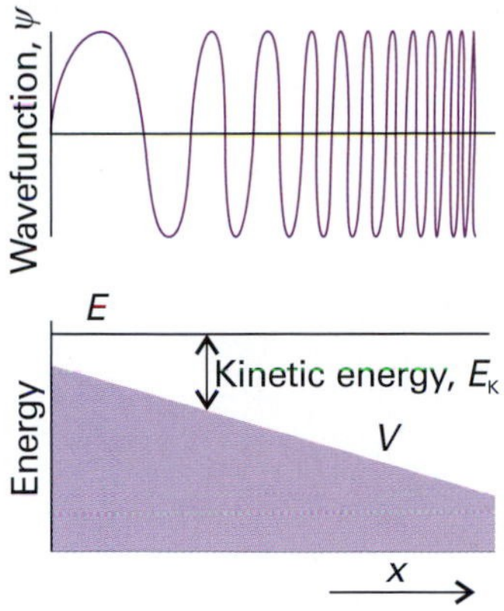

Fig. 1.28 The wavefunction of a particle in a potential decreasing towards the right and hence subjected to a constant force to the right. Only the real part of the wavefunction is shown, the imaginary part is similar, but displaced to the right.

Justification 1.3 *The reality of eigenvalues*

For a wavefunction ψ that is normalized and is an eigenfunction of an hermitian operator $\hat{\Omega}$ with eigenvalue ω, we can write

$$\int \psi^* \hat{\Omega} \psi \mathrm{d}\tau = \int \psi^* \omega \psi \mathrm{d}\tau = \omega \int \psi^* \psi \mathrm{d}\tau = \omega$$

However, by taking the complex conjugate we can write

$$\omega^* = \left\{ \int \psi^* \hat{\Omega} \psi \mathrm{d}\tau \right\}^* = \int \psi^* \hat{\Omega} \psi \mathrm{d}\tau = \omega$$

where in the second equality we have used the hermiticity of $\hat{\Omega}$. The conclusion that $\omega^* = \omega$ confirms that ω is real.

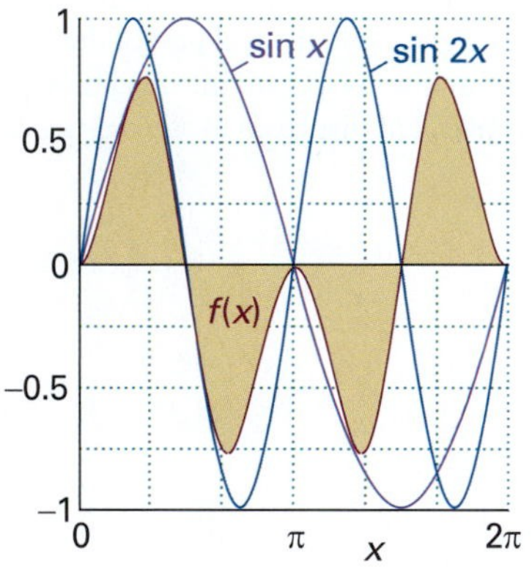

Fig. 1.29 The integral of the function $f(x) = \sin x \sin 2x$ is equal to the area (tinted) below the brown curve, and is zero, as can be inferred by symmetry. The function—and the value of the integral—repeats itself for all replications of the section between 0 and 2π, so the integral from $-\infty$ to ∞ is zero.

Illustration 1.2 *Confirming orthogonality*

The wavefunctions $\sin x$ and $\sin 2x$ are eigenfunctions of the hermitian operator $\mathrm{d}^2/\mathrm{d}x^2$, with eigenvalues -1 and -4, respectively. To verify that the two wavefunctions are mutually orthogonal, we integrate the product $(\sin x)(\sin 2x)$ over all space, which we may take to span from $x = 0$ to $x = 2\pi$, because both functions repeat themselves outside that range. Hence proving that the integral of their product is zero within that range implies that the integral over the whole of space is also integral (Fig. 1.29). A useful integral for this calculation is

$$\int \sin ax \sin bx \,\mathrm{d}x = \frac{\sin(a-b)x}{2(a-b)} - \frac{\sin(a+b)x}{2(a+b)} + \text{constant}, \qquad \text{if } a^2 \neq b^2$$

It follows that, for $a = 1$ and $b = 2$, and given the fact that $\sin 0 = 0$, $\sin 2\pi = 0$, and $\sin 6\pi = 0$,

$$\int_0^{2\pi} (\sin x)(\sin 2x)\mathrm{d}x = 0$$

and the two functions are mutually orthogonal.

Self-test 1.8 Confirm that the functions $\sin x$ and $\sin 3x$ are mutually orthogonal.

$$\left[\int_{-\infty}^{\infty} \sin x \sin 3x \,\mathrm{d}x = 0\right]$$

(d) Superpositions and expectation values

Suppose now that the wavefunction is the one given in eqn 1.19 (with $A = B$). What is the linear momentum of the particle it describes? We quickly run into trouble if we use the operator technique. When we operate with p_x, we find

$$\frac{\hbar}{\mathrm{i}} \frac{\mathrm{d}\psi}{\mathrm{d}x} = \frac{2\hbar}{\mathrm{i}} A \frac{\mathrm{d}\cos kx}{\mathrm{d}x} = -\frac{2k\hbar}{\mathrm{i}} A \sin kx \qquad (1.32)$$

This expression is not an eigenvalue equation, because the function on the right ($\sin kx$) is different from that on the left ($\cos kx$).

When the wavefunction of a particle is not an eigenfunction of an operator, the property to which the operator corresponds does not have a definite value. However, in the current example the momentum is not completely indefinite because the cosine wavefunction is a **linear combination**, or sum, of e^{ikx} and e^{-ikx}, and these two functions, as we have seen, individually correspond to definite momentum states. We say that the total wavefunction is a **superposition** of more than one wavefunction. Symbolically we can write the superposition as

$$\psi = \underbrace{\psi_{\rightarrow}}_{\substack{\text{Particle with}\\ \text{linear}\\ \text{momentum}\\ +k\hbar}} + \underbrace{\psi_{\leftarrow}}_{\substack{\text{Particle with}\\ \text{linear}\\ \text{momentum}\\ -k\hbar}}$$

The interpretation of this composite wavefunction is that, if the momentum of the particle is repeatedly measured in a long series of observations, then its magnitude will found to be $k\hbar$ in all the measurements (because that is the value for each component of the wavefunction). However, because the two component wavefunctions occur equally in the superposition, half the measurements will show that the particle is moving to the right ($p_x = +k\hbar$), and half the measurements will show that it is moving to the left ($p_x = -k\hbar$). According to quantum mechanics, we cannot predict in which direction the particle will in fact be found to be travelling; all we can say is that, in a long series of observations, if the particle is described by this wavefunction, then there are equal probabilities of finding the particle travelling to the right and to the left.

The same interpretation applies to any wavefunction written as a linear combination of eigenfunctions of an operator. Thus, suppose the wavefunction is known to be a superposition of many different linear momentum eigenfunctions and written as the linear combination

$$\psi = c_1\psi_1 + c_2\psi_2 + \cdots = \sum_k c_k\psi_k \qquad (1.33)$$

where the c_k are numerical (possibly complex) coefficients and the ψ_k correspond to different momentum states. The functions ψ_k are said to form a **complete set** in the sense that any arbitrary function can be expressed as a linear combination of them. Then according to quantum mechanics:

1 When the momentum is measured, in a single observation one of the eigenvalues corresponding to the ψ_k that contribute to the superposition will be found.

2 The probability of measuring a particular eigenvalue in a series of observations is proportional to the square modulus ($|c_k|^2$) of the corresponding coefficient in the linear combination.

3 The average value of a large number of observations is given by the expectation value, $\langle\Omega\rangle$, of the operator $\hat{\Omega}$ corresponding to the observable of interest.

The **expectation value** of an operator $\hat{\Omega}$ is defined as

$$\langle\Omega\rangle = \int \psi^*\hat{\Omega}\psi\,\mathrm{d}\tau \qquad [1.34]$$

This formula is valid only for normalized wavefunctions. As we see in the *Justification* below, an expectation value is the weighted average of a large number of observations of a property.

Comment 1.12

In general, a linear combination of two functions f and g is $c_1f + c_2g$, where c_1 and c_2 are numerical coefficients, so a linear combination is a more general term than 'sum'. In a sum, $c_1 = c_2 = 1$. A linear combination might have the form $0.567f + 1.234g$, for instance, so it is more general than the simple sum $f + g$.

Justification 1.4 *The expectation value of an operator*

If ψ is an eigenfunction of $\hat{\Omega}$ with eigenvalue ω, the expectation value of $\hat{\Omega}$ is

$$\langle\Omega\rangle = \int \psi^* \overbrace{\hat{\Omega}\psi}^{\omega\psi} \mathrm{d}\tau = \int \psi^* \omega\psi \mathrm{d}\tau = \omega \int \psi^* \psi \mathrm{d}\tau = \omega$$

because ω is a constant and may be taken outside the integral, and the resulting integral is equal to 1 for a normalized wavefunction. The interpretation of this expression is that, because every observation of the property Ω results in the value ω (because the wavefunction is an eigenfunction of $\hat{\Omega}$), the mean value of all the observations is also ω.

A wavefunction that is not an eigenfunction of the operator of interest can be written as a linear combination of eigenfunctions. For simplicity, suppose the wavefunction is the sum of two eigenfunctions (the general case, eqn 1.33, can easily be developed). Then

$$\begin{aligned}
\langle\Omega\rangle &= \int (c_1\psi_1 + c_2\psi_2)^* \hat{\Omega}(c_1\psi_1 + c_2\psi_2)\mathrm{d}\tau \\
&= \int (c_1\psi_1 + c_2\psi_2)^* (c_1\hat{\Omega}\psi_1 + c_2\hat{\Omega}\psi_2)\mathrm{d}\tau \\
&= \int (c_1\psi_1 + c_2\psi_2)^* (c_1\omega_1\psi_1 + c_2\omega_2\psi_2)\mathrm{d}\tau \\
&= c_1^*c_1\omega_1 \overbrace{\int \psi_1^*\psi_1\mathrm{d}\tau}^{1} + c_2^*c_2\omega_2 \overbrace{\int \psi_2^*\psi_2\mathrm{d}\tau}^{1} \\
&\quad + c_2^*c_1\omega_1 \overbrace{\int \psi_2^*\psi_1\mathrm{d}\tau}^{0} + c_1^*c_2\omega_2 \overbrace{\int \psi_1^*\psi_2\mathrm{d}\tau}^{0}
\end{aligned}$$

The first two integrals on the right are both equal to 1 because the wavefunctions are individually normalized. Because ψ_1 and ψ_2 correspond to different eigenvalues of an hermitian operator, they are orthogonal, so the third and fourth integrals on the right are zero. We can conclude that

$$\langle\Omega\rangle = |c_1|^2\omega_1 + |c_2|^2\omega_2$$

This expression shows that the expectation value is the sum of the two eigenvalues weighted by the probabilities that each one will be found in a series of measurements. Hence, the expectation value is the weighted mean of a series of observations.

Example 1.7 *Calculating an expectation value*

Calculate the average value of the distance of an electron from the nucleus in the hydrogen atom in its state of lowest energy.

Method The average radius is the expectation value of the operator corresponding to the distance from the nucleus, which is multiplication by r. To evaluate $\langle r\rangle$, we need to know the normalized wavefunction (from Example 1.4) and then evaluate the integral in eqn 1.34.

Answer The average value is given by the expectation value

$$\langle r\rangle = \int \psi^* r \psi \, d\tau$$

which we evaluate by using spherical polar coordinates. Using the normalized function in Example 1.4 gives

$$\langle r\rangle = \frac{1}{\pi a_0^3} \overbrace{\int_0^\infty r^3 e^{-2r/a_0} dr}^{3!a_0^3/2^4} \overbrace{\int_0^\pi \sin\theta \, d\theta}^{2} \overbrace{\int_0^{2\pi} d\phi}^{2\pi} = \tfrac{3}{2}a_0$$

Because $a_0 = 52.9$ pm (see Section 3.1), $\langle r\rangle = 79.4$ pm. This result means that, if a very large number of measurements of the distance of the electron from the nucleus are made, then their mean value will be 79.4 pm. However, each different observation will give a different and unpredictable individual result because the wavefunction is not an eigenfunction of the operator corresponding to r.

Self-test 1.9 Evaluate the root mean square distance, $\langle r^2\rangle^{1/2}$, of the electron from the nucleus in the hydrogen atom. [$3^{1/2}a_0 = 91.6$ pm]

The mean kinetic energy of a particle in one dimension is the expectation value of the operator given in eqn 1.28. Therefore, we can write

$$\langle E_K\rangle = \int \psi^* \hat{E}_K \psi \, d\tau = -\frac{\hbar^2}{2m}\int \psi^* \frac{d^2\psi}{dx^2} d\tau \qquad (1.35)$$

This conclusion confirms the previous assertion that the kinetic energy is a kind of average over the curvature of the wavefunction: we get a large contribution to the observed value from regions where the wavefunction is sharply curved (so $d^2\psi/dx^2$ is large) and the wavefunction itself is large (so that ψ^* is large too).

1.6 The uncertainty principle

We have seen that, if the wavefunction is Ae^{ikx}, then the particle it describes has a definite state of linear momentum, namely travelling to the right with momentum $p_x = +k\hbar$. However, we have also seen that the position of the particle described by this wavefunction is completely unpredictable. In other words, if the momentum is specified precisely, it is impossible to predict the location of the particle. This statement is one-half of a special case of the **Heisenberg uncertainty principle**, one of the most celebrated results of quantum mechanics:

> It is impossible to specify simultaneously, with arbitrary precision, both the momentum and the position of a particle.

Before discussing the principle further, we must establish its other half: that, if we know the position of a particle exactly, then we can say nothing about its momentum. The argument draws on the idea of regarding a wavefunction as a superposition of eigenfunctions, and runs as follows.

If we know that the particle is at a definite location, its wavefunction must be large there and zero everywhere else (Fig. 1.30). Such a wavefunction can be created by superimposing a large number of harmonic (sine and cosine) functions, or, equivalently, a number of e^{ikx} functions. In other words, we can create a sharply localized

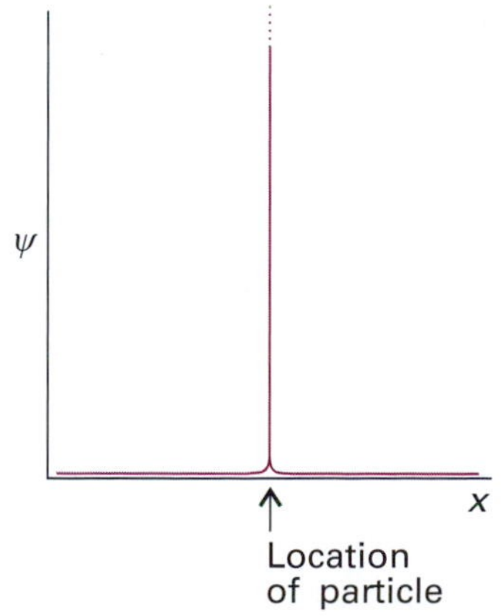

Fig. 1.30 The wavefunction for a particle at a well-defined location is a sharply spiked function that has zero amplitude everywhere except at the particle's position.

wavefunction, called a **wave packet**, by forming a linear combination of wavefunctions that correspond to many different linear momenta. The superposition of a few harmonic functions gives a wavefunction that spreads over a range of locations (Fig. 1.31). However, as the number of wavefunctions in the superposition increases, the wave packet becomes sharper on account of the more complete interference between the positive and negative regions of the individual waves. When an infinite number of components is used, the wave packet is a sharp, infinitely narrow spike, which corresponds to perfect localization of the particle. Now the particle is perfectly localized. However, we have lost all information about its momentum because, as we saw above, a measurement of the momentum will give a result corresponding to any one of the infinite number of waves in the superposition, and which one it will give is unpredictable. Hence, if we know the location of the particle precisely (implying that its wavefunction is a superposition of an infinite number of momentum eigenfunctions), then its momentum is completely unpredictable.

A quantitative version of this result is

$$\Delta p \Delta q \geq \tfrac{1}{2}\hbar \tag{1.36a}$$

In this expression Δp is the 'uncertainty' in the linear momentum parallel to the axis q, and Δq is the uncertainty in position along that axis. These 'uncertainties' are precisely defined, for they are the root mean square deviations of the properties from their mean values:

$$\Delta p = \{\langle p^2\rangle - \langle p\rangle^2\}^{1/2} \qquad \Delta q = \{\langle q^2\rangle - \langle q\rangle^2\}^{1/2} \tag{1.36b}$$

If there is complete certainty about the position of the particle ($\Delta q = 0$), then the only way that eqn 1.36a can be satisfied is for $\Delta p = \infty$, which implies complete uncertainty about the momentum. Conversely, if the momentum parallel to an axis is known exactly ($\Delta p = 0$), then the position along that axis must be completely uncertain ($\Delta q = \infty$).

The p and q that appear in eqn 1.36 refer to the same direction in space. Therefore, whereas simultaneous specification of the position on the x-axis and momentum parallel to the x-axis are restricted by the uncertainty relation, simultaneous location of position on x and motion parallel to y or z are not restricted. The restrictions that the uncertainty principle implies are summarized in Table 1.2.

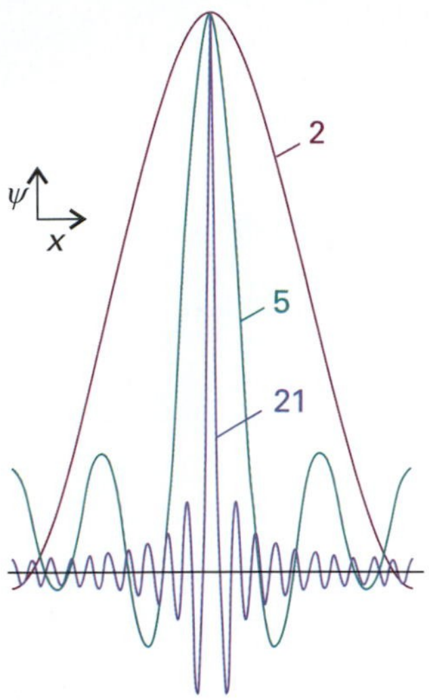

Fig. 1.31 The wavefunction for a particle with an ill-defined location can be regarded as the superposition of several wavefunctions of definite wavelength that interfere constructively in one place but destructively elsewhere. As more waves are used in the superposition (as given by the numbers attached to the curves), the location becomes more precise at the expense of uncertainty in the particle's momentum. An infinite number of waves is needed to construct the wavefunction of a perfectly localized particle.

Exploration Use mathematical software or an electronic spreadsheet to construct superpositions of cosine functions as $\psi(x) = \sum_{k=1}^{N} (1/N)\cos(k\pi x)$, where the constant $1/N$ is introduced to keep the superpositions with the same overall magnitude. Explore how the probability density $\psi^2(x)$ changes with the value of N.

Example 1.8 *Using the uncertainty principle*

Suppose the speed of a projectile of mass 1.0 g is known to within 1 μm s^{-1}. Calculate the minimum uncertainty in its position.

Method Estimate Δp from $m\Delta v$, where Δv is the uncertainty in the speed; then use eqn 1.36a to estimate the minimum uncertainty in position, Δq.

Answer The minimum uncertainty in position is

$$\Delta q = \frac{\hbar}{2m\Delta v}$$

$$= \frac{1.055 \times 10^{-34}\ \text{J s}}{2 \times (1.0 \times 10^{-3}\ \text{kg}) \times (1 \times 10^{-6}\ \text{m s}^{-1})} = 5 \times 10^{-26}\ \text{m}$$

where we have used 1 J = 1 kg m^2 s^{-2}. The uncertainty is completely negligible for all practical purposes concerning macroscopic objects. However, if the mass is that of an electron, then the same uncertainty in speed implies an uncertainty in position far larger than the diameter of an atom (the analogous calculation gives $\Delta q = 60$ m); so the concept of a trajectory, the simultaneous possession of a precise position and momentum, is untenable.

Self-test 1.10 Estimate the minimum uncertainty in the speed of an electron in a one-dimensional region of length $2a_0$. [$500\ \mathrm{km\ s^{-1}}$]

The Heisenberg uncertainty principle is more general than eqn 1.36 suggests. It applies to any pair of observables called **complementary observables**, which are defined in terms of the properties of their operators. Specifically, two observables Ω_1 and Ω_2 are complementary if

$$\hat{\Omega}_1(\hat{\Omega}_2\psi) \neq \hat{\Omega}_2(\hat{\Omega}_1\psi) \tag{1.37}$$

When the effect of two operators depends on their order (as this equation implies), we say that they do not **commute**. The different outcomes of the effect of applying $\hat{\Omega}_1$ and $\hat{\Omega}_2$ in a different order are expressed by introducing the **commutator** of the two operators, which is defined as

$$[\hat{\Omega}_1, \hat{\Omega}_2] = \hat{\Omega}_1\hat{\Omega}_2 - \hat{\Omega}_2\hat{\Omega}_1 \qquad [1.38]$$

We can conclude from *Illustration 1.3* that the commutator of the operators for position and linear momentum is

$$[\hat{x}, \hat{p}_x] = \mathrm{i}\hbar \tag{1.39}$$

Illustration 1.3 *Evaluating a commutator*

To show that the operators for position and momentum do not commute (and hence are complementary observables) we consider the effect of $\hat{x}\hat{p}_x$ (that is, the effect of $\hat{p}_x$ followed by the effect on the outcome of multiplication by x) on a wavefunction ψ:

$$\hat{x}\hat{p}_x\psi = x \times \frac{\hbar}{\mathrm{i}}\frac{\mathrm{d}\psi}{\mathrm{d}x}$$

Next, we consider the effect of $\hat{p}_x\hat{x}$ on the same function (that is, the effect of multiplication by x followed by the effect of $\hat{p}_x$ on the outcome):

$$\hat{p}_x\hat{x}\psi = \frac{\hbar}{\mathrm{i}}\frac{\mathrm{d}}{\mathrm{d}x}x\psi = \frac{\hbar}{\mathrm{i}}\left(\psi + x\frac{\mathrm{d}\psi}{\mathrm{d}x}\right)$$

For this step we have used the standard rule about differentiating a product of functions. The second expression is clearly different from the first, so the two operators do not commute. Subtraction of the second expression from the first gives eqn 1.39.

The commutator in eqn 1.39 is of such vital significance in quantum mechanics that it is taken as a fundamental distinction between classical mechanics and quantum mechanics. In fact, this commutator may be taken as a postulate of quantum mechanics, and is used to justify the choice of the operators for position and linear momentum given in eqn 1.26.

Table 1.2* Constraints of the uncertainty principle

	Variable 1					
Variable 2	x	y	z	p_x	p_y	p_z
x				□		
y					□	
z						□
p_x	□					
p_y		□				
p_z			□			

* Pairs of observables that cannot be determined simultaneously with arbitrary precision are marked with a white rectangle; all others are unrestricted.

Comment 1.13
For two functions f and g, $\mathrm{d}(fg) = f\mathrm{d}g + g\mathrm{d}f$.

Comment 1.14

The 'modulus' notation $|\ldots|$ means take the magnitude of the term the bars enclose: for a real quantity x, $|x|$ is the magnitude of x (its value without its sign); for an imaginary quantity iy, $|\mathrm{i}y|$ is the magnitude of y; and—most generally—for a complex quantity $z = x + \mathrm{i}y$, $|z|$ is the value of $(z^*z)^{1/2}$. For example, $|-2| = 2$, $|3\mathrm{i}| = 3$, and $|-2+3\mathrm{i}| = \{(-2-3\mathrm{i})(-2+3\mathrm{i})\}^{1/2} = 13^{1/2}$. Physically, the modulus on the right of eqn 1.40 ensures that the product of uncertainties has a real, non-negative value.

With the concept of commutator established, the Heisenberg uncertainty principle can be given its most general form. For *any* two pairs of observables, Ω_1 and Ω_2, the uncertainties (to be precise, the root mean square deviations of their values from the mean) in simultaneous determinations are related by

$$\Delta\Omega_1\Delta\Omega_2 \geq \tfrac{1}{2}|\langle[\hat{\Omega}_1,\hat{\Omega}_2]\rangle| \tag{1.40}$$

We obtain the special case of eqn 1.36a when we identify the observables with x and p_x and use eqn 1.39 for their commutator.

Complementary observables are observables with non-commuting operators. With the discovery that some pairs of observables are complementary (we meet more examples in the next chapter), we are at the heart of the difference between classical and quantum mechanics. Classical mechanics supposed, falsely as we now know, that the position and momentum of a particle could be specified simultaneously with arbitrary precision. However, quantum mechanics shows that position and momentum are complementary, and that we have to make a choice: we can specify position at the expense of momentum, or momentum at the expense of position.

The realization that some observables are complementary allows us to make considerable progress with the calculation of atomic and molecular properties, but it does away with some of the most cherished concepts of classical physics.

1.7 The postulates of quantum mechanics

For convenience, we collect here the postulates on which quantum mechanics is based and which have been introduced in the course of this chapter.

The wavefunction. All dynamical information is contained in the wavefunction ψ for the system, which is a mathematical function found by solving the Schrödinger equation for the system. In one dimension:

$$-\frac{\hbar^2}{2m}\frac{\mathrm{d}^2\psi}{\mathrm{d}x^2} + V(x)\psi = E\psi$$

The Born interpretation. If the wavefunction of a particle has the value ψ at some point $\boldsymbol{r}$, then the probability of finding the particle in an infinitesimal volume $\mathrm{d}\tau = \mathrm{d}x\mathrm{d}y\mathrm{d}z$ at that point is proportional to $|\psi|^2\mathrm{d}\tau$.

Acceptable wavefunctions. An acceptable wavefunction must be continuous, have a continuous first derivative, be single-valued, and be square-integrable.

Observables. Observables, Ω, are represented by operators, $\hat{\Omega}$, built from position and momentum operators of the form

$$\hat{x} = x\times \qquad \hat{p}_x = \frac{\hbar}{\mathrm{i}}\frac{\mathrm{d}}{\mathrm{d}x}$$

or, more generally, from operators that satisfy the commutation relation $[\hat{x}, \hat{p}_x] = \mathrm{i}\hbar$.

The Heisenberg uncertainty principle. It is impossible to specify simultaneously, with arbitrary precision, both the momentum and the position of a particle and, more generally, any pair of observable with operators that do not commute.

Checklist of key ideas

- ☐ 1. In classical physics, radiation is described in terms of an oscillating electromagnetic disturbance that travels through vacuum at a constant speed $c = \lambda\nu$.
- ☐ 2. A black body is an object that emits and absorbs all frequencies of radiation uniformly.
- ☐ 3. The variation of the energy output of a black body with wavelength is explained by invoking quantization of energy, the limitation of energies to discrete values, which in turn leads to the Planck distribution, eqn 1.5.
- ☐ 4. The variation of the molar heat capacity of a solid with temperature is explained by invoking quantization of energy, which leads to the Einstein and Debye formulas, eqns 1.7 and 1.9.
- ☐ 5. Spectroscopic transitions are changes in populations of quantized energy levels of a system involving the absorption, emission, or scattering of electromagnetic radiation, $\Delta E = h\nu$.
- ☐ 6. The photoelectric effect is the ejection of electrons from metals when they are exposed to ultraviolet radiation: $\frac{1}{2}m_e v^2 = h\nu - \Phi$, where Φ is the work function, the energy required to remove an electron from the metal to infinity.
- ☐ 7. The photoelectric effect and electron diffraction are phenomena that confirm wave–particle duality, the joint particle and wave character of matter and radiation.
- ☐ 8. The de Broglie relation, $\lambda = h/p$, relates the momentum of a particle with its wavelength.
- ☐ 9. A wavefunction is a mathematical function obtained by solving the Schrödinger equation and which contains all the dynamical information about a system.
- ☐ 10. The time-independent Schrödinger equation in one dimension is $-(\hbar^2/2m)(d^2\psi/dx^2) + V(x)\psi = E\psi$.
- ☐ 11. The Born interpretation of the wavefunction states that the value of $|\psi|^2$, the probability density, at a point is proportional to the probability of finding the particle at that point.
- ☐ 12. Quantization is the confinement of a dynamical observable to discrete values.
- ☐ 13. An acceptable wavefunction must be continuous, have a continuous first derivative, be single-valued, and be square-integrable.
- ☐ 14. An operator is something that carries out a mathematical operation on a function. The position and momentum operators are $\hat{x} = x\times$ and $\hat{p}_x = (\hbar/i)d/dx$, respectively.
- ☐ 15. The hamiltonian operator is the operator for the total energy of a system, $\hat{H}\psi = E\psi$ and is the sum of the operators for kinetic energy and potential energy.
- ☐ 16. An eigenvalue equation is an equation of the form $\hat{\Omega}\psi = \omega\psi$. The eigenvalue is the constant ω in the eigenvalue equation; the eigenfunction is the function ψ in the eigenvalue equation.
- ☐ 17. The expectation value of an operator is $\langle\Omega\rangle = \int\psi^*\hat{\Omega}\psi d\tau$.
- ☐ 18. An hermitian operator is one for which $\int\psi_i^*\hat{\Omega}\psi_j dx = (\int\psi_j^*\hat{\Omega}\psi_i dx)^*$. The eigenvalues of hermitian operators are real and correspond to observables, measurable properties of a system. The eigenfunctions of hermitian operations are orthogonal, meaning that $\int\psi_i^*\psi_j d\tau = 0$.
- ☐ 19. The Heisenberg uncertainty principle states that it is impossible to specify simultaneously, with arbitrary precision, both the momentum and the position of a particle; $\Delta p\Delta q \geq \frac{1}{2}\hbar$.
- ☐ 20. Two operators commute when $[\hat{\Omega}_1, \hat{\Omega}_2] = \hat{\Omega}_1\hat{\Omega}_2 - \hat{\Omega}_2\hat{\Omega}_1 = 0$.
- ☐ 21. Complementary observables are observables corresponding to non-commuting operators.
- ☐ 22. The general form of the Heisenberg uncertainty principle is $\Delta\Omega_1\Delta\Omega_2 \geq \frac{1}{2}|\langle[\hat{\Omega}_1, \hat{\Omega}_2]\rangle|$.

Further reading

Articles and texts

P.W. Atkins, *Quanta: A handbook of concepts*. Oxford University Press (1991).

P.W. Atkins and R.S. Friedman, *Molecular quantum mechanics*. Oxford University Press (2005).

D. Bohm, *Quantum theory*. Dover, New York (1989).

R.P. Feynman, R.B. Leighton, and M. Sands, *The Feynman lectures on physics*. Volume III. Addison–Wesley, Reading (1965).

C.S. Johnson, Jr. and L.G. Pedersen, *Problems and solutions in quantum chemistry and physics*. Dover, New York, 1986.

L. Pauling and E.B. Wilson, *Introduction to quantum mechanics with applications to chemistry*. Dover, New York (1985).

Each end-of-chapter discussion question, exercise, and problem for which a solution is available in the Student Solution Manual has a corresponding bracketed number referring to the Student Solutions Manual number for this problem, for example [SSM 8.1]. Each end-of-chapter exercise and problem for which a solution is available in the Instructors' Solution Manual has a corresponding bracketed number referring to the Instructor Solutions Manual number for this problem, for example [ISM 8.2].

Discussion questions

1.1 Summarize the evidence that led to the introduction of quantum mechanics. [SSM **8.1**]

1.2 Explain why Planck's introduction of quantization accounted for the properties of black-body radiation. [ISM **8.2**]

1.3 Explain why Einstein's introduction of quantization accounted for the properties of heat capacities at low temperatures. [SSM **8.3**]

1.4 Describe how a wavefunction determines the dynamical properties of a system and how those properties may be predicted. [ISM **8.4**]

1.5 Account for the uncertainty relation between position and linear momentum in terms of the shape of the wavefunction. [SSM **8.5**]

1.6 Suggest how the general shape of a wavefunction can be predicted without solving the Schrödinger equation explicitly. [ISM **8.6**]

Exercises

1.1a To what speed must an electron be accelerated for it to have a wavelength of 3.0 cm? [ISM **8.1a**]

1.1b To what speed must a proton be accelerated for it to have a wavelength of 3.0 cm? [SSM **8.1b**]

1.2a The fine-structure constant, α, plays a special role in the structure of matter; its approximate value is 1/137. What is the wavelength of an electron travelling at a speed αc, where c is the speed of light? (Note that the circumference of the first Bohr orbit in the hydrogen atom is 331 pm.) [ISM **8.2a**]

1.2b Calculate the linear momentum of photons of wavelength 350 nm. What speed does a hydrogen molecule need to travel to have the same linear momentum? [SSM **8.2b**]

1.3a The speed of a certain proton is 0.45 Mm s^{-1}. If the uncertainty in its momentum is to be reduced to 0.0100 per cent, what uncertainty in its location must be tolerated? [ISM **8.3a**]

1.3b The speed of a certain electron is 995 km s^{-1}. If the uncertainty in its momentum is to be reduced to 0.0010 per cent, what uncertainty in its location must be tolerated? [SSM **8.3b**]

1.4a Calculate the energy per photon and the energy per mole of photons for radiation of wavelength (a) 600 nm (red), (b) 550 nm (yellow), (c) 400 nm (blue). [ISM **8.4a**]

1.4b Calculate the energy per photon and the energy per mole of photons for radiation of wavelength (a) 200 nm (ultraviolet), (b) 150 pm (X-ray), (c) 1.00 cm (microwave). [SSM **8.4b**]

1.5a Calculate the speed to which a stationary H atom would be accelerated if it absorbed each of the photons used in Exercise 1.4a. [ISM **8.5a**]

1.5b Calculate the speed to which a stationary ^{4}He atom (mass 4.0026 u) would be accelerated if it absorbed each of the photons used in Exercise 1.4b. [SSM **8.5b**]

1.6a A glow-worm of mass 5.0 g emits red light (650 nm) with a power of 0.10 W entirely in the backward direction. To what speed will it have accelerated after 10 y if released into free space and assumed to live? [ISM **8.6a**]

1.6b A photon-powered spacecraft of mass 10.0 kg emits radiation of wavelength 225 nm with a power of 1.50 kW entirely in the backward direction. To what speed will it have accelerated after 10.0 y if released into free space? [SSM **8.6b**]

1.7a A sodium lamp emits yellow light (550 nm). How many photons does it emit each second if its power is (a) 1.0 W, (b) 100 W? [ISM **8.7a**]

1.7b A laser used to read CDs emits red light of wavelength 700 nm. How many photons does it emit each second if its power is (a) 0.10 W, (b) 1.0 W? [SSM **8.7b**]

1.8a The work function for metallic caesium is 2.14 eV. Calculate the kinetic energy and the speed of the electrons ejected by light of wavelength (a) 700 nm, (b) 300 nm. [ISM **8.8a**]

1.8b The work function for metallic rubidium is 2.09 eV. Calculate the kinetic energy and the speed of the electrons ejected by light of wavelength (a) 650 nm, (b) 195 nm. [SSM **8.8b**]

1.9a Calculate the size of the quantum involved in the excitation of (a) an electronic oscillation of period 1.0 fs, (b) a molecular vibration of period 10 fs, (c) a pendulum of period 1.0 s. Express the results in joules and kilojoules per mole. [ISM **8.9a**]

1.9b Calculate the size of the quantum involved in the excitation of (a) an electronic oscillation of period 2.50 fs, (b) a molecular vibration of period 2.21 fs, (c) a balance wheel of period 1.0 ms. Express the results in joules and kilojoules per mole. [SSM **8.9b**]

1.10a Calculate the de Broglie wavelength of (a) a mass of 1.0 g travelling at 1.0 cm s^{-1}, (b) the same, travelling at 100 km s^{-1}, (c) an He atom travelling at 1000 m s^{-1} (a typical speed at room temperature). [ISM **8.10a**]

1.10b Calculate the de Broglie wavelength of an electron accelerated from rest through a potential difference of (a) 100 V, (b) 1.0 kV, (c) 100 kV. [SSM **8.10b**]

1.11a Confirm that the operator $\hat{l}_z = (\hbar/\mathrm{i})\mathrm{d}/\mathrm{d}\phi$, where ϕ is an angle, is hermitian. [ISM **8.11a**]

1.11b Show that the linear combinations $\hat{A} + \mathrm{i}\hat{B}$ and $\hat{A} - \mathrm{i}\hat{B}$ are not hermitian if $\hat{A}$ and $\hat{B}$ are hermitian operators. [SSM **8.11b**]

1.12a Calculate the minimum uncertainty in the speed of a ball of mass 500 g that is known to be within 1.0 μm of a certain point on a bat. What is the minimum uncertainty in the position of a bullet of mass 5.0 g that is known to have a speed somewhere between 350.000 01 m s^{-1} and 350.000 00 m s^{-1}? [ISM **8.12a**]

1.12b An electron is confined to a linear region with a length of the same order as the diameter of an atom (about 100 pm). Calculate the minimum uncertainties in its position and speed. [SSM **8.12b**]

1.13a In an X-ray photoelectron experiment, a photon of wavelength 150 pm ejects an electron from the inner shell of an atom and it emerges with a speed of 21.4 Mm s^{-1}. Calculate the binding energy of the electron. [ISM **8.13a**]

1.13b In an X-ray photoelectron experiment, a photon of wavelength 121 pm ejects an electron from the inner shell of an atom and it emerges with a speed of 56.9 Mm s^{-1}. Calculate the binding energy of the electron. [SSM **8.13b**]

1.14a Determine the commutators of the operators (a) $\mathrm{d}/\mathrm{d}x$ and $1/x$, (b) $\mathrm{d}/\mathrm{d}x$ and x^2. [ISM **8.14a**]

1.14b Determine the commutators of the operators a and $a^\dagger$, where $a = (\hat{x} + \mathrm{i}\hat{p})/2^{1/2}$ and $a^\dagger = (\hat{x} - \mathrm{i}\hat{p})/2^{1/2}$. [SSM **8.14b**]

Problems*

Numerical problems

1.1 The Planck distribution gives the energy in the wavelength range $d\lambda$ at the wavelength λ. Calculate the energy density in the range 650 nm to 655 nm inside a cavity of volume 100 cm^3 when its temperature is (a) 25°C, (b) 3000°C. [SSM **8.1**]

1.2 For a black body, the temperature and the wavelength of emission maximum, λ_{max}, are related by Wien's law, $\lambda_{max}T = \frac{1}{5}c_2$, where $c_2 = hc/k$ (see Problem 8.10). Values of λ_{max} from a small pinhole in an electrically heated container were determined at a series of temperatures, and the results are given below. Deduce a value for Planck's constant.

θ/°C	1000	1500	2000	2500	3000	3500
λ_{max}/nm	2181	1600	1240	1035	878	763

[ISM **8.2**]

1.3 The Einstein frequency is often expressed in terms of an equivalent temperature θ_E, where $\theta_E = h\nu/k$. Confirm that θ_E has the dimensions of temperature, and express the criterion for the validity of the high-temperature form of the Einstein equation in terms of it. Evaluate θ_E for (a) diamond, for which $\nu = 46.5$ THz and (b) for copper, for which $\nu = 7.15$ THz. What fraction of the Dulong and Petit value of the heat capacity does each substance reach at 25°C? [SSM **8.3**]

1.4 The ground-state wavefunction for a particle confined to a one-dimensional box of length L is

$$\psi = \left(\frac{2}{L}\right)^{1/2} \sin\left(\frac{\pi x}{L}\right)$$

Suppose the box is 10.0 nm long. Calculate the probability that the particle is (a) between $x = 4.95$ nm and 5.05 nm, (b) between $x = 1.95$ nm and 2.05 nm, (c) between $x = 9.90$ nm and 10.00 nm, (d) in the right half of the box, (e) in the central third of the box. [ISM **8.4**]

1.5 The ground-state wavefunction of a hydrogen atom is

$$\psi = \left(\frac{1}{\pi a_0^3}\right)^{1/2} e^{-r/a_0}$$

where $a_0 = 53$ pm (the Bohr radius). (a) Calculate the probability that the electron will be found somewhere within a small sphere of radius 1.0 pm centred on the nucleus. (b) Now suppose that the same sphere is located at $r = a_0$. What is the probability that the electron is inside it? [SSM **8.5**]

1.6 The normalized wavefunctions for a particle confined to move on a circle are $\psi(\phi) = (1/2\pi)^{1/2}e^{-im\phi}$, where $m = 0, \pm1, \pm2, \pm3, \ldots$ and $0 \le \phi \le 2\pi$. Determine $\langle\phi\rangle$. [ISM **8.6**]

1.7 A particle is in a state described by the wavefunction $\psi(x) = (2a/\pi)^{1/4}e^{-ax^2}$, where a is a constant and $-\infty \le x \le \infty$. Verify that the value of the product $\Delta p\Delta x$ is consistent with the predictions from the uncertainty principle. [SSM **8.7**]

1.8 A particle is in a state described by the wavefunction $\psi(x) = a^{1/2}e^{-ax}$, where a is a constant and $0 \le x \le \infty$. Determine the expectation value of the commutator of the position and momentum operators. [ISM **8.8**]

Theoretical problems

1.9 Demonstrate that the Planck distribution reduces to the Rayleigh–Jeans law at long wavelengths. [SSM **8.9**]

1.10 Derive *Wien's law*, that $\lambda_{max}T$ is a constant, where λ_{max} is the wavelength corresponding to maximum in the Planck distribution at the temperature T, and deduce an expression for the constant as a multiple of the *second radiation constant*, $c_2 = hc/k$. [ISM **8.10**]

1.11 Use the Planck distribution to deduce the *Stefan–Boltzmann law* that the total energy density of black-body radiation is proportional to T^4, and find the constant of proportionality. [SSM **8.11**]

1.12‡ Prior to Planck's derivation of the distribution law for black-body radiation, Wien found empirically a closely related distribution function which is very nearly but not exactly in agreement with the experimental results, namely, $\rho = (a/\lambda^5)e^{-b/\lambda kT}$. This formula shows small deviations from Planck's at long wavelengths. (a) By fitting Wien's empirical formula to Planck's at short wavelengths determine the constants a and b. (b) Demonstrate that Wien's formula is consistent with Wien's law (Problem 1.10) and with the Stefan–Boltzmann law (Problem 1.11). [ISM **8.12**]

1.13 Normalize the following wavefunctions: (a) $\sin(n\pi x/L)$ in the range $0 \le x \le L$, where $n = 1, 2, 3, \ldots$, (b) a constant in the range $-L \le x \le L$, (c) $e^{-r/a}$ in three-dimensional space, (d) $xe^{-r/2a}$ in three-dimensional space. *Hint*: The volume element in three dimensions is $d\tau = r^2dr \sin\theta\, d\theta\, d\phi$, with $0 \le r < \infty$, $0 \le \theta \le \pi$, $0 \le \phi \le 2\pi$. Use the integral in Example 1.4. [SSM **8.13**]

1.14 (a) Two (unnormalized) excited state wavefunctions of the H atom are

$$\text{(i) } \psi = \left(2 - \frac{r}{a_0}\right)e^{-r/a_0} \qquad \text{(ii) } \psi = r\sin\theta\cos\phi\, e^{-r/2a_0}$$

Normalize both functions to 1. (b) Confirm that these two functions are mutually orthogonal. [ISM **8.14**]

1.15 Identify which of the following functions are eigenfunctions of the operator d/dx: (a) e^{ikx}, (b) $\cos kx$, (c) k, (d) kx, (e) $e^{-\alpha x^2}$. Give the corresponding eigenvalue where appropriate. [SSM **8.15**]

1.16 Determine which of the following functions are eigenfunctions of the inversion operator $\hat{\imath}$ (which has the effect of making the replacement $x \to -x$): (a) $x^3 - kx$, (b) $\cos kx$, (c) $x^2 + 3x - 1$. State the eigenvalue of $\hat{\imath}$ when relevant. [ISM **8.16**]

1.17 Which of the functions in Problem 1.15 are (a) also eigenfunctions of d^2/dx^2 and (b) only eigenfunctions of d^2/dx^2? Give the eigenvalues where appropriate. [SSM **8.17**]

1.18 A particle is in a state described by the wavefunction $\psi = (\cos\chi)e^{ikx} + (\sin\chi)e^{-ikx}$, where χ (chi) is a parameter. What is the probability that the particle will be found with a linear momentum (a) $+k\hbar$, (b) $-k\hbar$? What form would the wavefunction have if it were 90 per cent certain that the particle had linear momentum $+k\hbar$? [ISM **8.18**]

1.19 Evaluate the kinetic energy of the particle with wavefunction given in Problem 1.18. [SSM **8.19**]

* Problems denoted with the symbol ‡ were supplied by Charles Trapp, Carmen Giunta, and Marshall Cady.

1.20 Calculate the average linear momentum of a particle described by the following wavefunctions: (a) e^{ikx}, (b) $\cos kx$, (c) $e^{-\alpha x^2}$, where in each one x ranges from $-\infty$ to $+\infty$. [ISM **8.20**]

1.21 Evaluate the expectation values of r and r^2 for a hydrogen atom with wavefunctions given in Problem 1.14. [SSM **8.21**]

1.22 Calculate (a) the mean potential energy and (b) the mean kinetic energy of an electron in the ground state of a hydrogenic atom. [ISM **8.22**]

1.23 Use mathematical software to construct superpositions of cosine functions and determine the probability that a given momentum will be observed. If you plot the superposition (which you should), set $x = 0$ at the centre of the screen and build the superposition there. Evaluate the root mean square location of the packet, $\langle x^2\rangle^{1/2}$. [SSM **8.23**]

1.24 Show that the expectation value of an operator that can be written as the square of an hermitian operator is positive. [ISM **8.24**]

1.25 (a) Given that any operators used to represent observables must satisfy the commutation relation in eqn 1.38, what would be the operator for position if the choice had been made to represent linear momentum parallel to the x-axis by multiplication by the linear momentum? These different choices are all valid 'representations' of quantum mechanics. (b) With the identification of $\hat{x}$ in this representation, what would be the operator for $1/x$? *Hint.* Think of $1/x$ as x^{-1}. [SSM **8.25**]

Applications: to biology, environmental science, and astrophysics

1.26‡ The temperature of the Sun's surface is approximately 5800 K. On the assumption that the human eye evolved to be most sensitive at the wavelength of light corresponding to the maximum in the Sun's radiant energy distribution, determine the colour of light to which the eye is most sensitive. *Hint*: See Problem 1.10. [ISM **8.26**]

1.27 We saw in *Impact I1.1* that electron microscopes can obtain images with several hundredfold higher resolution than optical microscopes because of the short wavelength obtainable from a beam of electrons. For electrons moving at speeds close to c, the speed of light, the expression for the de Broglie wavelength (eqn 1.12) needs to be corrected for relativistic effects:

$$\lambda = \frac{h}{\left\{2m_e eV\left(1+\dfrac{eV}{2m_e c^2}\right)\right\}^{1/2}}$$

where c is the speed of light in vacuum and V is the potential difference through which the electrons are accelerated. (a) Use the expression above to calculate the de Broglie wavelength of electrons accelerated through 50 kV. (b) Is the relativistic correction important? [SSM **8.27**]

1.28‡ Solar energy strikes the top of the Earth's atmosphere at a rate of 343 W m^{-2}. About 30 per cent of this energy is reflected directly back into space by the Earth or the atmosphere. The Earth–atmosphere system absorbs the remaining energy and re-radiates it into space as black-body radiation. What is the average black-body temperature of the Earth? What is the wavelength of the most plentiful of the Earth's black-body radiation? *Hint.* Use Wien's law, Problem 1.10. [ISM **8.28**]

1.29‡ A star too small and cold to shine has been found by S. Kulkarni, K. Matthews, B.R. Oppenheimer, and T. Nakajima (*Science* **270**, 1478 (1995)). The spectrum of the object shows the presence of methane, which, according to the authors, would not exist at temperatures much above 1000 K. The mass of the star, as determined from its gravitational effect on a companion star, is roughly 20 times the mass of Jupiter. The star is considered to be a brown dwarf, the coolest ever found. (a) From available thermodynamic data, test the stability of methane at temperatures above 1000 K. (b) What is λ_{max} for this star? (c) What is the energy density of the star relative to that of the Sun (6000 K)? (d) To determine whether the star will shine, estimate the fraction of the energy density of the star in the visible region of the spectrum. [SSM **8.29**]

Quantum theory: techniques and applications

2

To find the properties of systems according to quantum mechanics we need to solve the appropriate Schrödinger equation. This chapter presents the essentials of the solutions for three basic types of motion: translation, vibration, and rotation. We shall see that only certain wavefunctions and their corresponding energies are acceptable. Hence, quantization emerges as a natural consequence of the equation and the conditions imposed on it. The solutions bring to light a number of highly nonclassical, and therefore surprising, features of particles, especially their ability to tunnel into and through regions where classical physics would forbid them to be found. We also encounter a property of the electron, its spin, that has no classical counterpart. The chapter concludes with an introduction to the experimental techniques used to probe the quantization of energy in molecules.

The three basic modes of motion—translation (motion through space), vibration, and rotation—all play an important role in chemistry because they are ways in which molecules store energy. Gas-phase molecules, for instance, undergo translational motion and their kinetic energy is a contribution to the total internal energy of a sample. Molecules can also store energy as rotational kinetic energy and transitions between their rotational energy states can be observed spectroscopically. Energy is also stored as molecular vibration and transitions between vibrational states are responsible for the appearance of infrared and Raman spectra.

Translational motion

Section 1.5 introduced the quantum mechanical description of free motion in one dimension. We saw there that the Schrödinger equation is

$$-\frac{\hbar^2}{2m}\frac{\mathrm{d}^2\psi}{\mathrm{d}x^2}=E\psi \tag{2.1a}$$

or more succinctly

$$\hat{H}\psi=E\psi \qquad \hat{H}=-\frac{\hbar^2}{2m}\frac{\mathrm{d}^2}{\mathrm{d}x^2} \tag{2.1b}$$

The general solutions of eqn 2.1 are

$$\psi_k=A\mathrm{e}^{\mathrm{i}kx}+B\mathrm{e}^{-\mathrm{i}kx} \qquad E_k=\frac{k^2\hbar^2}{2m} \tag{2.2}$$

Note that we are now labelling both the wavefunctions and the energies (that is, the eigenfunctions and eigenvalues of $\hat{H}$) with the index k. We can verify that these functions are solutions by substituting ψ_k into the left-hand side of eqn 2.1a and showing that the result is equal to $E_k\psi_k$. In this case, all values of k, and therefore all values of the energy, are permitted. It follows that the translational energy of a free particle is not quantized.

We saw in Section 1.5b that a wavefunction of the form e^{ikx} describes a particle with linear momentum $p_x = +k\hbar$, corresponding to motion towards positive x (to the right), and that a wavefunction of the form e^{-ikx} describes a particle with the same magnitude of linear momentum but travelling towards negative x (to the left). That is, e^{ikx} is an eigenfunction of the operator $\hat{p}_x$ with eigenvalue $+k\hbar$, and e^{-ikx} is an eigenfunction with eigenvalue $-k\hbar$. In either state, $|\psi|^2$ is independent of x, which implies that the position of the particle is completely unpredictable. This conclusion is consistent with the uncertainty principle because, if the momentum is certain, then the position cannot be specified (the operators $\hat{x}$ and $\hat{p}_x$ do not commute, Section 8.6).

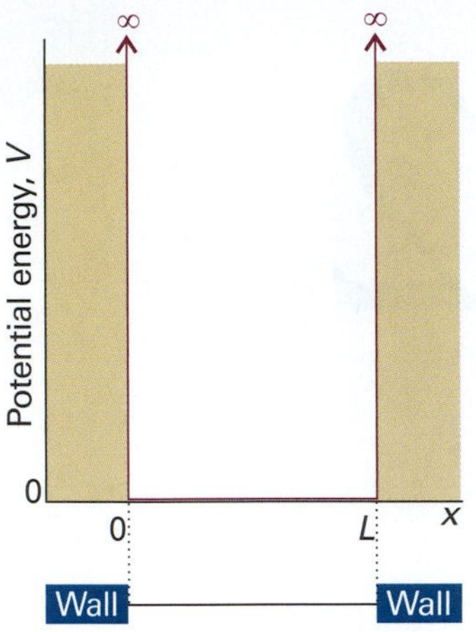

Fig. 2.1 A particle in a one-dimensional region with impenetrable walls. Its potential energy is zero between $x = 0$ and $x = L$, and rises abruptly to infinity as soon as it touches the walls.

2.1 A particle in a box

In this section, we consider a **particle in a box**, in which a particle of mass m is confined between two walls at $x = 0$ and $x = L$: the potential energy is zero inside the box but rises abruptly to infinity at the walls (Fig. 2.1). This model is an idealization of the potential energy of a gas-phase molecule that is free to move in a one-dimensional container. However, it is also the basis of the treatment of the electronic structure of metals and of a primitive treatment of conjugated molecules. The particle in a box is also used in statistical thermodynamics in assessing the contribution of the translational motion of molecules to their thermodynamic properties (Chapter 9).

(a) The acceptable solutions

The Schrödinger equation for the region between the walls (where $V = 0$) is the same as for a free particle (eqn 2.1), so the general solutions given in eqn 2.2 are also the same. However, we can us $e^{\pm ix} = \cos x \pm i \sin x$ to write

$$\begin{aligned}\psi_k &= Ae^{ikx} + Be^{-ikx} = A(\cos kx + i\sin kx) + B(\cos kx - i\sin kx)\\ &= (A+B)\cos kx + (A-B)i\sin kx\end{aligned}$$

If we absorb all numerical factors into two new coefficients C and D, then the general solutions take the form

$$\psi_k(x) = C\sin kx + D\cos kx \qquad E_k = \frac{k^2\hbar^2}{2m} \tag{2.3}$$

For a free particle, any value of E_k corresponds to an acceptable solution. However, when the particle is confined within a region, the acceptable wavefunctions must satisfy certain **boundary conditions**, or constraints on the function at certain locations. As we shall see when we discuss penetration into barriers, a wavefunction decays exponentially with distance inside a barrier, such as a wall, and the decay is infinitely fast when the potential energy is infinite. This behaviour is consistent with the fact that it is physically impossible for the particle to be found with an infinite potential energy. We conclude that the wavefunction must be zero where V is infinite, at $x < 0$ and $x > L$. The continuity of the wavefunction then requires it to vanish just inside the well at $x = 0$ and $x = L$. That is, the boundary conditions are $\psi_k(0) = 0$ and $\psi_k(L) = 0$. These boundary conditions imply quantization, as we show in the following *Justification*.

Justification 2.1 *The energy levels and wavefunctions of a particle in a one-dimensional box*

For an informal demonstration of quantization, we consider each wavefunction to be a de Broglie wave that must fit within the container. The permitted wavelengths satisfy

$$L = n \times \tfrac{1}{2}\lambda \qquad n = 1, 2, \ldots$$

and therefore

$$\lambda = \frac{2L}{n} \qquad \text{with } n = 1, 2, \ldots$$

According to the de Broglie relation, these wavelengths correspond to the momenta

$$p = \frac{h}{\lambda} = \frac{nh}{2L}$$

The particle has only kinetic energy inside the box (where $V = 0$), so the permitted energies are

$$E = \frac{p^2}{2m} = \frac{n^2h^2}{8mL^2} \qquad \text{with } n = 1, 2, \ldots$$

A more formal and widely applicable approach is as follows. Consider the wall at $x = 0$. According to eqn 2.3, $\psi(0) = D$ (because $\sin 0 = 0$ and $\cos 0 = 1$). But because $\psi(0) = 0$ we must have $D = 0$. It follows that the wavefunction must be of the form $\psi_k(x) = C \sin kx$. The value of ψ at the other wall (at $x = L$) is $\psi_k(L) = C \sin kL$, which must also be zero. Taking $C = 0$ would give $\psi_k(x) = 0$ for all x, which would conflict with the Born interpretation (the particle must be somewhere). Therefore, kL must be chosen so that $\sin kL = 0$, which is satisfied by

$$kL = n\pi \qquad n = 1, 2, \ldots$$

The value $n = 0$ is ruled out, because it implies $k = 0$ and $\psi_k(x) = 0$ everywhere (because $\sin 0 = 0$), which is unacceptable. Negative values of n merely change the sign of $\sin kL$ (because $\sin(-x) = -\sin x$). The wavefunctions are therefore

$$\psi_n(x) = C \sin(n\pi x/L) \qquad n = 1, 2, \ldots$$

(At this point we have started to label the solutions with the index n instead of k.) Because k and E_k are related by eqn 2.3, and k and n are related by $kL = n\pi$, it follows that the energy of the particle is limited to $E_n = n^2h^2/8mL^2$, the values obtained by the informal procedure.

We conclude that the energy of the particle in a one-dimensional box is quantized and that this quantization arises from the boundary conditions that ψ must satisfy if it is to be an acceptable wavefunction. This is a general conclusion: *the need to satisfy boundary conditions implies that only certain wavefunctions are acceptable, and hence restricts observables to discrete values.* So far, only energy has been quantized; shortly we shall see that other physical observables may also be quantized.

(b) Normalization

Before discussing the solution in more detail, we shall complete the derivation of the wavefunctions by finding the normalization constant (here written C and regarded as real, that is, does not contain i). To do so, we look for the value of C that ensures that the integral of ψ^2 over all the space available to the particle (that is, from $x = 0$ to $x = L$) is equal to 1:

$$\int_0^L \psi^2 \, dx = C^2 \int_0^L \sin^2 \frac{n\pi x}{L} = C^2 \times \frac{L}{2} = 1, \qquad \text{so } C = \left(\frac{2}{L}\right)^{1/2}$$

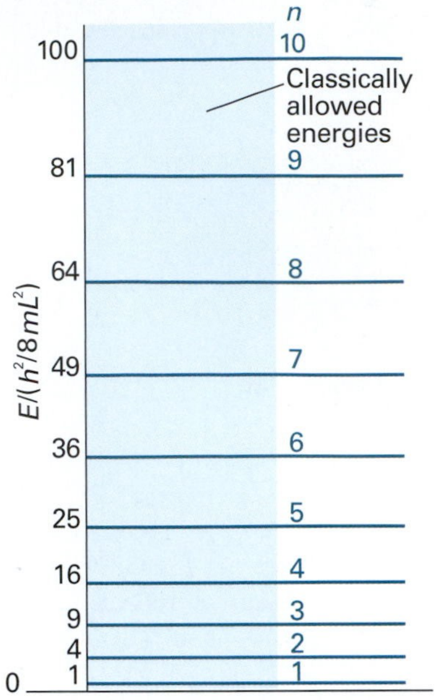

Fig. 2.2 The allowed energy levels for a particle in a box. Note that the energy levels increase as n^2, and that their separation increases as the quantum number increases.

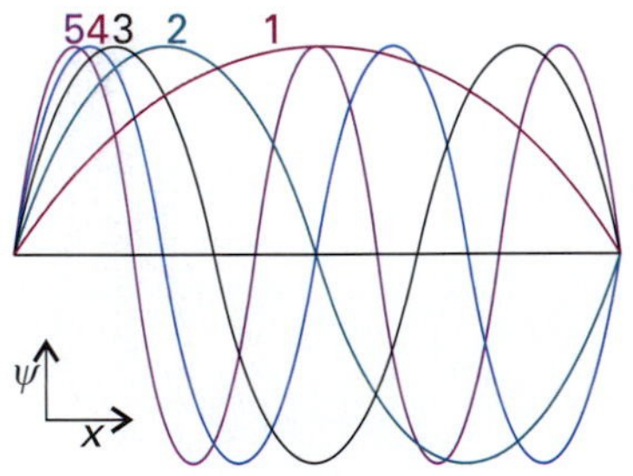

Fig. 2.3 The first five normalized wavefunctions of a particle in a box. Each wavefunction is a standing wave, and successive functions possess one more half wave and a correspondingly shorter wavelength.

Exploration Plot the probability density for a particle in a box with $n = 1, 2, \ldots 5$ and $n = 50$. How do your plots illustrate the correspondence principle?

Comment 2.1

It is sometimes useful to write

$\cos x = (e^{ix} + e^{-ix})/2 \quad \sin x = (e^{ix} - e^{-ix})/2i$

for all n. Therefore, the complete solution to the problem is

$$E_n = \frac{n^2h^2}{8mL^2} \qquad n = 1, 2, \ldots \tag{2.4a}$$

$$\psi_n(x) = \left(\frac{2}{L}\right)^{1/2} \sin\left(\frac{n\pi x}{L}\right) \qquad \text{for } 0 \le x \le L \tag{2.4b}$$

Self-test 2.1 Provide the intermediate steps for the determination of the normalization constant *C*. *Hint.* Use the standard integral $\int \sin^2 ax\, dx = \frac{1}{2}x - (\frac{1}{4}a) \sin 2ax +$ constant and the fact that $\sin 2m\pi = 0$, with $m = 0, 1, 2, \ldots$.

The energies and wavefunctions are labelled with the 'quantum number' n. A **quantum number** is an integer (in some cases, as we shall see, a half-integer) that labels the state of the system. For a particle in a box there is an infinite number of acceptable solutions, and the quantum number n specifies the one of interest (Fig. 2.2). As well as acting as a label, a quantum number can often be used to calculate the energy corresponding to the state and to write down the wavefunction explicitly (in the present example, by using eqn 2.4).

(c) The properties of the solutions

Figure 2.3 shows some of the wavefunctions of a particle in a box: they are all sine functions with the same amplitude but different wavelengths. Shortening the wavelength results in a sharper average curvature of the wavefunction and therefore an increase in the kinetic energy of the particle. Note that the number of nodes (points where the wavefunction passes *through* zero) also increases as n increases, and that the wavefunction ψ_n has $n - 1$ nodes. Increasing the number of nodes between walls of a given separation increases the average curvature of the wavefunction and hence the kinetic energy of the particle.

The linear momentum of a particle in a box is not well defined because the wavefunction $\sin kx$ is a standing wave and, like the example of $\cos kx$ treated in Section 1.5d, not an eigenfunction of the linear momentum operator. However, each wavefunction is a superposition of momentum eigenfunctions:

$$\psi_n = \left(\frac{2}{L}\right)^{1/2} \sin\frac{n\pi x}{L} = \frac{1}{2i}\left(\frac{2}{L}\right)^{1/2} (e^{ikx} - e^{-ikx}) \qquad k = \frac{n\pi}{L} \tag{2.5}$$

It follows that measurement of the linear momentum will give the value $+k\hbar$ for half the measurements of momentum and $-k\hbar$ for the other half. This detection of opposite directions of travel with equal probability is the quantum mechanical version of the classical picture that a particle in a box rattles from wall to wall, and in any given period spends half its time travelling to the left and half travelling to the right.

Self-test 2.2 What is (a) the average value of the linear momentum of a particle in a box with quantum number n, (b) the average value of p^2?

[(a) $\langle p\rangle = 0$, (b) $\langle p^2\rangle = n^2h^2/4L^2$]

Because n cannot be zero, the lowest energy that the particle may possess is not zero (as would be allowed by classical mechanics, corresponding to a stationary particle) but

$$E_1 = \frac{h^2}{8mL^2} \quad (2.6)$$

This lowest, irremovable energy is called the **zero-point energy**. The physical origin of the zero-point energy can be explained in two ways. First, the uncertainty principle requires a particle to possess kinetic energy if it is confined to a finite region: the location of the particle is not completely indefinite, so its momentum cannot be precisely zero. Hence it has nonzero kinetic energy. Second, if the wavefunction is to be zero at the walls, but smooth, continuous, and not zero everywhere, then it must be curved, and curvature in a wavefunction implies the possession of kinetic energy.

The separation between adjacent energy levels with quantum numbers n and $n+1$ is

$$E_{n+1} - E_n = \frac{(n+1)^2h^2}{8mL^2} - \frac{n^2h^2}{8mL^2} = (2n+1)\frac{h^2}{8mL^2} \quad (2.7)$$

This separation decreases as the length of the container increases, and is very small when the container has macroscopic dimensions. The separation of adjacent levels becomes zero when the walls are infinitely far apart. Atoms and molecules free to move in normal laboratory-sized vessels may therefore be treated as though their translational energy is not quantized. The translational energy of completely free particles (those not confined by walls) is not quantized.

Illustration 2.1 *Accounting for the electronic absorption spectra of polyenes*

β-Carotene (**1**) is a linear polyene in which 10 single and 11 double bonds alternate along a chain of 22 carbon atoms. If we take each CC bond length to be about 140 pm, then the length L of the molecular box in β-carotene is $L = 0.294$ nm. For reasons that will be familiar from introductory chemistry, each C atom contributes one p-electron to the π orbitals and, in the lowest energy state of the molecule, each level up to $n = 11$ is occupied by two electrons. From eqn 2.7 it follows that the separation in energy between the ground state and the state in which one electron is promoted from $n = 11$ to $n = 12$ is

$$\Delta E = E_{12} - E_{11} = (2\times 11+1)\frac{(6.626\times 10^{-34}\,\text{J s})^2}{8\times(9.110\times 10^{-31}\,\text{kg})\times(2.94\times 10^{-10}\,\text{m})^2}$$
$$= 1.60\times 10^{-19}\,\text{J}$$

It follows from the Bohr frequency condition (eqn 1.10, $\Delta E = h\nu$) that the frequency of radiation required to cause this transition is

$$\nu = \frac{\Delta E}{h} = \frac{1.60\times 10^{-19}\,\text{J}}{6.626\times 10^{-34}\,\text{J s}} = 2.41\times 10^{14}\,\text{s}^{-1}$$

The experimental value is $\nu = 6.03\times 10^{14}\,\text{s}^{-1}$ ($\lambda = 497$ nm), corresponding to radiation in the visible range of the electromagnetic spectrum.

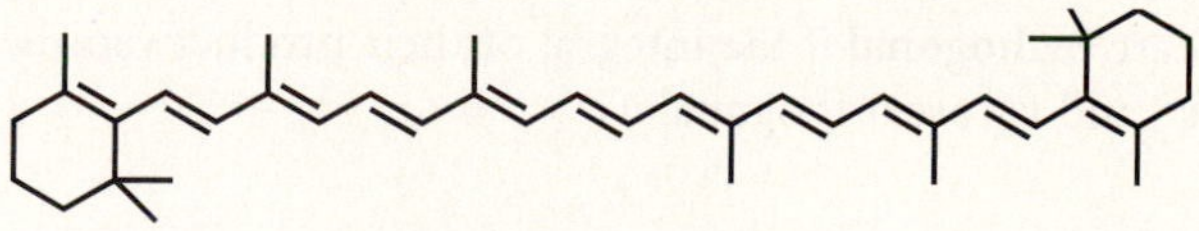

1 β-Carotene

Self-test 2.3 Estimate a typical nuclear excitation energy by calculating the first excitation energy of a proton confined to a square well with a length equal to the diameter of a nucleus (approximately 1 fm). [0.6 GeV]

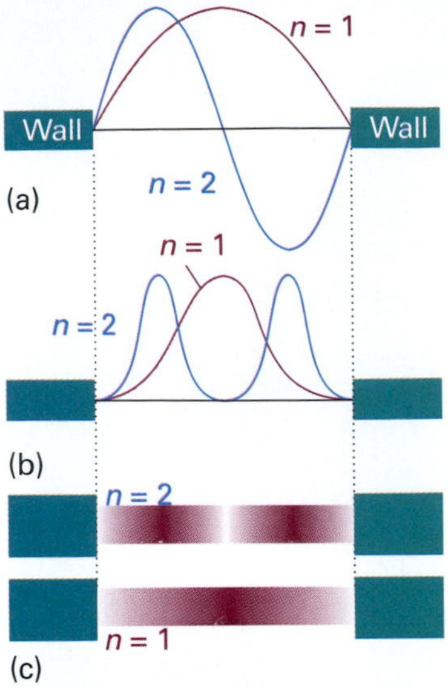

Fig. 2.4 (a) The first two wavefunctions, (b) the corresponding probability distributions, and (c) a representation of the probability distribution in terms of the darkness of shading.

The probability density for a particle in a box is

$$\psi^2(x) = \frac{2}{L}\sin^2\frac{n\pi x}{L} \tag{2.8}$$

and varies with position. The nonuniformity is pronounced when n is small (Fig. 2.4), but—provided we ignore the increasingly rapid oscillations—$\psi^2(x)$ becomes more uniform as n increases. The distribution at high quantum numbers reflects the classical result that a particle bouncing between the walls spends, on the average, equal times at all points. That the quantum result corresponds to the classical prediction at high quantum numbers is an illustration of the **correspondence principle**, which states that classical mechanics emerges from quantum mechanics as high quantum numbers are reached.

Example 2.1 *Using the particle in a box solutions*

The wavefunctions of an electron in a conjugated polyene can be approximated by particle-in-a-box wavefunctions. What is the probability, P, of locating the electron between $x = 0$ (the left-hand end of a molecule) and $x = 0.2$ nm in its lowest energy state in a conjugated molecule of length 1.0 nm?

Method The value of $\psi^2\mathrm{d}x$ is the probability of finding the particle in the small region $\mathrm{d}x$ located at x; therefore, the total probability of finding the electron in the specified region is the integral of $\psi^2\mathrm{d}x$ over that region. The wavefunction of the electron is given in eqn 2.4b with $n = 1$.

Answer The probability of finding the particle in a region between $x = 0$ and $x = l$ is

$$P = \int_0^l \psi_n^2\,\mathrm{d}x = \frac{2}{L}\int_0^l \sin^2\frac{n\pi x}{L}\,\mathrm{d}x = \frac{l}{L} - \frac{1}{2n\pi}\sin\frac{2\pi nl}{L}$$

We then set $n = 1$ and $l = 0.2$ nm, which gives $P = 0.05$. The result corresponds to a chance of 1 in 20 of finding the electron in the region. As n becomes infinite, the sine term, which is multiplied by $1/n$, makes no contribution to P and the classical result, $P = l/L$, is obtained.

Self-test 2.4 Calculate the probability that an electron in the state with $n = 1$ will be found between $x = 0.25L$ and $x = 0.75L$ in a conjugated molecule of length L (with $x = 0$ at the left-hand end of the molecule). [0.82]

(d) Orthogonality

We can now illustrate a property of wavefunctions first mentioned in Section 1.5. Two wavefunctions are **orthogonal** if the integral of their product vanishes. Specifically, the functions ψ_n and $\psi_{n'}$ are orthogonal if

$$\int \psi_n^* \psi_{n'}\,\mathrm{d}\tau = 0 \tag{2.9}$$

where the integration is over all space. A general feature of quantum mechanics, which we prove in the *Justification* below, is that *wavefunctions corresponding to different energies are orthogonal*; therefore, we can be confident that all the wavefunctions of a particle in a box are mutually orthogonal. A more compact notation for integrals of this kind is described in *Further information 2.1*.

Justification 2.2 *The orthogonality of wavefunctions*

Suppose we have two wavefunctions ψ_n and ψ_m corresponding to two different energies E_n and E_m, respectively. Then we can write

$$\hat{H}\psi_n = E_n\psi_n \qquad \hat{H}\psi_m = E_m\psi_m$$

Now multiply the first of these two Schrödinger equations by ψ_m^* and the second by ψ_n^* and integrate over all space:

$$\int \psi_m^* \hat{H}\psi_n \mathrm{d}\tau = E_n \int \psi_m^*\psi_n \mathrm{d}\tau \qquad \int \psi_n^* \hat{H}\psi_m \mathrm{d}\tau = E_m \int \psi_n^*\psi_m \mathrm{d}\tau$$

Next, noting that the energies themselves are real, form the complex conjugate of the second expression (for the state m) and subtract it from the first expression (for the state n):

$$\int \psi_m^* \hat{H}\psi_n \mathrm{d}\tau - \left(\int \psi_n^* \hat{H}\psi_m \mathrm{d}\tau\right)^* = E_n \int \psi_m^*\psi_n \mathrm{d}\tau - E_m \int \psi_n\psi_m^* \mathrm{d}\tau$$

By the hermiticity of the hamiltonian (Section 1.5c), the two terms on the left are equal, so they cancel and we are left with

$$0 = (E_n - E_m)\int \psi_m^*\psi_n \mathrm{d}\tau$$

However, the two energies are different; therefore the integral on the right must be zero, which confirms that two wavefunctions belonging to different energies are orthogonal.

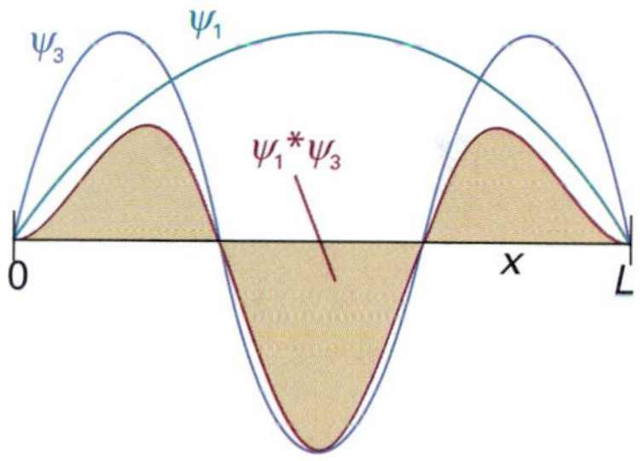

Fig. 2.5 Two functions are orthogonal if the integral of their product is zero. Here the calculation of the integral is illustrated graphically for two wavefunctions of a particle in a square well. The integral is equal to the total area beneath the graph of the product, and is zero.

Illustration 2.2 *Verifying the orthogonality of the wavefunctions for a particle in a box*

We can verify the orthogonality of wavefunctions of a particle in a box with $n = 1$ and $n = 3$ (Fig. 2.5):

$$\int_0^L \psi_1^*\psi_3 \mathrm{d}x = \frac{2}{L}\int_0^L \sin\frac{\pi x}{L}\sin\frac{3\pi x}{L}\mathrm{d}x = 0$$

We have used the standard integral given in *Illustration 1.2*.

The property of orthogonality is of great importance in quantum mechanics because it enables us to eliminate a large number of integrals from calculations. Orthogonality plays a central role in the theory of chemical bonding (Chapter 4) and spectroscopy (Chapter 7). Sets of functions that are normalized and mutually orthogonal are called **orthonormal**. The wavefunctions in eqn 2.4b are orthonormal.

2.2 Motion in two and more dimensions

Next, we consider a two-dimensional version of the particle in a box. Now the particle is confined to a rectangular surface of length L_1 in the x-direction and L_2 in the y-direction; the potential energy is zero everywhere except at the walls, where it is infinite (Fig. 2.6). The wavefunction is now a function of both x and y and the Schrödinger equation is

$$-\frac{\hbar^2}{2m}\left(\frac{\partial^2\psi}{\partial x^2} + \frac{\partial^2\psi}{\partial y^2}\right) = E\psi \qquad (2.10)$$

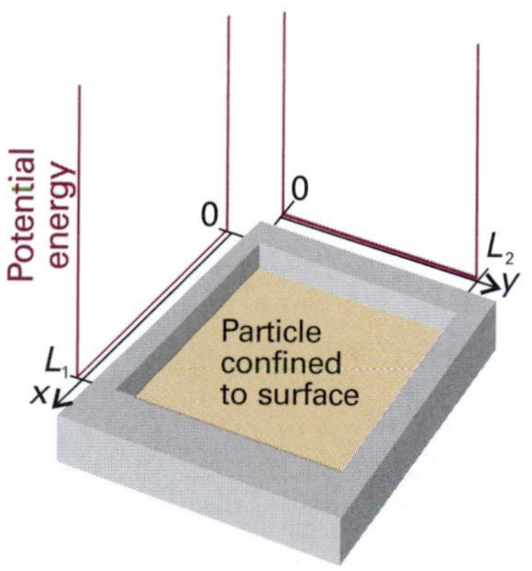

Fig. 2.6 A two-dimensional square well. The particle is confined to the plane bounded by impenetrable walls. As soon as it touches the walls, its potential energy rises to infinity.

We need to see how to solve this *partial* differential equation, an equation in more than one variable.

(a) Separation of variables

Some partial differential equations can be simplified by the **separation of variables technique**, which divides the equation into two or more ordinary differential equations, one for each variable. An important application of this procedure, as we shall see, is the separation of the Schrödinger equation for the hydrogen atom into equations that describe the radial and angular variation of the wavefunction. The technique is particularly simple for a two-dimensional square well, as can be seen by testing whether a solution of eqn 2.10 can be found by writing the wavefunction as a product of functions, one depending only on x and the other only on y:

$$\psi(x,y) = X(x)Y(y)$$

With this substitution, we show in the *Justification* below that eqn 2.10 separates into two ordinary differential equations, one for each coordinate:

$$-\frac{\hbar^2}{2m}\frac{d^2X}{dx^2} = E_X X \qquad -\frac{\hbar^2}{2m}\frac{d^2Y}{dy^2} = E_Y Y \qquad E = E_X + E_Y \tag{2.11}$$

The quantity E_X is the energy associated with the motion of the particle parallel to the x-axis, and likewise for E_Y and motion parallel to the y-axis.

Justification 2.3 *The separation of variables technique applied to the particle in a two-dimensional box*

The first step in the justification of the separability of the wavefunction into the product of two functions X and Y is to note that, because X is independent of y and Y is independent of x, we can write

$$\frac{\partial^2\psi}{\partial x^2} = \frac{\partial^2 XY}{\partial x^2} = Y\frac{d^2X}{dx^2} \qquad \frac{\partial^2\psi}{\partial y^2} = \frac{\partial^2 XY}{\partial y^2} = X\frac{d^2Y}{dy^2}$$

Then eqn 2.10 becomes

$$-\frac{\hbar^2}{2m}\left(Y\frac{d^2X}{dx^2} + X\frac{d^2Y}{dy^2}\right) = EXY$$

When both sides are divided by XY, we can rearrange the resulting equation into

$$\frac{1}{X}\frac{d^2X}{dx^2} + \frac{1}{Y}\frac{d^2Y}{dy^2} = -\frac{2mE}{\hbar^2}$$

The first term on the left is independent of y, so if y is varied only the second term can change. But the sum of these two terms is a constant given by the right-hand side of the equation; therefore, even the second term cannot change when y is changed. In other words, the second term is a constant, which we write $-2mE_Y/\hbar^2$. By a similar argument, the first term is a constant when x changes, and we write it $-2mE_X/\hbar^2$, and $E = E_X + E_Y$. Therefore, we can write

$$\frac{1}{X}\frac{d^2X}{dx^2} = -\frac{2mE_X}{\hbar^2} \qquad \frac{1}{Y}\frac{d^2Y}{dy^2} = -\frac{2mE_Y}{\hbar^2}$$

which rearrange into the two ordinary (that is, single variable) differential equations in eqn 2.11.

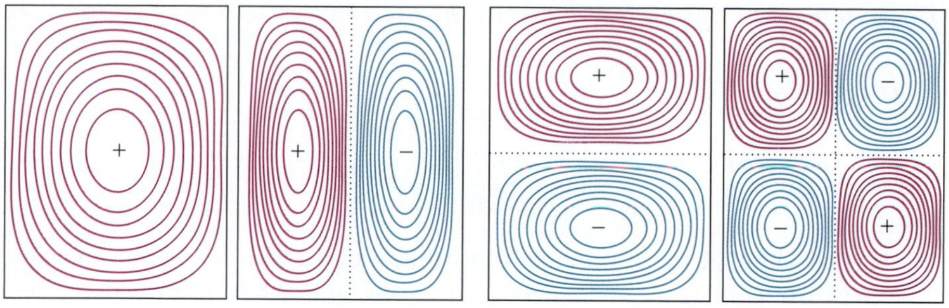

Fig. 2.7 The wavefunctions for a particle confined to a rectangular surface depicted as contours of equal amplitude. (a) $n_1 = 1$, $n_2 = 1$, the state of lowest energy, (b) $n_1 = 1$, $n_2 = 2$, (c) $n_1 = 2$, $n_2 = 1$, and (d) $n_1 = 2$, $n_2 = 2$.

Exploration Use mathematical software to generate three-dimensional plots of the functions in this illustration. Deduce a rule for the number of nodal lines in a wavefunction as a function of the values of n_x and n_y.

Each of the two ordinary differential equations in eqn 2.11 is the same as the one-dimensional square-well Schrödinger equation. We can therefore adapt the results in eqn 2.4 without further calculation:

$$X_{n_1}(x) = \left(\frac{2}{L_1}\right)^{1/2} \sin\frac{n_1\pi x}{L_1} \qquad Y_{n_2}(y) = \left(\frac{2}{L_2}\right)^{1/2} \sin\frac{n_2\pi y}{L_2}$$

Then, because $\psi = XY$ and $E = E_X + E_Y$, we obtain

$$\psi_{n_1,n_2}(x,y) = \frac{2}{(L_1L_2)^{1/2}} \sin\frac{n_1\pi x}{L_1} \sin\frac{n_2\pi y}{L_2} \qquad 0 \le x \le L_1, 0 \le y \le L_2$$

$$E_{n_1n_2} = \left(\frac{n_1^2}{L_1^2} + \frac{n_2^2}{L_2^2}\right)\frac{h^2}{8m} \tag{2.12a}$$

with the quantum numbers taking the values $n_1 = 1, 2, \ldots$ and $n_2 = 1, 2, \ldots$ independently. Some of these functions are plotted in Fig. 2.7. They are the two-dimensional versions of the wavefunctions shown in Fig. 2.3. Note that two quantum numbers are needed in this two-dimensional problem.

We treat a particle in a three-dimensional box in the same way. The wavefunctions have another factor (for the z-dependence), and the energy has an additional term in n_3^2/L_3^2. Solution of the Schrödinger equation by the separation of variables technique then gives

$$\psi_{n_1,n_2,n_3}(x,y,z) = \left(\frac{8}{L_1L_2L_3}\right)^{1/2} \sin\frac{n_1\pi x}{L_1} \sin\frac{n_2\pi y}{L_2} \sin\frac{n_3\pi z}{L_3}$$

$$0 \le x \le L_1, 0 \le y \le L_2, 0 \le z \le L_3 \tag{2.12b}$$

$$E_{n_1n_2n_3} = \left(\frac{n_1^2}{L_1^2} + \frac{n_2^2}{L_2^2} + \frac{n_3^2}{L_3^2}\right)\frac{h^2}{8m}$$

(b) Degeneracy

An interesting feature of the solutions for a particle in a two-dimensional box is obtained when the plane surface is square, with $L_1 = L_2 = L$. Then eqn 2.12a becomes

$$\psi_{n_1,n_2}(x,y) = \frac{2}{L} \sin\frac{n_1\pi x}{L} \sin\frac{n_2\pi y}{L} \qquad E_{n_1n_2} = (n_1^2 + n_2^2)\frac{h^2}{8mL^2} \tag{2.13}$$

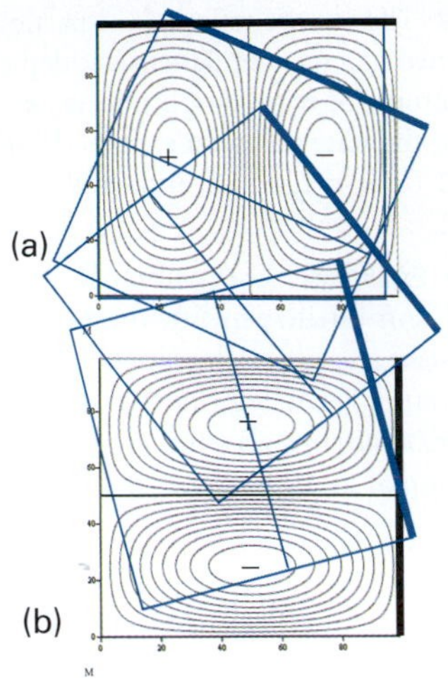

Fig. 2.8 The wavefunctions for a particle confined to a square surface. Note that one wavefunction can be converted into the other by a rotation of the box by 90°. The two functions correspond to the same energy. Degeneracy and symmetry are closely related.

Consider the cases $n_1 = 1, n_2 = 2$ and $n_1 = 2, n_2 = 1$:

$$\psi_{1,2} = \frac{2}{L}\sin\frac{\pi x}{L}\sin\frac{2\pi y}{L} \qquad E_{1,2} = \frac{5h^2}{8mL^2}$$

$$\psi_{2,1} = \frac{2}{L}\sin\frac{2\pi x}{L}\sin\frac{\pi y}{L} \qquad E_{2,1} = \frac{5h^2}{8mL^2}$$

We see that, although the wavefunctions are different, they are **degenerate**, meaning that they correspond to the same energy. In this case, in which there are two degenerate wavefunctions, we say that the energy level $5(h^2/8mL^2)$ is 'doubly degenerate'.

The occurrence of degeneracy is related to the symmetry of the system. Figure 2.8 shows contour diagrams of the two degenerate functions $\psi_{1,2}$ and $\psi_{2,1}$. Because the box is square, we can convert one wavefunction into the other simply by rotating the plane by 90°. Interconversion by rotation through 90° is not possible when the plane is not square, and $\psi_{1,2}$ and $\psi_{2,1}$ are then not degenerate. Similar arguments account for the degeneracy of states in a cubic box. We shall see many other examples of degeneracy in the pages that follow (for instance, in the hydrogen atom), and all of them can be traced to the symmetry properties of the system (see Section 5.4b).

2.3 Tunnelling

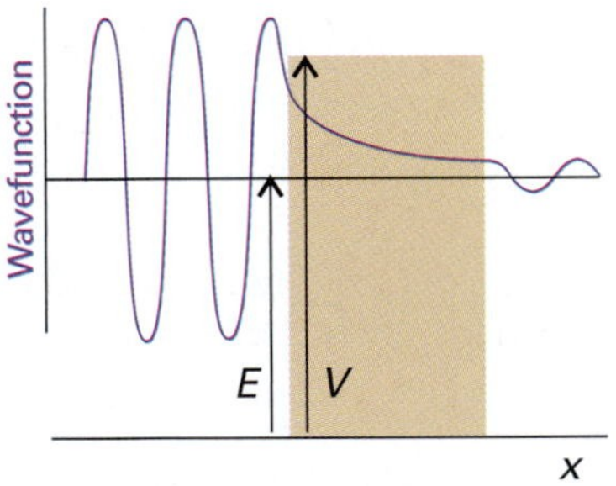

Fig. 2.9 A particle incident on a barrier from the left has an oscillating wave function, but inside the barrier there are no oscillations (for $E < V$). If the barrier is not too thick, the wavefunction is nonzero at its opposite face, and so oscillates begin again there. (Only the real component of the wavefunction is shown.)

If the potential energy of a particle does not rise to infinity when it is in the walls of the container, and $E < V$, the wavefunction does not decay abruptly to zero. If the walls are thin (so that the potential energy falls to zero again after a finite distance), then the wavefunction oscillates inside the box, varies smoothly inside the region representing the wall, and oscillates again on the other side of the wall outside the box (Fig. 2.9). Hence the particle might be found on the outside of a container even though according to classical mechanics it has insufficient energy to escape. Such leakage by penetration through a classically forbidden region is called **tunnelling**.

The Schrödinger equation can be used to calculate the probability of tunnelling of a particle of mass m incident on a finite barrier from the left. On the left of the barrier (for $x < 0$) the wavefunctions are those of a particle with $V = 0$, so from eqn 2.2 we can write

$$\psi = Ae^{ikx} + Be^{-ikx} \qquad k\hbar = (2mE)^{1/2} \tag{2.14}$$

The Schrödinger equation for the region representing the barrier (for $0 \le x \le L$), where the potential energy is the constant V, is

$$-\frac{\hbar^2}{2m}\frac{d^2\psi}{dx^2} + V\psi = E\psi \tag{2.15}$$

We shall consider particles that have $E < V$ (so, according to classical physics, the particle has insufficient energy to pass over the barrier), and therefore $V - E$ is positive. The general solutions of this equation are

$$\psi = Ce^{\kappa x} + De^{-\kappa x} \qquad \kappa\hbar = \{2m(V - E)\}^{1/2} \tag{2.16}$$

as we can readily verify by differentiating ψ twice with respect to x. The important feature to note is that the two exponentials are now real functions, as distinct from the complex, oscillating functions for the region where $V = 0$ (oscillating functions would be obtained if $E > V$). To the right of the barrier ($x > L$), where $V = 0$ again, the wavefunctions are

$$\psi = A'e^{ikx} + B'e^{-ikx} \qquad k\hbar = (2mE)^{1/2} \tag{2.17}$$

The complete wavefunction for a particle incident from the left consists of an incident wave, a wave reflected from the barrier, the exponentially changing amplitudes inside the barrier, and an oscillating wave representing the propagation of the particle to the right after tunnelling through the barrier successfully (Fig. 2.10). The acceptable wavefunctions must obey the conditions set out in Section 1.4b. In particular, they must be continuous at the edges of the barrier (at $x = 0$ and $x = L$, remembering that $e^0 = 1$):

$$A + B = C + D \qquad Ce^{\kappa L} + De^{-\kappa L} = A'e^{ikL} + B'e^{-ikL} \tag{2.18}$$

Their slopes (their first derivatives) must also be continuous there (Fig. 2.11):

$$ikA - ikB = \kappa C - \kappa D \qquad \kappa Ce^{\kappa L} - \kappa De^{-\kappa L} = ikA'e^{ikL} - ikB'e^{-ikL} \tag{2.19}$$

At this stage, we have four equations for the six unknown coefficients. If the particles are shot towards the barrier from the left, there can be no particles travelling to the left on the right of the barrier. Therefore, we can set $B' = 0$, which removes one more unknown. We cannot set $B = 0$ because some particles may be reflected back from the barrier toward negative x.

The probability that a particle is travelling towards positive x (to the right) on the left of the barrier is proportional to $|A|^2$, and the probability that it is travelling to the right on the right of the barrier is $|A'|^2$. The ratio of these two probabilities is called the **transmission probability**, T. After some algebra (see Problem 2.9) we find

$$T = \left\{1 + \frac{(e^{\kappa L} - e^{-\kappa L})^2}{16\varepsilon(1-\varepsilon)}\right\}^{-1} \tag{2.20a}$$

where $\varepsilon = E/V$. This function is plotted in Fig. 2.12; the transmission coefficient for $E > V$ is shown there too. For high, wide barriers (in the sense that $\kappa L \gg 1$), eqn 2.20a simplifies to

$$T \approx 16\varepsilon(1-\varepsilon)e^{-2\kappa L} \tag{2.20b}$$

The transmission probability decreases exponentially with the thickness of the barrier and with $m^{1/2}$. It follows that particles of low mass are more able to tunnel through barriers than heavy ones (Fig. 2.13). Tunnelling is very important for electrons and muons, and moderately important for protons; for heavier particles it is less important.

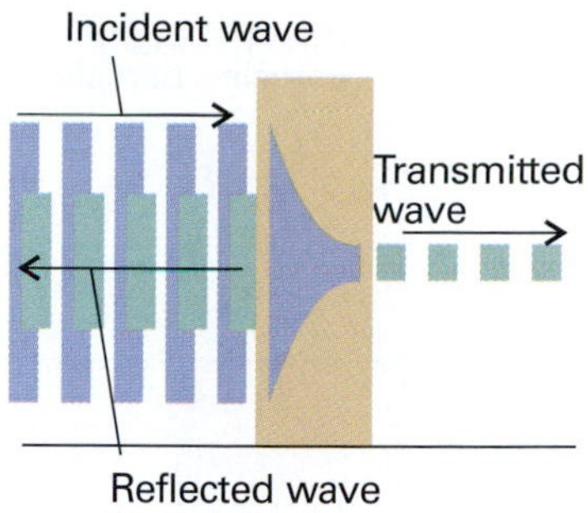

Fig. 2.10 When a particle is incident on a barrier from the left, the wavefunction consists of a wave representing linear momentum to the right, a reflected component representing momentum to the left, a varying but not oscillating component inside the barrier, and a (weak) wave representing motion to the right on the far side of the barrier.

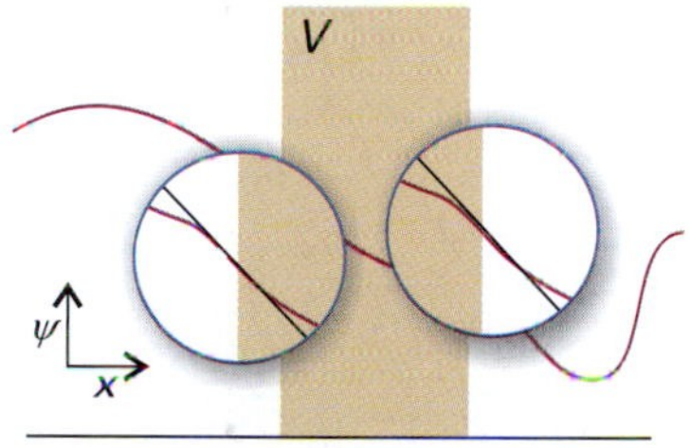

Fig. 2.11 The wavefunction and its slope must be continuous at the edges of the barrier. The conditions for continuity enable us to connect the wavefunctions in the three zones and hence to obtain relations between the coefficients that appear in the solutions of the Schrödinger equation.

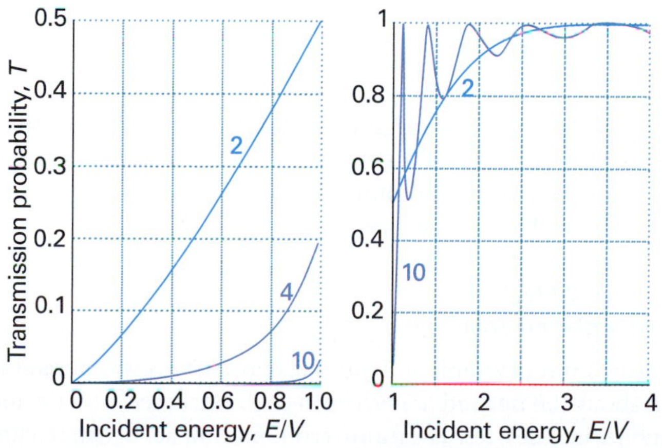

Fig. 2.12 The transition probabilities for passage through a barrier. The horizontal axis is the energy of the incident particle expressed as a multiple of the barrier height. The curves are labelled with the value of $L(2mV)^{1/2}/\hbar$. The graph on the left is for $E < V$ and that on the right for $E > V$. Note that $T > 0$ for $E < V$ whereas classically T would be zero. However, $T < 1$ for $E > V$, whereas classically T would be 1.

Exploration Plot T against ε for a hydrogen molecule, a proton and an electron.

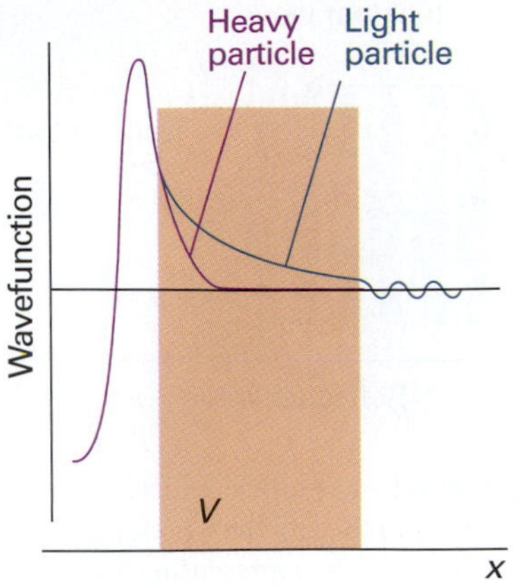

Fig. 2.13 The wavefunction of a heavy particle decays more rapidly inside a barrier than that of a light particle. Consequently, a light particle has a greater probability of tunnelling through the barrier.

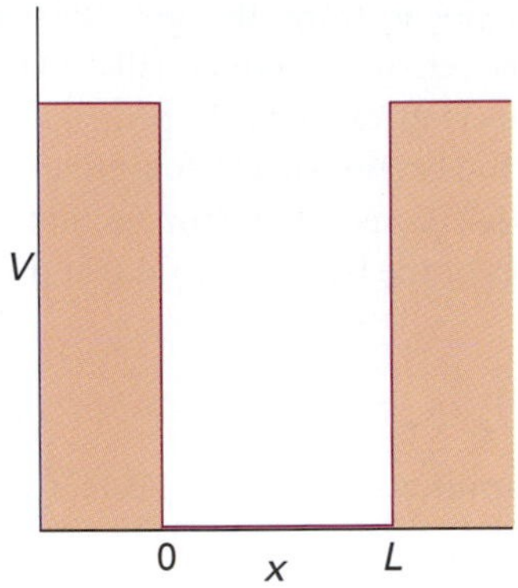

Fig. 2.14 A potential well with a finite depth.

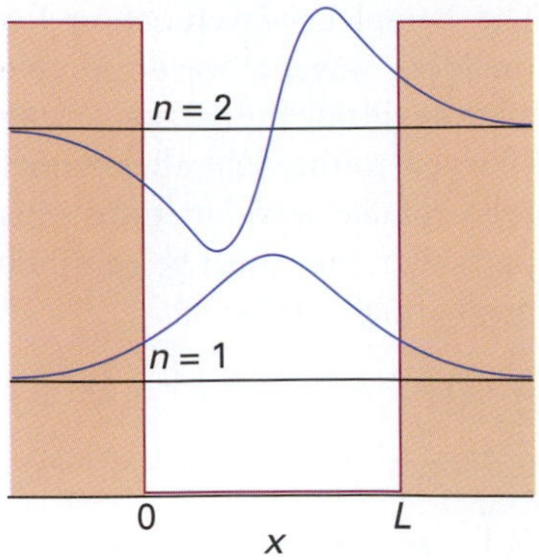

Fig. 2.15 The lowest two bound-state wavefunctions for a particle in the well shown in Fig. 2.14 and one of the wavefunctions corresponding to an unbound state ($E > V$).

A number of effects in chemistry (for example, the isotope-dependence of some reaction rates) depend on the ability of the proton to tunnel more readily than the deuteron. The very rapid equilibration of proton transfer reactions is also a manifestation of the ability of protons to tunnel through barriers and transfer quickly from an acid to a base. Tunnelling of protons between acidic and basic groups is also an important feature of the mechanism of some enzyme-catalysed reactions. Electron tunnelling is one of the factors that determine the rates of electron transfer reactions at electrodes and in biological systems.

A problem related to the one just considered is that of a particle in a square-well potential of finite depth (Fig. 2.14). In this kind of potential, the wavefunction penetrates into the walls, where it decays exponentially towards zero, and oscillates within the well. The wavefunctions are found by ensuring, as in the discussion of tunnelling, that they and their slopes are continuous at the edges of the potential. Some of the lowest energy solutions are shown in Fig. 2.15. A further difference from the solutions for an infinitely deep well is that there is only a finite number of bound states. Regardless of the depth and length of the well, there is always at least one bound state. Detailed consideration of the Schrödinger equation for the problem shows that in general the number of levels is equal to N, with

$$N - 1 < \frac{(8mVL)^{1/2}}{h} < N \tag{2.21}$$

where V is the depth of the well and L is its length (for a derivation of this expression, see *Further reading*). We see that the deeper and wider the well, the greater the number of bound states. As the depth becomes infinite, so the number of bound states also becomes infinite, as we have already seen.

IMPACT ON NANOSCIENCE
I2.1 Scanning probe microscopy

Nanoscience is the study of atomic and molecular assemblies with dimensions ranging from 1 nm to about 100 nm and *nanotechnology* is concerned with the incorporation of such assemblies into devices. The future economic impact of nanotechnology could be very significant. For example, increased demand for very small digital electronic devices has driven the design of ever smaller and more powerful microprocessors. However, there is an upper limit on the density of electronic circuits that can be incorporated into silicon-based chips with current fabrication technologies. As the ability to process data increases with the number of circuits in a chip, it follows that soon chips and the devices that use them will have to become bigger if processing power is to increase indefinitely. One way to circumvent this problem is to fabricate devices from nanometre-sized components.

We will explore several concepts of nanoscience throughout the text. We begin with the description of *scanning probe microscopy* (SPM), a collection of techniques that can be used to visualize and manipulate objects as small as atoms on surfaces. Consequently, SPM has far better resolution than electron microscopy (*Impact I1.1*).

One modality of SPM is *scanning tunnelling microscopy* (STM), in which a platinum–rhodium or tungsten needle is scanned across the surface of a conducting solid. When the tip of the needle is brought very close to the surface, electrons tunnel across the intervening space (Fig. 2.16). In the constant-current mode of operation, the stylus moves up and down corresponding to the form of the surface, and the topography of the surface, including any adsorbates, can be mapped on an atomic scale. The vertical motion of the stylus is achieved by fixing it to a piezoelectric cylinder, which contracts or expands according to the potential difference it experiences. In the constant-z mode, the vertical position of the stylus is held constant and the current is monitored. Because the tunnelling probability is very sensitive to the size of the gap, the microscope can detect tiny, atom-scale variations in the height of the surface.

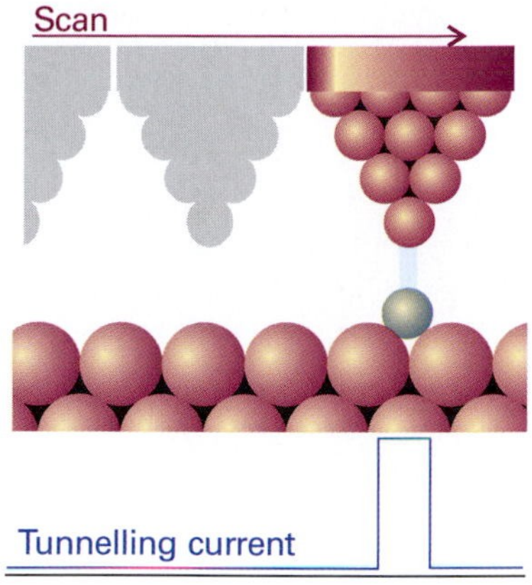

Fig. 2.16 A scanning tunnelling microscope makes use of the current of electrons that tunnel between the surface and the tip. That current is very sensitive to the distance of the tip above the surface.

Figure 2.17 shows an example of the kind of image obtained with a surface, in this case of gallium arsenide, that has been modified by addition of atoms, in this case caesium atoms. Each 'bump' on the surface corresponds to an atom. In a further variation of the STM technique, the tip may be used to nudge single atoms around on the surface, making possible the fabrication of complex and yet very tiny nanometre-sized structures.

In *atomic force microscopy* (AFM) a sharpened stylus attached to a cantilever is scanned across the surface. The force exerted by the surface and any bound species pushes or pulls on the stylus and deflects the cantilever (Fig. 2.18). The deflection is monitored either by interferometry or by using a laser beam. Because no current is needed between the sample and the probe, the technique can be applied to non-conducting surfaces too. A spectacular demonstration of the power of AFM is given in Fig. 2.19, which shows individual DNA molecules on a solid surface.

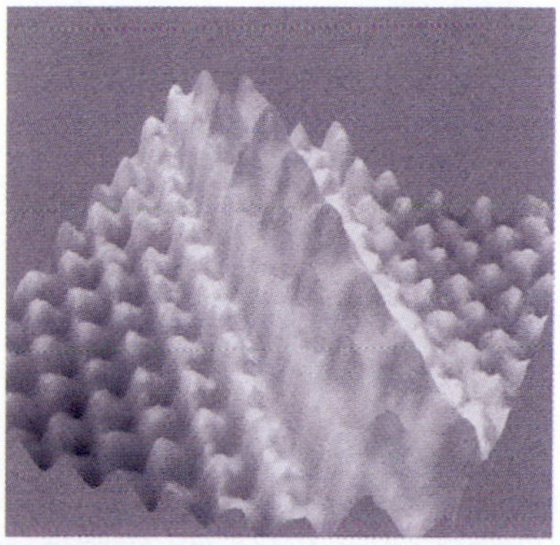

Fig. 2.17 An STM image of caesium atoms on a gallium arsenide surface.

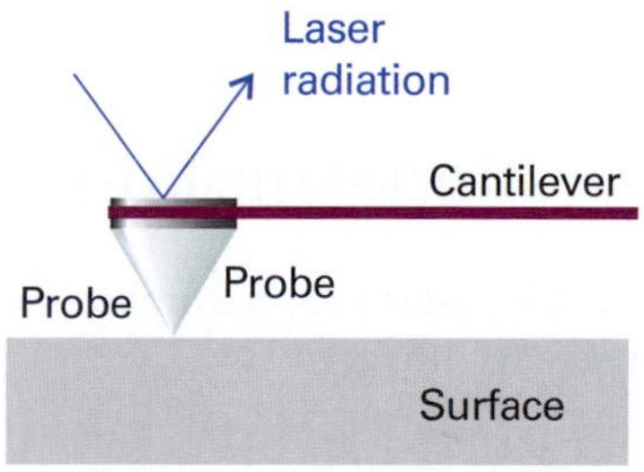

Fig. 2.18 In atomic force microscopy, a laser beam is used to monitor the tiny changes in the position of a probe as it is attracted to or repelled from atoms on a surface.

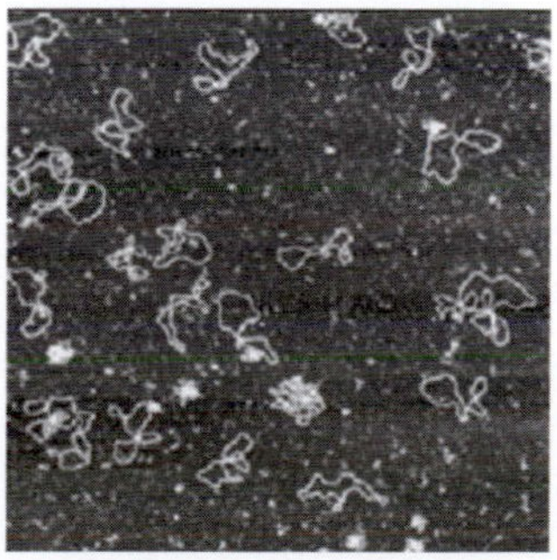

Fig. 2.19 An AFM image of bacterial DNA plasmids on a mica surface. (Courtesy of Veeco Instruments.)

Example 2.2 *Exploring the origin of the current in scanning tunnelling microscopy*

To get an idea of the distance dependence of the tunnelling current in STM, suppose that the wavefunction of the electron in the gap between sample and needle is given by $\psi = Be^{-\kappa x}$, where $\kappa = \{2m_e(V-E)/\hbar^2\}^{1/2}$; take $V - E = 2.0$ eV. By what factor would the current drop if the needle is moved from $L_1 = 0.50$ nm to $L_2 = 0.60$ nm from the surface?

Method We regard the tunnelling current to be proportional to the transmission probability T, so the ratio of the currents is equal to the ratio of the transmission probabilities. To choose between eqn 2.20a or 2.20b for the calculation of T, first calculate κL for the shortest distance L_1: if $\kappa L_1 > 1$, then use eqn 2.20b.

Answer When $L = L_1 = 0.50$ nm and $V - E = 2.0\ \text{eV} = 3.20 \times 10^{-19}$ J the value of κL is

$$\begin{aligned}\kappa L_1 &= \left\{\frac{2m_e(V-E)}{\hbar^2}\right\}^{1/2} L_1 \\ &= \left\{\frac{2 \times (9.109 \times 10^{-31}\ \text{kg}) \times (3.20 \times 10^{-19}\ \text{J})}{(1.054 \times 10^{-34}\ \text{J s})^2}\right\}^{1/2} \times (5.0 \times 10^{-10}\ \text{m}) \\ &= (7.25 \times 10^9\ \text{m}^{-1}) \times (5.0 \times 10^{-10}\ \text{m}) = 3.6\end{aligned}$$

Because $\kappa L_1 > 1$, we use eqn 2.20b to calculate the transmission probabilities at the two distances. It follows that

$$\begin{aligned}\frac{\text{current at } L_2}{\text{current at } L_1} &= \frac{T(L_2)}{T(L_1)} = \frac{16\varepsilon(1-\varepsilon)e^{-2\kappa L_2}}{16\varepsilon(1-\varepsilon)e^{-2\kappa L_1}} = e^{-2\kappa(L_2-L_1)} \\ &= e^{-2\times(7.25\times10^{-9}\ \text{m}^{-1})\times(1.0\times10^{-10}\ \text{m})} = 0.23\end{aligned}$$

We conclude that, at a distance of 0.60 nm between the surface and the needle, the current is 23 per cent of the value measured when the distance is 0.50 nm.

Self-test 2.5 The ability of a proton to tunnel through a barrier contributes to the rapidity of proton transfer reactions in solution and therefore to the properties of acids and bases. Estimate the relative probabilities that a proton and a deuteron can tunnel through the same barrier of height 1.0 eV (1.6×10^{-19} J) and length 100 pm when their energy is 0.9 eV. Any comment?

[$T_H/T_D = 3.7 \times 10^2$; we expect proton transfer reactions to be much faster than deuteron transfer reactions.]

Vibrational motion

A particle undergoes **harmonic motion** if it experiences a restoring force proportional to its displacement:

$$F = -kx \tag{2.22}$$

where k is the **force constant**: the stiffer the 'spring', the greater the value of k. Because force is related to potential energy by $F = -\mathrm{d}V/\mathrm{d}x$ (see *Appendix 3*), the force in eqn 2.22 corresponds to a potential energy

$$V = \tfrac{1}{2}kx^2 \tag{2.23}$$

This expression, which is the equation of a parabola (Fig. 2.20), is the origin of the term 'parabolic potential energy' for the potential energy characteristic of a harmonic oscillator. The Schrödinger equation for the particle is therefore

$$-\frac{\hbar^2}{2m}\frac{d^2\psi}{dx^2}+\tfrac{1}{2}kx^2\psi=E\psi \tag{2.24}$$

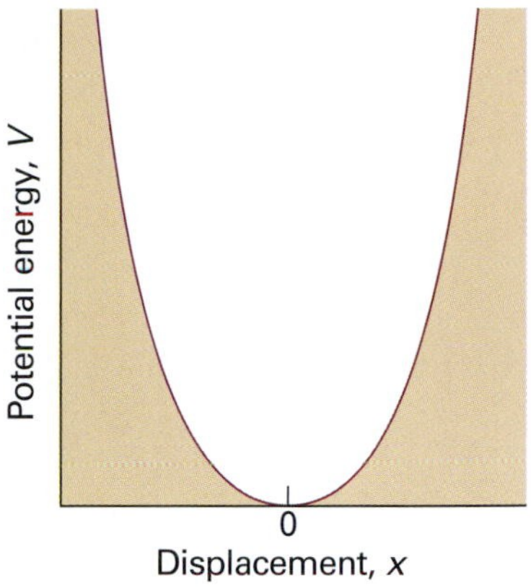

Fig. 2.20 The parabolic potential energy $V=\frac{1}{2}kx^2$ of a harmonic oscillator, where x is the displacement from equilibrium. The narrowness of the curve depends on the force constant k: the larger the value of k, the narrower the well.

2.4 The energy levels

Equation 2.24 is a standard equation in the theory of differential equations and its solutions are well known to mathematicians (for details, see *Further reading*). Quantization of energy levels arises from the boundary conditions: the oscillator will not be found with infinitely large compressions or extensions, so the only allowed solutions are those for which $\psi=0$ at $x=\pm\infty$. The permitted energy levels are

$$E_v=(v+\tfrac{1}{2})\hbar\omega \qquad \omega=\left(\frac{k}{m}\right)^{1/2} \qquad v=0,1,2,\ldots \tag{2.25}$$

Note that ω (omega) increases with increasing force constant and decreasing mass. It follows that the separation between adjacent levels is

$$E_{v+1}-E_v=\hbar\omega \tag{2.26}$$

which is the same for all v. Therefore, the energy levels form a uniform ladder of spacing $\hbar\omega$ (Fig. 2.21). The energy separation $\hbar\omega$ is negligibly small for macroscopic objects (with large mass), but is of great importance for objects with mass similar to that of atoms.

Because the smallest permitted value of v is 0, it follows from eqn 2.26 that a harmonic oscillator has a zero-point energy

$$E_0=\tfrac{1}{2}\hbar\omega \tag{2.27}$$

The mathematical reason for the zero-point energy is that v cannot take negative values, for if it did the wavefunction would be ill-behaved. The physical reason is the same as for the particle in a square well: the particle is confined, its position is not completely uncertain, and therefore its momentum, and hence its kinetic energy, cannot be exactly zero. We can picture this zero-point state as one in which the particle fluctuates incessantly around its equilibrium position; classical mechanics would allow the particle to be perfectly still.

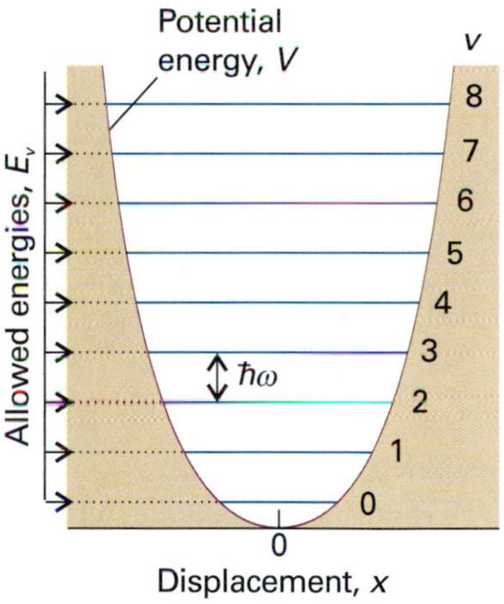

Fig. 2.21 The energy levels of a harmonic oscillator are evenly spaced with separation $\hbar\omega$, with $\omega=(k/m)^{1/2}$. Even in its lowest state, an oscillator has an energy greater than zero.

Illustration 2.3 *Calculating a molecular vibrational absorption frequency*

Atoms vibrate relative to one another in molecules with the bond acting like a spring. Consider an X—H chemical bond, where a heavy X atom forms a stationary anchor for the very light H atom. That is, only the H atom moves, vibrating as a simple harmonic oscillator. Therefore, eqn 2.25 describes the allowed vibrational energy levels of a X—H bond. The force constant of a typical X—H chemical bond is around 500 N m^{-1}. For example $k=516.3$ N m^{-1} for the $^{1}H^{35}Cl$ bond. Because the mass of a proton is about 1.7×10^{-27} kg, using $k=500$ N m^{-1} in eqn 2.25 gives $\omega\approx5.4\times10^{14}$ s^{-1} (5.4×10^{2} THz). It follows from eqn 9.26 that the separation of adjacent levels is $\hbar\omega\approx5.7\times10^{-20}$ J (57 zJ, about 0.36 eV). This energy separation corresponds to 34 kJ mol^{-1}, which is chemically significant. From eqn 2.27, the zero-point energy of this molecular oscillator is about 3 zJ, which corresponds to 0.2 eV, or 15 kJ mol^{-1}.

The excitation of the vibration of the bond from one level to the level immediately above requires 57 zJ. Therefore, if it is caused by a photon, the excitation requires radiation of frequency $\nu = \Delta E/h = 86$ THz and wavelength $\lambda = c/\nu = 3.5\ \mu$m. It follows that transitions between adjacent vibrational energy levels of molecules are stimulated by or emit infrared radiation. We shall see in Chapter 6 that the concepts just described represent the starting point for the interpretation of vibrational spectroscopy, an important technique for the characterization of small and large molecules in the gas phase or in condensed phases.

2.5 The wavefunctions

It is helpful at the outset to identify the similarities between the harmonic oscillator and the particle in a box, for then we shall be able to anticipate the form of the oscillator wavefunctions without detailed calculation. Like the particle in a box, a particle undergoing harmonic motion is trapped in a symmetrical well in which the potential energy rises to large values (and ultimately to infinity) for sufficiently large displacements (compare Figs. 2.1 and 2.20). However, there are two important differences. First, because the potential energy climbs towards infinity only as x^2 and not abruptly, the wavefunction approaches zero more slowly at large displacements than for the particle in a box. Second, as the kinetic energy of the oscillator depends on the displacement in a more complex way (on account of the variation of the potential energy), the curvature of the wavefunction also varies in a more complex way.

(a) The form of the wavefunctions

The detailed solution of eqn 2.24 shows that the wavefunction for a harmonic oscillator has the form

$$\psi(x) = N \times (\text{polynomial in } x) \times (\text{bell-shaped Gaussian function})$$

where N is a normalization constant. A Gaussian function is a function of the form e^{-x^2} (Fig. 2.22). The precise form of the wavefunctions are

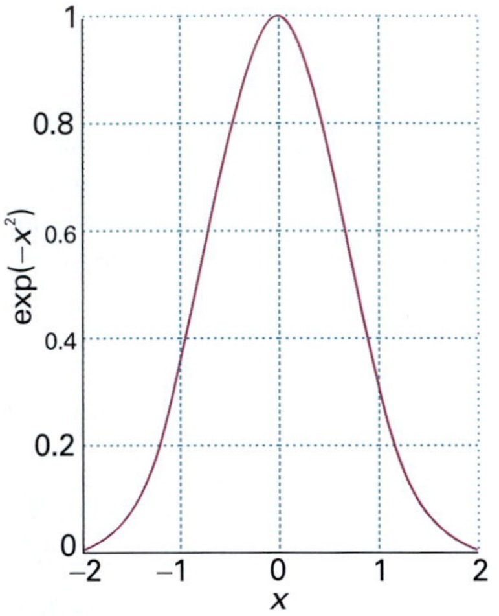

Fig. 2.22 The graph of the Gaussian function, $f(x) = e^{-x^2}$.

$$\psi_v(x) = N_v H_v(y) e^{-y^2/2} \qquad y = \frac{x}{\alpha} \qquad \alpha = \left(\frac{\hbar^2}{mk}\right)^{1/4} \tag{2.28}$$

The factor $H_v(y)$ is a **Hermite polynomial** (Table 2.1). For instance, because $H_0(y) = 1$, the wavefunction for the ground state (the lowest energy state, with $v = 0$) of the harmonic oscillator is

$$\psi_0(x) = N_0 e^{-y^2/2} = N_0 e^{-x^2/2\alpha^2} \tag{2.29a}$$

It follows that the probability density is the bell-shaped Gaussian function

$$\psi_0^2(x) = N_0^2 e^{-x^2/\alpha^2} \tag{2.29b}$$

The wavefunction and the probability distribution are shown in Fig. 2.23. Both curves have their largest values at zero displacement (at $x = 0$), so they capture the classical picture of the zero-point energy as arising from the ceaseless fluctuation of the particle about its equilibrium position.

The wavefunction for the first excited state of the oscillator, the state with $v = 1$, is obtained by noting that $H_1(y) = 2y$ (note that some of the Hermite polynomials are very simple functions!):

$$\psi_1(x) = N_1 \times 2y e^{-y^2/2} \tag{2.30}$$

Comment 2.2

The Hermite polynomials are solutions of the differential equation

$$H''_v - 2yH'_v + 2vH_v = 0$$

where primes denote differentiation. They satisfy the recursion relation

$$H_{v+1} - 2yH_v + 2vH_{v-1} = 0$$

An important integral is

$$\int_{-\infty}^{\infty} H_{v'}H_v e^{-y^2} dy = \begin{cases} 0 & \text{if } v' \neq v \\ \pi^{1/2}2^v v! & \text{if } v' = v \end{cases}$$

Hermite polynomials are members of a class of functions called *orthogonal polynomials.* These polynomials have a wide range of important properties that allow a number of quantum mechanical calculations to be done with relative ease. See *Further reading* for a reference to their properties.

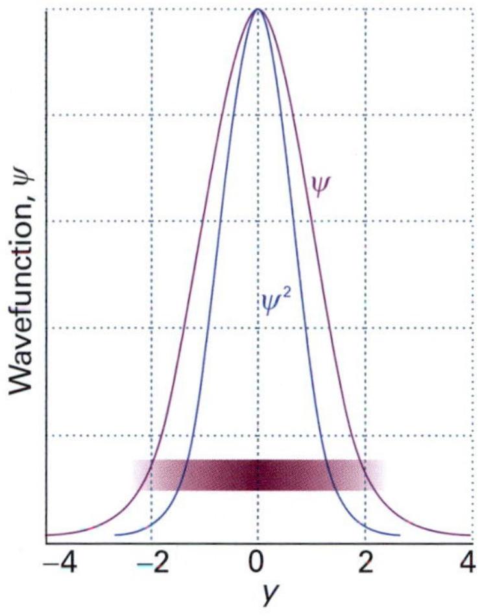

Fig. 2.23 The normalized wavefunction and probability distribution (shown also by shading) for the lowest energy state of a harmonic oscillator.

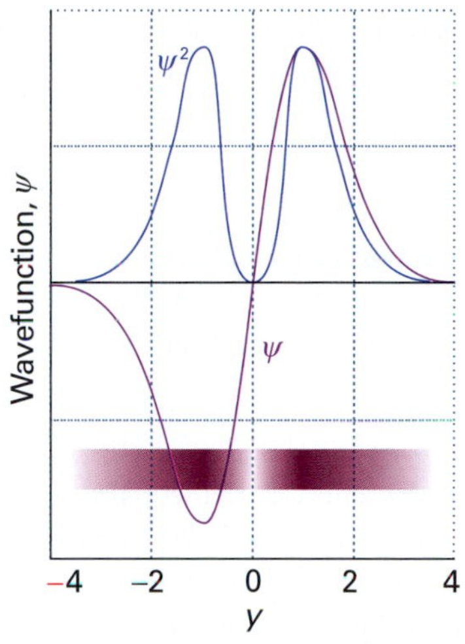

Fig. 2.24 The normalized wavefunction and probability distribution (shown also by shading) for the first excited state of a harmonic oscillator.

Table 2.1 The Hermite polynomials $H_v(y)$

v	$H_v(y)$
0	1
1	$2y$
2	$4y^2 - 2$
3	$8y^3 - 12y$
4	$16y^4 - 48y^2 + 12$
5	$32y^5 - 160y^3 + 120y$
6	$64y^6 - 480y^4 + 720y^2 - 120$

This function has a node at zero displacement ($x = 0$), and the probability density has maxima at $x = \pm\alpha$, corresponding to $y = \pm 1$ (Fig. 2.24).

Once again, we should interpret the mathematical expressions we have derived. In the case of the harmonic oscillator wavefunctions in eqn 2.28, we should note the following:

1. The Gaussian function goes very strongly to zero as the displacement increases (in either direction), so all the wavefunctions approach zero at large displacements.

2. The exponent y^2 is proportional to $x^2 \times (mk)^{1/2}$, so the wavefunctions decay more rapidly for large masses and stiff springs.

3. As v increases, the Hermite polynomials become larger at large displacements (as x^v), so the wavefunctions grow large before the Gaussian function damps them down to zero: as a result, the wavefunctions spread over a wider range as v increases.

The shapes of several wavefunctions are shown in Fig. 2.25. The shading in Fig. 2.26 that represents the probability density is based on the squares of these functions. At high quantum numbers, harmonic oscillator wavefunctions have their largest amplitudes near the turning points of the classical motion (the locations at which $V = E$, so

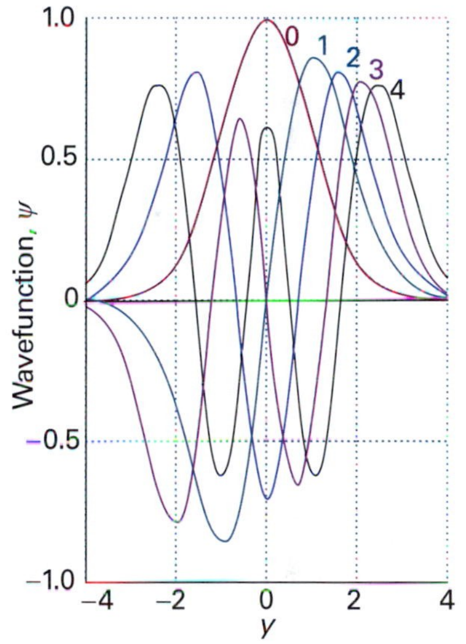

Fig. 2.25 The normalized wavefunctions for the first five states of a harmonic oscillator. Note that the number of nodes is equal to v and that alternate wavefunctions are symmetrical or antisymmetrical about $y = 0$ (zero displacement).

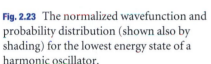

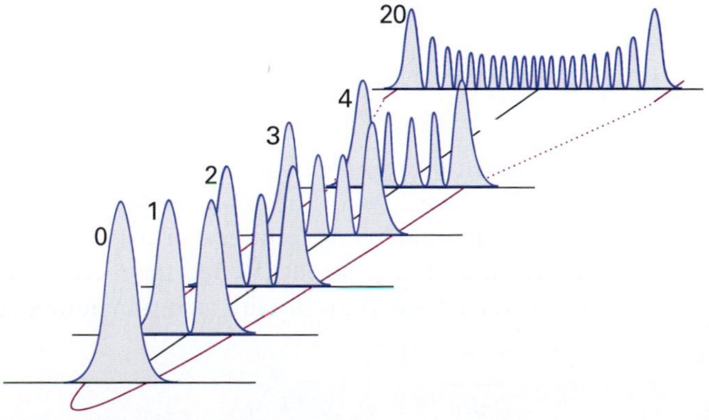

Fig. 2.26 The probability distributions for the first five states of a harmonic oscillator and the state with $v = 20$. Note how the regions of highest probability move towards the turning points of the classical motion as v increases.

Exploration To gain some insight into the origins of the nodes in the harmonic oscillator wavefunctions, plot the Hermite polynomials $H_v(y)$ for $v = 0$ through 5.

the kinetic energy is zero). We see classical properties emerging in the correspondence limit of high quantum numbers, for a classical particle is most likely to be found at the turning points (where it travels most slowly) and is least likely to be found at zero displacement (where it travels most rapidly).

Example 2.3 *Normalizing a harmonic oscillator wavefunction*

Find the normalization constant for the harmonic oscillator wavefunctions.

Method Normalization is always carried out by evaluating the integral of $|\psi|^2$ over all space and then finding the normalization factor from eqn 1.16. The normalized wavefunction is then equal to $N\psi$. In this one-dimensional problem, the volume element is dx and the integration is from $-\infty$ to $+\infty$. The wavefunctions are expressed in terms of the dimensionless variable $y = x/\alpha$, so begin by expressing the integral in terms of y by using $\mathrm{d}x = \alpha\mathrm{d}y$. The integrals required are given in *Comment 2.2*.

Answer The unnormalized wavefunction is

$$\psi_v(x) = H_v(y)\mathrm{e}^{-y^2/2}$$

It follows from the integrals given in *Comment 2.2* that

$$\int_{-\infty}^{\infty} \psi_v^* \psi_v \mathrm{d}x = \alpha \int_{-\infty}^{\infty} \psi_v^* \psi_v \mathrm{d}y = \alpha \int_{-\infty}^{\infty} H_v^2(y)\mathrm{e}^{-y^2}\mathrm{d}y = \alpha\pi^{1/2}2^v v!$$

where $v! = v(v-1)(v-2)\ldots 1$. Therefore,

$$N_v = \left(\frac{1}{\alpha\pi^{1/2}2^v v!}\right)^{1/2}$$

Note that for a harmonic oscillator N_v is different for each value of v.

Self-test 2.6 Confirm, by explicit evaluation of the integral, that ψ_0 and ψ_1 are orthogonal.

[Evaluate the integral $\int_{-\infty}^{\infty} \psi_0^* \psi_1 \,\mathrm{d}x$ by using the information in *Comment 2.2*.]

(b) The properties of oscillators

With the wavefunctions that are available, we can start calculating the properties of a harmonic oscillator. For instance, we can calculate the expectation values of an observable Ω by evaluating integrals of the type

$$\langle\Omega\rangle = \int_{-\infty}^{\infty} \psi_v^* \hat{\Omega} \psi_v \mathrm{d}x \tag{2.31}$$

(Here and henceforth, the wavefunctions are all taken to be normalized to 1.) When the explicit wavefunctions are substituted, the integrals look fearsome, but the Hermite polynomials have many simplifying features. For instance, we show in the following example that the mean displacement, $\langle x\rangle$, and the mean square displacement, $\langle x^2\rangle$, of the oscillator when it is in the state with quantum number v are

$$\langle x\rangle = 0 \qquad \langle x^2\rangle = (v + \tfrac{1}{2})\frac{\hbar}{(mk)^{1/2}} \tag{2.32}$$

The result for $\langle x \rangle$ shows that the oscillator is equally likely to be found on either side of $x = 0$ (like a classical oscillator). The result for $\langle x^2 \rangle$ shows that the mean square displacement increases with v. This increase is apparent from the probability densities in Fig. 2.26, and corresponds to the classical amplitude of swing increasing as the oscillator becomes more highly excited.

Example 2.4 *Calculating properties of a harmonic oscillator*

We can imagine the bending motion of a CO_2 molecule as a harmonic oscillation relative to the linear conformation of the molecule. We may be interested in the extent to which the molecule bends. Calculate the mean displacement of the oscillator when it is in a quantum state v.

Method Normalized wavefunctions must be used to calculate the expectation value. The operator for position along x is multiplication by the value of x (Section 1.5b). The resulting integral can be evaluated either by inspection (the integrand is the product of an odd and an even function), or by explicit evaluation using the formulas in *Comment 2.2*. To give practice in this type of calculation, we illustrate the latter procedure. We shall need the relation $x = \alpha y$, which implies that $dx = \alpha dy$.

Answer The integral we require is

$$\begin{aligned}\langle x \rangle &= \int_{-\infty}^{\infty} \psi_v^* x \psi_v \mathrm{d}x = N_v^2 \int_{-\infty}^{\infty} (H_v \mathrm{e}^{-y^2/2}) x (H_v \mathrm{e}^{-y^2/2}) \mathrm{d}x \\ &= \alpha^2 N_v^2 \int_{-\infty}^{\infty} (H_v \mathrm{e}^{-y^2/2}) y (H_v \mathrm{e}^{-y^2/2}) \mathrm{d}y \\ &= \alpha^2 N_v^2 \int_{-\infty}^{\infty} H_v y H_v \mathrm{e}^{-y^2} \mathrm{d}y\end{aligned}$$

Now use the recursion relation (see *Comment 2.2*) to form

$$yH_v = vH_{v-1} + \tfrac{1}{2}H_{v+1}$$

which turns the integral into

$$\int_{-\infty}^{\infty} H_v y H_v \mathrm{e}^{-y^2} \mathrm{d}y = v \int_{-\infty}^{\infty} H_{v-1} H_v \mathrm{e}^{-y^2} \mathrm{d}y + \tfrac{1}{2} \int_{-\infty}^{\infty} H_{v+1} H_v \mathrm{e}^{-y^2} \mathrm{d}y$$

Both integrals are zero, so $\langle x \rangle = 0$. As remarked in the text, the mean displacement is zero because the displacement occurs equally on either side of the equilibrium position. The following Self-test extends this calculation by examining the mean square displacement, which we can expect to be non-zero and to increase with increasing v.

Self-test 2.7 Calculate the mean square displacement $\langle x^2 \rangle$ of the particle from its equilibrium position. (Use the recursion relation twice.) [eqn 2.32]

Comment 2.3
An even function is one for which $f(-x) = f(x)$; an odd function is one for which $f(-x) = -f(x)$. The product of an odd and even function is itself odd, and the integral of an odd function over a symmetrical range about $x = 0$ is zero.

The mean potential energy of an oscillator, the expectation value of $V = \tfrac{1}{2}kx^2$, can now be calculated very easily:

$$\langle V \rangle = \langle \tfrac{1}{2}kx^2 \rangle = \tfrac{1}{2}(v + \tfrac{1}{2})\hbar \left(\frac{k}{m}\right)^{1/2} = \tfrac{1}{2}(v + \tfrac{1}{2})\hbar\omega \qquad (2.33)$$

Because the total energy in the state with quantum number v is $(v+\frac{1}{2})\hbar\omega$, it follows that

$$\langle V\rangle = \tfrac{1}{2}E_v \tag{2.34a}$$

The total energy is the sum of the potential and kinetic energies, so it follows at once that the mean kinetic energy of the oscillator is

$$\langle E_K\rangle = \tfrac{1}{2}E_v \tag{2.34b}$$

The result that the mean potential and kinetic energies of a harmonic oscillator are equal (and therefore that both are equal to half the total energy) is a special case of the **virial theorem**:

> If the potential energy of a particle has the form $V = ax^b$, then its mean potential and kinetic energies are related by
>
> $$2\langle E_K\rangle = b\langle V\rangle \tag{2.35}$$

For a harmonic oscillator $b = 2$, so $\langle E_K\rangle = \langle V\rangle$, as we have found. The virial theorem is a short cut to the establishment of a number of useful results, and we shall use it again.

An oscillator may be found at extensions with $V > E$ that are forbidden by classical physics, for they correspond to negative kinetic energy. For example, it follows from the shape of the wavefunction (see the *Justification* below) that in its lowest energy state there is about an 8 per cent chance of finding an oscillator stretched beyond its classical limit and an 8 per cent chance of finding it with a classically forbidden compression. These tunnelling probabilities are independent of the force constant and mass of the oscillator. The probability of being found in classically forbidden regions decreases quickly with increasing v, and vanishes entirely as v approaches infinity, as we would expect from the correspondence principle. Macroscopic oscillators (such as pendulums) are in states with very high quantum numbers, so the probability that they will be found in a classically forbidden region is wholly negligible. Molecules, however, are normally in their vibrational ground states, and for them the probability is very significant.

Justification 2.4 *Tunnelling in the quantum mechanical harmonic oscillator*

According to classical mechanics, the turning point, x_{tp}, of an oscillator occurs when its kinetic energy is zero, which is when its potential energy $\frac{1}{2}kx^2$ is equal to its total energy E. This equality occurs when

$$x_{tp}^2 = \frac{2E}{k} \qquad \text{or, } x_{tp} = \pm\left(\frac{2E}{k}\right)^{1/2}$$

with E given by eqn 2.25. The probability of finding the oscillator stretched beyond a displacement x_{tp} is the sum of the probabilities $\psi^2\mathrm{d}x$ of finding it in any of the intervals $\mathrm{d}x$ lying between x_{tp} and infinity:

$$P = \int_{x_{tp}}^{\infty} \psi_v^2\,\mathrm{d}x$$

The variable of integration is best expressed in terms of $y = x/\alpha$ with $\alpha = (\hbar^2/mk)^{1/2}$, and then the turning point on the right lies at

$$y_{tp} = \frac{x_{tp}}{\alpha} = \left\{\frac{2(v+\frac{1}{2})\hbar\omega}{\alpha^2 k}\right\}^{1/2} = (2v+1)^{1/2}$$

For the state of lowest energy ($v = 0$), $y_{tp} = 1$ and the probability is

$$P=\int_{x_{tp}}^{\infty}\psi_0^2\mathrm{d}x=\alpha N_0^2\int_{1}^{\infty}\mathrm{e}^{-y^2}\mathrm{d}y$$

The integral is a special case of the *error function*, erf *z*, which is defined as follows:

$$\mathrm{erf}\,z=1-\frac{2}{\pi^{1/2}}\int_{z}^{\infty}\mathrm{e}^{-y^2}\mathrm{d}y$$

The values of this function are tabulated and available in mathematical software packages, and a small selection of values is given in Table 2.2. In the present case

$$P=\tfrac{1}{2}(1-\mathrm{erf}\,1)=\tfrac{1}{2}(1-0.843)=0.079$$

It follows that, in 7.9 per cent of a large number of observations, any oscillator in the state $v=0$ will be found stretched to a classically forbidden extent. There is the same probability of finding the oscillator with a classically forbidden compression. The total probability of finding the oscillator tunnelled into a classically forbidden region (stretched or compressed) is about 16 per cent. A similar calculation for the state with $v=6$ shows that the probability of finding the oscillator outside the classical turning points has fallen to about 7 per cent.

Table 2.2 The error function

z	erf *z*
0	0
0.01	0.0113
0.05	0.0564
0.10	0.1125
0.50	0.5205
1.00	0.8427
1.50	0.9661
2.00	0.9953

Rotational motion

The treatment of rotational motion can be broken down into two parts. The first deals with motion in two dimensions and the second with rotation in three dimensions. It may be helpful to review the classical description of rotational motion given in *Appendix 3*, particularly the concepts of moment of inertia and angular momentum.

2.6 Rotation in two dimensions: a particle on a ring

We consider a particle of mass m constrained to move in a circular path of radius r in the xy-plane (Fig. 2.27). The total energy is equal to the kinetic energy, because $V=0$ everywhere. We can therefore write $E=p^2/2m$. According to classical mechanics, the **angular momentum**, J_z, around the z-axis (which lies perpendicular to the xy-plane) is $J_z=\pm pr$, so the energy can be expressed as $J_z^2/2mr^2$. Because mr^2 is the **moment of inertia**, I, of the mass on its path, it follows that

$$E=\frac{J_z^2}{2I} \tag{2.36}$$

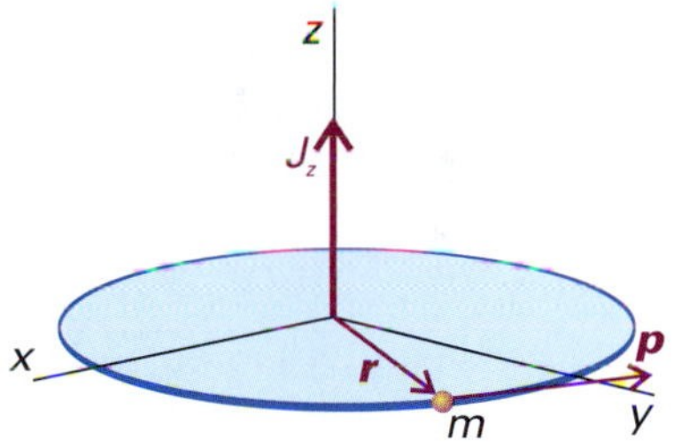

Fig. 2.27 The angular momentum of a particle of mass m on a circular path of radius r in the xy-plane is represented by a vector $\boldsymbol{J}$ with the single non-zero component J_z of magnitude pr perpendicular to the plane.

We shall now see that not all the values of the angular momentum are permitted in quantum mechanics, and therefore that both angular momentum and rotational energy are quantized.

(a) The qualitative origin of quantized rotation

Because $J_z=\pm pr$, and, from the de Broglie relation, $p=h/\lambda$, the angular momentum about the z-axis is

$$J_z=\pm\frac{hr}{\lambda}$$

Opposite signs correspond to opposite directions of travel. This equation shows that the shorter the wavelength of the particle on a circular path of given radius, the greater the angular momentum of the particle. It follows that, if we can see why the

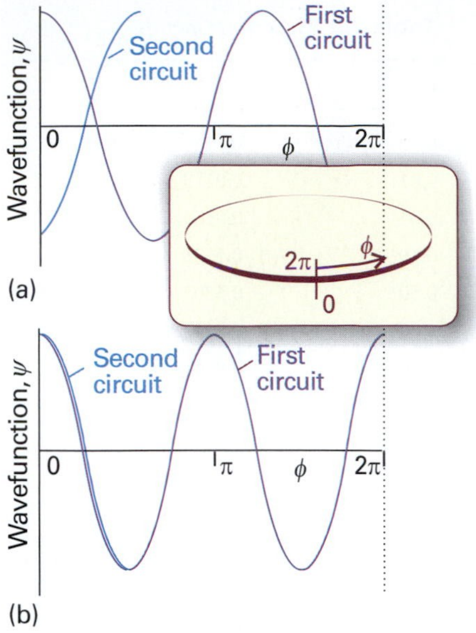

Fig. 2.28 Two solutions of the Schrödinger equation for a particle on a ring. The circumference has been opened out into a straight line; the points at $\phi = 0$ and 2π are identical. The solution in (a) is unacceptable because it is not single-valued. Moreover, on successive circuits it interferes destructively with itself, and does not survive. The solution in (b) is acceptable: it is single-valued, and on successive circuits it reproduces itself.

wavelength is restricted to discrete values, then we shall understand why the angular momentum is quantized.

Suppose for the moment that λ can take an arbitrary value. In that case, the wavefunction depends on the azimuthal angle ϕ as shown in Fig. 2.28a. When ϕ increases beyond 2π, the wavefunction continues to change, but for an arbitrary wavelength it gives rise to a different value at each point, which is unacceptable (Section 1.4b). An acceptable solution is obtained only if the wavefunction reproduces itself on successive circuits, as in Fig. 2.28b. Because only some wavefunctions have this property, it follows that only some angular momenta are acceptable, and therefore that only certain rotational energies exist. Hence, the energy of the particle is quantized. Specifically, the only allowed wavelengths are

$$\lambda = \frac{2\pi r}{m_l}$$

with m_l, the conventional notation for this quantum number, taking integral values including 0. The value $m_l = 0$ corresponds to $\lambda = \infty$; a 'wave' of infinite wavelength has a constant height at all values of ϕ. The angular momentum is therefore limited to the values

$$J_z = \pm\frac{hr}{\lambda} = \frac{m_l hr}{2\pi r} = \frac{m_l h}{2\pi}$$

where we have allowed m_l to have positive or negative values. That is,

$$J_z = m_l\hbar \qquad m_l = 0, \pm 1, \pm 2, \ldots \tag{2.37}$$

Positive values of m_l correspond to rotation in a clockwise sense around the z-axis (as viewed in the direction of z, Fig. 2.29) and negative values of m_l correspond to counter-clockwise rotation around z. It then follows from eqn 2.36 that the energy is limited to the values

$$E = \frac{J_z^2}{2I} = \frac{m_l^2\hbar^2}{2I} \tag{2.38a}$$

We shall see shortly that the corresponding normalized wavefunctions are

$$\psi_{m_l}(\phi) = \frac{e^{im_l\phi}}{(2\pi)^{1/2}} \tag{2.38b}$$

The wavefunction with $m_l = 0$ is $\psi_0(\phi) = 1/(2\pi)^{1/2}$, and has the same value at all points on the circle.

We have arrived at a number of conclusions about rotational motion by combining some classical notions with the de Broglie relation. Such a procedure can be very useful for establishing the general form (and, as in this case, the exact energies) for a quantum mechanical system. However, to be sure that the correct solutions have been obtained, and to obtain practice for more complex problems where this less formal approach is inadequate, we need to solve the Schrödinger equation explicitly. The formal solution is described in the *Justification* that follows.

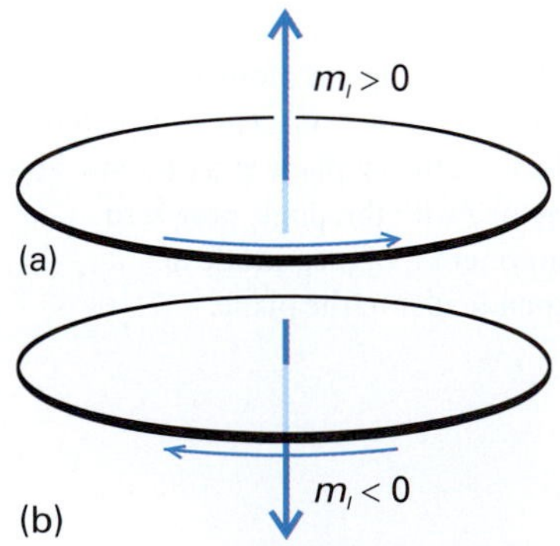

Fig. 2.29 The angular momentum of a particle confined to a plane can be represented by a vector of length $|m_l|$ units along the z-axis and with an orientation that indicates the direction of motion of the particle. The direction is given by the right-hand screw rule.

Justification 2.5 *The energies and wavefunctions of a particle on a ring*

The hamiltonian for a particle of mass m in a plane (with $V = 0$) is the same as that given in eqn 2.10:

$$\hat{H} = -\frac{\hbar^2}{2m}\left(\frac{\partial^2}{\partial x^2} + \frac{\partial^2}{\partial y^2}\right)$$

and the Schrödinger equation is $H\psi = E\psi$, with the wavefunction a function of the angle ϕ. It is always a good idea to use coordinates that reflect the full symmetry of the system, so we introduce the coordinates r and ϕ (Fig. 2.30), where $x = r\cos\phi$ and $y = r\sin\phi$. By standard manipulations (see *Further reading*) we can write

$$\frac{\partial^2}{\partial x^2}+\frac{\partial^2}{\partial y^2}=\frac{\partial^2}{\partial r^2}+\frac{1}{r}\frac{\partial}{\partial r}+\frac{1}{r^2}\frac{\partial^2}{\partial\phi^2} \tag{2.39}$$

However, because the radius of the path is fixed, the derivatives with respect to r can be discarded. The hamiltonian then becomes

$$\hat{H}=-\frac{\hbar^2}{2mr^2}\frac{d^2}{d\phi^2}$$

The moment of inertia $I = mr^2$ has appeared automatically, so H may be written

$$\hat{H}=-\frac{\hbar^2}{2I}\frac{d^2}{d\phi^2} \tag{2.40}$$

and the Schrödinger equation is

$$\frac{d^2\psi}{d\phi^2}=-\frac{2IE}{\hbar^2}\psi \tag{2.41}$$

The normalized general solutions of the equation are

$$\psi_{m_l}(\phi)=\frac{e^{im_l\phi}}{(2\pi)^{1/2}} \qquad m_l=\pm\frac{(2IE)^{1/2}}{\hbar} \tag{2.42}$$

The quantity m_l is just a dimensionless number at this stage.

We now select the acceptable solutions from among these general solutions by imposing the condition that the wavefunction should be single-valued. That is, the wavefunction ψ must satisfy a **cyclic boundary condition**, and match at points separated by a complete revolution: $\psi(\phi+2\pi)=\psi(\phi)$. On substituting the general wavefunction into this condition, we find

$$\psi_{m_l}(\phi+2\pi)=\frac{e^{im_l(\phi+2\pi)}}{(2\pi)^{1/2}}=\frac{e^{im_l\phi}e^{2\pi im_l}}{(2\pi)^{1/2}}=\psi_{m_l}(\phi)e^{2\pi im_l}$$

As $e^{i\pi}=-1$, this relation is equivalent to

$$\psi_{m_l}(\phi+2\pi)=(-1)^{2m_l}\psi(\phi) \tag{2.43}$$

Because we require $(-1)^{2m_l}=1$, $2m_l$ must be a positive or a negative even integer (including 0), and therefore m_l must be an integer: $m_l = 0, \pm1, \pm2, \ldots$.

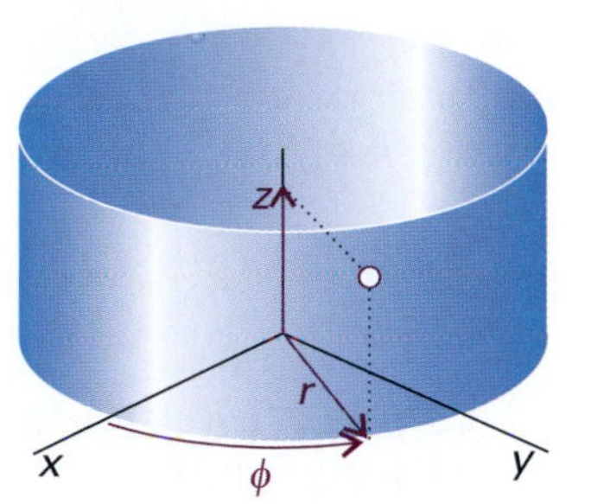

Fig. 2.30 The cylindrical coordinates z, r, and ϕ for discussing systems with axial (cylindrical) symmetry. For a particle confined to the xy-plane, only r and ϕ can change.

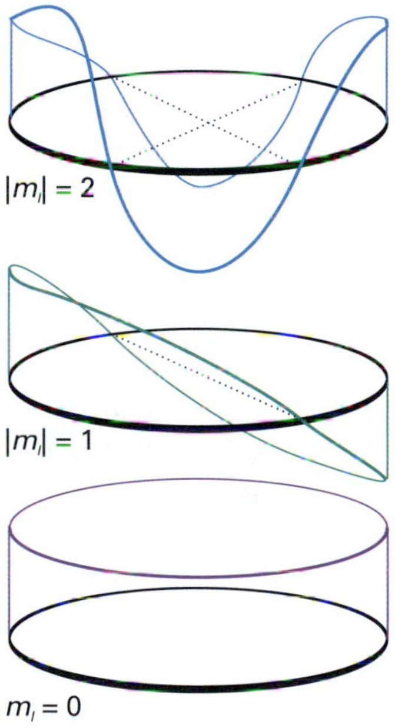

Fig. 2.31 The real parts of the wavefunctions of a particle on a ring. As shorter wavelengths are achieved, the magnitude of the angular momentum around the z-axis grows in steps of $\hbar$.

(b) Quantization of rotation

We can summarize the conclusions so far as follows. The energy is quantized and restricted to the values given in eqn 2.38a ($E = m_l^2\hbar^2/2I$). The occurrence of m_l as its square means that the energy of rotation is independent of the sense of rotation (the sign of m_l), as we expect physically. In other words, states with a given value of $|m_l|$ are doubly degenerate, except for $m_l = 0$, which is non-degenerate. Although the result has been derived for the rotation of a single mass point, it also applies to any body of moment of inertia I constrained to rotate about one axis.

We have also seen that the angular momentum is quantized and confined to the values given in eqn 2.37 ($J_z = m_l\hbar$). The increasing angular momentum is associated with the increasing number of nodes in the real and imaginary parts of the wavefunction: the wavelength decreases stepwise as $|m_l|$ increases, so the momentum with which the particle travels round the ring increases (Fig. 2.31). As shown in the following

Comment 2.4

The complex function $e^{im_l\phi}$ does not have nodes; however, it may be written as $\cos m_l\phi + i\sin m_l\phi$, and the real ($\cos m_l\phi$) and imaginary ($\sin m_l\phi$) components do have nodes.

Comment 2.5

The angular momentum in three dimensions is defined as

$$\boldsymbol{l}=\boldsymbol{r}\times\boldsymbol{p}=\begin{vmatrix} \boldsymbol{i} & \boldsymbol{j} & \boldsymbol{k} \\ x & y & z \\ p_x & p_y & p_z \end{vmatrix}$$
$$=(yp_z-zp_y)\boldsymbol{i}-(xp_z-zp_x)\boldsymbol{j}+(xp_y-yp_x)\boldsymbol{k}$$

where $\boldsymbol{i}$, $\boldsymbol{j}$, and $\boldsymbol{k}$ are unit vectors pointing along the positive directions on the x-, y-, and z-axes. It follows that the z-component of the angular momentum has a magnitude given by eqn 2.44. For more information on vectors, see *Appendix 2*.

Justification, we can come to the same conclusion more formally by using the argument about the relation between eigenvalues and the values of observables established in Section 1.5.

Justification 2.6 *The quantization of angular momentum*

In the discussion of translational motion in one dimension, we saw that the opposite signs in the wavefunctions e^{ikx} and e^{-ikx} correspond to opposite directions of travel, and that the linear momentum is given by the eigenvalue of the linear momentum operator. The same conclusions can be drawn here, but now we need the eigenvalues of the angular momentum operator. In classical mechanics the orbital angular momentum l_z about the z-axis is defined as

$$l_z=xp_y-yp_x \qquad [2.44]$$

where p_x is the component of linear motion parallel to the x-axis and p_y is the component parallel to the y-axis. The operators for the two linear momentum components are given in eqn 1.26, so the operator for angular momentum about the z-axis, which we denote $\hat{l}_z$, is

$$\hat{l}_z=\frac{\hbar}{\mathrm{i}}\left(x\frac{\partial}{\partial y}-y\frac{\partial}{\partial x}\right) \qquad (2.45)$$

When expressed in terms of the coordinates r and ϕ, by standard manipulations this equation becomes

$$\hat{l}_z=\frac{\hbar}{\mathrm{i}}\frac{\partial}{\partial\phi} \qquad (2.46)$$

With the angular momentum operator available, we can test the wavefunction in eqn 2.38b. Disregarding the normalization constant, we find

$$\hat{l}_z\psi_{m_l}=\frac{\hbar}{\mathrm{i}}\frac{\mathrm{d}\psi_{m_l}}{\mathrm{d}\phi}=\mathrm{i}m_l\frac{\hbar}{\mathrm{i}}e^{\mathrm{i}m_l\phi}=m_l\hbar\psi_{m_l} \qquad (2.47)$$

That is, ψ_{m_l} is an eigenfunction of $\hat{l}_z$, and corresponds to an angular momentum $m_l\hbar$. When m_l is positive, the angular momentum is positive (clockwise when seen from below); when m_l is negative, the angular momentum is negative (counterclockwise when seen from below). These features are the origin of the vector representation of angular momentum, in which the magnitude is represented by the length of a vector and the direction of motion by its orientation (Fig. 2.32).

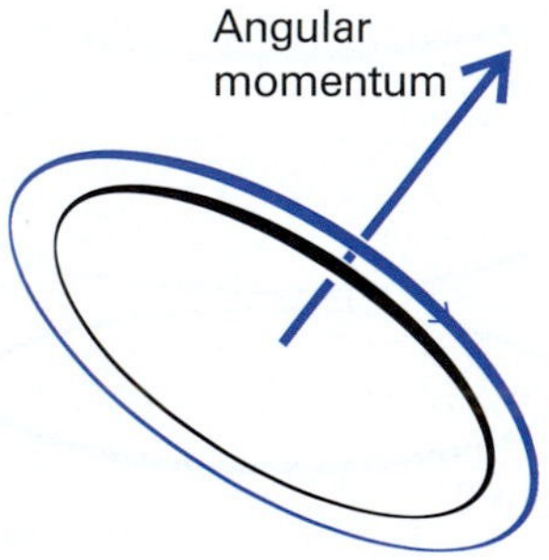

Fig. 2.32 The basic ideas of the vector representation of angular momentum: the magnitude of the angular momentum is represented by the length of the vector, and the orientation of the motion in space by the orientation of the vector (using the right-hand screw rule).

To locate the particle given its wavefunction in eqn 2.38b, we form the probability density:

$$\psi^{\star}_{m_l}\psi_{m_l}=\left(\frac{e^{\mathrm{i}m_l\phi}}{(2\pi)^{1/2}}\right)^{\star}\left(\frac{e^{\mathrm{i}m_l\phi}}{(2\pi)^{1/2}}\right)=\left(\frac{e^{-\mathrm{i}m_l\phi}}{(2\pi)^{1/2}}\right)\left(\frac{e^{\mathrm{i}m_l\phi}}{(2\pi)^{1/2}}\right)=\frac{1}{2\pi}$$

Because this probability density is independent of ϕ, the probability of locating the particle somewhere on the ring is also independent of ϕ (Fig. 2.33). Hence the location of the particle is completely indefinite, and knowing the angular momentum precisely eliminates the possibility of specifying the particle's location. Angular momentum and angle are a pair of complementary observables (in the sense defined in Section 1.6), and the inability to specify them simultaneously with arbitrary precision is another example of the uncertainty principle.

2.7 Rotation in three dimensions: the particle on a sphere

We now consider a particle of mass m that is free to move anywhere on the surface of a sphere of radius r. We shall need the results of this calculation when we come to describe rotating molecules and the states of electrons in atoms and in small clusters of atoms. The requirement that the wavefunction should match as a path is traced over the poles as well as round the equator of the sphere surrounding the central point introduces a second cyclic boundary condition and therefore a second quantum number (Fig. 2.34).

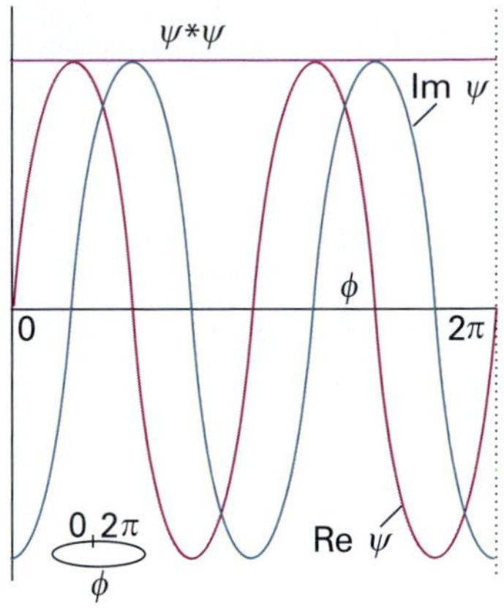

Fig. 2.33 The probability density for a particle in a definite state of angular momentum is uniform, so there is an equal probability of finding the particle anywhere on the ring.

(a) The Schrödinger equation

The hamiltonian for motion in three dimensions (Table 1.1) is

$$\hat{H} = -\frac{\hbar^2}{2m}\nabla^2 + V \qquad \nabla^2 = \frac{\partial^2}{\partial x^2} + \frac{\partial^2}{\partial y^2} + \frac{\partial^2}{\partial z^2} \tag{2.48}$$

The symbol ∇^2 is a convenient abbreviation for the sum of the three second derivatives; it is called the **laplacian**, and read either 'del squared' or 'nabla squared'. For the particle confined to a spherical surface, $V = 0$ wherever it is free to travel, and the radius r is a constant. The wavefunction is therefore a function of the **colatitude**, θ, and the **azimuth**, ϕ (Fig. 2.35), and we write it $\psi(\theta,\phi)$. The Schrödinger equation is

$$-\frac{\hbar^2}{2m}\nabla^2\psi = E\psi \tag{2.49}$$

As shown in the following *Justification*, this partial differential equation can be simplified by the separation of variables procedure by expressing the wavefunction (for constant r) as the product

$$\psi(\theta,\phi) = \Theta(\theta)\Phi(\phi) \tag{2.50}$$

where Θ is a function only of θ and Φ is a function only of ϕ.

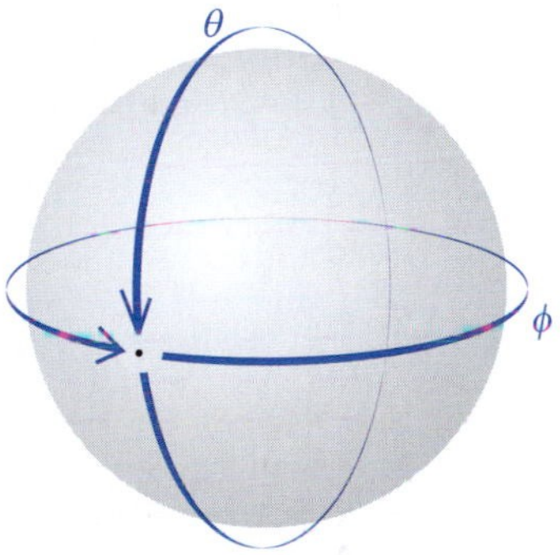

Fig. 2.34 The wavefunction of a particle on the surface of a sphere must satisfy two cyclic boundary conditions; this requirement leads to two quantum numbers for its state of angular momentum.

Justification 2.7 *The separation of variables technique applied to the particle on a sphere*

The laplacian in spherical polar coordinates is (see *Further reading*)

$$\nabla^2 = \frac{\partial^2}{\partial r^2} + \frac{2}{r}\frac{\partial}{\partial r} + \frac{1}{r^2}\Lambda^2 \tag{2.51a}$$

where the **legendrian**, Λ^2, is

$$\Lambda^2 = \frac{1}{\sin^2\theta}\frac{\partial^2}{\partial\phi^2} + \frac{1}{\sin\theta}\frac{\partial}{\partial\theta}\sin\theta\frac{\partial}{\partial\theta} \tag{2.51b}$$

Because r is constant, we can discard the part of the laplacian that involves differentiation with respect to r, and so write the Schrödinger equation as

$$\frac{1}{r^2}\Lambda^2\psi = -\frac{2mE}{\hbar^2}\psi$$

or, because $I = mr^2$, as

$$\Lambda^2\psi = -\varepsilon\psi \qquad \varepsilon = \frac{2IE}{\hbar^2}$$

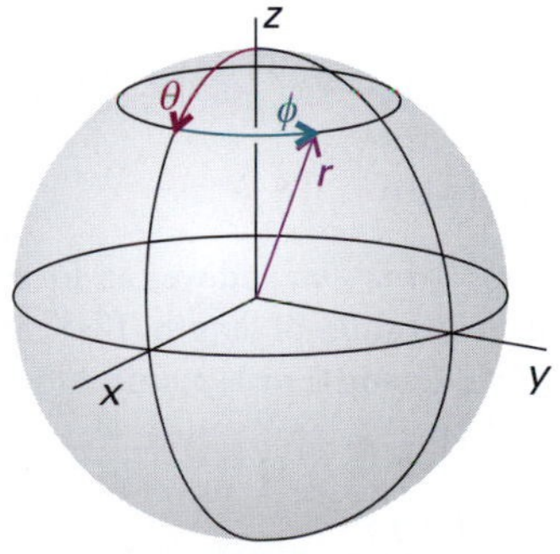

Fig. 2.35 Spherical polar coordinates. For a particle confined to the surface of a sphere, only the colatitude, θ, and the azimuth, ϕ, can change.

Table 2.3 The spherical harmonics

l	m_l	$Y_{l,m_l}(\theta,\varphi)$
0	0	$\left(\frac{1}{4\pi}\right)^{1/2}$
1	0	$\left(\frac{3}{4\pi}\right)^{1/2}\cos\theta$
	±1	$\mp\left(\frac{3}{8\pi}\right)^{1/2}\sin\theta\,e^{\pm i\phi}$
2	0	$\left(\frac{5}{16\pi}\right)^{1/2}(3\cos^2\theta-1)$
	±1	$\mp\left(\frac{15}{8\pi}\right)^{1/2}\cos\theta\sin\theta\,e^{\pm i\phi}$
	±2	$\left(\frac{15}{32\pi}\right)^{1/2}\sin^2\theta\,e^{\pm 2i\phi}$
3	0	$\left(\frac{7}{16\pi}\right)^{1/2}(5\cos^3\theta-3\cos\theta)$
	±1	$\mp\left(\frac{21}{64\pi}\right)^{1/2}(5\cos^2\theta-1)\sin\theta\,e^{\pm i\phi}$
	±2	$\left(\frac{105}{32\pi}\right)^{1/2}\sin^2\theta\cos\theta\,e^{\pm 2i\phi}$
	±3	$\mp\left(\frac{35}{64\pi}\right)^{1/2}\sin^3\theta\,e^{\pm 3i\phi}$

To verify that this expression is separable, we substitute $\psi = \Theta\Phi$:

$$\frac{1}{\sin^2\theta}\frac{\partial^2(\Theta\Phi)}{\partial\phi^2}+\frac{1}{\sin\theta}\frac{\partial}{\partial\theta}\sin\theta\frac{\partial(\Phi\Theta)}{\partial\theta}=-\varepsilon\Theta\Phi$$

We now use the fact that Θ and Φ are each functions of one variable, so the partial derivatives become complete derivatives:

$$\frac{\Theta}{\sin^2\theta}\frac{d^2\Phi}{d\phi^2}+\frac{\Phi}{\sin\theta}\frac{d}{d\theta}\sin\theta\frac{d\Theta}{d\theta}=-\varepsilon\Theta\Phi$$

Division through by $\Theta\Phi$, multiplication by $\sin^2\theta$, and minor rearrangement gives

$$\Phi\frac{d^2\Phi}{d\phi^2}+\frac{\sin\theta}{\Theta}\frac{d}{d\theta}\sin\theta\frac{d\Theta}{d\theta}+\varepsilon\sin^2\theta=0$$

The first term on the left depends only on ϕ and the remaining two terms depend only on θ. We met a similar situation when discussing a particle on a rectangular surface (*Justification 2.3*), and by the same argument, the complete equation can be separated. Thus, if we set the first term equal to the numerical constant $-m_l^2$ (using a notation chosen with an eye to the future), the separated equations are

$$\frac{1}{\Phi}\frac{d^2\Phi}{d\phi^2}=-m_l^2 \qquad \frac{\sin\theta}{\Theta}\frac{d}{d\theta}\sin\theta\frac{d\Theta}{d\theta}+\varepsilon\sin^2\theta=m_l^2$$

The first of these two equations is the same as that in *Justification 2.5*, so it has the same solutions (eqn 2.38b). The second is much more complicated to solve, but the solutions are tabulated as the *associated Legendre functions*. The cyclic boundary conditions on Θ result in the introduction of a second quantum number, l, which identifies the acceptable solutions. The presence of the quantum number m_l in the second equation implies, as we see below, that the range of acceptable values of m_l is restricted by the value of l.

As indicated in *Justification 2.7*, solution of the Schrödinger equation shows that the acceptable wavefunctions are specified by two quantum numbers l and m_l that are restricted to the values

$$l=0, 1, 2, \ldots \qquad m_l=l, l-1, \ldots, -l \tag{2.52}$$

Note that the **orbital angular momentum quantum number** l is non-negative and that, for a given value of l, there are $2l+1$ permitted values of the **magnetic quantum number**, m_l. The normalized wavefunctions are usually denoted $Y_{l,m_l}(\theta,\phi)$ and are called the **spherical harmonics** (Table 2.3).

Figure 2.36 is a representation of the spherical harmonics for $l=0$ to 4 and $m_l=0$ which emphasizes how the number of angular nodes (the positions at which the wavefunction passes through zero) increases as the value of l increases. There are no angular nodes around the z-axis for functions with $m_l = 0$, which corresponds to there being no component of orbital angular momentum about that axis. Figure 2.37 shows the distribution of the particle of a given angular momentum in more detail. In this representation, the value of $|Y_{l,m_l}|^2$ at each value of θ and ϕ is proportional to the distance of the surface from the origin. Note how, for a given value of l, the most probable location of the particle migrates towards the xy-plane as the value of $|m_l|$ increases.

It also follows from the solution of the Schrödinger equation that the energy E of the particle is restricted to the values

$$E=l(l+1)\frac{\hbar^2}{2I} \qquad l=0, 1, 2, \ldots \tag{2.53}$$

Comment 2.6

The spherical harmonics are orthogonal and normalized in the following sense:

$$\int_0^\pi\int_0^{2\pi} Y_{l',m_l'}(\theta,\phi)^* Y_{l,m_l}(\theta,\phi)\sin\theta\,d\theta\,d\phi = \delta_{l'l}\delta_{m_l'm_l}$$

An important 'triple integral' is

$$\int_0^\pi\int_0^{2\pi} Y_{l'',m_l''}(\theta,\phi)^* Y_{l',m_l'}(\theta,\phi) Y_{l,m_l}(\theta,\phi)\sin\theta\,d\theta\,d\phi=0$$

unless $m_l''=m_l'+m_l$ and we can form a triangle with sides of lengths l'', l', and l (such as 1, 2, and 3 or 1, 1, and 1, but not 1, 2, and 4).

Comment 2.7

The real and imaginary components of the Φ component of the wavefunctions, $e^{im_l\phi}=\cos m_l\phi + i\sin m_l\phi$, each have $|m_l|$ angular nodes, but these nodes are not seen when we plot the probability density, because $|e^{im_l\phi}|^2=1$.

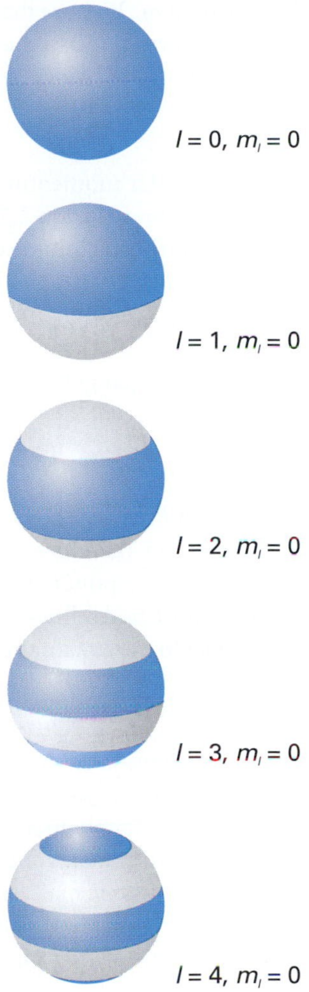

Fig. 2.36 A representation of the wavefunctions of a particle on the surface of a sphere which emphasizes the location of angular nodes: dark and light shading correspond to different signs of the wavefunction. Note that the number of nodes increases as the value of l increases. All these wavefunctions correspond to $m_l = 0$; a path round the vertical z-axis of the sphere does not cut through any nodes.

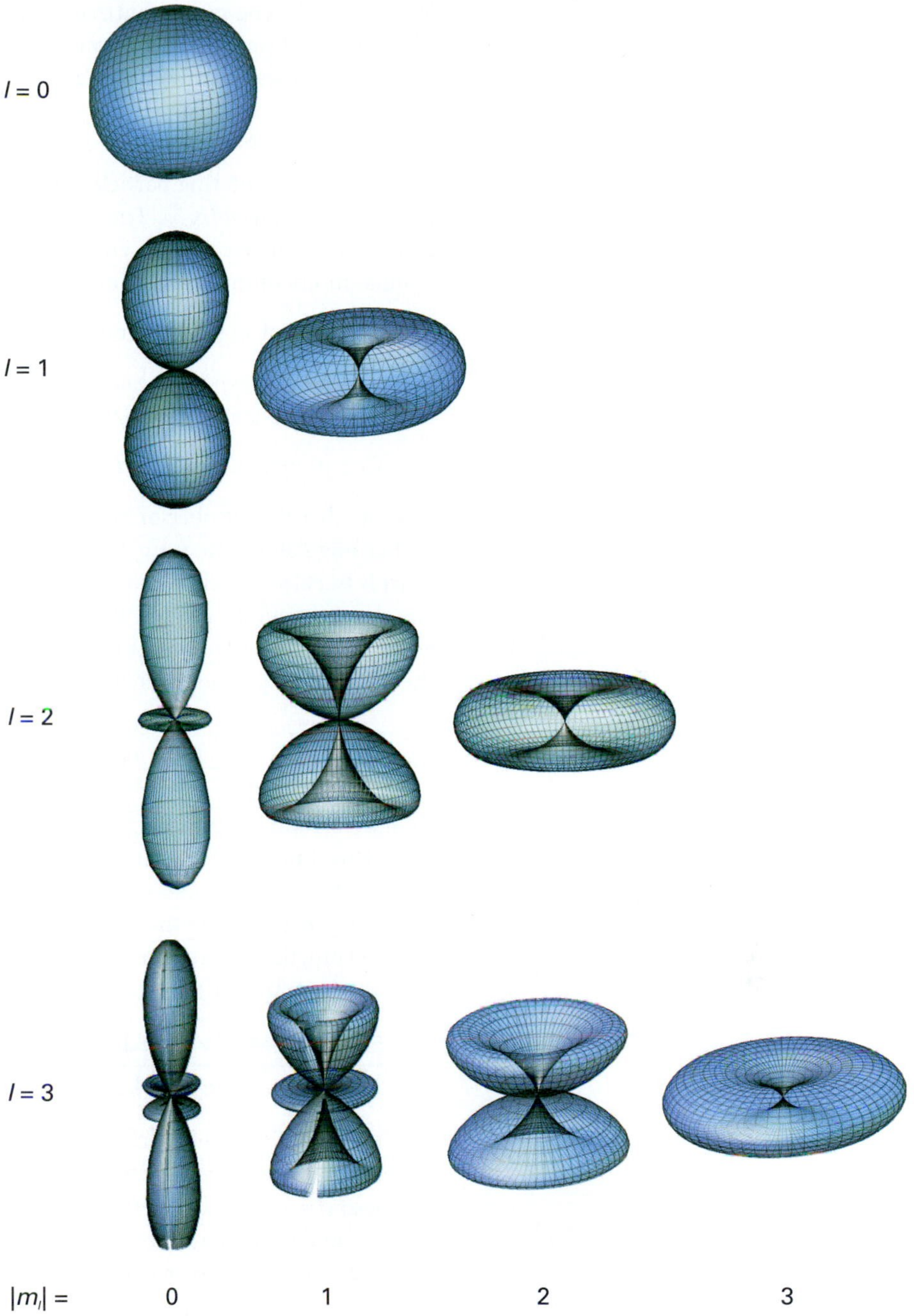

Fig. 2.37 A more complete representation of the wavefunctions for $l = 0$, 1, 2, and 3. The distance of a point on the surface from the origin is proportional to the square modulus of the amplitude of the wavefunction at that point.

Exploration Plot the variation with the radius r of the first ten energy levels of a particle on a sphere. Which of the following statements are true: (a) for a given value of r, the energy separation between adjacent levels decreases with increasing l, (b) increasing r leads to an decrease in the value of the energy for each level, (c) the energy difference between adjacent levels increases as r increases.

We see that the energy is quantized, and that it is independent of m_l. Because there are $2l + 1$ different wavefunctions (one for each value of m_l) that correspond to the same energy, it follows that a level with quantum number l is $(2l + 1)$-fold degenerate.

(b) Angular momentum

The energy of a rotating particle is related classically to its angular momentum J by $E = J^2/2I$ (see *Appendix 3*). Therefore, by comparing this equation with eqn 2.53, we can deduce that, because the energy is quantized, then so too is the magnitude of the angular momentum, and confined to the values

$$\text{Magnitude of angular momentum} = \{l(l+1)\}^{1/2}\hbar \qquad l = 0, 1, 2 \ldots \qquad (2.54a)$$

We have already seen (in the context of rotation in a plane) that the angular momentum about the z-axis is quantized, and that it has the values

$$z\text{-Component of angular momentum} = m_l\hbar \qquad m_l = l, l-1, \ldots, -l \qquad (2.54b)$$

The fact that the number of nodes in $\psi_{l,m_l}(\theta,\phi)$ increases with l reflects the fact that higher angular momentum implies higher kinetic energy, and therefore a more sharply buckled wavefunction. We can also see that the states corresponding to high angular momentum around the z-axis are those in which most nodal lines cut the equator: a high kinetic energy now arises from motion parallel to the equator because the curvature is greatest in that direction.

Illustration 2.4 *Calculating the frequency of a molecular rotational transition*

Under certain circumstances, the particle on a sphere is a reasonable model for the description of the rotation of diatomic molecules. Consider, for example, the rotation of a $^1H^{127}I$ molecule: because of the large difference in atomic masses, it is appropriate to picture the ^1H atom as orbiting a stationary ^{127}I atom at a distance $r = 160$ pm, the equilibrium bond distance. The moment of inertia of $^1H^{127}I$ is then $I = m_H r^2 = 4.288 \times 10^{-47}$ kg m^2. It follows that

$$\frac{\hbar^2}{2I} = \frac{(1.054\,57 \times 10^{-34}\ \text{J s})^2}{2 \times (4.288 \times 10^{-47}\ \text{kg m}^2)} = 1.297 \times 10^{-22}\ \text{J}$$

or 0.1297 zJ. This energy corresponds to 78.09 J mol^{-1}. From eqn 2.53, the first few rotational energy levels are therefore 0 ($l = 0$), 0.2594 zJ ($l = 1$), 0.7782 zJ ($l = 2$), and 1.556 zJ ($l = 3$). The degeneracies of these levels are 1, 3, 5, and 7, respectively (from $2l + 1$), and the magnitudes of the angular momentum of the molecule are 0, $2^{1/2}\hbar$, $6^{1/2}\hbar$, and $(12)^{1/2}\hbar$ (from eqn 2.54a). It follows from our calculations that the $l = 0$ and $l = 1$ levels are separated by $\Delta E = 0.2594$ zJ. A transition between these two rotational levels of the molecule can be brought about by the emission or absorption of a photon with a frequency given by the Bohr frequency condition (eqn 1.10):

$$\nu = \frac{\Delta E}{h} = \frac{2.594 \times 10^{-22}\ \text{J}}{6.626 \times 10^{-34}\ \text{J s}} = 3.915 \times 10^{11}\ \text{Hz} = 391.5\ \text{GHz}$$

Radiation with this frequency belongs to the microwave region of the electromagnetic spectrum, so microwave spectroscopy is a convenient method for the study of molecular rotations. Because the transition energies depend on the moment of inertia, microwave spectroscopy is a very accurate technique for the determination of bond lengths. We discuss rotational spectra further in Chapter 6.

Self-test 2.8 Repeat the calculation for a $^2H^{127}I$ molecule (same bond length as $^1H^{127}I$).

[Energies are smaller by a factor of two; same angular momenta and numbers of components]

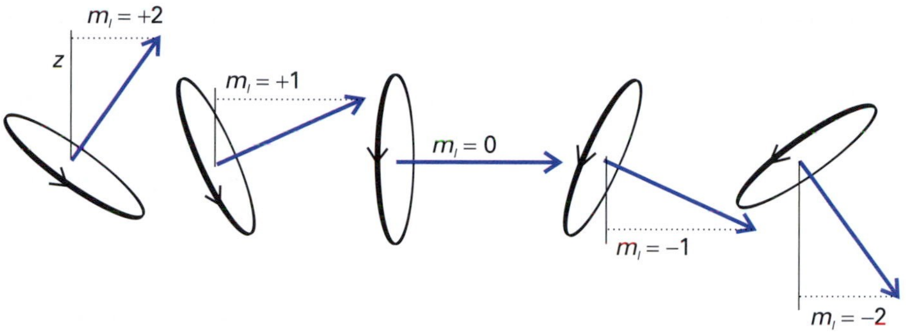

Fig. 2.38 The permitted orientations of angular momentum when $l = 2$. We shall see soon that this representation is too specific because the azimuthal orientation of the vector (its angle around z) is indeterminate.

(c) Space quantization

The result that m_l is confined to the values $l, l-1, \ldots, -l$ for a given value of l means that the component of angular momentum about the z-axis may take only $2l+1$ values. If the angular momentum is represented by a vector of length proportional to its magnitude (that is, of length $\{l(l+1)\}^{1/2}$ units), then to represent correctly the value of the component of angular momentum, the vector must be oriented so that its projection on the z-axis is of length m_l units. In classical terms, this restriction means that the plane of rotation of the particle can take only a discrete range of orientations (Fig. 2.38). The remarkable implication is that *the orientation of a rotating body is quantized*.

The quantum mechanical result that a rotating body may not take up an arbitrary orientation with respect to some specified axis (for example, an axis defined by the direction of an externally applied electric or magnetic field) is called **space quantization**. It was confirmed by an experiment first performed by Otto Stern and Walther Gerlach in 1921, who shot a beam of silver atoms through an inhomogeneous magnetic field (Fig. 2.39). The idea behind the experiment was that a rotating, charged body behaves like a magnet and interacts with the applied field. According to classical mechanics, because the orientation of the angular momentum can take any value, the associated magnet can take any orientation. Because the direction in which the magnet is driven by the inhomogeneous field depends on the magnet's orientation, it follows that a broad band of atoms is expected to emerge from the region where the magnetic field acts. According to quantum mechanics, however, because the angular momentum is quantized, the associated magnet lies in a number of discrete orientations, so several sharp bands of atoms are expected.

In their first experiment, Stern and Gerlach appeared to confirm the classical prediction. However, the experiment is difficult because collisions between the atoms in the beam blur the bands. When the experiment was repeated with a beam of very low intensity (so that collisions were less frequent) they observed discrete bands, and so confirmed the quantum prediction.

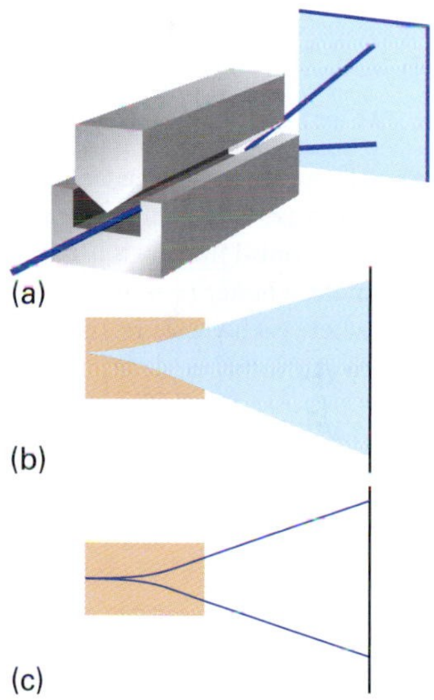

Fig. 2.39 (a) The experimental arrangement for the Stern–Gerlach experiment: the magnet provides an inhomogeneous field. (b) The classically expected result. (c) The observed outcome using silver atoms.

(d) The vector model

Throughout the preceding discussion, we have referred to the z-component of angular momentum (the component about an arbitrary axis, which is conventionally denoted z), and have made no reference to the x- and y-components (the components

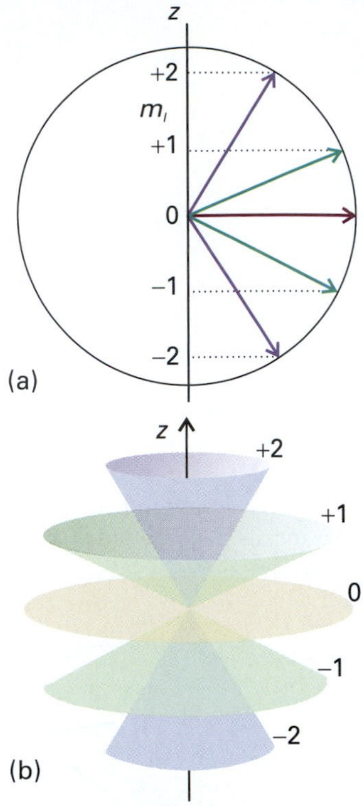

Fig. 2.40 (a) A summary of Fig. 2.38. However, because the azimuthal angle of the vector around the *z*-axis is indeterminate, a better representation is as in (b), where each vector lies at an unspecified azimuthal angle on its cone.

about the two axes perpendicular to *z*). The reason for this omission is found by examining the operators for the three components, each one being given by a term like that in eqn 2.45:

$$\hat{l}_x = \frac{\hbar}{\mathrm{i}}\left(y\frac{\partial}{\partial z} - z\frac{\partial}{\partial y}\right) \qquad \hat{l}_y = \frac{\hbar}{\mathrm{i}}\left(z\frac{\partial}{\partial x} - x\frac{\partial}{\partial z}\right) \qquad \hat{l}_z = \frac{\hbar}{\mathrm{i}}\left(x\frac{\partial}{\partial y} - y\frac{\partial}{\partial x}\right) \tag{2.55}$$

As you are invited to show in Problem 2.27, these three operators do not commute with one another:

$$[\hat{l}_x,\hat{l}_y] = \mathrm{i}\hbar\hat{l}_z \qquad [\hat{l}_y,\hat{l}_z] = \mathrm{i}\hbar\hat{l}_x \qquad [\hat{l}_z,\hat{l}_x] = \mathrm{i}\hbar\hat{l}_y \tag{2.56a}$$

Therefore, we cannot specify more than one component (unless $l = 0$). In other words, l_x, l_y, and l_z are complementary observables. On the other hand, the operator for the square of the magnitude of the angular momentum is

$$\hat{l}^2 = \hat{l}_x^2 + \hat{l}_y^2 + \hat{l}_z^2 = \hbar^2\Lambda^2 \tag{2.56b}$$

where Λ^2 is the legendrian in eqn 2.51b. This operator does commute with all three components:

$$[\hat{l}^2,\hat{l}_q] = 0 \qquad q = x, y, \text{and } z \tag{2.56c}$$

(See Problem 2.29.) Therefore, although we may specify the magnitude of the angular momentum and any of its components, if l_z is known, then it is impossible to ascribe values to the other two components. It follows that the illustration in Fig. 2.38, which is summarized in Fig. 2.40a, gives a false impression of the state of the system, because it suggests definite values for the *x*- and *y*-components. A better picture must reflect the impossibility of specifying l_x and l_y if l_z is known.

The **vector model** of angular momentum uses pictures like that in Fig. 2.40b. The cones are drawn with side $\{l(l+1)\}^{1/2}$ units, and represent the magnitude of the angular momentum. Each cone has a definite projection (of m_l units) on the *z*-axis, representing the system's precise value of l_z. The l_x and l_y projections, however, are indefinite. The vector representing the state of angular momentum can be thought of as lying with its tip on any point on the mouth of the cone. At this stage it should not be thought of as sweeping round the cone; that aspect of the model will be added later when we allow the picture to convey more information.

IMPACT ON NANOSCIENCE
I2.2 Quantum dots

In *Impact I2.1* we outlined some advantages of working in the nanometre regime. Another is the possibility of using quantum mechanical effects that render the properties of an assembly dependent on its size. Here we focus on the origins and consequences of these quantum mechanical effects.

Consider a sample of a metal, such as copper or gold. It carries an electrical current because the electrons are delocalized over all the atomic nuclei. That is, we may treat the movement of electrons in metals with a particle in a box model, though it is necessary to imagine that the electrons move independently of each other. Immediately, we predict from eqn 2.6 that the energy levels of the electrons in a large box, such as a copper wire commonly used to make electrical connections, form a continuum so we are justified in neglecting quantum mechanical effects on the properties of the material. However, consider a *nanocrystal*, a small cluster of atoms with dimensions in the nanometre scale. Again using eqn 2.6, we predict that the electronic energies are quantized and that the separation between energy levels decreases with increasing size of the cluster. This quantum mechanical effect can be observed in 'boxes' of any shape. For example, you are invited to show in Problem 2.39 that the energy levels of an electron *in* a sphere of radius R are given by

$$E_n = \frac{n^2h^2}{8m_eR^2} \tag{2.57}$$

The quantization of energy in nanocrystals has important technological implications when the material is a semiconductor, in which the electrical conductivity increases with increasing temperature or upon excitation by light. Transfer of energy to a semiconductor increases the mobility of electrons in the material. However, for every electron that moves to a different site in the sample, a unit of positive charge, called a *hole*, is left behind. The holes are also mobile, so to describe electrical conductivity in semiconductors we need to consider the movement of electron–hole pairs, also called **excitons**, in the material.

The electrons and holes may be regarded as particles trapped in a box, so eqn 2.6 can give us qualitative insight into the origins of conductivity in semiconductors. We conclude as before that only in nanocrystals are the energies of the charge carriers quantized. Now we explore the impact of energy quantization on the optical and electronic properties of semiconducting nanocrystals.

Three-dimensional nanocrystals of semiconducting materials containing 10^3 to 10^5 atoms are called **quantum dots.** They can be made in solution or by depositing atoms on a surface, with the size of the nanocrystal being determined by the details of the synthesis. A quantitative but approximate treatment that leads to the energy of the exciton begins with the following hamiltonian for a spherical quantum dot of radius R:

$$\hat{H} = -\frac{\hbar^2}{2m_e}\nabla_e^2 - \frac{\hbar^2}{2m_h}\nabla_h^2 + V(\boldsymbol{r}_e,\boldsymbol{r}_h) \tag{2.58}$$

where the first two terms are the kinetic energy operators for the electron and hole (with masses m_e and m_h, respectively), and the third term is the potential energy of interaction between electron and hole, which are located at positions $\boldsymbol{r}_e$ and $\boldsymbol{r}_h$ from the centre of the sphere. Taking into account only the Coulomb attraction between the hole, with charge $+e$, and the electron, with charge $-e$, we write (see Chapter 2 and *Appendix 3* for details):

$$V(\boldsymbol{r}_e,\boldsymbol{r}_h) = -\frac{e^2}{4\pi\varepsilon|\boldsymbol{r}_e - \boldsymbol{r}_h|} \tag{2.59}$$

where $|\boldsymbol{r}_e - \boldsymbol{r}_h|$ is the distance between the electron and hole and ε is the permittivity of the medium (we are ignoring the effect of polarization of the medium due to the presence of charges). Solving the Schrödinger equation in this case is not a trivial task, but the final expression for the energy of the exciton, E_{ex}, is relatively simple (see *Further reading* for details):

$$E_{ex} = \frac{h^2}{8R^2}\left(\frac{1}{m_e} + \frac{1}{m_h}\right) - \frac{1.8e^2}{4\pi\varepsilon R} \tag{2.60}$$

As expected, we see that the energy of the exciton decreases with increasing radius of the quantum dot. Moreover, for small R, the second term on the right of the preceding equation is smaller than the first term and the energy of the exciton is largely kinetic, with the resulting expression resembling the case for a particle in a sphere.

The expression for E_{ex} has important consequences for the optical properties of quantum dots. First, we see that the energy required to create mobile charge carriers and to induce electrical conductivity depends on the size of the quantum dot. The electrical properties of large, macroscopic samples of semiconductors cannot be tuned in this way. Second, in many quantum dots, such as the nearly spherical nanocrystals of cadmium selenide (CdSe), the exciton can be generated by absorption of visible light. Therefore, we predict that, as the radius of the quantum dot decreases, the excitation wavelength increases. That is, as the size of the quantum dot varies, so does the colour of the material. This phenomenon is indeed observed in suspensions of CdSe quantum dots of different sizes.

Because quantum dots are semiconductors with tunable electrical properties, it is easy to imagine uses for these materials in the manufacture of transistors. But the special optical properties of quantum dots can also be exploited. Just as the generation of an electron–hole pair requires absorption of light of a specific wavelength, so does recombination of the pair result in the emission of light of a specific wavelength. This property forms the basis for the use of quantum dots in the visualization of biological cells at work. For example, a CdSe quantum dot can be modified by covalent attachment of an organic spacer to its surface. When the other end of the spacer reacts specifically with a cellular component, such as a protein, nucleic acid, or membrane, the cell becomes labelled with a light-emitting quantum dot. The spatial distribution of emission intensity and, consequently, of the labelled molecule can then be measured with a microscope. Though this technique has been used extensively with organic molecules as labels, quantum dots are more stable and are stronger light emitters.

2.8 Spin

Stern and Gerlach observed *two* bands of Ag atoms in their experiment. This observation seems to conflict with one of the predictions of quantum mechanics, because an angular momentum l gives rise to $2l+1$ orientations, which is equal to 2 only if $l=\frac{1}{2}$, contrary to the conclusion that l must be an integer. The conflict was resolved by the suggestion that the angular momentum they were observing was not due to orbital angular momentum (the motion of an electron around the atomic nucleus) but arose instead from the motion of the electron about its own axis. This intrinsic angular momentum of the electron is called its **spin**. The explanation of the existence of spin emerged when Dirac combined quantum mechanics with special relativity and established the theory of relativistic quantum mechanics.

The spin of an electron about its own axis does not have to satisfy the same boundary conditions as those for a particle circulating around a central point, so the quantum number for spin angular momentum is subject to different restrictions. To distinguish this spin angular momentum from orbital angular momentum we use the **spin quantum number** s (in place of l; like l, s is a non-negative number) and m_s, the **spin magnetic quantum number**, for the projection on the z-axis. The magnitude of the spin angular momentum is $\{s(s+1)\}^{1/2}\hbar$ and the component $m_s\hbar$ is restricted to the $2s+1$ values

$$m_s = s, s-1, \ldots, -s \tag{2.61}$$

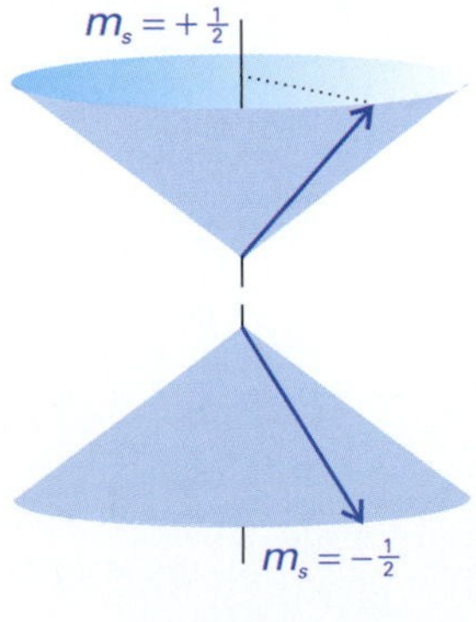

Fig. 2.41 An electron spin ($s=\frac{1}{2}$) can take only two orientations with respect to a specified axis. An α electron (top) is an electron with $m_s=+\frac{1}{2}$; a β electron (bottom) is an electron with $m_s=-\frac{1}{2}$. The vector representing the spin angular momentum lies at an angle of 55° to the z-axis (more precisely, the half-angle of the cones is $\arccos(1/3^{1/2})$).

The detailed analysis of the spin of a particle is sophisticated and shows that the property should not be taken to be an actual spinning motion. It is better to regard 'spin' as an intrinsic property like mass and charge. However, the picture of an actual spinning motion can be very useful when used with care. For an electron it turns out that only one value of s is allowed, namely $s=\frac{1}{2}$, corresponding to an angular momentum of magnitude $(\frac{3}{4})^{1/2}\hbar = 0.866\hbar$. This spin angular momentum is an intrinsic property of the electron, like its rest mass and its charge, and every electron has exactly the same value: the magnitude of the spin angular momentum of an electron cannot be changed. The spin may lie in $2s+1=2$ different orientations (Fig. 2.41).

One orientation corresponds to $m_s = +\frac{1}{2}$ (this state is often denoted α or $\uparrow$); the other orientation corresponds to $m_s = -\frac{1}{2}$ (this state is denoted β or $\downarrow$).

The outcome of the Stern–Gerlach experiment can now be explained if we suppose that each Ag atom possesses an angular momentum due to the spin of a single electron, because the two bands of atoms then correspond to the two spin orientations. Why the atoms behave like this is explained in Chapter 3 (but it is already probably familiar from introductory chemistry that the ground-state configuration of a silver atom is [Kr]$4d^{10}5s^1$, a single unpaired electron outside a closed shell).

Like the electron, other elementary particles have characteristic spin. For example, protons and neutrons are spin-$\frac{1}{2}$ particles (that is, $s = \frac{1}{2}$) and invariably spin with angular momentum $(\frac{3}{4})^{1/2}\hbar = 0.866\hbar$. Because the masses of a proton and a neutron are so much greater than the mass of an electron, yet they all have the same spin angular momentum, the classical picture would be of these two particles spinning much more slowly than an electron. Some elementary particles have $s = 1$, and so have an intrinsic angular momentum of magnitude $2^{1/2}\hbar$. Some mesons are spin-1 particles (as are some atomic nuclei), but for our purposes the most important spin-1 particle is the photon. From the discussion in this chapter, we see that the photon has zero rest mass, zero charge, an energy $h\nu$, a linear momentum h/λ or $h\nu/c$, an intrinsic angular momentum of $2^{1/2}\hbar$, and travels at the speed c. We shall see the importance of photon spin in the next chapter.

Particles with half-integral spin are called **fermions** and those with integral spin (including 0) are called **bosons**. Thus, electrons and protons are fermions and photons are bosons. It is a very deep feature of nature that all the elementary particles that constitute matter are fermions whereas the elementary particles that are responsible for the forces that bind fermions together are all bosons. Photons, for example, transmit the electromagnetic force that binds together electrically charged particles. Matter, therefore, is an assembly of fermions held together by forces conveyed by bosons.

The properties of angular momentum that we have developed are set out in Table 2.4. As mentioned there, when we use the quantum numbers l and m_l we shall mean orbital angular momentum (circulation in space). When we use s and m_s we shall mean spin angular momentum (intrinsic angular momentum). When we use j and m_j we shall mean either (or, in some contexts to be described in Chapter 3, a combination of orbital and spin momenta).

Table 2.4 Properties of angular momentum

Quantum number	Symbol	Values*	Specifies
Orbital angular momentum	l	0, 1, 2, . . .	Magnitude, $\{l(l+1)\}^{1/2}\hbar$
Magnetic	m_l	$l, l-1, \ldots, -l$	Component on z-axis, $m_l\hbar$
Spin	s	$\frac{1}{2}$	Magnitude, $\{s(s+1)\}^{1/2}\hbar$
Spin magnetic	m_s	$\pm\frac{1}{2}$	Component on z-axis, $m_s\hbar$
Total	j	$l+s, l+s-1, \ldots, \lvert l-s\rvert$	Magnitude, $\{j(j+1)\}^{1/2}\hbar$
Total magnetic	m_j	$j, j-1, \ldots, -j$	Component on z-axis, $m_j\hbar$

To combine two angular momenta, use the Clebsch–Gordan series:

$$j = j_1 + j_2, j_1 + j_2 - 1, \ldots, |j_1 - j_2|$$

For many-electron systems, the quantum numbers are designated by uppercase letters (L, M_L, S, M_S, etc.).

*Note that the quantum numbers for magnitude (l, s, j, etc.) are never negative.

Techniques of approximation

All the applications treated so far have had exact solutions. However, many problems—and almost all the problems of interest in chemistry—do not have exact solutions. To make progress with these problems we need to develop techniques of approximation. There are two major approaches, *variation theory* and *perturbation theory*. Variation theory is most commonly encountered in the context of molecular orbital theory, and we consider it there (Chapter 4). Here, we concentrate on perturbation theory.

2.9 Time-independent perturbation theory

In **perturbation theory**, we suppose that the hamiltonian for the problem we are trying to solve, $\hat{H}$, can be expressed as the sum of a simple hamiltonian, $\hat{H}^{(0)}$, which has known eigenvalues and eigenfunctions, and a contribution, $\hat{H}^{(1)}$, which represents the extent to which the true hamiltonian differs from the 'model' hamiltonian:

$$\hat{H}=\hat{H}^{(0)}+\hat{H}^{(1)} \tag{2.62}$$

In **time-independent perturbation theory**, the perturbation is always present and unvarying. For example, it might represent a dip in the potential energy of a particle in a box in some region along the length of the box.

In time-independent perturbation theory, we suppose that the true energy of the system differs from the energy of the simple system, and that we can write

$$E=E^{(0)}+E^{(1)}+E^{(2)}+\dots \tag{2.63}$$

where $E^{(1)}$ is the 'first-order correction' to the energy, a contribution proportional to $\hat{H}^{(1)}$, and $E^{(2)}$ is the 'second-order correction' to the energy, a contribution proportional to $\hat{H}^{(1)2}$, and so on. The true wavefunction also differs from the 'simple' wavefunction, and we write

$$\psi=\psi^{(0)}+\psi^{(1)}+\psi^{(2)}+\dots \tag{2.64}$$

In practice, we need to consider only the 'first-order correction' to the wavefunction, $\psi^{(1)}$. As we show in *Further information 2.2*, the first- and second-order corrections to the energy of the ground state (with the wavefunction ψ_0 and energy E_0), are

$$E_0^{(1)}=\int\psi_0^{(0)*}\hat{H}^{(1)}\psi_0^{(0)}\mathrm{d}\tau=H_{00}^{(1)} \tag{2.65a}$$

and

$$E_0^{(2)}=\sum_{n\neq0}\frac{\left|\int\psi_0^{(0)*}\hat{H}^{(1)}\psi_0^{(0)}\mathrm{d}\tau\right|^2}{E_0^{(0)}-E_n^{(0)}}=\sum_{n\neq0}\frac{|H_{n0}^{(1)}|^2}{E_0^{(0)}-E_n^{(0)}} \tag{2.65b}$$

where we have introduced the **matrix element**

$$\Omega_{nm}=\int\psi_n^*\hat{\Omega}\psi_m\mathrm{d}\tau \qquad [2.65c]$$

in a convenient compact notation for integrals that we shall use frequently.

As usual, it is important to be able to interpret these equations physically. We can interpret $E^{(1)}$ as the average value of the perturbation, calculated by using the unperturbed wavefunction. An analogy is the shift in energy of vibration of a violin string when small weights are hung along its length. The weights hanging close to the nodes have little effect on its energy of vibration. Those hanging at the antinodes, however, have a pronounced effect (Fig. 2.42a). The second-order energy represents a similar average of the perturbation, but now the average is taken over the *perturbed* wavefunctions. In terms of the violin analogy, the average is now taken over the distorted waveform of the vibrating string, in which the nodes and antinodes are slightly shifted (Fig. 2.42b).

We should note the following three features of eqn 2.65b:

1. Because $E_n(0) > E_0(0)$, all the terms in the denominator are negative and, because the numerators are all positive, the second-order correction is negative, which represents a *lowering* of the energy of the ground state.

2. The perturbation appears (as its square) in the numerator; so the stronger the perturbation, the greater the lowering of the ground-state energy.

3. If the energy levels of the system are widely spaced, all the denominators are large, so the sum is likely to be small; in which case the perturbation has little effect on the energy of the system: the system is 'stiff', and unresponsive to perturbations. The opposite is true when the energy levels lie close together.

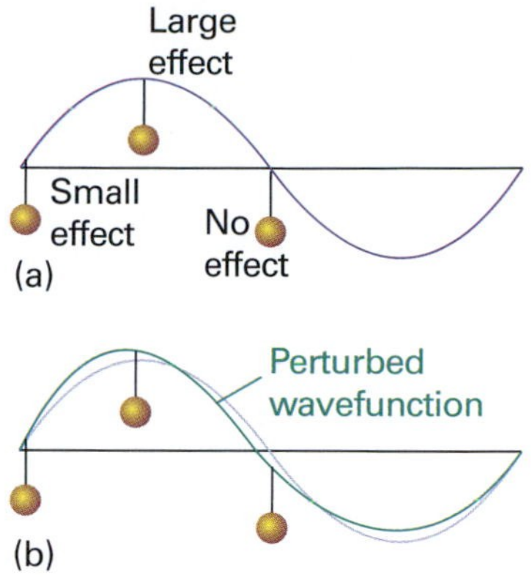

Fig. 2.42 (a) The first-order energy is an average of the perturbation (represented by the hanging weights) over the unperturbed wavefunction. (b) The second-order energy is a similar average, but over the distortion induced by the perturbation.

Example 2.5 *Using perturbation theory*

Find the first-order correction to the ground-state energy for a particle in a well with a variation in the potential of the form $V = -\varepsilon\sin(\pi x/L)$, as in Fig. 2.43.

Method Identify the first-order perturbation hamiltonian and evaluate $E_0^{(1)}$ from eqn 2.65a. We can expect a small lowering of the energy because the average potential energy of the particle is lower in the distorted box.

Answer The perturbation hamiltonian is

$$\hat{H}^{(1)} = -\varepsilon\sin(\pi x/L)$$

Therefore, the first-order correction to the energy is

$$E_0^{(1)} = \int_0^L \psi_1 \hat{H}^{(1)} \psi_1 \mathrm{d}x = -\frac{2\varepsilon}{L}\overbrace{\int_0^L \sin^3\frac{\pi x}{L}\mathrm{d}x}^{4L/3\pi} = -\frac{8\varepsilon}{3\pi}$$

Note that the energy is lowered by the perturbation, as would be expected for the shape shown in Fig. 2.43.

Self-test 2.9 Suppose that only ψ_3 contributes to the distortion of the wavefunction: calculate the coefficient c_3 and the second-order correction to the energy by using eqn 2.65b and eqn 2.76 in *Further information 2.2*.

$[c_3 = -8\varepsilon mL^2/15\pi h^2, E_0^{(2)} = -64\varepsilon^2 mL^2/225\pi^2 h^2]$

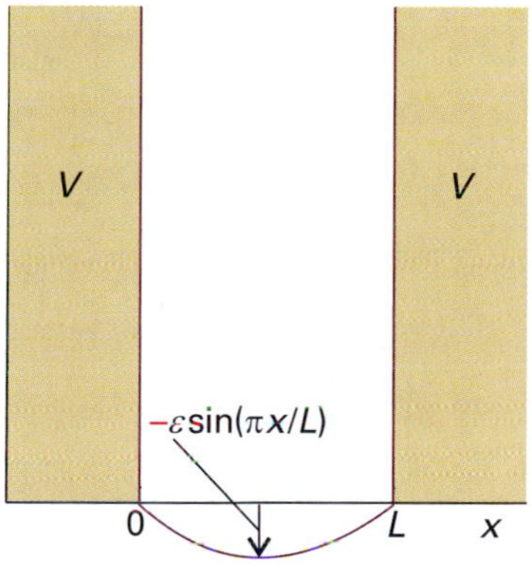

Fig. 2.43 The potential energy for a particle in a box with a potential that varies as $-\varepsilon\sin(\pi x/L)$ across the floor of the box. We can expect the particle to accumulate more in the centre of the box (in the ground state at least) than in the unperturbed box.

2.10 Time-dependent perturbation theory

In **time-dependent perturbation theory**, the perturbation is either switched on and allowed to rise to its final value or is varying in time. Many of the perturbations encountered in chemistry are time-dependent. The most important is the effect of an oscillating electromagnetic field, which is responsible for spectroscopic transitions between quantized energy levels in atoms and molecules.

Classically, for a molecule to be able to interact with the electromagnetic field and absorb or emit a photon of frequency ν, it must possess, at least transiently, a dipole oscillating at that frequency. In this section, we develop the quantum mechanical view and begin by writing the hamiltonian for the system as

$$\hat{H} = \hat{H}^{(0)} + \hat{H}^{(1)}(t) \tag{2.66}$$

where $\hat{H}^{(1)}(t)$ is the time-dependent perturbation. Because the perturbation arises from the effect of an oscillating electric field with the electric dipole moment, we write

$$\hat{H}^{(1)}(t) = -\mu_z \mathcal{E} \cos \omega t \tag{2.67}$$

where ω is the frequency of the field and $\mathcal{E}$ is its amplitude. We suppose that the perturbation is absent until $t = 0$, and then it is turned on.

Comment 2.8

An electric dipole consists of two electric charges $+q$ and $-q$ separated by a distance R. The electric dipole moment vector $\boldsymbol{\mu}$ has a magnitude $\mu = qR$.

We show in *Further information 2.2* that the rate of change of population of the state ψ_f due to transitions from state ψ_i, $w_{f\leftarrow i}$, is proportional to the square modulus of the matrix element of the perturbation between the two states:

$$w_{f\leftarrow i} \propto |H_{fi}^{(1)}|^2 \tag{2.68}$$

Because in our case the perturbation is that of the interaction of the electromagnetic field with a molecule (eqn 2.67), we conclude that

$$w_{f\leftarrow i} \propto |\mu_{z,fi}|^2 \mathcal{E}^2 \tag{2.69}$$

Therefore, the rate of transition, and hence the intensity of absorption of the incident radiation, is proportional to the square of the **transition dipole moment**:

$$\mu_{z,fi} = \int \psi_f^* \mu_z \psi_i d\tau \qquad [2.70]$$

The size of the transition dipole can be regarded as a measure of the charge redistribution that accompanies a transition.

The rate of transition is also proportional to $\mathcal{E}^2$, and therefore the intensity of the incident radiation (because the intensity is proportional to $\mathcal{E}^2$; see *Appendix 3*). This result will be the basis of most of our subsequent discussion of spectroscopy in Chapters 3 and 6–8 and of the kinetics of electron transfer in Chapter 11 of Volume 1.

Checklist of key ideas

☐ 1. The wavefunction of a free particle is $\psi_k = Ae^{ikx} + Be^{-ikx}$, $E_k = k^2\hbar^2/2m$.

☐ 2. The wavefunctions and energies of a particle in a one-dimensional box of length L are, respectively, $\psi_n(x) = (2/L)^{1/2} \sin(n\pi x/L)$ and $E_n = n^2h^2/8mL^2$, $n = 1,2,\ldots$. The zero-point energy, the lowest possible energy is $E_1 = h^2/8mL^2$.

☐ 3. The correspondence principle states that classical mechanics emerges from quantum mechanics as high quantum numbers are reached.

☐ 4. The functions ψ_n and $\psi_{n'}$ are orthogonal if $\int \psi_n^* \psi_{n'} d\tau = 0$; all wavefunctions corresponding to different energies of a system are orthogonal. Orthonormal functions are sets of functions that are normalized and mutually orthogonal.

☐ 5. The wavefunctions and energies of a particle in a two-dimensional box are given by eqn 2.12a.

☐ 6. Degenerate wavefunctions are different wavefunctions corresponding to the same energy.

☐ 7. Tunnelling is the penetration into or through classically forbidden regions. The transmission probability is given by eqn 2.20a.

☐ 8. Harmonic motion is the motion in the presence of a restoring force proportional to the displacement, $F = -kx$, where k is the force constant. As a consequence, $V = \frac{1}{2}kx^2$.

☐ 9. The wavefunctions and energy levels of a quantum mechanical harmonic oscillator are given by eqns 2.28 and 2.25, respectively.

☐ 10. The virial theorem states that, if the potential energy of a particle has the form $V = ax^b$, then its mean potential and kinetic energies are related by $2\langle E_K\rangle = b\langle V\rangle$.

☐ 11. Angular momentum is the moment of linear momentum around a point.

☐ 12. The wavefunctions and energies of a particle on a ring are, respectively, $\psi_{m_l}(\phi) = (1/2\pi)^{1/2}e^{im_l\phi}$ and $E = m_l^2\hbar^2/2I$, with $I = mr^2$ and $m_l = 0, \pm1, \pm2, \ldots$.

☐ 13. The wavefunctions of a particle on a sphere are the spherical harmonics, the functions $Y_{l,m_l}(\theta,\phi)$ (Table 2.3). The energies are $E = l(l+1)\hbar^2/2I$, $l = 0, 1, 2, \ldots$.

☐ 14. For a particle on a sphere, the magnitude of the angular momentum is $\{l(l+1)\}^{1/2}\hbar$ and the z-component of the angular momentum is $m_l\hbar$, $m_l = l, l-1, \ldots, -l$.

☐ 15. Space quantization is the restriction of the component of angular momentum around an axis to discrete values.

☐ 16. Spin is an intrinsic angular momentum of a fundamental particle. A fermion is a particle with a half-integral spin quantum number; a boson is a particle with an integral spin quantum number.

☐ 17. For an electron, the spin quantum number is $s = \frac{1}{2}$.

☐ 18. The spin magnetic quantum number is $m_s = s, s-1, \ldots, -s$; for an electron, $m_s = +\frac{1}{2}, -\frac{1}{2}$.

☐ 19. Perturbation theory is a technique that supplies approximate solutions to the Schrödinger equation and in which the hamiltonian for the problem is expressed as a sum of simpler hamiltonians.

☐ 20. In time-independent perturbation theory, the perturbation is always present and unvarying. The first- and second-order corrections to the energy are given by eqns 2.65a and 2.65b, respectively. In time-dependent perturbation theory, the perturbation is either switched on and allowed to rise to its final value or is varying in time.

☐ 21. The rate of change of population of the state ψ_f due to transitions from state ψ_i is $w_{f\leftarrow i} \propto |\mu_{z,fi}|^2\mathcal{E}^2$, where $\mu_{z,fi} = \int\psi_f^*\mu_z\psi_i\mathrm{d}\tau$ is the transition dipole moment.

Further reading

Articles and texts

P.W. Atkins and R.S. Friedman, *Molecular quantum mechanics*. Oxford University Press (2005).

C.S. Johnson, Jr. and L.G. Pedersen, *Problems and solutions in quantum chemistry and physics*. Dover, New York (1986).

I.N. Levine, *Quantum chemistry*. Prentice–Hall, Upper Saddle River (2000).

D.A. McQuarrie, *Mathematical methods for scientists and engineers*. University Science Books, Mill Valley (2003).

J.J.C. Mulder, Closed-form spherical harmonics: explicit polynomial expression for the associated legendre functions. *J. Chem. Educ.* **77**, 244 (2000).

L. Pauling and E.B. Wilson, *Introduction to quantum mechanics with applications to chemistry*. Dover, New York (1985).

Further information

Further information 2.1 *Dirac notation*

The integral in eqn 2.9 is often written

$$\langle n|n'\rangle = 0 \qquad (n' \neq n)$$

This **Dirac bracket notation** is much more succinct than writing out the integral in full. It also introduces the words 'bra' and 'ket' into the language of quantum mechanics. Thus, the **bra** $\langle n|$ corresponds to ψ_n^* and the **ket** $|n'\rangle$ corresponds to the wavefunction $\psi_{n'}$. When the bra and ket are put together as in this expression, the integration over all space is understood. Similarly, the normalization condition in eqn 1.17c becomes simply

$$\langle n|n\rangle = 1$$

in bracket notation. These two expressions can be combined into one:

$$\langle n|n'\rangle = \delta_{nn'} \tag{2.71}$$

Here $\delta_{nn'}$, which is called the **Kronecker delta**, is 1 when $n' = n$ and 0 when $n' \neq n$.

Integrals of the form $\int\psi_n^*\hat{\Omega}\psi_m\mathrm{d}\tau$, which we first encounter in connection with perturbation theory (Section 2.9) and which are commonly called 'matrix elements', are incorporated into the bracket notation by writing

$$\langle n|\hat{\Omega}|m\rangle = \int\psi_n^*\hat{\Omega}\psi_m\mathrm{d}\tau \qquad [2.72]$$

Note how the operator stands between the bra and the ket (which may denote different states), in the place of the c in ⟨bra|c|ket⟩. An integration is implied whenever a complete bracket is written. In this notation, an expectation value is

$$\langle\Omega\rangle = \langle n|\hat{\Omega}|n\rangle \tag{2.73}$$

with the bra and the ket corresponding to the same state (with quantum number n and wavefunction ψ_n). In this notation, an operator is hermitian (eqn 1.30) if

$$\langle n|\hat{\Omega}|m\rangle = \langle m|\hat{\Omega}|n\rangle^* \tag{2.74}$$

Further information 2.2 *Perturbation theory*

Here we treat perturbation theory in detail. Our first task is to develop the results of time-independent perturbation theory, in which a system is subjected to a perturbation that does not vary with time. Then, we go on to discuss time-dependent perturbation theory, in which a perturbation is turned on at a specific time and the system is allowed to evolve.

1 Time-independent perturbation theory

To develop expressions for the corrections to the wavefunction and energy of a system subjected to a time-independent perturbation, we write

$$\psi = \psi^{(0)} + \lambda\psi^{(1)} + \lambda^2\psi^{(2)} + \ldots$$

where the power of λ indicates the order of the correction. Likewise, we write

$$\hat{H} = \hat{H}^{(0)} + \lambda\hat{H}^{(1)}$$

and

$$E = E^{(0)} + \lambda E^{(1)} + \lambda^2 E^{(2)} + \ldots$$

When these expressions are inserted into the Schrödinger equation, $\hat{H}\psi = E\psi$, we obtain

$$\begin{aligned}(\hat{H}^{(0)} + \lambda\hat{H}^{(1)})(\psi^{(0)} + \lambda\psi^{(1)} + \lambda^2\psi^{(2)} + \ldots)\\ = (E^{(0)} + \lambda E^{(1)} + \lambda^2E^{(2)} + \ldots)(\psi^{(0)} + \lambda\psi^{(1)} + \lambda^2\psi^{(2)} + \ldots)\end{aligned}$$

which we can rewrite as

$$\begin{aligned}\hat{H}^{(0)}\psi^{(0)} + \lambda(\hat{H}^{(1)}\psi^{(0)} + \hat{H}^{(0)}\psi^{(1)}) + \lambda^2(\hat{H}^{(0)}\psi^{(2)} + \hat{H}^{(1)}\psi^{(1)}) + \ldots\\ = E^{(0)}\psi^{(0)} + \lambda(E^{(0)}\psi^{(1)} + E^{(1)}\psi^{(0)}) + \lambda^2(E^{(2)}\psi^{(0)} + E^{(1)}\psi^{(1)}\\ + E^{(0)}\psi^{(2)}) + \ldots\end{aligned}$$

By comparing powers of λ, we find

$$\begin{aligned}&\text{Terms in } \lambda^0: && \hat{H}^{(0)}\psi^{(0)} = E^{(0)}\psi^{(0)}\\ &\text{Terms in } \lambda: && \hat{H}^{(1)}\psi^{(0)} + \hat{H}^{(0)}\psi^{(1)} = E^{(0)}\psi^{(1)} + E^{(1)}\psi^{(0)}\\ &\text{Terms in } \lambda^2: && \hat{H}^{(0)}\psi^{(2)} + \hat{H}^{(1)}\psi^{(1)} = E^{(2)}\psi^{(0)} + E^{(1)}\psi^{(1)} + E^{(0)}\psi^{(2)}\end{aligned}$$

and so on.

The equations we have derived are applicable to any state of the system. From now on we shall consider only the ground state ψ_0 with energy E_0. The first equation, which we now write

$$\hat{H}^{(0)}\psi_0^{(0)} = E_0^{(0)}\psi_0^{(0)}$$

is the Schrödinger equation for the ground state of the unperturbed system, which we assume we can solve (for instance, it might be the equation for the ground state of the particle in a box, with the solutions given in eqn 2.7). To solve the next equation, which is now written

$$\hat{H}^{(1)}\psi_0^{(0)} + \hat{H}^{(0)}\psi_0^{(1)} = E_0^{(0)}\psi^{(1)} + E_0^{(1)}\psi_0^{(0)}$$

we suppose that the first-order correction to the wavefunction can be expressed as a linear combination of the wavefunctions of the unperturbed system, and write

$$\psi_0^{(1)} = \sum_n c_n\psi_n^{(0)} \tag{2.75}$$

Substitution of this expression gives

$$\hat{H}^{(1)}\psi_0^{(0)} + \sum_n c_n\hat{H}^{(0)}\psi_n^{(0)} = \sum_n c_nE_0^{(0)}\psi_n^{(0)} + E_0^{(1)}\psi_0^{(0)}$$

We can isolate the term in $E_0^{(1)}$ by making use of the fact that the $\psi_n^{(0)}$ form a complete orthogonal and normalized set in the sense that

$$\int \psi_0^{(0)*}\psi_n^{(0)}\mathrm{d}\tau = 0 \qquad \text{if } n \neq 0, \qquad \text{but 1 if } n = 0$$

Therefore, when we multiply through by $\psi_0^{(0)*}$ and integrate over all space, we get

$$\begin{aligned}&\int \psi_0^{(0)*}\hat{H}^{(1)}\psi_0^{(0)}\mathrm{d}\tau + \sum_n c_n\overbrace{\int \psi_0^{(0)*}\hat{H}^{(0)}\psi_n^{(0)}\mathrm{d}\tau}^{E_0^{(0)}\text{ if } n=0,\ 0 \text{ otherwise}}\\ &= \sum_n c_nE_0^{(0)}\overbrace{\int \psi_0^{(0)*}\psi_n^{(0)}\mathrm{d}\tau}^{1 \text{ if } n=0,\ 0\text{ otherwise}} + E_0^{(1)}\overbrace{\int \psi_0^{(0)*}\psi_0^{(0)}\mathrm{d}\tau}^{1}\end{aligned}$$

That is,

$$\int \psi_0^{(0)*}\hat{H}^{(1)}\psi_0^{(0)}\mathrm{d}\tau = E_0^{(1)}$$

which is eqn 2.65a.

To find the coefficients c_n, we multiply the same expression through by $\psi_k^{(0)*}$, where now $k \neq 0$, which gives

$$\begin{aligned}&\int \psi_k^{(0)*}\hat{H}^{(1)}\psi_0^{(0)}\mathrm{d}\tau + \sum_n c_n\overbrace{\int \psi_k^{(0)*}\hat{H}^{(0)}\psi_n^{(0)}\mathrm{d}\tau}^{E_k^{(0)}\delta_{kn}}\\ &= \sum_n c_nE_0^{(0)}\overbrace{\int \psi_k^{(0)*}\psi_n^{(0)}\mathrm{d}\tau}^{1 \text{ if } n=k,\ 0\text{ otherwise}} + E_0^{(1)}\overbrace{\int \psi_k^{(0)*}\psi_0^{(0)}\mathrm{d}\tau}^{0}\end{aligned}$$

That is,

$$\int \psi_k^{(0)*}\hat{H}^{(1)}\psi_0^{(0)}\mathrm{d}\tau + c_kE_k^{(0)} = c_kE_0^{(0)}$$

which we can rearrange into

$$c_k = -\frac{\int \psi_k^{(0)*}\hat{H}^{(1)}\psi_0^{(0)}\mathrm{d}\tau}{E_k^{(0)} - E_0^{(0)}} \tag{2.76}$$

The second-order energy is obtained starting from the second-order expression, which for the ground state is

$$\hat{H}^{(0)}\psi_0^{(2)} + \hat{H}^{(1)}\psi_0^{(1)} = E_0^{(2)}\psi_0^{(0)} + E_0^{(1)}\psi_0^{(1)} + E_0^{(0)}\psi_0^{(2)}$$

To isolate the term $E_0^{(2)}$ we multiply both sides by $\psi_0^{(0)*}$, integrate over all space, and obtain

$$\begin{aligned}&\overbrace{\int \psi_0^{(0)*}\hat{H}^{(0)}\psi_0^{(2)}\mathrm{d}\tau}^{E_0^{(0)}\int\psi_0^{(0)*}\psi_0^{(2)}\mathrm{d}\tau} + \int \psi_0^{(0)*}\hat{H}^{(1)}\psi_0^{(1)}\mathrm{d}\tau\\ &= \underbrace{E_0^{(2)}\int \psi_0^{(0)*}\psi_0^{(0)}\mathrm{d}\tau}_{1} + E_0^{(1)}\int \psi_0^{(0)*}\psi_0^{(1)}\mathrm{d}\tau + E_0^{(0)}\int \psi_0^{(0)*}\psi_0^{(2)}\mathrm{d}\tau\end{aligned}$$

The first and last terms cancel, and we are left with

$$E_0^{(2)} = \int \psi_0^{(0)*}\hat{H}^{(1)}\psi_0^{(1)}\mathrm{d}\tau - E_0^{(1)}\int \psi_0^{(0)*}\psi_0^{(1)}\mathrm{d}\tau$$

We have already found the first-order corrections to the energy and the wavefunction, so this expression could be regarded as an explicit expression for the second-order energy. However, we can go one step further by substituting eqn 2.75:

$$E_0^{(2)} = \sum_n c_n \overbrace{\int \psi_0^{(0)*}\hat{H}^{(1)}\psi_n^{(0)}\mathrm{d}\tau}^{H_{0n}^{(1)}} - \sum_n c_n E_0^{(1)} \overbrace{\int \psi_0^{(0)*}\psi_n^{(0)}\mathrm{d}\tau}^{\delta_{0n}}$$

$$= \sum_n c_n H_{0n}^{(1)} - c_0 E_0^{(1)}$$

The final term cancels the term $c_0 H_{00}^{(1)}$ in the sum, and we are left with

$$E_0^{(2)} = \sum_{n\neq 0} c_n H_{0n}^{(1)}$$

Substitution of the expression for c_n in eqn 2.76 now produces the final result, eqn 2.65b.

2 Time-dependent perturbation theory

To cope with a perturbed wavefunction that evolves with time, we need to solve the time-dependent Schrödinger equation,

$$\hat{H}\Psi = \mathrm{i}\hbar\frac{\partial \Psi}{\partial t} \qquad (2.77)$$

We confirm below, that if we write the first-order correction to the wavefunction as

$$\Psi_0^{(1)}(t) = \sum_n c_n(t)\Psi_n(t) = \sum_n c_n(t)\psi_n^{(0)}\mathrm{e}^{-\mathrm{i}E_n^{(0)}t/\hbar} \qquad (2.78a)$$

then the coefficients in this expansion are given by

$$c_n(t) = \frac{1}{\mathrm{i}\hbar}\int_0^t H_{n0}^{(1)}(t)\mathrm{e}^{\mathrm{i}\omega_{n0}t}\mathrm{d}t \qquad (2.78b)$$

The formal demonstration of eqn 2.78 is quite lengthy (see *Further reading*). Here we shall show that, given eqn 2.78b, a perturbation that is switched on very slowly to a constant value gives the same expression for the coefficients as we obtained for time-independent perturbation theory. For such a perturbation, we write

$$\hat{H}^{(1)}(t) = \hat{H}^{(1)}(1 - \mathrm{e}^{-t/\tau})$$

and take the time constant τ to be very long (Fig. 2.44). Substitution of this expression into eqn 2.78b gives

$$c_n(t) = \frac{1}{\mathrm{i}\hbar}H_{n0}^{(1)}\int_0^t (1 - \mathrm{e}^{-t/\tau})\mathrm{e}^{\mathrm{i}\omega_{n0}t}\mathrm{d}t$$

$$= \frac{1}{\mathrm{i}\hbar}H_{n0}^{(1)}\left\{\frac{\mathrm{e}^{\mathrm{i}\omega_{n0}t} - 1}{\mathrm{i}\omega_{n0}} - \frac{\mathrm{e}^{(\mathrm{i}\omega_{n0} - 1/\tau)t} - 1}{\mathrm{i}\omega_{n0} - 1/\tau}\right\}$$

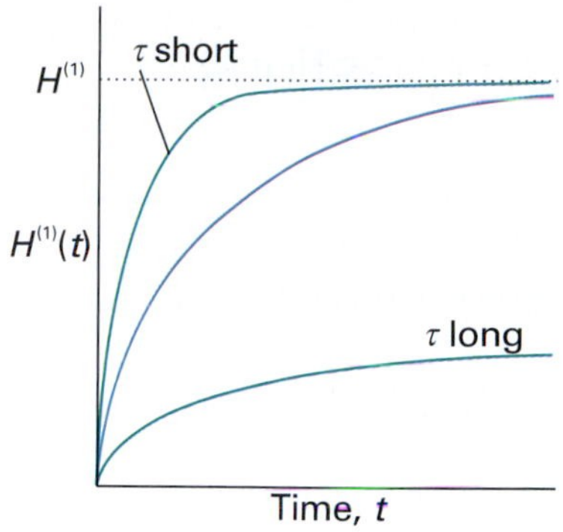

Fig. 2.44 The time-dependence of a slowly switched perturbation. A large value of τ corresponds to very slow switching.

At this point we suppose that the perturbation is switched slowly, in the sense that $\tau \gg 1/\omega_{n0}$ (so that the $1/\tau$ in the second denominator can be ignored). We also suppose that we are interested in the coefficients long after the perturbation has settled down into its final value, when $t \gg \tau$ (so that the exponential in the second numerator is close to zero and can be ignored). Under these conditions,

$$c_n(t) = -\frac{H_{n0}^{(0)}}{\hbar\omega_{n0}}\mathrm{e}^{\mathrm{i}\omega_{n0}t}$$

Now we recognize that $\hbar\omega_{n0} = E_n^{(0)} - E_0^{(0)}$, which gives

$$c_n(t) = -\frac{H_{n0}^{(0)}}{E_n^{(0)} - E_0^{(0)}}\mathrm{e}^{\mathrm{i}E_n^{(0)}t}\mathrm{e}^{-\mathrm{i}E_0^{(0)}t}$$

When this expression is substituted into eqn 2.78a, we obtain the time-independent expression, eqn 2.76 (apart from an irrelevant overall phase factor).

In accord with the general rules for the interpretation of wavefunctions, the probability that the system will be found in the state n is proportional to the square modulus of the coefficient of the state, $|c_n(t)|^2$. Therefore, the rate of change of population of a final state ψ_f due to transitions from an initial state ψ_i is

$$w_{\mathrm{f}\leftarrow\mathrm{i}} = \frac{\mathrm{d}|c_\mathrm{f}|^2}{\mathrm{d}t} = \frac{\mathrm{d}c_\mathrm{f}^{*}c_\mathrm{f}}{\mathrm{d}t} = c_\mathrm{f}\frac{\mathrm{d}c_\mathrm{f}^{*}}{\mathrm{d}t} + c_\mathrm{f}^{*}\frac{\mathrm{d}c_\mathrm{f}}{\mathrm{d}t}$$

Because the coefficients are proportional to the matrix elements of the perturbation, $w_{\mathrm{f}\leftarrow\mathrm{i}}$ is proportional to the square modulus of the matrix element of the perturbation between the two states:

$$w_{\mathrm{f}\leftarrow\mathrm{i}} \propto |H_{\mathrm{fi}}^{(1)}|^2$$

which is eqn 2.68.

Each end-of-chapter discussion question, exercise, and problem for which a solution is available in the Student Solution Manual has a corresponding bracketed number referring to the Student Solutions Manual number for this problem, for example [SSM 9.1]. Each end-of-chapter exercise and problem for which a solution is available in the Instructors' Solution Manual has a corresponding bracketed number referring to the Instructor Solutions Manual number for this problem, for example [ISM 9.2].

Discussion questions

2.1 Discuss the physical origin of quantization energy for a particle confined to moving inside a one-dimensional box or on a ring. [SSM **9.1**]

2.2 Discuss the correspondence principle and provide two examples. [ISM **9.2**]

2.3 Define, justify, and provide examples of zero-point energy. [SSM **9.3**]

2.4 Discuss the physical origins of quantum mechanical tunnelling. Why is tunnelling more likely to contribute to the mechanisms of electron transfer and proton transfer processes than to mechanisms of group transfer reactions, such as AB + C → A + BC (where A, B, and C are large molecular groups)? [ISM **9.4**]

2.5 Distinguish between a fermion and a boson. Provide examples of each type of particle. [SSM **9.5**]

2.6 Describe the features that stem from nanometre-scale dimensions that are not found in macroscopic objects. [ISM **9.6**]

Exercises

2.1a Calculate the energy separations in joules, kilojoules per mole, electronvolts, and reciprocal centimetres between the levels (a) $n=2$ and $n=1$, (b) $n=6$ and $n=5$ of an electron in a box of length 1.0 nm. [ISM **9.1a**]

2.1b Calculate the energy separations in joules, kilojoules per mole, electronvolts, and reciprocal centimetres between the levels (a) $n=3$ and $n=1$, (b) $n=7$ and $n=6$ of an electron in a box of length 1.50 nm. [SSM **9.1b**]

2.2a Calculate the probability that a particle will be found between $0.49L$ and $0.51L$ in a box of length L when it has (a) $n=1$, (b) $n=2$. Take the wavefunction to be a constant in this range. [ISM **9.2a**]

2.2b Calculate the probability that a particle will be found between $0.65L$ and $0.67L$ in a box of length L when it has (a) $n=1$, (b) $n=2$. Take the wavefunction to be a constant in this range. [SSM **9.2b**]

2.3a Calculate the expectation values of p and p^2 for a particle in the state $n=1$ in a square-well potential. [ISM **9.3a**]

2.3b Calculate the expectation values of p and p^2 for a particle in the state $n=2$ in a square-well potential. [SSM **9.3b**]

2.4a An electron is confined to a a square well of length L. What would be the length of the box such that the zero-point energy of the electron is equal to its rest mass energy, m_ec^2? Express your answer in terms of the parameter $\lambda_C = h/m_ec$, the 'Compton wavelength' of the electron. [ISM **9.4a**]

2.4b Repeat Exercise 2.4a for a general particle of mass m in a cubic box. [SSM **9.4b**]

2.5a What are the most likely locations of a particle in a box of length L in the state $n=3$? [ISM **9.5a**]

2.5b What are the most likely locations of a particle in a box of length L in the state $n=5$? [SSM **9.5b**]

2.6a Consider a particle in a cubic box. What is the degeneracy of the level that has an energy three times that of the lowest level? [ISM **9.6a**]

2.6b Consider a particle in a cubic box. What is the degeneracy of the level that has an energy $\frac{14}{3}$ times that of the lowest level? [SSM **9.6b**]

2.7a Calculate the percentage change in a given energy level of a particle in a cubic box when the length of the edge of the cube is decreased by 10 per cent in each direction. [ISM **9.7a**]

2.7b A nitrogen molecule is confined in a cubic box of volume 1.00 m^3. Assuming that the molecule has an energy equal to $\frac{3}{2}kT$ at $T=300$ K, what is the value of $n=(n_x^2+n_y^2+n_z^2)^{1/2}$ for this molecule? What is the energy separation between the levels n and $n+1$? What is its de Broglie wavelength? Would it be appropriate to describe this particle as behaving classically? [SSM **9.7b**]

2.8a Calculate the zero-point energy of a harmonic oscillator consisting of a particle of mass 2.33×10^{-26} kg and force constant 155 N m^{-1}. [ISM **9.8a**]

2.8b Calculate the zero-point energy of a harmonic oscillator consisting of a particle of mass 5.16×10^{-26} kg and force constant 285 N m^{-1}. [SSM **9.8b**]

2.9a For a harmonic oscillator of effective mass 1.33×10^{-25} kg, the difference in adjacent energy levels is 4.82 zJ. Calculate the force constant of the oscillator. [ISM **9.9a**]

2.9b For a harmonic oscillator of effective mass 2.88×10^{-25} kg, the difference in adjacent energy levels is 3.17 zJ. Calculate the force constant of the oscillator. [SSM **9.9b**]

2.10a Calculate the wavelength of a photon needed to excite a transition between neighbouring energy levels of a harmonic oscillator of effective mass equal to that of a proton (1.0078 u) and force constant 855 N m^{-1}. [ISM **9.10a**]

2.10b Calculate the wavelength of a photon needed to excite a transition between neighbouring energy levels of a harmonic oscillator of effective mass equal to that of an oxygen atom (15.9949 u) and force constant 544 N m^{-1}. [SSM **9.10b**]

2.11a Refer to Exercise 2.10a and calculate the wavelength that would result from doubling the effective mass of the oscillator. [ISM **9.11a**]

2.11b Refer to Exercise 2.10b and calculate the wavelength that would result from doubling the effective mass of the oscillator. [SSM **9.11b**]

2.12a Calculate the minimum excitation energies of (a) a pendulum of length 1.0 m on the surface of the Earth, (b) the balance-wheel of a clockwork watch ($\nu=5$ Hz). [ISM **9.12a**]

2.12b Calculate the minimum excitation energies of (a) the 33 kHz quartz crystal of a watch, (b) the bond between two O atoms in O_2, for which $k=1177$ N m^{-1}. [SSM **9.12b**]

2.13a Confirm that the wavefunction for the ground state of a one-dimensional linear harmonic oscillator given in Table 2.1 is a solution of the Schrödinger equation for the oscillator and that its energy is $\frac{1}{2}\hbar\omega$. [ISM **9.13a**]

2.13b Confirm that the wavefunction for the first excited state of a one-dimensional linear harmonic oscillator given in Table 2.1 is a solution of the Schrödinger equation for the oscillator and that its energy is $\frac{3}{2}\hbar\omega$. [SSM **9.13b**]

2.14a Locate the nodes of the harmonic oscillator wavefunction with $v = 4$. [ISM **9.14a**]

2.14b Locate the nodes of the harmonic oscillator wavefunction with $v = 5$. [SSM **9.14b**]

2.15a Assuming that the vibrations of a $^{35}Cl_2$ molecule are equivalent to those of a harmonic oscillator with a force constant $k = 329\ \mathrm{N\ m^{-1}}$, what is the zero-point energy of vibration of this molecule? The mass of a ^{35}Cl atom is 34.9688 u. [ISM **9.15a**]

2.15b Assuming that the vibrations of a $^{14}N_2$ molecule are equivalent to those of a harmonic oscillator with a force constant $k = 2293.8\ \mathrm{N\ m^{-1}}$, what is the zero-point energy of vibration of this molecule? The mass of a ^{14}N atom is 14.0031 u. [SSM **9.15b**]

2.16a The wavefunction, $\psi(\phi)$, for the motion of a particle in a ring is of the form $\psi = Ne^{im_l\phi}$. Determine the normalization constant, N. [ISM **9.16a**]

2.16b Confirm that wavefunctions for a particle in a ring with different values of the quantum number m_l are mutually orthogonal. [SSM **9.16b**]

2.17a A point mass rotates in a circle with $l = 1$, Calculate the magnitude of its angular momentum and the possible projections of the angular momentum on an arbitrary axis. [ISM **9.17a**]

2.17b A point mass rotates in a circle with $l = 2$, Calculate the magnitude of its angular momentum and the possible projections of the angular momentum on an arbitrary axis. [SSM **9.17b**]

2.18a Draw scale vector diagrams to represent the states (a) $s = \frac{1}{2}$, $m_s = +\frac{1}{2}$, (b) $l = 1$, $m_l = +1$, (c) $l = 2$, $m_l = 0$. [ISM **9.18a**]

2.18b Draw the vector diagram for all the permitted states of a particle with $l = 6$. [SSM **9.18b**]

Problems*

Numerical problems

2.1 Calculate the separation between the two lowest levels for an O_2 molecule in a one-dimensional container of length 5.0 cm. At what value of n does the energy of the molecule reach $\frac{1}{2}kT$ at 300 K, and what is the separation of this level from the one immediately below? [SSM **9.1**]

2.2 The mass to use in the expression for the vibrational frequency of a diatomic molecule is the effective mass $\mu = m_A m_B/(m_A + m_B)$, where m_A and m_B are the masses of the individual atoms. The following data on the infrared absorption wavenumbers (in $\mathrm{cm^{-1}}$) of molecules are taken from *Spectra of diatomic molecules*, G. Herzberg, van Nostrand (1950):

$H^{35}Cl$	$H^{81}Br$	HI	CO	NO
2990	2650	2310	2170	1904

Calculate the force constants of the bonds and arrange them in order of increasing stiffness. [ISM **9.2**]

2.3 The rotation of an $^{1}H^{127}I$ molecule can be pictured as the orbital motion of an H atom at a distance 160 pm from a stationary I atom. (This picture is quite good; to be precise, both atoms rotate around their common centre of mass, which is very close to the I nucleus.) Suppose that the molecule rotates only in a plane. Calculate the energy needed to excite the molecule into rotation. What, apart from 0, is the minimum angular momentum of the molecule? [SSM **9.3**]

2.4 Calculate the energies of the first four rotational levels of $^{1}H^{127}I$ free to rotate in three dimensions, using for its moment of inertia $I = \mu R^2$, with $\mu = m_H m_I/(m_H + m_I)$ and $R = 160$ pm. [ISM **9.4**]

2.5 A small step in the potential energy is introduced into the one-dimensional square-well problem as in Fig. 2.45. (a) Write a general expression for the first-order correction to the ground-state energy, $E_0^{(1)}$. (b) Evaluate the energy correction for $a = L/10$ (so the blip in the potential occupies the central 10 per cent of the well), with $n = 1$. [SSM **9.5**]

2.6 We normally think of the one-dimensional well as being horizontal. Suppose it is vertical; then the potential energy of the particle depends on x because of the presence of the gravitational field. Calculate the first-order correction to the zero-point energy, and evaluate it for an electron in a box on the surface of the Earth. Account for the result. *Hint.* The energy of the particle depends on its height as mgh, where $g = 9.81\ \mathrm{m\ s^{-2}}$. Because g is so small, the energy correction is small; but it would be significant if the box were near a very massive star. [ISM **9.6**]

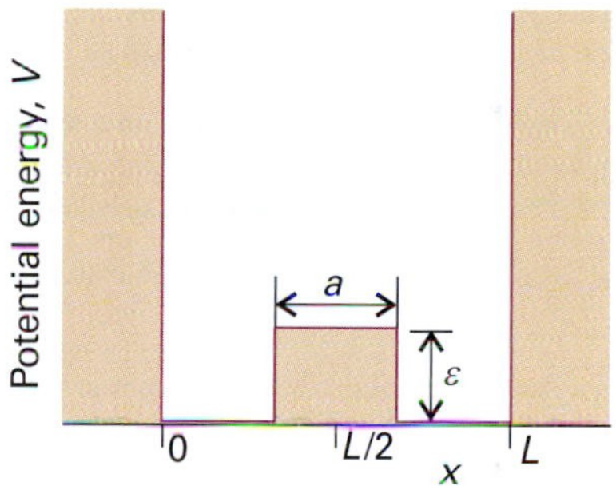

Fig. 2.45

2.7 Calculate the second-order correction to the energy for the system described in Problem 2.6 and calculate the ground-state wavefunction. Account for the shape of the distortion caused by the perturbation. *Hint.* The following integrals are useful

$$\int x \sin ax \sin bx\,dx = -\frac{d}{da}\int \cos ax \sin bx\,dx$$

$$\int \cos ax \sin bx\,dx = \frac{\cos(a-b)x}{2(a-b)} - \frac{\cos(a+b)x}{2(a+b)} + \text{constant}$$ [SSM **9.7**]

Theoretical problems

2.8 Suppose that 1.0 mol perfect gas molecules all occupy the lowest energy level of a cubic box. How much work must be done to change the volume of the box by ΔV? Would the work be different if the molecules all occupied a state $n \neq 1$? What is the relevance of this discussion to the discussion of

* Problems denoted with the symbol ‡ were supplied by Charles Trapp, Carmen Giunta, and Marshall Cady.

expansion work? Can you identify a distinction between adiabatic and isothermal expansion? [ISM 9.8]

2.9 Derive eqn 2.20a, the expression for the transmission probability. [SSM 9.9]

2.10‡ Consider the one-dimensional space in which a particle can experience one of three potentials depending upon its position. They are: $V = 0$ for $-\infty < x \leq 0$, 0, $V = V_2$ for $0 \leq x \leq L$, and $V = V_3$ for $L \leq x < \infty$. The particle wavefunction is to have both a component e^{ik_1x} that is incident upon the barrier V_2 and a reflected component e^{-ik_1x} in region 1 ($-\infty < x \leq 0$). In region 3 the wavefunction has only a forward component, e^{ik_3x}, which represents a particle that has traversed the barrier. The energy of the particle, E, is somewhere in the range of the $V_2 > E > V_3$. The transmission probability, T, is the ratio of the square modulus of the region 3 amplitude to the square modulus of the incident amplitude. (a) Base your calculation on the continuity of the amplitudes and the slope of the wavefunction at the locations of the zone boundaries and derive a general equation for T. (b) Show that the general equation for T reduces to eqn 2.20b in the high, wide barrier limit when $V_1 = V_3 = 0$. (c) Draw a graph of the probability of proton tunnelling when $V_3 = 0$, $L = 50$ pm, and $E = 10$ kJ mol^{-1} in the barrier range $E < V_2 < 2E$. [ISM 9.10]

2.11 The wavefunction inside a long barrier of height V is $\psi = Ne^{-\kappa x}$. Calculate (a) the probability that the particle is inside the barrier and (b) the average penetration depth of the particle into the barrier. [SSM 9.11]

2.12 Confirm that a function of the form e^{-gx^2} is a solution of the Schrödinger equation for the ground state of a harmonic oscillator and find an expression for g in terms of the mass and force constant of the oscillator. [ISM 9.12]

2.13 Calculate the mean kinetic energy of a harmonic oscillator by using the relations in Table 2.1. [SSM 9.13]

2.14 Calculate the values of $\langle x^3 \rangle$ and $\langle x^4 \rangle$ for a harmonic oscillator by using the relations in Table 2.1. [ISM 9.14]

2.15 Determine the values of $\delta x = (\langle x^2 \rangle - \langle x \rangle^2)^{1/2}$ and $\delta p = (\langle p^2 \rangle - \langle p \rangle^2)^{1/2}$ for (a) a particle in a box of length L and (b) a harmonic oscillator. Discuss these quantities with reference to the uncertainty principle. [SSM 9.15]

2.16 We shall see in Chapter 6 that the intensities of spectroscopic transitions between the vibrational states of a molecule are proportional to the square of the integral $\int \psi_{v'} x \psi_v \mathrm{d}x$ over all space. Use the relations between Hermite polynomials given in Table 2.1 to show that the only permitted transitions are those for which $v' = v \pm 1$ and evaluate the integral in these cases. [ISM 9.16]

2.17 The potential energy of the rotation of one CH_3 group relative to its neighbour in ethane can be expressed as $V(\varphi) = V_0 \cos 3\varphi$. Show that for small displacements the motion of the group is harmonic and calculate the energy of excitation from $v = 0$ to $v = 1$. What do you expect to happen to the energy levels and wavefunctions as the excitation increases? [SSM 9.17]

2.18 Show that, whatever superposition of harmonic oscillator states is used to construct a wavepacket, it is localized at the same place at the times 0, T, $2T$, . . . , where T is the classical period of the oscillator. [ISM 9.18]

2.19 Use the virial theorem to obtain an expression for the relation between the mean kinetic and potential energies of an electron in a hydrogen atom. [SSM 9.19]

2.20 Evaluate the z-component of the angular momentum and the kinetic energy of a particle on a ring that is described by the (unnormalized) wavefunctions (a) $e^{i\phi}$, (b) $e^{-2i\phi}$, (c) $\cos \phi$, and (d) $(\cos \chi)e^{i\phi} + (\sin \chi)e^{-i\phi}$. [ISM 9.20]

2.21 Is the Schrödinger equation for a particle on an elliptical ring of semimajor axes a and b separable? *Hint.* Although r varies with angle φ, the two are related by $r^2 = a^2 \sin^2\phi + b^2 \cos^2\phi$. [SSM 9.21]

2.22 Use mathematical software to construct a wavepacket of the form

$$\Psi(\phi,t) = \sum_{m_l=0}^{m_{l,\max}} c_{m_l} e^{i(m_l\phi - E_{m_l}t/\hbar)} \qquad E_{m_l} = m_l^2\hbar^2/2I$$

with coefficients c of your choice (for example, all equal). Explore how the wavepacket migrates on the ring but spreads with time. [ISM 9.22]

2.23 Confirm that the spherical harmonics (a) $Y_{0,0}$, (b) $Y_{2,-1}$, and (c) $Y_{3,+3}$ satisfy the Schrödinger equation for a particle free to rotate in three dimensions, and find its energy and angular momentum in each case. [SSM 9.23]

2.24 Confirm that $Y_{3,+3}$ is normalized to 1. (The integration required is over the surface of a sphere.) [ISM 9.24]

2.25 Derive an expression in terms of l and m_l for the half-angle of the apex of the cone used to represent an angular momentum according to the vector model. Evaluate the expression for an α spin. Show that the minimum possible angle approaches 0 as $l \to \infty$. [SSM 9.25]

2.26 Show that the function $f = \cos ax \cos \cos cz$ is an eigenfunction of ∇^2, and determine its eigenvalue. [ISM 9.26]

2.27 Derive (in Cartesian coordinates) the quantum mechanical operators for the three components of angular momentum starting from the classical definition of angular momentum, $\boldsymbol{l} = \boldsymbol{r} \times \boldsymbol{p}$. Show that any two of the components do not mutually commute, and find their commutator. [SSM 9.27]

2.28 Starting from the operator $l_z = xp_y - yp_x$, prove that in spherical polar coordinates $l_z = -i\hbar\partial/\partial\phi$. [ISM 9.28]

2.29 Show that the commutator $[l^2, l_z] = 0$, and then, without further calculation, justify the remark that $[l^2, l_q] = 0$ for all $q = x$, y, and z. [SSM 9.29]

2.30‡ A particle is confined to move in a one-dimensional box of length L. (a) If the particle is classical, show that the average value of x is $\tfrac{1}{2}L$ and that the root-mean square value is $L/3^{1/2}$. (b) Show that for large values of n, a quantum particle approaches the classical values. This result is an example of the correspondence principle, which states that, for very large values of the quantum numbers, the predictions of quantum mechanics approach those of classical mechanics. [ISM 9.30]

Applications: to biology and nanotechnology

2.31 When β-carotene is oxidized *in vivo*, it breaks in half and forms two molecules of retinal (vitamin A), which is a precursor to the pigment in the retina responsible for vision (*Impact I7.1*). The conjugated system of retinal consists of 11 C atoms and one O atom. In the ground state of retinal, each level up to $n = 6$ is occupied by two electrons. Assuming an average internuclear distance of 140 pm, calculate (a) the separation in energy between the ground state and the first excited state in which one electron occupies the state with $n = 7$, and (b) the frequency of the radiation required to produce a transition between these two states. (c) Using your results and *Illustration 9.1*, choose among the words in parentheses to generate a rule for the prediction of frequency shifts in the absorption spectra of linear polyenes:

> The absorption spectrum of a linear polyene shifts to (higher/lower) frequency as the number of conjugated atoms (increases/decreases).

[SSM 9.31]

2.32 Many biological electron transfer reactions, such as those associated with biological energy conversion, may be visualized as arising from electron tunnelling between protein-bound co-factors, such as cytochromes, quinones, flavins, and chlorophylls. This tunnelling occurs over distances that are often greater than 1.0 nm, with sections of protein separating electron donor from acceptor. For a specific combination of donor and acceptor, the rate of electron tunnelling is proportional to the transmission probability, with $\kappa \approx 7\ \text{nm}^{-1}$ (eqn 2.20). By what factor does the rate of electron tunnelling between two co-factors increase as the distance between them changes from 2.0 nm to 1.0 nm? [ISM **9.32**]

2.33 Carbon monoxide binds strongly to the Fe^{2+} ion of the haem group of the protein myoglobin. Estimate the vibrational frequency of CO bound to myoglobin by using the data in Problem 2.2 and by making the following assumptions: the atom that binds to the haem group is immobilized, the protein is infinitely more massive than either the C or O atom, the C atom binds to the Fe^{2+} ion, and binding of CO to the protein does not alter the force constant of the C≡O bond. [SSM **9.33**]

2.34 Of the four assumptions made in Problem 2.33, the last two are questionable. Suppose that the first two assumptions are still reasonable and that you have at your disposal a supply of myoglobin, a suitable buffer in which to suspend the protein, $^{12}C^{16}O$, $^{13}C^{16}O$, $^{12}C^{18}O$, $^{13}C^{18}O$, and an infrared spectrometer. Assuming that isotopic substitution does not affect the force constant of the C≡O bond, describe a set of experiments that: (a) proves which atom, C or O, binds to the haem group of myoglobin, and (b) allows for the determination of the force constant of the C≡O bond for myoglobin-bound carbon monoxide. [ISM **9.34**]

2.35 The particle on a ring is a useful model for the motion of electrons around the porphine ring (**2**), the conjugated macrocycle that forms the structural basis of the haem group and the chlorophylls. We may treat the group as a circular ring of radius 440 pm, with 22 electrons in the conjugated system moving along the perimeter of the ring. As in *Illustration 2.1*, we assume that in the ground state of the molecule quantized each state is occupied by two electrons. (a) Calculate the energy and angular momentum of an electron in the highest occupied level. (b) Calculate the frequency of radiation that can induce a transition between the highest occupied and lowest unoccupied levels. [SSM **9.35**]

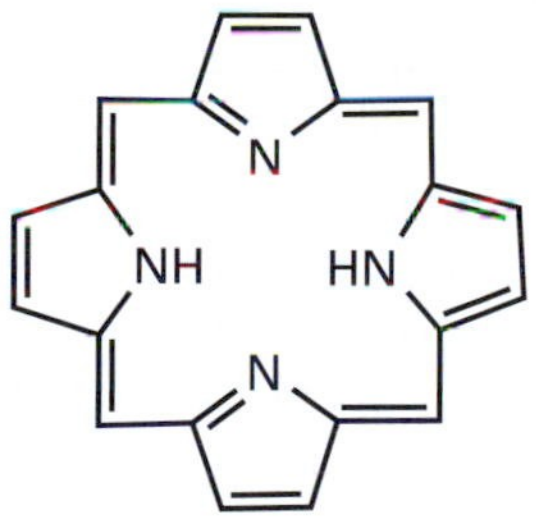

2 Porphine (free base form)

2.36 Some synthetic polymers and biological macromolecules adopt a configuration known as a *random coil*. For a one-dimensional random coil of N units, the restoring force at small displacements and at a temperature T is

$$F = -\frac{kT}{2l}\ln\left(\frac{N+n}{N-n}\right)$$

where l is the length of each monomer unit and nl is the distance between the ends of the chain. Show that for small extensions ($n \ll N$) the restoring force is proportional to n and therefore the coil undergoes harmonic oscillation with force constant kT/Nl^2. Suppose that the mass to use for the vibrating chain is its total mass Nm, where m is the mass of one monomer unit, and deduce the root mean square separation of the ends of the chain due to quantum fluctuations in its vibrational ground state. [ISM **9.36**]

2.37 The forces measured by AFM arise primarily from interactions between electrons of the stylus and on the surface. To get an idea of the magnitudes of these forces, calculate the force acting between two electrons separated by 2.0 nm. *Hints.* The Coulombic potential energy of a charge q_1 at a distance r from another charge q_2 is

$$V = \frac{q_1 q_2}{4\pi\varepsilon_0 r}$$

where $\varepsilon_0 = 8.854 \times 10^{-12}\ \text{C}^2\ \text{J}^{-1}\ \text{m}^{-1}$ is the vacuum permittivity. To calculate the force between the electrons, note that $F = -dV/dr$. [SSM **9.37**]

2.38 Here we explore further the idea introduced in *Impact I2.2* that quantum mechanical effects need to be invoked in the description of the electronic properties of metallic nanocrystals, here modelled as three-dimensional boxes. (a) Set up the Schrödinger equation for a particle of mass m in a three-dimensional rectangular box with sides L_1, L_2, and L_3. Show that the Schrödinger equation is separable. (b) Show that the wavefunction and the energy are defined by three quantum numbers. (c) Specialize the result from part (b) to an electron moving in a cubic box of side $L = 5$ nm and draw an energy diagram resembling Fig. 2.2 and showing the first 15 energy levels. Note that each energy level may consist of degenerate energy states. (d) Compare the energy level diagram from part (c) with the energy level diagram for an electron in a one-dimensional box of length $L = 5$ nm. Are the energy levels become more or less sparsely distributed in the cubic box than in the one-dimensional box? [ISM **9.38**]

2.39 We remarked in *Impact I2.2* that the particle *in* a sphere is a reasonable starting point for the discussion of the electronic properties of spherical metal nanoparticles. Here, we justify eqn 2.54, which shows that the energy of an electron in a sphere is quantized. (a) The Hamiltonian for a particle free to move inside a sphere of radius R is

$$\hat{H} = -\frac{\hbar}{2m}\nabla^2$$

Show that the Schrödinger equation is separable into radial and angular components. That is, begin by writing $\psi(r,\theta,\phi) = X(r)Y(\theta,\phi)$, where $X(r)$ depends only on the distance of the particle away from the centre of the sphere, and $Y(\theta,\phi)$ is a spherical harmonic. Then show that the Schrödinger equation can be separated into two equations, one for X, the radial equation, and the other for Y, the angular equation:

$$-\frac{\hbar^2}{2m}\left(\frac{d^2X(r)}{dr^2} + \frac{2}{r}\frac{dX(r)}{dr}\right) + \frac{l(l+1)\hbar^2}{2mr^2}X(r) = EX(r)$$

$$\Lambda^2 Y = -l(l+1)Y$$

You may wish to consult *Further information 10.1* for additional help. (c) Consider the case $l = 0$. Show by differentiation that the solution of the radial equation has the form

$$X(r) = (2\pi R)^{-1/2}\,\frac{\sin(n\pi r/R)}{r}$$

(e) Now go on to show that the allowed energies are given by:

$$E_n = \frac{n^2h^2}{8mR^2}$$

This result for the energy (which is eqn 2.54 after substituting m_e for m) also applies when $l \neq 0$. [SSM **9.39**]

3 Atomic structure and atomic spectra

We now use the principles of quantum mechanics introduced in the preceding two chapters to describe the internal structures of atoms. We see what experimental information is available from a study of the spectrum of atomic hydrogen. Then we set up the Schrödinger equation for an electron in an atom and separate it into angular and radial parts. The wavefunctions obtained are the 'atomic orbitals' of hydrogenic atoms. Next, we use these hydrogenic atomic orbitals to describe the structures of many-electron atoms. In conjunction with the Pauli exclusion principle, we account for the periodicity of atomic properties. The spectra of many-electron atoms are more complicated than those of hydrogen, but the same principles apply. We see in the closing sections of the chapter how such spectra are described by using term symbols, and the origin of the finer details of their appearance.

In this chapter we see how to use quantum mechanics to describe the **electronic structure** of an atom, the arrangement of electrons around a nucleus. The concepts we meet are of central importance for understanding the structures and reactions of atoms and molecules, and hence have extensive chemical applications. We need to distinguish between two types of atoms. A **hydrogenic atom** is a one-electron atom or ion of general atomic number Z; examples of hydrogenic atoms are H, He^+, Li^{2+}, O^{7+}, and even U^{91+}. A **many-electron atom** (or *polyelectronic atom*) is an atom or ion with more than one electron; examples include all neutral atoms other than H. So even He, with only two electrons, is a many-electron atom. Hydrogenic atoms are important because their Schrödinger equations can be solved exactly. They also provide a set of concepts that are used to describe the structures of many-electron atoms and, as we shall see in the next chapter, the structures of molecules too.

The structure and spectra of hydrogenic atoms

When an electric discharge is passed through gaseous hydrogen, the H_2 molecules are dissociated and the energetically excited H atoms that are produced emit light of discrete frequencies, producting a spectrum of a series of 'lines' (Fig. 3.1). The Swedish spectroscopist Johannes Rydberg noted (in 1890) that all of them are described by the expression

$$\tilde{\nu} = R_H\left(\frac{1}{n_1^2}-\frac{1}{n_2^2}\right) \qquad R_H = 109\ 677\ \text{cm}^{-1} \tag{3.1}$$

with $n_1 = 1$ (the *Lyman series*), 2 (the *Balmer series*), and 3 (the *Paschen series*), and that in each case $n_2 = n_1 + 1, n_1 + 2, \ldots$. The constant R_H is now called the **Rydberg constant** for the hydrogen atom.

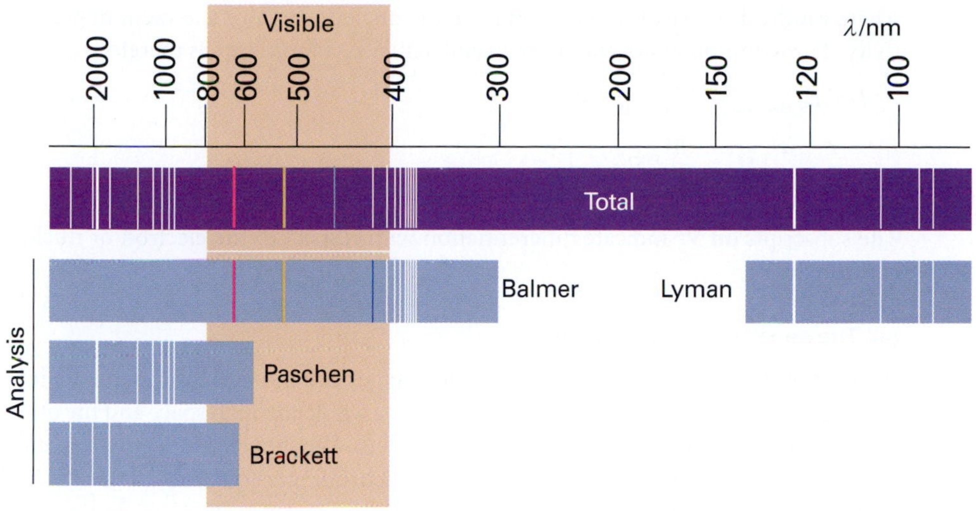

Fig. 3.1 The spectrum of atomic hydrogen. Both the observed spectrum and its resolution into overlapping series are shown. Note that the Balmer series lies in the visible region.

Self-test 3.1 Calculate the shortest wavelength line in the Paschen series.

[821 nm]

The form of eqn 3.1 strongly suggests that the wavenumber of each spectral line can be written as the difference of two **terms**, each of the form

$$T_n = \frac{R_{\mathrm{H}}}{n^2} \tag{3.2}$$

The **Ritz combination principle** states that *the wavenumber of any spectral line is the difference between two terms*. We say that two terms T_1 and T_2 'combine' to produce a spectral line of wavenumber

$$\tilde{\nu} = T_1 - T_2 \tag{3.3}$$

Thus, if each spectroscopic term represents an energy hcT, the difference in energy when the atom undergoes a transition between two terms is $\Delta E = hcT_1 - hcT_2$ and, according to the Bohr frequency conditions (Section 1.1d), the frequency of the radiation emitted is given by $\nu = cT_1 - cT_2$. This expression rearranges into the Ritz formula when expressed in terms of wavenumbers (on division by c). The Ritz combination principle applies to all types of atoms and molecules, but only for hydrogenic atoms do the terms have the simple form (constant)/n^2.

Because spectroscopic observations show that electromagnetic radiation is absorbed and emitted by atoms only at certain wavenumbers, it follows that only certain energy states of atoms are permitted. Our tasks in the first part of this chapter are to determine the origin of this energy quantization, to find the permitted energy levels, and to account for the value of R_{H}.

3.1 The structure of hydrogenic atoms

The Coulomb potential energy of an electron in a hydrogenic atom of atomic number Z (and nuclear charge Ze) is

$$V = -\frac{Ze^2}{4\pi\varepsilon_0 r} \tag{3.4}$$

where r is the distance of the electron from the nucleus and ε_0 is the vacuum permittivity. The hamiltonian for the electron and a nucleus of mass m_N is therefore

$$\hat{H} = \hat{E}_{K,\text{electron}} + \hat{E}_{K,\text{nucleus}} + \hat{V}$$
$$= -\frac{\hbar^2}{2m_e}\nabla_e^2 - \frac{\hbar^2}{2m_N}\nabla_N^2 - \frac{Ze^2}{4\pi\varepsilon_0 r} \tag{3.5}$$

The subscripts on ∇^2 indicate differentiation with respect to the electron or nuclear coordinates.

(a) The separation of variables

Physical intuition suggests that the full Schrödinger equation ought to separate into two equations, one for the motion of the atom as a whole through space and the other for the motion of the electron relative to the nucleus. We show in *Further information 3.1* how this separation is achieved, and that the Schrödinger equation for the internal motion of the electron relative to the nucleus is

$$-\frac{\hbar^2}{2\mu}\nabla^2\psi - \frac{Ze^2}{4\pi\varepsilon_0 r}\psi = E\psi \qquad \frac{1}{\mu} = \frac{1}{m_e} + \frac{1}{m_N} \tag{3.6}$$

where differentiation is now with respect to the coordinates of the electron relative to the nucleus. The quantity μ is called the **reduced mass**. The reduced mass is very similar to the electron mass because m_N, the mass of the nucleus, is much larger than the mass of an electron, so $1/\mu \approx 1/m_e$. In all except the most precise work, the reduced mass can be replaced by m_e.

Because the potential energy is centrosymmetric (independent of angle), we can suspect that the equation is separable into radial and angular components. Therefore, we write

$$\psi(r,\theta,\phi) = R(r)Y(\theta,\phi) \tag{3.7}$$

and examine whether the Schrödinger equation can be separated into two equations, one for R and the other for Y. As shown in *Further information 3.1*, the equation does separate, and the equations we have to solve are

$$\Lambda^2 Y = -l(l+1)Y \tag{3.8}$$

$$-\frac{\hbar^2}{2\mu}\frac{d^2u}{dr^2} + V_{\text{eff}}u = Eu \qquad u = rR \tag{3.9}$$

where

$$V_{\text{eff}} = -\frac{Ze^2}{4\pi\varepsilon_0 r} + \frac{l(l+1)\hbar^2}{2\mu r^2} \tag{3.10}$$

Equation 3.8 is the same as the Schrödinger equation for a particle free to move round a central point, and we considered it in Section 2.7. The solutions are the spherical harmonics (Table 2.3), and are specified by the quantum numbers l and m_l. We consider them in more detail shortly. Equation 3.9 is called the **radial wave equation**. The radial wave equation is the description of the motion of a particle of mass μ in a one-dimensional region $0 \le r < \infty$ where the potential energy is V_{eff}.

(b) The radial solutions

We can anticipate some features of the shapes of the radial wavefunctions by analysing the form of V_{eff}. The first term in eqn 3.10 is the Coulomb potential energy of the electron in the field of the nucleus. The second term stems from what in classical physics would be called the centrifugal force that arises from the angular momentum of the electron around the nucleus. When $l = 0$, the electron has no angular momentum, and the effective potential energy is purely Coulombic and attractive at all radii (Fig. 3.2). When $l \neq 0$, the centrifugal term gives a positive (repulsive) contribution to the effective potential energy. When the electron is close to the nucleus ($r \approx 0$), this repulsive term, which is proportional to $1/r^2$, dominates the attractive Coulombic component, which is proportional to $1/r$, and the net effect is an effective repulsion of the electron from the nucleus. The two effective potential energies, the one for $l = 0$ and the one for $l \neq 0$, are qualitatively very different close to the nucleus. However, they are similar at large distances because the centrifugal contribution tends to zero more rapidly (as $1/r^2$) than the Coulombic contribution (as $1/r$). Therefore, we can expect the solutions with $l = 0$ and $l \neq 0$ to be quite different near the nucleus but similar far away from it. We show in the *Justification* below that close to the nucleus the radial wavefunction is proportional to r^l, and the higher the orbital angular momentum, the less likely the electron is to be found (Fig. 3.3). We also show that far from the nucleus all wavefunctions approach zero exponentially.

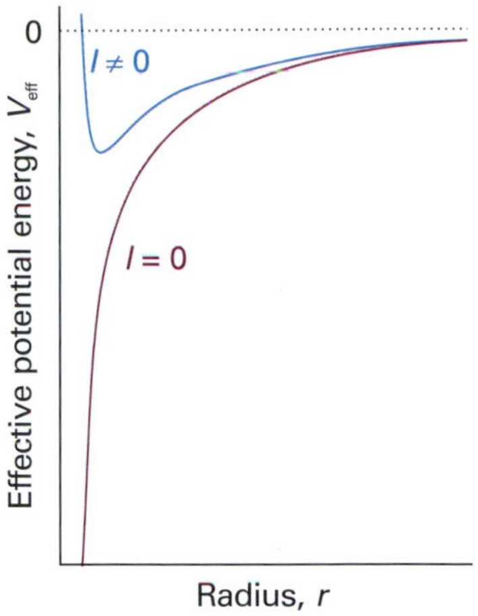

Fig. 3.2 The effective potential energy of an electron in the hydrogen atom. When the electron has zero orbital angular momentum, the effective potential energy is the Coulombic potential energy. When the electron has nonzero orbital angular momentum, the centrifugal effect gives rise to a positive contribution that is very large close to the nucleus. We can expect the $l = 0$ and $l \neq 0$ wavefunctions to be very different near the nucleus.

Exploration Plot the effective potential energy against r for several nonzero values of the orbital angular momentum l. How does the location of the minimum in the effective potential energy vary with l?

Justification 3.1 *The shape of the radial wavefunction*

When r is very small (close to the nucleus), $u \approx 0$, so the right-hand side of eqn 3.9 is zero; we can also ignore all but the largest terms (those depending on $1/r^2$) in eqn 3.9 and write

$$-\frac{d^2u}{dr^2} + \frac{l(l+1)}{r^2}u \approx 0$$

The solution of this equation (for $r \approx 0$) is

$$u \approx Ar^{l+1} + \frac{B}{r^l}$$

Because $R = u/r$, and R cannot be infinite at $r = 0$, we must set $B = 0$, and hence obtain $R \approx Ar^l$.

Far from the nucleus, when r is very large, we can ignore all terms in $1/r$ and eqn 3.9 becomes

$$-\frac{\hbar^2}{2\mu}\frac{d^2u}{dr^2} \simeq Eu$$

where $\simeq$ means 'asymptotically equal to'. Because

$$\frac{d^2u}{dr^2} = \frac{d^2(rR)}{dr^2} = r\frac{d^2R}{dr^2} + 2\frac{dR}{dr} \simeq r\frac{d^2R}{dr^2}$$

this equation has the form

$$-\frac{\hbar^2}{2\mu}\frac{d^2R}{dr^2} \simeq ER$$

The acceptable (finite) solution of this equation (for r large) is

$$R \simeq e^{-(2\mu|E|/\hbar^2)r}$$

and the wavefunction decays exponentially towards zero as r increases.

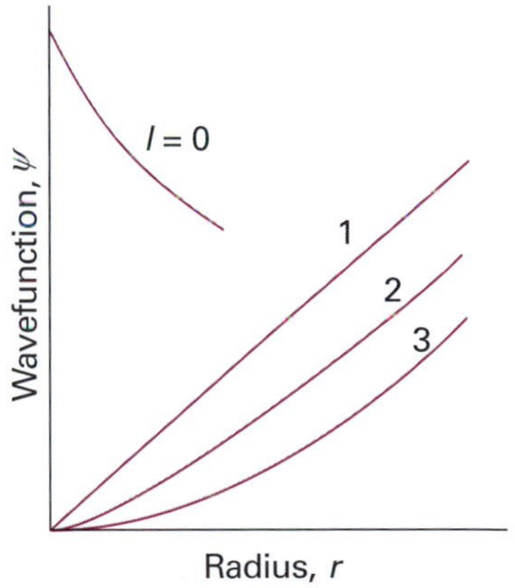

Fig. 3.3 Close to the nucleus, orbitals with $l = 1$ are proportional to r, orbitals with $l = 2$ are proportional to r^2, and orbitals with $l = 3$ are proportional to r^3. Electrons are progressively excluded from the neighbourhood of the nucleus as l increases. An orbital with $l = 0$ has a finite, nonzero value at the nucleus.

We shall not go through the technical steps of solving the radial equation for the full range of radii, and see how the form r^l close to the nucleus blends into the exponentially decaying form at great distances (see *Further reading*). It is sufficient to know that the two limits can be matched only for integral values of a quantum number n, and that the allowed energies corresponding to the allowed solutions are

$$E_n = -\frac{Z^2\mu e^4}{32\pi^2\varepsilon_0^2\hbar^2 n^2} \tag{3.11}$$

with $n = 1, 2, \ldots$ Likewise, the radial wavefunctions depend on the values of both n and l (but not on m_l), and all of them have the form

$$R(r) = (\text{polynomial in } r) \times (\text{decaying exponential in } r) \tag{3.12}$$

These functions are most simply written in terms of the dimensionless quantity ρ (rho), where

$$\rho = \frac{2Zr}{na_0} \qquad a_0 = \frac{4\pi\varepsilon_0\hbar^2}{m_e e^2} \tag{3.13}$$

The **Bohr radius**, a_0, has the value 52.9 pm; it is so called because the same quantity appeared in Bohr's early model of the hydrogen atom as the radius of the electron orbit of lowest energy. Specifically, the radial wavefunctions for an electron with quantum numbers n and l are the (real) functions

$$R_{n,l}(r) = N_{n,l}\rho^l L_{n+1}^{2l+1}(\rho)e^{-\rho/2} \tag{3.14}$$

where L is a polynomial in ρ called an *associated Laguerre polynomial*: it links the $r \approx 0$ solutions on its left (corresponding to $R \propto \rho^l$) to the exponentially decaying function on its right. The notation might look fearsome, but the polynomials have quite simple forms, such as 1, ρ, and $2 - \rho$ (they can be picked out in Table 3.1). Specifically, we can interpret the components of this expression as follows:

1 The exponential factor ensures that the wavefunction approaches zero far from the nucleus.

Table 3.1 Hydrogenic radial wavefunctions

Orbital	n	l	$R_{n,l}$
1*s*	1	0	$2\left(\frac{Z}{a}\right)^{3/2} e^{-\rho/2}$
2*s*	2	0	$\frac{1}{8^{1/2}}\left(\frac{Z}{a}\right)^{3/2}(2-\rho)e^{-\rho/2}$
2*p*	2	1	$\frac{1}{24^{1/2}}\left(\frac{Z}{a}\right)^{3/2}\rho e^{-\rho/2}$
3*s*	3	0	$\frac{1}{243^{1/2}}\left(\frac{Z}{a}\right)^{3/2}(6-6\rho+\rho^2)e^{-\rho/2}$
3*p*	3	1	$\frac{1}{486^{1/2}}\left(\frac{Z}{a}\right)^{3/2}(4-\rho)\rho e^{-\rho/2}$
3*d*	3	2	$\frac{1}{2430^{1/2}}\left(\frac{Z}{a}\right)^{3/2}\rho^2 e^{-\rho/2}$

$\rho = (2Z/na)r$ with $a = 4\pi\varepsilon_0\hbar^2/\mu e^2$. For an infinitely heavy nucleus (or one that may be assumed to be so), $\mu = m_e$ and $a = a_0$, the Bohr radius. The full wavefunction is obtained by multiplying R by the appropriate Y given in Table 2.3.

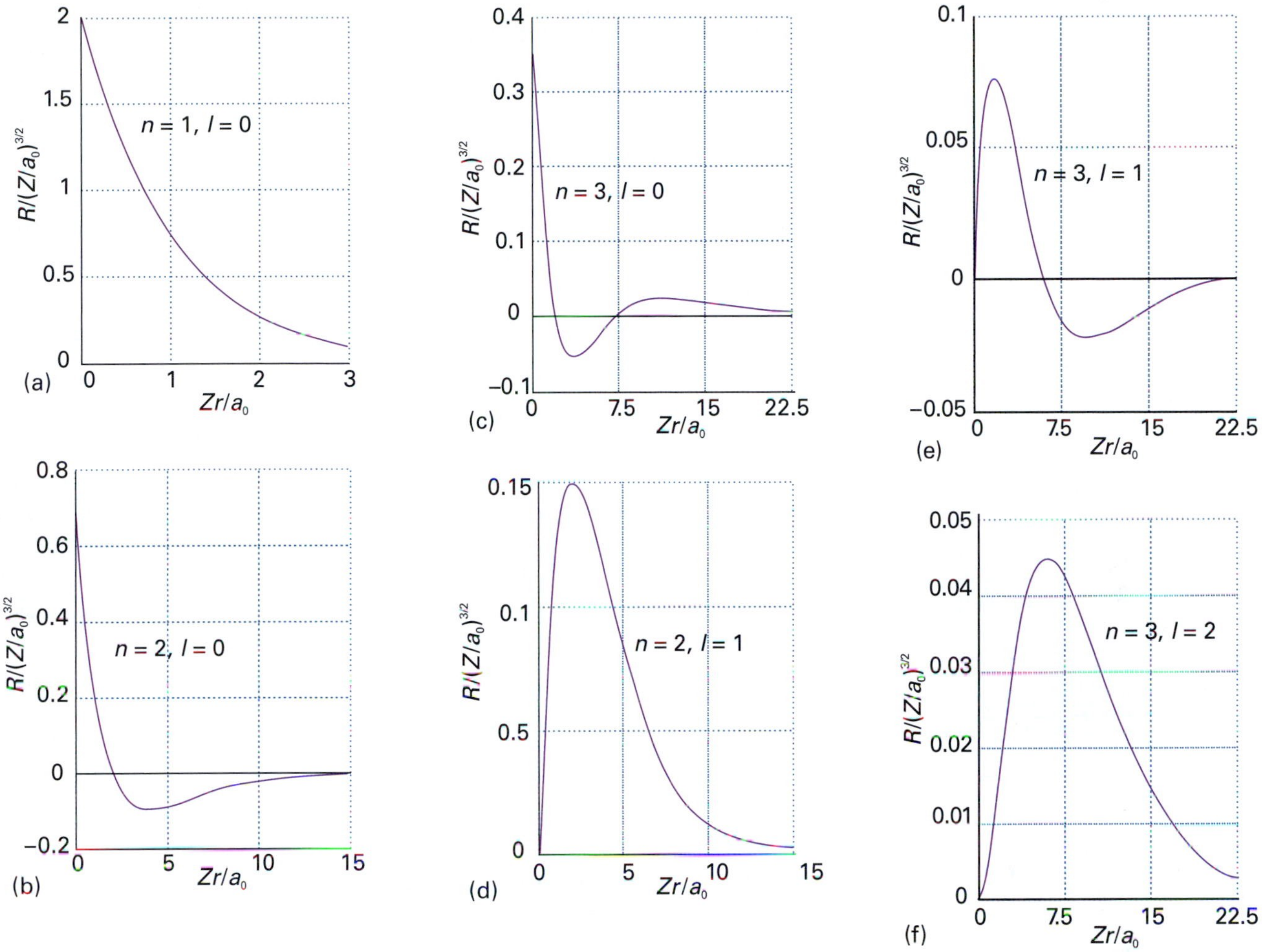

Fig. 3.4 The radial wavefunctions of the first few states of hydrogenic atoms of atomic number Z. Note that the orbitals with $l = 0$ have a nonzero and finite value at the nucleus. The horizontal scales are different in each case: orbitals with high principal quantum numbers are relatively distant from the nucleus.

Exploration Use mathematical software to find the locations of the radial nodes in hydrogenic wavefunctions with n up to 3.

2 The factor ρ^l ensures that (provided $l > 0$) the wavefunction vanishes at the nucleus.

3 The associated Laguerre polynomial is a function that oscillates from positive to negative values and accounts for the presence of radial nodes.

Expressions for some radial wavefunctions are given in Table 3.1 and illustrated in Fig. 3.4.

Comment 3.1

The zero at $r = 0$ is not a radial node because the radial wavefunction does not pass through zero at that point (because r cannot be negative). Nodes at the nucleus are all angular nodes.

Illustration 3.1 *Calculating a probability density*

To calculate the probability density at the nucleus for an electron with $n = 1$, $l = 0$, and $m_l = 0$, we evaluate ψ at $r = 0$:

$$\psi_{1,0,0}(0,\theta,\phi) = R_{1,0}(0)Y_{0,0}(\theta,\phi) = 2\left(\frac{Z}{a_0}\right)^{3/2}\left(\frac{1}{4\pi}\right)^{1/2}$$

The probability density is therefore

$$\psi_{1,0,0}(0,\theta,\phi)^2 = \frac{Z^3}{\pi a_0^3}$$

which evaluates to $2.15 \times 10^{-6}\ \text{pm}^{-3}$ when $Z = 1$.

Self-test 3.2 Evaluate the probability density at the nucleus of the electron for an electron with $n = 2$, $l = 0$, $m_l = 0$. $[(Z/a_0)^3/8\pi]$

3.2 Atomic orbitals and their energies

An **atomic orbital** is a one-electron wavefunction for an electron in an atom. Each hydrogenic atomic orbital is defined by three quantum numbers, designated n, l, and m_l. When an electron is described by one of these wavefunctions, we say that it 'occupies' that orbital. We could go on to say that the electron is in the state $|n,l,m_l\rangle$. For instance, an electron described by the wavefunction $\psi_{1,0,0}$ and in the state $|1,0,0\rangle$ is said to occupy the orbital with $n = 1$, $l = 0$, and $m_l = 0$.

The quantum number n is called the **principal quantum number**; it can take the values $n = 1, 2, 3, \ldots$ and determines the energy of the electron:

An electron in an orbital with quantum number n has an energy given by eqn 3.11.

The two other quantum numbers, l and m_l, come from the angular solutions, and specify the angular momentum of the electron around the nucleus:

An electron in an orbital with quantum number l has an angular momentum of magnitude $\{l(l+1)\}^{1/2}\hbar$, with $l = 0, 1, 2, \ldots, n-1$.

An electron in an orbital with quantum number m_l has a z-component of angular momentum $m_l\hbar$, with $m_l = 0, \pm1, \pm2, \ldots, \pm l$.

Note how the value of the principal quantum number, n, controls the maximum value of l and l controls the range of values of m_l.

To define the state of an electron in a hydrogenic atom fully we need to specify not only the orbital it occupies but also its spin state. We saw in Section 9.8 that an electron possesses an intrinsic angular momentum that is described by the two quantum numbers s and m_s (the analogues of l and m_l). The value of s is fixed at $\frac{1}{2}$ for an electron, so we do not need to consider it further at this stage. However, m_s may be either $+\frac{1}{2}$ or $-\frac{1}{2}$, and to specify the electron's state in a hydrogenic atom we need to specify which of these values describes it. It follows that, to specify the state of an electron in a hydrogenic atom, we need to give the values of four quantum numbers, namely n, l, m_l, and m_s.

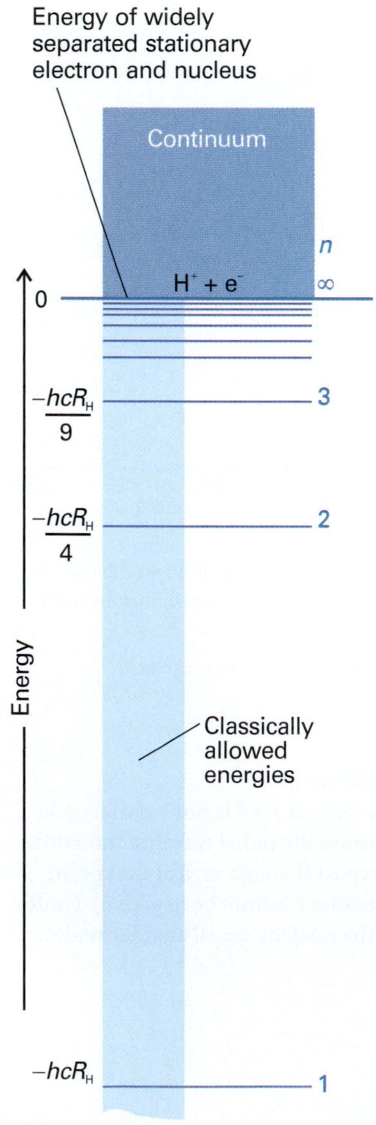

Fig. 3.5 The energy levels of a hydrogen atom. The values are relative to an infinitely separated, stationary electron and a proton.

(a) The energy levels

The energy levels predicted by eqn 3.11 are depicted in Fig. 3.5. The energies, and also the separation of neighbouring levels, are proportional to Z^2, so the levels are four times as wide apart (and the ground state four times deeper in energy) in He^+ ($Z = 2$) than in H ($Z = 1$). All the energies given by eqn 3.11 are negative. They refer to the **bound states** of the atom, in which the energy of the atom is lower than that of the infinitely separated, stationary electron and nucleus (which corresponds to the zero of energy). There are also solutions of the Schrödinger equation with positive energies. These solutions correspond to **unbound states** of the electron, the states to which an electron is raised when it is ejected from the atom by a high-energy collision or photon. The energies of the unbound electron are not quantized and form the continuum states of the atom.

Equation 3.11 is consistent with the spectroscopic result summarized by eqn 3.1, and we can identify the Rydberg constant for hydrogen ($Z = 1$) as

$$hcR_H = \frac{\mu_H e^4}{32\pi^2\varepsilon_0^2\hbar^2} \quad [3.15]$$

where μ_H is the reduced mass for hydrogen. The **Rydberg constant** itself, R, is defined by the same expression except for the replacement of μ_H by the mass of an electron, m_e, corresponding to a nucleus of infinite mass:

$$R_H = \frac{\mu_H}{m_e}R \qquad R = \frac{m_e e^4}{8\varepsilon_0^2 h^3 c} \quad [3.16]$$

Insertion of the values of the fundamental constants into the expression for R_H gives almost exact agreement with the experimental value. The only discrepancies arise from the neglect of relativistic corrections (in simple terms, the increase of mass with speed), which the non-relativistic Schrödinger equation ignores.

Comment 3.2

The particle in a finite well, discussed in Section 2.3, is a primitive but useful model that gives insight into the bound and unbound states of the electron in a hydrogenic atom. Figure 2.15 shows that the energies of a particle (for example, an electron in a hydrogenic atom) are quantized when its total energy, E, is lower than its potential energy, V (the Coulomb interaction energy between the electron and the nucleus). When $E > V$, the particle can escape from the well (the atom is ionized) and its energies are no longer quantized, forming a continuum.

(b) Ionization energies

The **ionization energy**, I, of an element is the minimum energy required to remove an electron from the ground state, the state of lowest energy, of one of its atoms. Because the ground state of hydrogen is the state with $n = 1$, with energy $E_1 = -hcR_H$ and the atom is ionized when the electron has been excited to the level corresponding to $n = \infty$ (see Fig. 3.5), the energy that must be supplied is

$$I = hcR_H \quad (3.17)$$

The value of I is 2.179 aJ (a, for atto, is the prefix that denotes 10^{-18}), which corresponds to 13.60 eV.

Example 3.1 *Measuring an ionization energy spectroscopically*

The emission spectrum of atomic hydrogen shows lines at 82 259, 97 492, 102 824, 105 292, 106 632, and 107 440 cm^{-1}, which correspond to transitions to the same lower state. Determine (a) the ionization energy of the lower state, (b) the value of the Rydberg constant.

Method The spectroscopic determination of ionization energies depends on the determination of the series limit, the wavenumber at which the series terminates and becomes a continuum. If the upper state lies at an energy $-hcR_H/n^2$, then, when the atom makes a transition to E_{lower}, a photon of wavenumber

$$\tilde{\nu} = -\frac{R_H}{n^2} - \frac{E_{lower}}{hc}$$

is emitted. However, because $I = -E_{lower}$, it follows that

$$\tilde{\nu} = \frac{I}{hc} - \frac{R_H}{n^2}$$

A plot of the wavenumbers against $1/n^2$ should give a straight line of slope $-R_H$ and intercept I/hc. Use a computer to make a least-squares fit of the data to get a result that reflects the precision of the data.

Answer The wavenumbers are plotted against $1/n^2$ in Fig. 3.6. The (least-squares) intercept lies at 109 679 cm^{-1}, so the ionization energy is 2.1788 aJ (1312.1 kJ mol^{-1}).

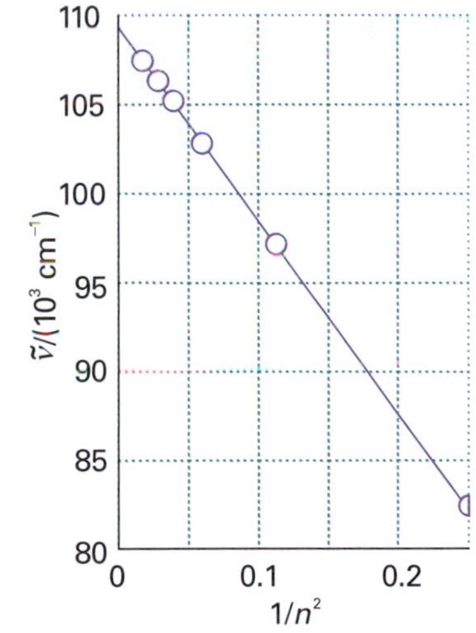

Fig. 3.6 The plot of the data in Example 3.1 used to determine the ionization energy of an atom (in this case, of H).

Exploration The initial value of n was not specified in Example 3.1. Show that the correct value can be determined by making several choices and selecting the one that leads to a straight line.

The slope is, in this instance, numerically the same, so $R_H = 109\,679\ cm^{-1}$. A similar extrapolation procedure can be used for many-electron atoms (see Section 3.5).

Self-test 3.3 The emission spectrum of atomic deuterium shows lines at 15 238, 20 571, 23 039, and 24 380 cm^{-1}, which correspond to transitions to the same lower state. Determine (a) the ionization energy of the lower state, (b) the ionization energy of the ground state, (c) the mass of the deuteron (by expressing the Rydberg constant in terms of the reduced mass of the electron and the deuteron, and solving for the mass of the deuteron).

[(a) 328.1 kJ mol^{-1}, (b) 1312.4 kJ mol^{-1}, (c) 2.8×10^{-27} kg, a result very sensitive to R_D]

(c) Shells and subshells

All the orbitals of a given value of n are said to form a single **shell** of the atom. In a hydrogenic atom, all orbitals of given n, and therefore belonging to the same shell, have the same energy. It is common to refer to successive shells by letters:

$n =$	1	2	3	4 . . .
	K	L	M	N . . .

Thus, all the orbitals of the shell with $n = 2$ form the L shell of the atom, and so on.

The orbitals with the same value of n but different values of l are said to form a **subshell** of a given shell. These subshells are generally referred to by letters:

$l =$	0	1	2	3	4	5	6 . . .
	s	p	d	f	g	h	i . . .

The letters then run alphabetically (j is not used). Figure 3.7 is a version of Fig. 3.5 which shows the subshells explicitly. Because l can range from 0 to $n - 1$, giving n values in all, it follows that there are n subshells of a shell with principal quantum number n. Thus, when $n = 1$, there is only one subshell, the one with $l = 0$. When $n = 2$, there are two subshells, the 2s subshell (with $l = 0$) and the 2p subshell (with $l = 1$).

When $n = 1$ there is only one subshell, that with $l = 0$, and that subshell contains only one orbital, with $m_l = 0$ (the only value of m_l permitted). When $n = 2$, there are four orbitals, one in the s subshell with $l = 0$ and $m_l = 0$, and three in the $l = 1$ subshell with $m_l = +1, 0, -1$. When $n = 3$ there are nine orbitals (one with $l = 0$, three with $l = 1$, and five with $l = 2$). The organization of orbitals in the shells is summarized in Fig. 3.8. In general, the number of orbitals in a shell of principal quantum number n is n^2, so in a hydrogenic atom each energy level is n^2-fold degenerate.

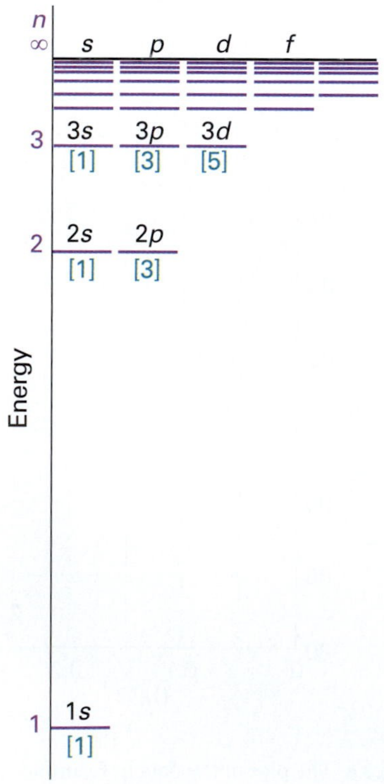

Fig. 3.7 The energy levels of the hydrogen atom showing the subshells and (in square brackets) the numbers of orbitals in each subshell. In hydrogenic atoms, all orbitals of a given shell have the same energy.

(d) Atomic orbitals

The orbital occupied in the ground state is the one with $n = 1$ (and therefore with $l = 0$ and $m_l = 0$, the only possible values of these quantum numbers when $n = 1$). From Table 3.1 we can write (for $Z = 1$):

$$\psi = \frac{1}{(\pi a_0^3)^{1/2}} e^{-r/a_0} \tag{3.18}$$

This wavefunction is independent of angle and has the same value at all points of constant radius; that is, the 1s orbital is *spherically symmetrical.* The wavefunction decays exponentially from a maximum value of $1/(\pi a_0^3)^{1/2}$ at the nucleus (at $r = 0$). It follows that the most probable point at which the electron will be found is at the nucleus itself.

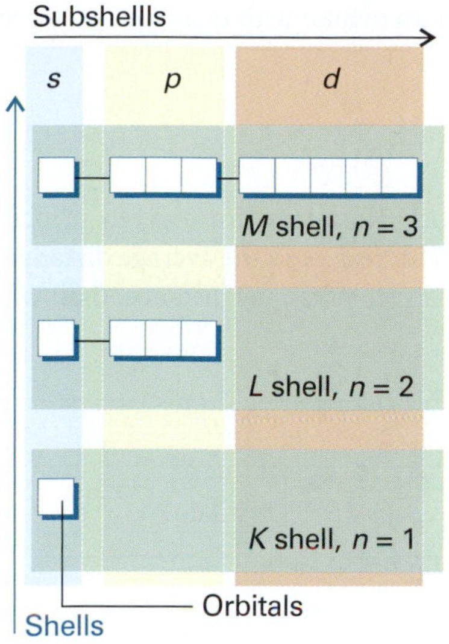

Fig. 3.8 The organization of orbitals (white squares) into subshells (characterized by l) and shells (characterized by n).

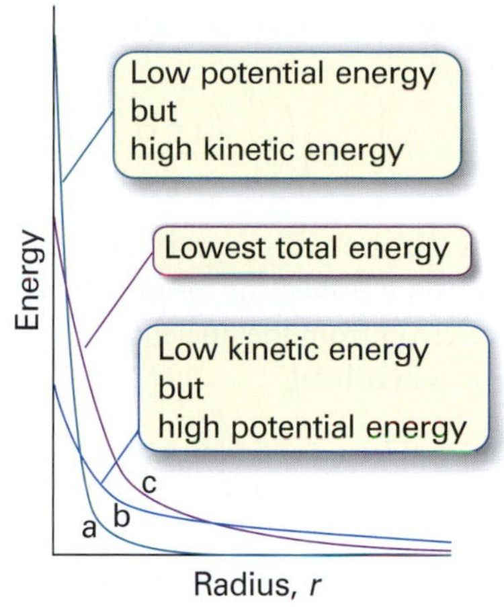

Fig. 3.9 The balance of kinetic and potential energies that accounts for the structure of the ground state of hydrogen (and similar atoms). (a) The sharply curved but localized orbital has high mean kinetic energy, but low mean potential energy;
(b) the mean kinetic energy is low, but the potential energy is not very favourable;
(c) the compromise of moderate kinetic energy and moderately favourable potential energy.

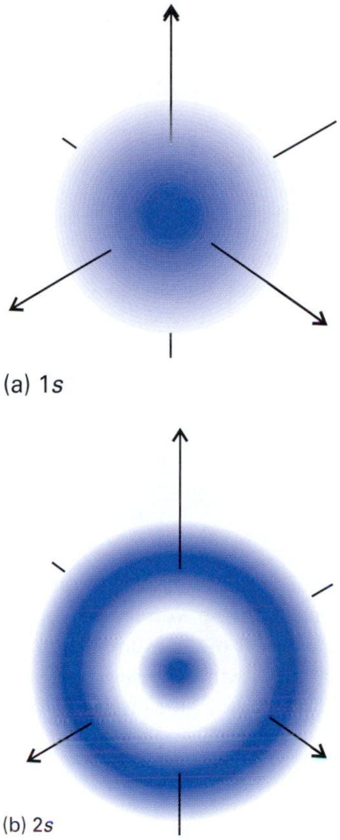

Fig. 3.10 Representations of the 1*s* and 2*s* hydrogenic atomic orbitals in terms of their electron densities (as represented by the density of shading)

We can understand the general form of the ground-state wavefunction by considering the contributions of the potential and kinetic energies to the total energy of the atom. The closer the electron is to the nucleus on average, the lower its average potential energy. This dependence suggests that the lowest potential energy should be obtained with a sharply peaked wavefunction that has a large amplitude at the nucleus and is zero everywhere else (Fig. 3.9). However, this shape implies a high kinetic energy, because such a wavefunction has a very high average curvature. The electron would have very low kinetic energy if its wavefunction had only a very low average curvature. However, such a wavefunction spreads to great distances from the nucleus and the average potential energy of the electron will be correspondingly high. The actual ground-state wavefunction is a compromise between these two extremes: the wavefunction spreads away from the nucleus (so the expectation value of the potential energy is not as low as in the first example, but nor is it very high) and has a reasonably low average curvature (so the expectation of the kinetic energy is not very low, but nor is it as high as in the first example).

The energies of *ns* orbitals increase (become less negative; the electron becomes less tightly bound) as n increases because the average distance of the electron from the nucleus increases. By the virial theorem with $b=-1$ (eqn 2.35), $\langle E_K \rangle = -\frac{1}{2}\langle V \rangle$ so, even though the average kinetic energy decreases as n increases, the total energy is equal to $\frac{1}{2}\langle V \rangle$, which becomes less negative as n increases.

One way of depicting the probability density of the electron is to represent $|\psi|^2$ by the density of shading (Fig. 3.10). A simpler procedure is to show only the **boundary surface**, the surface that captures a high proportion (typically about 90 per cent) of the electron probability. For the 1*s* orbital, the boundary surface is a sphere centred on the nucleus (Fig. 3.11).

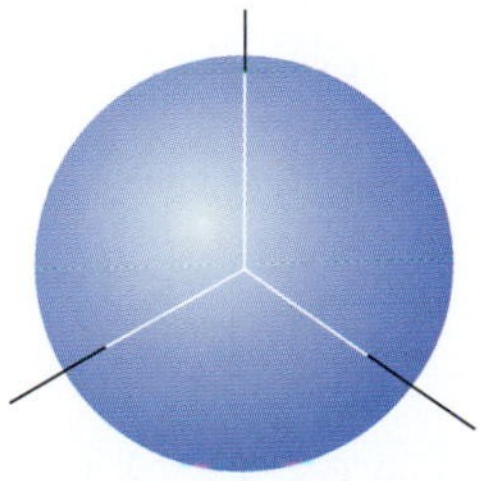

Fig. 3.11 The boundary surface of an *s* orbital, within which there is a 90 per cent probability of finding the electron.

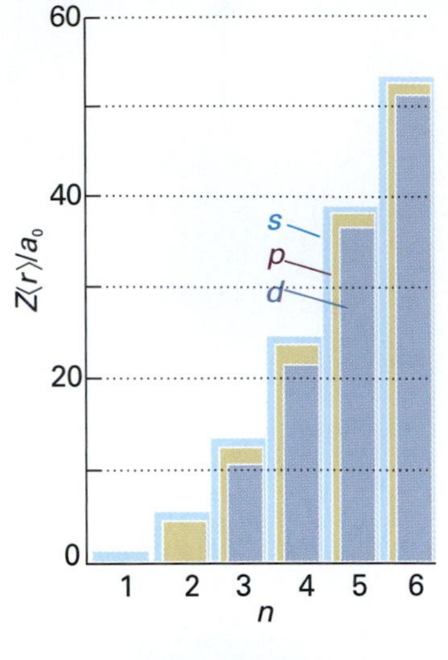

Fig. 3.12 The variation of the mean radius of a hydrogenic atom with the principal and orbital angular momentum quantum numbers. Note that the mean radius lies in the order $d < p < s$ for a given value of n.

The general expression for the mean radius of an orbital with quantum numbers l and n is

$$r_{n,l} = n^2\left\{1+\tfrac{1}{2}\left(1-\frac{l(l+1)}{n^2}\right)\right\}\frac{a_0}{Z} \tag{3.19}$$

The variation with n and l is shown in Fig. 3.12. Note that, for a given principal quantum number, the mean radius decreases as l increases, so the average distance of an electron from the nucleus is less when it is in a $2p$ orbital, for instance, than when it is in a $2s$ orbital.

Example 3.2 *Calculating the mean radius of an orbital*

Use hydrogenic orbitals to calculate the mean radius of a $1s$ orbital.

Method The mean radius is the expectation value

$$\langle r\rangle = \int \psi^* r\psi\,\mathrm{d}\tau = \int r|\psi|^2\,\mathrm{d}\tau$$

We therefore need to evaluate the integral using the wavefunctions given in Table 3.1 and $\mathrm{d}\tau = r^2\mathrm{d}r\sin\theta\,\mathrm{d}\theta\,\mathrm{d}\phi$. The angular parts of the wavefunction are normalized in the sense that

$$\int_0^{\pi}\int_0^{2\pi} |Y_{l,m_l}|^2 \sin\theta\mathrm{d}\theta\,\mathrm{d}\phi = 1$$

The integral over r required is given in Example 1.7.

Answer With the wavefunction written in the form $\psi = RY$, the integration is

$$\langle r\rangle = \int_0^{\infty}\int_0^{\pi}\int_0^{2\pi} rR_{n,l}^2|Y_{l,m_l}|^2 r^2\,\mathrm{d}r\sin\theta\,\mathrm{d}\theta\,\mathrm{d}\phi = \int_0^{\infty} r^3 R_{n,l}^2\mathrm{d}r$$

For a $1s$ orbital,

$$R_{1,0} = 2\left(\frac{Z}{a_0^3}\right)^{1/2}\mathrm{e}^{-Zr/a_0}$$

Hence

$$\langle r\rangle = \frac{4Z}{a_0^3}\int_0^{\infty} r^3\mathrm{e}^{-2Zr/a_0}\mathrm{d}r = \frac{3a_0}{2Z}$$

Self-test 3.4 Evaluate the mean radius (a) of a $3s$ orbital by integration, and (b) of a $3p$ orbital by using the general formula, eqn 3.19. [(a) $27a_0/2Z$; (b) $25a_0/2Z$]

All s-orbitals are spherically symmetric, but differ in the number of radial nodes. For example, the $1s$, $2s$, and $3s$ orbitals have 0, 1, and 2 radial nodes, respectively. In general, an ns orbital has $n-1$ radial nodes.

Self-test 3.5 (a) Use the fact that a $2s$ orbital has radial nodes where the polynomial factor (Table 3.1) is equal to zero, and locate the radial node at $2a_0/Z$ (see Fig. 3.4). (b) Similarly, locate the two nodes of a $3s$ orbital.

[(a) $2a_0/Z$; (b) $1.90a_0/Z$ and $7.10a_0/Z$]

(e) Radial distribution functions

The wavefunction tells us, through the value of $|\psi|^2$, the probability of finding an electron in any region. We can imagine a probe with a volume $d\tau$ and sensitive to electrons, and which we can move around near the nucleus of a hydrogen atom. Because the probability density in the ground state of the atom is $|\psi|^2 \propto e^{-2Zr/a_0}$, the reading from the detector decreases exponentially as the probe is moved out along any radius but is constant if the probe is moved on a circle of constant radius (Fig. 3.13).

Now consider the probability of finding the electron *anywhere* between the two walls of a spherical shell of thickness dr at a radius r. The sensitive volume of the probe is now the volume of the shell (Fig. 3.14), which is $4\pi r^2 dr$ (the product of its surface area, $4\pi r^2$, and its thickness, dr). The probability that the electron will be found between the inner and outer surfaces of this shell is the probability density at the radius r multiplied by the volume of the probe, or $|\psi|^2 \times 4\pi r^2 dr$. This expression has the form $P(r)dr$, where

$$P(r) = 4\pi r^2 \psi^2 \tag{3.20}$$

The more general expression, which also applies to orbitals that are not spherically symmetrical, is

$$P(r) = r^2 R(r)^2 \tag{3.21}$$

where $R(r)$ is the radial wavefunction for the orbital in question.

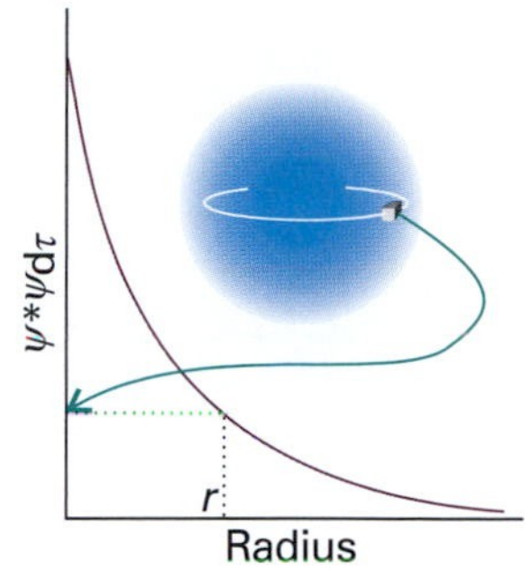

Fig. 3.13 A constant-volume electron-sensitive detector (the small cube) gives its greatest reading at the nucleus, and a smaller reading elsewhere. The same reading is obtained anywhere on a circle of given radius: the *s* orbital is spherically symmetrical.

Justification 3.2 *The general form of the radial distribution function*

The probability of finding an electron in a volume element $d\tau$ when its wavefunction is $\psi = RY$ is $|RY|^2 d\tau$ with $d\tau = r^2 dr \sin\theta\, d\theta d\phi$. The total probability of finding the electron at any angle at a constant radius is the integral of this probability over the surface of a sphere of radius r, and is written $P(r)dr$; so

$$P(r)dr = \int_0^{\pi}\int_0^{2\pi} \overbrace{R(r)^2|Y(\theta,\phi)|^2}^{\psi^2}\, \overbrace{r^2 dr \sin\theta d\theta d\phi}^{d\tau}$$

$$= r^2 R(r)^2 dr \overbrace{\int_0^{\pi}\int_0^{2\pi} |Y(\theta,\phi)|^2 \sin\theta d\theta d\phi}^{1} = r^2 R(r)^2 dr$$

The last equality follows from the fact that the spherical harmonics are normalized to 1 (see Example 3.2). It follows that $P(r) = r^2 R(r)^2$, as stated in the text.

The **radial distribution function**, $P(r)$, is a probability density in the sense that, when it is multiplied by dr, it gives the probability of finding the electron anywhere between the two walls of a spherical shell of thickness dr at the radius r. For a 1*s* orbital,

$$P(r) = \frac{4Z^3}{a_0^3} r^2 e^{-2Zr/a_0} \tag{3.22}$$

Let's interpret this expression:

1 Because $r^2 = 0$ at the nucleus, at the nucleus $P(0) = 0$.

2 As $r \to \infty$, $P(r) \to 0$ on account of the exponential term.

3 The increase in r^2 and the decrease in the exponential factor means that P passes through a maximum at an intermediate radius (see Fig. 3.14).

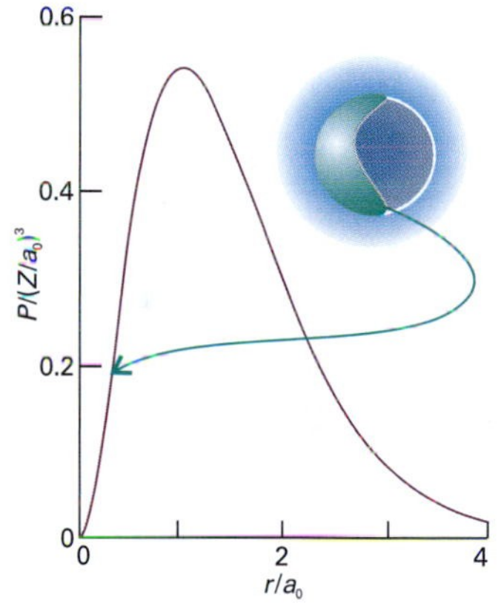

Fig. 3.14 The radial distribution function P gives the probability that the electron will be found anywhere in a shell of radius r. For a 1*s* electron in hydrogen, P is a maximum when r is equal to the Bohr radius a_0. The value of P is equivalent to the reading that a detector shaped like a spherical shell would give as its radius is varied.

The maximum of $P(r)$, which can be found by differentiation, marks the most probable radius at which the electron will be found, and for a 1*s* orbital in hydrogen occurs at $r = a_0$, the Bohr radius. When we carry through the same calculation for the radial distribution function of the 2*s* orbital in hydrogen, we find that the most probable radius is $5.2a_0 = 275$ pm. This larger value reflects the expansion of the atom as its energy increases.

Example 3.3 *Calculating the most probable radius*

Calculate the most probable radius, r^*, at which an electron will be found when it occupies a 1*s* orbital of a hydrogenic atom of atomic number Z, and tabulate the values for the one-electron species from H to Ne^{9+}.

Method We find the radius at which the radial distribution function of the hydrogenic 1*s* orbital has a maximum value by solving $\mathrm{d}P/\mathrm{d}r = 0$. If there are several maxima, then we choose the one corresponding to the greatest amplitude (the outermost one).

Answer The radial distribution function is given in eqn 3.22. It follows that

$$\frac{\mathrm{d}P}{\mathrm{d}r} = \frac{4Z^3}{a_0^3}\left(2r - \frac{2Zr^2}{a_0}\right)\mathrm{e}^{-2Zr/a_0}$$

This function is zero where the term in parentheses is zero, which is at

$$r^* = \frac{a_0}{Z}$$

Then, with $a_0 = 52.9$ pm, the radial node lies at

	H	He^+	Li^{2+}	Be^{3+}	B^{4+}	C^{5+}	N^{6+}	O^{7+}	F^{8+}	Ne^{9+}
r^*/pm	52.9	26.5	17.6	13.2	10.6	8.82	7.56	6.61	5.88	5.29

Notice how the 1*s* orbital is drawn towards the nucleus as the nuclear charge increases. At uranium the most probable radius is only 0.58 pm, almost 100 times closer than for hydrogen. (On a scale where $r^* = 10$ cm for H, $r^* = 1$ mm for U.) The electron then experiences strong accelerations and relativistic effects are important.

Self-test 3.6 Find the most probable distance of a 2*s* electron from the nucleus in a hydrogenic atom. $[(3 + 5^{1/2})a_0/Z]$

(f) *p* Orbitals

The three 2*p* orbitals are distinguished by the three different values that m_l can take when $l = 1$. Because the quantum number m_l tells us the orbital angular momentum around an axis, these different values of m_l denote orbitals in which the electron has different orbital angular momenta around an arbitrary *z*-axis but the same magnitude of that momentum (because l is the same for all three). The orbital with $m_l = 0$, for instance, has zero angular momentum around the *z*-axis. Its angular variation is proportional to $\cos\theta$, so the probability density, which is proportional to $\cos^2\theta$, has its maximum value on either side of the nucleus along the *z*-axis (at $\theta = 0$ and 180°). The wavefunction of a 2*p*-orbital with $m_l = 0$ is

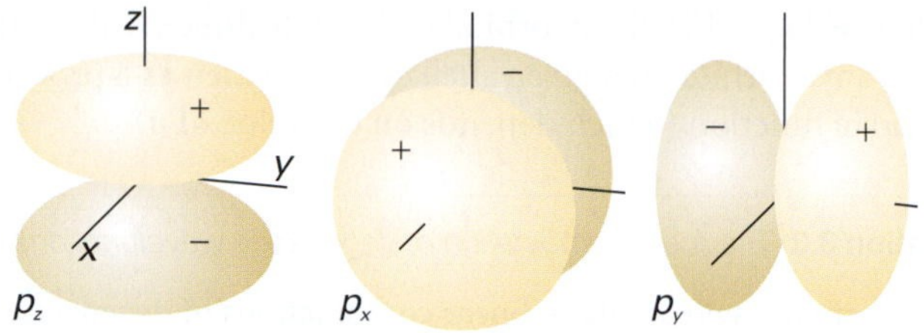

Fig. 3.15 The boundary surfaces of p orbitals. A nodal plane passes through the nucleus and separates the two lobes of each orbital. The dark and light areas denote regions of opposite sign of the wavefunction.

Exploration Use mathematical software to plot the boundary surfaces of the real parts of the spherical harmonics $Y_{1,m_l}(\theta,\phi)$. The resulting plots are not strictly the p orbital boundary surfaces, but sufficiently close to be reasonable representations of the shapes of hydrogenic orbitals.

$$\psi_{p_0} = R_{2,1}(r)Y_{1,0}(\theta,\phi) = \frac{1}{4(2\pi)^{1/2}}\left(\frac{Z}{a_0}\right)^{5/2} r\cos\theta e^{-Zr/2a_0}$$
$$= r\cos\theta f(r)$$

where $f(r)$ is a function only of r. Because in spherical polar coordinates $z = r\cos\theta$, this wavefunction may also be written

$$\psi_{p_z} = zf(r) \tag{3.23}$$

All p orbitals with $m_l = 0$ have wavefunctions of this form regardless of the value of n. This way of writing the orbital is the origin of the name 'p_z orbital': its boundary surface is shown in Fig. 3.15. The wavefunction is zero everywhere in the xy-plane, where $z = 0$, so the xy-plane is a **nodal plane** of the orbital: the wavefunction changes sign on going from one side of the plane to the other.

The wavefunctions of $2p$ orbitals with $m_l = \pm 1$ have the following form:

$$\psi_{p_{\pm 1}} = R_{2,1}(r)Y_{1,\pm 1}(\theta,\phi) = \mp\frac{1}{8\pi^{1/2}}\left(\frac{Z}{a_0}\right)^{5/2} re^{-Zr/2a_0}\sin\theta e^{\pm i\phi}$$
$$= \mp\frac{1}{2^{1/2}} r\sin\theta e^{\pm i\phi} f(r)$$

We saw in Chapter 1 that a moving particle can be described by a complex wavefunction. In the present case, the functions correspond to non-zero angular momentum about the z-axis: $e^{+i\phi}$ corresponds to clockwise rotation when viewed from below, and $e^{-i\phi}$ corresponds to counter-clockwise rotation (from the same viewpoint). They have zero amplitude where $\theta = 0$ and 180° (along the z-axis) and maximum amplitude at 90°, which is in the xy-plane. To draw the functions it is usual to represent them as standing waves. To do so, we take the real linear combinations

$$\psi_{p_x} = -\frac{1}{2^{1/2}}(p_{+1} - p_{-1}) = r\sin\theta\cos\phi f(r) = xf(r)$$
$$\psi_{p_y} = \frac{i}{2^{1/2}}(p_{+1} + p_{-1}) = r\sin\theta\sin\phi f(r) = yf(r) \tag{3.24}$$

These linear combinations are indeed standing waves with no net orbital angular momentum around the z-axis, as they are superpositions of states with equal and opposite values of m_l. The p_x orbital has the same shape as a p_z orbital, but it is directed along the x-axis (see Fig. 3.15); the p_y orbital is similarly directed along the y-axis. The wavefunction of any p orbital of a given shell can be written as a product of x, y, or z and the same radial function (which depends on the value of n).

Justification 3.3 *The linear combination of degenerate wavefunctions*

We justify here the step of taking linear combinations of degenerate orbitals when we want to indicate a particular point. The freedom to do so rests on the fact that, whenever two or more wavefunctions correspond to the same energy, any linear combination of them is an equally valid solution of the Schrödinger equation.

Suppose ψ_1 and ψ_2 are both solutions of the Schrödinger equation with energy E; then we know that

$$H\psi_1 = E\psi_1 \qquad H\psi_2 = E\psi_2$$

Now consider the linear combination

$$\psi = c_1\psi_1 + c_2\psi_2$$

where c_1 and c_2 are arbitrary coefficients. Then it follows that

$$H\psi = H(c_1\psi_1 + c_2\psi_2) = c_1H\psi_1 + c_2H\psi_2 = c_1E\psi_1 + c_2E\psi_2 = E\psi$$

Hence, the linear combination is also a solution corresponding to the same energy E.

(g) *d* Orbitals

When $n=3$, l can be 0, 1, or 2. As a result, this shell consists of one 3s orbital, three 3p orbitals, and five 3d orbitals. The five d orbitals have $m_l = +2, +1, 0, -1, -2$ and correspond to five different angular momenta around the z-axis (but the same *magnitude* of angular momentum, because $l=2$ in each case). As for the p orbitals, d orbitals with opposite values of m_l (and hence opposite senses of motion around the z-axis) may be combined in pairs to give real standing waves, and the boundary surfaces of the resulting shapes are shown in Fig. 3.16. The real combinations have the following forms:

$$d_{xy} = xyf(r) \qquad d_{yz} = yzf(r) \qquad d_{zx} = zxf(r)$$
$$d_{x^2-y^2} = \tfrac{1}{2}(x^2-y^2)f(r) \qquad d_{z^2} = (\tfrac{1}{2}\sqrt{3})(3z^2-r^2)f(r) \qquad (3.25)$$

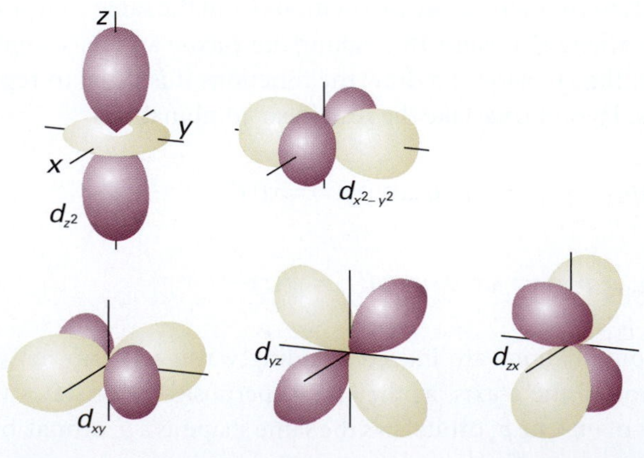

Fig. 3.16 The boundary surfaces of d orbitals. Two nodal planes in each orbital intersect at the nucleus and separate the lobes of each orbital. The dark and light areas denote regions of opposite sign of the wavefunction.

Exploration To gain insight into the shapes of the f orbitals, use mathematical software to plot the boundary surfaces of the spherical harmonics $Y_{3,m_l}(\theta,\phi)$.

3.3 Spectroscopic transitions and selection rules

The energies of the hydrogenic atoms are given by eqn 3.11. When the electron undergoes a **transition**, a change of state, from an orbital with quantum numbers n_1, l_1, m_{l1} to another (lower energy) orbital with quantum numbers n_2, l_2, m_{l2}, it undergoes a change of energy ΔE and discards the excess energy as a photon of electromagnetic radiation with a frequency ν given by the Bohr frequency condition (eqn 1.10).

It is tempting to think that all possible transitions are permissible, and that a spectrum arises from the transition of an electron from any initial orbital to any other orbital. However, this is not so, because a photon has an intrinsic spin angular momentum corresponding to $s = 1$ (Section 2.8). The change in angular momentum of the electron must compensate for the angular momentum carried away by the photon. Thus, an electron in a *d* orbital ($l = 2$) cannot make a transition into an *s* orbital ($l = 0$) because the photon cannot carry away enough angular momentum. Similarly, an *s* electron cannot make a transition to another *s* orbital, because there would then be no change in the electron's angular momentum to make up for the angular momentum carried away by the photon. It follows that some spectroscopic transitions are **allowed**, meaning that they can occur, whereas others are **forbidden**, meaning that they cannot occur.

A **selection rule** is a statement about which transitions are allowed. They are derived (for atoms) by identifying the transitions that conserve angular momentum when a photon is emitted or absorbed. The selection rules for hydrogenic atoms are

$$\Delta l = \pm 1 \qquad \Delta m_l = 0, \pm 1 \tag{3.26}$$

The principal quantum number n can change by any amount consistent with the Δl for the transition, because it does not relate directly to the angular momentum.

Justification 3.4 *The identification of selection rules*

We saw in Section 2.10 that the rate of transition between two states is proportional to the square of the transition dipole moment, $\boldsymbol{\mu}_{\mathrm{fi}}$, between the initial and final states, where (using the notation introduced in *Further information 2.1*)

$$\boldsymbol{\mu}_{\mathrm{fi}} = \langle \mathrm{f} | \boldsymbol{\mu} | \mathrm{i} \rangle \qquad [3.27]$$

and $\boldsymbol{\mu}$ is the electric dipole moment operator. For a one-electron atom $\boldsymbol{\mu}$ is multiplication by $-e\boldsymbol{r}$ with components $\mu_x = -ex$, $\mu_y = -ey$, and $\mu_z = -ez$. If the transition dipole moment is zero, the transition is forbidden; the transition is allowed if the transition moment is non-zero. Physically, the transition dipole moment is a measure of the dipolar 'kick' that the electron gives to or receives from the electromagnetic field.

To evaluate a transition dipole moment, we consider each component in turn. For example, for the z-component,

$$\mu_{z,\mathrm{fi}} = -e\langle \mathrm{f} | z | \mathrm{i} \rangle = -e\int \psi_{\mathrm{f}}^{\star} z \psi_{\mathrm{i}} \,\mathrm{d}\tau \tag{3.28}$$

To evaluate the integral, we note from Table 2.3 that $z = (4\pi/3)^{1/2} rY_{1,0}$, so

$$\int \psi_{\mathrm{f}}^{\star} z \psi_{\mathrm{i}} \mathrm{d}\tau = \int_0^{\infty}\int_0^{\pi}\int_0^{2\pi} \overbrace{R_{n_{\mathrm{f}},l_{\mathrm{f}}} Y_{l_{\mathrm{f}},m_{l,\mathrm{f}}}^{\star}}^{\psi_{\mathrm{f}}} \overbrace{\left(\frac{4\pi}{3}\right)^{1/2} rY_{1,0}}^{z} \overbrace{R_{n_{\mathrm{i}},l_{\mathrm{i}}} Y_{l_{\mathrm{i}},m_{l,\mathrm{i}}}}^{\psi_{\mathrm{i}}} \overbrace{r^2 \mathrm{d}r \sin\theta \mathrm{d}\theta \mathrm{d}\phi}^{\mathrm{d}\tau}$$

This multiple integral is the product of three factors, an integral over r and two integrals over the angles, so the factors on the right can be grouped as follows:

$$\int \psi_{\mathrm{f}}^{\star} z \psi_{\mathrm{i}} \mathrm{d}\tau = \left(\frac{4\pi}{3}\right)^{1/2} \int_0^{\infty} R_{n_{\mathrm{f}},l_{\mathrm{f}}} r R_{n_{\mathrm{i}},l_{\mathrm{i}}} r^2 \mathrm{d}r \int_0^{\pi}\int_0^{2\pi} Y_{l_{\mathrm{f}},m_{l,\mathrm{f}}}^{\star} Y_{1,0} Y_{l_{\mathrm{i}},m_{l,\mathrm{i}}} \sin\theta \,\mathrm{d}\theta \mathrm{d}\phi$$

It follows from the properties of the spherical harmonics (Comment 2.6) that the integral

$$\int_0^{\pi}\int_0^{2\pi} Y^{\star}_{l_f m_{l_f}} Y_{1,m} Y_{l_i m_{l_i}} \sin\theta \mathrm{d}\theta\mathrm{d}\phi$$

is zero unless $l_f = l_i \pm 1$ and $m_{l,f} = m_{l,i} + m$. Because $m = 0$ in the present case, the angular integral, and hence the z-component of the transition dipole moment, is zero unless $\Delta l = \pm 1$ and $\Delta m_l = 0$, which is a part of the set of selection rules. The same procedure, but considering the x- and y-components, results in the complete set of rules.

Illustration 3.2 *Applying selection rules*

To identify the orbitals to which a 4*d* electron may make radiative transitions, we first identify the value of l and then apply the selection rule for this quantum number. Because $l = 2$, the final orbital must have $l = 1$ or 3. Thus, an electron may make a transition from a 4*d* orbital to any *np* orbital (subject to $\Delta m_l = 0, \pm 1$) and to any *nf* orbital (subject to the same rule). However, it cannot undergo a transition to any other orbital, so a transition to any *ns* orbital or to another *nd* orbital is forbidden.

Self-test 3.7 To what orbitals may a 4*s* electron make electric-dipole allowed radiative transitions? [to *np* orbitals only]

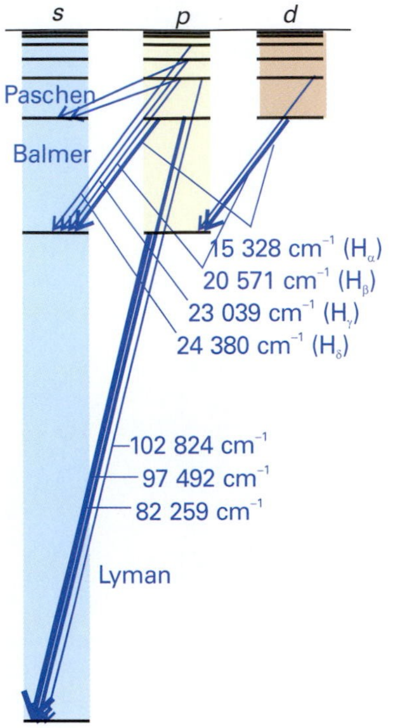

Fig. 3.17 A Grotrian diagram that summarizes the appearance and analysis of the spectrum of atomic hydrogen. The thicker the line, the more intense the transition.

The selection rules and the atomic energy levels jointly account for the structure of a **Grotrian diagram** (Fig. 3.17), which summarizes the energies of the states and the transitions between them. The thicknesses of the transition lines in the diagram denote their relative intensities in the spectrum; we see how to determine transition intensities in Section 6.2.

The structures of many-electron atoms

The Schrödinger equation for a many-electron atom is highly complicated because all the electrons interact with one another. Even for a helium atom, with its two electrons, no analytical expression for the orbitals and energies can be given, and we are forced to make approximations. We shall adopt a simple approach based on what we already know about the structure of hydrogenic atoms. Later we shall see the kind of numerical computations that are currently used to obtain accurate wavefunctions and energies.

3.4 The orbital approximation

The wavefunction of a many-electron atom is a very complicated function of the coordinates of all the electrons, and we should write it $\psi(\boldsymbol{r}_1,\boldsymbol{r}_2,\ldots)$, where $\boldsymbol{r}_i$ is the vector from the nucleus to electron i. However, in the **orbital approximation** we suppose that a reasonable first approximation to this exact wavefunction is obtained by thinking of each electron as occupying its 'own' orbital, and write

$$\psi(\boldsymbol{r}_1,\boldsymbol{r}_2,\ldots) = \psi(\boldsymbol{r}_1)\psi(\boldsymbol{r}_2)\ldots \tag{3.29}$$

We can think of the individual orbitals as resembling the hydrogenic orbitals, but corresponding to nuclear charges modified by the presence of all the other electrons in the atom. This description is only approximate, but it is a useful model for discussing the chemical properties of atoms, and is the starting point for more sophisticated descriptions of atomic structure.

Justification 3.5 *The orbital approximation*

The orbital approximation would be exact if there were no interactions between electrons. To demonstrate the validity of this remark, we need to consider a system in which the hamiltonian for the energy is the sum of two contributions, one for electron 1 and the other for electron 2:

$$\hat{H} = \hat{H}_1 + \hat{H}_2$$

In an actual atom (such as helium atom), there is an additional term corresponding to the interaction of the two electrons, but we are ignoring that term. We shall now show that if $\psi(\boldsymbol{r}_1)$ is an eigenfunction of $\hat{H}_1$ with energy E_1, and $\psi(\boldsymbol{r}_2)$ is an eigenfunction of $\hat{H}_2$ with energy E_2, then the product $\psi(\boldsymbol{r}_1,\boldsymbol{r}_2) = \psi(\boldsymbol{r}_1)\psi(\boldsymbol{r}_2)$ is an eigenfunction of the combined hamiltonian $\hat{H}$. To do so we write

$$\begin{aligned}\hat{H}\psi(\boldsymbol{r}_1,\boldsymbol{r}_2) &= (\hat{H}_1 + \hat{H}_2)\psi(\boldsymbol{r}_1)\psi(\boldsymbol{r}_2) = \hat{H}_1\psi(\boldsymbol{r}_1)\psi(\boldsymbol{r}_2) + \psi(\boldsymbol{r}_1)\hat{H}_2\psi(\boldsymbol{r}_2)\\ &= E_1\psi(\boldsymbol{r}_1)\psi(\boldsymbol{r}_2) + \psi(\boldsymbol{r}_1)E_2\psi(\boldsymbol{r}_2) = (E_1 + E_2)\psi(\boldsymbol{r}_1)\psi(\boldsymbol{r}_2)\\ &= E\psi(\boldsymbol{r}_1,\boldsymbol{r}_2)\end{aligned}$$

where $E = E_1 + E_2$. This is the result we need to prove. However, if the electrons interact (as they do in fact), then the proof fails.

(a) The helium atom

The orbital approximation allows us to express the electronic structure of an atom by reporting its **configuration**, the list of occupied orbitals (usually, but not necessarily, in its ground state). Thus, as the ground state of a hydrogenic atom consists of the single electron in a 1*s* orbital, we report its configuration as $1s^1$.

The He atom has two electrons. We can imagine forming the atom by adding the electrons in succession to the orbitals of the bare nucleus (of charge 2*e*). The first electron occupies a 1*s* hydrogenic orbital, but because $Z = 2$ that orbital is more compact than in H itself. The second electron joins the first in the 1*s* orbital, so the electron configuration of the ground state of He is $1s^2$.

(b) The Pauli principle

Lithium, with $Z = 3$, has three electrons. The first two occupy a 1*s* orbital drawn even more closely than in He around the more highly charged nucleus. The third electron, however, does not join the first two in the 1*s* orbital because that configuration is forbidden by the **Pauli exclusion principle**:

> No more than two electrons may occupy any given orbital, and if two do occupy one orbital, then their spins must be paired.

Electrons with paired spins, denoted ↑↓, have zero net spin angular momentum because the spin of one electron is cancelled by the spin of the other. Specifically, one electron has $m_s = +\frac{1}{2}$, the other has $m_s = -\frac{1}{2}$ and they are orientated on their respective cones so that the resultant spin is zero (Fig. 3.18). The exclusion principle is the key to the structure of complex atoms, to chemical periodicity, and to molecular structure. It was proposed by Wolfgang Pauli in 1924 when he was trying to account for the

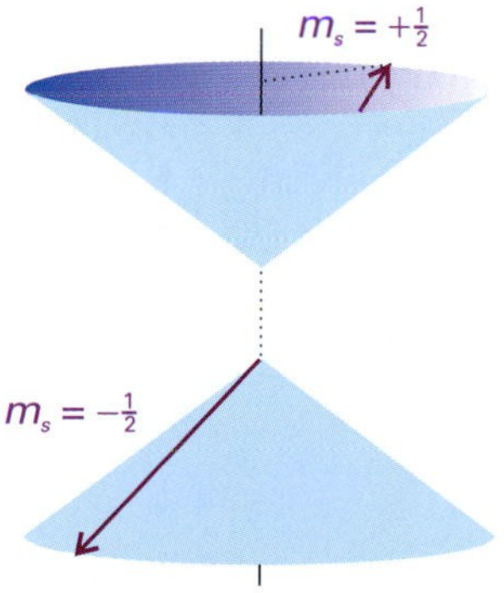

Fig. 3.18 Electrons with paired spins have zero resultant spin angular momentum. They can be represented by two vectors that lie at an indeterminate position on the cones shown here, but wherever one lies on its cone, the other points in the opposite direction; their resultant is zero.

absence of some lines in the spectrum of helium. Later he was able to derive a very general form of the principle from theoretical considerations.

The Pauli exclusion principle in fact applies to any pair of identical fermions (particles with half integral spin). Thus it applies to protons, neutrons, and ^{13}C nuclei (all of which have spin $\frac{1}{2}$) and to ^{35}Cl nuclei (which have spin $\frac{3}{2}$). It does not apply to identical bosons (particles with integral spin), which include photons (spin 1), ^{12}C nuclei (spin 0). Any number of identical bosons may occupy the same state (that is, be described by the same wavefunction).

The Pauli exclusion principle is a special case of a general statement called the **Pauli principle**:

> When the labels of any two identical fermions are exchanged, the total wavefunction changes sign; when the labels of any two identical bosons are exchanged, the total wavefunction retains the same sign.

By 'total wavefunction' is meant the entire wavefunction, including the spin of the particles. To see that the Pauli principle implies the Pauli exclusion principle, we consider the wavefunction for two electrons $\psi(1,2)$. The Pauli principle implies that it is a fact of nature (which has its roots in the theory of relativity) that the wavefunction must change sign if we interchange the labels 1 and 2 wherever they occur in the function:

$$\psi(2,1) = -\psi(1,2) \tag{3.30}$$

Suppose the two electrons in an atom occupy an orbital ψ, then in the orbital approximation the overall wavefunction is $\psi(1)\psi(2)$. To apply the Pauli principle, we must deal with the total wavefunction, the wavefunction including spin. There are several possibilities for two spins: both α, denoted $\alpha(1)\alpha(2)$, both β, denoted $\beta(1)\beta(2)$, and one α the other β, denoted either $\alpha(1)\beta(2)$ or $\alpha(2)\beta(1)$. Because we cannot tell which electron is α and which is β, in the last case it is appropriate to express the spin states as the (normalized) linear combinations

$$\begin{aligned}\sigma_+(1,2) &= (1/2^{1/2})\{\alpha(1)\beta(2) + \beta(1)\alpha(2)\}\\ \sigma_-(1,2) &= (1/2^{1/2})\{\alpha(1)\beta(2) - \beta(1)\alpha(2)\}\end{aligned} \tag{3.31}$$

Comment 3.3

A stronger justification for taking linear combinations in eqn 3.31 is that they correspond to eigenfunctions of the total spin operators S^2 and S_z, with $M_S = 0$ and, respectively, $S = 1$ and 0. See Section 3.7.

because these combinations allow one spin to be α and the other β with equal probability. The total wavefunction of the system is therefore the product of the orbital part and one of the four spin states:

$$\psi(1)\psi(2)\alpha(1)\alpha(2) \quad \psi(1)\psi(2)\beta(1)\beta(2) \quad \psi(1)\psi(2)\sigma_+(1,2) \quad \psi(1)\psi(2)\sigma_-(1,2)$$

The Pauli principle says that for a wavefunction to be acceptable (for electrons), it must change sign when the electrons are exchanged. In each case, exchanging the labels 1 and 2 converts the factor $\psi(1)\psi(2)$ into $\psi(2)\psi(1)$, which is the same, because the order of multiplying the functions does not change the value of the product. The same is true of $\alpha(1)\alpha(2)$ and $\beta(1)\beta(2)$. Therefore, the first two overall products are not allowed, because they do not change sign. The combination $\sigma_+(1,2)$ changes to

$$\sigma_+(2,1) = (1/2^{1/2})\{\alpha(2)\beta(1) + \beta(2)\alpha(1)\} = \sigma_+(1,2)$$

because it is simply the original function written in a different order. The third overall product is therefore also disallowed. Finally, consider $\sigma_-(1,2)$:

$$\begin{aligned}\sigma_-(2,1) &= (1/2^{1/2})\{\alpha(2)\beta(1) - \beta(2)\alpha(1)\}\\ &= -(1/2^{1/2})\{\alpha(1)\beta(2) - \beta(1)\alpha(2)\} = -\sigma_-(1,2)\end{aligned}$$

This combination does change sign (it is 'antisymmetric'). The product $\psi(1)\psi(2)$ $\sigma_-(1,2)$ also changes sign under particle exchange, and therefore it is acceptable.

Now we see that only one of the four possible states is allowed by the Pauli principle, and the one that survives has paired α and β spins. This is the content of the Pauli exclusion principle. The exclusion principle is irrelevant when the orbitals occupied by the electrons are different, and both electrons may then have (but need not have) the same spin state. Nevertheless, even then the overall wavefunction must still be antisymmetric overall, and must still satisfy the Pauli principle itself.

A final point in this connection is that the acceptable product wavefunction $\psi(1)\psi(2)\sigma_-(1,2)$ can be expressed as a determinant:

$$\frac{1}{2^{1/2}}\begin{vmatrix}\psi(1)\alpha(1) & \psi(2)\alpha(2)\\ \psi(1)\beta(1) & \psi(2)\beta(2)\end{vmatrix} = \frac{1}{2^{1/2}}\{\psi(1)\alpha(1)\psi(2)\beta(2)-\psi(2)\alpha(2)\psi(1)\beta(1)\}$$
$$=\psi(1)\psi(2)\sigma_-(1,2)$$

Any acceptable wavefunction for a closed-shell species can be expressed as a **Slater determinant**, as such determinants are known. In general, for N electrons in orbitals $\psi_a, \psi_b, \ldots$

$$\psi(1,2,\ldots,N)=\frac{1}{(N!)^{1/2}}\begin{vmatrix}\psi_a(1)\alpha(1) & \psi_a(2)\alpha(2) & \psi_a(3)\alpha(3) & \ldots & \psi_a(N)\alpha(N)\\ \psi_a(1)\beta(1) & \psi_a(2)\beta(2) & \psi_a(3)\beta(3) & \ldots & \psi_a(N)\beta(N)\\ \psi_b(1)\alpha(1) & \psi_b(2)\alpha(2) & \psi_b(3)\alpha(3) & \ldots & \psi_b(N)\alpha(N)\\ \vdots & \vdots & \vdots & \vdots & \vdots\\ \psi_z(1)\beta(1) & \psi_z(2)\beta(2) & \psi_z(3)\beta(3) & \ldots & \psi_z(N)\beta(N)\end{vmatrix} \quad [3.32]$$

Writing a many-electron wavefunction in this way ensures that it is antisymmetric under the interchange of any pair of electrons, as is explored in Problem 3.23.

Now we can return to lithium. In Li ($Z=3$), the third electron cannot enter the 1*s* orbital because that orbital is already full: we say the *K* shell is *complete* and that the two electrons form a **closed shell**. Because a similar closed shell is characteristic of the He atom, we denote it [He]. The third electron is excluded from the *K* shell and must occupy the next available orbital, which is one with $n=2$ and hence belonging to the *L* shell. However, we now have to decide whether the next available orbital is the 2*s* orbital or a 2*p* orbital, and therefore whether the lowest energy configuration of the atom is [He]$2s^1$ or [He]$2p^1$.

(c) Penetration and shielding

Unlike in hydrogenic atoms, the 2*s* and 2*p* orbitals (and, in general, all subshells of a given shell) are not degenerate in many-electron atoms. As will be familiar from introductory chemistry, an electron in a many-electron atom experiences a Coulombic repulsion from all the other electrons present. If it is at a distance *r* from the nucleus, it experiences an average repulsion that can be represented by a point negative charge located at the nucleus and equal in magnitude to the total charge of the electrons within a sphere of radius *r* (Fig. 3.19). The effect of this point negative charge, when averaged over all the locations of the electron, is to reduce the full charge of the nucleus from Ze to $Z_{\text{eff}}e$, the **effective nuclear charge**. In everyday parlance, Z_{eff} itself is commonly referred to as the 'effective nuclear charge'. We say that the electron experiences a **shielded** nuclear charge, and the difference between Z and Z_{eff} is called the **shielding constant**, σ:

$$Z_{\text{eff}}=Z-\sigma \quad [3.33]$$

The electrons do not actually 'block' the full Coulombic attraction of the nucleus: the shielding constant is simply a way of expressing the net outcome of the nuclear

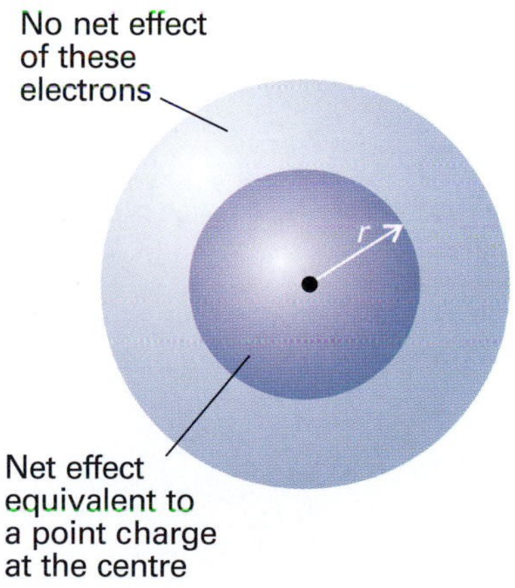

Fig. 3.19 An electron at a distance *r* from the nucleus experiences a Coulombic repulsion from all the electrons within a sphere of radius *r* and which is equivalent to a point negative charge located on the nucleus. The negative charge reduces the effective nuclear charge of the nucleus from Ze to $Z_{\text{eff}}e$.

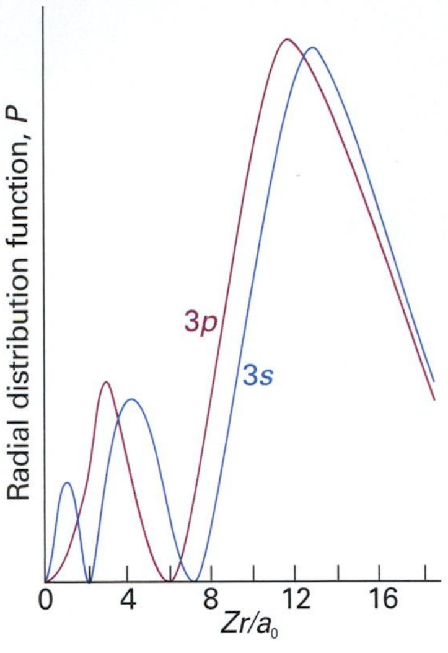

Fig. 3.20 An electron in an *s* orbital (here a 3*s* orbital) is more likely to be found close to the nucleus than an electron in a *p* orbital of the same shell (note the closeness of the innermost peak of the 3*s* orbital to the nucleus at $r = 0$). Hence an *s* electron experiences less shielding and is more tightly bound than a *p* electron.

Exploration Calculate and plot the graphs given above for $n = 4$.

Synoptic table 3.2* Effective nuclear charge, $Z_{\text{eff}} = Z - \sigma$

Element	Z	Orbital	Z_{eff}
He	2	1*s*	1.6875
C	6	1*s*	5.6727
		2*s*	3.2166
		2*p*	3.1358

* More values are given in the *Data section*.

attraction and the electronic repulsions in terms of a single equivalent charge at the centre of the atom.

The shielding constant is different for *s* and *p* electrons because they have different radial distributions (Fig. 3.20). An *s* electron has a greater **penetration** through inner shells than a *p* electron, in the sense that it is more likely to be found close to the nucleus than a *p* electron of the same shell (the wavefunction of a *p* orbital, remember, is zero at the nucleus). Because only electrons inside the sphere defined by the location of the electron (in effect, the core electrons) contribute to shielding, an *s* electron experiences less shielding than a *p* electron. Consequently, by the combined effects of penetration and shielding, an *s* electron is more tightly bound than a *p* electron of the same shell. Similarly, a *d* electron penetrates less than a *p* electron of the same shell (recall that the wavefunction of a *d* orbital varies as r^2 close to the nucleus, whereas a *p* orbital varies as *r*), and therefore experiences more shielding.

Shielding constants for different types of electrons in atoms have been calculated from their wavefunctions obtained by numerical solution of the Schrödinger equation for the atom (Table 3.2). We see that, in general, valence-shell *s* electrons do experience higher effective nuclear charges than *p* electrons, although there are some discrepancies. We return to this point shortly.

The consequence of penetration and shielding is that the energies of subshells of a shell in a many-electron atom in general lie in the order

$$s < p < d < f$$

The individual orbitals of a given subshell remain degenerate because they all have the same radial characteristics and so experience the same effective nuclear charge.

We can now complete the Li story. Because the shell with $n = 2$ consists of two non-degenerate subshells, with the 2*s* orbital lower in energy than the three 2*p* orbitals, the third electron occupies the 2*s* orbital. This occupation results in the ground-state configuration $1s^22s^1$, with the central nucleus surrounded by a complete helium-like shell of two 1*s* electrons, and around that a more diffuse 2*s* electron. The electrons in the outermost shell of an atom in its ground state are called the **valence electrons** because they are largely responsible for the chemical bonds that the atom forms. Thus, the valence electron in Li is a 2*s* electron and its other two electrons belong to its core.

(d) The building-up principle

The extension of this argument is called the **building-up principle**, or the *Aufbau principle*, from the German word for building up, which will be familiar from introductory courses. In brief, we imagine the bare nucleus of atomic number *Z*, and then feed into the orbitals *Z* electrons in succession. The order of occupation is

$$1s \quad 2s \quad 2p \quad 3s \quad 3p \quad 4s \quad 3d \quad 4p \quad 5s \quad 4d \quad 5p \quad 6s$$

and each orbital may accommodate up to two electrons. As an example, consider the carbon atom, for which $Z = 6$ and there are six electrons to accommodate. Two electrons enter and fill the 1*s* orbital, two enter and fill the 2*s* orbital, leaving two electrons to occupy the orbitals of the 2*p* subshell. Hence the ground-state configuration of C is $1s^22s^22p^2$, or more succinctly $[\text{He}]2s^22p^2$, with [He] the helium-like $1s^2$ core. However, we can be more precise: we can expect the last two electrons to occupy different 2*p* orbitals because they will then be further apart on average and repel each other less than if they were in the same orbital. Thus, one electron can be thought of as occupying the $2p_x$ orbital and the other the $2p_y$ orbital (the *x*, *y*, *z* designation is arbitrary, and it would be equally valid to use the complex forms of these orbitals), and the lowest energy configuration of the atom is $[\text{He}]2s^22p_x^12p_y^1$. The same rule applies whenever degenerate orbitals of a subshell are available for occupation. Thus, another rule of the building-up principle is:

> Electrons occupy different orbitals of a given subshell before doubly occupying any one of them.

For instance, nitrogen ($Z=7$) has the configuration $[He]2s^22p_x^12p_y^12p_z^1$, and only when we get to oxygen ($Z=8$) is a $2p$ orbital doubly occupied, giving $[He]2s^22p_x^22p_y^12p_z^1$. When electrons occupy orbitals singly we invoke **Hund's maximum multiplicity rule**:

> An atom in its ground state adopts a configuration with the greatest number of unpaired electrons.

The explanation of Hund's rule is subtle, but it reflects the quantum mechanical property of **spin correlation**, that electrons with parallel spins behave as if they have a tendency to stay well apart, and hence repel each other less. In essence, the effect of spin correlation is to allow the atom to shrink slightly, so the electron–nucleus interaction is improved when the spins are parallel. We can now conclude that, in the ground state of the carbon atom, the two $2p$ electrons have the same spin, that all three $2p$ electrons in the N atoms have the same spin, and that the two $2p$ electrons in different orbitals in the O atom have the same spin (the two in the $2p_x$ orbital are necessarily paired).

Justification 3.6 *Spin correlation*

Suppose electron 1 is described by a wavefunction $\psi_a(\boldsymbol{r}_1)$ and electron 2 is described by a wavefunction $\psi_b(\boldsymbol{r}_2)$; then, in the orbital approximation, the joint wavefunction of the electrons is the product $\psi=\psi_a(\boldsymbol{r}_1)\psi_b(\boldsymbol{r}_2)$. However, this wavefunction is not acceptable, because it suggests that we know which electron is in which orbital, whereas we cannot keep track of electrons. According to quantum mechanics, the correct description is either of the two following wavefunctions:

$$\psi_\pm=(1/2^{1/2})\{\psi_a(\boldsymbol{r}_1)\psi_b(\boldsymbol{r}_2)\pm\psi_b(\boldsymbol{r}_1)\psi_a(\boldsymbol{r}_2)\}$$

According to the Pauli principle, because ψ_+ is symmetrical under particle interchange, it must be multiplied by an antisymmetric spin function (the one denoted σ_-). That combination corresponds to a spin-paired state. Conversely, ψ_- is antisymmetric, so it must be multiplied by one of the three symmetric spin states. These three symmetric states correspond to electrons with parallel spins (see Section 3.7 for an explanation).

Now consider the values of the two combinations when one electron approaches another, and $\boldsymbol{r}_1=\boldsymbol{r}_2$. We see that ψ_- vanishes, which means that there is zero probability of finding the two electrons at the same point in space when they have parallel spins. The other combination does not vanish when the two electrons are at the same point in space. Because the two electrons have different relative spatial distributions depending on whether their spins are parallel or not, it follows that their Coulombic interaction is different, and hence that the two states have different energies.

Neon, with $Z=10$, has the configuration $[He]2s^22p^6$, which completes the L shell. This closed-shell configuration is denoted [Ne], and acts as a core for subsequent elements. The next electron must enter the $3s$ orbital and begin a new shell, so an Na atom, with $Z=11$, has the configuration $[Ne]3s^1$. Like lithium with the configuration $[He]2s^1$, sodium has a single s electron outside a complete core. This analysis has brought us to the origin of chemical periodicity. The L shell is completed by eight electrons, so the element with $Z=3$ (Li) should have similar properties to the element with $Z=11$ (Na). Likewise, Be ($Z=4$) should be similar to $Z=12$ (Mg), and so on, up to the noble gases He ($Z=2$), Ne ($Z=10$), and Ar ($Z=18$).

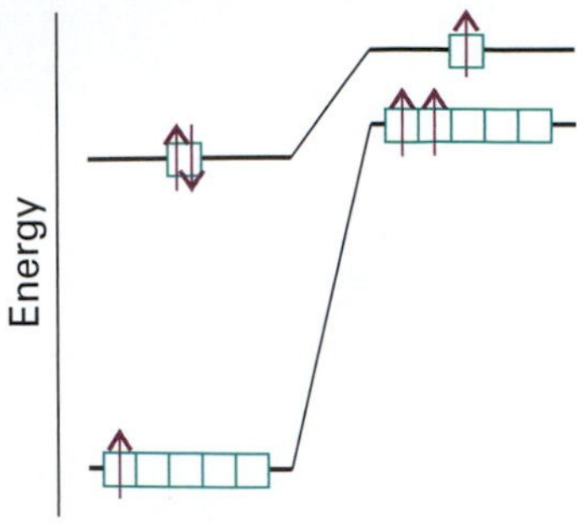

Fig. 3.21 Strong electron–electron repulsions in the $3d$ orbitals are minimized in the ground state of Sc if the atom has the configuration $[Ar]3d^14s^2$ (shown on the left) instead of $[Ar]3d^24s^1$ (shown on the right). The total energy of the atom is lower when it has the $[Ar]3d^14s^2$ configuration despite the cost of populating the high energy $4s$ orbital.

Ten electrons can be accommodated in the five $3d$ orbitals, which accounts for the electron configurations of scandium to zinc. Calculations of the type discussed in Section 3.5 show that for these atoms the energies of the $3d$ orbitals are always lower than the energy of the $4s$ orbital. However, spectroscopic results show that Sc has the configuration $[Ar]3d^14s^2$, instead of $[Ar]3d^3$ or $[Ar]3d^24s^1$. To understand this observation, we have to consider the nature of electron–electron repulsions in $3d$ and $4s$ orbitals. The most probable distance of a $3d$ electron from the nucleus is less than that for a $4s$ electron, so two $3d$ electrons repel each other more strongly than two $4s$ electrons. As a result, Sc has the configuration $[Ar]3d^14s^2$ rather than the two alternatives, for then the strong electron–electron repulsions in the $3d$ orbitals are minimized. The total energy of the atom is least despite the cost of allowing electrons to populate the high energy $4s$ orbital (Fig. 3.21). The effect just described is generally true for scandium through zinc, so their electron configurations are of the form $[Ar]3d^n4s^2$, where $n = 1$ for scandium and $n = 10$ for zinc. Two notable exceptions, which are observed experimentally, are Cr, with electron configuration $[Ar]3d^54s^1$, and Cu, with electron configuration $[Ar]3d^{10}4s^1$ (see *Further reading* for a discussion of the theoretical basis for these exceptions).

At gallium, the building-up principle is used in the same way as in preceding periods. Now the $4s$ and $4p$ subshells constitute the valence shell, and the period terminates with krypton. Because 18 electrons have intervened since argon, this period is the first 'long period' of the periodic table. The existence of the d-block elements (the 'transition metals') reflects the stepwise occupation of the $3d$ orbitals, and the subtle shades of energy differences and effects of electron–electron repulsion along this series gives rise to the rich complexity of inorganic d-metal chemistry. A similar intrusion of the f orbitals in Periods 6 and 7 accounts for the existence of the f block of the periodic table (the lanthanoids and actinoids).

We derive the configurations of cations of elements in the s, p, and d blocks of the periodic table by removing electrons from the ground-state configuration of the neutral atom in a specific order. First, we remove valence p electrons, then valence s electrons, and then as many d electrons as are necessary to achieve the specified charge. For instance, because the configuration of V is $[Ar]3d^34s^2$, the V^{2+} cation has the configuration $[Ar]3d^3$. It is reasonable that we remove the more energetic $4s$ electrons in order to form the cation, but it is not obvious why the $[Ar]3d^3$ configuration is preferred in V^{2+} over the $[Ar]3d^14s^2$ configuration, which is found in the isoelectronic Sc atom. Calculations show that the energy difference between $[Ar]3d^3$ and $[Ar]3d^14s^2$ depends on Z_{eff}. As Z_{eff} increases, transfer of a $4s$ electron to a $3d$ orbital becomes more favourable because the electron–electron repulsions are compensated by attractive interactions between the nucleus and the electrons in the spatially compact $3d$ orbital. Indeed, calculations reveal that, for a sufficiently large Z_{eff}, $[Ar]3d^3$ is lower in energy than $[Ar]3d^14s^2$. This conclusion explains why V^{2+} has a $[Ar]3d^3$ configuration and also accounts for the observed $[Ar]4s^03d^n$ configurations of the M^{2+} cations of Sc through Zn.

The configurations of anions of the p-block elements are derived by continuing the building-up procedure and adding electrons to the neutral atom until the configuration of the next noble gas has been reached. Thus, the configuration of the O^{2-} ion is achieved by adding two electrons to $[He]2s^22p^4$, giving $[He]2s^22p^6$, the same as the configuration of neon.

Comment 3.4
The web site for this text contains links to databases of atomic properties.

(e) Ionization energies and electron affinities

The minimum energy necessary to remove an electron from a many-electron atom in the gas phase is the **first ionization energy**, I_1, of the element. The **second ionization energy**, I_2, is the minimum energy needed to remove a second electron (from the

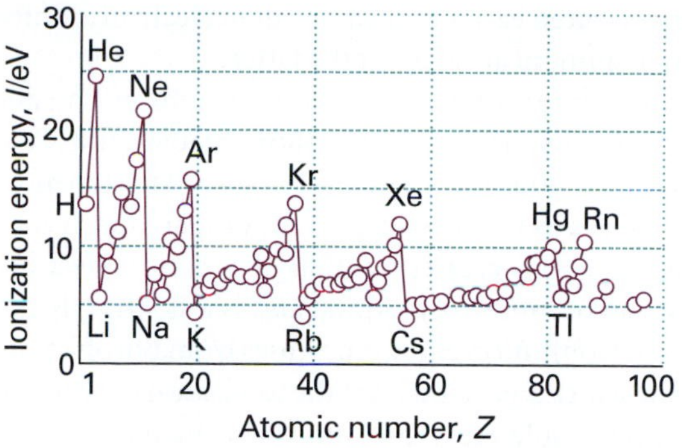

Fig. 3.22 The first ionization energies of the elements plotted against atomic number.

singly charged cation). The variation of the first ionization energy through the periodic table is shown in Fig. 3.22 and some numerical values are given in Table 3.3. In thermodynamic calculations we often need the **standard enthalpy of ionization**, $\Delta_{\text{ion}}H^{\ominus}$. As shown in the *Justification* below, the two are related by

$$\Delta_{\text{ion}}H^{\ominus}(T) = I + \tfrac{5}{2}RT \qquad (3.34)$$

At 298 K, the difference between the ionization enthalpy and the corresponding ionization energy is 6.20 kJ mol^{-1}.

Justification 3.7 *The ionization enthalpy and the ionization energy*

It follows from Kirchhoff's law (Volume 1, Section 2.9 and eqn 2.36) that the reaction enthalpy for

$$M(g) \rightarrow M^{+}(g) + e^{-}(g)$$

at a temperature T is related to the value at $T = 0$ by

$$\Delta_r H^{\ominus}(T) = \Delta_r H^{\ominus}(0) + \int_0^T \Delta_r C_p^{\ominus} \mathrm{d}T$$

The molar constant-pressure heat capacity of each species in the reaction is $\tfrac{5}{2}R$, so $\Delta_r C_p^{\ominus} = +\tfrac{5}{2}R$. The integral in this expression therefore evaluates to $+\tfrac{5}{2}RT$. The reaction enthalpy at $T = 0$ is the same as the (molar) ionization energy, I. Equation 3.33 then follows. The same expression applies to each successive ionization step, so the overall ionization enthalpy for the formation of M^{2+} is

$$\Delta_r H^{\ominus}(T) = I_1 + I_2 + 5RT$$

Synoptic table 3.3* First and second ionization energies

Element	I_1/(kJ mol^{-1})	I_2/(kJ mol^{-1})
H	1312	
He	2372	5251
Mg	738	1451
Na	496	4562

* More values are given in the *Data section*.

Synoptic table 3.4* Electron affinities, E_a/(kJ mol^{-1})

Cl	349		
F	322		
H	73		
O	141	O^-	−844

* More values are given in the *Data section*.

The **electron affinity**, E_{ea}, is the energy released when an electron attaches to a gas-phase atom (Table 3.4). In a common, logical, but not universal convention (which we adopt), the electron affinity is positive if energy is released when the electron attaches to the atom (that is, $E_{\text{ea}} > 0$ implies that electron attachment is exothermic). It follows from a similar argument to that given in the *Justification* above that the **standard enthalpy of electron gain**, $\Delta_{\text{eg}}H^{\ominus}$, at a temperature T is related to the electron affinity by

$$\Delta_{\text{eg}}H^{\ominus}(T) = -E_{\text{ea}} - \tfrac{5}{2}RT \qquad (3.35)$$

Note the change of sign. In typical thermodynamic cycles the $\tfrac{5}{2}RT$ that appears in eqn 3.35 cancels that in eqn 3.34, so ionization energies and electron affinities can be used directly. A final preliminary point is that the electron-gain enthalpy of a species X is the negative of the ionization enthalpy of its negative ion:

$$\Delta_{\text{eg}}H^{\ominus}(X) = -\Delta_{\text{ion}}H^{\ominus}(X^{-}) \qquad (3.36)$$

As ionization energy is often easier to measure than electron affinity, this relation can be used to determine numerical values of the latter.

As will be familiar from introductory chemistry, ionization energies and electron affinities show periodicities. The former is more regular and we concentrate on it. Lithium has a low first ionization energy because its outermost electron is well shielded from the nucleus by the core ($Z_{\text{eff}} = 1.3$, compared with $Z = 3$). The ionization energy of beryllium ($Z = 4$) is greater but that of boron is lower because in the latter the outermost electron occupies a $2p$ orbital and is less strongly bound than if it had been a $2s$ electron. The ionization energy increases from boron to nitrogen on account of the increasing nuclear charge. However, the ionization energy of oxygen is less than would be expected by simple extrapolation. The explanation is that at oxygen a $2p$ orbital must become doubly occupied, and the electron–electron repulsions are increased above what would be expected by simple extrapolation along the row. In addition, the loss of a $2p$ electron results in a configuration with a half-filled subshell (like that of N), which is an arrangement of low energy, so the energy of $O^+ + e^-$ is lower than might be expected, and the ionization energy is correspondingly low too. (The kink is less pronounced in the next row, between phosphorus and sulfur because their orbitals are more diffuse.) The values for oxygen, fluorine, and neon fall roughly on the same line, the increase of their ionization energies reflecting the increasing attraction of the more highly charged nuclei for the outermost electrons.

The outermost electron in sodium is $3s$. It is far from the nucleus, and the latter's charge is shielded by the compact, complete neon-like core. As a result, the ionization energy of sodium is substantially lower than that of neon. The periodic cycle starts again along this row, and the variation of the ionization energy can be traced to similar reasons.

Electron affinities are greatest close to fluorine, for the incoming electron enters a vacancy in a compact valence shell and can interact strongly with the nucleus. The attachment of an electron to an anion (as in the formation of O^{2-} from O^-) is invariably endothermic, so E_{ea} is negative. The incoming electron is repelled by the charge already present. Electron affinities are also small, and may be negative, when an electron enters an orbital that is far from the nucleus (as in the heavier alkali metal atoms) or is forced by the Pauli principle to occupy a new shell (as in the noble gas atoms).

3.5 Self-consistent field orbitals

The central difficulty of the Schrödinger equation is the presence of the electron–electron interaction terms. The potential energy of the electrons is

$$V = -\sum_i \frac{Ze^2}{4\pi\varepsilon_0 r_i} + \tfrac{1}{2}\sum_{i,j}{}' \frac{e^2}{4\pi\varepsilon_0 r_{ij}} \tag{3.37}$$

The prime on the second sum indicates that $i \neq j$, and the factor of one-half prevents double-counting of electron pair repulsions (1 with 2 is the same as 2 with 1). The first term is the total attractive interaction between the electrons and the nucleus. The second term is the total repulsive interaction between the electrons; r_{ij} is the distance between electrons i and j. It is hopeless to expect to find analytical solutions of a Schrödinger equation with such a complicated potential energy term, but computational techniques are available that give very detailed and reliable numerical solutions for the wavefunctions and energies. The techniques were originally introduced by D.R. Hartree (before computers were available) and then modified by V. Fock to take into account the Pauli principle correctly. In broad outline, the **Hartree–Fock self-consistent field** (HF-SCF) procedure is as follows.

Imagine that we have a rough idea of the structure of the atom. In the Ne atom, for instance, the orbital approximation suggests the configuration $1s^22s^22p^6$ with the orbitals approximated by hydrogenic atomic orbitals. Now consider one of the $2p$ electrons. A Schrödinger equation can be written for this electron by ascribing to it a potential energy due to the nuclear attraction and the repulsion from the other electrons. This equation has the form

$$H(1)\psi_{2p}(1) + V(\text{other electrons})\psi_{2p}(1) - V(\text{exchange correction})\psi_{2p}(1) = E_{2p}\psi_{2p}(1) \qquad (3.38)$$

A similar equation can be written for the $1s$ and $2s$ orbitals in the atom. The various terms are as follows:

1 The first term on the left is the contribution of the kinetic energy and the attraction of the electron to the nucleus, just as in a hydrogenic atom.

2 The second takes into account the potential energy of the electron of interest due to the electrons in the other occupied orbitals.

3 The third term takes into account the spin correlation effects discussed earlier.

Although the equation is for the $2p$ orbital in neon, it depends on the wavefunctions of all the other occupied orbitals in the atom.

There is no hope of solving eqn 3.38 analytically. However, it can be solved numerically if we guess an approximate form of the wavefunctions of all the orbitals except $2p$. The procedure is then repeated for the other orbitals in the atom, the $1s$ and $2s$ orbitals. This sequence of calculations gives the form of the $2p$, $2s$, and $1s$ orbitals, and in general they will differ from the set used initially to start the calculation. These improved orbitals can be used in another cycle of calculation, and a second improved set of orbitals is obtained. The recycling continues until the orbitals and energies obtained are insignificantly different from those used at the start of the current cycle. The solutions are then self-consistent and accepted as solutions of the problem.

Figure 3.23 shows plots of some of the HF-SCF radial distribution functions for sodium. They show the grouping of electron density into shells, as was anticipated by the early chemists, and the differences of penetration as discussed above. These SCF calculations therefore support the qualitative discussions that are used to explain chemical periodicity. They also considerably extend that discussion by providing detailed wavefunctions and precise energies.

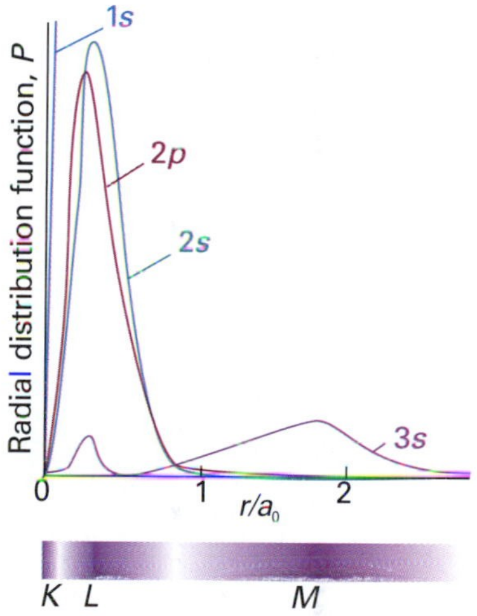

Fig. 3.23 The radial distribution functions for the orbitals of Na based on SCF calculations. Note the shell-like structure, with the $3s$ orbital outside the inner K and L shells.

The spectra of complex atoms

The spectra of atoms rapidly become very complicated as the number of electrons increases, but there are some important and moderately simple features that make atomic spectroscopy useful in the study of the composition of samples as large and as complex as stars (*Impact I3.1*). The general idea is straightforward: lines in the spectrum (in either emission or absorption) occur when the atom undergoes a transition with a change of energy $|\Delta E|$, and emits or absorbs a photon of frequency $\nu = |\Delta E|/h$ and $\tilde{\nu} = |\Delta E|/hc$. Hence, we can expect the spectrum to give information about the energies of electrons in atoms. However, the actual energy levels are not given solely by the energies of the orbitals, because the electrons interact with one another in various ways, and there are contributions to the energy in addition to those we have already considered.

Comment 3.5
The web site for this text contains links to databases of atomic spectra.

IMPACT ON ASTROPHYSICS
I3.1 Spectroscopy of stars

The bulk of stellar material consists of neutral and ionized forms of hydrogen and helium atoms, with helium being the product of 'hydrogen burning' by nuclear fusion. However, nuclear fusion also makes heavier elements. It is generally accepted that the outer layers of stars are composed of lighter elements, such as H, He, C, N, O, and Ne in both neutral and ionized forms. Heavier elements, including neutral and ionized forms of Si, Mg, Ca, S, and Ar, are found closer to the stellar core. The core itself contains the heaviest elements and ^{56}Fe is particularly abundant because it is a very stable nuclide. All these elements are in the gas phase on account of the very high temperatures in stellar interiors. For example, the temperature is estimated to be 3.6 MK half way to the centre of the Sun.

Astronomers use spectroscopic techniques to determine the chemical composition of stars because each element, and indeed each isotope of an element, has a characteristic spectral signature that is transmitted through space by the star's light. To understand the spectra of stars, we must first know why they shine. Nuclear reactions in the dense stellar interior generate radiation that travels to less dense outer layers. Absorption and re-emission of photons by the atoms and ions in the interior give rise to a quasi-continuum of radiation energy that is emitted into space by a thin layer of gas called the *photosphere*. To a good approximation, the distribution of energy emitted from a star's photosphere resembles the Planck distribution for a very hot black body (Section 2.1). For example, the energy distribution of our Sun's photosphere may be modelled by a Planck distribution with an effective temperature of 5.8 kK. Superimposed on the black-body radiation continuum are sharp absorption and emission lines from neutral atoms and ions present in the photosphere. Analysis of stellar radiation with a spectrometer mounted on to a telescope yields the chemical composition of the star's photosphere by comparison with known spectra of the elements. The data can also reveal the presence of small molecules, such as CN, C_2, TiO, and ZrO, in certain 'cold' stars, which are stars with relatively low effective temperatures.

The two outermost layers of a star are the *chromosphere*, a region just above the photosphere, and the *corona*, a region above the chromosphere that can be seen (with proper care) during eclipses. The photosphere, chromosphere, and corona comprise a star's 'atmosphere'. Our Sun's chromosphere is much less dense than its photosphere and its temperature is much higher, rising to about 10 kK. The reasons for this increase in temperature are not fully understood. The temperature of our Sun's corona is very high, rising up to 1.5 MK, so black-body emission is strong from the X-ray to the radiofrequency region of the spectrum. The spectrum of the Sun's corona is dominated by emission lines from electronically excited species, such as neutral atoms and a number of highly ionized species . The most intense emission lines in the visible range are from the Fe^{13+} ion at 530.3 nm, the Fe^{9+} ion at 637.4 nm, and the Ca^{4+} ion at 569.4 nm.

Because only light from the photosphere reaches our telescopes, the overall chemical composition of a star must be inferred from theoretical work on its interior and from spectral analysis of its atmosphere. Data on the Sun indicate that it is 92 per cent hydrogen and 7.8 per cent helium. The remaining 0.2 per cent is due to heavier elements, among which C, N, O, Ne, and Fe are the most abundant. More advanced analysis of spectra also permit the determination of other properties of stars, such as their relative speeds (Problem 3.27) and their effective temperatures (Problem 6.29).

3.6 Quantum defects and ionization limits

One application of atomic spectroscopy is to the determination of ionization energies. However, we cannot use the procedure illustrated in Example 10.1 indiscriminately because the energy levels of a many-electron atom do not in general vary as $1/n^2$. If we confine attention to the outermost electrons, then we know that, as a result of penetration and shielding, they experience a nuclear charge of slightly more than $1e$ because in a neutral atom the other $Z-1$ electrons cancel all but about one unit of nuclear charge. Typical values of Z_{eff} are a little more than 1, so we expect binding energies to be given by a term of the form $-hcR/n^2$, but lying slightly lower in energy than this formula predicts. We therefore introduce a **quantum defect**, δ, and write the energy as $-hcR/(n-\delta)^2$. The quantum defect is best regarded as a purely empirical quantity.

There are some excited states that are so diffuse that the $1/n^2$ variation is valid: these states are called **Rydberg states**. In such cases we can write

$$\tilde{\nu} = \frac{I}{hc} - \frac{R}{n^2} \tag{3.39}$$

and a plot of wavenumber against $1/n^2$ can be used to obtain I by extrapolation; in practice, one would use a linear regression fit using a computer. If the lower state is not the ground state (a possibility if we wish to generalize the concept of ionization energy), the ionization energy of the ground state can be determined by adding the appropriate energy difference to the ionization energy obtained as described here.

3.7 Singlet and triplet states

Suppose we were interested in the energy levels of a He atom, with its two electrons. We know that the ground-state configuration is $1s^2$, and can anticipate that an excited configuration will be one in which one of the electrons has been promoted into a $2s$ orbital, giving the configuration $1s^12s^1$. The two electrons need not be paired because they occupy different orbitals. According to Hund's maximum multiplicity rule, the state of the atom with the spins parallel lies lower in energy than the state in which they are paired. Both states are permissible, and can contribute to the spectrum of the atom.

Parallel and antiparallel (paired) spins differ in their overall spin angular momentum. In the paired case, the two spin momenta cancel each other, and there is zero net spin (as was depicted in Fig. 3.18). The paired-spin arrangement is called a **singlet**. Its spin state is the one we denoted σ_- in the discussion of the Pauli principle:

$$\sigma_-(1,2) = (1/2^{1/2})\{\alpha(1)\beta(2) - \beta(1)\alpha(2)\} \tag{3.40a}$$

The angular momenta of two parallel spins add together to give a nonzero total spin, and the resulting state is called a **triplet**. As illustrated in Fig. 3.24, there are three ways of achieving a nonzero total spin, but only one way to achieve zero spin. The three spin states are the symmetric combinations introduced earlier:

$$\alpha(1)\alpha(2) \qquad \sigma_+(1,2) = (1/2^{1/2})\{\alpha(1)\beta(2) + \beta(1)\alpha(2)\} \qquad \beta(1)\beta(2) \tag{3.40b}$$

The fact that the parallel arrangement of spins in the $1s^12s^1$ configuration of the He atom lies lower in energy than the antiparallel arrangement can now be expressed by saying that the triplet state of the $1s^12s^1$ configuration of He lies lower in energy than the singlet state. This is a general conclusion that applies to other atoms (and molecules) and, *for states arising from the same configuration, the triplet state generally lies lower than the singlet state*. The origin of the energy difference lies in the effect of spin correlation on the Coulombic interactions between electrons, as we saw in the case of Hund's rule for ground-state configurations. Because the Coulombic interaction between electrons in an atom is strong, the difference in energies between singlet and triplet states of the same configuration can be large. The two states of $1s^12s^1$ He, for instance, differ by 6421 cm^{-1} (corresponding to 0.80 eV).

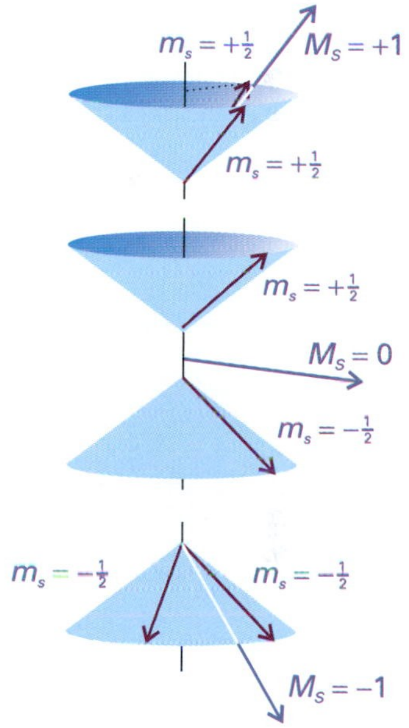

Fig. 3.24 When two electrons have parallel spins, they have a nonzero total spin angular momentum. There are three ways of achieving this resultant, which are shown by these vector representations. Note that, although we cannot know the orientation of the spin vectors on the cones, the angle between the vectors is the same in all three cases, for all three arrangements have the same total spin angular momentum (that is, the resultant of the two vectors has the same length in each case, but points in different directions). Compare this diagram with Fig. 3.18, which shows the antiparallel case. Note that, whereas two paired spins are precisely antiparallel, two 'parallel' spins are not strictly parallel.

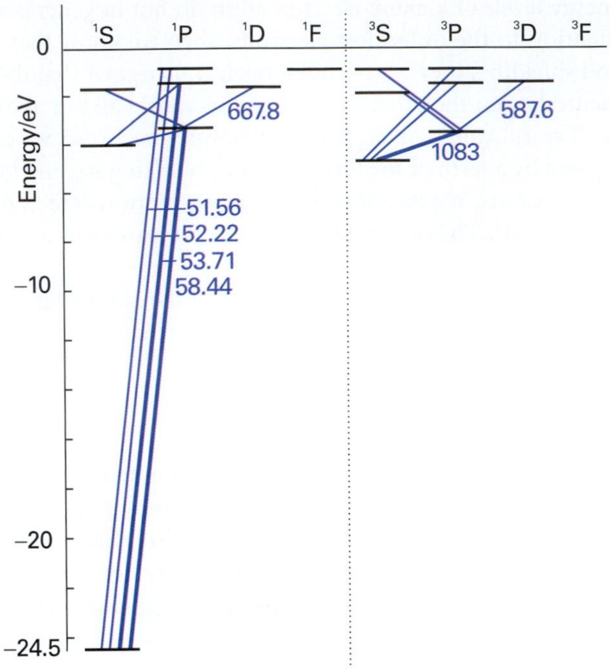

Fig. 3.25 Part of the Grotrian diagram for a helium atom. Note that there are no transitions between the singlet and triplet levels.

The spectrum of atomic helium is more complicated than that of atomic hydrogen, but there are two simplifying features. One is that the only excited configurations it is necessary to consider are of the form $1s^1nl^1$: that is, only one electron is excited. Excitation of two electrons requires an energy greater than the ionization energy of the atom, so the He^+ ion is formed instead of the doubly excited atom. Second, no radiative transitions take place between singlet and triplet states because the relative orientation of the two electron spins cannot change during a transition. Thus, there is a spectrum arising from transitions between singlet states (including the ground state) and between triplet states, but not between the two. Spectroscopically, helium behaves like two distinct species, and the early spectroscopists actually thought of helium as consisting of 'parahelium' and 'orthohelium'. The Grotrian diagram for helium in Fig. 3.25 shows the two sets of transitions.

Comment 3.6

We have already remarked that the electron's spin is a purely quantum mechanical phenomenon that has no classical counterpart. However a classical model can give us partial insight into the origin of an electron's magnetic moment. Namely, the magnetic field generated by a spinning electron, regarded classically as a moving charge, induces a magnetic moment. This model is merely a visualization aid and cannot be used to explain the magnitude of the magnetic moment of the electron or the origin of spin magnetic moments in electrically neutral particles, such as the neutron.

3.8 Spin–orbit coupling

An electron has a magnetic moment that arises from its spin (Fig. 3.26). Similarly, an electron with orbital angular momentum (that is, an electron in an orbital with $l > 0$) is in effect a circulating current, and possesses a magnetic moment that arises from its orbital momentum. The interaction of the spin magnetic moment with the magnetic field arising from the orbital angular momentum is called **spin–orbit coupling**. The strength of the coupling, and its effect on the energy levels of the atom, depend on the relative orientations of the spin and orbital magnetic moments, and therefore on the relative orientations of the two angular momenta (Fig. 3.27).

(a) The total angular momentum

One way of expressing the dependence of the spin–orbit interaction on the relative orientation of the spin and orbital momenta is to say that it depends on the total angular momentum of the electron, the vector sum of its spin and orbital momenta. Thus, when the spin and orbital angular momenta are nearly parallel, the total angular

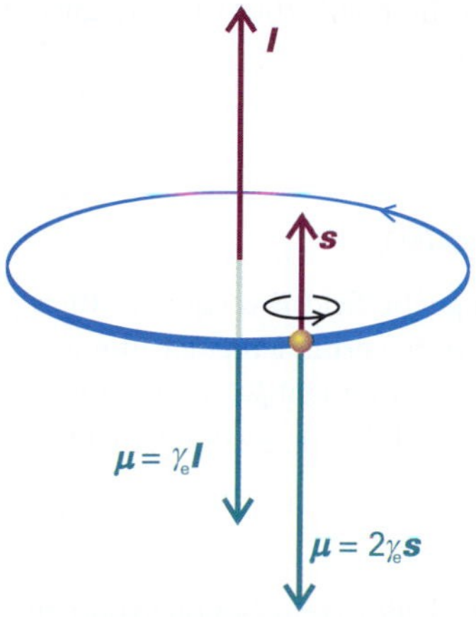

Fig. 3.26 Angular momentum gives rise to a magnetic moment ($\boldsymbol{\mu}$). For an electron, the magnetic moment is antiparallel to the orbital angular momentum, but proportional to it. For spin angular momentum, there is a factor 2, which increases the magnetic moment to twice its expected value (see Section 3.10).

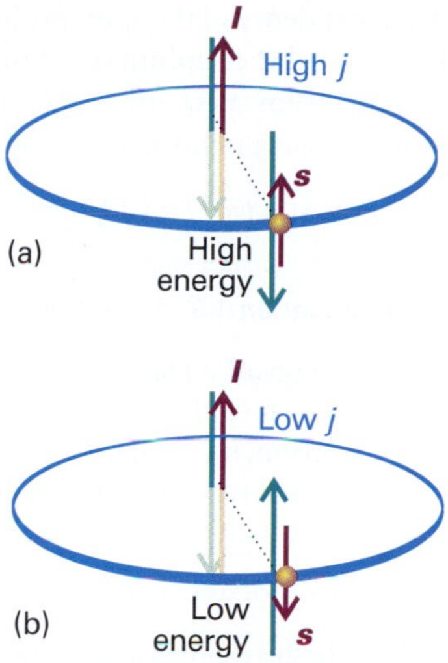

Fig. 3.27 Spin–orbit coupling is a magnetic interaction between spin and orbital magnetic moments. When the angular momenta are parallel, as in (a), the magnetic moments are aligned unfavourably; when they are opposed, as in (b), the interaction is favourable. This magnetic coupling is the cause of the splitting of a configuration into levels.

momentum is high; when the two angular momenta are opposed, the total angular momentum is low.

The total angular momentum of an electron is described by the quantum numbers j and m_j, with $j = l + \frac{1}{2}$ (when the two angular momenta are in the same direction) or $j = l - \frac{1}{2}$ (when they are opposed, Fig. 3.28). The different values of j that can arise for a given value of l label **levels** of a term. For $l = 0$, the only permitted value is $j = \frac{1}{2}$ (the total angular momentum is the same as the spin angular momentum because there is no other source of angular momentum in the atom). When $l = 1$, j may be either $\frac{3}{2}$ (the spin and orbital angular momenta are in the same sense) or $\frac{1}{2}$ (the spin and angular momenta are in opposite senses).

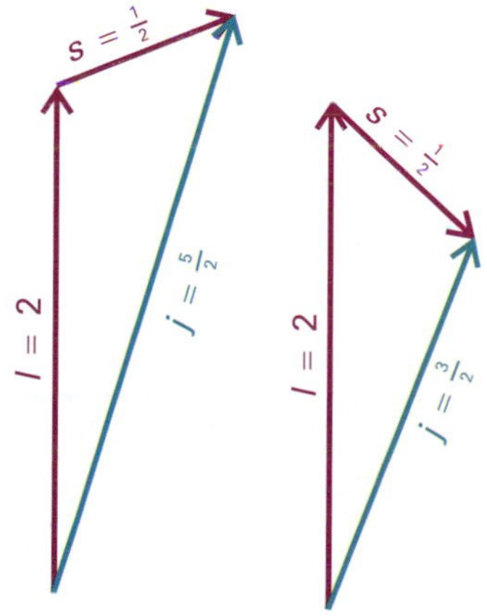

Fig. 3.28 The coupling of the spin and orbital angular momenta of a d electron ($l = 2$) gives two possible values of j depending on the relative orientations of the spin and orbital angular momenta of the electron.

Example 3.4 *Identifying the levels of a configuration*

Identify the levels that may arise from the configurations (a) d^1, (b) s^1.

Method In each case, identify the value of l and then the possible values of j. For these one-electron systems, the total angular momentum is the sum and difference of the orbital and spin momenta.

Answer (a) For a d electron, $l = 2$ and there are two levels in the configuration, one with $j = 2 + \frac{1}{2} = \frac{5}{2}$ and the other with $j = 2 - \frac{1}{2} = \frac{3}{2}$. (b) For an s electron $l = 0$, so only one level is possible, and $j = \frac{1}{2}$.

Self-test 3.8 Identify the levels of the configurations (a) p^1 and (b) f^1.
[(a) $\frac{3}{2}, \frac{1}{2}$; (b) $\frac{7}{2}, \frac{5}{2}$]

The dependence of the spin–orbit interaction on the value of j is expressed in terms of the **spin–orbit coupling constant**, A (which is typically expressed as a wavenumber). A quantum mechanical calculation leads to the result that the energies of the levels with quantum numbers s, l, and j are given by

$$E_{l,s,j} = \tfrac{1}{2}hcA\{j(j+1) - l(l+1) - s(s+1)\} \tag{3.41}$$

Comment 3.7

The scalar product (or dot product) $\boldsymbol{u}\cdot\boldsymbol{v}$ of two vectors $\boldsymbol{u}$ and $\boldsymbol{v}$ with magnitudes u and v is $\boldsymbol{u}\cdot\boldsymbol{v} = uv\cos\theta$, where θ is the angle between the two vectors.

Justification 3.8 *The energy of spin–orbit interaction*

The energy of a magnetic moment $\boldsymbol{\mu}$ in a magnetic field $\boldsymbol{B}$ is equal to their scalar product $-\boldsymbol{\mu}\cdot\boldsymbol{B}$. If the magnetic field arises from the orbital angular momentum of the electron, it is proportional to $\boldsymbol{l}$; if the magnetic moment $\boldsymbol{\mu}$ is that of the electron spin, then it is proportional to $\boldsymbol{s}$. It then follows that the energy of interaction is proportional to the scalar product $\boldsymbol{s}\cdot\boldsymbol{l}$:

$$\text{Energy of interaction} = -\boldsymbol{\mu}\cdot\boldsymbol{B} \propto \boldsymbol{s}\cdot\boldsymbol{l}$$

We take this expression to be the first-order perturbation contribution to the hamiltonian. Next, we note that the total angular momentum is the vector sum of the spin and orbital momenta: $\boldsymbol{j} = \boldsymbol{l} + \boldsymbol{s}$. The magnitude of the vector $\boldsymbol{j}$ is calculated by evaluating

$$\boldsymbol{j}\cdot\boldsymbol{j} = (\boldsymbol{l}+\boldsymbol{s})\cdot(\boldsymbol{l}+\boldsymbol{s}) = \boldsymbol{l}\cdot\boldsymbol{l} + \boldsymbol{s}\cdot\boldsymbol{s} + 2\boldsymbol{s}\cdot\boldsymbol{l}$$

That is,

$$\boldsymbol{s}\cdot\boldsymbol{l} = \tfrac{1}{2}\{j^2 - l^2 - s^2\}$$

where we have used the fact that the scalar product of two vectors $\boldsymbol{u}$ and v is $\boldsymbol{u}\cdot v = uv\cos\theta$, from which it follows that $\boldsymbol{u}\cdot\boldsymbol{u} = u^2$.

The preceding equation is a classical result. To make the transition to quantum mechanics, we treat all the quantities as operators, and write

$$\hat{\boldsymbol{s}}\cdot\hat{\boldsymbol{l}} = \tfrac{1}{2}\{\hat{j}^2 - \hat{l}^2 - \hat{s}^2\} \tag{3.42}$$

At this point, we calculate the first-order correction to the energy by evaluating the expectation value:

$$\langle j,l,s|\hat{\boldsymbol{s}}\cdot\hat{\boldsymbol{l}}|j,l,s\rangle = \tfrac{1}{2}\langle j,l,s|\hat{j}^2 - \hat{l}^2 - \hat{s}^2|j,l,s\rangle = \tfrac{1}{2}\{j(j+1) - l(l+1) - s(s+1)\}\hbar^2 \tag{3.43}$$

Then, by inserting this expression into the formula for the energy of interaction ($E \propto \boldsymbol{s}\cdot\boldsymbol{l}$), and writing the constant of proportionality as $hcA/\hbar^2$, we obtain eqn 3.42. The calculation of A is much more complicated: see *Further reading*.

Illustration 3.3 *Calculating the energies of levels*

The unpaired electron in the ground state of an alkali metal atom has $l = 0$, so $j = \tfrac{1}{2}$. Because the orbital angular momentum is zero in this state, the spin–orbit coupling energy is zero (as is confirmed by setting $j = s$ and $l = 0$ in eqn 3.41). When the electron is excited to an orbital with $l = 1$, it has orbital angular momentum and can give rise to a magnetic field that interacts with its spin. In this configuration the electron can have $j = \tfrac{3}{2}$ or $j = \tfrac{1}{2}$, and the energies of these levels are

$$E_{3/2} = \tfrac{1}{2}hcA\{\tfrac{3}{2}\times\tfrac{5}{2} - 1\times 2 - \tfrac{1}{2}\times\tfrac{3}{2}\} = \tfrac{1}{2}hcA$$
$$E_{1/2} = \tfrac{1}{2}hcA\{\tfrac{1}{2}\times\tfrac{3}{2} - 1\times 2 - \tfrac{1}{2}\times\tfrac{3}{2}\} = -hcA$$

The corresponding energies are shown in Fig. 3.29. Note that the baricentre (the 'centre of gravity') of the levels is unchanged, because there are four states of energy $\tfrac{1}{2}hcA$ and two of energy $-hcA$.

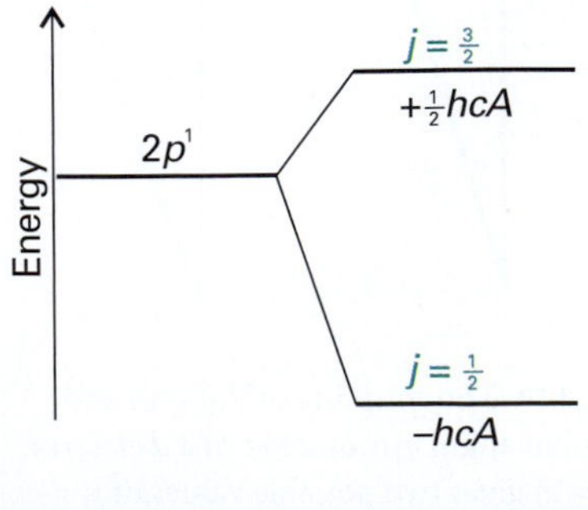

Fig. 3.29 The levels of a ^{2}P term arising from spin–orbit coupling. Note that the low-j level lies below the high-j level in energy.

The strength of the spin–orbit coupling depends on the nuclear charge. To understand why this is so, imagine riding on the orbiting electron and seeing a charged nucleus apparently orbiting around us (like the Sun rising and setting). As a result, we find ourselves at the centre of a ring of current. The greater the nuclear charge, the greater this current, and therefore the stronger the magnetic field we detect. Because the spin magnetic moment of the electron interacts with this orbital magnetic field, it follows that the greater the nuclear charge, the stronger the spin–orbit interaction. The coupling increases sharply with atomic number (as Z^4). Whereas it is only small in H (giving rise to shifts of energy levels of no more than about 0.4 $\mathrm{cm^{-1}}$), in heavy atoms like Pb it is very large (giving shifts of the order of thousands of reciprocal centimetres).

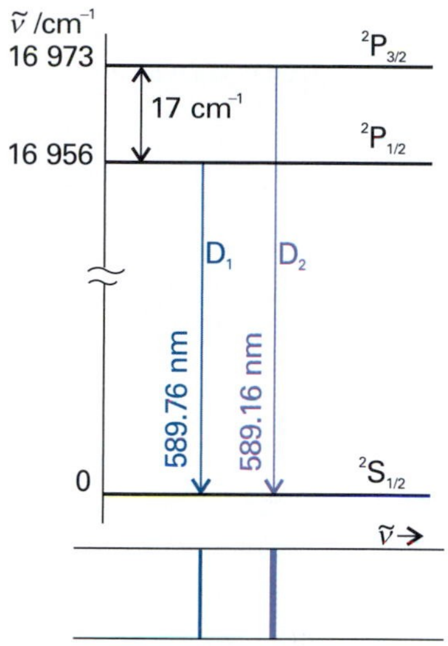

Fig. 3.30 The energy-level diagram for the formation of the sodium D lines. The splitting of the spectral lines (by 17 $\mathrm{cm^{-1}}$) reflects the splitting of the levels of the ^{2}P term.

(b) Fine structure

Two spectral lines are observed when the *p* electron of an electronically excited alkali metal atom undergoes a transition and falls into a lower *s* orbital. One line is due to a transition starting in a $j=\frac{3}{2}$ level and the other line is due to a transition starting in the $j=\frac{1}{2}$ level of the same configuration. The two lines are an example of the **fine structure** of a spectrum, the structure in a spectrum due to spin–orbit coupling. Fine structure can be clearly seen in the emission spectrum from sodium vapour excited by an electric discharge (for example, in one kind of street lighting). The yellow line at 589 nm (close to 17 000 $\mathrm{cm^{-1}}$) is actually a doublet composed of one line at 589.76 nm (16 956.2 $\mathrm{cm^{-1}}$) and another at 589.16 nm (16 973.4 $\mathrm{cm^{-1}}$); the components of this doublet are the 'D lines' of the spectrum (Fig. 3.30). Therefore, in Na, the spin–orbit coupling affects the energies by about 17 $\mathrm{cm^{-1}}$.

Example 3.5 *Analysing a spectrum for the spin–orbit coupling constant*

The origin of the D lines in the spectrum of atomic sodium is shown in Fig. 3.30. Calculate the spin–orbit coupling constant for the upper configuration of the Na atom.

Method We see from Fig. 3.30 that the splitting of the lines is equal to the energy separation of the $j=\frac{3}{2}$ and $\frac{1}{2}$ levels of the excited configuration. This separation can be expressed in terms of *A* by using eqn 3.40. Therefore, set the observed splitting equal to the energy separation calculated from eqn 3.40 and solve the equation for *A*.

Answer The two levels are split by

$$\Delta\tilde{\nu}=A\,\tfrac{1}{2}\{\tfrac{3}{2}(\tfrac{3}{2}+1)-\tfrac{1}{2}(\tfrac{1}{2}+1)\}=\tfrac{3}{2}A$$

The experimental value is 17.2 $\mathrm{cm^{-1}}$; therefore

$$A=\tfrac{2}{3}\times(17.2\ \mathrm{cm^{-1}})=11.5\ \mathrm{cm^{-1}}$$

The same calculation repeated for the other alkali metal atoms gives Li: 0.23 $\mathrm{cm^{-1}}$, K: 38.5 $\mathrm{cm^{-1}}$, Rb: 158 $\mathrm{cm^{-1}}$, Cs: 370 $\mathrm{cm^{-1}}$. Note the increase of *A* with atomic number (but more slowly than Z^4 for these many-electron atoms).

Self-test 3.9 The configuration . . . $4p^65d^1$ of rubidium has two levels at 25 700.56 $\mathrm{cm^{-1}}$ and 25 703.52 $\mathrm{cm^{-1}}$ above the ground state. What is the spin–orbit coupling constant in this excited state? [1.18 $\mathrm{cm^{-1}}$]

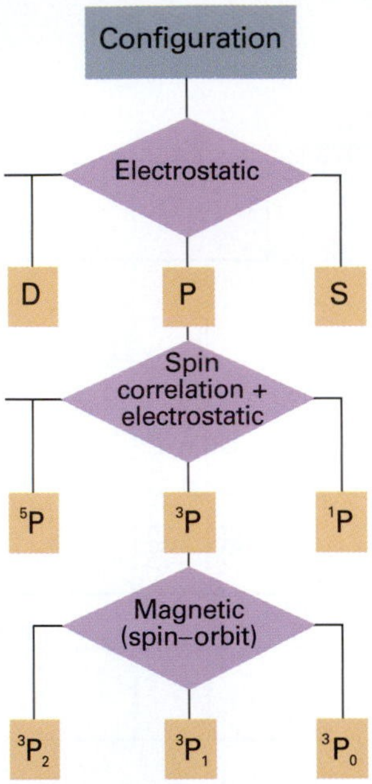

Fig. 3.31 A summary of the types of interaction that are responsible for the various kinds of splitting of energy levels in atoms. For light atoms, magnetic interactions are small, but in heavy atoms they may dominate the electrostatic (charge–charge) interactions.

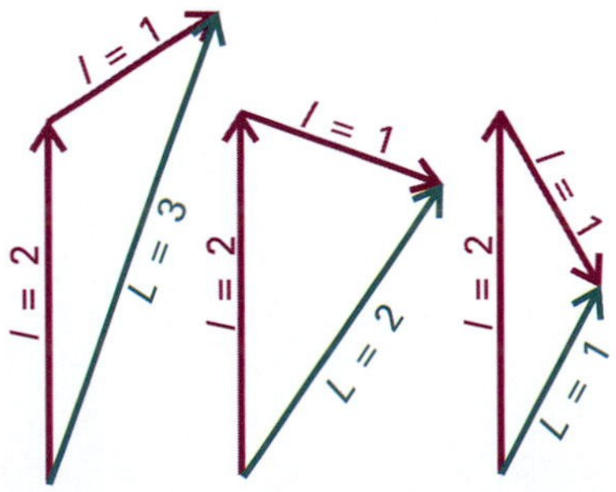

Fig. 3.32 The total angular orbital momenta of a *p* electron and a *d* electron correspond to $L = 3$, 2, and 1 and reflect the different relative orientations of the two momenta.

3.9 Term symbols and selection rules

We have used expressions such as 'the $j = \frac{3}{2}$ level of a configuration'. A **term symbol**, which is a symbol looking like $^2P_{3/2}$ or $^3D_2$, conveys this information much more succinctly. The convention of using lowercase letters to label orbitals and uppercase letters to label overall states applies throughout spectroscopy, not just to atoms.

A term symbol gives three pieces of information:

1 The letter (P or D in the examples) indicates the total orbital angular momentum quantum number, L.

2 The left superscript in the term symbol (the 2 in $^2P_{3/2}$) gives the multiplicity of the term.

3 The right subscript on the term symbol (the $\frac{3}{2}$ in $^2P_{3/2}$) is the value of the total angular momentum quantum number, J.

We shall now say what each of these statements means; the contributions to the energies which we are about to discuss are summarized in Fig. 3.31.

(a) The total orbital angular momentum

When several electrons are present, it is necessary to judge how their individual orbital angular momenta add together or oppose each other. The **total orbital angular momentum quantum number**, L, tells us the magnitude of the angular momentum through $\{L(L+1)\}^{1/2}\hbar$. It has $2L + 1$ orientations distinguished by the quantum number M_L, which can take the values $L, L - 1, \ldots, -L$. Similar remarks apply to the **total spin quantum number**, S, and the quantum number M_S, and the **total angular momentum quantum number**, J, and the quantum number M_J.

The value of L (a non-negative integer) is obtained by coupling the individual orbital angular momenta by using the **Clebsch–Gordan series:**

$$L = l_1 + l_2, l_1 + l_2 - 1, \ldots, |l_1 - l_2| \qquad (3.44)$$

The modulus signs are attached to $l_1 - l_2$ because L is non-negative. The maximum value, $L = l_1 + l_2$, is obtained when the two orbital angular momenta are in the same direction; the lowest value, $|l_1 - l_2|$, is obtained when they are in opposite directions. The intermediate values represent possible intermediate relative orientations of the two momenta (Fig. 3.32). For two *p* electrons (for which $l_1 = l_2 = 1$), $L = 2, 1, 0$. The code for converting the value of L into a letter is the same as for the *s, p, d, f*, . . . designation of orbitals, but uses uppercase Roman letters:

L:	0	1	2	3	4	5	6 . . .
	S	P	D	F	G	H	I . . .

Thus, a p^2 configuration can give rise to D, P, and S terms. The terms differ in energy on account of the different spatial distribution of the electrons and the consequent differences in repulsion between them.

A closed shell has zero orbital angular momentum because all the individual orbital angular momenta sum to zero. Therefore, when working out term symbols, we need consider only the electrons of the unfilled shell. In the case of a single electron outside a closed shell, the value of L is the same as the value of l; so the configuration $[Ne]3s^1$ has only an S term.

Example 3.6 *Deriving the total orbital angular momentum of a configuration*

Find the terms that can arise from the configurations (a) d^2, (b) p^3.

Method Use the Clebsch–Gordan series and begin by finding the minimum value of L (so that we know where the series terminates). When there are more than two electrons to couple together, use two series in succession: first couple two electrons, and then couple the third to each combined state, and so on.

Answer (a) The minimum value is $|l_1 - l_2| = |2 - 2| = 0$. Therefore,

$$L = 2 + 2, 2 + 2 - 1, \ldots, 0 = 4, 3, 2, 1, 0$$

corresponding to G, F, D, P, S terms, respectively. (b) Coupling two electrons gives a minimum value of $|1 - 1| = 0$. Therefore,

$$L' = 1 + 1, 1 + 1 - 1, \ldots, 0 = 2, 1, 0$$

Now couple l_3 with $L' = 2$, to give $L = 3, 2, 1$; with $L' = 1$, to give $L = 2, 1, 0$; and with $L' = 0$, to give $L = 1$. The overall result is

$$L = 3, 2, 2, 1, 1, 1, 0$$

giving one F, two D, three P, and one S term.

Self-test 3.10 Repeat the question for the configurations (a) f^1d^1 and (b) d^3.

[(a) H, G, F, D, P; (b) I, 2H, 3G, 4F, 5D, 3P, S]

(b) The multiplicity

When there are several electrons to be taken into account, we must assess their total spin angular momentum quantum number, S (a non-negative integer or half integer). Once again, we use the Clebsch–Gordan series in the form

$$S = s_1 + s_2, s_1 + s_2 - 1, \ldots, |s_1 - s_2| \tag{3.45}$$

to decide on the value of S, noting that each electron has $s = \frac{1}{2}$, which gives $S = 1, 0$ for two electrons (Fig. 3.33). If there are three electrons, the total spin angular momentum is obtained by coupling the third spin to each of the values of S for the first two spins, which results in $S = \frac{3}{2}$, and $S = \frac{1}{2}$.

The **multiplicity** of a term is the value of $2S + 1$. When $S = 0$ (as for a closed shell, like $1s^2$) the electrons are all paired and there is no net spin: this arrangement gives a singlet term, ^1S. A single electron has $S = s = \frac{1}{2}$, so a configuration such as [Ne]$3s^1$ can give rise to a doublet term, ^2S. Likewise, the configuration [Ne]$3p^1$ is a doublet, ^2P. When there are two unpaired electrons $S = 1$, so $2S + 1 = 3$, giving a triplet term, such as ^3D. We discussed the relative energies of singlets and triplets in Section 3.7 and saw that their energies differ on account of the different effects of spin correlation.

(c) The total angular momentum

As we have seen, the quantum number j tells us the relative orientation of the spin and orbital angular momenta of a single electron. The **total angular momentum quantum number**, J (a non-negative integer or half integer), does the same for several electrons. If there is a single electron outside a closed shell, $J = j$, with j either $l + \frac{1}{2}$ or $|l - \frac{1}{2}|$. The [Ne]$3s^1$ configuration has $j = \frac{1}{2}$ (because $l = 0$ and $s = \frac{1}{2}$), so the $2s$ term has a single level, which we denote $^2S_{1/2}$. The [Ne]$3p^1$ configuration has $l = 1$; therefore

Comment 3.8

Throughout our discussion of atomic spectroscopy, distinguish italic S, the total spin quantum number, from roman S, the term label.

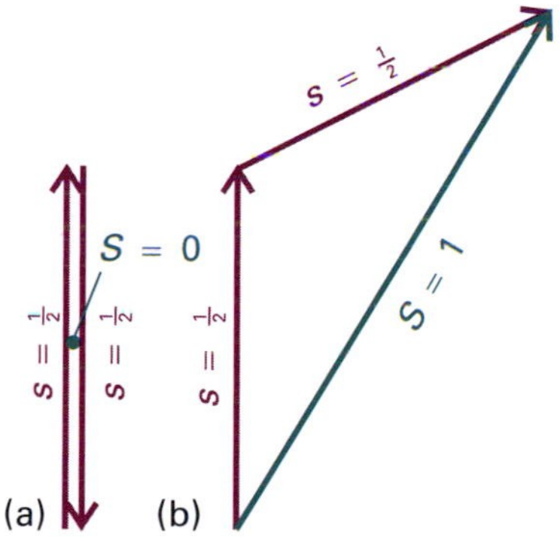

Fig. 3.33 For two electrons (each of which has $s = \frac{1}{2}$), only two total spin states are permitted ($S = 0, 1$). The state with $S = 0$ can have only one value of M_S ($M_S = 0$) and is a singlet; the state with $S = 1$ can have any of three values of M_S (+1, 0, −1) and is a triplet. The vector representation of the singlet and triplet states are shown in Figs. 3.18 and 3.24, respectively.

$j = \frac{3}{2}$ and $\frac{1}{2}$; the ^{2}P term therefore has two levels, $^2P_{3/2}$ and $^2P_{1/2}$. These levels lie at different energies on account of the magnetic spin–orbit interaction.

If there are several electrons outside a closed shell we have to consider the coupling of all the spins and all the orbital angular momenta. This complicated problem can be simplified when the spin–orbit coupling is weak (for atoms of low atomic number), for then we can use the **Russell–Saunders coupling** scheme. This scheme is based on the view that, if spin–orbit coupling is weak, then it is effective only when all the orbital momenta are operating cooperatively. We therefore imagine that all the orbital angular momenta of the electrons couple to give a total L, and that all the spins are similarly coupled to give a total S. Only at this stage do we imagine the two kinds of momenta coupling through the spin–orbit interaction to give a total J. The permitted values of J are given by the Clebsch–Gordan series

$$J = L + S, L + S - 1, \ldots, |L - S| \qquad (3.46)$$

For example, in the case of the ^{3}D term of the configuration [Ne]$2p^13p^1$, the permitted values of J are 3, 2, 1 (because ^{3}D has $L = 2$ and $S = 1$), so the term has three levels, $^3D_3$, $^3D_2$, and $^3D_1$.

When $L \geq S$, the multiplicity is equal to the number of levels. For example, a ^{2}P term has the two levels $^2P_{3/2}$ and $^2P_{1/2}$, and ^{3}D has the three levels $^3D_3$, $^3D_2$, and $^3D_1$. However, this is not the case when $L < S$: the term ^{2}S, for example, has only the one level $^2S_{1/2}$.

Example 3.7 *Deriving term symbols*

Write the term symbols arising from the ground-state configurations of (a) Na and (b) F, and (c) the excited configuration $1s^22s^22p^13p^1$ of C.

Method Begin by writing the configurations, but ignore inner closed shells. Then couple the orbital momenta to find L and the spins to find S. Next, couple L and S to find J. Finally, express the term as $^{2S+1}\{L\}_J$, where $\{L\}$ is the appropriate letter. For F, for which the valence configuration is $2p^5$, treat the single gap in the closed-shell $2p^6$ configuration as a single particle.

Answer (a) For Na, the configuration is [Ne]$3s^1$, and we consider the single 3s electron. Because $L = l = 0$ and $S = s = \frac{1}{2}$, it is possible for $J = j = s = \frac{1}{2}$ only. Hence the term symbol is $^2S_{1/2}$. (b) For F, the configuration is [He]$2s^22p^5$, which we can treat as [Ne]$2p^{-1}$ (where the notation $2p^{-1}$ signifies the absence of a 2p electron). Hence $L = 1$, and $S = s = \frac{1}{2}$. Two values of $J = j$ are allowed: $J = \frac{3}{2}, \frac{1}{2}$. Hence, the term symbols for the two levels are $^2P_{3/2}$, $^2P_{1/2}$. (c) We are treating an excited configuration of carbon because, in the ground configuration, $2p^2$, the Pauli principle forbids some terms, and deciding which survive (^{1}D, ^{3}P, ^{1}S, in fact) is quite complicated. That is, there is a distinction between 'equivalent electrons', which are electrons that occupy the same orbitals, and 'inequivalent electrons', which are electrons that occupy different orbitals. For information about how to deal with equivalent electrons, see *Further reading*. The excited configuration of C under consideration is effectively $2p^13p^1$. This is a two-electron problem, and $l_1 = l_2 = 1$, $s_1 = s_2 = \frac{1}{2}$. It follows that $L = 2, 1, 0$ and $S = 1, 0$. The terms are therefore ^{3}D and ^{1}D, ^{3}P and ^{1}P, and ^{3}S and ^{1}S. For ^{3}D, $L = 2$ and $S = 1$; hence $J = 3, 2, 1$ and the levels are $^3D_3$, $^3D_2$, and $^3D_1$. For ^{1}D, $L = 2$ and $S = 0$, so the single level is $^1D_2$. The triplet of levels of ^{3}P is $^3P_2$, $^3P_1$, and $^3P_0$, and the singlet is $^1P_1$. For the ^{3}S term there is only one level, $^3S_1$ (because $J = 1$ only), and the singlet term is $^1S_0$.

Self-test 3.11 Write down the terms arising from the configurations (a) $2s^12p^1$, (b) $2p^13d^1$.

[(a) $^3P_2$, $^3P_1$, $^3P_0$, $^1P_1$;
(b) $^3F_4$, $^3F_3$, $^3F_2$, $^1F_3$, $^3D_3$, $^3D_2$, $^3D_1$, $^1D_2$, $^3P_1$, $^3P_0$, $^1P_1$]

Russell–Saunders coupling fails when the spin–orbit coupling is large (in heavy atoms). In that case, the individual spin and orbital momenta of the electrons are coupled into individual j values; then these momenta are combined into a grand total, J. This scheme is called ***jj*-coupling**. For example, in a p^2 configuration, the individual values of j are $\frac{3}{2}$ and $\frac{1}{2}$ for each electron. If the spin and the orbital angular momentum of each electron are coupled together strongly, it is best to consider each electron as a particle with angular momentum $j = \frac{3}{2}$ or $\frac{1}{2}$. These individual total momenta then couple as follows:

$$j_1 = \tfrac{3}{2} \text{ and } j_2 = \tfrac{3}{2} \qquad J = 3, 2, 1, 0$$
$$j_1 = \tfrac{3}{2} \text{ and } j_2 = \tfrac{1}{2} \qquad J = 2, 1$$
$$j_1 = \tfrac{1}{2} \text{ and } j_2 = \tfrac{3}{2} \qquad J = 2, 1$$
$$j_1 = \tfrac{1}{2} \text{ and } j_2 = \tfrac{1}{2} \qquad J = 1, 0$$

For heavy atoms, in which *jj*-coupling is appropriate, it is best to discuss their energies using these quantum numbers.

Although *jj*-coupling should be used for assessing the energies of heavy atoms, the term symbols derived from Russell–Saunders coupling can still be used as labels. To see why this procedure is valid, we need to examine how the energies of the atomic states change as the spin–orbit coupling increases in strength. Such a **correlation diagram** is shown in Fig. 3.34. It shows that there is a correspondence between the low spin–orbit coupling (Russell–Saunders coupling) and high spin–orbit coupling (*jj*-coupling) schemes, so the labels derived by using the Russell–Saunders scheme can be used to label the states of the *jj*-coupling scheme.

(d) Selection rules

Any state of the atom, and any spectral transition, can be specified by using term symbols. For example, the transitions giving rise to the yellow sodium doublet (which were shown in Fig. 3.30) are

$$3p^1\ {}^2P_{3/2} \rightarrow 3s^1\ {}^2S_{1/2} \qquad 3p^1\ {}^2P_{1/2} \rightarrow 3s^1\ {}^2S_{1/2}$$

By convention, the upper term precedes the lower. The corresponding absorptions are therefore denoted

$${}^2P_{3/2} \leftarrow {}^2S_{1/2} \qquad {}^2P_{1/2} \leftarrow {}^2S_{1/2}$$

(The configurations have been omitted.)

We have seen that selection rules arise from the conservation of angular momentum during a transition and from the fact that a photon has a spin of 1. They can therefore be expressed in terms of the term symbols, because the latter carry information about angular momentum. A detailed analysis leads to the following rules:

$$\Delta S = 0 \qquad \Delta L = 0, \pm 1 \qquad \Delta l = \pm 1 \qquad \Delta J = 0, \pm 1, \text{ but } J = 0 \leftarrow\!|\!\rightarrow J = 0 \tag{3.47}$$

where the symbol $\leftarrow\!|\!\rightarrow$ denotes a forbidden transition. The rule about ΔS (no change of overall spin) stems from the fact that the light does not affect the spin directly. The rules about ΔL and Δl express the fact that the orbital angular momentum of an

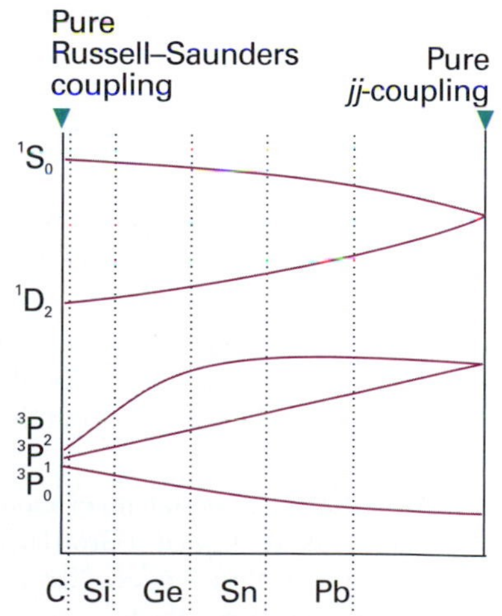

Fig. 3.34 The correlation diagram for some of the states of a two-electron system. All atoms lie between the two extremes, but the heavier the atom, the closer it lies to the pure *jj*-coupling case.

individual electron must change (so $\Delta l = \pm 1$), but whether or not this results in an overall change of orbital momentum depends on the coupling.

The selection rules given above apply when Russell–Saunders coupling is valid (in light atoms). If we insist on labelling the terms of heavy atoms with symbols like ^{3}D, then we shall find that the selection rules progressively fail as the atomic number increases because the quantum numbers S and L become ill defined as *jj*-coupling becomes more appropriate. As explained above, Russell–Saunders term symbols are only a convenient way of labelling the terms of heavy atoms: they do not bear any direct relation to the actual angular momenta of the electrons in a heavy atom. For this reason, transitions between singlet and triplet states (for which $\Delta S = \pm 1$), while forbidden in light atoms, are allowed in heavy atoms.

Checklist of key ideas

- ☐ 1. A hydrogenic atom is a one-electron atom or ion of general atomic number Z. A many-electron (polyelectronic) atom is an atom or ion with more than one electron.
- ☐ 2. The Lyman, Balmer, and Paschen series in the spectrum of atomic hydrogen arise, respectively, from the transitions $n \rightarrow 1$, $n \rightarrow 2$, and $n \rightarrow 3$.
- ☐ 3. The wavenumbers of all the spectral lines of a hydrogen atom can be expressed as $\tilde{\nu} = R_{\mathrm{H}}(1/n_1^2 - 1/n_2^2)$, where R_{H} is the Rydberg constant for hydrogen.
- ☐ 4. The Ritz combination principle states that the wavenumber of any spectral line is the difference between two spectroscopic energy levels, or terms: $\tilde{\nu} = T_1 - T_2$.
- ☐ 5. The wavefunction of the hydrogen atom is the product of a radial wavefunction and a radial wavefunction: $\psi(r,\theta,\phi) = R(r)Y(\theta,\phi)$.
- ☐ 6. An atomic orbital is a one-electron wavefunction for an electron in an atom.
- ☐ 7. The energies of an electron in a hydrogen atom are given by $E_n = -Z^2\mu e^4/32\pi^2\varepsilon_0^2\hbar^2 n^2$, where n is the principal quantum number, $n = 1, 2, \ldots$.
- ☐ 8. All the orbitals of a given value of n belong to a given shell; orbitals with the same value of n but different values of l belong to different subshells.
- ☐ 9. The radial distribution function is a probability density that, when it is multiplied by dr, gives the probability of finding the electron anywhere in a shell of thickness dr at the radius r; $P(r) = r^2R(r)^2$.
- ☐ 10. A selection rule is a statement about which spectroscopic transitions are allowed; a Grotrian diagram is a diagram summarizing the energies of the states and atom and the transitions between them.
- ☐ 11. In the orbital approximation it is supposed that each electron occupies its 'own' orbital, $\psi(\boldsymbol{r}_1,\boldsymbol{r}_2,\ldots) = \psi(\boldsymbol{r}_1)\psi(\boldsymbol{r}_2)\ldots$.
- ☐ 12. The Pauli exclusion principle states that no more than two electrons may occupy any given orbital and, if two do occupy one orbital, then their spins must be paired.
- ☐ 13. The Pauli principle states that, when the labels of any two identical fermions are exchanged, the total wavefunction changes sign; when the labels of any two identical bosons are exchanged, the total wavefunction retains the same sign.
- ☐ 14. The effective nuclear charge Z_{eff} is the net charge experienced by an electron allowing for electron–electron repulsions.
- ☐ 15. Shielding is the effective reduction in charge of a nucleus by surrounding electrons; the shielding constant σ is given by $Z_{\mathrm{eff}} = Z - \sigma$.
- ☐ 16. Penetration is the ability of an electron to be found inside inner shells and close to the nucleus.
- ☐ 17. The building-up (*Aufbau*) principle is the procedure for filling atomic orbitals that leads to the ground-state configuration of an atom.
- ☐ 18. Hund's maximum multiplicity rule states that an atom in its ground state adopts a configuration with the greatest number of unpaired electrons.
- ☐ 19. The first ionization energy I_1 is the minimum energy necessary to remove an electron from a many-electron atom in the gas phase; the second ionization energy I_2 is the minimum energy necessary to remove an electron from an ionized many-electron atom in the gas phase.
- ☐ 20. The electron affinity E_{ea} is the energy released when an electron attaches to a gas-phase atom.
- ☐ 21. A singlet term has $S = 0$; a triplet term has $S = 1$.
- ☐ 22. Spin–orbit coupling is the interaction of the spin magnetic moment with the magnetic field arising from the orbital angular momentum.
- ☐ 23. Fine structure is the structure in a spectrum due to spin–orbit coupling.
- ☐ 24. A term symbol is a symbolic specification of the state of an atom, $^{2S+1}\{L\}_J$.
- ☐ 25. The allowed values of a combined angular momenta are obtained by using the Clebsch–Gordan series: $J = j_1 + j_2, j_1 + j_2 - 1, \ldots |j_1 - j_2|$.

☐ 26. The multiplicity of a term is the value of $2S+1$; provided $L \geq S$, the multiplicity is the number of levels of the term.

☐ 28. A level is a group of states with a common value of J.

☐ 29. Russell–Saunders coupling is a coupling scheme based on the view that, if spin–orbit coupling is weak, then it is effective only when all the orbital momenta are operating cooperatively.

☐ 30. *jj*-Coupling is a coupling scheme based on the view that the individual spin and orbital momenta of the electrons are coupled into individual j values and these momenta are combined into a grand total, J.

☐ 31. The selection rules for spectroscopic transitions in polyelectronic atoms are: $\Delta S=0$, $\Delta L=0, \pm1$, $\Delta l=\pm1$, $\Delta J=0, \pm1$, but $J=0 \leftarrow\!|\!\rightarrow J=0$.

Further reading

Articles and texts

P.W. Atkins, *Quanta: a handbook of concepts*. Oxford University Press (1991).

P.F. Bernath, *Spectra of atoms and molecules*. Oxford University Press (1995).

K. Bonin and W. Happer, Atomic spectroscopy. In *Encyclopedia of applied physics* (ed. G.L. Trigg), **2**, 245. VCH, New York (1991).

E.U. Condon and H. Odabaçi, *Atomic structure*. Cambridge University Press (1980).

C.W. Haigh, The theory of atomic spectroscopy: *jj* coupling, intermediate coupling, and configuration interaction. *J. Chem. Educ.* **72**, 206 (1995).

C.S. Johnson, Jr. and L.G. Pedersen, *Problems and solutions in quantum chemistry and physics*. Dover, New York (1986).

J.C. Morrison, A.W. Weiss, K. Kirby, and D. Cooper, Electronic structure of atoms and molecules. In *Encyclopedia of applied physics* (ed. G.L. Trigg), **6**, 45. VCH, New York (1993).

N. Shenkuan, The physical basis of Hund's rule: orbital contraction effects. *J. Chem. Educ.* **69**, 800 (1992).

L.G. Vanquickenborne, K. Pierloot, and D. Devoghel, Transition metals and the *Aufbau* principle. *J. Chem. Educ.* **71**, 469 (1994).

Sources of data and information

S. Bashkin and J.O. Stonor, Jr., *Atomic energy levels and Grotrian diagrams*. North-Holland, Amsterdam (1975–1982).

D.R. Lide (ed.), *CRC handbook of chemistry and physics*, Section 10, CRC Press, Boca Raton (2000).

Further information

Further information 3.1 *The separation of motion*

The separation of internal and external motion

Consider a one-dimensional system in which the potential energy depends only on the separation of the two particles. The total energy is

$$E=\frac{p_1^2}{2m_1}+\frac{p_2^2}{2m_2}+V$$

where $p_1=m_1\dot{x}_1$ and $p_2=m_2\dot{x}_2$, the dot signifying differentiation with respect to time. The centre of mass (Fig. 3.35) is located at

$$X=\frac{m_1}{m}x_1+\frac{m_2}{m}x_2 \qquad m=m_1+m_2$$

and the separation of the particles is $x=x_1-x_2$. It follows that

$$x_1=X+\frac{m_2}{m}x \qquad x_2=X-\frac{m_1}{m}x$$

The linear momenta of the particles can be expressed in terms of the rates of change of x and X:

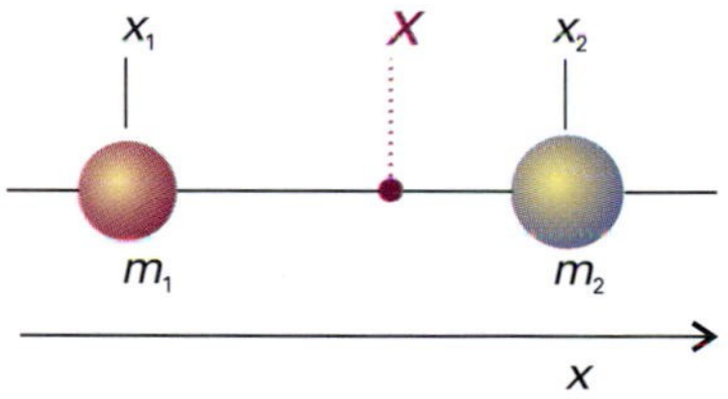

Fig. 3.35 The coordinates used for discussing the separation of the relative motion of two particles from the motion of the centre of mass.

$$p_1=m_1\dot{x}_1=m_1\dot{X}+\frac{m_1m_2}{m}\dot{x} \qquad p_2=m_2\dot{x}_2=m_2\dot{X}-\frac{m_1m_2}{m}\dot{x}$$

Then it follows that

$$\frac{p_1^2}{2m_1}+\frac{p_2^2}{2m_2}=\tfrac{1}{2}m\dot{X}^2+\tfrac{1}{2}\mu\dot{x}^2$$

where μ is given in eqn 3.6. By writing $P=m\dot{X}$ for the linear momentum of the system as a whole and defining p as $\mu\dot{x}$, we find

$$E=\frac{P^2}{2m}+\frac{p^2}{2\mu}+V$$

The corresponding hamiltonian (generalized to three dimensions) is therefore

$$\hat{H}=-\frac{\hbar^2}{2m}\nabla^2_{\text{c.m.}}-\frac{\hbar^2}{2\mu}\nabla^2+V$$

where the first term differentiates with respect to the centre of mass coordinates and the second with respect to the relative coordinates.

Now we write the overall wavefunction as the product $\psi_{\text{total}}=\psi_{\text{c.m.}}\psi$, where the first factor is a function of only the centre of mass coordinates and the second is a function of only the relative coordinates. The overall Schrödinger equation, $H\psi_{\text{total}}=E_{\text{total}}\psi_{\text{total}}$, then separates by the argument that we have used in Sections 2.2a and 2.7, with $E_{\text{total}}=E_{\text{c.m.}}+E$.

The separation of angular and radial motion

The laplacian in three dimensions is given in eqn 2.51a. It follows that the Schrödinger equation in eqn 3.6 is

$$-\frac{\hbar^2}{2\mu}\left(\frac{\partial^2}{\partial r^2}+\frac{2}{r}\frac{\partial}{\partial r}+\frac{1}{r^2}\Lambda^2\right)RY+VRY=ERY$$

Because R depends only on r and Y depends only on the angular coordinates, this equation becomes

$$-\frac{\hbar^2}{2\mu}\left(Y\frac{\mathrm{d}^2R}{\mathrm{d}r^2}+\frac{2Y}{r}\frac{\mathrm{d}R}{\mathrm{d}r}+\frac{R}{r^2}\Lambda^2Y\right)+VRY=ERY$$

If we multiply through by r^2/RY, we obtain

$$-\frac{\hbar^2}{2\mu R}\left(r^2\frac{\mathrm{d}^2R}{\mathrm{d}r^2}+2r\frac{\mathrm{d}R}{\mathrm{d}r}\right)+Vr^2-\frac{\hbar^2}{2\mu Y}\Lambda^2Y=Er^2$$

At this point we employ the usual argument. The term in Y is the only one that depends on the angular variables, so it must be a constant. When we write this constant as $\hbar^2 l(l+1)/2\mu$, eqn 3.10 follows immediately.

Each end-of-chapter discussion question, exercise, and problem for which a solution is available in the Student Solution Manual has a corresponding bracketed number referring to the Student Solutions Manual number for this problem, for example [SSM 10.1]. Each end-of-chapter exercise and problem for which a solution is available in the Instructors' Solution Manual has a corresponding bracketed number referring to the Instructor Solutions Manual number for this problem, for example [ISM 10.2].

Discussion questions

3.1 Describe the separation of variables procedure as it is applied to simplify the description of a hydrogenic atom free to move through space. [SSM 10.1]

3.2 List and describe the significance of the quantum numbers needed to specify the internal state of a hydrogenic atom. [ISM 10.2]

3.3 Specify and account for the selection rules for transitions in hydrogenic atoms. [SSM 10.3]

3.4 Explain the significance of (a) a boundary surface and (b) the radial distribution function for hydrogenic orbitals. [ISM 10.4]

3.5 Outline the electron configurations of many-electron atoms in terms of their location in the periodic table. [SSM 10.5]

3.6 Describe and account for the variation of first ionization energies along Period 2 of the periodic table. [ISM 10.6]

3.7 Describe the orbital approximation for the wavefunction of a many-electron atom. What are the limitations of the approximation? [SSM 10.7]

3.8 Explain the origin of spin–orbit coupling and how it affects the appearance of a spectrum. [ISM 10.8]

Exercises

3.1a When ultraviolet radiation of wavelength 58.4 nm from a helium lamp is directed on to a sample of krypton, electrons are ejected with a speed of 1.59 Mm s^{-1}. Calculate the ionization energy of krypton. [ISM 10.1a]

3.1b When ultraviolet radiation of wavelength 58.4 nm from a helium lamp is directed on to a sample of xenon, electrons are ejected with a speed of 1.79 Mm s^{-1}. Calculate the ionization energy of xenon. [SSM 10.1b]

3.2a By differentiation of the 2*s* radial wavefunction, show that it has two extrema in its amplitude, and locate them. [ISM 10.2a]

3.2b By differentiation of the 3*s* radial wavefunction, show that it has three extrema in its amplitude, and locate them. [SSM 10.2b]

3.3a Locate the radial nodes in the 3*s* orbital of an H atom. [ISM 10.3a]

3.3b Locate the radial nodes in the 4*p* orbital of an H atom where, in the notation of Table 2.1, the radial wavefunction is proportional to $20-10\rho+\rho^2$. [SSM 10.3b]

3.4a The wavefunction for the ground state of a hydrogen atom is Ne^{-r/a_0}. Determine the normalization constant N. [ISM 10.4a]

3.4b The wavefunction for the 2*s* orbital of a hydrogen atom is $N(2-r/a_0)e^{-r/2a_0}$. Determine the normalization constant N. [SSM 10.4b]

3.5a Calculate the average kinetic and potential energies of an electron in the ground state of a hydrogen atom. [ISM 10.5a]

3.5b Calculate the average kinetic and potential energies of a 2*s* electron in a hydrogenic atom of atomic number Z. [SSM 10.5b]

3.6a Write down the expression for the radial distribution function of a 2*s* electron in a hydrogenic atom and determine the radius at which the electron is most likely to be found. [ISM **10.6a**]

3.6b Write down the expression for the radial distribution function of a 3*s* electron in a hydrogenic atom and determine the radius at which the electron is most likely to be found. [SSM **10.6b**]

3.7a Write down the expression for the radial distribution function of a 2*p* electron in a hydrogenic atom and determine the radius at which the electron is most likely to be found. [ISM **10.7a**]

3.7b Write down the expression for the radial distribution function of a 3*p* electron in a hydrogenic atom and determine the radius at which the electron is most likely to be found. [SSM **10.7b**]

3.8a What is the orbital angular momentum of an electron in the orbitals (a) 1*s*, (b) 3*s*, (c) 3*d*? Give the numbers of angular and radial nodes in each case. [ISM **10.8a**]

10.8b What is the orbital angular momentum of an electron in the orbitals (a) 4*d*, (b) 2*p*, (c) 3*p*? Give the numbers of angular and radial nodes in each case. [SSM **10.8b**]

3.9a Calculate the permitted values of *j* for (a) a *d* electron, (b) an *f* electron. [ISM **10.9a**]

3.9b Calculate the permitted values of *j* for (a) a *p* electron, (b) an *h* electron. [SSM **10.9b**]

3.10a An electron in two different states of an atom is known to have $j=\frac{3}{2}$ and $\frac{1}{2}$. What is its orbital angular momentum quantum number in each case? [ISM **10.10a**]

3.10b What are the allowed total angular momentum quantum numbers of a composite system in which $j_1 = 5$ and $j_2 = 3$? [SSM **10.10b**]

3.11a State the orbital degeneracy of the levels in a hydrogen atom that have energy (a) $-hcR_H$; (b) $-\frac{1}{9}hcR_H$; (c) $-\frac{1}{25}hcR_H$. [ISM **10.11a**]

3.11b State the orbital degeneracy of the levels in a hydrogenic atom (*Z* in parentheses) that have energy (a) $-4hcR_{atom}$ (2); (b) $-\frac{1}{4}hcR_{atom}$ (4), and (c) $-hcR_{atom}$ (5). [SSM **10.11b**]

3.12a What information does the term symbol $^1D_2$ provide about the angular momentum of an atom? [ISM **10.12a**]

3.12b What information does the term symbol $^3F_4$ provide about the angular momentum of an atom? [SSM **10.12b**]

3.13a At what radius does the probability of finding an electron at a point in the H atom fall to 50 per cent of its maximum value? [ISM **10.13a**]

3.13b At what radius in the H atom does the radial distribution function of the ground state have (a) 50 per cent, (b) 75 per cent of its maximum value? [SSM **10.13b**]

3.14a Which of the following transitions are allowed in the normal electronic emission spectrum of an atom: (a) $2s \rightarrow 1s$, (b) $2p \rightarrow 1s$, (c) $3d \rightarrow 2p$? [ISM **10.14a**]

3.14b Which of the following transitions are allowed in the normal electronic emission spectrum of an atom: (a) $5d \rightarrow 2s$, (b) $5p \rightarrow 3s$, (c) $6p \rightarrow 4f$? [SSM **10.14b**]

3.15a (a) Write the electronic configuration of the Ni^{2+} ion. (b) What are the possible values of the total spin quantum numbers *S* and M_S for this ion? [ISM **10.15a**]

3.15b (a) Write the electronic configuration of the V^{2+} ion. (b) What are the possible values of the total spin quantum numbers *S* and M_S for this ion? [SSM **10.15b**]

3.16a Suppose that an atom has (a) 2, (b) 3 electrons in different orbitals. What are the possible values of the total spin quantum number *S*? What is the multiplicity in each case? [ISM **10.16a**]

3.16b Suppose that an atom has (a) 4, (b) 5 electrons in different orbitals. What are the possible values of the total spin quantum number *S*? What is the multiplicity in each case? [SSM **10.16b**]

3.17a What atomic terms are possible for the electron configuration ns^1nd^1? Which term is likely to lie lowest in energy? [ISM **10.17a**]

3.17b What atomic terms are possible for the electron configuration np^1nd^1? Which term is likely to lie lowest in energy? [SSM **10.17b**]

3.18a What values of *J* may occur in the terms (a) ^1S, (b) ^2P, (c) ^3P? How many states (distinguished by the quantum number M_J) belong to each level? [ISM **10.18a**]

3.18b What values of *J* may occur in the terms (a) ^3D, (b) ^4D, (c) ^2G? How many states (distinguished by the quantum number M_J) belong to each level? [SSM **10.18b**]

3.19a Give the possible term symbols for (a) Li [He]$2s^1$, (b) Na [Ne]$3p^1$. [ISM **10.19a**]

3.19b Give the possible term symbols for (a) Sc [Ar]$3d^14s^2$, (b) Br [Ar]$3d^{10}4s^24p^5$. [SSM **10.19b**]

Problems*

Numerical problems

3.1 The *Humphreys series* is a group of lines in the spectrum of atomic hydrogen. It begins at 12 368 nm and has been traced to 3281.4 nm. What are the transitions involved? What are the wavelengths of the intermediate transitions? [SSM **10.1**]

3.2 A series of lines in the spectrum of atomic hydrogen lies at 656.46 nm, 486.27 nm, 434.17 nm, and 410.29 nm. What is the wavelength of the next line in the series? What is the ionization energy of the atom when it is in the lower state of the transitions? [ISM **10.2**]

3.3 The Li^{2+} ion is hydrogenic and has a Lyman series at 740 747 cm^{-1}, 877 924 cm^{-1}, 925 933 cm^{-1}, and beyond. Show that the energy levels are of the form $-hcR/n^2$ and find the value of *R* for this ion. Go on to predict the wavenumbers of the two longest-wavelength transitions of the Balmer series of the ion and find the ionization energy of the ion. [SSM **10.3**]

3.4 A series of lines in the spectrum of neutral Li atoms rise from combinations of $1s^22p^1\ ^2P$ with $1s^2nd^1\ ^2D$ and occur at 610.36 nm, 460.29 nm, and 413.23 nm. The *d* orbitals are hydrogenic. It is known that the ^2P term lies at 670.78 nm above the ground state, which is $1s^22s^1\ ^2S$. Calculate the ionization energy of the ground-state atom. [ISM **10.4**]

* Problems denoted with the symbol ‡ were supplied by Charles Trapp, Carmen Giunta, and Marshall Cady.

3.5‡ W.P. Wijesundera, S.H. Vosko, and F.A. Parpia (*Phys. Rev.* A **51**, 278 (1995)) attempted to determine the electron configuration of the ground state of lawrencium, element 103. The two contending configurations are $[Rn]5f^{14}7s^27p^1$ and $[Rn]5f^{14}6d7s^2$. Write down the term symbols for each of these configurations, and identify the lowest level within each configuration. Which level would be lowest according to a simple estimate of spin–orbit coupling? [SSM **10.5**]

3.6 The characteristic emission from K atoms when heated is purple and lies at 770 nm. On close inspection, the line is found to have two closely spaced components, one at 766.70 nm and the other at 770.11 nm. Account for this observation, and deduce what information you can. [ISM **10.6**]

3.7 Calculate the mass of the deuteron given that the first line in the Lyman series of H lies at 82 259.098 cm^{-1} whereas that of D lies at 82 281.476 cm^{-1}. Calculate the ratio of the ionization energies of H and D. [SSM **10.7**]

3.8 Positronium consists of an electron and a positron (same mass, opposite charge) orbiting round their common centre of mass. The broad features of the spectrum are therefore expected to be hydrogen-like, the differences arising largely from the mass differences. Predict the wavenumbers of the first three lines of the Balmer series of positronium. What is the binding energy of the ground state of positronium? [ISM **10.8**]

3.9 The *Zeeman effect* is the modification of an atomic spectrum by the application of a strong magnetic field. It arises from the interaction between applied magnetic fields and the magnetic moments due to orbital and spin angular momenta (recall the evidence provided for electron spin by the Stern–Gerlach experiment, Section 2.8). To gain some appreciation for the so-called *normal Zeeman effect*, which is observed in transitions involving singlet states, consider a *p* electron, with $l = 1$ and $m_l = 0, \pm 1$. In the absence of a magnetic field, these three states are degenerate. When a field of magnitude B is present, the degeneracy is removed and it is observed that the state with $m_l = +1$ moves up in energy by $\mu_B B$, the state with $m_l = 0$ is unchanged, and the state with $m_l = -1$ moves down in energy by $\mu_B B$, where $\mu_B = e\hbar/2m_e = 9.274 \times 10^{-24}$ J T^{-1} is the Bohr magneton (see Section 8.1). Therefore, a transition between a $^1S_0$ term and a $^1P_1$ term consists of three spectral lines in the presence of a magnetic field where, in the absence of the magnetic field, there is only one. (a) Calculate the splitting in reciprocal centimetres between the three spectral lines of a transition between a $^1S_0$ term and a $^1P_1$ term in the presence of a magnetic field of 2 T (where 1 T = 1 kg s^{-2} A^{-1}). (b) Compare the value you calculated in (a) with typical optical transition wavenumbers, such as those for the Balmer series of the H atom. Is the line splitting caused by the normal Zeeman effect relatively small or relatively large? [SSM **10.9**]

3.10 In 1976 it was mistakenly believed that the first of the 'superheavy' elements had been discovered in a sample of mica. Its atomic number was believed to be 126. What is the most probable distance of the innermost electrons from the nucleus of an atom of this element? (In such elements, relativistic effects are very important, but ignore them here.) [ISM **10.10**]

Theoretical problems

3.11 What is the most probable point (not radius) at which a 2*p* electron will be found in the hydrogen atom? [SSM **10.11**]

3.12 Show by explicit integration that (a) hydrogenic 1*s* and 2*s* orbitals, (b) $2p_x$ and $2p_y$ orbitals are mutually orthogonal. [ISM **10.12**]

3.13‡ Explicit expressions for hydrogenic orbitals are given in Tables 3.1 and 2.3. (a) Verify both that the $3p_x$ orbital is normalized (to 1) and that $3p_x$ and $3d_{xy}$ are mutually orthogonal. (b) Determine the positions of both the radial nodes and nodal planes of the 3*s*, $3p_x$, and $3d_{xy}$ orbitals. (c) Determine the mean radius of the 3*s* orbital. (d) Draw a graph of the radial distribution function for the three orbitals (of part (b)) and discuss the significance of the graphs for interpreting the properties of many-electron atoms. (e) Create both *xy*-plane polar plots and boundary surface plots for these orbitals. Construct the boundary plots so that the distance from the origin to the surface is the absolute value of the angular part of the wavefunction. Compare the *s*, *p*, and *d* boundary surface plots with that of an *f*-orbital; e.g. $\psi_f \propto x(5z^2 - r^2) \propto \sin\theta\,(5\cos^2\theta - 1)\cos\phi$. [SSM **10.13**]

3.14 Determine whether the p_x and p_y orbitals are eigenfunctions of l_z. If not, does a linear combination exist that is an eigenfunction of l_z? [ISM **10.14**]

3.15 Show that l_z and l^2 both commute with the hamiltonian for a hydrogen atom. What is the significance of this result? [SSM **10.15**]

3.16 The 'size' of an atom is sometimes considered to be measured by the radius of a sphere that contains 90 per cent of the charge density of the electrons in the outermost occupied orbital. Calculate the 'size' of a hydrogen atom in its ground state according to this definition. Go on to explore how the 'size' varies as the definition is changed to other percentages, and plot your conclusion. [ISM **10.16**]

3.17 Some atomic properties depend on the average value of $1/r$ rather than the average value of r itself. Evaluate the expectation value of $1/r$ for (a) a hydrogen 1*s* orbital, (b) a hydrogenic 2*s* orbital, (c) a hydrogenic 2*p* orbital. [SSM **10.17**]

3.18 One of the most famous of the obsolete theories of the hydrogen atom was proposed by Bohr. It has been replaced by quantum mechanics but, by a remarkable coincidence (not the only one where the Coulomb potential is concerned), the energies it predicts agree exactly with those obtained from the Schrödinger equation. In the Bohr atom, an electron travels in a circle around the nucleus. The Coulombic force of attraction ($Ze^2/4\pi\varepsilon_0 r^2$) is balanced by the centrifugal effect of the orbital motion. Bohr proposed that the angular momentum is limited to integral values of $\hbar$. When the two forces are balanced, the atom remains in a stationary state until it makes a spectral transition. Calculate the energies of a hydrogenic atom using the Bohr model. [ISM **10.18**]

3.19 The Bohr model of the atom is specified in Problem 3.18. What features of it are untenable according to quantum mechanics? How does the Bohr ground state differ from the actual ground state? Is there an experimental distinction between the Bohr and quantum mechanical models of the ground state? [SSM **10.19**]

3.20 Atomic units of length and energy may be based on the properties of a particular atom. The usual choice is that of a hydrogen atom, with the unit of length being the Bohr radius, a_0, and the unit of energy being the (negative of the) energy of the 1*s* orbital. If the positronium atom (e^+,e^-) were used instead, with analogous definitions of units of length and energy, what would be the relation between these two sets of atomic units? [ISM **10.20**]

3.21 Some of the selection rules for hydrogenic atoms were derived in *Justification* 3.4. Complete the derivation by considering the *x*- and *y*-components of the electric dipole moment operator. [SSM **10.21**]

3.22‡ Stern–Gerlach splittings of atomic beams are small and require either large magnetic field gradients or long magnets for their observation. For a beam of atoms with zero orbital angular momentum, such as H or Ag, the deflection is given by $x = \pm(\mu_B L^2/4E_K)\mathrm{d}B/\mathrm{d}z$, where μ_B is the Bohr magneton (Problem 3.9), L is the length of the magnet, E_K is the average kinetic energy of the atoms in the beam, and $\mathrm{d}B/\mathrm{d}z$ is the magnetic field gradient across the beam. (a) Use the Maxwell–Boltzmann velocity distribution to show that the average translational kinetic energy of the atoms emerging as a beam from a pinhole in an oven at temperature T is $2kT$. (b) Calculate the magnetic field gradient required to produce a splitting of 1.00 mm in a beam of Ag atoms from an oven at 1000 K with a magnet of length 50 cm. [ISM **10.22**]

3.23 The wavefunction of a many-electron closed-shell atom can expressed as a Slater determinant (Section 3.4b). A useful property of determinants is that interchanging any two rows or columns changes their sign and therefore, if any two rows or columns are identical, then the determinant vanishes.

Use this property to show that (a) the wavefunction is antisymmetric under particle exchange, (b) no two electrons can occupy the same orbital with the same spin. [SSM **10.23**]

Applications: to astrophysics and biochemistry

3.24 Hydrogen is the most abundant element in all stars. However, neither absorption nor emission lines due to neutral hydrogen are found in the spectra of stars with effective temperatures higher than 25 000 K. Account for this observation. [ISM **10.24**]

3.25 The distribution of isotopes of an element may yield clues about the nuclear reactions that occur in the interior of a star. Show that it is possible to use spectroscopy to confirm the presence of both $^4He^+$ and $^3He^+$ in a star by calculating the wavenumbers of the $n = 3 \rightarrow n = 2$ and of the $n = 2 \rightarrow n = 1$ transitions for each isotope. [SSM **10.25**]

3.26‡ Highly excited atoms have electrons with large principal quantum numbers. Such *Rydberg atoms* have unique properties and are of interest to astrophysicists . For hydrogen atoms with large *n*, derive a relation for the separation of energy levels. Calculate this separation for $n = 100$; also calculate the average radius, the geometric cross-section, and the ionization energy. Could a thermal collision with another hydrogen atom ionize this Rydberg atom? What minimum velocity of the second atom is required? Could a normal-sized neutral H atom simply pass through the Rydberg atom leaving it undisturbed? What might the radial wavefunction for a 100*s* orbital be like? [ISM **10.26**]

3.27 The spectrum of a star is used to measure its *radial velocity* with respect to the Sun, the component of the star's velocity vector that is parallel to a vector connecting the star's centre to the centre of the Sun. The measurement relies on the *Doppler effect*, in which radiation is shifted in frequency when the source is moving towards or away from the observer. When a star emitting electromagnetic radiation of frequency ν moves with a speed s relative to an observer, the observer detects radiation of frequency $\nu_{receding} = \nu f$ or $\nu_{approaching} = \nu/f$, where $f = \{(1 - s/c)/(1 + s/c)\}^{1/2}$ and c is the speed of light. It is easy to see that $\nu_{receding} < \nu$ and a receding star is characterized by a *red shift* of its spectrum with respect to the spectrum of an identical, but stationary source. Furthermore, $\nu_{approaching} > \nu$ and an approaching star is characterized by a *blue shift* of its spectrum with respect to the spectrum of an identical, but stationary source. In a typical experiment, ν is the frequency of a spectral line of an element measured in a stationary Earth-bound laboratory from a calibration source, such as an arc lamp. Measurement of the same spectral line in a star gives ν_{star} and the speed of recession or approach may be calculated from the value of ν and the equations above. (a) Three Fe I lines of the star HDE 271 182, which belongs to the Large Magellanic Cloud, occur at 438.882 nm, 441.000 nm, and 442.020 nm. The same lines occur at 438.392 nm, 440.510 nm, and 441.510 nm in the spectrum of an Earth-bound iron arc. Determine whether HDE 271 182 is receding from or approaching the Earth and estimate the star's radial speed with respect to the Earth. (b) What additional information would you need to calculate the radial velocity of HDE 271 182 with respect to the Sun? [SSM **10.27**]

3.28 The *d*-metals iron, copper, and manganese form cations with different oxidation states. For this reason, they are found in many oxidoreductases and in several proteins of oxidative phosphorylation and photosynthesis. Explain why many *d*-metals form cations with different oxidation states. [ISM **10.28**]

3.29 Thallium, a neurotoxin, is the heaviest member of Group 13 of the periodic table and is found most usually in the +1 oxidation state. Aluminium, which causes anaemia and dementia, is also a member of the group but its chemical properties are dominated by the +3 oxidation state. Examine this issue by plotting the first, second, and third ionization energies for the Group 13 elements against atomic number. Explain the trends you observe. *Hints.* The third ionization energy, I_3, is the minimum energy needed to remove an electron from the doubly charged cation: $E^{2+}(g) \rightarrow E^{3+}(g) + e^-(g)$, $I_3 = E(E^{3+}) - E(E^{2+})$. For data, see the links to databases of atomic properties provided in the text's web site. [SSM **10.29**]

4 Molecular structure

The concepts developed in Chapter 3, particularly those of orbitals, can be extended to a description of the electronic structures of molecules. There are two principal quantum mechanical theories of molecular electronic structure. In valence-bond theory, the starting point is the concept of the shared electron pair. We see how to write the wavefunction for such a pair, and how it may be extended to account for the structures of a wide variety of molecules. The theory introduces the concepts of σ and π bonds, promotion, and hybridization that are used widely in chemistry. In molecular orbital theory (with which the bulk of the chapter is concerned), the concept of atomic orbital is extended to that of molecular orbital, which is a wavefunction that spreads over all the atoms in a molecule.

In this chapter we consider the origin of the strengths, numbers, and three-dimensional arrangement of chemical bonds between atoms. The quantum mechanical description of chemical bonding has become highly developed through the use of computers, and it is now possible to consider the structures of molecules of almost any complexity. We shall concentrate on the quantum mechanical description of the **covalent bond**, which was identified by G.N. Lewis (in 1916, before quantum mechanics was fully established) as an electron pair shared between two neighbouring atoms. We shall see, however, that the other principal type of bond, an **ionic bond**, in which the cohesion arises from the Coulombic attraction between ions of opposite charge, is also captured as a limiting case of a covalent bond between dissimilar atoms. In fact, although the Schrödinger equation might shroud the fact in mystery, all chemical bonding can be traced to the interplay between the attraction of opposite charges, the repulsion of like charges, and the effect of changing kinetic energy as the electrons are confined to various regions when bonds form.

There are two major approaches to the calculation of molecular structure, **valence-bond theory** (VB theory) and **molecular orbital theory** (MO theory). Almost all modern computational work makes use of MO theory, and we concentrate on that theory in this chapter. Valence-bond theory, though, has left its imprint on the language of chemistry, and it is important to know the significance of terms that chemists use every day. Therefore, our discussion is organized as follows. First, we set out the concepts common to all levels of description. Then we present VB theory, which gives us a simple qualitative understanding of bond formation. Next, we present the basic ideas of MO theory. Finally, we see how computational techniques pervade all current discussions of molecular structure, including the prediction of chemical reactivity.

The Born–Oppenheimer approximation

All theories of molecular structure make the same simplification at the outset. Whereas the Schrödinger equation for a hydrogen atom can be solved exactly, an exact solution is not possible for any molecule because the simplest molecule consists of three particles (two nuclei and one electron). We therefore adopt the **Born–Oppenheimer approximation** in which it is supposed that the nuclei, being so much heavier than an electron, move relatively slowly and may be treated as stationary while the electrons move in their field. We can therefore think of the nuclei as being fixed at arbitrary locations, and then solve the Schrödinger equation for the wavefunction of the electrons alone.

The approximation is quite good for ground-state molecules, for calculations suggest that the nuclei in H_2 move through only about 1 pm while the electron speeds through 1000 pm, so the error of assuming that the nuclei are stationary is small. Exceptions to the approximation's validity include certain excited states of polyatomic molecules and the ground states of cations; both types of species are important when considering photoelectron spectroscopy (Section 4.4) and mass spectrometry.

The Born–Oppenheimer approximation allows us to select an internuclear separation in a diatomic molecule and then to solve the Schrödinger equation for the electrons at that nuclear separation. Then we choose a different separation and repeat the calculation, and so on. In this way we can explore how the energy of the molecule varies with bond length (in polyatomic molecules, with angles too) and obtain a **molecular potential energy curve** (Fig. 4.1). When more than one molecular parameter is changed in a polyatomic molecule, we obtain a potential energy surface. It is called a *potential* energy curve because the kinetic energy of the stationary nuclei is zero. Once the curve has been calculated or determined experimentally (by using the spectroscopic techniques described in Chapters 6 and 7), we can identify the **equilibrium bond length**, R_e, the internuclear separation at the minimum of the curve, and the **bond dissociation energy**, D_0, which is closely related to the depth, D_e, of the minimum below the energy of the infinitely widely separated and stationary atoms.

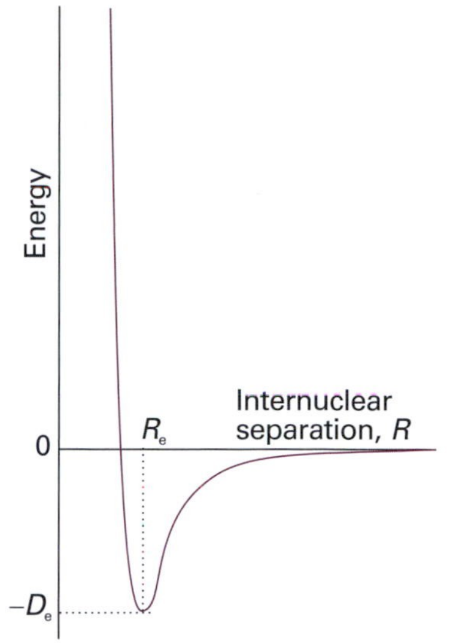

Fig. 4.1 A molecular potential energy curve. The equilibrium bond length corresponds to the energy minimum.

Comment 4.1

The dissociation energy differs from the depth of the well by an energy equal to the zero-point vibrational energy of the bonded atoms: $D_0 = D_e - \frac{1}{2}\hbar\omega$, where ω is the vibrational frequency of the bond (Section 6.9).

Valence-bond theory

Valence-bond theory was the first quantum mechanical theory of bonding to be developed. The language it introduced, which includes concepts such as spin pairing, orbital overlap, σ and π bonds, and hybridization, is widely used throughout chemistry, especially in the description of the properties and reactions of organic compounds. Here we summarize essential topics of VB theory that are familiar from introductory chemistry and set the stage for the development of MO theory.

4.1 Homonuclear diatomic molecules

In VB theory, a bond is regarded as forming when an electron in an atomic orbital on one atom pairs its spin with that of an electron in an atomic orbital on another atom. To understand why this pairing leads to bonding, we have to examine the wavefunction for the two electrons that form the bond. We begin by considering the simplest possible chemical bond, the one in molecular hydrogen, H_2.

The spatial wavefunction for an electron on each of two widely separated H atoms is

$$\psi = \chi_{H1s_A}(\boldsymbol{r}_1)\chi_{H1s_B}(\boldsymbol{r}_2)$$

if electron 1 is on atom A and electron 2 is on atom B; in this chapter we use χ (chi) to denote atomic orbitals. For simplicity, we shall write this wavefunction as $\psi = A(1)B(2)$. When the atoms are close, it is not possible to know whether it is electron 1 that is on A or electron 2. An equally valid description is therefore $\psi = A(2)B(1)$, in which electron 2 is on A and electron 1 is on B. When two outcomes are equally probable,

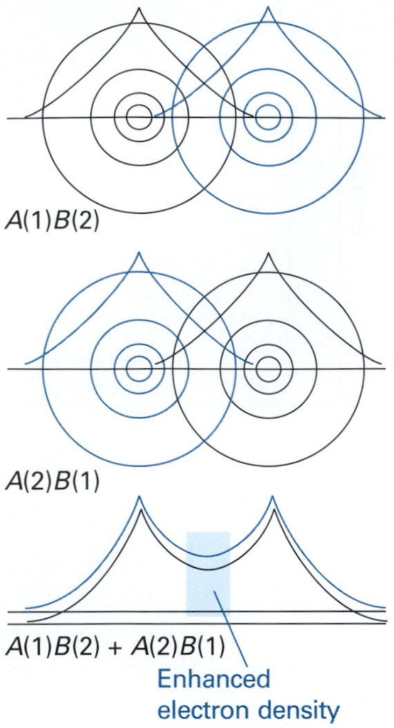

Fig. 4.2 It is very difficult to represent valence-bond wavefunctions because they refer to two electrons simultaneously. However, this illustration is an attempt. The atomic orbital for electron 1 is represented by the black contours, and that of electron 2 is represented by the blue contours. The top illustration represents $A(1)B(2)$, and the middle illustration represents the contribution $A(2)B(1)$. When the two contributions are superimposed, there is interference between the black contributions and between the blue contributions, resulting in an enhanced (two-electron) density in the internuclear region.

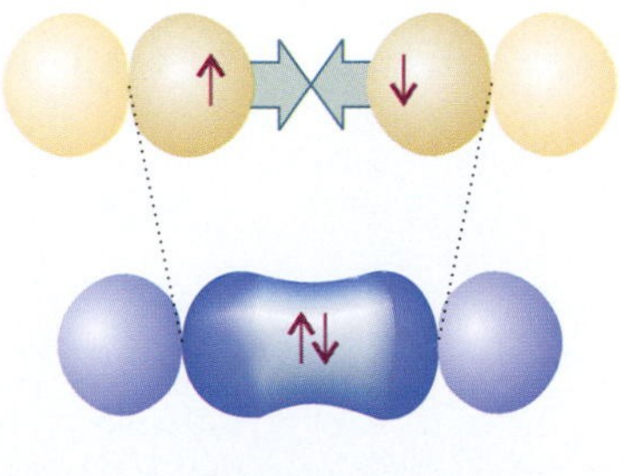

Fig. 4.3 The orbital overlap and spin pairing between electrons in two collinear *p* orbitals that results in the formation of a σ bond.

quantum mechanics instructs us to describe the true state of the system as a superposition of the wavefunctions for each possibility (Section 1.5d), so a better description of the molecule than either wavefunction alone is the (unnormalized) linear combination

$$\psi = A(1)B(2) \pm A(2)B(1) \qquad (4.1)$$

It turns out that the combination with lower energy is the one with a + sign, so the valence-bond wavefunction of the H_2 molecule is

$$\psi = A(1)B(2) + A(2)B(1) \qquad (4.2)$$

The formation of the bond in H_2 can be pictured as due to the high probability that the two electrons will be found between the two nuclei and hence will bind them together. More formally, the wave pattern represented by the term $A(1)B(2)$ interferes constructively with the wave pattern represented by the contribution $A(2)B(1)$, and there is an enhancement in the value of the wavefunction in the internuclear region (Fig. 4.2).

The electron distribution described by the wavefunction in eqn 4.2 is called a **σ bond**. A σ bond has cylindrical symmetry around the internuclear axis, and is so called because, when viewed along the internuclear axis, it resembles a pair of electrons in an *s* orbital (and σ is the Greek equivalent of s).

A chemist's picture of a covalent bond is one in which the spins of two electrons pair as the atomic orbitals overlap. The origin of the role of spin is that the wavefunction given in eqn 4.2 can be formed only by a pair of electrons with opposed spins. Spin pairing is not an end in itself: it is a means of achieving a wavefunction (and the probability distribution it implies) that corresponds to a low energy.

Justification 4.1 *Electron pairing in VB theory*

The Pauli principle requires the wavefunction of two electrons to change sign when the labels of the electrons are interchanged (see Section 3.4b). The total VB wavefunction for two electrons is

$$\psi(1,2) = \{A(1)B(2) + A(2)B(1)\}\sigma(1,2)$$

where σ represents the spin component of the wavefunction. When the labels 1 and 2 are interchanged, this wavefunction becomes

$$\psi(2,1) = \{A(2)B(1) + A(1)B(2)\}\sigma(2,1) = \{A(1)B(2) + A(2)B(1)\}\sigma(2,1)$$

The Pauli principle requires that $\psi(2,1) = -\psi(1,2)$, which is satisfied only if $\sigma(2,1) = -\sigma(1,2)$. The combination of two spins that has this property is

$$\sigma_-(1,2) = (1/2^{1/2})\{\alpha(1)\beta(2) - \alpha(2)\beta(1)\}$$

which corresponds to paired electron spins (Section 3.7). Therefore, we conclude that the state of lower energy (and hence the formation of a chemical bond) is achieved if the electron spins are paired.

The VB description of H_2 can be applied to other homonuclear diatomic molecules, such as nitrogen, N_2. To construct the valence bond description of N_2, we consider the valence electron configuration of each atom, which is $2s^2 2p_x^1 2p_y^1 2p_z^1$. It is conventional to take the *z*-axis to be the internuclear axis, so we can imagine each atom as having a $2p_z$ orbital pointing towards a $2p_z$ orbital on the other atom (Fig. 4.3), with the $2p_x$ and $2p_y$ orbitals perpendicular to the axis. A σ bond is then formed by spin pairing between the two electrons in the two $2p_z$ orbitals. Its spatial wavefunction is given by eqn 4.2, but now *A* and *B* stand for the two $2p_z$ orbitals.

The remaining $2p$ orbitals cannot merge to give σ bonds as they do not have cylindrical symmetry around the internuclear axis. Instead, they merge to form two π bonds. A **π bond** arises from the spin pairing of electrons in two p orbitals that approach side-by-side (Fig. 4.4). It is so called because, viewed along the internuclear axis, a π bond resembles a pair of electrons in a p orbital (and π is the Greek equivalent of p).

There are two π bonds in N_2, one formed by spin pairing in two neighbouring $2p_x$ orbitals and the other by spin pairing in two neighbouring $2p_y$ orbitals. The overall bonding pattern in N_2 is therefore a σ bond plus two π bonds (Fig. 4.5), which is consistent with the Lewis structure :N≡N: for nitrogen.

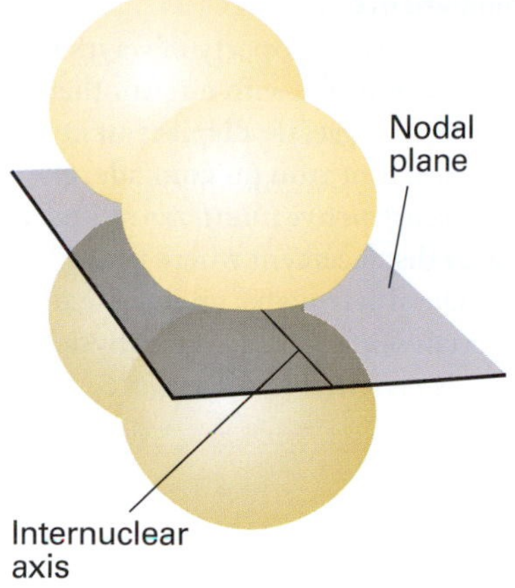

Fig. 4.4 A π bond results from orbital overlap and spin pairing between electrons in p orbitals with their axes perpendicular to the internuclear axis. The bond has two lobes of electron density separated by a nodal plane.

4.2 Polyatomic molecules

Each σ bond in a polyatomic molecule is formed by the spin pairing of electrons in atomic orbitals with cylindrical symmetry about the relevant internuclear axis. Likewise, π bonds are formed by pairing electrons that occupy atomic orbitals of the appropriate symmetry.

The VB description of H_2O will make this clear. The valence electron configuration of an O atom is $2s^22p_x^22p_y^12p_z^1$. The two unpaired electrons in the O$2p$ orbitals can each pair with an electron in an H$1s$ orbital, and each combination results in the formation of a σ bond (each bond has cylindrical symmetry about the respective O—H internuclear axis). Because the $2p_y$ and $2p_z$ orbitals lie at 90° to each other, the two σ bonds also lie at 90° to each other (Fig. 4.6). We can predict, therefore, that H_2O should be an angular molecule, which it is. However, the theory predicts a bond angle of 90°, whereas the actual bond angle is 104.5°.

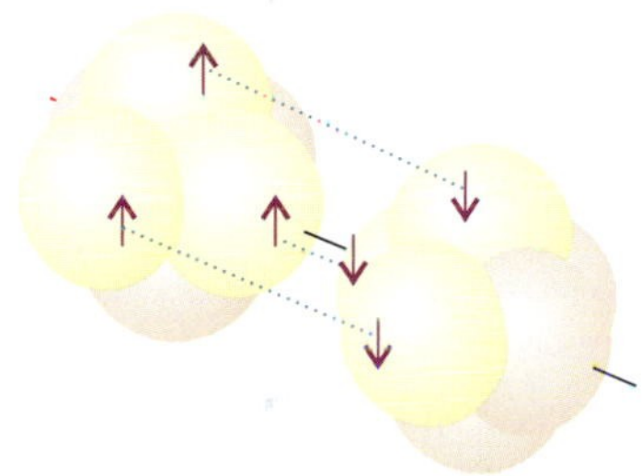

Fig. 4.5 The structure of bonds in a nitrogen molecule: there is one σ bond and two π bonds. As explained later, the overall electron density has cylindrical symmetry around the internuclear axis.

Self-test 4.1 Use valence-bond theory to suggest a shape for the ammonia molecule, NH_3.

[A trigonal pyramidal molecule with each N—H bond 90°; experimental: 107°]

Another deficiency of VB theory is its inability to account for carbon's tetravalence (its ability to form four bonds). The ground-state configuration of C is $2s^22p_x^12p_y^1$, which suggests that a carbon atom should be capable of forming only two bonds, not four. This deficiency is overcome by allowing for **promotion**, the excitation of an electron to an orbital of higher energy. In carbon, for example, the promotion of a $2s$ electron to a $2p$ orbital can be thought of as leading to the configuration $2s^12p_x^12p_y^12p_z^1$, with four unpaired electrons in separate orbitals. These electrons may pair with four electrons in orbitals provided by four other atoms (such as four H$1s$ orbitals if the molecule is CH_4), and hence form four σ bonds. Although energy was required to promote the electron, it is more than recovered by the promoted atom's ability to form four bonds in place of the two bonds of the unpromoted atom. Promotion, and the formation of four bonds, is a characteristic feature of carbon because the promotion energy is quite small: the promoted electron leaves a doubly occupied $2s$ orbital and enters a vacant $2p$ orbital, hence significantly relieving the electron–electron repulsion it experiences in the former. However, we need to remember that promotion is not a 'real' process in which an atom somehow becomes excited and then forms bonds: it is a notional contribution to the overall energy change that occurs when bonds form.

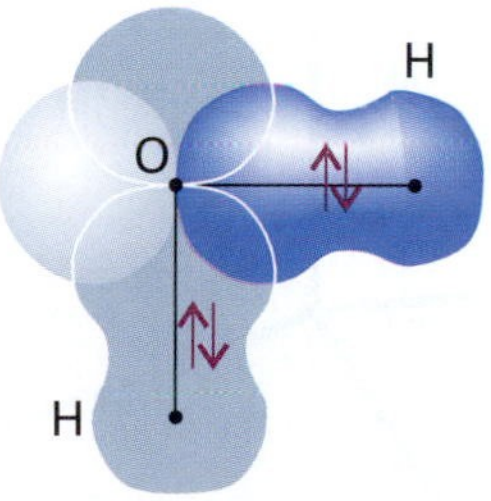

Fig. 4.6 A first approximation to the valence-bond description of bonding in an H_2O molecule. Each σ bond arises from the overlap of an H$1s$ orbital with one of the O$2p$ orbitals. This model suggests that the bond angle should be 90°, which is significantly different from the experimental value.

The description of the bonding in CH_4 (and other alkanes) is still incomplete because it implies the presence of three σ bonds of one type (formed from H$1s$ and C$2p$

Comment 4.2

A characteristic property of waves is that they interfere with one another, resulting in a greater displacement where peaks or troughs coincide, giving rise to *constructive interference*, and a smaller displacement where peaks coincide with troughs, giving rise to *destructive interference*. The physics of waves is reviewed in *Appendix 3*.

orbitals) and a fourth σ bond of a distinctly different character (formed from H1*s* and C2*s*). This problem is overcome by realizing that the electron density distribution in the promoted atom is equivalent to the electron density in which each electron occupies a **hybrid orbital** formed by interference between the C2*s* and C2*p* orbitals. The origin of the hybridization can be appreciated by thinking of the four atomic orbitals centred on a nucleus as waves that interfere destructively and constructively in different regions, and give rise to four new shapes.

The specific linear combinations that give rise to four equivalent hybrid orbitals are

$$h_1 = s + p_x + p_y + p_z \qquad h_2 = s - p_x - p_y + p_z$$
$$h_3 = s - p_x + p_y - p_z \qquad h_4 = s + p_x - p_y - p_z \tag{4.3}$$

As a result of the interference between the component orbitals, each hybrid orbital consists of a large lobe pointing in the direction of one corner of a regular tetrahedron (Fig. 4.7). The angle between the axes of the hybrid orbitals is the tetrahedral angle, 109.47°. Because each hybrid is built from one *s* orbital and three *p* orbitals, it is called an sp^3 **hybrid orbital**.

It is now easy to see how the valence-bond description of the CH_4 molecule leads to a tetrahedral molecule containing four equivalent C—H bonds. Each hybrid orbital of the promoted C atom contains a single unpaired electron; an H1*s* electron can pair with each one, giving rise to a σ bond pointing in a tetrahedral direction. For example, the (un-normalized) wavefunction for the bond formed by the hybrid orbital h_1 and the $1s_A$ orbital (with wavefunction that we shall denote A) is

$$\psi = h_1(1)A(2) + h_1(2)A(1)$$

Because each sp^3 hybrid orbital has the same composition, all four σ bonds are identical apart from their orientation in space (Fig. 4.8).

A hybrid orbital has enhanced amplitude in the internuclear region, which arises from the constructive interference between the *s* orbital and the positive lobes of the *p* orbitals (Fig. 4.9). As a result, the bond strength is greater than for a bond formed

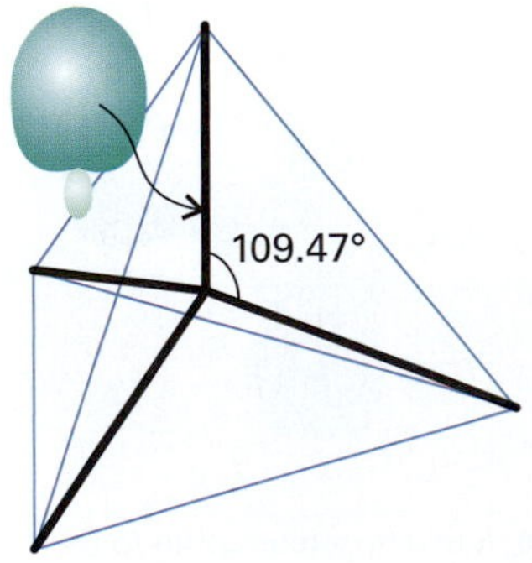

Fig. 4.7 An sp^3 hybrid orbital formed from the superposition of *s* and *p* orbitals on the same atom. There are four such hybrids: each one points towards the corner of a regular tetrahedron. The overall electron density remains spherically symmetrical.

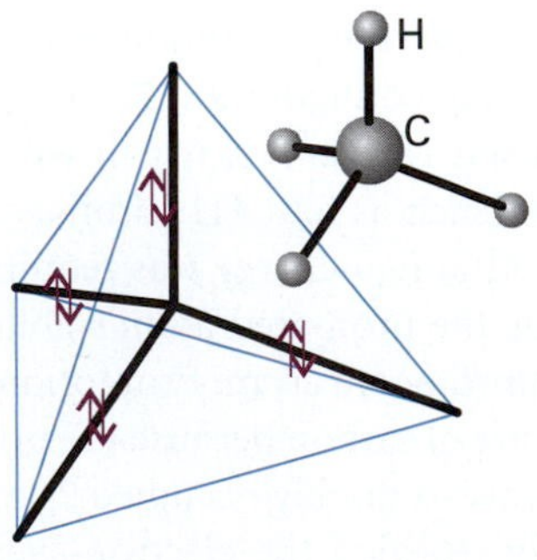

Fig. 4.8 Each sp^3 hybrid orbital forms a σ bond by overlap with an H1*s* orbital located at the corner of the tetrahedron. This model accounts for the equivalence of the four bonds in CH_4.

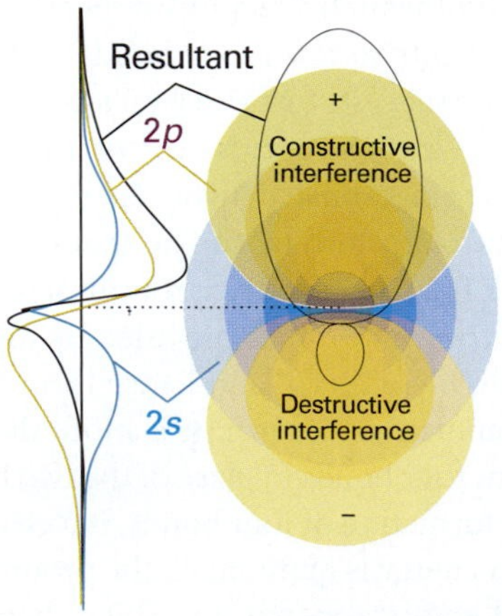

Fig. 4.9 A more detailed representation of the formation of an sp^3 hybrid by interference between wavefunctions centred on the same atomic nucleus. (To simplify the representation, we have ignored the radial node of the 2*s* orbital.)

from an s or p orbital alone. This increased bond strength is another factor that helps to repay the promotion energy.

Hybridization can also be used to describe the structure of an ethene molecule, $H_2C{=}CH_2$, and the torsional rigidity of double bonds. An ethene molecule is planar, with HCH and HCC bond angles close to 120°. To reproduce the σ bonding structure, we promote each C atom to a $2s^12p^3$ configuration. However, instead of using all four orbitals to form hybrids, we form ***sp*² hybrid orbitals**:

$$h_1 = s + 2^{1/2}p_y \qquad h_2 = s + (\tfrac{3}{2})^{1/2}p_x - (\tfrac{1}{2})^{1/2}p_y \qquad h_3 = s - (\tfrac{3}{2})^{1/2}p_x - (\tfrac{1}{2})^{1/2}p_y \qquad (4.4)$$

that lie in a plane and point towards the corners of an equilateral triangle (Fig. 4.10). The third $2p$ orbital ($2p_z$) is not included in the hybridization; its axis is perpendicular to the plane in which the hybrids lie. As always in superpositions, the proportion of each orbital in the mixture is given by the *square* of the corresponding coefficient. Thus, in the first of these hybrids the ratio of s to p contributions is 1:2. Similarly, the total p contribution in each of h_2 and h_3 is $\frac{3}{2} + \frac{1}{2} = 2$, so the ratio for these orbitals is also 1:2. The different signs of the coefficients ensure that constructive interference takes place in different regions of space, so giving the patterns in the illustration.

We can now describe the structure of $CH_2{=}CH_2$ as follows. The sp^2-hybridized C atoms each form three σ bonds by spin pairing with either the h_1 hybrid of the other C atom or with H1s orbitals. The σ framework therefore consists of C—H and C—C σ bonds at 120° to each other. When the two CH_2 groups lie in the same plane, the two electrons in the unhybridized p orbitals can pair and form a π bond (Fig. 4.11). The formation of this π bond locks the framework into the planar arrangement, for any rotation of one CH_2 group relative to the other leads to a weakening of the π bond (and consequently an increase in energy of the molecule).

A similar description applies to ethyne, HC≡CH, a linear molecule. Now the C atoms are ***sp* hybridized**, and the σ bonds are formed using hybrid atomic orbitals of the form

$$h_1 = s + p_z \qquad h_2 = s - p_z \qquad (4.5)$$

These two orbitals lie along the internuclear axis. The electrons in them pair either with an electron in the corresponding hybrid orbital on the other C atom or with an electron in one of the H1s orbitals. Electrons in the two remaining p orbitals on each atom, which are perpendicular to the molecular axis, pair to form two perpendicular π bonds (Fig. 4.12).

Self-test 4.2 Hybrid orbitals do not always form bonds. They may also contain lone pairs of electrons. Use valence-bond theory to suggest possible shapes for the hydrogen peroxide molecule, H_2O_2.

[Each H—O—O bond angle is predicted to be approximately 109° (experimental: 94.8°); rotation around the O—O bond is possible, so the molecule interconverts between planar and non-planar geometries at high temperatures.]

Other hybridization schemes, particularly those involving d orbitals, are often invoked in elementary work to be consistent with other molecular geometries (Table 4.1). The hybridization of N atomic orbitals always results in the formation of N hybrid orbitals, which may either form bonds or may contain lone pairs of electrons. For example, sp^3d^2 hybridization results in six equivalent hybrid orbitals pointing towards the corners of a regular octahedron and is sometimes invoked to account for the structure of octahedral molecules, such as SF_6.

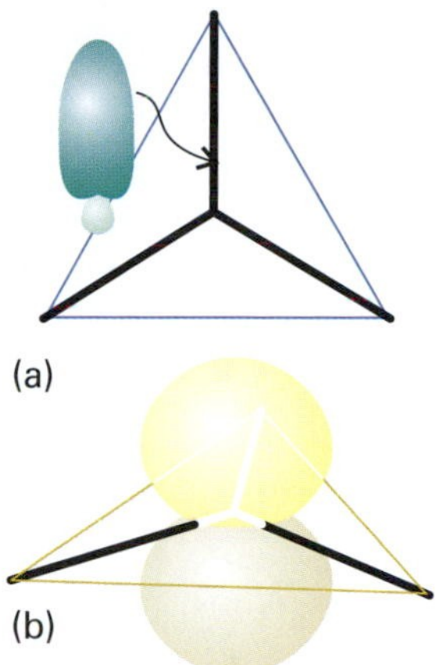

Fig. 4.10 (a) An s orbital and two p orbitals can be hybridized to form three equivalent orbitals that point towards the corners of an equilateral triangle. (b) The remaining, unhybridized p orbital is perpendicular to the plane.

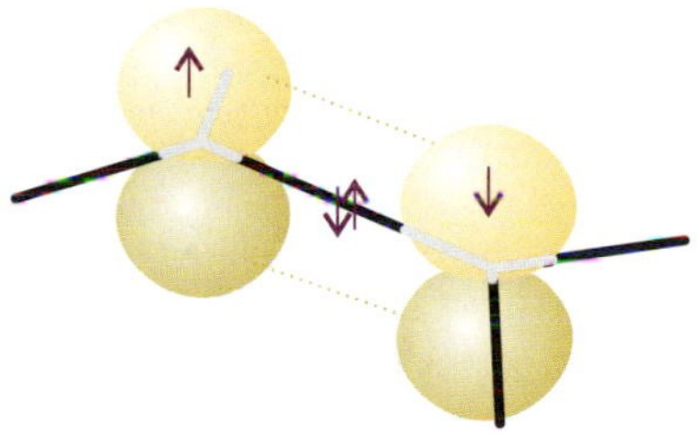

Fig. 4.11 A representation of the structure of a double bond in ethene; only the π bond is shown explicitly.

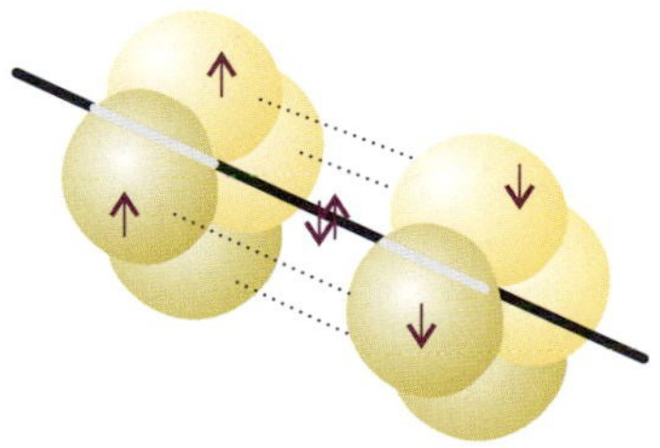

Fig. 4.12 A representation of the structure of a triple bond in ethyne; only the π bonds are shown explicitly. The overall electron density has cylindrical symmetry around the axis of the molecule.

Table 4.1* Some hybridization schemes

Coordination number	Arrangement	Composition
2	Linear	*sp*, *pd*, *sd*
	Angular	*sd*
3	Trigonal planar	sp^2, p^2d
	Unsymmetrical planar	*spd*
	Trigonal pyramidal	pd^2
4	Tetrahedral	sp^3, sd^3
	Irregular tetrahedral	spd^2, p^3d, dp^3
	Square planar	p^2d^2, sp^2d
5	Trigonal bipyramidal	sp^3d, spd^2
	Tetragonal pyramidal	sp^2d^2, sd^4, pd^4, p^3d^2
	Pentagonal planar	p^2d^3
6	Octahedral	sp^3d^2
	Trigonal prismatic	spd^4, pd^5
	Trigonal antiprismatic	p^3d^2

* Source: H. Eyring, J. Walter, and G.E. Kimball, *Quantum chemistry*, Wiley (1944).

Molecular orbital theory

In MO theory, it is accepted that electrons should not be regarded as belonging to particular bonds but should be treated as spreading throughout the entire molecule. This theory has been more fully developed than VB theory and provides the language that is widely used in modern discussions of bonding. To introduce it, we follow the same strategy as in Chapter 3, where the one-electron H atom was taken as the fundamental species for discussing atomic structure and then developed into a description of many-electron atoms. In this chapter we use the simplest molecular species of all, the hydrogen molecule-ion, H_2^+, to introduce the essential features of bonding, and then use it as a guide to the structures of more complex systems. To that end, we will progress to **homonuclear diatomic molecules**, which, like the H_2^+ molecule-ion, are formed from two atoms of the same element, then describe **heteronuclear diatomic molecules**, which are diatomic molecules formed from atoms of two different elements (such as CO and HCl), and end with a treatment of polyatomic molecules that forms the basis for modern computational models of molecular structure and chemical reactivity.

4.3 The hydrogen molecule-ion

The hamiltonian for the single electron in H_2^+ is

$$H = -\frac{\hbar^2}{2m_e}\nabla_1^2 + V \qquad V = -\frac{e^2}{4\pi\varepsilon_0}\left(\frac{1}{r_{A1}} + \frac{1}{r_{B1}} - \frac{1}{R}\right) \tag{4.6}$$

where r_{A1} and r_{B1} are the distances of the electron from the two nuclei (**1**) and R is the distance between the two nuclei. In the expression for V, the first two terms in parentheses are the attractive contribution from the interaction between the electron and the nuclei; the remaining term is the repulsive interaction between the nuclei.

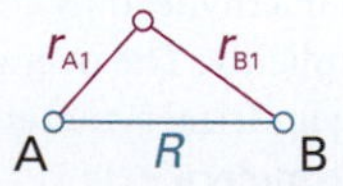

1

The one-electron wavefunctions obtained by solving the Schrödinger equation $H\psi = E\psi$ are called **molecular orbitals** (MO). A molecular orbital ψ gives, through the value of $|\psi|^2$, the distribution of the electron in the molecule. A molecular orbital is like an atomic orbital, but spreads throughout the molecule.

The Schrödinger equation can be solved analytically for H_2^+ (within the Born–Oppenheimer approximation), but the wavefunctions are very complicated functions; moreover, the solution cannot be extended to polyatomic systems. Therefore, we adopt a simpler procedure that, while more approximate, can be extended readily to other molecules.

(a) Linear combinations of atomic orbitals

If an electron can be found in an atomic orbital belonging to atom A and also in an atomic orbital belonging to atom B, then the overall wavefunction is a superposition of the two atomic orbitals:

$$\psi_{\pm} = N(A \pm B) \tag{4.7}$$

where, for H_2^+, A denotes χ_{H1s_A}, B denotes χ_{H1s_B}, and N is a normalization factor. The technical term for the superposition in eqn 4.7 is a **linear combination of atomic orbitals** (LCAO). An approximate molecular orbital formed from a linear combination of atomic orbitals is called an **LCAO-MO**. A molecular orbital that has cylindrical symmetry around the internuclear axis, such as the one we are discussing, is called a **σ orbital** because it resembles an *s* orbital when viewed along the axis and, more precisely, because it has zero orbital angular momentum around the internuclear axis.

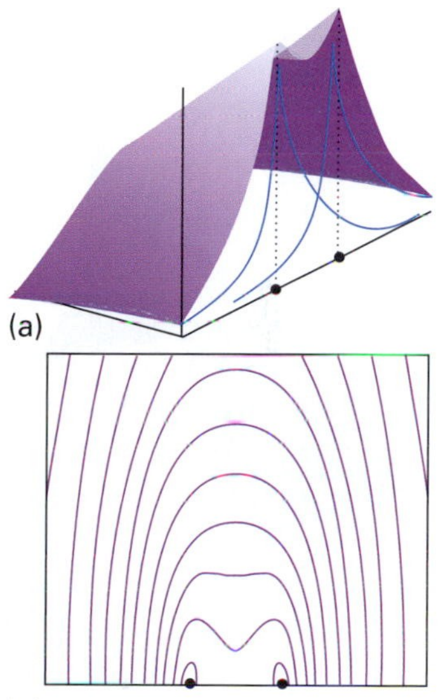

Fig. 4.13 (a) The amplitude of the bonding molecular orbital in a hydrogen molecule-ion in a plane containing the two nuclei and (b) a contour representation of the amplitude.

Exploration Plot the 1σ orbital for different values of the internuclear distance. Point to the features of the 1σ orbital that lead to bonding.

Example 4.1 *Normalizing a molecular orbital*

Normalize the molecular orbital ψ_+ in eqn 4.7.

Method We need to find the factor N such that

$$\int \psi^* \psi \, d\tau = 1$$

To proceed, substitute the LCAO into this integral, and make use of the fact that the atomic orbitals are individually normalized.

Answer When we substitute the wavefunction, we find

$$\int \psi^* \psi \, d\tau = N^2 \left\{ \int A^2 d\tau + \int B^2 d\tau + 2\int AB \, d\tau \right\} = N^2(1+1+2S)$$

where $S = \int AB \, d\tau$. For the integral to be equal to 1, we require

$$N = \frac{1}{\{2(1+S)\}^{1/2}}$$

In H_2^+, $S \approx 0.59$, so $N = 0.56$.

Self-test 4.3 Normalize the orbital ψ_- in eqn 4.7.

[$N = 1/\{2(1-S)\}^{1/2}$, so $N = 1.10$]

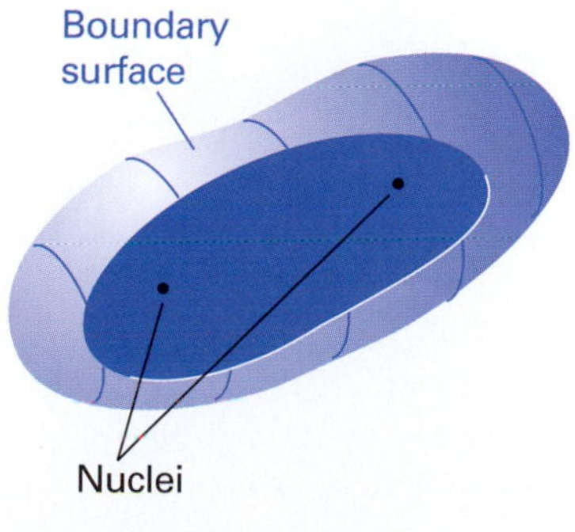

Fig. 4.14 A general indication of the shape of the boundary surface of a σ orbital.

Figure 4.13 shows the contours of constant amplitude for the two molecular orbitals in eqn 4.7, and Fig. 4.14 shows their boundary surfaces. Plots like these are readily obtained using commercially available software. The calculation is quite straightforward, because all we need do is feed in the mathematical forms of the two atomic orbitals and then let the program do the rest. In this case, we use

$$A = \frac{e^{-r_A/a_0}}{(\pi a_0^3)^{1/2}} \qquad B = \frac{e^{-r_B/a_0}}{(\pi a_0^3)^{1/2}} \tag{4.8}$$

and note that r_A and r_B are not independent (**2**), but related by the *law of cosines* (see *Comment 4.3*):

$$r_B = \{r_A^2 + R^2 - 2r_AR\cos\theta\}^{1/2} \tag{4.9}$$

To make this plot, we have taken $N^2 = 0.31$ (Example 4.1).

Comment 4.3

The *law of cosines* states that for a triangle such as that shown in (2) with sides r_A, r_B, and R, and angle θ facing side r_B we may write: $r_B^2 = r_A^2 + R^2 - 2r_AR\cos\theta$.

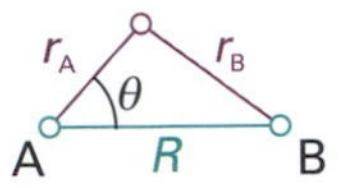

2

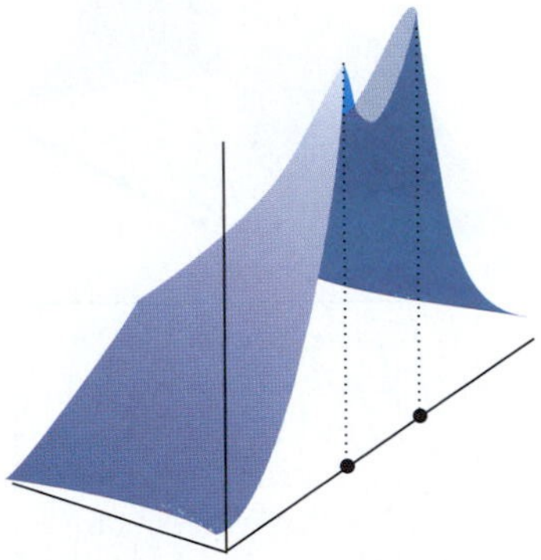

Fig. 4.15 The electron density calculated by forming the square of the wavefunction used to construct Fig. 4.13. Note the accumulation of electron density in the internuclear region.

(b) Bonding orbitals

According to the Born interpretation, the probability density of the electron in H_2^+ is proportional to the square modulus of its wavefunction. The probability density corresponding to the (real) wavefunction ψ_+ in eqn 4.7 is

$$\psi_+^2 = N^2(A^2 + B^2 + 2AB) \tag{4.10}$$

This probability density is plotted in Fig. 4.15.

An important feature of the probability density becomes apparent when we examine the internuclear region, where both atomic orbitals have similar amplitudes. According to eqn 4.10, the total probability density is proportional to the sum of

1 A^2, the probability density if the electron were confined to the atomic orbital A.

2 B^2, the probability density if the electron were confined to the atomic orbital B.

3 $2AB$, an extra contribution to the density.

This last contribution, the **overlap density**, is crucial, because it represents an enhancement of the probability of finding the electron in the internuclear region. The enhancement can be traced to the constructive interference of the two atomic orbitals: each has a positive amplitude in the internuclear region, so the total amplitude is greater there than if the electron were confined to a single atomic orbital.

We shall frequently make use of the result that *electrons accumulate in regions where atomic orbitals overlap and interfere constructively*. The conventional explanation is based on the notion that accumulation of electron density between the nuclei puts the electron in a position where it interacts strongly with both nuclei. Hence, the energy of the molecule is lower than that of the separate atoms, where each electron can interact strongly with only one nucleus. This conventional explanation, however, has been called into question, because shifting an electron away from a nucleus into the internuclear region *raises* its potential energy. The modern (and still controversial) explanation does not emerge from the simple LCAO treatment given here. It seems that, at the same time as the electron shifts into the internuclear region, the atomic orbitals shrink. This orbital shrinkage improves the electron–nucleus attraction more than it is decreased by the migration to the internuclear region, so there is a net lowering of potential energy. The kinetic energy of the electron is also modified because the curvature of the wavefunction is changed, but the change in kinetic energy is dominated by the change in potential energy. Throughout the following discussion we ascribe the strength of chemical bonds to the accumulation of electron density in the internuclear region. We leave open the question whether in molecules more complicated than H_2^+ the true source of energy lowering is that accumulation itself or some indirect but related effect.

The σ orbital we have described is an example of a **bonding orbital**, an orbital which, if occupied, helps to bind two atoms together. Specifically, we label it 1σ as it is the σ orbital of lowest energy. An electron that occupies a σ orbital is called a **σ electron**, and if that is the only electron present in the molecule (as in the ground state of H_2^+), then we report the configuration of the molecule as $1\sigma^1$.

The energy $E_{1\sigma}$ of the 1σ orbital is (see Problem 4.23):

$$E_{1\sigma} = E_{H1s} + \frac{e^2}{4\pi\varepsilon_0 R} - \frac{j+k}{1+S} \tag{4.11}$$

where

$$S = \int AB\,d\tau = \left\{1 + \frac{R}{a_0} + \frac{1}{3}\left(\frac{R}{a_0}\right)^2\right\} e^{-R/a_0} \tag{4.12a}$$

$$j = \frac{e^2}{4\pi\varepsilon_0}\int \frac{A^2}{r_B}\,d\tau = \frac{e^2}{4\pi\varepsilon_0 R}\left\{1 - \left(1 + \frac{R}{a_0}\right)e^{-2R/a_0}\right\} \tag{4.12b}$$

$$k = \frac{e^2}{4\pi\varepsilon_0}\int \frac{AB}{r_B}\,d\tau = \frac{e^2}{4\pi\varepsilon_0 a_0}\left(1 + \frac{R}{a_0}\right)e^{-R/a_0} \tag{4.12c}$$

We can interpret the preceding integrals as follows:

1 All three integrals are positive and decline towards zero at large internuclear separations (S and k on account of the exponential term, j on account of the factor $1/R$).

2 The integral j is a measure of the interaction between a nucleus and electron density centred on the other nucleus.

3 The integral k is a measure of the interaction between a nucleus and the excess probability in the internuclear region arising from overlap.

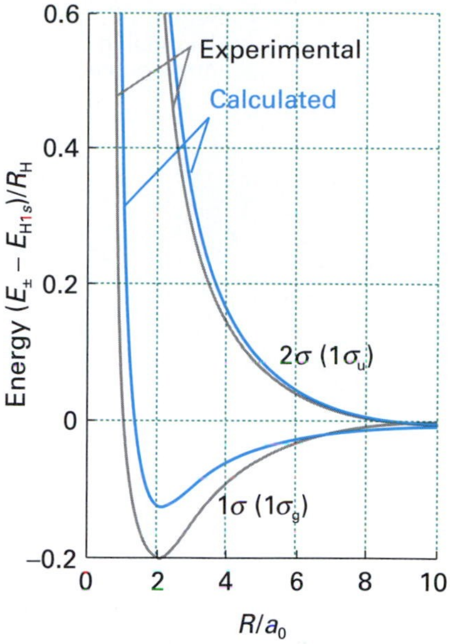

Fig. 4.16 The calculated and experimental molecular potential energy curves for a hydrogen molecule-ion showing the variation of the energy of the molecule as the bond length is changed. The alternative g,u notation is introduced in Section 4.3c.

Figure 4.16 is a plot of $E_{1\sigma}$ against R relative to the energy of the separated atoms. The energy of the 1σ orbital decreases as the internuclear separation decreases from large values because electron density accumulates in the internuclear region as the constructive interference between the atomic orbitals increases (Fig. 4.17). However, at small separations there is too little space between the nuclei for significant accumulation of electron density there. In addition, the nucleus–nucleus repulsion (which is proportional to $1/R$) becomes large. As a result, the energy of the molecule rises at short distances, and there is a minimum in the potential energy curve. Calculations on H_2^+ give $R_e = 130$ pm and $D_e = 1.77$ eV (171 kJ mol^{-1}); the experimental values are 106 pm and 2.6 eV, so this simple LCAO-MO description of the molecule, while inaccurate, is not absurdly wrong.

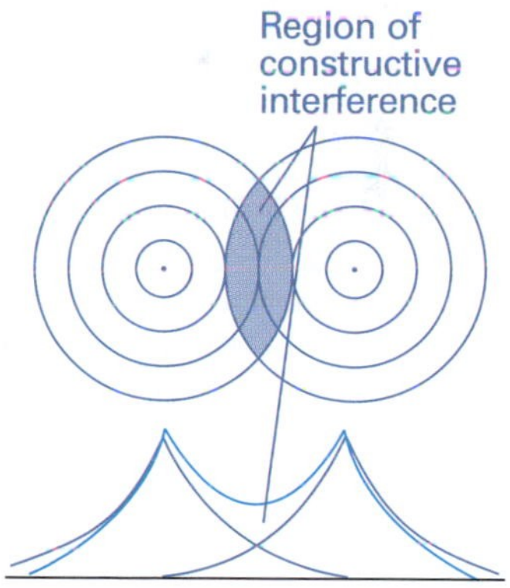

Fig. 4.17 A representation of the constructive interference that occurs when two H1s orbitals overlap and form a bonding σ orbital.

(c) Antibonding orbitals

The linear combination ψ_- in eqn 4.7 corresponds to a higher energy than that of ψ_+. Because it is also a σ orbital we label it 2σ. This orbital has an internuclear nodal plane where A and B cancel exactly (Figs. 4.18 and 4.19). The probability density is

$$\psi_-^2 = N^2(A^2 + B^2 - 2AB) \tag{4.13}$$

There is a reduction in probability density between the nuclei due to the $-2AB$ term (Fig. 4.20); in physical terms, there is destructive interference where the two atomic orbitals overlap. The 2σ orbital is an example of an **antibonding orbital**, an orbital that, if occupied, contributes to a reduction in the cohesion between two atoms and helps to raise the energy of the molecule relative to the separated atoms.

The energy $E_{2\sigma}$ of the 2σ antibonding orbital is given by (see Problem 4.23)

$$E_{2\sigma} = E_{H1s} + \frac{e^2}{4\pi\varepsilon_0 R} - \frac{j-k}{1-S} \tag{4.14}$$

where the integrals S, j, and k are given by eqn 4.12. The variation of $E_{2\sigma}$ with R is shown in Fig. 4.16, where we see the destabilizing effect of an antibonding electron.

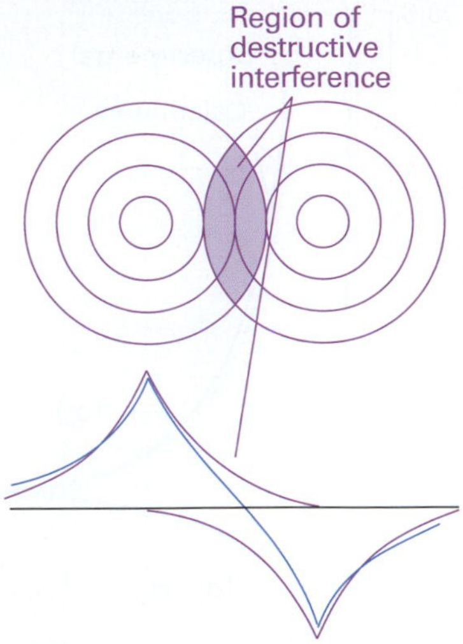

Fig. 4.18 A representation of the destructive interference that occurs when two H1*s* orbitals overlap and form an antibonding 2σ orbital.

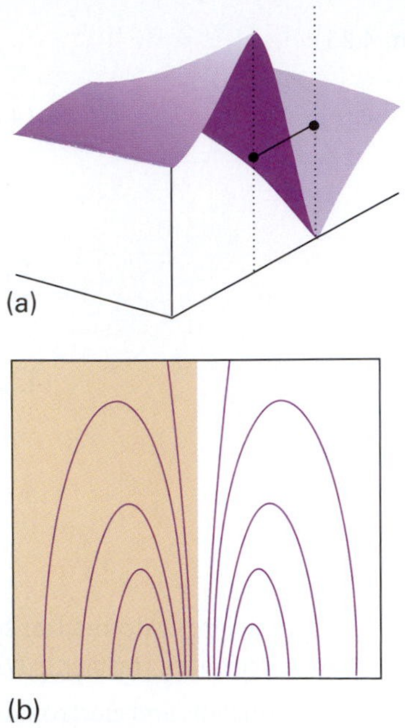

Fig. 4.19 (a) The amplitude of the antibonding molecular orbital in a hydrogen molecule-ion in a plane containing the two nuclei and (b) a contour representation of the amplitude. Note the internuclear node.

Exploration Plot the 2σ orbital for different values of the internuclear distance. Point to the features of the 2σ orbital that lead to antibonding.

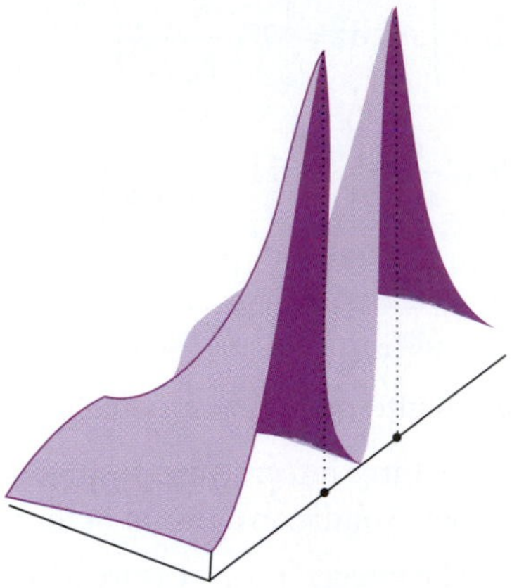

Fig. 4.20 The electron density calculated by forming the square of the wavefunction used to construct Fig. 4.19. Note the elimination of electron density from the internuclear region.

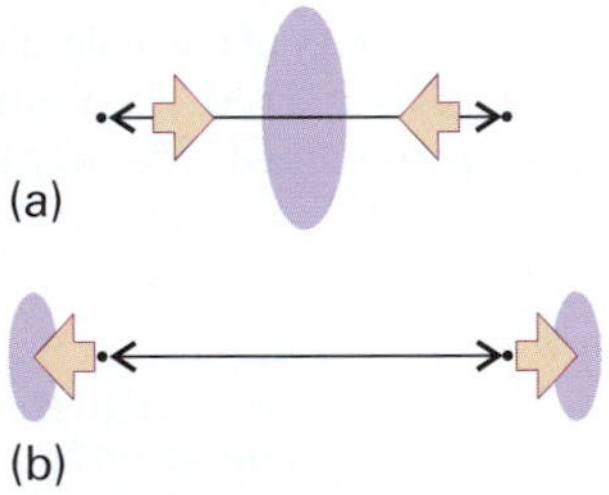

Fig. 4.21 A partial explanation of the origin of bonding and antibonding effects. (a) In a bonding orbital, the nuclei are attracted to the accumulation of electron density in the internuclear region. (b) In an antibonding orbital, the nuclei are attracted to an accumulation of electron density outside the internuclear region.

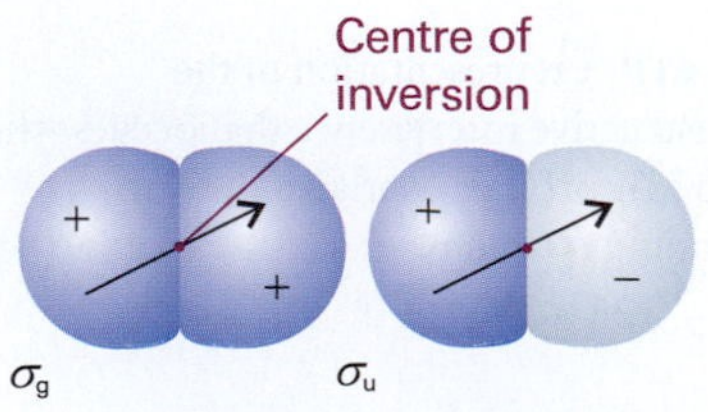

Fig. 4.22 The parity of an orbital is even (g) if its wavefunction is unchanged under inversion through the centre of symmetry of the molecule, but odd (u) if the wavefunction changes sign. Heteronuclear diatomic molecules do not have a centre of inversion, so for them the g, u classification is irrelevant.

The effect is partly due to the fact that an antibonding electron is excluded from the internuclear region, and hence is distributed largely outside the bonding region. In effect, whereas a bonding electron pulls two nuclei together, an antibonding electron pulls the nuclei apart (Fig. 4.21). Figure 4.16 also shows another feature that we draw on later: $|E_- - E_{H1s}| > |E_+ - E_{H1s}|$, which indicates that *the antibonding orbital is more antibonding than the bonding orbital is bonding*. This important conclusion stems in part from the presence of the nucleus–nucleus repulsion ($e^2/4\pi\varepsilon_0 R$): this contribution raises the energy of both molecular orbitals. Antibonding orbitals are often labelled with an asterisk (*), so the 2σ orbital could also be denoted 2σ* (and read '2 sigma star').

For homonuclear diatomic molecules, it is helpful to describe a molecular orbital by identifying its **inversion symmetry**, the behaviour of the wavefunction when it is inverted through the centre (more formally, the centre of inversion) of the molecule. Thus, if we consider any point on the bonding σ orbital, and then project it through the centre of the molecule and out an equal distance on the other side, then we arrive at an identical value of the wavefunction (Fig. 4.22). This so-called **gerade symmetry** (from the German word for 'even') is denoted by a subscript g, as in σ_g. On the other hand, the same procedure applied to the antibonding 2σ orbital results in the same size but opposite sign of the wavefunction. This **ungerade symmetry** ('odd symmetry') is denoted by a subscript u, as in σ_u. This inversion symmetry classification is not applicable to diatomic molecules formed by atoms from two different elements (such as CO) because these molecules do not have a centre of inversion. When using the g,u notation, each set of orbitals of the same inversion symmetry are labelled separately so, whereas 1σ becomes $1\sigma_g$, its antibonding partner, which so far we have called 2σ, is the first orbital of a different symmetry, and is denoted $1\sigma_u$. The general rule is that *each set of orbitals of the same symmetry designation is labelled separately*.

Comment 4.4

When treating homonuclear diatomic molecules, we shall favour the more modern notation that focuses attention on the symmetry properties of the orbital. For all other molecules, we shall use asterisks from time to time to denote antibonding orbitals.

4.4 Homonuclear diatomic molecules

In Chapter 3 we used the hydrogenic atomic orbitals and the building-up principle to deduce the ground electronic configurations of many-electron atoms. We now do the same for many-electron diatomic molecules by using the H_2^+ molecular orbitals. The general procedure is to construct molecular orbitals by combining the available atomic orbitals. The electrons supplied by the atoms are then accommodated in the orbitals so as to achieve the lowest overall energy subject to the constraint of the Pauli exclusion principle, that no more than two electrons may occupy a single orbital (and then must be paired). As in the case of atoms, if several degenerate molecular orbitals are available, we add the electrons singly to each individual orbital before doubly occupying any one orbital (because that minimizes electron–electron repulsions). We also take note of Hund's maximum multiplicity rule (Section 3.4) that, if electrons do occupy different degenerate orbitals, then a lower energy is obtained if they do so with parallel spins.

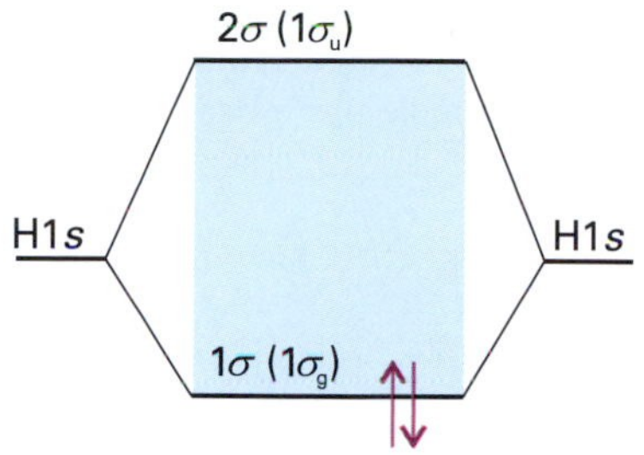

Fig. 4.23 A molecular orbital energy level diagram for orbitals constructed from the overlap of H1*s* orbitals; the separation of the levels corresponds to that found at the equilibrium bond length. The ground electronic configuration of H_2 is obtained by accommodating the two electrons in the lowest available orbital (the bonding orbital).

(a) σ orbitals

Consider H_2, the simplest many-electron diatomic molecule. Each H atom contributes a 1*s* orbital (as in H_2^+), so we can form the $1\sigma_g$ and $1\sigma_u$ orbitals from them, as we have seen already. At the experimental internuclear separation these orbitals will have the energies shown in Fig. 4.23, which is called a **molecular orbital energy level diagram**. Note that from two atomic orbitals we can build two molecular orbitals. In general, from *N* atomic orbitals we can build *N* molecular orbitals.

There are two electrons to accommodate, and both can enter $1\sigma_g$ by pairing their spins, as required by the Pauli principle (see the following *Justification*). The ground-state configuration is therefore $1\sigma_g^2$ and the atoms are joined by a bond consisting of an electron pair in a bonding σ orbital. This approach shows that an electron pair, which was the focus of Lewis's account of chemical bonding, represents the maximum number of electrons that can enter a bonding molecular orbital.

Justification 4.2 *Electron pairing in MO theory*

The spatial wavefunction for two electrons in a bonding molecular orbital ψ such as the bonding orbital in eqn 4.7, is $\psi(1)\psi(2)$. This two-electron wavefunction is obviously symmetric under interchange of the electron labels. To satisfy the Pauli principle, it must be multiplied by the antisymmetric spin state $\alpha(1)\beta(2)-\beta(1)\alpha(2)$ to give the overall antisymmetric state

$$\psi(1,2)=\psi(1)\psi(2)\{\alpha(1)\beta(2)-\beta(1)\alpha(2)\}$$

Because $\alpha(1)\beta(2)-\beta(1)\alpha(2)$ corresponds to paired electron spins, we see that two electrons can occupy the same molecular orbital (in this case, the bonding orbital) only if their spins are paired.

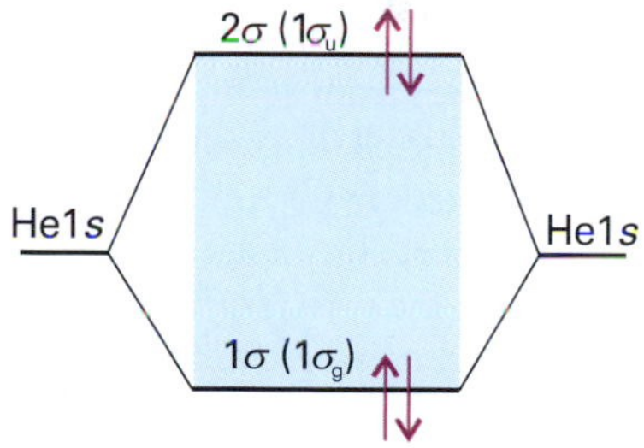

Fig. 4.24 The ground electronic configuration of the hypothetical four-electron molecule He_2 has two bonding electrons and two antibonding electrons. It has a higher energy than the separated atoms, and so is unstable.

The same argument shows why He does not form diatomic molecules. Each He atom contributes a 1*s* orbital, so $1\sigma_g$ and $1\sigma_u$ molecular orbitals can be constructed. Although these orbitals differ in detail from those in H_2, the general shape is the same, and we can use the same qualitative energy level diagram in the discussion. There are four electrons to accommodate. Two can enter the $1\sigma_g$ orbital, but then it is full, and the next two must enter the $1\sigma_u$ orbital (Fig. 4.24). The ground electronic configuration

Comment 4.5

Diatomic helium 'molecules' have been prepared: they consist of pairs of atoms held together by weak van der Waals forces.

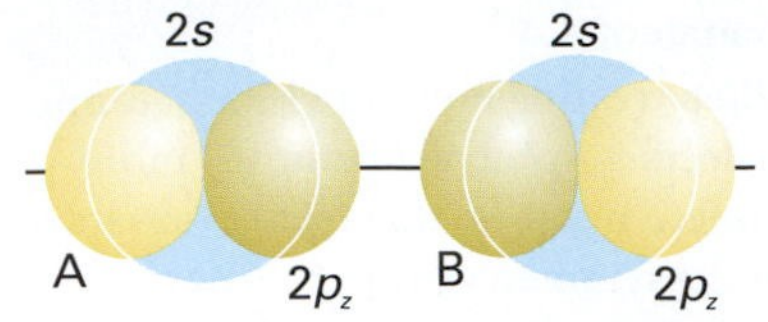

Fig. 4.25 According to molecular orbital theory, σ orbitals are built from all orbitals that have the appropriate symmetry. In homonuclear diatomic molecules of Period 2, that means that two 2*s* and two $2p_z$ orbitals should be used. From these four orbitals, four molecular orbitals can be built.

of He_2 is therefore $1\sigma_g^2 1\sigma_u^2$. We see that there is one bond and one antibond. Because an antibond is slightly more antibonding than a bond is bonding, an He_2 molecule has a higher energy than the separated atoms, so it is unstable relative to the individual atoms.

We shall now see how the concepts we have introduced apply to homonuclear diatomic molecules in general. In elementary treatments, only the orbitals of the valence shell are used to form molecular orbitals so, for molecules formed with atoms from Period 2 elements, only the 2*s* and 2*p* atomic orbitals are considered.

A general principle of molecular orbital theory is that *all orbitals of the appropriate symmetry contribute to a molecular orbital.* Thus, to build σ orbitals, we form linear combinations of all atomic orbitals that have cylindrical symmetry about the internuclear axis. These orbitals include the 2*s* orbitals on each atom and the $2p_z$ orbitals on the two atoms (Fig. 4.25). The general form of the σ orbitals that may be formed is therefore

$$\psi = c_{A2s}\chi_{A2s} + c_{B2s}\chi_{B2s} + c_{A2p_z}\chi_{A2p_z} + c_{B2p_z}\chi_{B2p_z} \tag{4.15}$$

From these four atomic orbitals we can form four molecular orbitals of σ symmetry by an appropriate choice of the coefficients *c*.

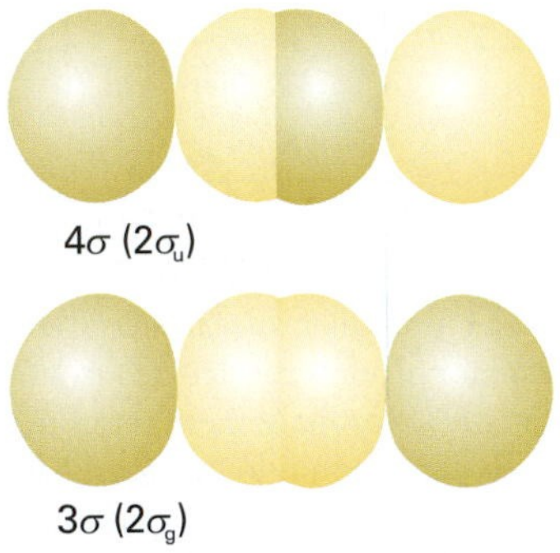

Fig. 4.26 A representation of the composition of bonding and antibonding σ orbitals built from the overlap of *p* orbitals. These illustrations are schematic.

The procedure for calculating the coefficients will be described in Section 4.6. At this stage we adopt a simpler route, and suppose that, because the 2*s* and $2p_z$ orbitals have distinctly different energies, they may be treated separately. That is, the four σ orbitals fall approximately into two sets, one consisting of two molecular orbitals of the form

$$\psi = c_{A2s}\chi_{A2s} + c_{B2s}\chi_{B2s} \tag{4.16a}$$

and another consisting of two orbitals of the form

$$\psi = c_{A2p_z}\chi_{A2p_z} + c_{B2p_z}\chi_{B2p_z} \tag{4.16b}$$

Comment 4.6

Note that we number only the molecular orbitals formed from atomic orbitals in the valence shell. In an alternative system of notation, $1\sigma_g$ and $1\sigma_u$ are used to designate the molecular orbitals formed from the core 1*s* orbitals of the atoms; the orbitals we are considering would then be labelled starting from 2.

Because atoms A and B are identical, the energies of their 2*s* orbitals are the same, so the coefficients are equal (apart from a possible difference in sign); the same is true of the $2p_z$ orbitals. Therefore, the two sets of orbitals have the form $\chi_{A2s} \pm \chi_{B2s}$ and $\chi_{A2p_z} \pm \chi_{B2p_z}$.

The 2*s* orbitals on the two atoms overlap to give a bonding and an antibonding σ orbital ($1\sigma_g$ and $1\sigma_u$, respectively) in exactly the same way as we have already seen for 1*s* orbitals. The two $2p_z$ orbitals directed along the internuclear axis overlap strongly. They may interfere either constructively or destructively, and give a bonding or antibonding σ orbital, respectively (Fig. 4.26). These two σ orbitals are labelled $2\sigma_g$ and $2\sigma_u$, respectively. In general, note how the numbering follows the order of increasing energy.

(b) π orbitals

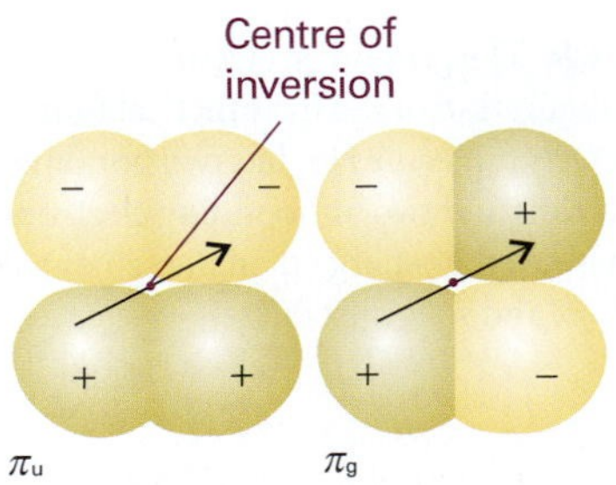

Fig. 4.27 A schematic representation of the structure of π bonding and antibonding molecular orbitals. The figure also shows that the bonding π orbital has odd parity, whereas the antiboding π orbital has even parity.

Now consider the $2p_x$ and $2p_y$ orbitals of each atom. These orbitals are perpendicular to the internuclear axis and may overlap broadside-on. This overlap may be constructive or destructive, and results in a bonding or an antibonding **π orbital** (Fig. 4.27). The notation π is the analogue of *p* in atoms, for when viewed along the axis of the molecule, a π orbital looks like a *p* orbital, and has one unit of orbital angular momentum around the internuclear axis. The two $2p_x$ orbitals overlap to give a bonding and antibonding π_x orbital, and the two $2p_y$ orbitals overlap to give two π_y orbitals. The π_x and π_y bonding orbitals are degenerate; so too are their antibonding partners. We also see from Fig. 4.27 that a bonding π orbital has odd parity and is denoted π_u and an antibonding π orbital has even parity, denoted π_g.

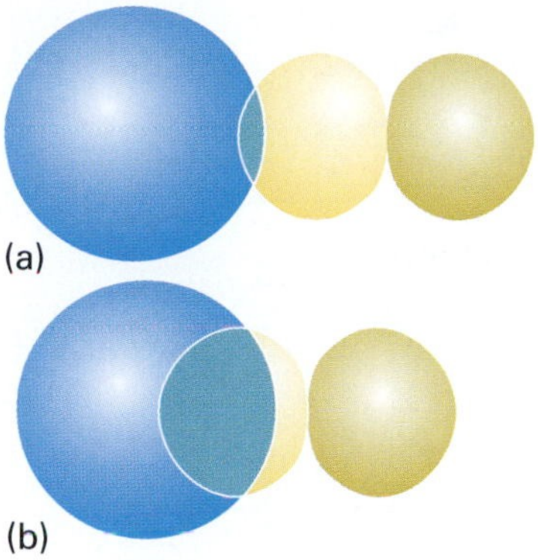

Fig. 4.28 (a) When two orbitals are on atoms that are far apart, the wavefunctions are small where they overlap, so S is small. (b) When the atoms are closer, both orbitals have significant amplitudes where they overlap, and S may approach 1. Note that S will decrease again as the two atoms approach more closely than shown here, because the region of negative amplitude of the p orbital starts to overlap the positive overlap of the s orbital. When the centres of the atoms coincide, $S = 0$.

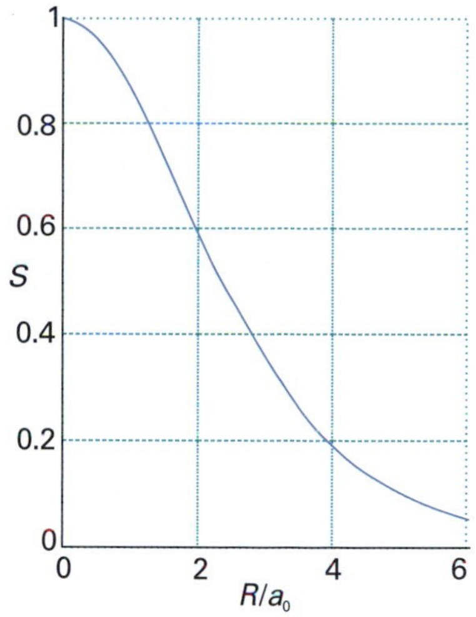

Fig. 4.29 The overlap integral, S, between two H1s orbitals as a function of their separation R.

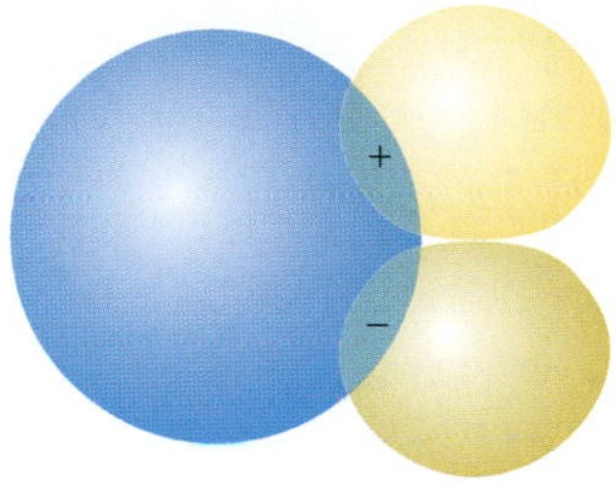

Fig. 4.30 A p orbital in the orientation shown here has zero net overlap ($S = 0$) with the s orbital at all internuclear separations.

(c) The overlap integral

The extent to which two atomic orbitals on different atoms overlap is measured by the **overlap integral**, S:

$$S = \int \chi_A^* \chi_B \, d\tau \qquad [4.17]$$

If the atomic orbital χ_A on A is small wherever the orbital χ_B on B is large, or vice versa, then the product of their amplitudes is everywhere small and the integral—the sum of these products—is small (Fig. 4.28). If χ_A and χ_B are simultaneously large in some region of space, then S may be large. If the two normalized atomic orbitals are identical (for instance, 1s orbitals on the same nucleus), then $S = 1$. In some cases, simple formulas can be given for overlap integrals and the variation of S with bond length plotted (Fig. 4.29). It follows that $S = 0.59$ for two H1s orbitals at the equilibrium bond length in H_2^+, which is an unusually large value. Typical values for orbitals with $n = 2$ are in the range 0.2 to 0.3.

Now consider the arrangement in which an s orbital is superimposed on a p_x orbital of a different atom (Fig. 4.30). The integral over the region where the product of orbitals is positive exactly cancels the integral over the region where the product of orbitals is negative, so overall $S = 0$ exactly. Therefore, there is no net overlap between the s and p orbitals in this arrangement.

(d) The electronic structures of homonuclear diatomic molecules

To construct the molecular orbital energy level diagram for Period 2 homonuclear diatomic molecules, we form eight molecular orbitals from the eight valence shell orbitals (four from each atom). In some cases, π orbitals are less strongly bonding than σ orbitals because their maximum overlap occurs off-axis. This relative weakness suggests that the molecular orbital energy level diagram ought to be as shown in Fig. 4.31. However, we must remember that we have assumed that 2s and 2p_z orbitals

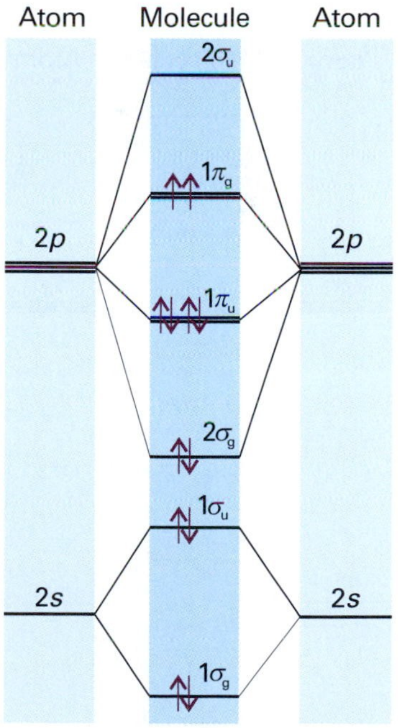

Fig. 4.31 The molecular orbital energy level diagram for homonuclear diatomic molecules. The lines in the middle are an indication of the energies of the molecular orbitals that can be formed by overlap of atomic orbitals. As remarked in the text, this diagram should be used for O_2 (the configuration shown) and F_2.

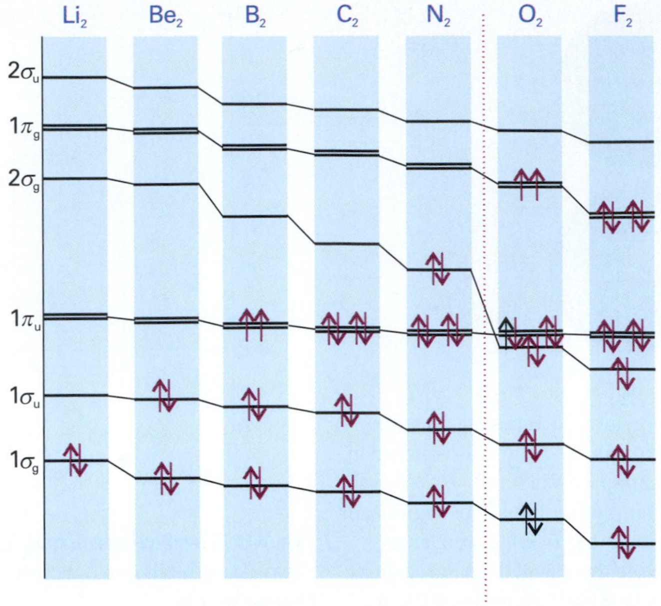

Fig. 4.32 The variation of the orbital energies of Period 2 homonuclear diatomics.

contribute to different sets of molecular orbitals whereas in fact all four atomic orbitals contribute jointly to the four σ orbitals. Hence, there is no guarantee that this order of energies should prevail, and it is found experimentally (by spectroscopy) and by detailed calculation that the order varies along Period 2 (Fig. 4.32). The order shown in Fig. 4.33 is appropriate as far as N_2, and Fig. 4.31 applies for O_2 and F_2. The relative order is controlled by the separation of the 2*s* and 2*p* orbitals in the atoms, which increases across the group. The consequent switch in order occurs at about N_2.

With the orbitals established, we can deduce the ground configurations of the molecules by adding the appropriate number of electrons to the orbitals and following the building-up rules. Anionic species (such as the peroxide ion, O_2^{2-}) need more electrons than the parent neutral molecules; cationic species (such as O_2^+) need fewer.

Consider N_2, which has 10 valence electrons. Two electrons pair, occupy, and fill the $1\sigma_g$ orbital; the next two occupy and fill the $1\sigma_u$ orbital. Six electrons remain. There are two $1\pi_u$ orbitals, so four electrons can be accommodated in them. The last two enter the $2\sigma_g$ orbital. Therefore, the ground-state configuration of N_2 is $1\sigma_g^2 1\sigma_u^2 1\pi_u^4 2\sigma_g^2$.

A measure of the net bonding in a diatomic molecule is its **bond order**, b:

$$b = \tfrac{1}{2}(n - n^{*}) \qquad [4.18]$$

where n is the number of electrons in bonding orbitals and n^{*} is the number of electrons in antibonding orbitals. Thus each electron pair in a bonding orbital increases the bond order by 1 and each pair in an antibonding orbital decreases b by 1. For H_2, $b = 1$, corresponding to a single bond, H—H, between the two atoms. In He_2, $b = 0$, and there is no bond. In N_2, $b = \tfrac{1}{2}(8 - 2) = 3$. This bond order accords with the Lewis structure of the molecule (:N≡N:).

The ground-state electron configuration of O_2, with 12 valence electrons, is based on Fig. 4.31, and is $1\sigma_g^2 1\sigma_u^2 2\sigma_g^2 1\pi_u^4 1\pi_g^2$. Its bond order is 2. According to the building-up principle, however, the two $1\pi_g$ electrons occupy different orbitals: one will enter $1\pi_{u,x}$ and the other will enter $1\pi_{u,y}$. Because the electrons are in different orbitals, they will have parallel spins. Therefore, we can predict that an O_2 molecule will have a net spin angular momentum $S = 1$ and, in the language introduced in Section 3.7, be in a triplet state. Because electron spin is the source of a magnetic moment, we can go on to predict that oxygen should be paramagnetic. This prediction, which VB theory does not make, is confirmed by experiment.

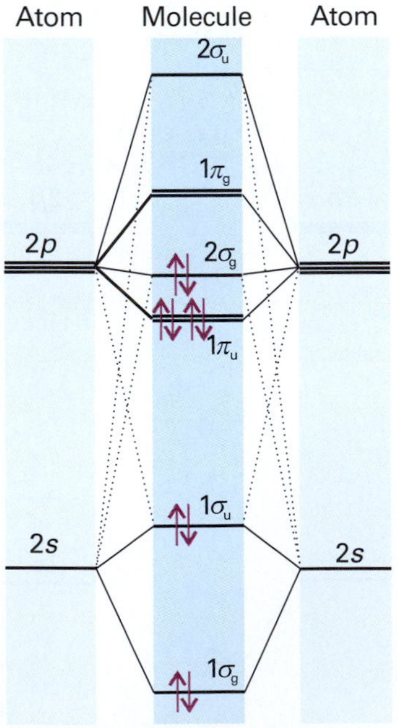

Fig. 4.33 An alternative molecular orbital energy level diagram for homonuclear diatomic molecules. As remarked in the text, this diagram should be used for diatomics up to and including N_2 (the configuration shown).

Synoptic table 4.2* Bond lengths

Bond	Order	R_e/pm
HH	1	74.14
NN	3	109.76
HCl	1	127.45
CH	1	114
CC	1	*154*
CC	2	*134*
CC	3	*120*

* More values will be found in the *Data section*. Numbers in italics are mean values for polyatomic molecules.

Synoptic table 4.3* Bond dissociation energies

Bond	Order	D_0/(kJ mol^{-1})
HH	1	432.1
NN	3	941.7
HCl	1	427.7
CH	1	*435*
CC	1	*368*
CC	2	*720*
CC	3	*962*

* More values will be found in the *Data section*. Numbers in italics are mean values for polyatomic molecules.

Comment 4.7

A paramagnetic substance tends to move into a magnetic field; a diamagnetic substance tends to move out of one. Paramagnetism, the rarer property, arises when the molecules have unpaired electron spins.

An F_2 molecule has two more electrons than an O_2 molecule. Its configuration is therefore $1\sigma_g^2 1\sigma_u^2 2\sigma_g^2 1\pi_u^4 1\pi_g^4$ and $b = 1$. We conclude that F_2 is a singly-bonded molecule, in agreement with its Lewis structure. The hypothetical molecule dineon, Ne_2, has two further electrons: its configuration is $1\sigma_g^2 1\sigma_u^2 2\sigma_g^2 1\pi_u^4 1\pi_g^2 2\sigma_u^2$ and $b = 0$. The zero bond order is consistent with the monatomic nature of Ne.

The bond order is a useful parameter for discussing the characteristics of bonds, because it correlates with bond length and bond strength. For bonds between atoms of a given pair of elements:

1 The greater the bond order, the shorter the bond.

2 The greater the bond order, the greater the bond strength.

Table 4.2 lists some typical bond lengths in diatomic and polyatomic molecules. The strength of a bond is measured by its bond dissociation energy, D_e, the energy required to separate the atoms to infinity. Table 4.3 lists some experimental values of dissociation energies.

Comment 4.8

Bond dissociation energies are commonly used in thermodynamic cycles, where bond enthalpies, $\Delta_{bond}H^{\ominus}$, should be used instead. It follows from the same kind of argument used in *Justification 3.7* concerning ionization enthalpies, that

$$X_2(g) \rightarrow 2\,X(g) \qquad \Delta_{bond}H^{\ominus}(T) = D_e + \tfrac{3}{2}RT$$

To derive this relation, we have supposed that the molar constant-pressure heat capacity of X_2 is $\tfrac{7}{2}R$ (Volume 1, *Molecular interpretation 2.2*) for there is a contribution from two rotational modes as well as three translational modes.

Example 4.2 *Judging the relative bond strengths of molecules and ions*

Judge whether N_2^+ is likely to have a larger or smaller dissociation energy than N_2.

Method Because the molecule with the larger bond order is likely to have the larger dissociation energy, compare their electronic configurations and assess their bond orders.

Answer From Fig. 4.33, the electron configurations and bond orders are

$$N_2 \qquad 1\sigma_g^2 1\sigma_u^2 1\pi_u^4 2\sigma_g^2 \qquad b = 3$$
$$N_2^+ \qquad 1\sigma_g^2 1\sigma_u^2 1\pi_u^4 2\sigma_g^1 \qquad b = 2\tfrac{1}{2}$$

Because the cation has the smaller bond order, we expect it to have the smaller dissociation energy. The experimental dissociation energies are 945 kJ mol^{-1} for N_2 and 842 kJ mol^{-1} for N_2^+.

Self-test 4.4 Which can be expected to have the higher dissociation energy, F_2 or F_2^+? [F_2^+]

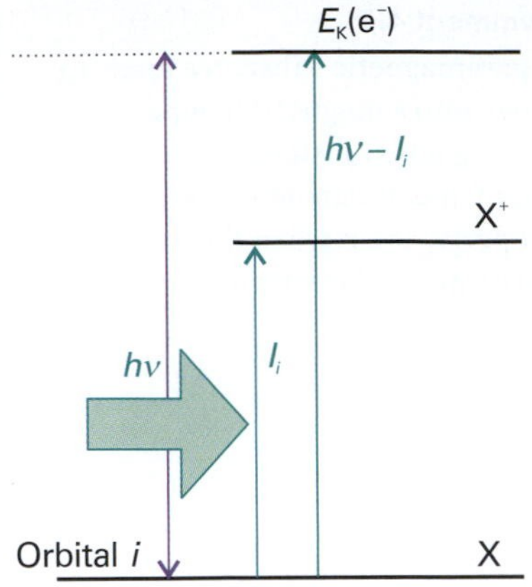

Fig. 4.34 An incoming photon carries an energy $h\nu$; an energy I_i is needed to remove an electron from an orbital i, and the difference appears as the kinetic energy of the electron.

(e) Photoelectron spectroscopy

So far we have treated molecular orbitals as purely theoretical constructs, but is there experimental evidence for their existence? **Photoelectron spectroscopy** (PES) measures the ionization energies of molecules when electrons are ejected from different orbitals by absorption of a photon of the proper energy, and uses the information to infer the energies of molecular orbitals. The technique is also used to study solids and their surfaces.

Because energy is conserved when a photon ionizes a sample, the energy of the incident photon $h\nu$ must be equal to the sum of the ionization energy, I, of the sample and the kinetic energy of the **photoelectron**, the ejected electron (Fig. 4.34):

$$h\nu = \tfrac{1}{2}m_e v^2 + I \qquad (4.19)$$

This equation (which is like the one used for the photoelectric effect, Section 1.2a) can be refined in two ways. First, photoelectrons may originate from one of a number of different orbitals, and each one has a different ionization energy. Hence, a series of different kinetic energies of the photoelectrons will be obtained, each one satisfying

$$h\nu = \tfrac{1}{2}m_e v^2 + I_i \qquad (4.20)$$

where I_i is the ionization energy for ejection of an electron from an orbital i. Therefore, by measuring the kinetic energies of the photoelectrons, and knowing ν, these ionization energies can be determined. Photoelectron spectra are interpreted in terms of an approximation called **Koopmans' theorem**, which states that the ionization energy I_i is equal to the orbital energy of the ejected electron (formally: $I_i = -\varepsilon_i$). That is, we can identify the ionization energy with the energy of the orbital from which it is ejected. Similarly, the energy of unfilled ('virtual orbitals') is related to the electron affinity. The theorem is only an approximation because it ignores the fact that the remaining electrons adjust their distributions when ionization occurs.

The ionization energies of molecules are several electronvolts even for valence electrons, so it is essential to work in at least the ultraviolet region of the spectrum and with wavelengths of less than about 200 nm. Much work has been done with radiation generated by a discharge through helium: the He(I) line ($1s^1 2p^1 \rightarrow 1s^2$) lies at 58.43 nm, corresponding to a photon energy of 21.22 eV. Its use gives rise to the technique of **ultraviolet photoelectron spectroscopy** (UPS). When core electrons are being studied, photons of even higher energy are needed to expel them: X–rays are used, and the technique is denoted XPS.

The kinetic energies of the photoelectrons are measured using an electrostatic deflector that produces different deflections in the paths of the photoelectrons as they pass between charged plates (Fig. 4.35). As the field strength is increased, electrons of different speeds, and therefore kinetic energies, reach the detector. The electron flux can be recorded and plotted against kinetic energy to obtain the photoelectron spectrum.

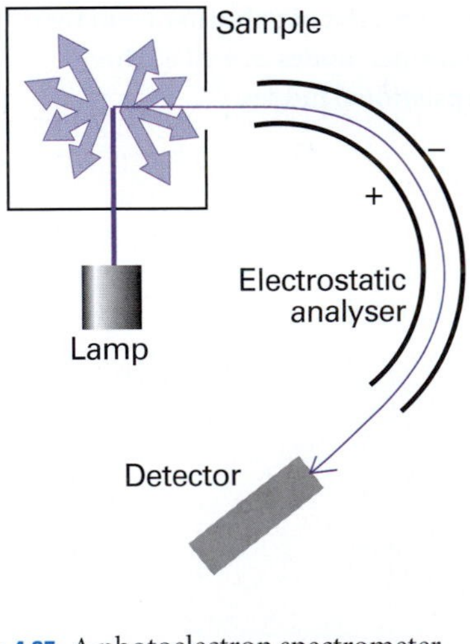

Fig. 4.35 A photoelectron spectrometer consists of a source of ionizing radiation (such as a helium discharge lamp for UPS and an X-ray source for XPS), an electrostatic analyser, and an electron detector. The deflection of the electron path caused by the analyser depends on their speed.

Illustration 4.1 *Interpreting a photoelectron spectrum*

Photoelectrons ejected from N_2 with He(I) radiation had kinetic energies of 5.63 eV (1 eV = 8065.5 cm^{-1}). Helium(I) radiation of wavelength 58.43 nm has wavenumber 1.711×10^5 cm^{-1} and therefore corresponds to an energy of 21.22 eV. Then, from eqn 4.20, 21.22 eV = 5.63 eV + I_i, so I_i = 15.59 eV. This ionization energy is the energy needed to remove an electron from the occupied molecular orbital with the highest energy of the N_2 molecule, the $2\sigma_g$ bonding orbital (see Fig. 4.33).

Self-test 4.5 Under the same circumstances, photoelectrons are also detected at 4.53 eV. To what ionization energy does that correspond? Suggest an origin.

[16.7 eV, $1\pi_u$]

4.5 Heteronuclear diatomic molecules

The electron distribution in the covalent bond between the atoms in a heteronuclear diatomic molecule is not shared evenly because it is energetically favourable for the electron pair to be found closer to one atom than the other. This imbalance results in a **polar bond**, a covalent bond in which the electron pair is shared unequally by the two atoms. The bond in HF, for instance, is polar, with the electron pair closer to the F atom. The accumulation of the electron pair near the F atom results in that atom having a net negative charge, which is called a **partial negative charge** and denoted $\delta-$. There is a matching **partial positive charge**, $\delta+$, on the H atom.

(a) Polar bonds

A polar bond consists of two electrons in an orbital of the form

$$\psi = c_A A + c_B B \tag{4.21}$$

with unequal coefficients. The proportion of the atomic orbital A in the bond is $|c_A|^2$ and that of B is $|c_B|^2$. A nonpolar bond has $|c_A|^2 = |c_B|^2$ and a pure ionic bond has one coefficient zero (so the species A^+B^- would have $c_A = 0$ and $c_B = 1$). The atomic orbital with the lower energy makes the larger contribution to the bonding molecular orbital. The opposite is true of the antibonding orbital, for which the dominant component comes from the atomic orbital with higher energy.

These points can be illustrated by considering HF, and judging the energies of the atomic orbitals from the ionization energies of the atoms. The general form of the molecular orbitals is

$$\psi = c_H \chi_H + c_F \chi_F \tag{4.22}$$

where χ_H is an H1s orbital and χ_F is an F2p orbital. The H1s orbital lies 13.6 eV below the zero of energy (the separated proton and electron) and the F2p orbital lies at 18.6 eV (Fig. 4.36). Hence, the bonding σ orbital in HF is mainly F2p and the antibonding σ orbital is mainly H1s orbital in character. The two electrons in the bonding orbital are most likely to be found in the F2p orbital, so there is a partial negative charge on the F atom and a partial positive charge on the H atom.

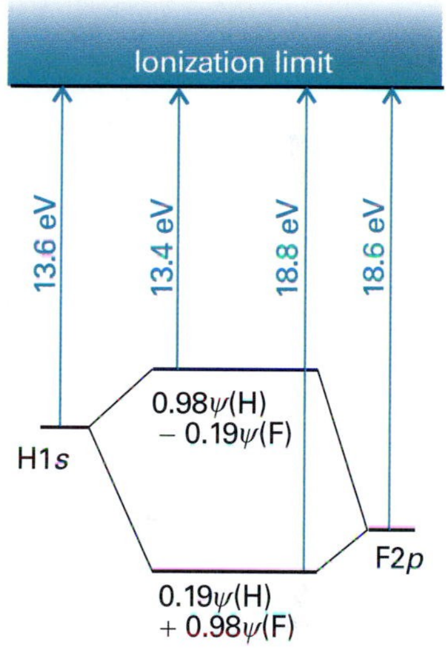

Fig. 4.36 The atomic orbital energy levels of H and F atoms and the molecular orbitals they form.

(b) Electronegativity

The charge distribution in bonds is commonly discussed in terms of the **electronegativity**, χ, of the elements involved (there should be little danger of confusing this use of χ with its use to denote an atomic orbital, which is another common convention). The electronegativity is a parameter introduced by Linus Pauling as a measure of the power of an atom to attract electrons to itself when it is part of a compound. Pauling used valence-bond arguments to suggest that an appropriate numerical scale of electronegativities could be defined in terms of bond dissociation energies, D, in electronvolts and proposed that the difference in electronegativities could be expressed as

$$|\chi_A - \chi_B| = 0.102\{D(\mathrm{A{-}B}) - \tfrac{1}{2}[D(\mathrm{A{-}A}) + D(\mathrm{B{-}B})]\}^{1/2} \tag{4.23}$$

Electronegativities based on this definition are called **Pauling electronegativities**. A list of Pauling electronegativities is given in Table 4.4. The most electronegative

Synoptic table 4.4* Pauling electronegativities

Element	χ_P
H	2.2
C	2.6
N	3.0
O	3.4
F	4.0
Cl	3.2
Cs	0.79

* More values will be found in the *Data section*.

elements are those close to fluorine; the least are those close to caesium. It is found that the greater the difference in electronegativities, the greater the polar character of the bond. The difference for HF, for instance, is 1.78; a C—H bond, which is commonly regarded as almost nonpolar, has an electronegativity difference of 0.51.

The spectroscopist Robert Mulliken proposed an alternative definition of electronegativity. He argued that an element is likely to be highly electronegative if it has a high ionization energy (so it will not release electrons readily) and a high electron affinity (so it is energetically favorable to acquire electrons). The **Mulliken electronegativity scale** is therefore based on the definition

$$\chi_M = \tfrac{1}{2}(I + E_{ea}) \qquad [4.24]$$

where I is the ionization energy of the element and E_{ea} is its electron affinity (both in electronvolts, Section 3.4e). The Mulliken and Pauling scales are approximately in line with each other. A reasonably reliable conversion between the two is $\chi_P = 1.35\chi_M^{1/2} - 1.37$.

(c) The variation principle

A more systematic way of discussing bond polarity and finding the coefficients in the linear combinations used to build molecular orbitals is provided by the **variation principle**:

> If an arbitrary wavefunction is used to calculate the energy, the value calculated is never less than the true energy.

This principle is the basis of all modern molecular structure calculations (Section 4.7). The arbitrary wavefunction is called the **trial wavefunction**. The principle implies that, if we vary the coefficients in the trial wavefunction until the lowest energy is achieved (by evaluating the expectation value of the hamiltonian for each wavefunction), then those coefficients will be the best. We might get a lower energy if we use a more complicated wavefunction (for example, by taking a linear combination of several atomic orbitals on each atom), but we shall have the optimum (minimum energy) molecular orbital that can be built from the chosen **basis set**, the given set of atomic orbitals.

Comment 4.9

The name 'secular' is derived from the Latin word for age or generation. The term comes via astronomy, where the same equations appear in connection with slowly accumulating modifications of planetary orbits.

The method can be illustrated by the trial wavefunction in eqn 4.21. We show in the *Justification* below that the coefficients are given by the solutions of the two **secular equations**

$$(\alpha_A - E)c_A + (\beta - ES)c_B = 0 \qquad (4.25a)$$

$$(\beta - ES)c_A + (\alpha_B - E)c_B = 0 \qquad (4.25b)$$

The parameter α is called a **Coulomb integral**. It is negative and can be interpreted as the energy of the electron when it occupies A (for α_A) or B (for α_B). In a homonuclear diatomic molecule, $\alpha_A = \alpha_B$. The parameter β is called a **resonance integral** (for classical reasons). It vanishes when the orbitals do not overlap, and at equilibrium bond lengths it is normally negative.

Justification 4.3 *The variation principle applied to a heteronuclear diatomic molecule*

The trial wavefunction in eqn 4.21 is real but not normalized because at this stage the coefficients can take arbitrary values. Therefore, we can write $\psi^* = \psi$ but do not assume that $\int \psi^2 d\tau = 1$. The energy of the trial wavefunction is the expectation value of the energy operator (the hamiltonian, $\hat{H}$, Section 1.5):

$$E = \frac{\int \psi^* \hat{H} \psi \, d\tau}{\int \psi^* \psi \, d\tau} \tag{4.26}$$

We must search for values of the coefficients in the trial function that minimize the value of E. This is a standard problem in calculus, and is solved by finding the coefficients for which

$$\frac{\partial E}{\partial c_A} = 0 \qquad \frac{\partial E}{\partial c_B} = 0$$

The first step is to express the two integrals in terms of the coefficients. The denominator is

$$\begin{aligned}\int \psi^2 \, d\tau &= \int (c_A A + c_B B)^2 \, d\tau \\ &= c_A^2 \int A^2 \, d\tau + c_B^2 \int B^2 \, d\tau + 2c_A c_B \int AB \, d\tau \\ &= c_A^2 + c_B^2 + 2c_A c_B S\end{aligned}$$

because the individual atomic orbitals are normalized and the third integral is the overlap integral S (eqn 4.17). The numerator is

$$\begin{aligned}\int \psi \hat{H} \psi \, d\tau &= \int (c_A A + c_B B) \hat{H} (c_A A + c_B B) \, d\tau \\ &= c_A^2 \int A \hat{H} A \, d\tau + c_B^2 \int B \hat{H} B \, d\tau + c_A c_B \int A \hat{H} B \, d\tau + c_A c_B \int B \hat{H} A \, d\tau\end{aligned}$$

There are some complicated integrals in this expression, but we can combine them all into the parameters

$$\alpha_A = \int A \hat{H} A \, d\tau \qquad \alpha_B = \int B \hat{H} B \, d\tau \tag{4.27}$$

$$\beta = \int A \hat{H} B \, d\tau = \int B \hat{H} A \, d\tau \, (\text{by the hermiticity of } \hat{H})$$

Then

$$\int \psi \hat{H} \psi \, d\tau = c_A^2 \alpha_A + c_B^2 \alpha_B + 2c_A c_B \beta$$

The complete expression for E is

$$E = \frac{c_A^2 \alpha_A + c_B^2 \alpha_B + 2c_A c_B \beta}{c_A^2 + c_B^2 + 2c_A c_B S} \tag{4.28}$$

Its minimum is found by differentiation with respect to the two coefficients and setting the results equal to 0. After a bit of work, we obtain

$$\frac{\partial E}{\partial c_A} = \frac{2 \times (c_A \alpha_A - c_A E + c_B \beta - c_B S E)}{c_A^2 + c_B^2 + 2c_A c_B S} = 0$$

$$\frac{\partial E}{\partial c_B} = \frac{2 \times (c_B \alpha_B - c_B E + c_A \beta - c_A S E)}{c_A^2 + c_B^2 + 2c_A c_B S} = 0$$

For the derivatives to vanish, the numerators of the expressions above must vanish. That is, we must find values of c_A and c_B that satisfy the conditions

$$c_A\alpha_A - c_AE + c_B\beta - c_BSE = (\alpha_A - E)c_A + (\beta - ES)c_B = 0$$
$$c_A\beta - c_ASE + c_B\alpha_B - c_BE = (\beta - ES)c_A + (\alpha_B - E)c_B = 0$$

which are the secular equations (eqn 4.25).

To solve the secular equations for the coefficients we need to know the energy E of the orbital. As for any set of simultaneous equations, the secular equations have a solution if the **secular determinant**, the determinant of the coefficients, is zero; that is, if

$$\begin{vmatrix} \alpha_A - E & \beta - ES \\ \beta - ES & \alpha_B - E \end{vmatrix} = 0 \qquad (4.29)$$

Comment 4.10
We need to know that a 2×2 determinant expands as follows:

$$\begin{vmatrix} a & b \\ c & d \end{vmatrix} = ad - bc$$

This determinant expands to a quadratic equation in E (see *Illustration 4.2*). Its two roots give the energies of the bonding and antibonding molecular orbitals formed from the atomic orbitals and, according to the variation principle, the lower root is the best energy achievable with the given basis set.

Illustration 4.2 *Using the variation principle (1)*

To find the energies E of the bonding and antibonding orbitals of a homonuclear diatomic molecule set with $\alpha_A = \alpha_B = \alpha$ in eqn 4.29 and get

$$\begin{vmatrix} \alpha - E & \beta - ES \\ \beta - ES & \alpha - E \end{vmatrix} = (\alpha - E)^2 - (\beta - ES)^2 = 0$$

The solutions of this equation are

$$E_\pm = \frac{\alpha \pm \beta}{1 \pm S}$$

The values of the coefficients in the linear combination are obtained by solving the secular equations using the two energies obtained from the secular determinant. The lower energy (E_+ in the *Illustration*) gives the coefficients for the bonding molecular orbital, the upper energy (E_-) the coefficients for the antibonding molecular orbital. The secular equations give expressions for the ratio of the coefficients in each case, so we need a further equation in order to find their individual values. This equation is obtained by demanding that the best wavefunction should also be normalized. This condition means that, at this final stage, we must also ensure that

$$\int \psi^2\, d\tau = c_A^2 + c_B^2 + 2c_Ac_BS = 1 \qquad (4.30)$$

Illustration 4.3 *Using the variation principle (2)*

To find the values of the coefficients c_A and c_B in the linear combination that corresponds to the energy E_+ from *Illustration 4.2*, we use eqn 4.28 (with $\alpha_A = \alpha_B = \alpha$) to write

$$E_+ = \frac{\alpha + \beta}{1 + S} = \frac{c_A^2\alpha + c_B^2\alpha + 2c_Ac_B\beta}{c_A^2 + c_B^2 + 2c_Ac_BS}$$

Now we use the normalization condition, eqn 4.30, to set $c_A^2 + c_B^2 + 2c_Ac_BS = 1$, and so write

$$\frac{\alpha+\beta}{1+S} = (c_A^2 + c_B^2)\alpha + 2c_Ac_B\beta$$

This expression implies that

$$c_A^2 + c_B^2 = 2c_Ac_B = \frac{1}{1+S} \quad \text{and} \quad |c_A| = \frac{1}{\{2(1+S)\}^{1/2}} \quad c_B = c_A$$

Proceeding in a similar way to find the coefficients in the linear combination that corresponds to the energy E_-,we write

$$E_- = \frac{\alpha-\beta}{1-S} = (c_A^2 + c_B^2)\alpha + 2c_Ac_B\beta$$

which implies that

$$c_A^2 + c_B^2 = -2c_Ac_B = \frac{1}{1-S} \quad \text{and} \quad |c_A| = \frac{1}{\{2(1-S)\}^{1/2}} \quad c_B = -c_A$$

(d) Two simple cases

The complete solutions of the secular equations are very cumbersome, even for 2×2 determinants, but there are two cases where the roots can be written down very simply.

We saw in *Illustrations* 4.2 and 4.3 that, when the two atoms are the same, and we can write $\alpha_A = \alpha_B = \alpha$, the solutions are

$$E_+ = \frac{\alpha+\beta}{1+S} \qquad c_A = \frac{1}{\{2(1+S)\}^{1/2}} \qquad c_B = c_A \tag{4.31a}$$

$$E_- = \frac{\alpha-\beta}{1-S} \qquad c_A = \frac{1}{\{2(1-S)\}^{1/2}} \qquad c_B = -c_A \tag{4.31b}$$

In this case, the bonding orbital has the form

$$\psi_+ = \frac{A+B}{\{2(1+S)\}^{1/2}} \tag{4.32a}$$

and the corresponding antibonding orbital is

$$\psi_- = \frac{A-B}{\{2(1-S)\}^{1/2}} \tag{4.32b}$$

in agreement with the discussion of homonuclear diatomics we have already given, but now with the normalization constant in place.

The second simple case is for a heteronuclear diatomic molecule but with $S = 0$ (a common approximation in elementary work). The secular determinant is then

$$\begin{vmatrix} \alpha_A - E & \beta \\ \beta & \alpha_B - E \end{vmatrix} = (\alpha_A - E)(\alpha_B - E) - \beta^2 = 0$$

The solutions can be expressed in terms of the parameter ζ (zeta), with

$$\zeta = \tfrac{1}{2}\arctan\frac{2|\beta|}{\alpha_B - \alpha_A} \tag{4.33}$$

and are

$$E_- = \alpha_B - \beta \tan \zeta \qquad \psi_- = -A \sin \zeta + B \cos \zeta \tag{4.34a}$$

$$E_+ = \alpha_A + \beta \tan \zeta \qquad \psi_+ = A \cos \zeta + B \sin \zeta \tag{4.34b}$$

An important feature revealed by these solutions is that as the energy difference $|\alpha_B - \alpha_A|$ between the interacting atomic orbitals increases, the value of ζ decreases. We show in the following *Justification* that, when the energy difference is very large, in the sense that $|\alpha_B - \alpha_A| \gg 2|\beta|$, the energies of the resulting molecular orbitals differ only slightly from those of the atomic orbitals, which implies in turn that the bonding and antibonding effects are small. That is, the strongest bonding and antibonding effects are obtained when the two contributing orbitals have closely similar energies. The difference in energy between core and valence orbitals is the justification for neglecting the contribution of core orbitals to bonding. The core orbitals of one atom have a similar energy to the core orbitals of the other atom; but core–core interaction is largely negligible because the overlap between them (and hence the value of β) is so small.

Justification 4.4 *Bonding and antibonding effects in heteronuclear diatomic molecules*

When $|\alpha_B - \alpha_A| \gg 2|\beta|$ and $2|\beta|/|\alpha_B - \alpha_A| \ll 1$, we can write $\arctan 2|\beta|/|\alpha_B - \alpha_A| \approx 2|\beta|/|\alpha_B - \alpha_A|$ and, from eqn 4.33, $\zeta \approx |\beta|/(\alpha_B - \alpha_A)$. It follows that $\tan \zeta \approx |\beta|/(\alpha_B - \alpha_A)$. Noting that β is normally a negative number, so that $\beta/|\beta| = -1$, we can use eqn 4.34 to write

$$E_- = \alpha_B + \frac{\beta^2}{\alpha_B - \alpha_A} \qquad E_+ = \alpha_A - \frac{\beta^2}{\alpha_B - \alpha_A}$$

Comment 4.11

For $x \ll 1$, we can write: $\sin x \approx x$, $\cos x \approx 1$, $\tan x \approx x$, and $\arctan x = \tan^{-1} x \approx x$.

(In Problem 4.25 you are invited to derive these expressions via a different route.) It follows that, when the energy difference between the atomic orbitals is so large that $|\alpha_B - \alpha_A| \gg 2|\beta|$, the energies of the two molecular orbitals are $E_- \approx \alpha_B$ and $E_+ \approx \alpha_A$.

Now we consider the behaviour of the wavefunctions in the limit of large $|\alpha_B - \alpha_A|$, when $\zeta \ll 1$. In this case, $\sin \zeta \approx \zeta$ and $\cos \zeta \approx 1$ and, from eqn 4.34, we write $\psi_- \approx B$ and $\psi_+ \approx A$. That is, the molecular orbitals are respectively almost pure B and almost pure A.

Example 4.3 *Calculating the molecular orbitals of HF*

Calculate the wavefunctions and energies of the σ orbitals in the HF molecule, taking $\beta = -1.0$ eV and the following ionization energies: H1*s*: 13.6 eV, F2*s*: 40.2 eV, F2*p*: 17.4 eV.

Method Because the F2*p* and H1*s* orbitals are much closer in energy than the F2*s* and H1*s* orbitals, to a first approximation neglect the contribution of the F2*s* orbital. To use eqn 4.34, we need to know the values of the Coulomb integrals α_H and α_F. Because these integrals represent the energies of the H1*s* and F2*p* electrons, respectively, they are approximately equal to (the negative of) the ionization energies of the atoms. Calculate ζ from eqn 4.33 (with A identified as F and B as H), and then write the wavefunctions by using eqn 4.34.

Answer Setting $\alpha_H = -13.6$ eV and $\alpha_F = -17.4$ eV gives $\tan 2\zeta = 0.58$; so $\zeta = 13.9°$. Then

$E_- = -13.4$ eV $\quad \psi_- = 0.97\chi_H - 0.24\chi_F$

$E_+ = -17.6$ eV $\quad \psi_+ = 0.24\chi_H + 0.97\chi_F$

Notice how the lower energy orbital (the one with energy −17.6 eV) has a composition that is more F2*p* orbital than H1*s*, and that the opposite is true of the higher energy, antibonding orbital.

Self-test 4.6 The ionization energy of Cl is 13.1 eV; find the form and energies of the σ orbitals in the HCl molecule using $\beta = -1.0$ eV.

[$E_- = -12.8$ eV, $\psi_- = -0.62\chi_H + 0.79\chi_{Cl}$; $E_+ = -13.9$ eV, $\psi_+ = 0.79\chi_H + 0.62\chi_{Cl}$]

IMPACT ON BIOCHEMISTRY

I4.1 The biochemical reactivity of O_2, N_2, and NO

We can now see how some of these concepts are applied to diatomic molecules that play a vital biochemical role. At sea level, air contains approximately 23.1 per cent O_2 and 75.5 per cent N_2 by mass. Molecular orbital theory predicts—correctly—that O_2 has unpaired electron spins and, consequently, is a reactive component of the Earth's atmosphere; its most important biological role is as an oxidizing agent. By contrast N_2, the major component of the air we breathe, is so stable (on account of the triple bond connecting the atoms) and unreactive that *nitrogen fixation*, the reduction of atmospheric N_2 to NH_3, is among the most thermodynamically demanding of biochemical reactions, in the sense that it requires a great deal of energy derived from metabolism. So taxing is the process that only certain bacteria and archaea are capable of carrying it out, making nitrogen available first to plants and other microorganisms in the form of ammonia. Only after incorporation into amino acids by plants does nitrogen adopt a chemical form that, when consumed, can be used by animals in the synthesis of proteins and other nitrogen-containing molecules.

The reactivity of O_2, while important for biological energy conversion, also poses serious physiological problems. During the course of metabolism, some electrons escape from complexes I, II, and III of the respiratory chain and reduce O_2 to superoxide ion, O_2^-. The ground-state electronic configuration of O_2^- is $1\sigma_g^2 1\sigma_u^2 2\sigma_g^2 1\pi_u^4 1\pi_g^3$, so the ion is a radical with a bond order $b = \frac{3}{2}$. We predict that the superoxide ion is a reactive species that must be scavenged to prevent damage to cellular components. The enzyme superoxide dismutase protects cells by catalysing the disproportionation (or dismutation) of O_2^- into O_2 and H_2O_2:

$$2\,O_2^- + 2\,H^+ \rightarrow H_2O_2 + O_2$$

However, H_2O_2 (hydrogen peroxide), formed by the reaction above and by leakage of electrons out of the respiratory chain, is a powerful oxidizing agent and also harmful to cells. It is metabolized further by catalases and peroxidases. A catalase catalyses the reaction

$$2\,H_2O_2 \rightarrow 2\,H_2O + O_2$$

and a peroxidase reduces hydrogen peroxide to water by oxidizing an organic molecule. For example, the enzyme glutathione peroxidase catalyses the reaction

$$2\,\text{glutathione}_{\text{red}} + H_2O_2 \rightarrow \text{glutathione}_{\text{ox}} + 2\,H_2O$$

There is growing evidence for the involvement of the damage caused by reactive oxygen species (ROS), such as O_2^-, H_2O_2, and ·OH (the hydroxyl radical), in the mechanism of ageing and in the development of cardiovascular disease, cancer,

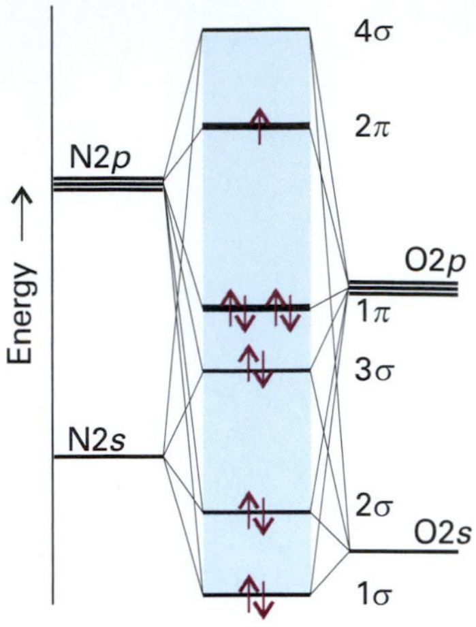

Fig. 4.37 The molecular orbital energy level diagram for NO.

stroke, inflammatory disease, and other conditions. For this reason, much effort has been expended on studies of the biochemistry of *antioxidants*, substances that can either deactivate ROS directly (as glutathione does) or halt the progress of cellular damage through reactions with radicals formed by processes initiated by ROS. Important examples of antioxidants are vitamin C (ascorbic acid), vitamin E (α-tocopherol), and uric acid.

Nitric oxide (nitrogen monoxide, NO) is a small molecule that diffuses quickly between cells, carrying chemical messages that help initiate a variety of processes, such as regulation of blood pressure, inhibition of platelet aggregation, and defence against inflammation and attacks to the immune system. The molecule is synthesized from the amino acid arginine in a series of reactions catalysed by nitric oxide synthase and requiring O_2 and NADPH.

Figure 4.37 shows the bonding scheme in NO and illustrates a number of points we have made about heteronuclear diatomic molecules. The ground configuration is $1\sigma^2 2\sigma^2 3\sigma^2 1\pi^4 2\pi^1$. The $3\sigma$and 1πorbitals are predominantly of O character as that is the more electronegative element. The highest-energy occupied orbital is 2π, contains one electron, and has more N character than O character. It follows that NO is a radical with an unpaired electron that can be regarded as localized more on the N atom than on the O atom. The lowest-energy occupied orbital is 4σ, which is also localized predominantly on N.

Because NO is a radical, we expect it to be reactive. Its half-life is estimated at approximately 1–5 s, so it needs to be synthesized often in the cell. As we saw above, there is a biochemical price to be paid for the reactivity of biological radicals. Like O_2, NO participates in some reactions that are not beneficial to the cell. Indeed, the radicals O_2^- and NO combine to form the peroxynitrite ion:

$$NO\cdot + O_2^-\cdot \rightarrow ONOO^-$$

where we have shown the unpaired electrons explicitly. The peroxynitrite ion is a reactive oxygen species that damages proteins, DNA, and lipids, possibly leading to heart disease, amyotrophic lateral sclerosis (Lou Gehrig's disease), Alzheimer's disease, and multiple sclerosis. Note that the structure of the ion is consistent with the bonding scheme in Fig. 4.37: because the unpaired electron in NO is slightly more localized on the N atom, we expect that atom to form a bond with an O atom from the O_2^- ion.

Molecular orbitals for polyatomic systems

The molecular orbitals of polyatomic molecules are built in the same way as in diatomic molecules, the only difference being that we use more atomic orbitals to construct them. As for diatomic molecules, polyatomic molecular orbitals spread over the entire molecule. A molecular orbital has the general form

$$\psi = \sum_i c_i \chi_i \qquad (4.35)$$

where χ_i is an atomic orbital and the sum extends over all the valence orbitals of all the atoms in the molecule. To find the coefficients, we set up the secular equations and the secular determinant, just as for diatomic molecules, solve the latter for the energies, and then use these energies in the secular equations to find the coefficients of the atomic orbitals for each molecular orbital.

The principal difference between diatomic and polyatomic molecules lies in the greater range of shapes that are possible: a diatomic molecule is necessarily linear, but a triatomic molecule, for instance, may be either linear or angular with a characteristic bond angle. The shape of a polyatomic molecule—the specification of its bond lengths and its bond angles—can be predicted by calculating the total energy of the molecule for a variety of nuclear positions, and then identifying the conformation that corresponds to the lowest energy.

4.6 The Hückel approximation

Molecular orbital theory takes large molecules and extended aggregates of atoms, such as solid materials, in its stride. First we shall consider conjugated molecules, in which there is an alternation of single and double bonds along a chain of carbon atoms. Although the classification of an orbital as σ or π is strictly valid only in linear molecules, as will be familiar from introductory chemistry courses, it is also used to denote the local symmetry with respect to a given A—B bond axis.

The π molecular orbital energy level diagrams of conjugated molecules can be constructed using a set of approximations suggested by Erich Hückel in 1931. In his approach, the π orbitals are treated separately from the σ orbitals, and the latter form a rigid framework that determines the general shape of the molecule. All the C atoms are treated identically, so all the Coulomb integrals α for the atomic orbitals that contribute to the π orbitals are set equal. For example, in ethene, we take the σ bonds as fixed, and concentrate on finding the energies of the single π bond and its companion antibond.

(a) Ethene and frontier orbitals

We express the π orbitals as LCAOs of the C2p orbitals that lie perpendicular to the molecular plane. In ethene, for instance, we would write

$$\psi = c_A A + c_B B \tag{4.36}$$

where the A is a C2p orbital on atom A, and so on. Next, the optimum coefficients and energies are found by the variation principle as explained in Section 4.5. That is, we have to solve the secular determinant, which in the case of ethene is eqn 4.29 with $\alpha_A = \alpha_B = \alpha$:

$$\begin{vmatrix} \alpha - E & \beta - ES \\ \beta - ES & \alpha - E \end{vmatrix} = 0 \tag{4.37}$$

The roots of this determinant can be found very easily (they are the same as those in *Illustration 4.2*). In a modern computation all the resonance integrals and overlap integrals would be included, but an indication of the molecular orbital energy level diagram can be obtained very readily if we make the following additional **Hückel approximations**:

1 All overlap integrals are set equal to zero.

2 All resonance integrals between non-neighbours are set equal to zero.

3 All remaining resonance integrals are set equal (to β).

These approximations are obviously very severe, but they let us calculate at least a general picture of the molecular orbital energy levels with very little work. The assumptions result in the following structure of the secular determinant:

1 All diagonal elements: $\alpha - E$.

2 Off-diagonal elements between neighbouring atoms: β.

3 All other elements: 0.

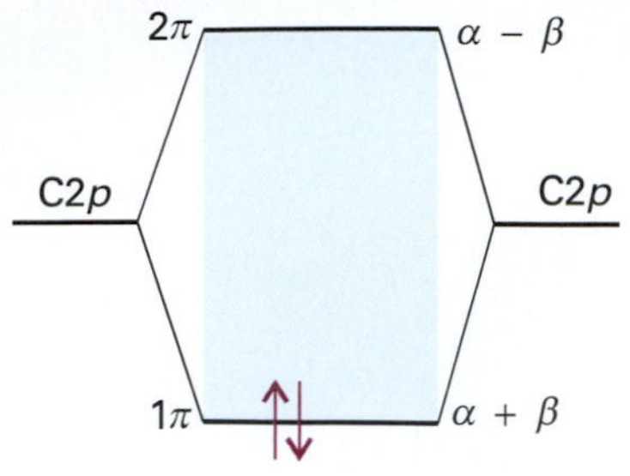

Fig. 4.38 The Hückel molecular orbital energy levels of ethene. Two electrons occupy the lower π orbital.

These approximations lead to

$$\begin{vmatrix} \alpha - E & \beta \\ \beta & \alpha - E \end{vmatrix} = (\alpha - E)^2 - \beta^2 = 0 \tag{4.38}$$

The roots of the equation are

$$E_{\pm} = \alpha \pm \beta \tag{4.39}$$

The + sign corresponds to the bonding combination (β is negative) and the − sign corresponds to the antibonding combination (Fig. 4.38). We see the effect of neglecting overlap by comparing this result with eqn 4.31.

The building-up principle leads to the configuration $1\pi^2$, because each carbon atom supplies one electron to the π system. The **highest occupied molecular orbital** in ethene, its HOMO, is the 1π orbital; the **lowest unfilled molecular orbital**, its LUMO, is the 2π orbital (or, as it is sometimes denoted, the $2\pi^*$ orbital). These two orbitals jointly form the **frontier orbitals** of the molecule. The frontier orbitals are important because they are largely responsible for many of the chemical and spectroscopic properties of the molecule. For example, we can estimate that $2|\beta|$ is the $\pi^* \leftarrow \pi$ excitation energy of ethene, the energy required to excite an electron from the 1π to the 2π orbital. The constant β is often left as an adjustable parameter; an approximate value for π bonds formed from overlap of two C2*p* atomic orbitals is about −2.4 eV (−230 kJ mol^{-1}).

(b) The matrix formulation of the Hückel method

In preparation for making Hückel theory more sophisticated and readily applicable to bigger molecules, we need to reformulate it in terms of matrices and vectors (see *Appendix 2*). We have seen that the secular equations that we have to solve for a two-atom system have the form

$$(H_{AA} - E_i S_{AA})c_{i,A} + (H_{AB} - E_i S_{AB})c_{i,B} = 0 \tag{4.40a}$$

$$(H_{BA} - E_i S_{BA})c_{i,A} + (H_{BB} - E_i S_{BB})c_{i,B} = 0 \tag{4.40b}$$

where the eigenvalue E_i corresponds to a wavefunction of the form $\psi_i = c_{i,A}A + c_{i,B}B$. (These expressions generalize eqn 4.25). There are two atomic orbitals, two eigenvalues, and two wavefunctions, so there are two pairs of secular equations, with the first corresponding to E_1 and ψ_1:

$$(H_{AA} - E_1 S_{AA})c_{1,A} + (H_{AB} - E_1 S_{AB})c_{1,B} = 0 \tag{4.41a}$$

$$(H_{BA} - E_1 S_{BA})c_{1,A} + (H_{BB} - E_1 S_{BB})c_{1,B} = 0 \tag{4.41b}$$

and another corresponding to E_2 and ψ_2:

$$(H_{AA} - E_2 S_{AA})c_{2,A} + (H_{AB} - E_2 S_{AB})c_{2,B} = 0 \tag{4.41c}$$

$$(H_{BA} - E_2 S_{BA})c_{2,A} + (H_{BB} - E_2 S_{BB})c_{2,B} = 0 \tag{4.41d}$$

If we introduce the following matrices and column vectors

$$\boldsymbol{H} = \begin{pmatrix} H_{AA} & H_{AB} \\ H_{BA} & H_{BB} \end{pmatrix} \qquad \boldsymbol{S} = \begin{pmatrix} S_{AA} & S_{AB} \\ S_{BA} & S_{BB} \end{pmatrix} \qquad \boldsymbol{c}_i = \begin{pmatrix} c_{i,A} \\ c_{i,B} \end{pmatrix} \tag{4.42}$$

then each pair of equations may be written more succinctly as

$$(\boldsymbol{H} - E_i \boldsymbol{S})\boldsymbol{c}_i = 0 \qquad \text{or} \qquad \boldsymbol{H}\boldsymbol{c}_i = \boldsymbol{S}\boldsymbol{c}_i E_i \tag{4.43}$$

where $\boldsymbol{H}$ is the hamiltonian matrix and $\boldsymbol{S}$ is the overlap matrix. To proceed with the calculation of the eigenvalues and coefficients, we introduce the matrices

$$\boldsymbol{C} = (\boldsymbol{c}_1 \quad \boldsymbol{c}_2) = \begin{pmatrix} c_{1,A} & c_{2,A} \\ c_{1,B} & c_{2,B} \end{pmatrix} \qquad \boldsymbol{E} = \begin{pmatrix} E_1 & 0 \\ 0 & E_2 \end{pmatrix} \tag*{[4.44]}$$

for then the entire set of equations we have to solve can be expressed as

$$HC = SCE \tag{4.45}$$

Self-test 4.7 Show by carrying out the necessary matrix operations that eqn 4.45 is a representation of the system of equations consisting of eqns 4.41(a)–(d).

In the Hückel approximation, $H_{AA} = H_{BB} = \alpha$, $H_{AB} = H_{BA} = \beta$, and we neglect overlap, setting $\boldsymbol{S} = \boldsymbol{1}$, the unit matrix (with 1 on the diagonal and 0 elsewhere). Then

$$HC = CE$$

At this point, we multiply from the left by the inverse matrix $\boldsymbol{C}^{-1}$, and find

$$C^{-1}HC = E \tag{4.46}$$

where we have used $\boldsymbol{C}^{-1}\boldsymbol{C} = \boldsymbol{1}$. In other words, to find the eigenvalues E_i, we have to find a transformation of $\boldsymbol{H}$ that makes it diagonal. This procedure is called **matrix diagonalization**. The diagonal elements then correspond to the eigenvalues E_i and the columns of the matrix $\boldsymbol{C}$ that brings about this diagonalization are the coefficients of the members of the **basis set**, the set of atomic orbitals used in the calculation, and hence give us the composition of the molecular orbitals. If there are N orbitals in the basis set (there are only two in our example), then there are N eigenvalues E_i and N corresponding column vectors $\boldsymbol{c}_i$. As a result, we have to solve N equations of the form $\boldsymbol{Hc}_i = \boldsymbol{Sc}_iE_i$ by diagonalization of the $N \times N$ matrix $\boldsymbol{H}$, as directed by eqn 4.46.

Example 4.4 *Finding the molecular orbitals by matrix diagonalization*

Set up and solve the matrix equations within the Hückel approximation for the π-orbitals of butadiene (**3**).

3

Method The matrices will be four-dimensional for this four-atom system. Ignore overlap, and construct the matrix $\boldsymbol{H}$ by using the Hückel values α and β. Find the matrix $\boldsymbol{C}$ that diagonalizes $\boldsymbol{H}$: for this step, use mathematical software. Full details are given in *Appendix 2*.

Solution

$$\boldsymbol{H} = \begin{pmatrix} H_{11} & H_{12} & H_{13} & H_{14} \\ H_{21} & H_{22} & H_{23} & H_{24} \\ H_{31} & H_{32} & H_{33} & H_{34} \\ H_{41} & H_{42} & H_{43} & H_{44} \end{pmatrix} = \begin{pmatrix} \alpha & \beta & 0 & 0 \\ \beta & \alpha & \beta & 0 \\ 0 & \beta & \alpha & \beta \\ 0 & 0 & \beta & \alpha \end{pmatrix}$$

Mathematical software then diagonalizes this matrix to

$$\boldsymbol{E} = \begin{pmatrix} \alpha + 1.62\beta & 0 & 0 & 0 \\ 0 & \alpha + 0.62\beta & 0 & 0 \\ 0 & 0 & \alpha - 0.62\beta & 0 \\ 0 & 0 & 0 & \alpha - 1.62\beta \end{pmatrix}$$

and the matrix that achieves the diagonalization is

$$\boldsymbol{C} = \begin{pmatrix} 0.372 & 0.602 & 0.602 & -0.372 \\ 0.602 & 0.372 & -0.372 & 0.602 \\ 0.602 & -0.372 & -0.372 & -0.602 \\ 0.372 & -0.602 & 0.602 & 0.372 \end{pmatrix}$$

We can conclude that the energies and molecular orbitals are

$$E_1 = \alpha + 1.62\beta \qquad \psi_1 = 0.372\chi_A + 0.602\chi_B + 0.602\chi_C + 0.372\chi_D$$
$$E_2 = \alpha + 0.62\beta \qquad \psi_2 = 0.602\chi_A + 0.372\chi_B - 0.372\chi_C - 0.602\chi_D$$
$$E_3 = \alpha - 0.62\beta \qquad \psi_3 = 0.602\chi_A - 0.372\chi_B - 0.372\chi_C + 0.602\chi_D$$
$$E_4 = \alpha - 1.62\beta \qquad \psi_4 = -0.372\chi_A + 0.602\chi_B - 0.602\chi_C - 0.372\chi_D$$

where the C2*p* atomic orbitals are denoted by $\chi_A, \ldots, \chi_D$. Note that the orbitals are mutually orthogonal and, with overlap neglected, normalized.

Self-test 4.8 Repeat the exercise for the allyl radical, $\cdot CH_2{-}CH{=}CH_2$.

$[E = \alpha + 2^{1/2}\beta, \alpha, \alpha - 2^{1/2}\beta;\ \psi_1 = \frac{1}{2}\chi_A + (\frac{1}{2})^{1/2}\chi_B + \frac{1}{2}\chi_C,$
$\psi_2 = (\frac{1}{2})^{1/2}\chi_A - (\frac{1}{2})^{1/2}\chi_C,\ \psi_3 = \frac{1}{2}\chi_A - (\frac{1}{2})^{1/2}\chi_B + \frac{1}{2}\chi_C]$

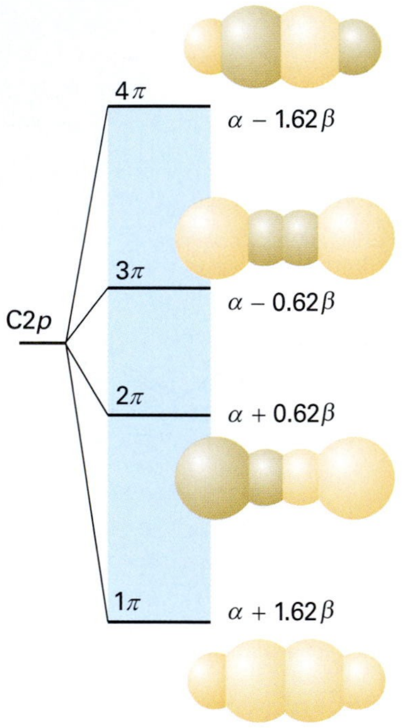

Fig. 4.39 The Hückel molecular orbital energy levels of butadiene and the top view of the corresponding π orbitals. The four *p* electrons (one supplied by each C) occupy the two lower π orbitals. Note that the orbitals are delocalized.

(c) Butadiene and π-electron binding energy

As we saw in the preceding example, the energies of the four LCAO-MOs for butadiene are

$$E = \alpha \pm 1.62\beta, \qquad \alpha \pm 0.62\beta \tag{4.47}$$

These orbitals and their energies are drawn in Fig. 4.39. Note that the greater the number of internuclear nodes, the higher the energy of the orbital. There are four electrons to accommodate, so the ground-state configuration is $1\pi^2 2\pi^2$. The frontier orbitals of butadiene are the 2π orbital (the HOMO, which is largely bonding) and the 3π orbital (the LUMO, which is largely antibonding). 'Largely' bonding means that an orbital has both bonding and antibonding interactions between various neighbours, but the bonding effects dominate. 'Largely antibonding' indicates that the antibonding effects dominate.

An important point emerges when we calculate the total **π-electron binding energy**, E_π, the sum of the energies of each π electron, and compare it with what we find in ethene. In ethene the total energy is

$$E_\pi = 2(\alpha + \beta) = 2\alpha + 2\beta$$

In butadiene it is

$$E_\pi = 2(\alpha + 1.62\beta) + 2(\alpha + 0.62\beta) = 4\alpha + 4.48\beta$$

Therefore, the energy of the butadiene molecule lies lower by 0.48β (about 110 kJ mol^{-1}) than the sum of two individual π bonds. This extra stabilization of a conjugated system is called the **delocalization energy**. A closely related quantity is the **π-bond formation energy**, the energy released when a π bond is formed. Because the contribution of α is the same in the molecule as in the atoms, we can find the π-bond formation energy from the π-electron binding energy by writing

$$E_{bf} = E_\pi - N\alpha \tag{4.48}$$

where N is the number of carbon atoms in the molecule. The π-bond formation energy in butadiene, for instance, is 4.48β.

Example 4.5 *Estimating the delocalization energy*

Use the Hückel approximation to find the energies of the π orbitals of cyclobutadiene, and estimate the delocalization energy.

Method Set up the secular determinant using the same basis as for butadiene, but note that atoms A and D are also now neighbours. Then solve for the roots of the secular equation and assess the total π-bond energy. For the delocalization energy, subtract from the total π-bond energy the energy of two π-bonds.

Answer The hamiltonian matrix is

$$\boldsymbol{H}=\begin{pmatrix}\alpha & \beta & 0 & \beta\\ \beta & \alpha & \beta & 0\\ 0 & \beta & \alpha & \beta\\ \beta & 0 & \beta & \alpha\end{pmatrix}$$

Diagonalization gives the energies of the orbitals as

$$E=\alpha+2\beta, \qquad \alpha, \qquad \alpha, \qquad \alpha-2\beta$$

Four electrons must be accommodated. Two occupy the lowest orbital (of energy $\alpha+2\beta$), and two occupy the doubly degenerate orbitals (of energy α). The total energy is therefore $4\alpha+4\beta$. Two isolated π bonds would have an energy $4\alpha+4\beta$; therefore, in this case, the delocalization energy is zero.

Self-test 4.9 Repeat the calculation for benzene. [See next subsection]

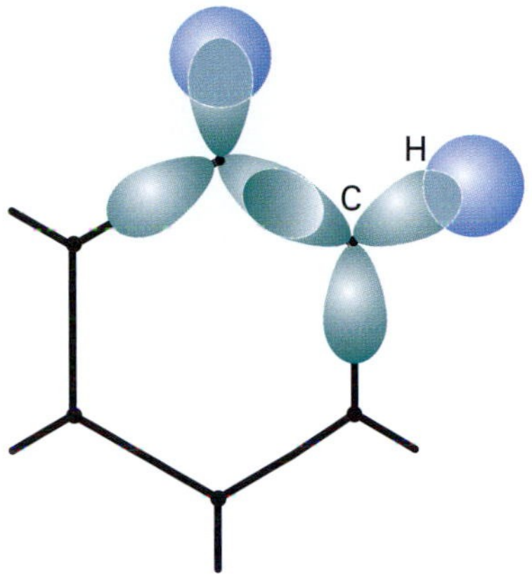

Fig. 4.40 The σ framework of benzene is formed by the overlap of Csp^2 hybrids, which fit without strain into a hexagonal arrangement.

(d) Benzene and aromatic stability

The most notable example of delocalization conferring extra stability is benzene and the aromatic molecules based on its structure. Benzene is often expressed in a mixture of valence-bond and molecular orbital terms, with typically valence-bond language used for its σ framework and molecular orbital language used to describe its π electrons.

First, the valence-bond component. The six C atoms are regarded as sp^2 hybridized, with a single unhydridized perpendicular $2p$ orbital. One H atom is bonded by (Csp^2,H1s) overlap to each C carbon, and the remaining hybrids overlap to give a regular hexagon of atoms (Fig. 4.40). The internal angle of a regular hexagon is 120°, so sp^2 hybridization is ideally suited for forming σ bonds. We see that benzene's hexagonal shape permits strain-free σ bonding.

Now consider the molecular orbital component of the description. The six C2p orbitals overlap to give six π orbitals that spread all round the ring. Their energies are calculated within the Hückel approximation by diagonalizing the hamiltonian matrix

$$\boldsymbol{H}=\begin{pmatrix}\alpha & \beta & 0 & 0 & 0 & \beta\\ \beta & \alpha & \beta & 0 & 0 & 0\\ 0 & \beta & \alpha & \beta & 0 & 0\\ 0 & 0 & \beta & \alpha & \beta & 0\\ 0 & 0 & 0 & \beta & \alpha & \beta\\ \beta & 0 & 0 & 0 & \beta & \alpha\end{pmatrix}$$

The MO energies, the eigenvalues of this matrix, are simply

$$E=\alpha\pm2\beta, \alpha\pm\beta, \alpha\pm\beta \tag{4.49}$$

as shown in Fig. 4.41. The orbitals there have been given symmetry labels that we explain in Chapter 5. Note that the lowest energy orbital is bonding between all neighbouring atoms, the highest energy orbital is antibonding between each pair of neighbours, and the intermediate orbitals are a mixture of bonding, nonbonding, and antibonding character between adjacent atoms.

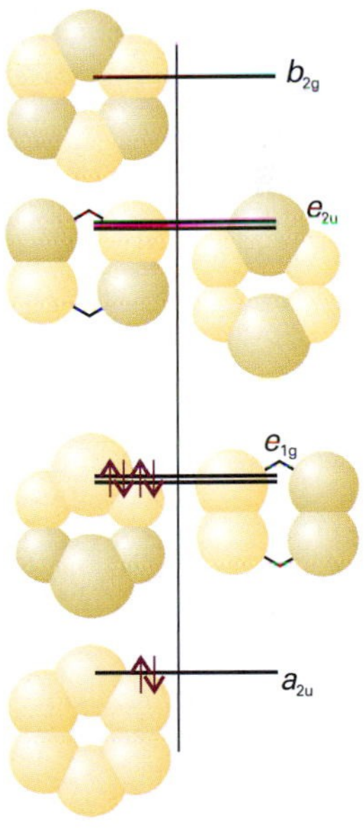

Fig. 4.41 The Hückel orbitals of benzene and the corresponding energy levels. The symmetry labels are explained in Chapter 5. The bonding and antibonding character of the delocalized orbitals reflects the numbers of nodes between the atoms. In the ground state, only the bonding orbitals are occupied.

We now apply the building-up principle to the π system. There are six electrons to accommodate (one from each C atom), so the three lowest orbitals (a_{2u} and the doubly-degenerate pair e_{1g}) are fully occupied, giving the ground-state configuration $a_{2u}^2 e_{1g}^4$. A significant point is that the only molecular orbitals occupied are those with net bonding character.

The π-electron energy of benzene is

$$E_\pi = 2(\alpha + 2\beta) + 4(\alpha + \beta) = 6\alpha + 8\beta$$

If we ignored delocalization and thought of the molecule as having three isolated π bonds, it would be ascribed a π-electron energy of only $3(2\alpha + 2\beta) = 6\alpha + 6\beta$. The delocalization energy is therefore $2\beta \approx -460$ kJ mol^{-1}, which is considerably more than for butadiene. The π-bond formation energy in benzene is 8β.

This discussion suggests that aromatic stability can be traced to two main contributions. First, the shape of the regular hexagon is ideal for the formation of strong σ bonds: the σ framework is relaxed and without strain. Second, the π orbitals are such as to be able to accommodate all the electrons in bonding orbitals, and the delocalization energy is large.

4.7 Computational chemistry

The difficulties arising from the severe assumptions of Hückel method have been overcome by more sophisticated theories that not only calculate the shapes and energies of molecular orbitals but also predict with reasonable accuracy the structure and reactivity of molecules. The full treatment of molecular electronic structure is quite easy to formulate but difficult to implement. However, it has received an enormous amount of attention by chemists, and has become a keystone of modern chemical research. John Pople and Walter Kohn were awarded the Nobel Prize in Chemistry for 1998 for their contributions to the development of computational techniques for the elucidation of molecular structure and reactivity.

Comment 4.12

The web site contains links to sites where you may perform semi-empirical and *ab initio* calculations on simple molecules directly from your web browser.

(a) The Hartree–Fock equations

The starting point is to write down the many-electron wavefunction as a product of one-electron wavefunctions:

$$\Psi = \psi_{a,\alpha}(1)\psi_{a,\beta}(2)\ldots\psi_{z,\beta}(N)$$

This is the wavefunction for an N-electron closed-shell molecule in which electron 1 occupies molecular orbital ψ_a with spin α, electron 2 occupies molecular orbital ψ_a with spin β, and so on. However, the wavefunction must satisfy the Pauli principle and change sign under the permutation of any pair of electrons. To achieve this behaviour, we write the wavefunction as a sum of all possible permutations with the appropriate sign:

$$\Psi = \psi_{a,\alpha}(1)\psi_{a,\beta}(2)\ldots\psi_{z,\beta}(N) - \psi_{a,\alpha}(2)\psi_{a,\beta}(1)\ldots\psi_{z,\beta}(N) + \cdots$$

There are $N!$ terms in this sum, and the entire sum can be written as a determinant:

$$\Psi = \frac{1}{\sqrt{N!}}\begin{vmatrix} \psi_{a,\alpha}(1) & \psi_{a,\beta}(1) & \cdots & \cdots & \psi_{z,\beta}(1) \\ \psi_{a,\alpha}(2) & \psi_{a,\beta}(2) & \cdots & \cdots & \psi_{z,\beta}(2) \\ \vdots & \vdots & & & \vdots \\ \vdots & \vdots & & & \vdots \\ \psi_{a,\alpha}(N) & \psi_{a,\beta}(N) & \cdots & \cdots & \psi_{z,\beta}(N) \end{vmatrix} \tag{4.50a}$$

The initial factor ensures that the wavefunction is normalized if the component molecular orbitals are normalized. To save the tedium of writing out large determinants, the wavefunction is normally written simply as

$$\Psi = (1/N!)^{1/2}\det|\psi_{a,\alpha}(1)\psi_{a,\beta}(2)\ldots\psi_{z,\beta}(N)| \quad (4.50b)$$

When the determinantal wavefunction is combined with the variation principle (Section 4.5c), the optimum wavefunctions, in the sense of corresponding to the lowest total energy, must satisfy the **Hartree–Fock equations:**

$$f_1\psi_{a,\sigma}(1) = \varepsilon\psi_{a,\sigma}(1) \quad (4.51)$$

where σ is either α or β. The **Fock operator** f_1 is

$$f_1 = h_1 + \Sigma_j\{2J_j(1) - K_j(1)\} \quad (4.52)$$

The three terms in this expression are the **core hamiltonian**

$$h_1 = -\frac{\hbar^2}{2m_e}\nabla_1^2 - \sum_n \frac{Z_n e^2}{4\pi\varepsilon_0 r_{ni}} \quad [4.53a]$$

the **Coulomb operator** J, where

$$J_j(1)\psi_a(1) = \int \psi_j^*(2)\psi_j(2)\left(\frac{e^2}{4\pi\varepsilon_0 r_{12}}\right)\psi_a(1)\mathrm{d}\tau_2 \quad [4.53b]$$

and the **exchange operator**, K, where

$$K_j(1)\psi_a(1) = \int \psi_j^*(2)\psi_a(2)\left(\frac{e^2}{4\pi\varepsilon_0 r_{12}}\right)\psi_j(1)\mathrm{d}\tau_2 \quad [4.53c]$$

Although the Hartree–Fock equations look deceptively simple, with the Fock operator looking like a hamiltonian, we see from these definitions that f actually depends on the wavefunctions of all the electrons. To proceed, we have to guess the initial form of the wavefunctions, use them in the definition of the Coulomb and exchange operators, and solve the Hartree–Fock equations. That process is then continued using the newly found wavefunctions until each cycle of calculation leaves the energies and wavefunctions unchanged to within a chosen criterion. This is the origin of the term **self-consistent field** (SCF) for this type of procedure.

The difficulty in this procedure is in the solution of the Hartree–Fock equations. To make progress, we have to express the wavefunctions as linear combinations of M atomic orbitals χ_i, and write

$$\psi_a = \sum_{i=1}^{M} c_{ia}\chi_i$$

As we show in the *Justification* below, the use of a linear combination like this leads to a set of equations that can be expressed in a matrix form known as the **Roothaan equations:**

$$\boldsymbol{FC} = \boldsymbol{SC\varepsilon} \quad (4.54)$$

where $\boldsymbol{F}$ is the matrix formed from the Fock operator:

$$F_{ij} = \int \chi_i^*(1)f_1\chi_j(1)\mathrm{d}\tau \quad (4.55a)$$

and $\boldsymbol{S}$ is the matrix of overlap integrals

$$S_{ij} = \int \chi_i^*(1)\chi_j(1)\mathrm{d}\tau \quad (4.55b)$$

Justification 4.5 *The Roothaan equations*

To construct the Roothaan equations we substitute the linear combination of atomic orbitals into eqn 4.51, which gives

$$f_1 \sum_{i=1}^{M} c_{i\alpha}\chi_i(1) = \varepsilon_\alpha \sum_{i=1}^{M} c_{i\alpha}\chi_i(1)$$

Now multiply from the left by $\chi_j^*(1)$ and integrate over the coordinates of electron 1:

$$\sum_{i=1}^{M} c_{i\alpha} \overbrace{\int \chi_j(1)^* f(1)\chi_i(1)\mathrm{d}\boldsymbol{r}_1}^{F_{ji}} = \varepsilon_\alpha \sum_{i=1}^{M} c_{i\alpha} \overbrace{\int \chi_j(1)^* \chi_i(1)\mathrm{d}\boldsymbol{r}_1}^{S_{ji}}$$

That is,

$$\sum_{i=1}^{M} F_{ji}c_{i\alpha} = \varepsilon_\alpha \sum_{i=1}^{M} S_{ji}c_{i\alpha}$$

This expression has the form of the matrix equation in eqn 4.54.

(b) Semi-empirical and *ab initio* methods

There are two main strategies for continuing the calculation from this point. In the **semi-empirical methods**, many of the integrals are estimated by appealing to spectroscopic data or physical properties such as ionization energies, and using a series of rules to set certain integrals equal to zero. In the ***ab initio* methods**, an attempt is made to calculate all the integrals that appear in the Fock and overlap matrices. Both procedures employ a great deal of computational effort and, along with cryptanalysts and meteorologists, theoretical chemists are among the heaviest users of the fastest computers.

The Fock matrix has elements that consist of integrals of the form

$$(AB|CD) = \int A(1)B(1)\left(\frac{e^2}{4\pi\varepsilon_0 r_{12}}\right)C(2)D(2)\mathrm{d}\tau_1\mathrm{d}\tau_2 \qquad (4.56)$$

where A, B, C, and D are atomic orbitals that in general may be centred on different nuclei. It can be appreciated that, if there are several dozen atomic orbitals used to build the molecular orbitals, then there will be tens of thousands of integrals of this form to evaluate (the number of integrals increases as the fourth power of the number of atomic orbitals in the basis). One severe approximation is called **complete neglect of differential overlap** (CNDO), in which all integrals are set to zero unless A and B are the same orbitals centred on the same nucleus, and likewise for C and D. The surviving integrals are then adjusted until the energy levels are in good agreement with experiment. The more recent semi-empirical methods make less draconian decisions about which integrals are to be ignored, but they are all descendants of the early CNDO technique. These procedures are now readily available in commercial software packages and can be used with very little detailed knowledge of their mode of calculation. The packages also have sophisticated graphical output procedures, which enable one to analyse the shapes of orbitals and the distribution of electric charge in molecules. The latter is important when assessing, for instance, the likelihood that a given molecule will bind to an active site in an enzyme.

Commercial packages are also available for *ab initio* calculations. Here the problem is to evaluate as efficiently as possible thousands of integrals. This task is greatly facilitated by expressing the atomic orbitals used in the LCAOs as linear combinations of

Gaussian orbitals. A **Gaussian type orbital** (GTO) is a function of the form $e^{-\zeta r^2}$. The advantage of GTOs over the correct orbitals (which for hydrogenic systems are proportional to $e^{-\zeta r}$) is that the product of two Gaussian functions is itself a Gaussian function that lies between the centres of the two contributing functions (Fig. 4.42). In this way, the four-centre integrals like that in eqn 4.56 become two-centre integrals of the form

$$(AB|CD) = \int X(1)\left(\frac{e^2}{4\pi\varepsilon_0 r_{12}}\right)Y(2)\mathrm{d}\tau_1\mathrm{d}\tau_2 \tag{4.57}$$

where X is the Gaussian corresponding to the product AB and Y is the corresponding Gaussian from CD. Integrals of this form are much easier and faster to evaluate numerically than the original four-centre integrals. Although more GTOs have to be used to simulate the atomic orbitals, there is an overall increase in speed of computation.

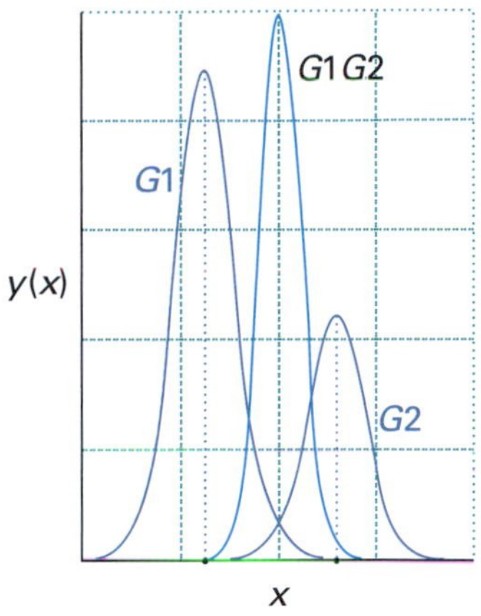

Fig. 4.42 The product of two Gaussian functions (the purple curves) is itself a Gaussian function located between the two contributing Gaussians.

(c) Density functional theory

A technique that has gained considerable ground in recent years to become one of the most widely used techniques for the calculation of molecular structure is **density functional theory** (DFT). Its advantages include less demanding computational effort, less computer time, and—in some cases (particularly *d*-metal complexes)—better agreement with experimental values than is obtained from Hartree–Fock procedures.

The central focus of DFT is the electron density, ρ, rather than the wavefunction ψ. The 'functional' part of the name comes from the fact that the energy of the molecule is a function of the electron density, written $E[\rho]$, and the electron density is itself a function of position, $\rho(\boldsymbol{r})$, and in mathematics a function of a function is called a *functional*. The exact ground-state energy of an n-electron molecule is

$$E[\rho] = E_{\mathrm{K}} + E_{\mathrm{P;e,N}} + E_{\mathrm{P;e,e}} + E_{\mathrm{XC}}[\rho] \tag{4.58}$$

where E_{K} is the total electron kinetic energy, $E_{\mathrm{P;e,N}}$ the electron–nucleus potential energy, $E_{\mathrm{P;e,e}}$ the electron–electron potential energy, and $E_{\mathrm{XC}}[\rho]$ the **exchange–correlation energy**, which takes into account all the effects due to spin. The orbitals used to construct the electron density from

$$\rho(\boldsymbol{r}) = \sum_{i=1}^{N} |\psi_i(\boldsymbol{r})|^2 \tag{4.59}$$

are calculated from the **Kohn–Sham equations**, which are found by applying the variation principle to the electron energy, and are like the Hartree–Fock equations except for a term V_{XC}, which is called the **exchange–correlation potential**:

$$\left\{\underbrace{-\frac{\hbar^2}{2m_{\mathrm{e}}}\nabla_1^2}_{\text{Kinetic energy}} - \underbrace{\sum_{j=1}^{N}\frac{Z_j e^2}{4\pi\varepsilon_0 r_{j1}}}_{\text{Electron–nucleus attraction}} + \underbrace{\int\frac{\rho(\boldsymbol{r}_2)e^2}{4\pi\varepsilon_0 r_{12}}\mathrm{d}\boldsymbol{r}_2}_{\text{Electron–electron repulsion}} + \underbrace{V_{\mathrm{XC}}(\boldsymbol{r}_1)}_{\text{Exchange–correlation}}\right\}\psi_i(\boldsymbol{r}_1) = \varepsilon_i\psi_i(\boldsymbol{r}_1) \tag{4.60}$$

The exchange–correlation potential is the 'functional derivative' of the exchange–correlation energy:

$$V_{\mathrm{XC}}[\rho] = \frac{\delta E_{\mathrm{XC}}[\rho]}{\delta\rho} \tag{4.61}$$

The Kohn–Sham equations are solved iteratively and self-consistently. First, we guess the electron density. For this step it is common to use a superposition of atomic

Comment 4.13

Consider the functional $G[f]$ where f is a function of x. When x changes to $x + \delta x$, the function changes to $f + \delta f$ and the functional changes to $G[f + \delta f]$. By analogy with the derivative of a function, the functional derivative is then defined as

$$\frac{\delta G}{\delta f} = \lim_{\delta f \to 0}\frac{G[f+\delta f] - G[f]}{\delta f}$$

where the manner in which δf goes to zero must be specified explicitly. See *Appendix 2* for more details and examples.

electron densities. Then the exchange–correlation potential is calculated by assuming an approximate form of the dependence of the exchange–correlation energy on the electron density and evaluating the functional derivative in eqn 4.61. For this step, the simplest approximation is the **local-density approximation** and to write

$$E_{XC}[\rho] = \int \rho(\boldsymbol{r})\varepsilon_{XC}[\rho(\boldsymbol{r})]\mathrm{d}\boldsymbol{r} \qquad (4.62)$$

where ε_{XC} is the exchange–correlation energy per electron in a homogeneous gas of constant density. Next, the Kohn–Sham equations are solved to obtain an initial set of orbitals. This set of orbitals is used to obtain a better approximation to the electron density (from eqn 4.59) and the process is repeated until the density and the exchange–correlation energy are constant to within some tolerance.

4.8 The prediction of molecular properties

The results of molecular orbital calculations are only approximate, with deviations from experimental values increasing with the size of the molecule. Therefore, one goal of computational chemistry is to gain insight into trends in properties of molecules, without necessarily striving for ultimate accuracy. In the next sections we give a brief summary of strategies used by computational chemists for the prediction of molecular properties.

(a) Electron density and the electrostatic potential surfaces

One of the most significant developments in computational chemistry has been the introduction of graphical representations of molecular orbitals and electron densities. The raw output of a molecular structure calculation is a list of the coefficients of the atomic orbitals in each molecular orbital and the energies of these orbitals. The graphical representation of a molecular orbital uses stylized shapes to represent the basis set, and then scales their size to indicate the coefficient in the linear combination. Different signs of the wavefunctions are represented by different colours.

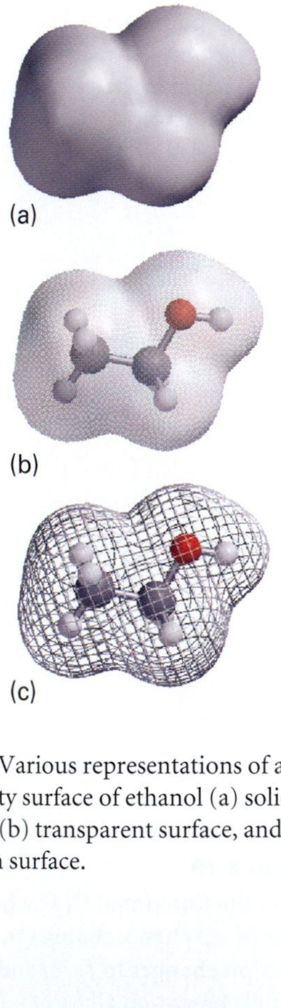

Fig. 4.43 Various representations of an isodensity surface of ethanol (a) solid surface, (b) transparent surface, and (c) mesh surface.

Once the coefficients are known, we can build up a representation of the electron density in the molecule by noting which orbitals are occupied and then forming the squares of those orbitals. The total electron density at any point is then the sum of the squares of the wavefunctions evaluated at that point. The outcome is commonly represented by a **isodensity surface**, a surface of constant total electron density (Fig. 4.43). As shown in the illustration, there are several styles of representing an isodensity surface, as a solid form, as a transparent form with a ball-and-stick representation of the molecule within, or as a mesh. A related representation is a **solvent-accessible surface** in which the shape represents the shape of the molecule by imagining a sphere representing a solvent molecule rolling across the surface and plotting the locations of the centre of that sphere.

One of the most important aspects of a molecule other than its geometrical shape is the distribution of charge over its surface. The net charge at each point on an isodensity surface can be calculated by subtracting the charge due to the electron density at that point form the charge due to the nuclei: the result is an **electrostatic potential surface** (an 'elpot surface') in which net positive charge is shown in one colour and net negative charge is shown in another, with intermediate gradations of colour (Fig. 4.44).

Fig. 4.44 An elpot diagram of ethanol.

Representations such as those we have illustrated are of critical importance in a number of fields. For instance, they may be used to identify an electron-poor region of a molecule that is susceptible to association with or chemical attack by an electron-rich region of another molecule. Such considerations are important for assessing the pharmacological activity of potential drugs.

(b) Thermodynamic and spectroscopic properties

Computational chemistry is becoming the technique of choice for estimating standard enthalpies of formation of molecules with complex three-dimensional structures. The computational approach also makes it possible to gain insight into the effect of solvation on the enthalpy of formation without conducting experiments. A calculation performed in the absence of solvent molecules estimates the properties of the molecule of interest in the gas phase. Computational methods are available that allow for the inclusion of several solvent molecules around a solute molecule, thereby taking into account the effect of molecular interactions with the solvent on the enthalpy of formation of the solute. Again, the numerical results are only estimates and the primary purpose of the calculation is to predict whether interactions with the solvent increase or decrease the enthalpy of formation. As an example, consider the amino acid glycine, which can exist in a neutral (**4**) or zwitterionic (**5**) form, in which the amino group is protonated and the carboxyl group is deprotonated. It is possible to show computationally that in the gas phase the neutral form has a lower enthalpy of formation than the zwitterionic form. However, in water the opposite is true because of strong interactions between the polar solvent and the charges in the zwitterion.

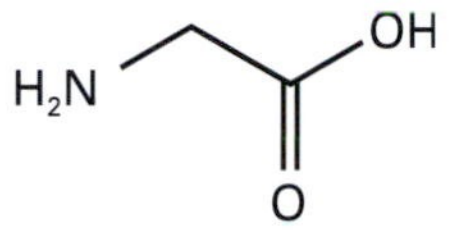

4

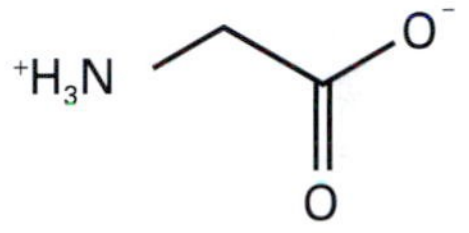

5

Molecular orbital calculations can also be used to predict trends in electrochemical properties, such as standard potentials. Several experimental and computational studies of aromatic hydrocarbons indicate that decreasing the energy of the LUMO enhances the ability of a molecule to accept an electron into the LUMO, with an attendant increase in the value of the molecule's standard potential. The effect is also observed in quinones and flavins, co-factors involved in biological electron transfer reactions. For example, stepwise substitution of the hydrogen atoms in *p*-benzoquinone by methyl groups ($-CH_3$) results in a systematic increase in the energy of the LUMO and a decrease in the standard potential for formation of the semiquinone radical (**6**):

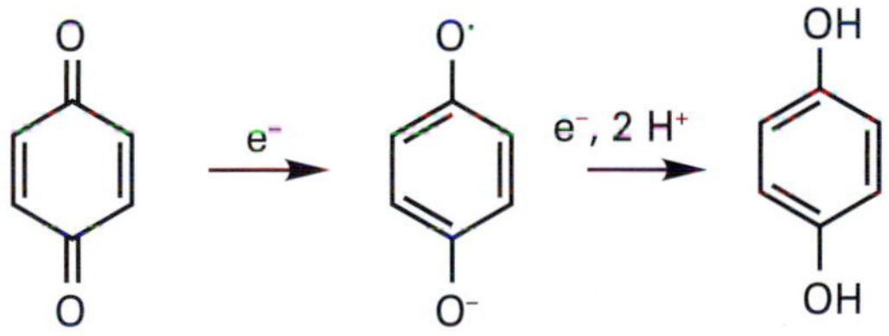

6

The standard potentials of naturally occurring quinones are also modified by the presence of different substituents, a strategy that imparts specific functions to specific quinones. For example, the substituents in coenzyme Q are largely responsible for poising its standard potential so that the molecule can function as an electron shuttle between specific electroactive proteins in the respiratory chain (*Impact* I10.2).

We remarked in Chapter 1 that a molecule can absorb or emit a photon of energy hc/λ, resulting in a transition between two quantized molecular energy levels. The transition of lowest energy (and longest wavelength) occurs between the HOMO and LUMO. We can use calculations based on semi-empirical, *ab initio*, and DFT methods to correlate the calculated HOMO–LUMO energy gap with the wavelength of absorption. For example, consider the linear polyenes shown in Table 4.5: ethene (C_2H_4), butadiene (C_4H_6), hexatriene (C_6H_8), and octatetraene (C_8H_{10}), all of which absorb in the ultraviolet region of the spectrum. The table also shows that, as expected, the wavelength of the lowest-energy electronic transition decreases as the energy separation between the HOMO and LUMO increases. We also see that the smallest HOMO–LUMO gap and longest transition wavelength correspond to octatetraene, the longest

Table 4.5 *Ab initio* calculations and spectroscopic data

Polyene	$\{E(\text{HOMO}) - E(\text{LUMO})\}$/eV	λ/nm
═ (C_2H_4)	18.1	163
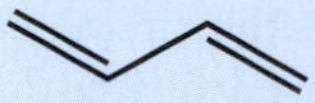	14.5	217
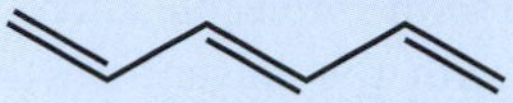	12.7	252
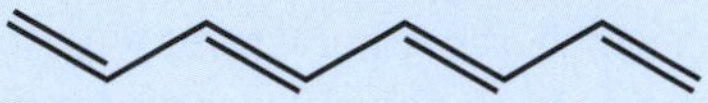	11.8	304

polyene in the group. It follows that the wavelength of the transition increases with increasing number of conjugated double bonds in linear polyenes. Extrapolation of the trend suggests that a sufficiently long linear polyene should absorb light in the visible region of the electromagnetic spectrum. This is indeed the case for β-carotene (7), which absorbs light with $\lambda \approx 450$ nm. The ability of β-carotene to absorb visible light is part of the strategy employed by plants to harvest solar energy for use in photosynthesis.

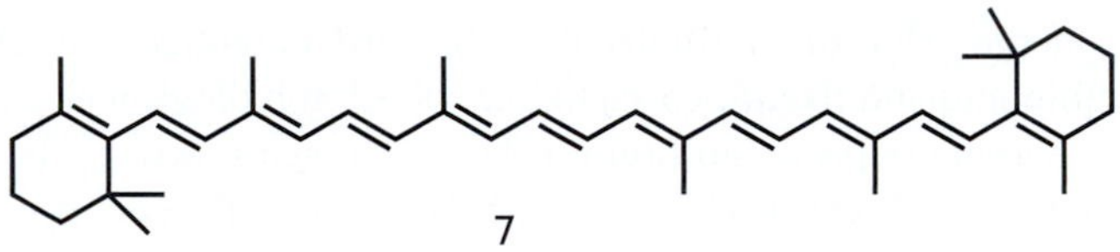

7

Checklist of key ideas

- ☐ 1. In the Born–Oppenheimer approximation, nuclei are treated as stationary while electrons move around them.
- ☐ 2. In valence-bond theory (VB theory), a bond is regarded as forming when an electron in an atomic orbital on one atoms pairs its spin with that of an electron in an atomic orbital on another atom.
- ☐ 3. A valence bond wavefunction with cylindrical symmetry around the internuclear axis is a σ bond. A π bond arises from the merging of two *p* orbitals that approach side-by-side and the pairing of electrons that they contain.
- ☐ 4. Hybrid orbitals are mixtures or atomic orbitals on the same atom and are invoked in VB theory to explain molecular geometries.
- ☐ 5. In molecular orbital theory (MO theory), electrons are treated as spreading throughout the entire molecule.
- ☐ 6. A bonding orbital is a molecular orbital that, if occupied, contributes to the strength of a bond between two atoms. An antibonding orbital is a molecular orbital that, if occupied, decreases the strength of a bond between two atoms.
- ☐ 7. A σ molecular orbital has zero orbital angular momentum about the internuclear axis. A π molecular orbital has one unit of angular momentum around the internuclear axis; in a nonlinear molecule, it has a nodal plane that includes the internucelar axis.
- ☐ 8. The electron configurations of homonuclear diatomic molecules are shown in Figs. 4.31 and 4.33.
- ☐ 9. When constructing molecular orbitals, we need to consider only combinations of atomic orbitals of similar energies and of the same symmetry around the internuclear axis.
- ☐ 10. The bond order of a diatomic molecule is $b = \frac{1}{2}(n - n^*)$, where n and n^* are the numbers of electrons in bonding and antibonding orbitals, respectively.
- ☐ 11. The electronegativity, χ, of an element is the power of its atoms to draw electrons to itself when it is part of a compound.
- ☐ 12. In a bond between dissimilar atoms, the atomic orbital belonging to the more electronegative atom makes the larger contribution to the molecular orbital with the lowest energy. For the molecular orbital with the highest energy, the principal contribution comes from the atomic orbital belonging to the less electronegative atom.

☐ 13. The hamiltonian matrix, $\boldsymbol{H}$, is formed of all integrals $H_{ij}=\int\psi_i^*\hat{H}\psi_j\mathrm{d}\tau$. The overlap matrix, $\boldsymbol{S}$, is formed of all $S_{ij}=\int\psi_i^*\psi_j\mathrm{d}\tau$.

☐ 14. The variation principle states that if an arbitrary wavefunction is used to calculate the energy, the value calculated is never less than the true energy.

☐ 15. In the Hückel method, all Coulomb integrals H_{ii} are set equal (to α), all overlap integrals are set equal to zero, all resonance integrals H_{ij} between non-neighbours are set equal to zero, and all remaining resonance integrals are set equal (to β).

☐ 16. The π-electron binding energy is the sum of the energies of each π electron. The π-bond formation energy is the energy released when a π bond is formed. The delocalization energy is the extra stabilization of a conjugated system.

☐ 17. In the self-consistent field procedure, an initial guess about the composition of the molecular orbitals is successively refined until the solution remains unchanged in a cycle of calculations.

☐ 18. In semi-empirical methods for the determination of electronic structure, the Schrödinger equation is written in terms of parameters chosen to agree with selected experimental quantities. In *ab initio* and density functional methods, the Schrödinger equation is solved numerically, without the need of parameters that appeal to experimental data.

Further reading

Articles and texts

T.A. Albright and J.K. Burdett, *Problems in molecular orbital theory*. Oxford University Press (1992).

P.W. Atkins and R.S. Friedman, *Molecular quantum mechanics*. Oxford University Press (2005).

I.N. Levine, *Quantum chemistry*. Prentice–Hall, Upper Saddle River (2000).

D.A. McQuarrie, *Mathematical methods for scientists and engineers*. University Science Books, Mill Valley (2003).

R.C. Mebane, S.A. Schanley, T.R. Rybolt, and C.D. Bruce, The correlation of physical properties of organic molecules with computed molecular surface areas. *J. Chem. Educ.* **76**, 688 (1999).

L. Pauling, *The nature of the chemical bond*. Cornell University Press, Ithaca (1960).

C.M. Quinn, *Computational quantum chemistry: an interactive guide to basis set theory*. Academic Press, San Diego (2002).

Sources of data and information

D.R. Lide (ed.), *CRC handbook of chemistry and physics*, Section 9, CRC Press, Boca Raton (2000).

P.R. Scott and W.G. Richards, *Energy levels in atoms and molecules*. Oxford Chemistry Primers, Oxford University Press (1994).

Each end-of-chapter discussion question, exercise, and problem for which a solution is available in the Student Solution Manual has a corresponding bracketed number referring to the Student Solutions Manual number for this problem, for example [SSM 11.1]. Each end-of-chapter exercise and problem for which a solution is available in the Instructors' Solution Manual has a corresponding bracketed number referring to the Instructor Solutions Manual number for this problem, for example [ISM 11.2].

Discussion questions

4.1 Compare the approximations built into valence-bond theory and molecular-orbital theory. [SSM **11.1**]

4.2 Discuss the steps involved in the construction of sp^3, sp^2, and sp hybrid orbitals. [ISM **11.2**]

4.3 Distinguish between the Pauling and Mulliken electronegativity scales. [SSM **11.3**]

4.4 Discuss the steps involved in the calculation of the energy of a system by using the variation principle. [ISM **11.4**]

4.5 Discuss the approximations built into the Hückel method. [SSM **11.5**]

4.6 Distinguish between delocalization energy, π-electron binding energy, and π-bond formation energy. [ISM **11.6**]

4.7 Use concepts of molecular orbital theory to describe the biochemical reactivity of O_2, N_2, and NO. [SSM **11.7**]

4.8 Distinguish between semi-empirical, *ab initio*, and density functional theory methods of electronic structure determination. [ISM **11.8**]

Exercises

4.1a Give the ground-state electron configurations and bond orders of (a) Li_2, (b) Be_2, and (c) C_2. [ISM **11.1a**]

4.1b Give the ground-state electron configurations of (a) H_2^-, (b) N_2, and (c) O_2. [SSM **11.1b**]

4.2a Give the ground-state electron configurations of (a) CO, (b) NO, and (c) CN^-. [ISM **11.2a**]

4.2b Give the ground-state electron configurations of (a) ClF, (b) CS, and (c) O_2^-. [SSM **11.2b**]

4.3a From the ground-state electron configurations of B_2 and C_2, predict which molecule should have the greater bond dissociation energy. [ISM **11.3a**]

4.3b Which of the molecules N_2, NO, O_2, C_2, F_2, and CN would you expect to be stabilized by (a) the addition of an electron to form AB^-, (b) the removal of an electron to form AB^+? [SSM **11.3b**]

4.4a Sketch the molecular orbital energy level diagram for XeF and deduce its ground-state electron configurations. Is XeF likely to have a shorter bond length than XeF^+? [ISM **11.4a**]

4.4b Sketch the molecular orbital energy level diagrams for BrCl and deduce its ground-state electron configurations. Is BrCl likely to have a shorter bond length than $BrCl^-$? [SSM **11.4b**]

4.5a Use the electron configurations of NO and N_2 to predict which is likely to have the shorter bond length. [ISM **11.5a**]

4.5b Arrange the species O_2^+, O_2, O_2^-, O_2^{2-} in order of increasing bond length. [SSM **11.5b**]

4.6a Show that the sp^2 hybrid orbital $(s + 2^{1/2}p)/3^{1/2}$ is normalized to 1 if the s and p orbitals are normalized to 1. [ISM **11.6a**]

4.6b Normalize the molecular orbital $\psi_A + \lambda\psi_B$ in terms of the parameter λ and the overlap integral S. [SSM **11.6b**]

4.7a Confirm that the bonding and antibonding combinations $\psi_A \pm \psi_B$ are mutually orthogonal in the sense that their mutual overlap is zero. [ISM **11.7a**]

4.7b Suppose that a molecular orbital has the form $N(0.145A + 0.844B)$. Find a linear combination of the orbitals A and B that is orthogonal to this combination. [SSM **11.7b**]

4.8a Can the function $\psi = x(L - x)$ be used as a trial wavefunction for the $n = 1$ state of a particle with mass m in a one-dimensional box of length L? If the answer is yes, then express the energy of this trial wavefunction in terms of h, m, and L and compare it with the exact result (eqn 2.4). If the answer is no, explain why this is not a suitable trial wavefunction. [ISM **11.8a**]

4.8b Can the function $\psi = x^2(L - 2x)$ be used as a trial wavefunction for the $n = 1$ state of a particle with mass m in a one-dimensional box of length L? If the answer is yes, then express the energy of this trial wavefunction in terms of h, m, and L and compare it with the exact result (eqn 2.4). If the answer is no, explain why this is not a suitable trial wavefunction. [SSM **11.8b**]

4.9a Suppose that the function $\psi = Ae^{-ar^2}$, with A being the normalization constant and a being an adjustable parameter, is used as a trial wavefunction for the 1s orbital of the hydrogen atom. Express the energy of this trial wavefunction as a function of the h, a, e, the electron charge, and μ, the effective mass of the H atom. [ISM **11.9a**]

4.9b Suppose that the function $\psi = Ae^{-ar^2}$, with A being the normalization constant and a being an adjustable parameter, is used as a trial wavefunction for the 1s orbital of the hydrogen atom. The energy of this trial wavefunction is

$$E = \frac{3a\hbar^2}{2\mu} - \frac{e^2}{\varepsilon_0}\left(\frac{a}{2\pi^3}\right)^{1/2}$$

where e is the electron charge, and μ is the effective mass of the H atom. What is the minimum energy associated with this trial wavefunction? [SSM **11.9b**]

4.10a What is the energy of an electron that has been ejected from an orbital of ionization energy 11.0 eV by a photon of radiation of wavelength 100 nm? [ISM **11.10a**]

4.10b What is the energy of an electron that has been ejected from an orbital of ionization energy 4.69 eV by a photon of radiation of wavelength 584 pm? [SSM **11.10b**]

4.11a Construct the molecular orbital energy level diagrams of ethene on the basis that the molecule is formed from the appropriately hybridized CH_2 or CH fragments. [ISM **11.11a**]

4.11b Construct the molecular orbital energy level diagrams of ethyne (acetylene) on the basis that the molecule is formed from the appropriately hybridized CH_2 or CH fragments. [SSM **11.11b**]

4.12a Write down the secular determinants for (a) linear H_3, (b) cyclic H_3 within the Hückel approximation. [ISM **11.12a**]

4.12b Predict the electronic configurations of (a) the benzene anion, (b) the benzene cation. Estimate the π-electron binding energy in each case. [SSM **11.12b**]

4.13a Write down the secular determinants (a) anthracene (**8**), (b) phenanthrene (**9**) within the Hückel approximation and using the C2p orbitals as the basis set. [ISM **11.13a**]

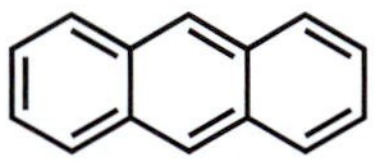

8

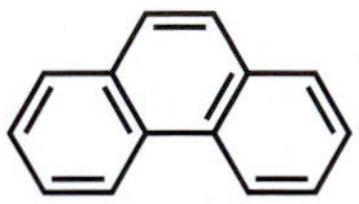

9

4.13b Use mathematical software to estimate the π-electron binding energy of (a) anthracene (**7**), (b) phenanthrene (**8**) within the Hückel approximation. [SSM **11.13b**]

Problems*

Numerical problems

4.1 Show that, if a wave cos kx centred on A (so that x is measured from A) interferes with a similar wave cos $k'x$ centred on B (with x measured from B) a distance R away, then constructive interference occurs in the intermediate region when $k = k' = \pi/2R$ and destructive interference if $kR = \frac{1}{2}\pi$ and $k'R = \frac{3}{2}\pi$. [SSM **11.1**]

4.2 The overlap integral between two H1s orbitals on nuclei separated by a distance R is $S = \{1 + (R/a_0) + \frac{1}{3}(R/a_0)^2\}e^{-R/a_0}$. Plot this function for $0 \le R < \infty$. [ISM **11.2**]

4.3 Before doing the calculation below, sketch how the overlap between a 1s orbital and a 2p orbital can be expected to depend on their separation. The overlap integral between an H1s orbital and an H2p orbital on nuclei separated by a distance R and forming a σ orbital is $S = (R/a_0)\{1 + (R/a_0) + \frac{1}{3}(R/a_0)^2\}e^{-R/a_0}$. Plot this function, and find the separation for which the overlap is a maximum. [SSM **11.3**]

4.4 Calculate the total amplitude of the normalized bonding and antibonding LCAO-MOs that may be formed from two H1s orbitals at a separation of 106 pm. Plot the two amplitudes for positions along the molecular axis both inside and outside the internuclear region. [ISM **11.4**]

4.5 Repeat the calculation in Problem 4.4 but plot the probability densities of the two orbitals. Then form the difference density, the difference between ψ^2 and $\frac{1}{2}\{\psi_A^2 + \psi_B^2\}$. [SSM **11.5**]

4.6‡ Use the $2p_x$ and $2p_z$ hydrogenic atomic orbitals to construct simple LCAO descriptions of $2p\sigma$ and $2p\pi$ molecular orbitals. (a) Make a probability density plot, and both surface and contour plots of the xz-plane amplitudes of the $2p_z\sigma$ and $2p_z\sigma^*$ molecular orbitals. (b) Make surface and contour plots of the xz-plane amplitudes of the $2p_x\pi$ and $2p_x\pi^*$ molecular orbitals. Include plots for both internuclear distances, R, of $10a_0$ and $3a_0$, where $a_0 = 52.9$ pm. Interpret the graphs, and describe why this graphical information is useful. [ISM **11.6**]

4.7 Imagine a small electron-sensitive probe of volume 1.00 pm^3 inserted into an H_2^+ molecule-ion in its ground state. Calculate the probability that it will register the presence of an electron at the following positions: (a) at nucleus A, (b) at nucleus B, (c) half-way between A and B, (c) at a point 20 pm along the bond from A and 10 pm perpendicularly. Do the same for the molecule-ion the instant after the electron has been excited into the antibonding LCAO-MO. [SSM **11.7**]

4.8 The energy of H_2^+ with internuclear separation R is given by the expression

$$E = E_H + \frac{e^2}{4\pi\varepsilon_0 R} - \frac{V_1 + V_2}{1 + S}$$

where E_H is the energy of an isolated H atom, V_1 is the attractive potential energy between the electron centred on one nucleus and the charge of the other nucleus, V_2 is the attraction between the overlap density and one of the nuclei, S is the overlap integral. The values are given below. Plot the molecular potential energy curve and find the bond dissociation energy (in electronvolts) and the equilibrium bond length.

R/a_0	0	1	2	3	4
V_1/E_h	11.000	10.729	10.473	10.330	10.250
V_2/E_h	11.000	10.736	10.406	10.199	10.092
S	1.000	0.858	0.587	0.349	0.189

where $E_h = 27.3$ eV and $a_0 = 52.9$ pm and $E_H = -\frac{1}{2}E_h$. [ISM **11.8**]

4.9 The same data as in Problem 4.8 may be used to calculate the molecular potential energy curve for the antibonding orbital, which is given by

$$E = E_H + \frac{e^2}{4\pi\varepsilon_0 R} - \frac{V_1 - V_2}{1 - S}$$

Plot the curve. [SSM **11.9**]

4.10‡ J.G. Dojahn, E.C.M. Chen, and W.E. Wentworth (*J. Phys. Chem.* **100**, 9649 (1996)) characterized the potential energy curves of homonuclear diatomic halogen molecules and molecular anions. Among the properties they report are the equilibrium internuclear distance R_e, the vibrational wavenumber, $\tilde{\nu}$, and the dissociation energy, D_e:

Species	R_e	$\tilde{\nu}$/cm^{-1}	D_e/eV
F_2	1.411	916.6	1.60
F_2^-	1.900	450.0	1.31

Rationalize these data in terms of molecular orbital configurations. [ISM **11.10**]

4.11‡ *Rydberg molecules* are molecules with an electron in an atomic orbital with principal quantum number n one higher than the valence shells of the constituent atoms. Speculate about the existence of 'hyper Rydberg' H_2 formed from two H atoms with 100s electrons. Make reasonable guesses about the binding energy, the equilibrium internuclear separation, the vibrational force constant, and the rotational constant. Is such a molecule likely to exist under any circumstances? [SSM **11.11**]

4.12 In a particular photoelectron spectrum using 21.21 eV photons, electrons were ejected with kinetic energies of 11.01 eV, 8.23 eV, and 5.22 eV. Sketch the molecular orbital energy level diagram for the species, showing the ionization energies of the three identifiable orbitals. [ISM **11.12**]

4.13‡ Set up and solve the Hückel secular equations for the π electrons of NO_3^-. Express the energies in terms of the Coulomb integrals α_O and α_N and the resonance integral β. Determine the delocalization energy of the ion. [SSM **11.13**]

4.14 In the 'free electron molecular orbital' (FEMO) theory, the electrons in a conjugated molecule are treated as independent particles in a box of length L. Sketch the form of the two occupied orbitals in butadiene predicted by this model and predict the minimum excitation energy of the molecule. The tetraene $CH_2{=}CHCH{=}CHCH{=}CHCH{=}CH_2$ can be treated as a box of length $8R$, where $R \approx 140$ pm (as in this case, an extra half bond-length is often added at each end of the box). Calculate the minimum excitation energy of the molecule and sketch the HOMO and LUMO. Estimate the colour a sample of the compound is likely to appear in white light. [ISM **11.14**]

4.15 The FEMO theory (Problem 4.14) of conjugated molecules is rather crude and better results are obtained with simple Hückel theory. (a) For a linear conjugated polyene with each of N carbon atoms contributing an electron in a 2p orbital, the energies E_k of the resulting π molecular orbitals are given by:

$$E_k = \alpha + 2\beta\cos\frac{k\pi}{N+1} \qquad k = 1, 2, 3, \ldots, N$$

Use this expression to determine a reasonable empirical estimate of the resonance integral β for the homologous series consisting of ethene,

* Problems denoted with the symbol ‡ were supplied by Charles Trapp, Carmen Giunta, and Marshall Cady.

butadiene, hexatriene, and octatetraene given that $\pi^* \leftarrow \pi$ ultraviolet absorptions from the HOMO to the LUMO occur at 61 500, 46 080, 39 750, and 32 900 cm^{-1}, respectively. (b) Calculate the π-electron delocalization energy, $E_{\text{deloc}} = E_\pi - n(\alpha + \beta)$, of octatetraene, where E_π is the total π-electron binding energy and n is the total number of π-electrons. (c) In the context of this Hückel model, the π molecular orbitals are written as linear combinations of the carbon 2p orbitals. The coefficient of the jth atomic orbital in the kth molecular orbital is given by:

$$c_{kj} = \left(\frac{2}{N+1}\right)^{1/2} \sin\frac{jk\pi}{N+1} \qquad j = 1, 2, 3, \ldots, N$$

Determine the values of the coefficients of each of the six 2p orbitals in each of the six π molecular orbitals of hexatriene. Match each set of coefficients (that is, each molecular orbital) with a value of the energy calculated with the expression given in part (a) of the molecular orbital. Comment on trends that relate the energy of a molecular orbital with its 'shape', which can be inferred from the magnitudes and signs of the coefficients in the linear combination that describes the molecular orbital. [SSM **11.15**]

4.16 For monocyclic conjugated polyenes (such as cyclobutadiene and benzene) with each of N carbon atoms contributing an electron in a 2p orbital, simple Hückel theory gives the following expression for the energies E_k of the resulting π molecular orbitals:

$$E_k = \alpha + 2\beta\cos\frac{2k\pi}{N} \qquad k = 0, \pm1, \pm2, \ldots, \pm N/2 \text{ (even } N)$$

$$k = 0, \pm1, \pm2, \ldots, \pm(N-1)/2 \text{ (odd } N)$$

(a) Calculate the energies of the π molecular orbitals of benzene and cyclooctatetraene. Comment on the presence or absence of degenerate energy levels. (b) Calculate and compare the delocalization energies of benzene (using the expression above) and hexatriene (see Problem 4.15a). What do you conclude from your results? (c) Calculate and compare the delocalization energies of cyclooctaene and octatetraene. Are your conclusions for this pair of molecules the same as for the pair of molecules investigated in part (b)? [ISM **11.16**]

4.17 If you have access to mathematical software that can perform matrix diagonalization, use it to solve Problems 4.15 and 4.16, disregarding the expressions for the energies and coefficients given there. [SSM **11.17**]

4.18 Molecular orbital calculations based on semi-empirical, *ab initio*, and DFT methods describe the spectroscopic properties of conjugated molecules a bit better than simple Hückel theory. (a) Using molecular modelling software[2] and the computational method of your choice (semi-empirical, *ab initio*, or density functioncal methods), calculate the energy separation between the HOMO and LUMO of ethene, butadiene, hexatriene, and octatetraene. (b) Plot the HOMO–LUMO energy separations against the experimental frequencies for $\pi^* \leftarrow \pi$ ultraviolet absorptions for these molecules (Problem 4.15). Use mathematical software to find the polynomial equation that best fits the data. (c) Use your polynomial fit from part (b) to estimate the frequency of the $\pi^* \leftarrow \pi$ ultraviolet absorption of decapentaene from the calculated HOMO–LUMO energy separation. (d) Discuss why the calibration procedure of part (b) is necessary. [ISM **11.18**]

4.19 Electronic excitation of a molecule may weaken or strengthen some bonds because bonding and antibonding characteristics differ between the HOMO and the LUMO. For example, a carbon–carbon bond in a linear polyene may have bonding character in the HOMO and antibonding character in the LUMO. Therefore, promotion of an electron from the HOMO to the LUMO weakens this carbon–carbon bond in the excited electronic state, relative to the ground electronic state. Display the HOMO and LUMO of each molecule in Problem 4.15 and discuss in detail any changes in bond order that accompany the $\pi^* \leftarrow \pi$ ultraviolet absorptions in these molecules. [SSM **11.19**]

4.20 Molecular electronic structure methods may be used to estimate the standard enthalpy of formation of molecules in the gas phase. (a) Using molecular modelling software and a semi-empirical method of your choice, calculate the standard enthalpy of formation of ethene, butadiene, hexatriene, and octatetraene in the gas phase. (b) Consult a database of thermochemical data, such as the online sources listed in this textbook's web site, and, for each molecule in part (a), calculate the relative error between the calculated and experimental values of the standard enthalpy of formation. (c) A good thermochemical database will also report the uncertainty in the experimental value of the standard enthalpy of formation. Compare experimental uncertainties with the relative errors calculated in part (b) and discuss the reliability of your chosen semi-empirical method for the estimation of thermochemical properties of linear polyenes. [ISM **11.20**]

Theoretical problems

4.21 An sp^2 hybrid orbital that lies in the xy-plane and makes an angle of 120° to the x-axis has the form

$$\psi = \frac{1}{3^{1/2}}\left(s - \frac{1}{2^{1/2}}p_x + \frac{3^{1/2}}{2^{1/2}}p_y\right)$$

Use hydrogenic atomic orbitals to write the explicit form of the hybrid orbital. Show that it has its maximum amplitude in the direction specified. [SSM **11.21**]

4.22 Use the expressions in Problems 4.8 and 4.9 to show that the antibonding orbital is more antibonding than the bonding orbital is bonding at most internuclear separations. [ISM **11.22**]

4.23 Derive eqns 4.11 and 4.14 by working with the normalized LCAO-MOs for the H_2^+ molecule-ion (Section 4.3a). Proceed by evaluating the expectation value of the hamiltonian for the ion. Make use of the fact that A and B each individually satisfy the Schrödinger equation for an isolated H atom. [SSM **11.23**]

4.24 Take as a trial function for the ground state of the hydrogen atom (a) e^{-kr}, (b) e^{-kr^2} and use the variation principle to find the optimum value of k in each case. Identify the better wavefunction. The only part of the laplacian that need be considered is the part that involves radial derivatives (eqn 2.5). [ISM **11.24**]

4.25 We saw in Section 4.5 that, to find the energies of the bonding and antibonding orbitals of a heteronuclear diatomic molecule, we need to solve the secular determinant

$$\begin{vmatrix} \alpha_A - E & B \\ \beta & \alpha_B - E \end{vmatrix} = 0$$

where $\alpha_A \neq \alpha_B$ and we have taken $S = 0$. Equations 4.34a and 4.34b give the general solution to this problem. Here, we shall develop the result for the case $(\alpha_B - \alpha_A)^2 \gg \beta^2$. (a) Begin by showing that

$$E_\pm = \frac{\alpha_A + \alpha_B}{2} \pm \frac{\alpha_A - \alpha_B}{2}\left[1 + \frac{4\beta^2}{(\alpha_A - \alpha_B)^2}\right]^{1/2}$$

where E_+ and E_- are the energies of the bonding and antibonding molecular orbitals, respectively. (b) Now use the expansion

$$(1+x)^{1/2} = 1 + \frac{x}{2} - \frac{x^3}{8} + \cdots$$

to show that

$$E_- = \alpha_B + \frac{\beta^2}{\alpha_B - \alpha_A} \qquad E_+ = \alpha_A - \frac{\beta^2}{\alpha_B - \alpha_A}$$

which is the limiting result used in *Justification 4.4*. [SSM **11.25**]

[2] The web site contains links to molecular modelling freeware and to other sites where you may perform molecular orbital calculations directly from your web browser.

Applications: to astrophysics and biology

4.26‡ In Exercise 4.12a you were invited to set up the Hückel secular determinant for linear and cyclic H_3. The same secular determinant applies to the molecular ions H_3^+ and D_3^+. The molecular ion H_3^+ was discovered as long ago as 1912 by J.J. Thomson, but only more recently has the equivalent equilateral triangular structure been confirmed by M.J. Gaillard *et al.* (*Phys. Rev.* **A17**, 1797 (1978)). The molecular ion H_3^+ is the simplest polyatomic species with a confirmed existence and plays an important role in chemical reactions occurring in interstellar clouds that may lead to the formation of water, carbon monoxide, and ethyl alcohol. The H_3^+ ion has also been found in the atmospheres of Jupiter, Saturn, and Uranus. (a) Solve the Hückel secular equations for the energies of the H_3 system in terms of the parameters α and β, draw an energy level diagram for the orbitals, and determine the binding energies of H_3^+, H_3, and H_3^-. (b) Accurate quantum mechanical calculations by G.D. Carney and R.N. Porter (*J. Chem. Phys.* **65**, 3547 (1976)) give the dissociation energy for the process $H_3^+ \rightarrow H + H + H^+$ as 849 kJ mol^{-1}. From this information and data in Table 4.3, calculate the enthalpy of the reaction $H^+(g) + H_2(g) \rightarrow H_3^+(g)$. (c) From your equations and the information given, calculate a value for the resonance integral β in H_3^+. Then go on to calculate the bind energies of the other H_3 species in (a). [ISM **11.26**]

4.27‡ There is some indication that other hydrogen ring compounds and ions in addition to H_3 and D_3 species may play a role in interstellar chemistry. According to J.S. Wright and G.A. DiLabio (*J. Phys. Chem.* **96**, 10793 (1992)), H_5^-, H_6, and H_7^+ are particularly stable whereas H_4 and H_5^+ are not. Confirm these statements by Hückel calculations. [SSM **11.27**]

4.28 Here we develop a molecular orbital theory treatment of the peptide group (**10**), which links amino acids in proteins. Specifically, we shall describe the factors that stabilize the planar conformation of the peptide group.

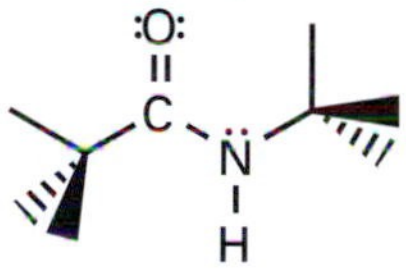

10

(a) It will be familiar from introductory chemistry that valence bond theory explains the planar conformation of the peptide group by invoking delocalization of the π bond between the oxygen, carbon, and nitrogen atoms (**11**, **11**):

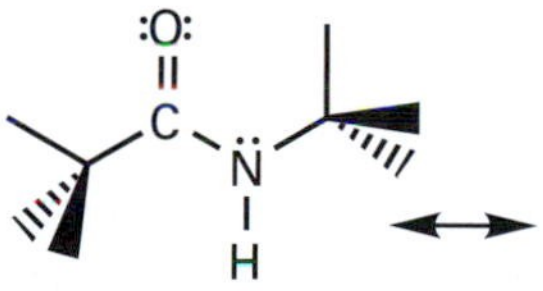

11

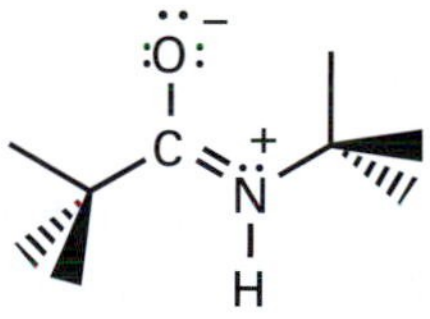

12

It follows that we can model the peptide group with molecular orbital theory by making LCAO-MOs from 2*p* orbitals perpendicular to the plane defined by the O, C, and N atoms. The three combinations have the form:

$$\psi_1 = a\psi_O + b\psi_C + c\psi_N \qquad \psi_2 = d\psi_O - e\psi_N \qquad \psi_3 = f\psi_O - g\psi_C + h\psi_N$$

where the coefficients *a* through *h* are all positive. Sketch the orbitals ψ_1, ψ_2, and ψ_3 and characterize them as bonding, non-bonding, or antibonding molecular orbitals. In a non-bonding molecular orbital, a pair of electrons resides in an orbital confined largely to one atom and not appreciably involved in bond formation. (b) Show that this treatment is consistent only with a planar conformation of the peptide link. (c) Draw a diagram showing the relative energies of these molecular orbitals and determine the occupancy of the orbitals. *Hint.* Convince yourself that there are four electrons to be distributed among the molecular orbitals. (d) Now consider a non-planar conformation of the peptide link, in which the O2*p* and C2*p* orbitals are perpendicular to the plane defined by the O, C, and N atoms, but the N2*p* orbital lies on that plane. The LCAO-MOs are given by

$$\psi_4 = a\psi_O + b\psi_C \qquad \psi_5 = e\psi_N \qquad \psi_6 = f\psi_O - g\psi_C$$

Just as before, sketch these molecular orbitals and characterize them as bonding, non-bonding, or antibonding. Also, draw an energy level diagram and determine the occupancy of the orbitals. (e) Why is this arrangement of atomic orbitals consistent with a non-planar conformation for the peptide link? (f) Does the bonding MO associated with the planar conformation have the same energy as the bonding MO associated with the non-planar conformation? If not, which bonding MO is lower in energy? Repeat the analysis for the non-bonding and anti-bonding molecular orbitals. (g) Use your results from parts (a)–(f) to construct arguments that support the planar model for the peptide link. [ISM **11.28**]

4.29 Molecular orbital calculations may be used to predict trends in the standard potentials of conjugated molecules, such as the quinones and flavins, that are involved in biological electron transfer reactions (*Impact I10.2*). It is commonly assumed that decreasing the energy of the LUMO enhances the ability of a molecule to accept an electron into the LUMO, with an attendant increase in the value of the molecule's standard potential. Furthermore, a number of studies indicate that there is a linear correlation between the LUMO energy and the reduction potential of aromatic hydrocarbons (see, for example, J.P. Lowe, *Quantum chemistry*, Chapter 8, Academic Press (1993)). (a) The standard potentials at pH = 7 for the one-electron reduction of methyl-substituted 1,4-benzoquinones (**13**) to their respective semiquinone radical anions are:

R_2	R_3	R_5	R_6	$E^\ominus$/V
H	H	H	H	0.078
CH_3	H	H	H	0.023
CH_3	H	CH_3	H	−0.067
CH_3	CH_3	CH_3	H	−0.165
CH_3	CH_3	CH_3	CH_3	−0.260

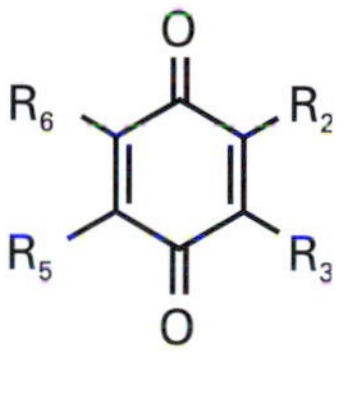

13

Using molecular modelling software and the computational method of your choice (semi-empirical, *ab initio*, or density functional theory methods), calculate E_{LUMO}, the energy of the LUMO of each substituted 1,4-benzoquinone, and plot E_{LUMO} against $E^\ominus$. Do your calculations support a linear relation between E_{LUMO} and $E^\ominus$? (b) The 1,4-benzoquinone for which $R_2 = R_3 = CH_3$ and $R_5 = R_6 = OCH_3$ is a suitable model of ubiquinone, a component of the respiratory electron transport chain. Determine E_{LUMO} of this quinone and then use your results from part (a) to estimate its standard potential. (c) The 1,4-benzoquinone for which $R_2 = R_3 = R_5 = CH_3$ and $R_6 = H$ is a suitable model of plastoquinone, a component of the photosynthetic electron transport chain. Determine E_{LUMO} of this quinone and then use your results from part (a) to estimate its standard potential. Is plastoquinone expected to be a better or worse oxidizing agent than ubiquinone? (d) Based on your predictions and on basic concepts of biological electron transport, suggest a reason why ubiquinone is used in respiration and plastoquinone is used in photosynthesis. [SSM **11.29**]

5 Molecular symmetry

In this chapter we sharpen the concept of 'shape' into a precise definition of 'symmetry', and show that symmetry may be discussed systematically. We see how to classify any molecule according to its symmetry and how to use this classification to discuss molecular properties. After describing the symmetry properties of molecules themselves, we turn to a consideration of the effect of symmetry transformations on orbitals and see that their transformation properties can be used to set up a labelling scheme. These symmetry labels are used to identify integrals that necessarily vanish. One important integral is the overlap integral between two orbitals. By knowing which atomic orbitals may have nonzero overlap, we can decide which ones can contribute to molecular orbitals. We also see how to select linear combinations of atomic orbitals that match the symmetry of the nuclear framework. Finally, by considering the symmetry properties of integrals, we see that it is possible to derive the selection rules that govern spectroscopic transitions.

The systematic discussion of symmetry is called **group theory**. Much of group theory is a summary of common sense about the symmetries of objects. However, because group theory is systematic, its rules can be applied in a straightforward, mechanical way. In most cases the theory gives a simple, direct method for arriving at useful conclusions with the minimum of calculation, and this is the aspect we stress here. In some cases, though, it leads to unexpected results.

The symmetry elements of objects

Some objects are 'more symmetrical' than others. A sphere is more symmetrical than a cube because it looks the same after it has been rotated through any angle about any diameter. A cube looks the same only if it is rotated through certain angles about specific axes, such as 90°, 180°, or 270° about an axis passing through the centres of any of its opposite faces (Fig. 5.1), or by 120° or 240° about an axis passing through any of its opposite corners. Similarly, an NH_3 molecule is 'more symmetrical' than an H_2O molecule because NH_3 looks the same after rotations of 120° or 240° about the axis shown in Fig. 5.2, whereas H_2O looks the same only after a rotation of 180°.

An action that leaves an object looking the same after it has been carried out is called a **symmetry operation**. Typical symmetry operations include rotations, reflections, and inversions. There is a corresponding **symmetry element** for each symmetry operation, which is the point, line, or plane with respect to which the symmetry operation is performed. For instance, a rotation (a symmetry operation) is carried out around an axis (the corresponding symmetry element). We shall see that we can classify molecules by identifying all their symmetry elements, and grouping together molecules that

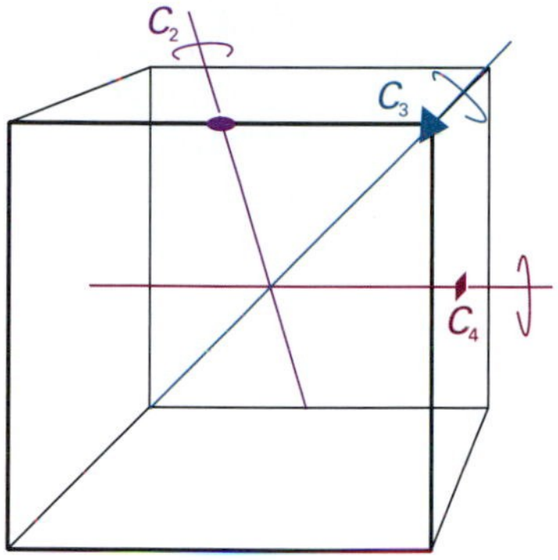

Fig. 5.1 Some of the symmetry elements of a cube. The twofold, threefold, and fourfold axes are labelled with the conventional symbols.

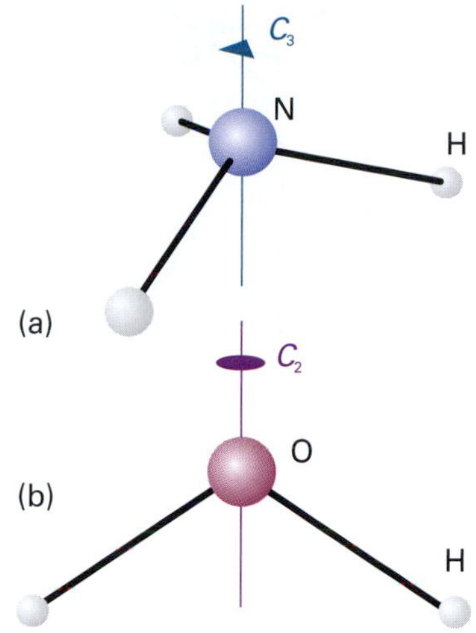

Fig. 5.2 (a) An NH_3 molecule has a threefold (C_3) axis and (b) an H_2O molecule has a twofold (C_2) axis. Both have other symmetry elements too.

possess the same set of symmetry elements. This procedure, for example, puts the trigonal pyramidal species NH_3 and SO_3^{2-} into one group and the angular species H_2O and SO_2 into another group.

5.1 Operations and symmetry elements

The classification of objects according to symmetry elements corresponding to operations that leave at least one common point unchanged gives rise to the **point groups**. There are five kinds of symmetry operation (and five kinds of symmetry element) of this kind. To apply group theory to solids we need to consider symmetries arising from translation through space. These more extensive groups are called **space groups**.

The **identity**, E, consists of doing nothing; the corresponding symmetry element is the entire object. Because every molecule is indistinguishable from itself if nothing is done to it, every object possesses at least the identity element. One reason for including the identity is that some molecules have only this symmetry element (**1**); another reason is technical and connected with the detailed formulation of group theory.

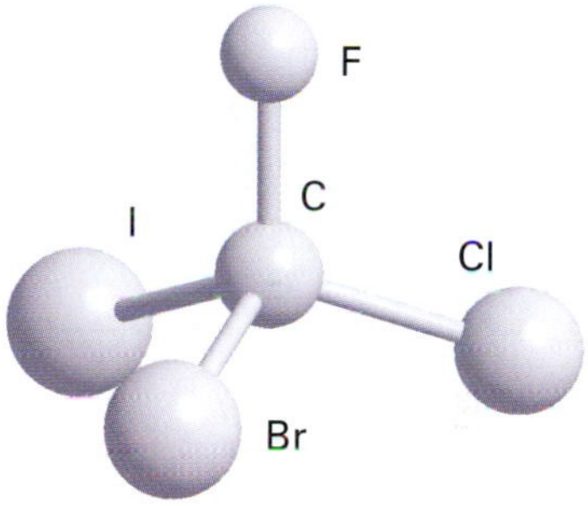

1 CBrClFI

An ***n*-fold rotation** (the operation) about an ***n*-fold axis of symmetry**, C_n (the corresponding element) is a rotation through 360°/n. The operation C_1 is a rotation through 360°, and is equivalent to the identity operation E. An H_2O molecule has one twofold axis, C_2. An NH_3 molecule has one threefold axis, C_3, with which is associated two symmetry operations, one being 120° rotation in a clockwise sense and the other 120° rotation in a counter-clockwise sense. A pentagon has a C_5 axis, with two (clockwise and counterclockwise) rotations through 72° associated with it. It also has an axis denoted C_5^2, corresponding to two successive C_5 rotations; there are two such operations, one through 144° in a clockwise sense and the other through 144° in a counter-clockwise sense. A cube has three C_4 axes, four C_3 axes, and six C_2 axes. However, even this high symmetry is exceeded by a sphere, which possesses an infinite number of symmetry axes (along any diameter) of all possible integral values of n. If a molecule possesses several rotation axes, then the one (or more) with the greatest value of n is called the **principal axis**. The principal axis of a benzene molecule is the sixfold axis perpendicular to the hexagonal ring (**2**).

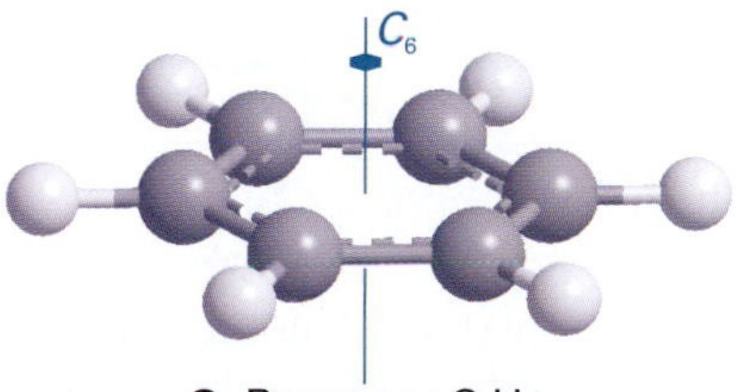

2 Benzene, C_6H_6

Comment 5.1

There is only one twofold rotation associated with a C_2 axis because clockwise and counter-clockwise 180° rotations are identical.

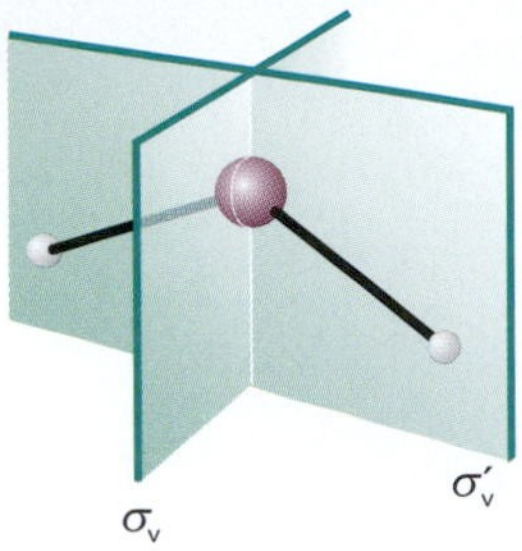

Fig. 5.3 An H_2O molecule has two mirror planes. They are both vertical (i.e. contain the principal axis), so are denoted σ_v and σ'_v.

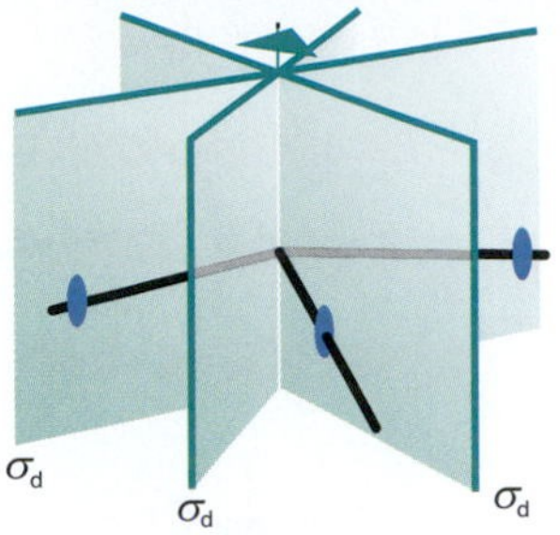

Fig. 5.4 Dihedral mirror planes (σ_d) bisect the C_2 axes perpendicular to the principal axis.

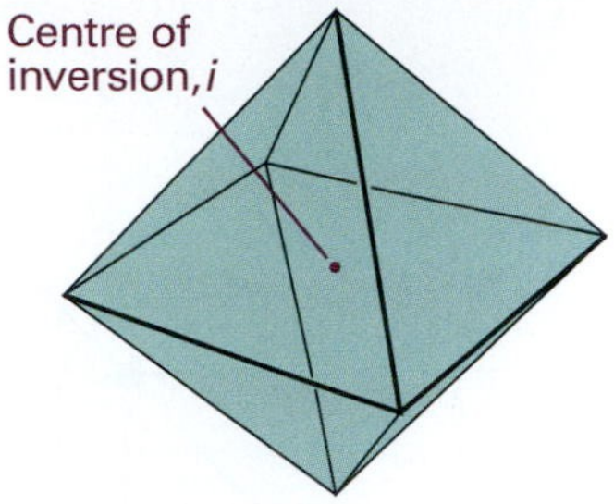

Fig. 5.5 A regular octahedron has a centre of inversion (i).

A **reflection** (the operation) in a **mirror plane,** σ (the element), may contain the principal axis of a molecule or be perpendicular to it. If the plane is parallel to the principal axis, it is called 'vertical' and denoted σ_v. An H_2O molecule has two vertical planes of symmetry (Fig. 5.3) and an NH_3 molecule has three. A vertical mirror plane that bisects the angle between two C_2 axes is called a 'dihedral plane' and is denoted σ_d (Fig. 5.4). When the plane of symmetry is perpendicular to the principal axis it is called 'horizontal' and denoted σ_h. A C_6H_6 molecule has a C_6 principal axis and a horizontal mirror plane (as well as several other symmetry elements).

In an **inversion** (the operation) through a **centre of symmetry,** i (the element), we imagine taking each point in a molecule, moving it to the centre of the molecule, and then moving it out the same distance on the other side; that is, the point (x, y, z) is taken into the point $(-x, -y, -z)$. Neither an H_2O molecule nor an NH_3 molecule has a centre of inversion, but a sphere and a cube do have one. A C_6H_6 molecule does have a centre of inversion, as does a regular octahedron (Fig. 5.5); a regular tetrahedron and a CH_4 molecule do not.

An ***n*-fold improper rotation** (the operation) about an ***n*-fold axis of improper rotation** or an ***n*-fold improper rotation axis,** S_n, (the symmetry element) is composed of two successive transformations. The first component is a rotation through $360°/n$, and the second is a reflection through a plane perpendicular to the axis of that rotation; neither operation alone needs to be a symmetry operation. A CH_4 molecule has three S_4 axes (Fig. 5.6).

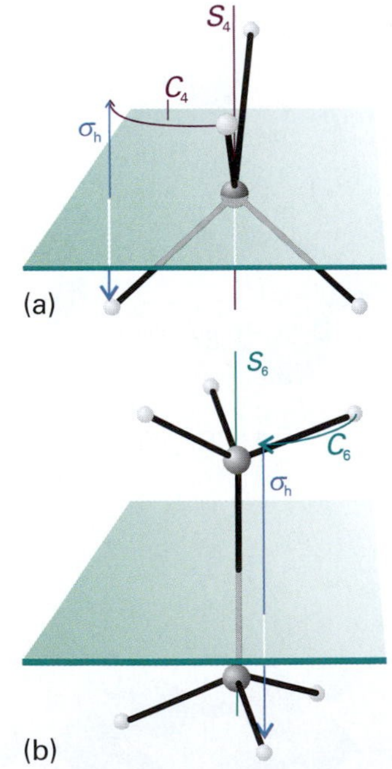

Fig. 5.6 (a) A CH_4 molecule has a fourfold improper rotation axis (S_4): the molecule is indistinguishable after a 90° rotation followed by a reflection across the horizontal plane, but neither operation alone is a symmetry operation. (b) The staggered form of ethane has an S_6 axis composed of a 60° rotation followed by a reflection.

5.2 The symmetry classification of molecules

To classify molecules according to their symmetries, we list their symmetry elements and collect together molecules with the same list of elements. This procedure puts CH_4 and CCl_4, which both possess the same symmetry elements as a regular tetrahedron, into the same group, and H_2O into another group.

The name of the group to which a molecule belongs is determined by the symmetry elements it possesses. There are two systems of notation (Table 5.1). The **Schoenflies system** (in which a name looks like C_{4v}) is more common for the discussion of individual molecules, and the **Hermann–Mauguin system,** or **International system** (in which a name looks like $4mm$), is used almost exclusively in the discussion of crystal symmetry. The identification of a molecule's point group according to the Schoenflies system is simplified by referring to the flow diagram in Fig. 5.7 and the shapes shown in Fig. 5.8.

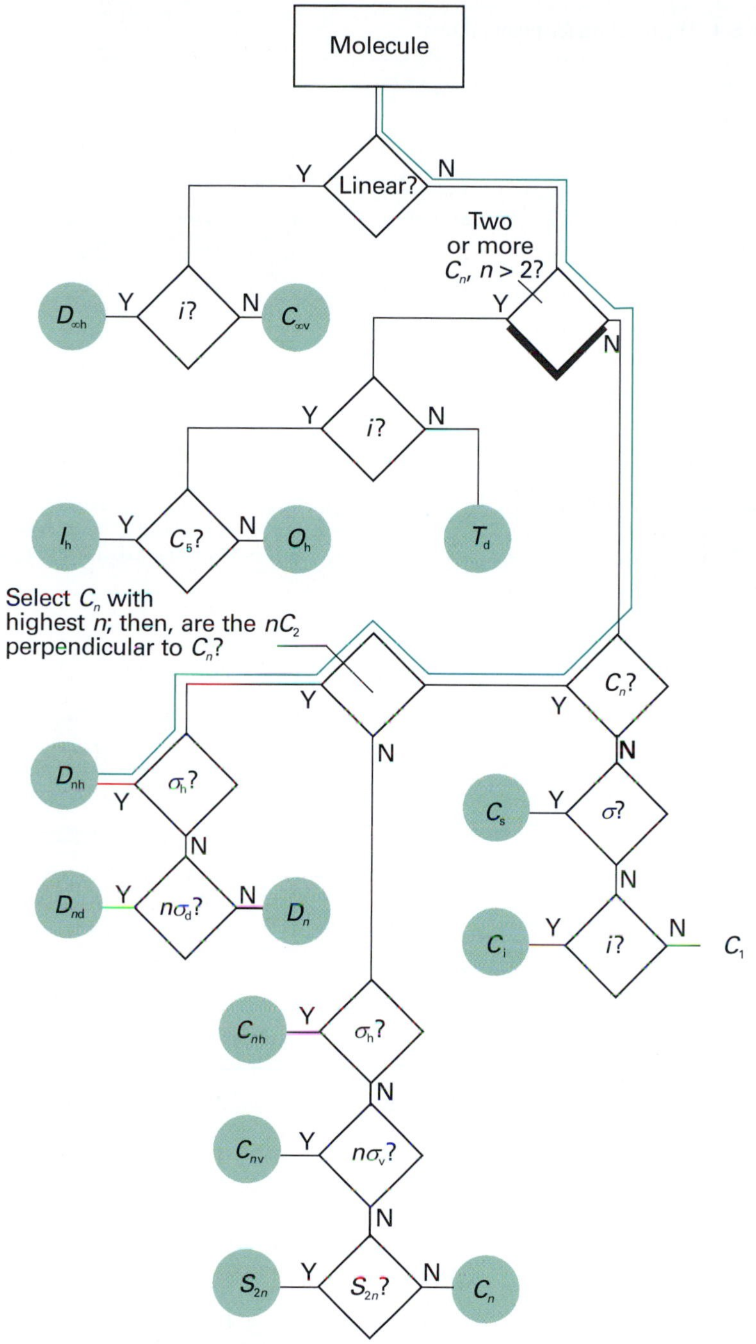

Fig. 5.7 A flow diagram for determining the point group of a molecule. Start at the top and answer the question posed in each diamond (Y = yes, N = no).

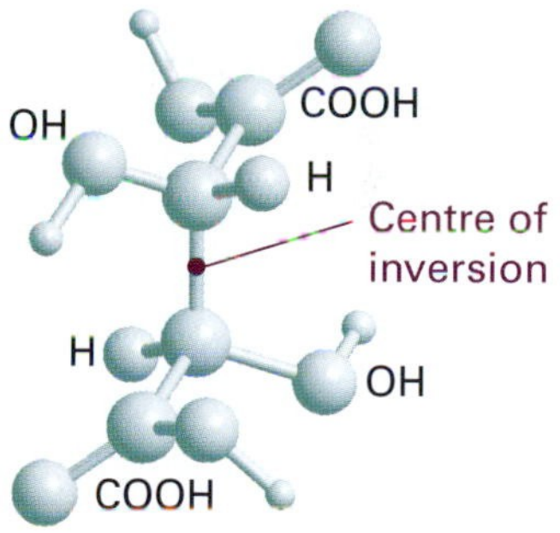

3 Meso-tartaric acid, HOOCCH(OH)CH(OH)COOH

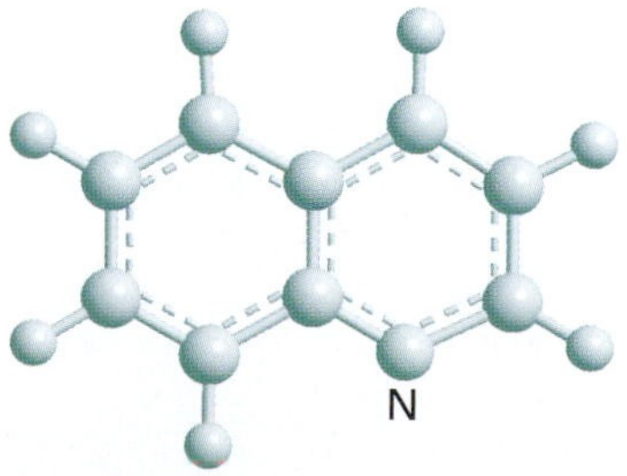

4 Quinoline, C_9H_7N

(a) The groups C_1, C_i, and C_s

A molecule belongs to the group C_1 if it has no element other than the identity, as in (**1**). It belongs to C_i if it has the identity and the inversion alone (**3**), and to C_s if it has the identity and a mirror plane alone (**4**).

Table 5.1 The notation for point groups*

C_i	$\bar{1}$								
C_s	m								
C_1	1	C_2	2	C_3	3	C_4	4	C_6	6
		C_{2v}	$2mm$	C_{3v}	$3m$	C_{4v}	$4mm$	C_{6v}	$6mm$
		C_{2h}	$2m$	C_{3h}	$\bar{6}$	C_{4h}	$4/m$	C_{6h}	$6/m$
		D_2	222	D_3	32	D_4	422	D_6	622
		D_{2h}	mmm	D_{3h}	$\bar{6}2m$	D_{4h}	$4/mmm$	D_{6h}	$6/mmm$
		D_{2d}	$\bar{4}2m$	D_{3d}	$\bar{3}m$	S_4	$\bar{4}/m$	S_6	$\bar{3}$
T	23	T_d	$\bar{4}3m$	T_h	$m3$				
O	432	O_h	$m3m$						

* In the International system (or Hermann–Mauguin system) for point groups, a number n denotes the presence of an n-fold axis and m denotes a mirror plane. A slash (/) indicates that the mirror plane is perpendicular to the symmetry axis. It is important to distinguish symmetry elements of the same type but of different classes, as in $4/mmm$, in which there are three classes of mirror plane. A bar over a number indicates that the element is combined with an inversion. The only groups listed here are the so-called 'crystallographic point groups'.

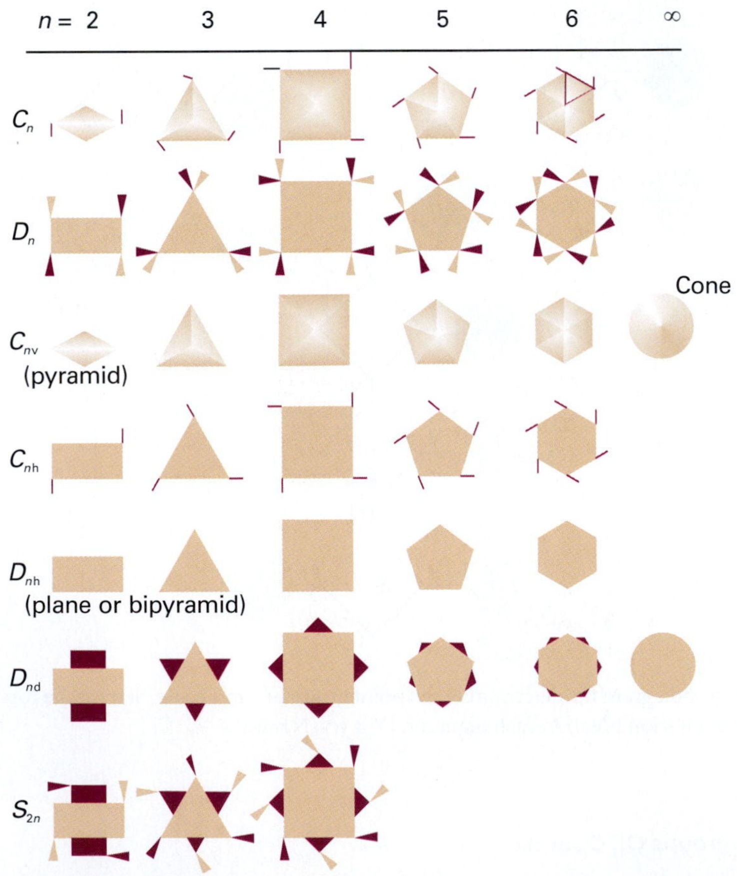

Fig. 5.8 A summary of the shapes corresponding to different point groups. The group to which a molecule belongs can often be identified from this diagram without going through the formal procedure in Fig. 5.7.

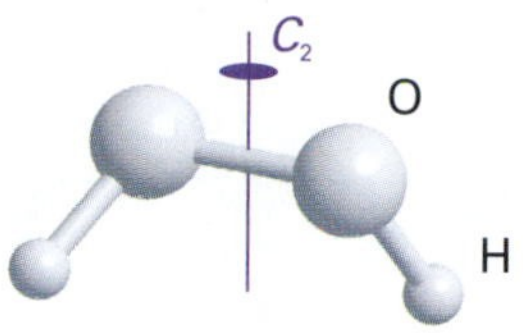

5 Hydrogen peroxide, H_2O_2

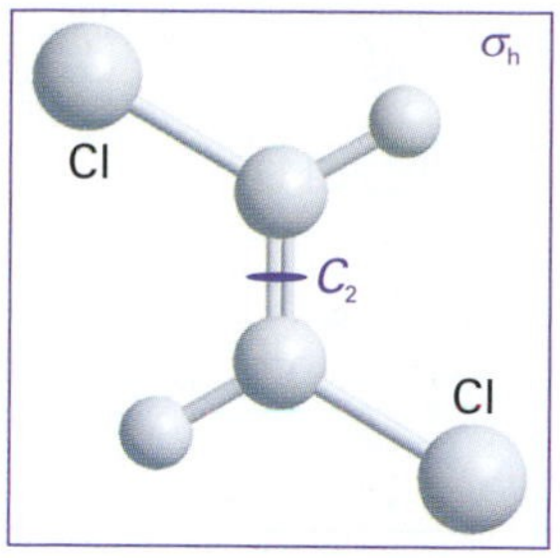

6 *trans*-CHCl=CHCl

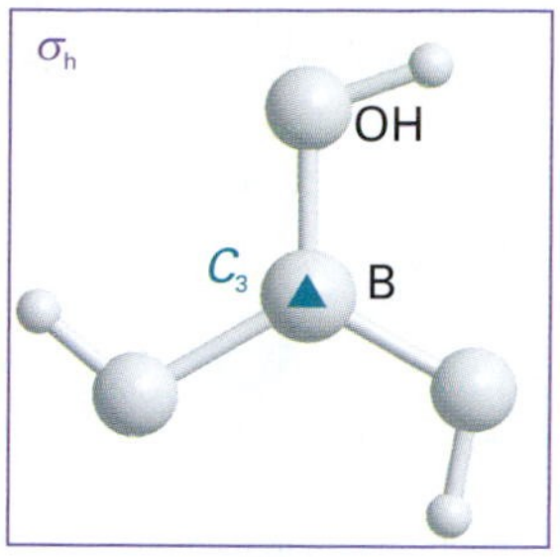

7 $B(OH)_3$

(b) The groups C_n, C_{nv}, and C_{nh}

A molecule belongs to the group C_n if it possesses an n-fold axis. Note that symbol C_n is now playing a triple role: as the label of a symmetry element, a symmetry operation, and a group. For example, an H_2O_2 molecule has the elements E and C_2 (**5**), so it belongs to the group C_2.

If in addition to the identity and a C_n axis a molecule has n vertical mirror planes σ_v, then it belongs to the group C_{nv}. An H_2O molecule, for example, has the symmetry elements E, C_2, and $2\sigma_v$, so it belongs to the group C_{2v}. An NH_3 molecule has the elements E, C_3, and $3\sigma_v$, so it belongs to the group C_{3v}. A heteronuclear diatomic molecule such as HCl belongs to the group $C_{\infty v}$ because all rotations around the axis and reflections across the axis are symmetry operations. Other members of the group $C_{\infty v}$ include the linear OCS molecule and a cone.

Objects that in addition to the identity and an n-fold principal axis also have a horizontal mirror plane σ_h belong to the groups C_{nh}. An example is *trans*-CHCl=CHCl (**6**), which has the elements E, C_2, and σ_h, so belongs to the group C_{2h}; the molecule $B(OH)_3$ in the conformation shown in (**7**) belongs to the group C_{3h}. The presence of certain symmetry elements may be implied by the presence of others: thus, in C_{2h} the operations C_2 and σ_h jointly imply the presence of a centre of inversion (Fig. 5.9).

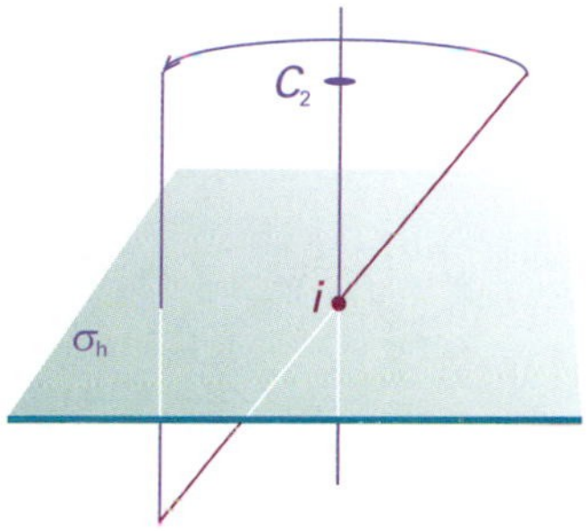

Fig. 5.9 The presence of a twofold axis and a horizontal mirror plane jointly imply the presence of a centre of inversion in the molecule.

(c) The groups D_n, D_{nh}, and D_{nd}

We see from Fig. 5.7 that a molecule that has an n-fold principal axis and n twofold axes perpendicular to C_n belongs to the group D_n. A molecule belongs to D_{nh} if it also possesses a horizontal mirror plane. The planar trigonal BF_3 molecule has the elements E, C_3, $3C_2$, and σ_h (with one C_2 axis along each B—F bond), so belongs to D_{3h} (**8**). The C_6H_6 molecule has the elements E, C_6, $3C_2$, $3C_2'$, and σ_h together with some others that these elements imply, so it belongs to D_{6h}. All homonuclear diatomic molecules, such as N_2, belong to the group $D_{\infty h}$ because all rotations around the axis are symmetry operations, as are end-to-end rotation and end-to-end reflection; $D_{\infty h}$ is also the group of the linear OCO and HCCH molecules and of a uniform cylinder. Other examples of D_{nh} molecules are shown in (**9**), (**10**), and (**11**).

A molecule belongs to the group D_{nd} if in addition to the elements of D_n it possesses n dihedral mirror planes σ_d. The twisted, 90° allene (**12**) belongs to D_{2d}, and the staggered conformation of ethane (**13**) belongs to D_{3d}.

Comment 5.2

The prime on $3C_2'$ indicates that the three C_2 axes are different from the other three C_2 axes. In benzene, three of the C_2 axes bisect C—C bonds and the other three pass through vertices of the hexagon formed by the carbon framework of the molecule.

(d) The groups S_n

Molecules that have not been classified into one of the groups mentioned so far, but that possess one S_n axis, belong to the group S_n. An example is tetraphenylmethane, which belongs to the point group S_4 (**14**). Molecules belonging to S_n with $n > 4$ are rare. Note that the group S_2 is the same as C_i, so such a molecule will already have been classified as C_i.

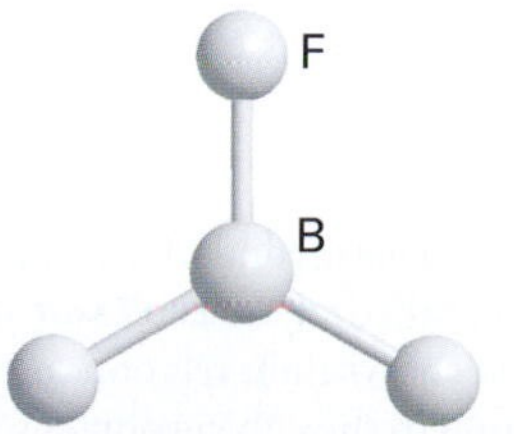

8 Boron trifluoride, BF_3

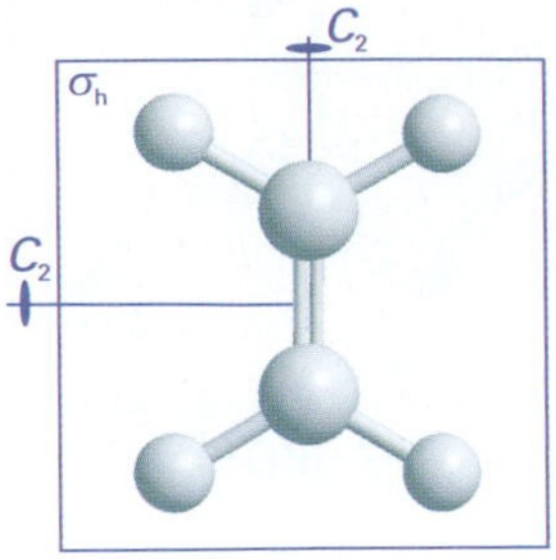

9 Ethene, $CH_2{=}CH_2$ (D_{2h})

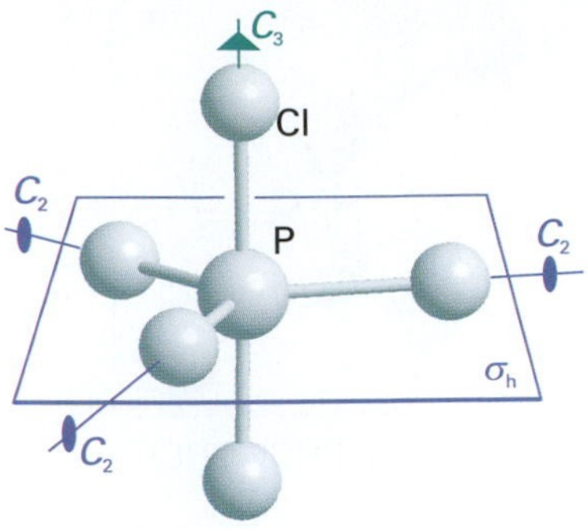

10 Phosphorus pentachloride, PCl_5 (D_{3h})

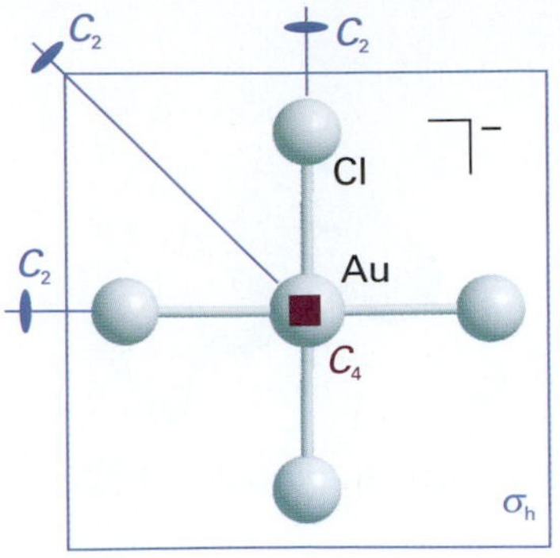

11 Tetrachloroaurate(III) ion, $[AuCl_4]^-$ (D_{4h})

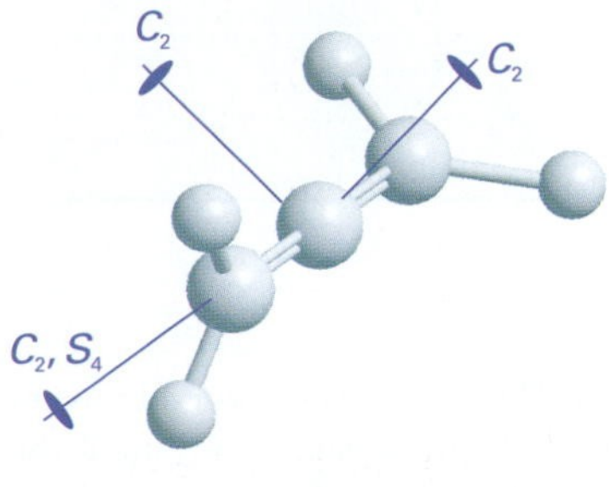

12 Allene, C_3H_4 (D_{2d})

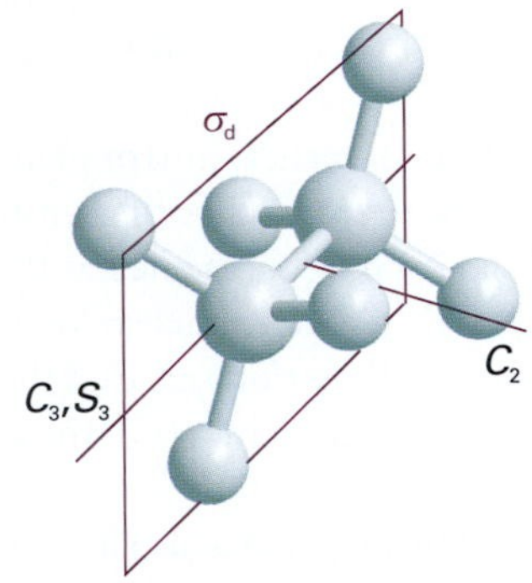

13 Ethane, C_2H_6 (D_{3d})

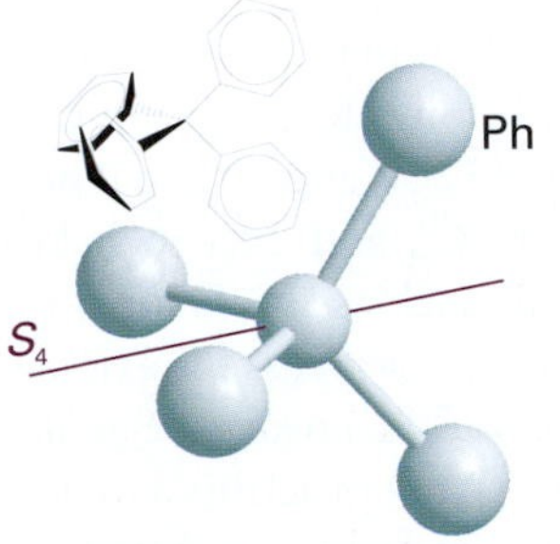

14 Tetraphenylmethane, $C(C_6H_5)_4$ (S_4)

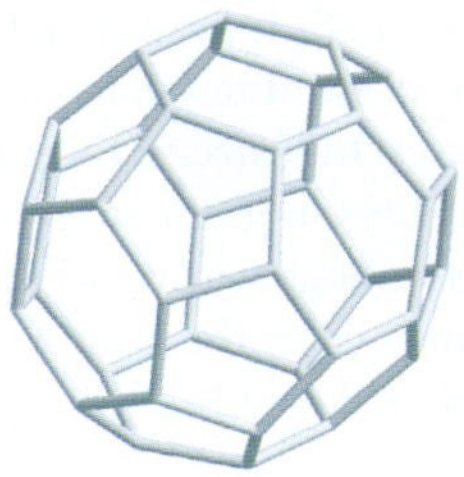

15 Buckminsterfullerene, C_{60} (I)

(e) The cubic groups

A number of very important molecules (e.g. CH_4 and SF_6) possess more than one principal axis. Most belong to the **cubic groups**, and in particular to the **tetrahedral groups** T, T_d, and T_h (Fig. 12.10a) or to the **octahedral groups** O and O_h (Fig. 5.10b). A few icosahedral (20-faced) molecules belonging to the **icosahedral group**, I (Fig. 5.10c), are also known: they include some of the boranes and buckminsterfullerene, C_{60} (**15**). The groups T_d and O_h are the groups of the regular tetrahedron (for instance, CH_4) and the regular octahedron (for instance, SF_6), respectively. If the object possesses the rotational symmetry of the tetrahedron or the octahedron, but none of their planes of reflection, then it belongs to the simpler groups T or O (Fig. 5.11). The group T_h is based on T but also contains a centre of inversion (Fig. 5.12).

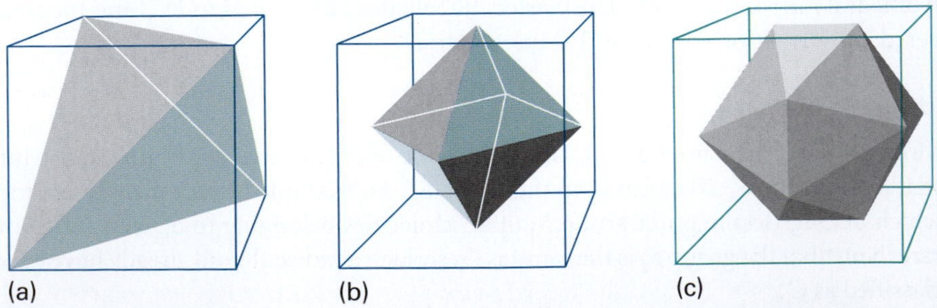

Fig. 5.10 (a) Tetrahedral, (b) octahedral, and (c) icosahedral molecules are drawn in a way that shows their relation to a cube: they belong to the cubicgroups T_d, O_h, and I_h, respectively.

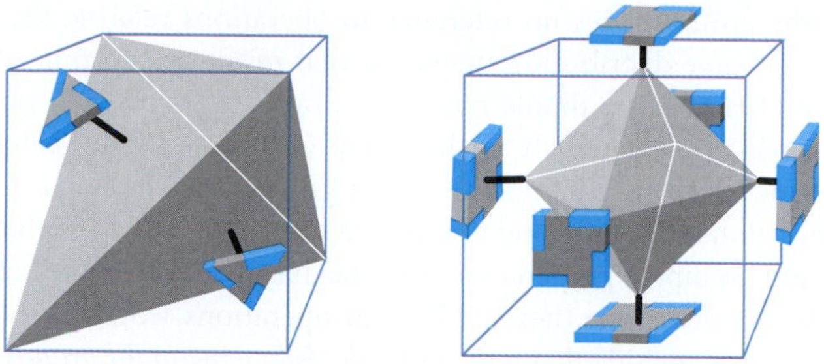

Fig. 5.11 Shapes corresponding to the point groups (a) T and (b) O. The presence of the decorated slabs reduces the symmetry of the object from T_d and O_h, respectively.

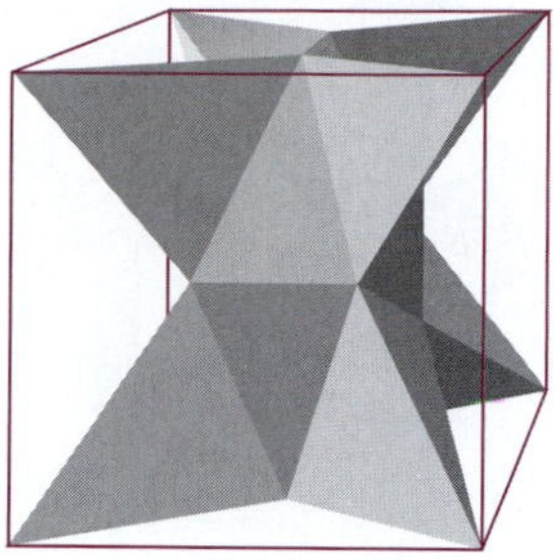

Fig. 5.12 The shape of an object belonging to the group T_h.

(f) The full rotation group

The **full rotation group**, R_3 (the 3 refers to rotation in three dimensions), consists of an infinite number of rotation axes with all possible values of n. A sphere and an atom belong to R_3, but no molecule does. Exploring the consequences of R_3 is a very important way of applying symmetry arguments to atoms, and is an alternative approach to the theory of orbital angular momentum.

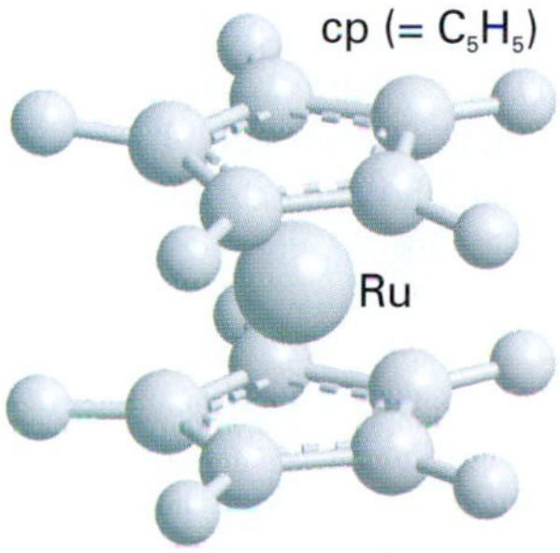

16 Ruthenocene, Ru(cp)$_2$

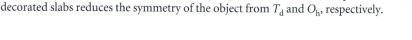

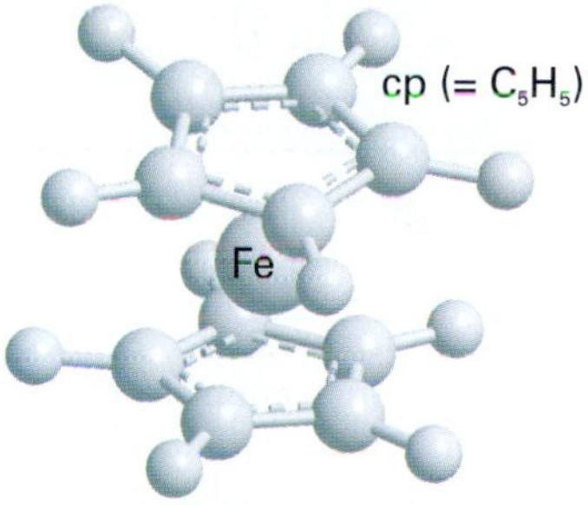

17 Ferrocene, Fe(cp)$_2$

Example 5.1 *Identifying a point group of a molecule*

Identify the point group to which a ruthenocene molecule (**16**) belongs.

Method Use the flow diagram in Fig. 5.7.

Answer The path to trace through the flow diagram in Fig. 5.7 is shown by a green line; it ends at D_{nh}. Because the molecule has a fivefold axis, it belongs to the group D_{5h}. If the rings were staggered, as they are in an excited state of ferrocene that lies 4 kJ mol^{-1} above the ground state (**17**), the horizontal reflection plane would be absent, but dihedral planes would be present.

Self-test 5.1 Classify the pentagonal antiprismatic excited state of ferrocene (**17**). [D_{5d}]

5.3 Some immediate consequences of symmetry

Some statements about the properties of a molecule can be made as soon as its point group has been identified.

Comment 5.3
The web site contains links to interactive tutorials, where you use your web browser to to assign point groups of molecules.

(a) Polarity

A **polar molecule** is one with a permanent electric dipole moment (HCl, O_3, and NH_3 are examples). If the molecule belongs to the group C_n with $n > 1$, it cannot possess a charge distribution with a dipole moment perpendicular to the symmetry axis because the symmetry of the molecule implies that any dipole that exists in one direction perpendicular to the axis is cancelled by an opposing dipole (Fig. 5.13a). For example, the perpendicular component of the dipole associated with one O—H bond in H_2O is cancelled by an equal but opposite component of the dipole of the second O—H bond, so any dipole that the molecule has must be parallel to the twofold symmetry axis.

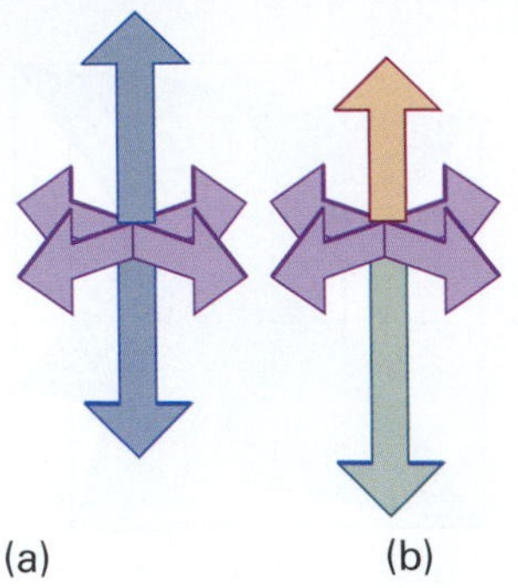

Fig. 5.13 (a) A molecule with a C_n axis cannot have a dipole perpendicular to the axis, but (b) it may have one parallel to the axis. The arrows represent local contributions to the overall electric dipole, such as may arise from bonds between pairs of neighbouring atoms with different electronegativities.

However, as the group makes no reference to operations relating the two ends of the molecule, a charge distribution may exist that results in a dipole along the axis (Fig. 5.13b), and H_2O has a dipole moment parallel to its twofold symmetry axis. The same remarks apply generally to the group C_{nv}, so molecules belonging to any of the C_{nv} groups may be polar. In all the other groups, such as C_{3h}, D, etc., there are symmetry operations that take one end of the molecule into the other. Therefore, as well as having no dipole perpendicular to the axis, such molecules can have none along the axis, for otherwise these additional operations would not be symmetry operations. We can conclude that *only molecules belonging to the groups* C_n, C_{nv}, *and* C_s *may have a permanent electric dipole moment.*

For C_n and C_{nv}, that dipole moment must lie along the symmetry axis. Thus ozone, O_3, which is angular and belongs to the group C_{2v}, may be polar (and is), but carbon dioxide, CO_2, which is linear and belongs to the group $D_{\infty h}$, is not.

(b) Chirality

A **chiral molecule** (from the Greek word for 'hand') is a molecule that cannot be superimposed on its mirror image. An **achiral molecule** is a molecule that can be superimposed on its mirror image. Chiral molecules are **optically active** in the sense that they rotate the plane of polarized light (a property discussed in more detail in Appendix 3). A chiral molecule and its mirror-image partner constitute an **enantiomeric pair** of isomers and rotate the plane of polarization in equal but opposite directions.

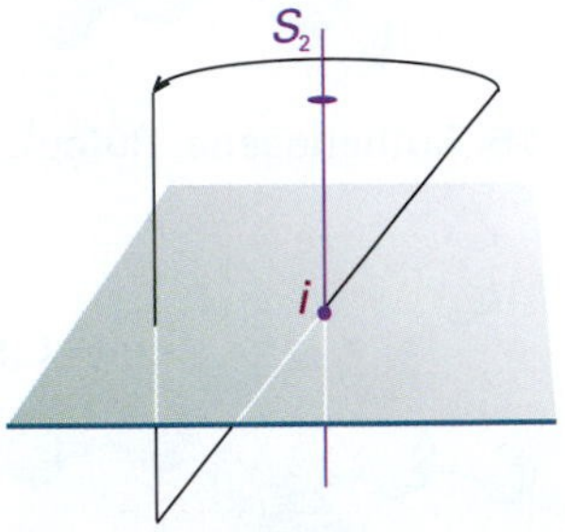

Fig. 5.14 Some symmetry elements are implied by the other symmetry elements in a group. Any molecule containing an inversion also possesses at least an S_2 element because i and S_2 are equivalent.

A molecule may be chiral, and therefore optically active, only if it does not possess an axis of improper rotation, S_n. However, we need to be aware that such an axis may be present under a different name, and be implied by other symmetry elements that are present. For example, molecules belonging to the groups C_{nh} possess an S_n axis implicitly because they possess both C_n and σ_h, which are the two components of an improper rotation axis. Any molecule containing a centre of inversion, i, also possesses an S_2 axis, because i is equivalent to C_2 in conjunction with σ_h, and that combination of elements is S_2 (Fig. 5.14). It follows that all molecules with centres of inversion are achiral and hence optically inactive. Similarly, because $S_1 = \sigma$, it follows that any molecule with a mirror plane is achiral.

A molecule may be chiral if it does not have a centre of inversion or a mirror plane, which is the case with the amino acid alanine (**18**), but not with glycine (**19**). However, a molecule may be achiral even though it does not have a centre of inversion. For example, the S_4 species (**20**) is achiral and optically inactive: though it lacks i (that is, S_2) it does have an S_4 axis.

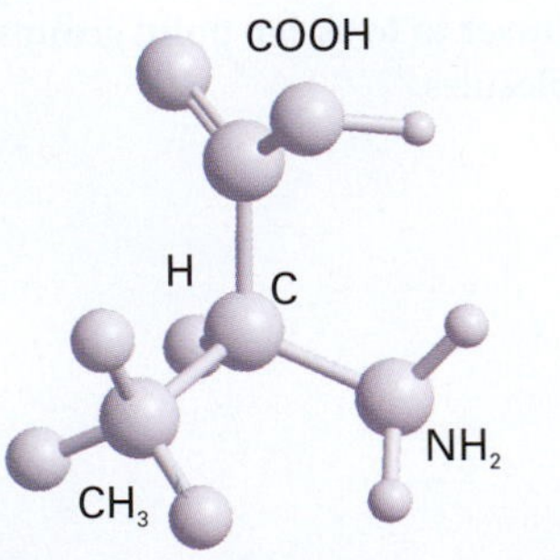

18 L-Alanine, $NH_2CH(CH_3)COOH$

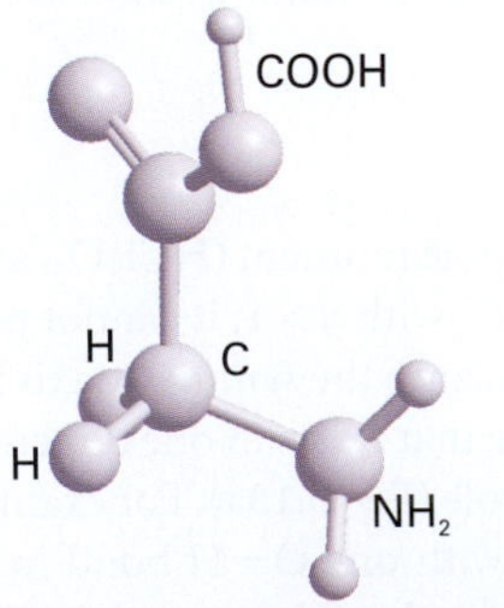

19 Glycine, NH_2CH_2COOH

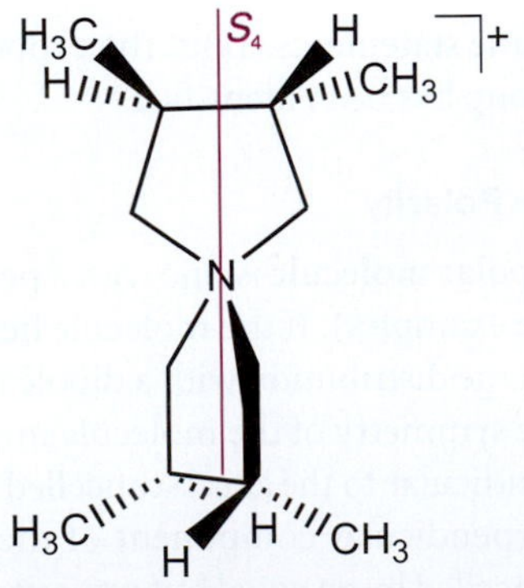

20 $N(CH_2CH(CH_3)CH(CH_3)CH_2)_2^+$

Applications to molecular orbital theory and spectroscopy

We shall now turn our attention away from the symmetries of molecules themselves and direct it towards the symmetry characteristics of orbitals that belong to the various atoms in a molecule. This material will enable us to discuss the formulation and labelling of molecular orbitals and selection rules in spectroscopy.

5.4 Character tables and symmetry labels

We saw in Chapter 4 that molecular orbitals of diatomic and linear polyatomic molecules are labelled σ, π, etc. These labels refer to the symmetries of the orbitals with respect to rotations around the principal symmetry axis of the molecule. Thus, a σ orbital does not change sign under a rotation through any angle, a π orbital changes sign when rotated by 180°, and so on (Fig. 5.15). The symmetry classifications σ and π can also be assigned to individual atomic orbitals in a linear molecule. For example, we can speak of an individual p_z orbital as having σ symmetry if the z-axis lies along the bond, because p_z is cylindrically symmetrical about the bond. This labelling of orbitals according to their behaviour under rotations can be generalized and extended to nonlinear polyatomic molecules, where there may be reflections and inversions to take into account as well as rotations.

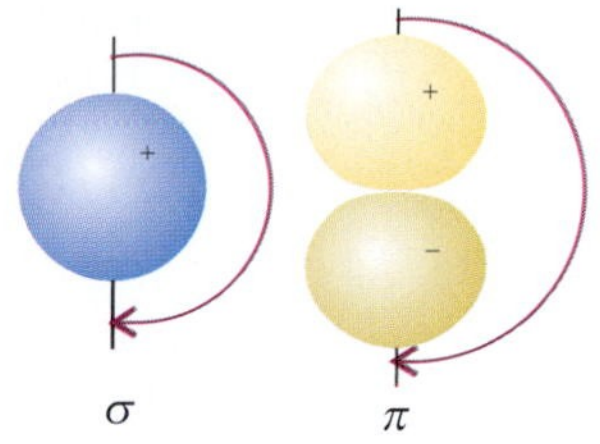

Fig. 5.15 A rotation through 180° about the internuclear axis (perpendicular to the page) leaves the sign of a σ orbital unchanged but the sign of a π orbital is changed. In the language introduced in this chapter, the characters of the C_2 rotation are +1 and −1 for the σ and π orbitals, respectively.

(a) Representations and characters

Labels analogous to σ and π are used to denote the symmetries of orbitals in polyatomic molecules. These labels look like a, a_1, e, e_g, and we first encountered them in Fig. 4.4 in connection with the molecular orbitals of benzene. As we shall see, these labels indicate the behaviour of the orbitals under the symmetry operations of the relevant point group of the molecule.

A label is assigned to an orbital by referring to the **character table** of the group, a table that characterizes the different symmetry types possible in the point group. Thus, to assign the labels σ and π, we use the table shown in the margin. This table is a fragment of the full character table for a linear molecule. The entry +1 shows that the orbital remains the same and the entry −1 shows that the orbital changes sign under the operation C_2 at the head of the column (as illustrated in Fig. 5.15). So, to assign the label σ or π to a particular orbital, we compare the orbital's behaviour with the information in the character table.

	C_2	(i.e. rotation by 180°)
σ	+1	(i.e. no change of sign)
π	−1	(i.e. change of sign)

The entries in a complete character table are derived by using the formal techniques of group theory and are called **characters**, χ (chi). These numbers characterize the essential features of each symmetry type in a way that we can illustrate by considering the C_{2v} molecule SO_2 and the valence p_x orbitals on each atom, which we shall denote p_S, p_A, and p_B (Fig. 5.16).

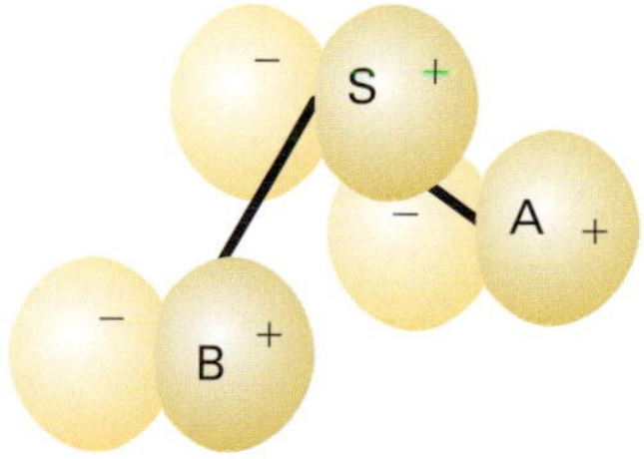

Fig. 5.16 The three p_x orbitals that are used to illustrate the construction of a matrix representation in a C_{2v} molecule (SO_2).

Under σ_v, the change $(p_S, p_B, p_A) \leftarrow (p_S, p_A, p_B)$ takes place. We can express this transformation by using matrix multiplication:

$$(p_S, p_B, p_A) = (p_S, p_A, p_B)\begin{pmatrix} 1 & 0 & 0 \\ 0 & 0 & 1 \\ 0 & 1 & 0 \end{pmatrix} = (p_S, p_A, p_B)\boldsymbol{D}(\sigma_v) \tag{5.1}$$

The matrix $\boldsymbol{D}(\sigma_v)$ is called a **representative** of the operation σ_v. Representatives take different forms according to the **basis**, the set of orbitals, that has been adopted.

Comment 5.4

See *Appendix 2* for a summary of the rules of matrix algebra.

We can use the same technique to find matrices that reproduce the other symmetry operations. For instance, C_2 has the effect $(-p_S, -p_B, -p_A) \leftarrow (p_S, p_A, p_B)$, and its representative is

$$\boldsymbol{D}(C_2) = \begin{pmatrix} -1 & 0 & 0 \\ 0 & 0 & -1 \\ 0 & -1 & 0 \end{pmatrix} \tag{5.2}$$

The effect of σ_v' is $(-p_S, -p_A, -p_B) \leftarrow (p_S, p_A, p_B)$, and its representative is

$$\boldsymbol{D}(\sigma_v') = \begin{pmatrix} -1 & 0 & 0 \\ 0 & -1 & 0 \\ 0 & 0 & -1 \end{pmatrix} \tag{5.3}$$

The identity operation has no effect on the basis, so its representative is the 3×3 unit matrix:

$$\boldsymbol{D}(E) = \begin{pmatrix} 1 & 0 & 0 \\ 0 & 1 & 0 \\ 0 & 0 & 1 \end{pmatrix} \tag{5.4}$$

The set of matrices that represents *all* the operations of the group is called a **matrix representation**, Γ (uppercase gamma), of the group for the particular basis we have chosen. We denote this three-dimensional representation by $\Gamma^{(3)}$. The discovery of a matrix representation of the group means that we have found a link between symbolic manipulations of operations and algebraic manipulations of numbers.

The character of an operation in a particular matrix representation is the sum of the diagonal elements of the representative of that operation. Thus, in the basis we are illustrating, the characters of the representatives are

$\boldsymbol{D}(E)$	$\boldsymbol{D}(C_2)$	$\boldsymbol{D}(\sigma_v)$	$\boldsymbol{D}(\sigma_v')$
3	−1	1	−3

The character of an operation depends on the basis.

Inspection of the representatives shows that they are all of **block-diagonal form**:

$$\boldsymbol{D} = \begin{pmatrix} [\blacksquare] & 0 & 0 \\ 0 & [\blacksquare] & [\blacksquare] \\ 0 & [\blacksquare] & [\blacksquare] \end{pmatrix}$$

The block-diagonal form of the representatives show us that the symmetry operations of C_{2v} never mix p_S with the other two functions. Consequently, the basis can be cut into two parts, one consisting of p_S alone and the other of (p_A, p_B). It is readily verified that the p_S orbital itself is a basis for the one-dimensional representation

$$\boldsymbol{D}(E) = 1 \qquad \boldsymbol{D}(C_2) = -1 \qquad \boldsymbol{D}(\sigma_v) = 1 \qquad \boldsymbol{D}(\sigma_v') = -1$$

which we shall call $\Gamma^{(1)}$. The remaining two basis functions are a basis for the two-dimensional representation $\Gamma^{(2)}$:

$$\boldsymbol{D}(E) = \begin{pmatrix} 1 & 0 \\ 0 & 1 \end{pmatrix} \quad \boldsymbol{D}(C_2) = \begin{pmatrix} 0 & -1 \\ -1 & 0 \end{pmatrix} \quad \boldsymbol{D}(\sigma_v) = \begin{pmatrix} 0 & 1 \\ 1 & 0 \end{pmatrix} \quad \boldsymbol{D}(\sigma_v') = \begin{pmatrix} -1 & 0 \\ 0 & -1 \end{pmatrix}$$

These matrices are the same as those of the original three-dimensional representation, except for the loss of the first row and column. We say that the original three-dimensional representation has been **reduced** to the 'direct sum' of a one-dimensional representation 'spanned' by p_S, and a two-dimensional representation spanned by (p_A, p_B). This reduction is consistent with the common sense view that the central orbital plays a role different from the other two. We denote the reduction symbolically by writing

$$\Gamma^{(3)} = \Gamma^{(1)} + \Gamma^{(2)} \tag{5.5}$$

The one-dimensional representation $\Gamma^{(1)}$ cannot be reduced any further, and is called an **irreducible representation** of the group (an 'irrep'). We can demonstrate that the two-dimensional representation $\Gamma^{(2)}$ is reducible (for this basis in this group) by switching attention to the linear combinations $p_1 = p_A + p_B$ and $p_2 = p_A - p_B$. These combinations are sketched in Fig. 5.17. The representatives in the new basis can be constructed from the old by noting, for example, that under σ_v, $(p_B, p_A) \leftarrow (p_A, p_B)$. In this way we find the following representation in the new basis:

$$\boldsymbol{D}(E) = \begin{pmatrix} 1 & 0 \\ 0 & 1 \end{pmatrix} \quad \boldsymbol{D}(C_2) = \begin{pmatrix} -1 & 0 \\ 0 & 1 \end{pmatrix} \quad \boldsymbol{D}(\sigma_v) = \begin{pmatrix} 1 & 0 \\ 0 & -1 \end{pmatrix} \quad \boldsymbol{D}(\sigma_v') = \begin{pmatrix} -1 & 0 \\ 0 & -1 \end{pmatrix}$$

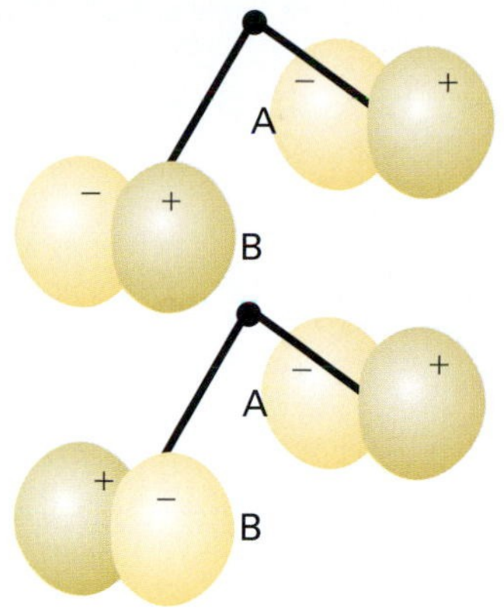

Fig. 5.17 Two symmetry-adapted linear combinations of the basis orbitals shown in Fig. 5.16. The two combinations each span a one-dimensional irreducible representation, and their symmetry species are different.

The new representatives are all in block-diagonal form, and the two combinations are not mixed with each other by any operation of the group. We have therefore achieved the reduction of $\Gamma^{(2)}$ to the sum of two one-dimensional representations. Thus, p_1 spans

$$\boldsymbol{D}(E) = 1 \quad \boldsymbol{D}(C_2) = -1 \quad \boldsymbol{D}(\sigma_v) = 1 \quad \boldsymbol{D}(\sigma_v') = -1$$

which is the same one-dimensional representation as that spanned by p_S, and p_2 spans

$$\boldsymbol{D}(E) = 1 \quad \boldsymbol{D}(C_2) = 1 \quad \boldsymbol{D}(\sigma_v) = -1 \quad \boldsymbol{D}(\sigma_v') = -1$$

which is a different one-dimensional representation; we shall denote it $\Gamma^{(1)\prime}$.

At this point we have found two irreducible representations of the group C_{2v} (Table 5.2). The two irreducible representations are normally labelled B_1 and A_2, respectively. An A or a B is used to denote a one-dimensional representation; A is used if the character under the principal rotation is +1, and B is used if the character is −1. Subscripts are used to distinguish the irreducible representations if there is more than one of the same type: A_1 is reserved for the representation with character 1 for all operations. When higher dimensional irreducible representations are permitted, E denotes a two-dimensional irreducible representation and T a three-dimensional irreducible representation; all the irreducible representations of C_{2v} are one-dimensional.

There are in fact only two more species of irreducible representations of this group, for a surprising theorem of group theory states that

$$\text{Number of symmetry species} = \text{number of classes} \tag{5.6}$$

Symmetry operations fall into the same **class** if they are of the same type (for example, rotations) and can be transformed into one another by a symmetry operation of the

Table 5.2* The C_{2v} character table

C_{2v}, 2*mm*	E	C_2	σ_v	σ_v'	$h = 4$	
A_1	1	1	1	1	z	z^2, y^2, x^2
A_2	1	1	−1	−1		xy
B_1	1	−1	1	−1	x	zx
B_2	1	−1	−1	1	y	yz

* More character tables are given at the end of the *Data section*.

Table 5.3* The C_{3v} character table

C_{3v}, $3m$	E	$2C_3$	$3\sigma_v$	$h=6$	
A_1	1	1	1	z	z^2, x^2+y^2
A_2	1	1	−1		
E	2	−1	0	(x, y)	$(xy, x^2-y^2), (yz, zx)$

* More character tables are given at the end of the *Data section*.

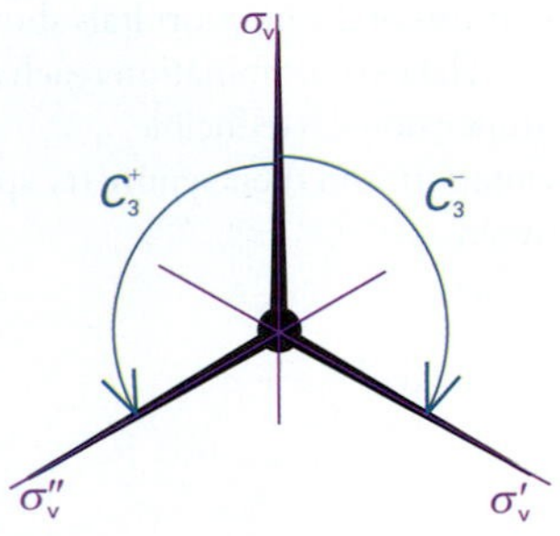

Fig. 5.18 Symmetry operations in the same class are related to one another by the symmetry operations of the group. Thus, the three mirror frames shown here are related by threefold rotations, and the two rotations shown here are related by reflection in σ_v.

group. In C_{2v}, for instance, there are four classes (four columns in the character table), so there are only four species of irreducible representation. The character table in Table 5.2 therefore shows the characters of all the irreducible representations of this group.

(b) The structure of character tables

In general, the columns in a character table are labelled with the symmetry operations of the group. For instance, for the group C_{3v} the columns are headed E, C_3, and σ_v (Table 5.3). The numbers multiplying each operation are the numbers of members of each class. In the C_{3v} character table we see that the two threefold rotations (clockwise and counter-clockwise rotations by 120°) belong to the same class: they are related by a reflection (Fig. 5.18). The three reflections (one through each of the three vertical mirror planes) also lie in the same class: they are related by the threefold rotations. The two reflections of the group C_{2v} fall into different classes: although they are both reflections, one cannot be transformed into the other by any symmetry operation of the group.

The total number of operations in a group is called the **order**, h, of the group. The order of the group C_{3v}, for instance, is 6.

The rows under the labels for the operations summarize the symmetry properties of the orbitals. They are labelled with the **symmetry species** (the analogues of the labels σ and π). More formally, the symmetry species label the irreducible representations of the group, which are the basic types of behaviour that orbitals may show when subjected to the symmetry operations of the group, as we have illustrated for the group C_{2v}. By convention, irreducible representations are labelled with upper case Roman letters (such as A_1 and E) but the orbitals to which they apply are labelled with the lower case italic equivalents (so an orbital of symmetry species A_1 is called an a_1 orbital). Examples of each type of orbital are shown in Fig. 5.19.

Comment 5.5

Note that care must be taken to distinguish the identity element E (italic, a column heading) from the symmetry label E (roman, a row label).

(c) Character tables and orbital degeneracy

The character of the identity operation E tells us the degeneracy of the orbitals. Thus, in a C_{3v} molecule, any orbital with a symmetry label a_1 or a_2 is nondegenerate. Any doubly degenerate pair of orbitals in C_{3v} must be labelled e because, in this group, only E symmetry species have characters greater than 1.

Because there are no characters greater than 2 in the column headed E in C_{3v}, we know that there can be no triply degenerate orbitals in a C_{3v} molecule. This last point is a powerful result of group theory, for it means that, with a glance at the character table of a molecule, we can state the maximum possible degeneracy of its orbitals.

Example 5.2 *Using a character table to judge degeneracy*

Can a trigonal planar molecule such as BF_3 have triply degenerate orbitals? What is the minimum number of atoms from which a molecule can be built that does display triple degeneracy?

Method First, identify the point group, and then refer to the corresponding character table in the *Data section*. The maximum number in the column headed by the identity E is the maximum orbital degeneracy possible in a molecule of that point group. For the second part, consider the shapes that can be built from two, three, etc. atoms, and decide which number can be used to form a molecule that can have orbitals of symmetry species T.

Answer Trigonal planar molecules belong to the point group D_{3h}. Reference to the character table for this group shows that the maximum degeneracy is 2, as no character exceeds 2 in the column headed E. Therefore, the orbitals cannot be triply degenerate. A tetrahedral molecule (symmetry group T) has an irreducible representation with a T symmetry species. The minimum number of atoms needed to build such a molecule is four (as in P_4, for instance).

Self-test 5.2 A buckminsterfullerene molecule, C_{60} (**15**), belongs to the icosahedral point group. What is the maximum possible degree of degeneracy of its orbitals? [5]

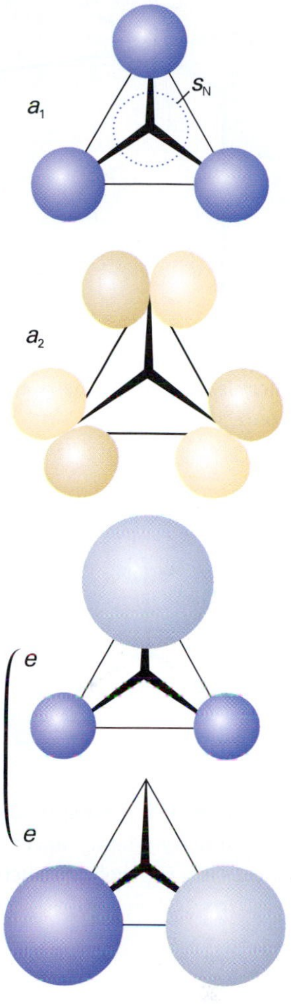

Fig. 5.19 Typical symmetry-adapted linear combinations of orbitals in a C_{3v} molecule.

(d) Characters and operations

The characters in the rows labelled A and B and in the columns headed by symmetry operations other than the identity E indicate the behaviour of an orbital under the corresponding operations: a +1 indicates that an orbital is unchanged, and a −1 indicates that it changes sign. It follows that we can identify the symmetry label of the orbital by comparing the changes that occur to an orbital under each operation, and then comparing the resulting +1 or −1 with the entries in a row of the character table for the point group concerned.

For the rows labelled E or T (which refer to the behaviour of sets of doubly and triply degenerate orbitals, respectively), the characters in a row of the table are the sums of the characters summarizing the behaviour of the individual orbitals in the basis. Thus, if one member of a doubly degenerate pair remains unchanged under a symmetry operation but the other changes sign (Fig. 5.20), then the entry is reported as $\chi = 1 - 1 = 0$. Care must be exercised with these characters because the transformations of orbitals can be quite complicated; nevertheless, the sums of the individual characters are integers.

As an example, consider the O2p_x orbital in H_2O. Because H_2O belongs to the point group C_{2v}, we know by referring to the C_{2v} character table (Table 5.2) that the labels available for the orbitals are a_1, a_2, b_1, and b_2. We can decide the appropriate label for O2p_x by noting that under a 180° rotation (C_2) the orbital changes sign (Fig. 5.21), so it must be either B_1 or B_2, as only these two symmetry types have character −1 under C_2. The O2p_x orbital also changes sign under the reflection σ_v', which identifies it as B_1. As we shall see, any molecular orbital built from this atomic orbital will also be a b_1 orbital. Similarly, O2p_y changes sign under C_2 but not under σ_v'; therefore, it can contribute to b_2 orbitals.

The behaviour of *s*, *p*, and *d* orbitals on a central atom under the symmetry operations of the molecule is so important that the symmetry species of these orbitals

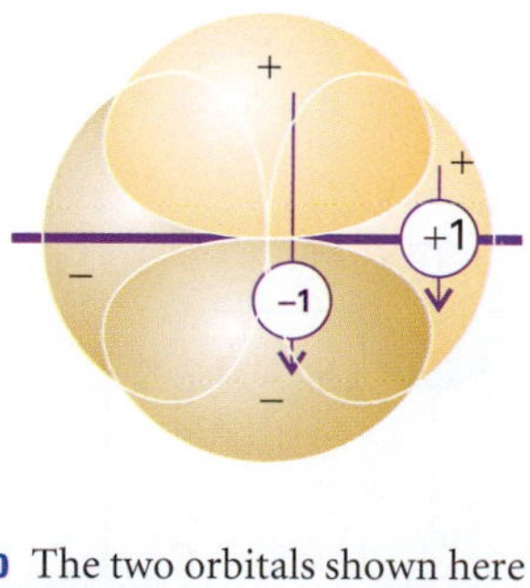

Fig. 5.20 The two orbitals shown here have different properties under reflection through the mirror plane: one changes sign (character −1), the other does not (character +1).

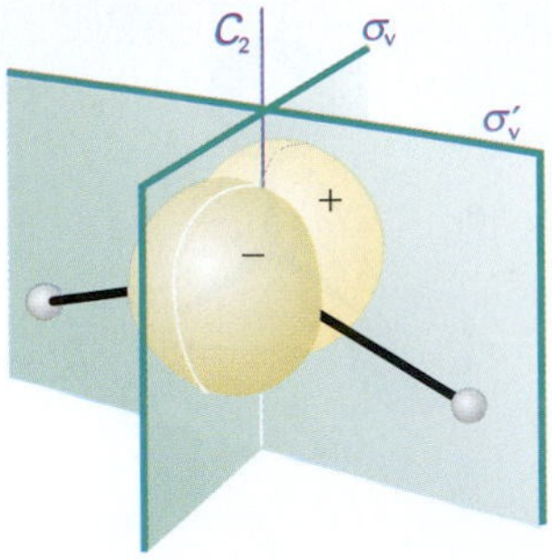

Fig. 5.21 A p_x orbital on the central atom of a C_{2v} molecule and the symmetry elements of the group.

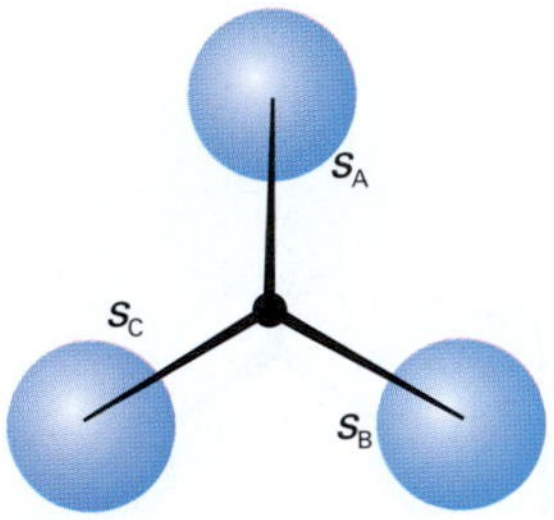

Fig. 5.22 The three H1*s* orbitals used to construct symmetry-adapted linear combinations in a C_{3v} molecule such as NH_3.

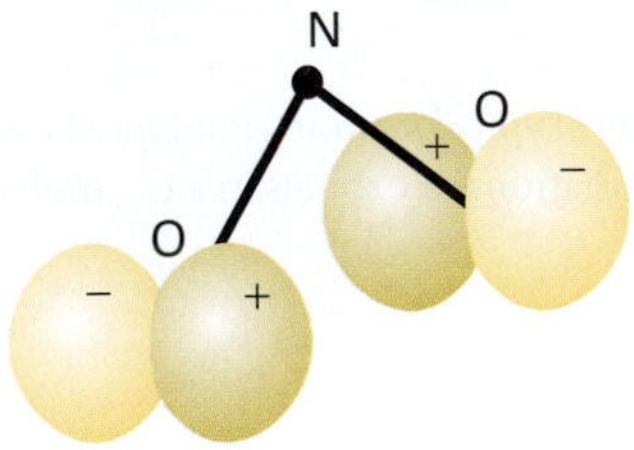

Fig. 5.23 One symmetry-adapted linear combination of O2p_x orbitals in the C_{2v} NO_2^- ion.

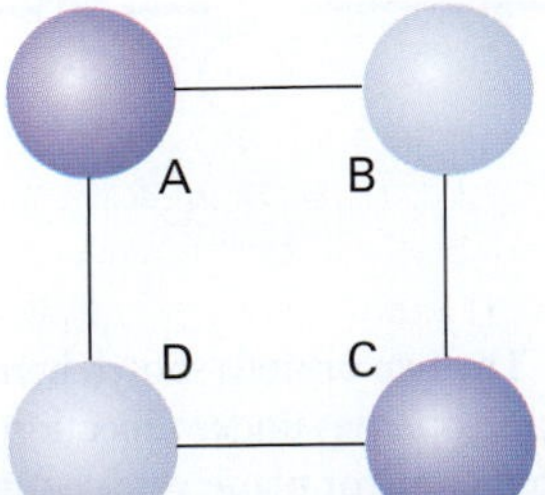

21

are generally indicated in a character table. To make these allocations, we look at the symmetry species of *x*, *y*, and *z*, which appear on the right-hand side of the character table. Thus, the position of *z* in Table 5.3 shows that p_z (which is proportional to $zf(r)$), has symmetry species A_1 in C_{3v}, whereas p_x and p_y (which are proportional to $xf(r)$ and $yf(r)$, respectively) are jointly of E symmetry. In technical terms, we say that p_x and p_y jointly **span** an irreducible representation of symmetry species E. An *s* orbital on the central atom always spans the fully symmetrical irreducible representation (typically labelled A_1 but sometimes A_1') of a group as it is unchanged under all symmetry operations.

The five *d* orbitals of a shell are represented by *xy* for d_{xy}, etc, and are also listed on the right of the character table. We can see at a glance that in C_{3v}, d_{xy} and $d_{x^2-y^2}$ on a central atom jointly belong to E and hence form a doubly degenerate pair.

(e) The classification of linear combinations of orbitals

So far, we have dealt with the symmetry classification of individual orbitals. The same technique may be applied to linear combinations of orbitals on atoms that are related by symmetry transformations of the molecule, such as the combination $\psi_1 = \psi_A + \psi_B + \psi_C$ of the three H1*s* orbitals in the C_{3v} molecule NH_3 (Fig. 5.22). This combination remains unchanged under a C_3 rotation and under any of the three vertical reflections of the group, so its characters are

$$\chi(E) = 1 \qquad \chi(C_3) = 1 \qquad \chi(\sigma_v) = 1$$

Comparison with the C_{3v} character table shows that ψ_1 is of symmetry species A_1, and therefore that it contributes to a_1 molecular orbitals in NH_3.

Example 5.3 *Identifying the symmetry species of orbitals*

Identify the symmetry species of the orbital $\psi = \psi_A - \psi_B$ in a C_{2v} NO_2 molecule, where ψ_A is an O2p_x orbital on one O atom and ψ_B that on the other O atom.

Method The negative sign in ψ indicates that the sign of ψ_B is opposite to that of ψ_A. We need to consider how the combination changes under each operation of the group, and then write the character as +1, −1, or 0 as specified above. Then we compare the resulting characters with each row in the character table for the point group, and hence identify the symmetry species.

Answer The combination is shown in Fig. 5.23. Under C_2, ψ changes into itself, implying a character of +1. Under the reflection σ_v, both orbitals change sign, so $\psi \rightarrow -\psi$, implying a character of −1. Under σ_v', $\psi \rightarrow -\psi$, so the character for this operation is also −1. The characters are therefore

$$\chi(E) = 1 \qquad \chi(C_2) = 1 \qquad \chi(\sigma_v) = -1 \qquad \chi(\sigma_v') = -1$$

These values match the characters of the A_2 symmetry species, so ψ can contribute to an a_2 orbital.

Self-test 5.3 Consider $PtCl_4^-$, in which the Cl ligands form a square planar array of point group D_{4h} (**21**). Identify the symmetry type of the combination $\psi_A - \psi_B + \psi_C - \psi_D$. [$B_{2g}$]

5.5 Vanishing integrals and orbital overlap

Suppose we had to evaluate the integral

$$I = \int f_1 f_2 \, d\tau \qquad (5.7)$$

where f_1 and f_2 are functions. For example, f_1 might be an atomic orbital A on one atom and f_2 an atomic orbital B on another atom, in which case I would be their overlap integral. If we knew that the integral is zero, we could say at once that a molecular orbital does not result from (A,B) overlap in that molecule. We shall now see that character tables provide a quick way of judging whether an integral is necessarily zero.

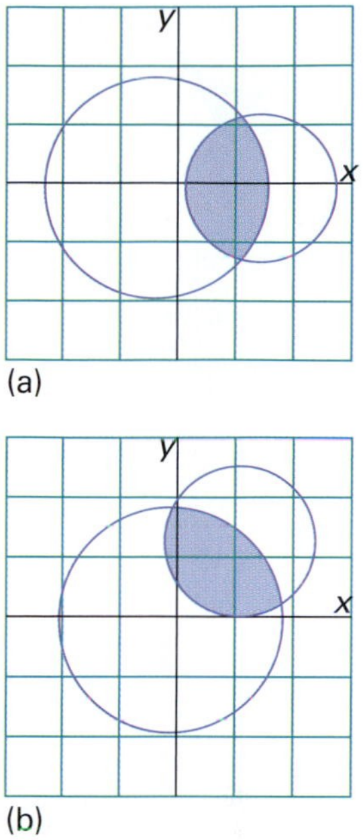

Fig. 5.24 The value of an integral I (for example, an area) is independent of the coordinate system used to evaluate it. That is, I is a basis of a representation of symmetry species A_1 (or its equivalent).

(a) The criteria for vanishing integrals

The key point in dealing with the integral I is that the value of any integral, and of an overlap integral in particular, is independent of the orientation of the molecule (Fig. 5.24). In group theory we express this point by saying that *I is invariant under any symmetry operation of the molecule*, and that each operation brings about the trivial transformation $I \rightarrow I$. Because the volume element $d\tau$ is invariant under any symmetry operation, it follows that the integral is nonzero only if the integrand itself, the product $f_1 f_2$, is unchanged by any symmetry operation of the molecular point group. If the integrand changed sign under a symmetry operation, the integral would be the sum of equal and opposite contributions, and hence would be zero. It follows that the only contribution to a nonzero integral comes from functions for which under any symmetry operation of the molecular point group $f_1 f_2 \rightarrow f_1 f_2$, and hence for which the characters of the operations are all equal to +1. Therefore, for I not to be zero, *the integrand $f_1 f_2$ must have symmetry species A_1* (or its equivalent in the specific molecular point group).

We use the following procedure to deduce the symmetry species spanned by the product $f_1 f_2$ and hence to see whether it does indeed span A_1.

1 Decide on the symmetry species of the individual functions f_1 and f_2 by reference to the character table, and write their characters in two rows in the same order as in the table.

2 Multiply the numbers in each column, writing the results in the same order.

3 Inspect the row so produced, and see if it can be expressed as a sum of characters from each column of the group. The integral must be zero if this sum does not contain A_1.

For example, if f_1 is the s_N orbital in NH_3 and f_2 is the linear combination $s_3 = s_B - s_C$ (Fig. 5.25), then, because s_N spans A_1 and s_3 is a member of the basis spanning E, we write

f_1:	1	1	1
f_2:	2	−1	0
$f_1 f_2$:	2	−1	0

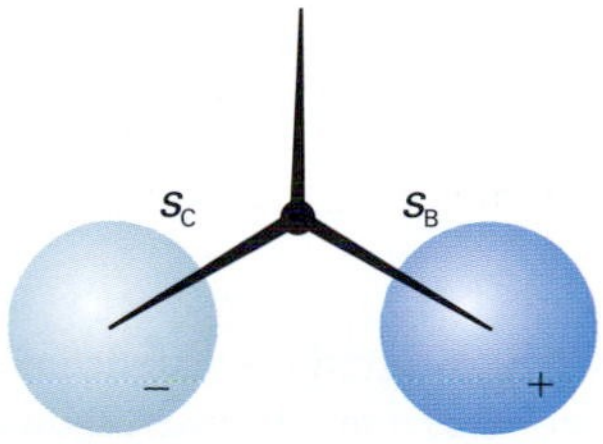

Fig. 5.25 A symmetry-adapted linear combination that belongs to the symmetry species E in a C_{3v} molecule such as NH_3. This combination can form a molecular orbital by overlapping with the p_x orbital on the central atom (the orbital with its axis parallel to the width of the page; see Fig. 5.28c).

The characters 2, −1, 0 are those of E alone, so the integrand does not span A_1. It follows that the integral must be zero. Inspection of the form of the functions (see Fig. 5.25) shows why this is so: s_3 has a node running through s_N. Had we taken $f_1 = s_N$ and $f_2 = s_1$ instead, where $s_1 = s_A + s_B + s_C$, then because each spans A_1 with characters 1,1,1:

f_1:	1	1	1
f_2:	1	1	1
$f_1 f_2$:	1	1	1

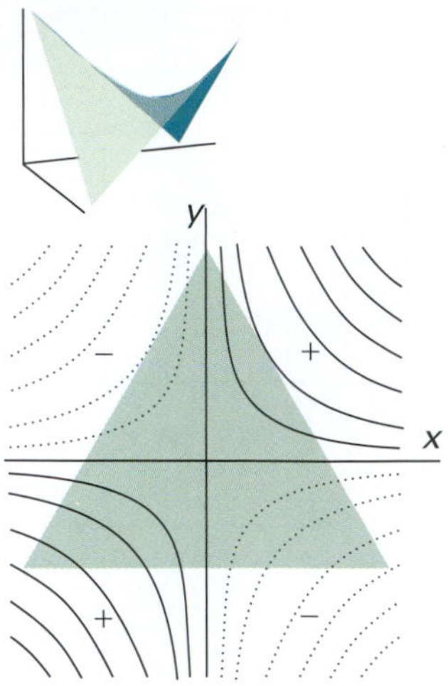

Fig. 5.26 The integral of the function $f = xy$ over the tinted region is zero. In this case, the result is obvious by inspection, but group theory can be used to establish similar results in less obvious cases. The insert shows the shape of the function in three dimensions.

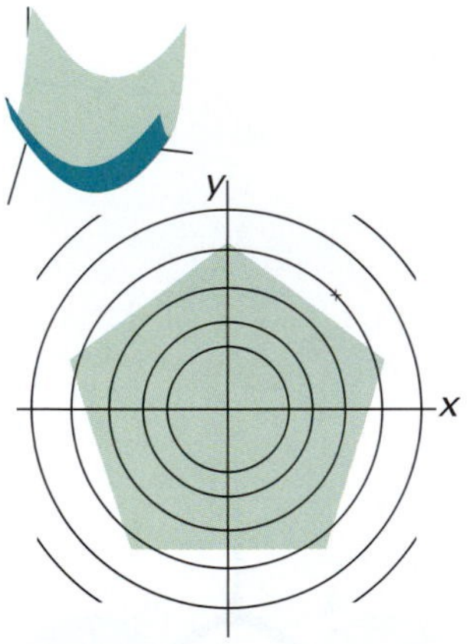

Fig. 5.27 The integration of a function over a pentagonal region. The insert shows the shape of the function in three dimensions.

The characters of the product are those of A_1 itself. Therefore, s_1 and s_N may have nonzero overlap. A shortcut that works when f_1 and f_2 are bases for irreducible representations of a group is to note their symmetry species: if they are different, then the integral of their product must vanish; if they are the same, then the integral may be nonzero.

It is important to note that group theory is specific about when an integral must be zero, but integrals that it allows to be nonzero may be zero for reasons unrelated to symmetry. For example, the N—H distance in ammonia may be so great that the (s_1, s_N) overlap integral is zero simply because the orbitals are so far apart.

Example 5.4 *Deciding if an integral must be zero (1)*

May the integral of the function $f = xy$ be nonzero when evaluated over a region the shape of an equilateral triangle centred on the origin (Fig. 5.26)?

Method First, note that an integral over a single function f is included in the previous discussion if we take $f_1 = f$ and $f_2 = 1$ in eqn 5.7. Therefore, we need to judge whether f alone belongs to the symmetry species A_1 (or its equivalent) in the point group of the system. To decide that, we identify the point group and then examine the character table to see whether f belongs to A_1 (or its equivalent).

Answer An equilateral triangle has the point-group symmetry D_{3h}. If we refer to the character table of the group, we see that xy is a member of a basis that spans the irreducible representation E′. Therefore, its integral must be zero, because the integrand has no component that spans A_1'.

Self-test 5.4 Can the function $x^2 + y^2$ have a nonzero integral when integrated over a regular pentagon centred on the origin? [Yes, Fig. 5.27]

In many cases, the product of functions f_1 and f_2 spans a sum of irreducible representations. For instance, in C_{2v} we may find the characters 2, 0, 0, −2 when we multiply the characters of f_1 and f_2 together. In this case, we note that these characters are the sum of the characters for A_2 and B_1:

	E	C_{2v}	σ_v	σ_v'
A_2	1	1	−1	−1
B_1	1	−1	1	−1
$A_2 + B_1$	2	0	0	−2

To summarize this result we write the symbolic expression $A_2 \times B_1 = A_2 + B_1$, which is called the **decomposition of a direct product**. This expression is symbolic. The × and + signs in this expression are not ordinary multiplication and addition signs: formally, they denote technical procedures with matrices called a 'direct product' and a 'direct sum'. Because the sum on the right does not include a component that is a basis for an irreducible representation of symmetry species A_1, we can conclude that the integral of $f_1 f_2$ over all space is zero in a C_{2v} molecule.

Whereas the decomposition of the characters 2, 0, 0, −2 can be done by inspection in this simple case, in other cases and more complex groups the decomposition is often far from obvious. For example, if we found the characters 8, −2, −6, 4, it would not be obvious that the sum contains A_1. Group theory, however, provides a systematic way of using the characters of the representation spanned by a product to find the symmetry species of the irreducible representations. The recipe is as follows:

1 Write down a table with columns headed by the symmetry operations of the group.

2 In the first row write down the characters of the symmetry species we want to analyse.

3 In the second row, write down the characters of the irreducible representation Γ we are interested in.

4 Multiply the two rows together, add the products together, and divide by the order of the group.

The resulting number is the number of times Γ occurs in the decomposition.

Illustration 5.1 *To find whether A_1 occurs in a direct product*

To find whether A_1 does indeed occur in the product with characters 8, −2, −6, 4 in C_{2v}, we draw up the following table:

	E	C_{2v}	σ_v	σ'_v	$h = 4$ (the order of the group)
f_1f_2	8	−2	−6	4	(the characters of the product)
A_1	1	1	1	1	(the symmetry species we are interested in)
	8	−2	−6	4	(the product of the two sets of characters)

The sum of the numbers in the last line is 4; when that number is divided by the order of the group, we get 1, so A_1 occurs once in the decomposition. When the procedure is repeated for all four symmetry species, we find that f_1f_2 spans $A_1 + 2A_2 + 5B_2$.

Self-test 5.5 Does A_2 occur among the symmetry species of the irreducible representations spanned by a product with characters 7, −3, −1, 5 in the group C_{2v}?

[No]

(b) Orbitals with nonzero overlap

The rules just given let us decide which atomic orbitals may have nonzero overlap in a molecule. We have seen that s_N may have nonzero overlap with s_1 (the combination $s_A + s_B + s_C$), so bonding and antibonding molecular orbitals can form from (s_N, s_1) overlap (Fig. 5.28). The general rule is that *only orbitals of the same symmetry species may have nonzero overlap*, so only orbitals of the same symmetry species form bonding and antibonding combinations. It should be recalled from Chapter 4 that the selection of atomic orbitals that had mutual nonzero overlap is the central and initial step in the construction of molecular orbitals by the LCAO procedure. We are therefore at the point of contact between group theory and the material introduced in that chapter. The molecular orbitals formed from a particular set of atomic orbitals with nonzero overlap are labelled with the lowercase letter corresponding to the symmetry species. Thus, the (s_N, s_1)-overlap orbitals are called a_1 orbitals (or a_1^*, if we wish to emphasize that they are antibonding).

The linear combinations $s_2 = 2s_a - s_b - s_c$ and $s_3 = s_b - s_c$ have symmetry species E. Does the N atom have orbitals that have nonzero overlap with them (and give rise to e molecular orbitals)? Intuition (as supported by Figs. 5.28b and c) suggests that N2p_x and N2p_y should be suitable. We can confirm this conclusion by noting that the character table shows that, in C_{3v}, the functions x and y jointly belong to the symmetry species E. Therefore, N2p_x and N2p_y also belong to E, so may have nonzero overlap with s_2 and s_3. This conclusion can be verified by multiplying the characters

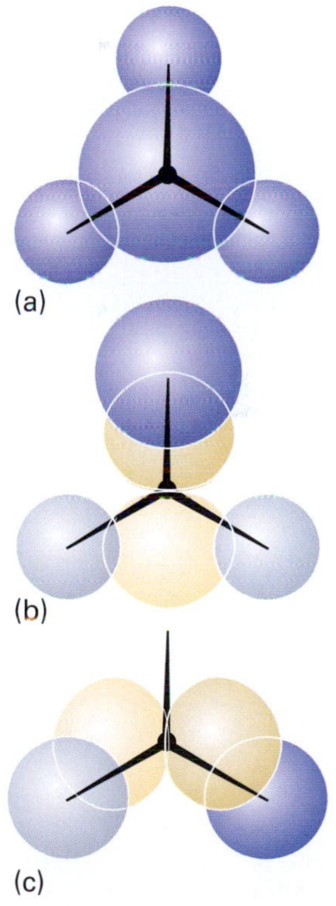

Fig. 5.28 Orbitals of the same symmetry species may have non-vanishing overlap. This diagram illustrates the three bonding orbitals that may be constructed from (N2s, H1s) and (N2p, H1s) overlap in a C_{3v} molecule. (a) a_1; (b) and (c) the two components of the doubly degenerate e orbitals. (There are also three antibonding orbitals of the same species.)

and finding that the product of characters can be expressed as the decomposition $E \times E = A_1 + A_2 + E$. The two *e* orbitals that result are shown in Fig. 5.28 (there are also two antibonding *e* orbitals).

We can see the power of the method by exploring whether any *d* orbitals on the central atom can take part in bonding. As explained earlier, reference to the C_{3v} character table shows that d_{z^2} has A_1 symmetry and that the pairs $(d_{x^2-y^2}, d_{xy})$ and (d_{yz}, d_{zx}) each transform as E. It follows that molecular orbitals may be formed by (s_1, d_{z^2}) overlap and by overlap of the s_2, s_3 combinations with the E *d* orbitals. Whether or not the *d* orbitals are in fact important is a question group theory cannot answer because the extent of their involvement depends on energy considerations, not symmetry.

Example 5.5 *Determining which orbitals can contribute to bonding*

The four H1*s* orbitals of methane span $A_1 + T_2$. With which of the C atom orbitals can they overlap? What bonding pattern would be possible if the C atom had *d* orbitals available?

Method Refer to the T_d character table (in the *Data section*) and look for *s*, *p*, and *d* orbitals spanning A_1 or T_2.

Answer An *s* orbital spans A_1, so it may have nonzero overlap with the A_1 combination of H1*s* orbitals. The C2*p* orbitals span T_2, so they may have nonzero overlap with the T_2 combination. The d_{xy}, d_{yz}, and d_{zx} orbitals span T_2, so they may overlap the same combination. Neither of the other two *d* orbitals span A_1 (they span E), so they remain nonbonding orbitals. It follows that in methane there are (C2*s*,H1*s*)-overlap a_1 orbitals and (C2*p*,H1*s*)-overlap t_2 orbitals. The C3*d* orbitals might contribute to the latter. The lowest energy configuration is probably $a_1^2 t_2^6$, with all bonding orbitals occupied.

Self-test 5.6 Consider the octahedral SF_6 molecule, with the bonding arising from overlap of S orbitals and a 2*p* orbital on each F directed towards the central S atom. The latter span $A_{1g} + E_g + T_{1u}$. What *s* orbitals have nonzero overlap? Suggest what the ground-state configuration is likely to be.

[3*s*(A_{1g}), 3*p*(T_{1u}), 3*d*(E_g); $a_{1g}^2 t_{1u}^6 e_g^4$]

(c) Symmetry-adapted linear combinations

So far, we have only asserted the forms of the linear combinations (such as s_1, etc.) that have a particular symmetry. Group theory also provides machinery that takes an arbitrary **basis**, or set of atomic orbitals (s_A, etc.), as input and generates combinations of the specified symmetry. Because these combinations are adapted to the symmetry of the molecule, they are called **symmetry-adapted linear combinations** (SALC). Symmetry-adapted linear combinations are the building blocks of LCAO molecular orbitals, for they include combinations such as those used to construct molecular orbitals in benzene. The construction of SALCs is the first step in any molecular orbital treatment of molecules.

The technique for building SALCs is derived by using the full power of group theory. We shall not show the derivation (see *Further reading*), which is very lengthy, but present the main conclusions as a set of rules:

1 Construct a table showing the effect of each operation on each orbital of the original basis.

2 To generate the combination of a specified symmetry species, take each column in turn and:

(i) Multiply each member of the column by the character of the corresponding operation.

(ii) Add together all the orbitals in each column with the factors as determined in (i).

(iii) Divide the sum by the order of the group.

	Original basis			
	s_N	s_A	s_B	s_C
Under E	s_N	s_A	s_B	s_C
C_3^+	s_N	s_B	s_C	s_A
C_3^-	s_N	s_C	s_A	s_B
σ_v	s_N	s_A	s_C	s_B
σ_v'	s_N	s_B	s_A	s_C
σ_v''	s_N	s_C	s_B	s_A

For example, from the (s_N,s_A,s_B,s_C) basis in NH_3 we form the table shown in the margin. To generate the A_1 combination, we take the characters for A_1 (1,1,1,1,1,1); then rules (i) and (ii) lead to

$$\psi \propto s_N + s_N + \cdots = 6s_N$$

The order of the group (the number of elements) is 6, so the combination of A_1 symmetry that can be generated from s_N is s_N itself. Applying the same technique to the column under s_A gives

$$\psi = \tfrac{1}{6}(s_A + s_B + s_C + s_A + s_B + s_C) = \tfrac{1}{3}(s_A + s_B + s_C)$$

The same combination is built from the other two columns, so they give no further information. The combination we have just formed is the s_1 combination we used before (apart from the numerical factor).

We now form the overall molecular orbital by forming a linear combination of all the SALCs of the specified symmetry species. In this case, therefore, the a_1 molecular orbital is

$$\psi = c_N s_N + c_1 s_1$$

This is as far as group theory can take us. The coefficients are found by solving the Schrödinger equation; they do not come directly from the symmetry of the system.

We run into a problem when we try to generate an SALC of symmetry species E, because, for representations of dimension 2 or more, the rules generate sums of SALCs. This problem can be illustrated as follows. In C_{3v}, the E characters are 2, −1, −1, 0, 0, 0, so the column under s_N gives

$$\psi = \tfrac{1}{6}(2s_N - s_N - s_N + 0 + 0 + 0) = 0$$

The other columns give

$$\tfrac{1}{6}(2s_A - s_B - s_C) \qquad \tfrac{1}{6}(2s_B - s_A - s_C) \qquad \tfrac{1}{6}(2s_C - s_B - s_A)$$

However, any one of these three expressions can be expressed as a sum of the other two (they are not 'linearly independent'). The difference of the second and third gives $\tfrac{1}{2}(s_B - s_C)$, and this combination and the first, $\tfrac{1}{6}(2s_A - s_B - s_C)$ are the two (now linearly independent) SALCs we have used in the discussion of e orbitals.

5.6 Vanishing integrals and selection rules

Integrals of the form

$$I = \int f_1 f_2 f_3 \, d\tau \tag{5.8}$$

are also common in quantum mechanics for they include matrix elements of operators (Section 1.5d), and it is important to know when they are necessarily zero. For the integral to be nonzero, *the product $f_1 f_2 f_3$ must span A_1 (or its equivalent) or contain a component that spans A_1*. To test whether this is so, the characters of all three functions are multiplied together in the same way as in the rules set out above.

Example 5.6 *Deciding if an integral must be zero (2)*

Does the integral $\int(3d_{z^2})x(3d_{xy})\mathrm{d}\tau$ vanish in a C_{2v} molecule?

Method We must refer to the C_{2v} character table (Table 5.2) and the characters of the irreducible representations spanned by $3z^2 - r^2$ (the form of the d_{z^2} orbital), x, and xy; then we can use the procedure set out above (with one more row of multiplication).

Answer We draw up the following table:

	E	C_2	σ_v	σ_v'	
$f_3 = d_{xy}$	1	1	−1	−1	A_2
$f_2 = x$	1	−1	1	−1	B_1
$f_1 = d_{z^2}$	1	1	1	1	A_1
$f_1 f_2 f_3$	1	−1	−1	1	

The characters are those of B_2. Therefore, the integral is necessarily zero.

Self-test 5.7 Does the integral $\int(2p_x)(2p_y)(2p_z)\mathrm{d}\tau$ necessarily vanish in an octahedral environment? [No]

We saw in Chapters 2 and 3, and will see in more detail in Chapters 6 and 7, that the intensity of a spectral line arising from a molecular transition between some initial state with wavefunction ψ_i and a final state with wavefunction ψ_f depends on the (electric) transition dipole moment, $\boldsymbol{\mu}_{fi}$. The z-component of this vector is defined through

$$\mu_{z,fi} = -e\int \psi_f^* z \psi_i \,\mathrm{d}\tau \qquad [5.9]$$

where $-e$ is the charge of the electron. The transition moment has the form of the integral in eqn 5.8, so, once we know the symmetry species of the states, we can use group theory to formulate the selection rules for the transitions.

As an example, we investigate whether an electron in an a_1 orbital in H_2O (which belongs to the group C_{2v}) can make an electric dipole transition to a b_1 orbital (Fig. 5.29). We must examine all three components of the transition dipole moment, and take f_2 in eqn 5.8 as x, y, and z in turn. Reference to the C_{2v} character table shows that these components transform as B_1, B_2, and A_1, respectively. The three calculations run as follows:

	x-component				y-component				z-component			
	E	C_2	σ_v	σ_v'	E	C_2	σ_v	σ_v'	E	C_2	σ_v	σ_v'
f_3	1	−1	1	−1	1	−1	1	−1	1	−1	1	−1 B_1
f_2	1	−1	1	−1	1	−1	−1	1	1	1	1	1
f_1	1	1	1	1	1	1	1	1	1	1	1	1 A_1
$f_1 f_2 f_3$	1	1	1	1	1	1	−1	−1	1	−1	1	−1

Only the first product (with $f_2 = x$) spans A_1, so only the x-component of the transition dipole moment may be nonzero. Therefore, we conclude that the electric dipole

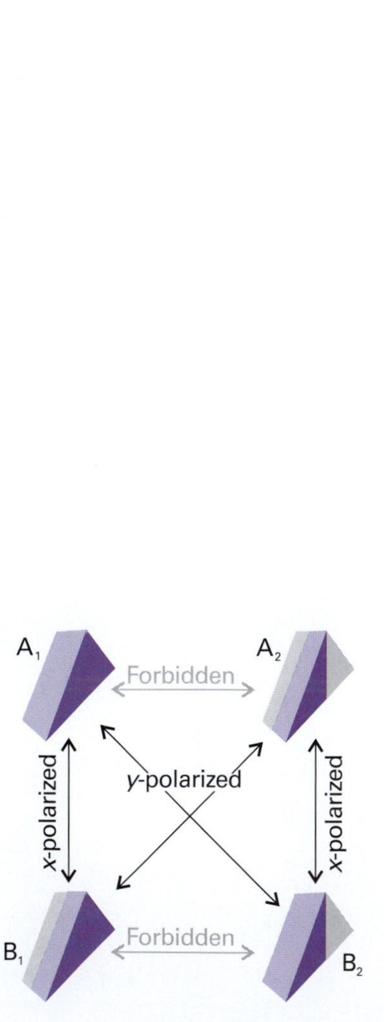

Fig. 5.29 The polarizations of the allowed transitions in a C_{2v} molecule. The shading indicates the structure of the orbitals of the specified symmetry species. The perspective view of the molecule makes it look rather like a door stop; however, from the side, each 'door stop' is in fact an isosceles triangle.

transitions between a_1 and b_1 are allowed. We can go on to state that the radiation emitted (or absorbed) is x-polarized and has its electric field vector in the x-direction, because that form of radiation couples with the x–component of a transition dipole.

Example 5.7 *Deducing a selection rule*

Is $p_x \rightarrow p_y$ an allowed transition in a tetrahedral environment?

Method We must decide whether the product $p_y q p_x$, with $q = x, y$, or z, spans A_1 by using the T_d character table.

Answer The procedure works out as follows:

	E	$8C_3$	$3C_2$	$6\sigma_d$	$6S_4$	
$f_3(p_y)$	3	0	−1	−1	1	T_2
$f_2(q)$	3	0	−1	−1	1	T_2
$f_1(p_x)$	3	0	−1	−1	1	T_2
$f_1f_2f_3$	27	0	−1	−1	1	

We can use the decomposition procedure described in Section 5.5a to deduce that A_1 occurs (once) in this set of characters, so $p_x \rightarrow p_y$ is allowed.

A more detailed analysis (using the matrix representatives rather than the characters) shows that only $q = z$ gives an integral that may be nonzero, so the transition is z-polarized. That is, the electromagnetic radiation involved in the transition has its electric vector aligned in the z-direction.

Self-test 5.8 What are the allowed transitions, and their polarizations, of a b_1 electron in a C_{4v} molecule? $[b_1 \rightarrow b_1(z);\ b_1 \rightarrow e(x,y)]$

The following chapters will show many more examples of the systematic use of symmetry. We shall see that the techniques of group theory greatly simplify the analysis of molecular structure and spectra.

Checklist of key ideas

☐ 1. A symmetry operation is an action that leaves an object looking the same after it has been carried out.

☐ 2. A symmetry element is a point, line, or plane with respect to which a symmetry operation is performed.

☐ 3. A point group is a group of symmetry operations that leaves at least one common point unchanged. A space group, a group of symmetry operations that includes translation through space.

☐ 4. The notation for point groups commonly used for molecules and solids is summarized in Table 5.1.

☐ 5. To be polar, a molecule must belong to C_n, C_{nv}, or C_s (and have no higher symmetry).

☐ 6. A molecule may be chiral only if it does not possess an axis of improper rotation, S_n.

☐ 7. A representative $\boldsymbol{D}(X)$ is a matrix that brings about the transformation of the basis under the operation X. The basis is the set of functions on which the representative acts.

☐ 8. A character, χ, is the sum of the diagonal elements of a matrix representative.

☐ 9. A character table characterizes the different symmetry types possible in the point group.

☐ 10. In a reduced representation all the matrices have block-diagonal form. An irreducible representation cannot be reduced further.

☐ 11. Symmetry species are the labels for the irreducible representations of a group.

☐ 12. Decomposition of the direct product is the reduction of a product of symmetry species to a sum of symmetry species, $\Gamma \times \Gamma' = \Gamma^{(1)} + \Gamma^{(2)} + \cdots$

☐ 13. For an integral $\int f_1 f_2\, d\tau$ to be nonzero, the integrand $f_1 f_2$ must have the symmetry species A_1 (or its equivalent in the specific molecular point group).

☐ 14. A symmetry-adapted linear combination (SALC) is a combination of atomic orbitals adapted to the symmetry of the molecule and used as the building blocks for LCAO molecular orbitals.

☐ 15. Allowed and forbidden spectroscopic transitions can be identified by considering the symmetry criteria for the non-vanishing of the transition moment between the initial and final states.

Further reading

Articles and texts

P.W. Atkins and R.S. Friedman, *Molecular quantum mechanics.* Oxford University Press (2005).

F.A. Cotton, *Chemical applications of group theory.* Wiley, New York (1990).

R. Drago, *Physical methods for chemists.* Saunders, Philadelphia (1992).

D.C. Harris and M.D. Bertollucci, *Symmetry and spectroscopy: an introduction to vibrational and electronic spectroscopy.* Dover, New York (1989).

S.F.A. Kettle, *Symmetry and structure: readable group theory for chemists.* Wiley, New York (1995).

Sources of data and information

G.L. Breneman, Crystallographic symmetry point group notation flow chart. *J. Chem. Educ.* **64**, 216 (1987).

P.W. Atkins, M.S. Child, and C.S.G. Phillips, *Tables for group theory.* Oxford University Press (1970).

Each end-of-chapter discussion question, exercise, and problem for which a solution is available in the Student Solution Manual has a corresponding bracketed number referring to the Student Solutions Manual number for this problem, for example [SSM 12.1]. Each end-of-chapter exercise and problem for which a solution is available in the Instructors' Solution Manual has a corresponding bracketed number referring to the Instructor Solutions Manual number for this problem, for example [ISM 12.2].

Discussion questions

5.1 Explain how a molecule is assigned to a point group. [SSM **12.1**]

5.2 List the symmetry operations and the corresponding symmetry elements of the point groups. [ISM **12.2**]

5.3 Explain the symmetry criteria that allow a molecule to be polar. [SSM **12.3**]

5.4 Explain the symmetry criteria that allow a molecule to be optically active. [ISM **12.4**]

5.5 Explain what is meant by (a) a representative and (b) a representation in the context of group theory. [SSM **12.5**]

5.6 Explain the construction and content of a character table. [ISM **12.6**]

5.7 Explain how spectroscopic selection rules arise and how they are formulated by using group theory. [SSM **12.7**]

5.8 Outline how a direct product is expressed as a direct sum and how to decide whether the totally symmetric irreducible representation is present in the direct product. [ISM **12.8**]

Exercises

5.1a The CH_3Cl molecule belongs to the point group C_{3v}. List the symmetry elements of the group and locate them in the molecule. [ISM **12.1a**]

5.1b The CCl_4 molecule belongs to the point group T_d. List the symmetry elements of the group and locate them in the molecule. [SSM **12.1b**]

5.2a Which of the following molecules may be polar? (a) pyridine (C_{2v}), (b) nitroethane (C_s), (c) gas-phase $HgBr_2$ ($D_{\infty h}$), (d) $B_3N_3N_6$ (D_{3h}). [ISM **12.2a**]

5.2b Which of the following molecules may be polar? (a) CH_3Cl (C_{3v}), (b) $HW_2(CO)_{10}$ (D_{4h}), (c) $SnCl_4$ (T_d). [SSM **12.2b**]

5.3a Use symmetry properties to determine whether or not the integral $\int p_x z p_z\, d\tau$ is necessarily zero in a molecule with symmetry C_{4v}. [ISM **12.3a**]

5.3b Use symmetry properties to determine whether or not the integral $\int p_x z p_z\, d\tau$ is necessarily zero in a molecule with symmetry D_{6h}. [SSM **12.3b**]

5.4a Show that the transition $A_1 \rightarrow A_2$ is forbidden for electric dipole transitions in a C_{3v} molecule. [ISM **12.4a**]

5.4b Is the transition $A_{1g} \rightarrow E_{2u}$ forbidden for electric dipole transitions in a D_{6h} molecule? [SSM **12.4b**]

5.5a Show that the function *xy* has symmetry species B_2 in the group C_{4v}. [ISM **12.5a**]

5.5b Show that the function *xyz* has symmetry species A_1 in the group D_2 [SSM **12.5b**]

5.6a Molecules belonging to the point groups D_{2h} or C_{3h} cannot be chiral. Which elements of these groups rule out chirality? [ISM **12.6a**]

5.6b Molecules belonging to the point groups T_h or T_d cannot be chiral. Which elements of these groups rule out chirality? [SSM **12.6b**]

5.7a The group D_2 consists of the elements E, C_2, C_2', and C_2'', where the three twofold rotations are around mutually perpendicular axes. Construct the group multiplication table. [ISM **12.7a**]

5.7b The group C_{4v} consists of the elements E, $2C_4$, C_2, and $2\sigma_v$, $2\sigma_d$. Construct the group multiplication table. [SSM **12.7b**]

5.8a Identify the point groups to which the following objects belong: (a) a sphere, (b) an isosceles triangle, (c) an equilateral triangle, (d) an unsharpened cylindrical pencil. [ISM **12.8a**]

5.8b Identify the point groups to which the following objects belong: (a) a sharpened cylindrical pencil, (b) a three-bladed propellor, (c) a four-legged table, (d) yourself (approximately). [SSM **12.8b**]

5.9a List the symmetry elements of the following molecules and name the point groups to which they belong: (a) NO_2, (b) N_2O, (c) $CHCl_3$, (d) $CH_2{=}CH_2$, (e) *cis*-CHBr=CHBr, (f) *trans*-CHCl=CHCl. [ISM **12.9a**]

5.9b List the symmetry elements of the following molecules and name the point groups to which they belong: (a) naphthalene, (b) anthracene, (c) the three dichlorobenzenes. [SSM **12.9b**]

5.10a Assign (a) *cis*-dichloroethene and (b) *trans*-dichloroethene to point groups. [ISM **12.10a**]

5.10b Assign the following molecules to point groups: (a) HF, (b) IF_7 (pentagonal bipyramid), (c) XeO_2F_2 (see-saw), (d) $Fe_2(CO)_9$ (**22**), (e) cubane, C_8H_8, (f) tetrafluorocubane, $C_8H_4F_4$ (**23**). [SSM **12.10b**]

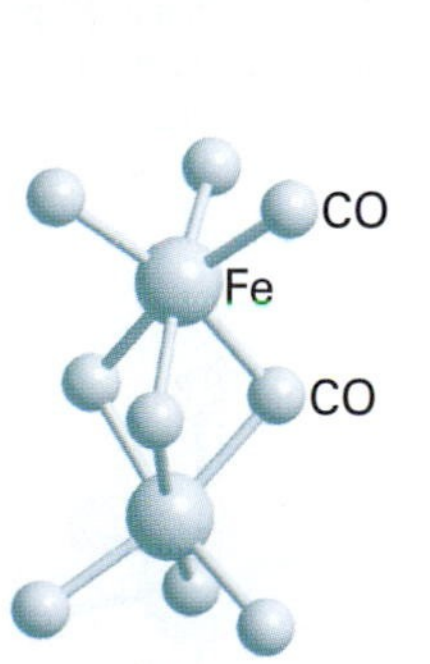

22

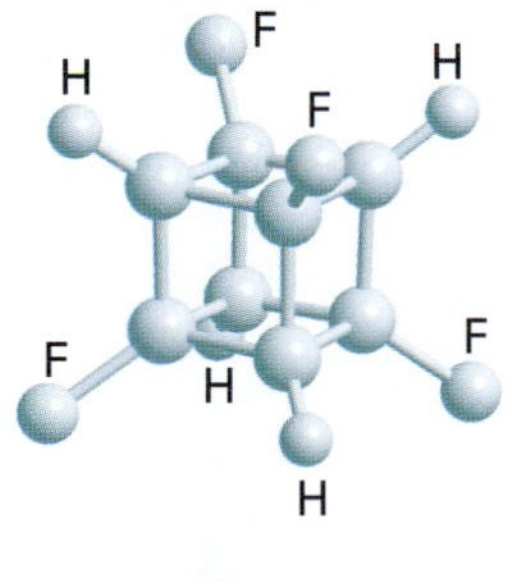

23

5.11a Which of the molecules in Exercises 5.9a and 5.10a can be (a) polar, (b) chiral? [ISM **12.11a**]

5.11b Which of the molecules in Exercises 5.9b and 5.10b can be (a) polar, (b) chiral? [SSM **12.11b**]

5.12a Consider the C_{2v} molecule NO_2. The combination $p_x(A) - p_x(B)$ of the two O atoms (with *x* perpendicular to the plane) spans A_2. Is there any orbital of the central N atom that can have a nonzero overlap with that combination of O orbitals? What would be the case in SO_2, where 3*d* orbitals might be available? [ISM **12.12a**]

5.12b Consider the C_{3v} ion NO_3^-. Is there any orbital of the central N atom that can have a nonzero overlap with the combination $2p_z(A) - p_z(B) - p_z(C)$ of the three O atoms (with *z* perpendicular to the plane). What would be the case in SO_3, where 3*d* orbitals might be available? [SSM **12.12b**]

5.13a The ground state of NO_2 is A_1 in the group C_{2v}. To what excited states may it be excited by electric dipole transitions, and what polarization of light is it necessary to use? [ISM **12.13a**]

5.13b The ClO_2 molecule (which belongs to the group C_{2v}) was trapped in a solid. Its ground state is known to be B_1. Light polarized parallel to the *y*-axis (parallel to the OO separation) excited the molecule to an upper state. What is the symmetry of that state? [SSM **12.13b**]

5.14a What states of (a) benzene, (b) naphthalene may be reached by electric dipole transitions from their (totally symmetrical) ground states? [ISM **12.14a**]

5.14b What states of (a) anthracene, (b) coronene (**24**) may be reached by electric dipole transitions from their (totally symmetrical) ground states? [SSM **12.14b**]

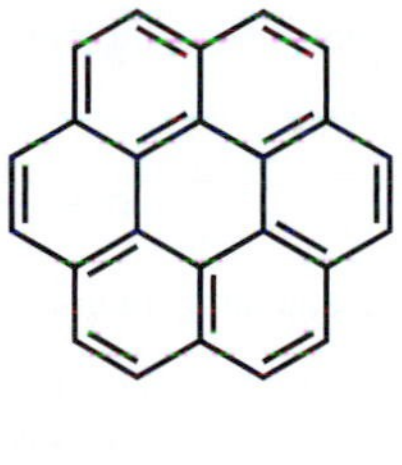

24 Coronene

5.15a Write $f_1 = \sin\theta$ and $f_2 = \cos\theta$, and show by symmetry arguments using the group C_s that the integral of their product over a symmetrical range around $\theta = 0$ is zero. [ISM **12.15a**]

5.15b Determine whether the integral over f_1 and f_2 in Exercise 5.15a is zero over a symmetrical range about $\theta = 0$ in the group C_{3v}. [SSM **12.5b**]

Problems*

5.1 List the symmetry elements of the following molecules and name the point groups to which they belong: (a) staggered CH_3CH_3, (b) chair and boat cyclohexane, (c) B_2H_6, (d) $[Co(en)_3]^{3+}$, where en is ethylenediamine (ignore its detailed structure), (e) crown-shaped S_8. Which of these molecules can be (i) polar, (ii) chiral? [SSM **12.1**]

5.2 The group C_{2h} consists of the elements E, C_2, σ_h, i. Construct the group multiplication table and find an example of a molecule that belongs to the group. [ISM **12.2**]

5.3 The group D_{2h} has a C_2 axis perpendicular to the principal axis and a horizontal mirror plane. Show that the group must therefore have a centre of inversion. [SSM **12.3**]

5.4 Consider the H_2O molecule, which belongs to the group C_{2v}. Take as a basis the two H1*s* orbitals and the four valence orbital of the O atom and set up the 6×6 matrices that represent the group in this basis. Confirm by explicit matrix multiplication that the group multiplications (a) $C_2\sigma_v = \sigma_v'$ and (b) $\sigma_v\sigma_v' = C_2$. Confirm, by calculating the traces of the matrices: (a) that symmetry elements in the same class have the same character, (b) that the representation is reducible, and (c) that the basis spans $3A_1 + B_1 + 2B_2$. [ISM **12.4**]

5.5 Confirm that the *z*-component of orbital angular momentum is a basis for an irreducible representation of A_2 symmetry in C_{3v}. [SSM **12.5**]

5.6 The (one-dimensional) matrices $\boldsymbol{D}(C_3) = \mathbf{1}$ and $\boldsymbol{D}(C_2) = \mathbf{1}$, and $\boldsymbol{D}(C_3) = \mathbf{1}$ and $\boldsymbol{D}(C_2) = \mathbf{-1}$ both represent the group multiplication $C_3C_2 = C_6$ in the group C_{6v} with $\boldsymbol{D}(C_6) = +1$ and -1, respectively. Use the character table to confirm these remarks. What are the representatives of σ_v and σ_d in each case? [ISM **12.6**]

5.7 Construct the multiplication table of the Pauli spin matrices, $\boldsymbol{\sigma}$, and the 2 × 2 unit matrix:

$$\sigma_x = \begin{pmatrix} 0 & 1 \\ 1 & 0 \end{pmatrix} \qquad \sigma_y = \begin{pmatrix} 0 & -i \\ i & 0 \end{pmatrix} \qquad \sigma_z = \begin{pmatrix} 1 & 0 \\ 0 & -1 \end{pmatrix} \qquad \sigma_0 = \begin{pmatrix} 1 & 0 \\ 0 & 1 \end{pmatrix}$$

Do the four matrices form a group under multiplication? [SSM **12.7**]

5.8 What irreducible representations do the four H1*s* orbitals of CH_4 span? Are there *s* and *p* orbitals of the central C atom that may form molecular orbitals with them? Could *d* orbitals, even if they were present on the C atom, play a role in orbital formation in CH_4? [ISM **12.8**]

5.9 Suppose that a methane molecule became distorted to (a) C_{3v} symmetry by the lengthening of one bond, (b) C_{2v} symmetry, by a kind of scissors action in which one bond angle opened and another closed slightly. Would more *d* orbitals become available for bonding? [SSM **12.9**]

5.10‡ B.A. Bovenzi and G.A. Pearse, Jr. (*J. Chem. Soc. Dalton Trans.*, 2763 (1997)) synthesized coordination compounds of the tridentate ligand pyridine-2,6-diamidoxime ($C_7H_9N_5O_2$, **25**). Reaction with $NiSO_4$ produced a complex in which two of the essentially planar ligands are bonded at right angles to a single Ni atom. Name the point group and the symmetry operations of the resulting $[Ni(C_7H_9N_5O_2)_2]^{2+}$ complex cation. [ISM **12.10**]

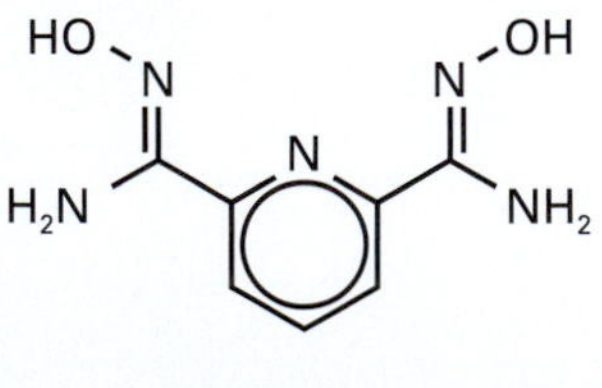

25

5.11‡ R. Eujen, B. Hoge, and D.J. Brauer (*Inorg. Chem.* **36**, 1464 (1997)) prepared and characterized several square-planar Ag(III) complex anions. In the complex anion $[trans\text{-}Ag(CF_3)_2(CN)_2]^-$, the Ag—CN groups are collinear. (a) Assuming free rotation of the CF_3 groups (that is, disregarding the AgCF and AgCH angles), name the point group of this complex anion. (b) Now suppose the CF_3 groups cannot rotate freely (because the ion was in a solid, for example). Structure (**26**) shows a plane that bisects the NC—Ag—CN axis and is perpendicular to it. Name the point group of the complex if each CF_3 group has a CF bond in that plane (so the CF_3 groups do not point to either CN group preferentially) and the CF_3 groups are (i) staggered, (ii) eclipsed. [SSM **12.11**]

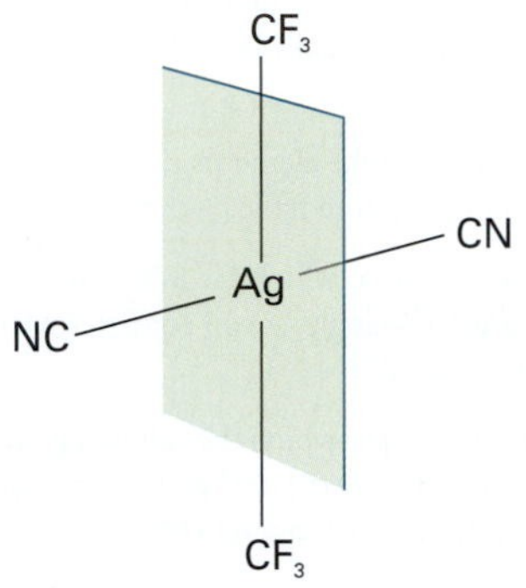

26

5.12‡ A computational study by C.J. Marsden (*Chem. Phys. Lett.* **245**, 475 (1995)) of AM_x compounds, where A is in Group 14 of the periodic table and M is an alkali metal, shows several deviations from the most symmetric structures for each formula. For example, most of the AM_4 structures were not tetrahedral but had two distinct values for MAM bond angles. They could be derived from a tetrahedron by a distortion shown in (**27**). (a) What is the point group of the distorted tetrahedron? (b) What is the symmetry species of the distortion considered as a vibration in the new, less symmetric group? Some AM_6 structures are not octahedral, but could be derived from an octahedron by translating a C—M—C axis as in (**28**). (c) What is the point group of the distorted octahedron? (d) What is the symmetry species of the distortion considered as a vibration in the new, less symmetric group? [ISM **12.12**]

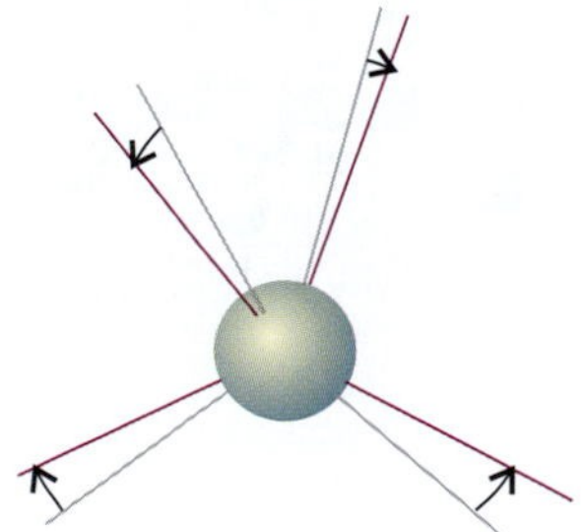

27

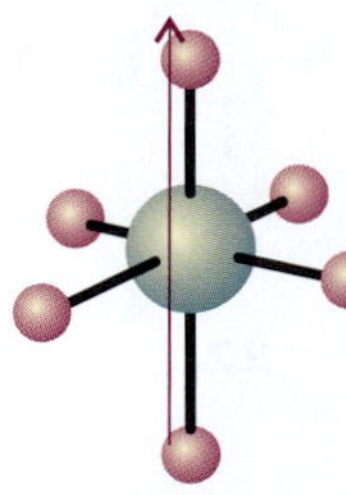

28

5.13 The algebraic forms of the *f* orbitals are a radial function multiplied by one of the factors (a) $z(5z^2 - 3r^2)$, (b) $y(5y^2 - 3r^2)$, (c) $x(5x^2 - 3r^2)$,

* Problems denoted with the symbol ‡ were supplied by Charles Trapp and Carmen Giunta.

(d) $z(x^2 - y^2)$, (e) $y(x^2 - z^2)$, (f) $x(z^2 - y^2)$, (g) xyz. Identify the irreducible representations spanned by these orbitals in (a) C_{2v}, (b) C_{3v}, (c) T_d, (d) O_h. Consider a lanthanide ion at the centre of (a) a tetrahedral complex, (b) an octahedral complex. What sets of orbitals do the seven *f* orbitals split into? [SSM **12.13**]

5.14 Does the product *xyz* necessarily vanish when integrated over (a) a cube, (b) a tetrahedron, (c) a hexagonal prism, each centred on the origin? [ISM **12.14**]

5.15 The NO_2 molecule belongs to the group C_{2v}, with the C_2 axis bisecting the ONO angle. Taking as a basis the N2*s*, N2*p*, and O2*p* orbitals, identify the irreducible representations they span, and construct the symmetry-adapted linear combinations. [SSM **12.15**]

5.16 Construct the symmetry-adapted linear combinations of C2p_z orbitals for benzene, and use them to calculate the Hückel secular determinant. This procedure leads to equations that are much easier to solve than using the original orbitals, and show that the Hückel orbitals are those specified in Section 4.6d. [ISM **12.16**]

5.17 The phenanthrene molecule (**29**) belongs to the group C_{2v} with the C_2 axis perpendicular to the molecular plane. (a) Classify the irreducible representations spanned by the carbon 2p_z orbitals and find their symmetry-adapted linear combinations. (b) Use your results from part (a) to calculate the Hückel secular determinant. (c) What states of phenanthrene may be reached by electric dipole transitions from its (totally symmetrical) ground state? [SSM **12.17**]

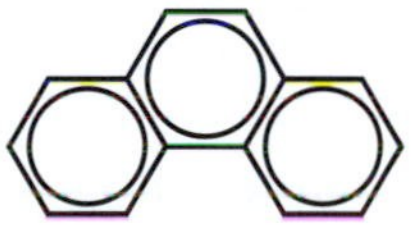

29 Phenanthrene

5.18‡ In a spectroscopic study of C_{60}, F. Negri, G. Orlandi, and F. Zerbetto (*J. Phys. Chem.* **100**, 10849 (1996)) assigned peaks in the fluorescence spectrum. The molecule has icosahedral symmetry (I_h). The ground electronic state is A_{1g}, and the lowest-lying excited states are T_{1g} and G_g. (a) Are photon-induced transitions allowed from the ground state to either of these excited states? Explain your answer. (b) What if the transition is accompanied by a vibration that breaks the parity? [ISM **12.18**]

Applications: to astrophysics and biology

5.19‡ The H_3^+ molecular ion, which plays an important role in chemical reactions occurring in interstellar clouds, is known to be equilateral triangular. (a) Identify the symmetry elements and determine the point group of this molecule. (b) Take as a basis for a representation of this molecule the three H1*s* orbitals and set up the matrices that group in this basis. (c) Obtain the group multiplication table by explicit multiplication of the matrices. (d) Determine if the representation is reducible and, if so, give the irreducible representations obtained. [SSM **12.19**]

5.20‡ The H_3^+ molecular ion has recently been found in the interstellar medium and in the atmospheres of Jupiter, Saturn, and Uranus. The H_4 analogues have not yet been found, and the square planar structure is thought to be unstable with respect to vibration. Take as a basis for a representation of the point group of this molecule the four H1*s* orbitals and determine if this representation is reducible. [ISM **12.20**]

5.21 Some linear polyenes, of which β-carotene is an example, are important biological co-factors that participate in processes as diverse as the absorption of solar energy in photosynthesis and protection against harmful biological oxidations. Use as a model of β-carotene a linear polyene containing 22 conjugated C atoms. (a) To what point group does this model of β-carotene belong? (b) Classify the irreducible representations spanned by the carbon 2p_z orbitals and find their symmetry-adapted linear combinations. (c) Use your results from part (b) to calculate the Hückel secular determinant. (d) What states of this model of β-carotene may be reached by electric dipole transitions from its (totally symmetrical) ground state? [SSM **12.21**]

5.22 The chlorophylls that participate in photosynthesis and the haem groups of cytochromes are derived from the porphine dianion group (**30**), which belongs to the D_{4h} point group. The ground electronic state is A_{1g} and the lowest-lying excited state is E_u. Is a photon-induced transition allowed from the ground state to the excited state? Explain your answer. [ISM **12.22**]

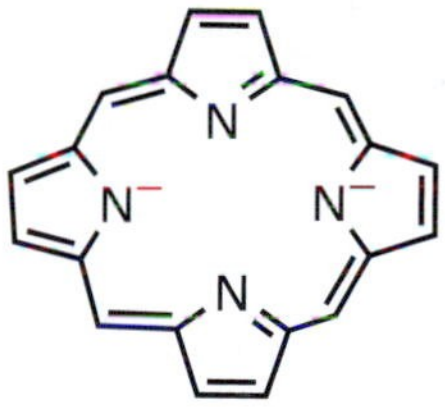

30

6 Molecular spectroscopy 1: rotational and vibrational spectra

The general strategy we adopt in the chapter is to set up expressions for the energy levels of molecules and then apply selection rules and considerations of populations to infer the form of spectra. Rotational energy levels are considered first, and we see how to derive expressions for their values and how to interpret rotational spectra in terms of molecular dimensions. Not all molecules can occupy all rotational states: we see the experimental evidence for this restriction and its explanation in terms of nuclear spin and the Pauli principle. Next, we consider the vibrational energy levels of diatomic molecules, and see that we can use the properties of harmonic oscillators developed in Chapter 2. Then we consider polyatomic molecules and find that their vibrations may be discussed as though they consisted of a set of independent harmonic oscillators, so the same approach as employed for diatomic molecules may be used. We also see that the symmetry properties of the vibrations of polyatomic molecules are helpful for deciding which modes of vibration can be studied spectroscopically.

The origin of spectral lines in molecular spectroscopy is the absorption, emission, or scattering of a photon when the energy of a molecule changes. The difference from atomic spectroscopy is that the energy of a molecule can change not only as a result of electronic transitions but also because it can undergo changes of rotational and vibrational state. Molecular spectra are therefore more complex than atomic spectra. However, they also contain information relating to more properties, and their analysis leads to values of bond strengths, lengths, and angles. They also provide a way of determining a variety of molecular properties, particularly molecular dimensions, shapes, and dipole moments. Molecular spectroscopy is also useful to astrophysicists and environmental scientists, for the chemical composition of interstellar space and of planetary atmospheres can be inferred from their rotational, vibrational, and electronic spectra.

Pure rotational spectra, in which only the rotational state of a molecule changes, can be observed in the gas phase. Vibrational spectra of gaseous samples show features that arise from rotational transitions that accompany the excitation of vibration. Electronic spectra, which are described in Chapter 7, show features arising from simultaneous vibrational and rotational transitions. The simplest way of dealing with these complexities is to tackle each type of transition in turn, and then to see how simultaneous changes affect the appearance of spectra.

General features of spectroscopy

All types of spectra have some features in common, and we examine these first. In **emission spectroscopy**, a molecule undergoes a transition from a state of high energy E_1 to a state of lower energy E_2 and emits the excess energy as a photon. In **absorption spectroscopy**, the net absorption of nearly monochromatic (single frequency) incident radiation is monitored as the radiation is swept over a range of frequencies. We say *net* absorption, because it will become clear that, when a sample is irradiated, both absorption and emission at a given frequency are stimulated, and the detector measures the difference, the net absorption.

The energy, $h\nu$, of the photon emitted or absorbed, and therefore the frequency ν of the radiation emitted or absorbed, is given by the Bohr frequency condition, $h\nu = |E_1 - E_2|$ (eqn 1.10). Emission and absorption spectroscopy give the same information about energy level separations, but practical considerations generally determine which technique is employed. We shall discuss emission spectroscopy in Chapter 14; here we focus on absorption spectroscopy, which is widely employed in studies of electronic transitions, molecular rotations, and molecular vibrations.

In Chapter 2 we saw that transitions between electronic energy levels are stimulated by or emit ultraviolet, visible, or near-infrared radiation. Vibrational and rotational transitions, the focus of the discussion in this chapter, can be induced in two ways. First, the direct absorption or emission of infrared radiation can cause changes in vibrational energy levels, whereas absorption or emission of microwave radiation gives information about rotational energy levels. Second, vibrational and rotational energy levels can be explored by examining the frequencies present in the radiation scattered by molecules in **Raman spectroscopy**. About 1 in 10^7 of the incident photons collide with the molecules, give up some of their energy, and emerge with a lower energy. These scattered photons constitute the lower-frequency **Stokes radiation** from the sample (Fig. 6.1). Other incident photons may collect energy from the molecules (if they are already excited), and emerge as higher-frequency **anti-Stokes radiation**. The component of radiation scattered without change of frequency is called **Rayleigh radiation**.

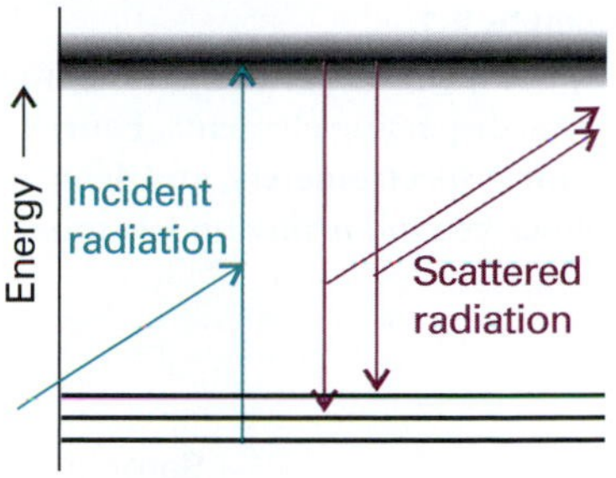

Fig. 6.1 In Raman spectroscopy, an incident photon is scattered from a molecule with either an increase in frequency (if the radiation collects energy from the molecule) or—as shown here for the case of scattered Stokes radiation—with a lower frequency if it loses energy to the molecule. The process can be regarded as taking place by an excitation of the molecule to a wide range of states (represented by the shaded band), and the subsequent return of the molecule to a lower state; the net energy change is then carried away by the photon.

6.1 Experimental techniques

A **spectrometer** is an instrument that detects the characteristics of light scattered, emitted, or absorbed by atoms and molecules. Figure 6.2 shows the general layouts of absorption and emission spectrometers operating in the ultraviolet and visible ranges. Radiation from an appropriate source is directed toward a sample. In most spectrom-

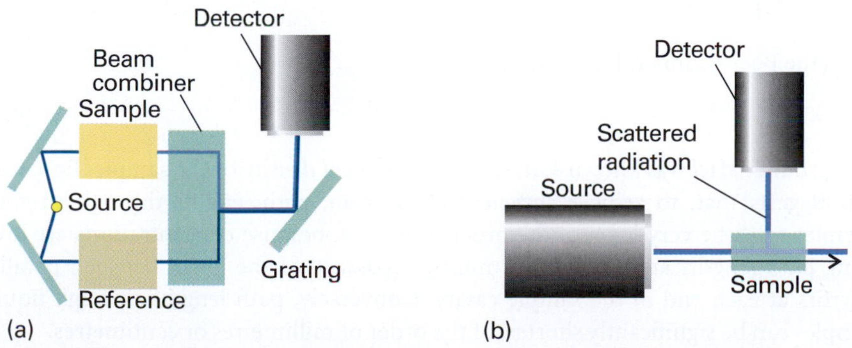

Fig. 6.2 Two examples of spectrometers: (a) the layout of an absorption spectrometer, used primarily for studies in the ultraviolet and visible ranges, in which the exciting beams of radiation pass alternately through a sample and a reference cell, and the detector is synchronized with them so that the relative absorption can be determined, and (b) a simple emission spectrometer, where light emitted or scattered by the sample is detected at right angles to the direction of propagation of an incident beam of radiation.

Comment 6.1

The principles of operation of radiation sources, dispersing elements, Fourier transform spectrometers, and detectors are described in *Further information 6.1*.

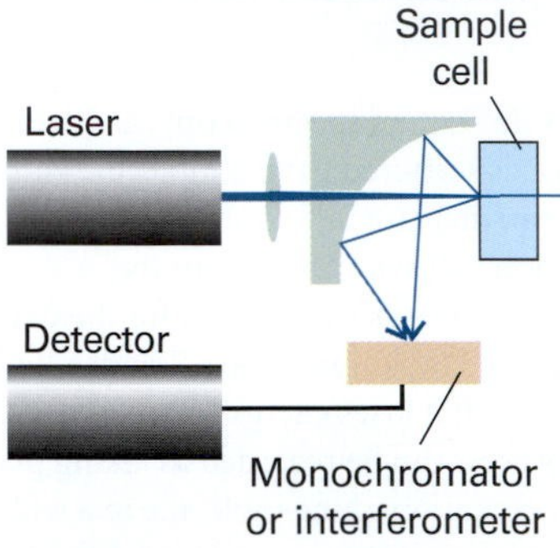

Fig. 6.3 A common arrangement adopted in Raman spectroscopy. A laser beam first passes through a lens and then through a small hole in a mirror with a curved reflecting surface. The focused beam strikes the sample and scattered light is both deflected and focused by the mirror. The spectrum is analysed by a monochromator or an interferometer.

eters, light transmitted, emitted, or scattered by the sample is collected by mirrors or lenses and strikes a **dispersing element** that separates radiation into different frequencies. The intensity of light at each frequency is then analysed by a suitable detector. In a typical Raman spectroscopy experiment, a monochromatic incident laser beam is passed through the sample and the radiation scattered from the front face of the sample is monitored (Fig. 6.3). This detection geometry allows for the study of gases, pure liquids, solutions, suspensions, and solids.

Modern spectrometers, particularly those operating in the infrared and near-infrared, now almost always use **Fourier transform techniques** of spectral detection and analysis. The heart of a Fourier transform spectrometer is a *Michelson interferometer*, a device for analysing the frequencies present in a composite signal. The total signal from a sample is like a chord played on a piano, and the Fourier transform of the signal is equivalent to the separation of the chord into its individual notes, its spectrum.

6.2 The intensities of spectral lines

The ratio of the transmitted intensity, I, to the incident intensity, I_0, at a given frequency is called the **transmittance**, T, of the sample at that frequency:

$$T = \frac{I}{I_0} \qquad [6.1]$$

It is found empirically that the transmitted intensity varies with the length, l, of the sample and the molar concentration, [J], of the absorbing species J in accord with the **Beer–Lambert law**:

$$I = I_0 10^{-\varepsilon[\mathrm{J}]l} \qquad (6.2)$$

The quantity ε is called the **molar absorption coefficient** (formerly, and still widely, the 'extinction coefficient'). The molar absorption coefficient depends on the frequency of the incident radiation and is greatest where the absorption is most intense. Its dimensions are 1/(concentration × length), and it is normally convenient to express it in cubic decimetres per mole per centimetre ($\mathrm{dm^3\,mol^{-1}\,cm^{-1}}$). Alternative units are square centimetres per mole ($\mathrm{cm^2\,mol^{-1}}$). This change of units demonstrates that ε may be regarded as a molar cross-section for absorption and, the greater the cross-sectional area of the molecule for absorption, the greater its ability to block the passage of the incident radiation.

To simplify eqn 6.2, we introduce the **absorbance**, A, of the sample at a given wavenumber as

$$A = \log\frac{I_0}{I} \quad \text{or} \quad A = -\log T \qquad [6.3]$$

Then the Beer–Lambert law becomes

$$A = \varepsilon[\mathrm{J}]l \qquad (6.4)$$

The product $\varepsilon[\mathrm{J}]l$ was known formerly as the *optical density* of the sample. Equation 6.4 suggests that, to achieve sufficient absorption, path lengths through gaseous samples must be very long, of the order of metres, because concentrations are low. Long path lengths are achieved by multiple passage of the beam between parallel mirrors at each end of the sample cavity. Conversely, path lengths through liquid samples can be significantly shorter, of the order of millimetres or centimetres.

Justification 6.1 *The Beer–Lambert law*

The Beer–Lambert law is an empirical result. However, it is simple to account for its form. The reduction in intensity, $\mathrm{d}I$, that occurs when light passes through a layer of thickness $\mathrm{d}l$ containing an absorbing species J at a molar concentration [J] is proportional to the thickness of the layer, the concentration of J, and the intensity, I, incident on the layer (because the rate of absorption is proportional to the intensity, see below). We can therefore write

$$\mathrm{d}I = -\kappa[\mathrm{J}]I\mathrm{d}l$$

where κ (kappa) is the proportionality coefficient, or equivalently

$$\frac{\mathrm{d}I}{I} = -\kappa[\mathrm{J}]\mathrm{d}l$$

This expression applies to each successive layer into which the sample can be regarded as being divided. Therefore, to obtain the intensity that emerges from a sample of thickness l when the intensity incident on one face of the sample is I_0, we sum all the successive changes:

$$\int_{I_0}^{I} \frac{\mathrm{d}I}{I} = -\kappa \int_0^l [\mathrm{J}]\mathrm{d}l$$

If the concentration is uniform, [J] is independent of location, and the expression integrates to

$$\ln \frac{I}{I_0} = -\kappa[\mathrm{J}]l$$

This expression gives the Beer–Lambert law when the logarithm is converted to base 10 by using $\ln x = (\ln 10)\log x$ and replacing κ by $\varepsilon \ln 10$.

Illustration 6.1 *Using the Beer–Lambert law*

The Beer–Lambert law implies that the intensity of electromagnetic radiation transmitted through a sample at a given wavenumber decreases exponentially with the sample thickness and the molar concentration. If the transmittance is 0.1 for a path length of 1 cm (corresponding to a 90 per cent reduction in intensity), then it would be $(0.1)^2 = 0.01$ for a path of double the length (corresponding to a 99 per cent reduction in intensity overall).

The maximum value of the molar absorption coefficient, $\varepsilon_{\mathrm{max}}$, is an indication of the intensity of a transition. However, as absorption bands generally spread over a range of wavenumbers, quoting the absorption coefficient at a single wavenumber might not give a true indication of the intensity of a transition. The **integrated absorption coefficient**, $\mathcal{A}$, is the sum of the absorption coefficients over the entire band (Fig. 6.4), and corresponds to the area under the plot of the molar absorption coefficient against wavenumber:

$$\mathcal{A} = \int_{\mathrm{band}} \varepsilon(\tilde{\nu})\,\mathrm{d}\tilde{\nu} \qquad [6.5]$$

For lines of similar widths, the integrated absorption coefficients are proportional to the heights of the lines.

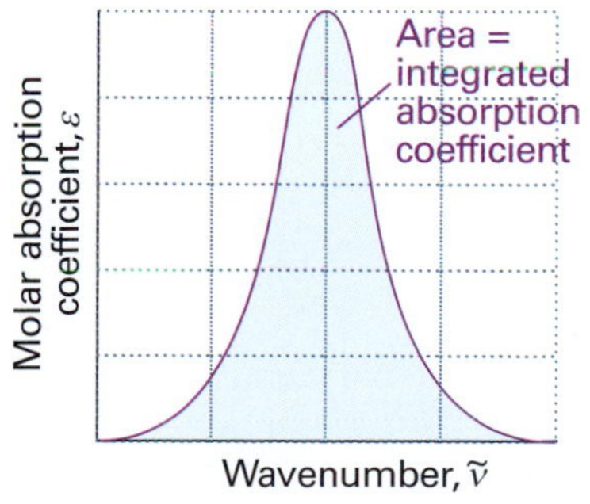

Fig. 6.4 The integrated absorption coefficient of a transition is the area under a plot of the molar absorption coefficient against the wavenumber of the incident radiation.

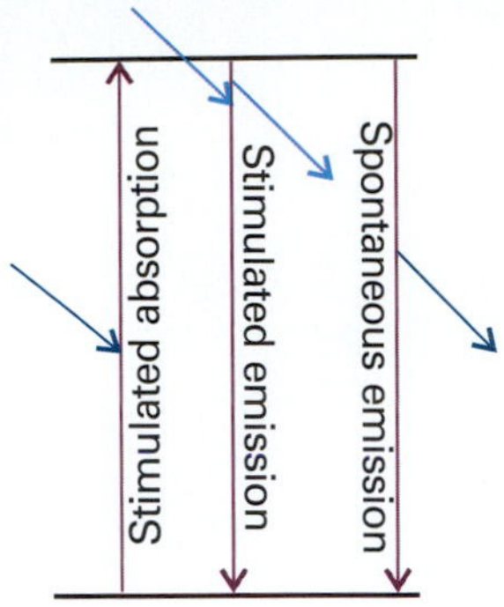

Fig. 6.5 The processes that account for absorption and emission of radiation and the attainment of thermal equilibrium. The excited state can return to the lower state spontaneously as well as by a process stimulated by radiation already present at the transition frequency.

Comment 6.2

The slight difference between the forms of the Planck distribution shown in eqns 1.5 and 6.7 stems from the fact that it is written here as $\rho d\nu$, and $d\lambda = (c/\nu^2)d\nu$.

(a) Absorption intensities

Einstein identified three contributions to the transitions between states. **Stimulated absorption** is the transition from a low energy state to one of higher energy that is driven by the electromagnetic field oscillating at the transition frequency. We saw in Section 2.10 that the transition rate, w, is the rate of change of probability of the molecule being found in the upper state. We also saw that the more intense the electromagnetic field (the more intense the incident radiation), the greater the rate at which transitions are induced and hence the stronger the absorption by the sample (Fig. 6.5). Einstein wrote the transition rate as

$$w = B\rho \tag{6.6}$$

The constant B is the **Einstein coefficient of stimulated absorption** and $\rho d\nu$ is the energy density of radiation in the frequency range ν to $\nu + d\nu$, where ν is the frequency of the transition. When the molecule is exposed to black-body radiation from a source of temperature T, ρ is given by the Planck distribution (eqn 1.5):

$$\rho = \frac{8\pi h\nu^3/c^3}{e^{h\nu/kT} - 1} \tag{6.7}$$

For the time being, we can treat B as an empirical parameter that characterizes the transition: if B is large, then a given intensity of incident radiation will induce transitions strongly and the sample will be strongly absorbing. The **total rate of absorption,** W, the number of molecules excited during an interval divided by the duration of the interval, is the transition rate of a single molecule multiplied by the number of molecules N in the lower state: $W = Nw$.

Einstein considered that the radiation was also able to induce the molecule in the upper state to undergo a transition to the lower state, and hence to generate a photon of frequency ν. Thus, he wrote the rate of this stimulated emission as

$$w' = B'\rho \tag{6.8}$$

where B' is the **Einstein coefficient of stimulated emission.** Note that only radiation of the same frequency as the transition can stimulate an excited state to fall to a lower state. However, he realized that stimulated emission was not the only means by which the excited state could generate radiation and return to the lower state, and suggested that an excited state could undergo **spontaneous emission** at a rate that was independent of the intensity of the radiation (of any frequency) that is already present. Einstein therefore wrote the total rate of transition from the upper to the lower state as

$$w' = A + B'\rho \tag{6.9}$$

The constant A is the **Einstein coefficient of spontaneous emission.** The overall rate of emission is

$$W' = N'(A + B'\rho) \tag{6.10}$$

where N' is the population of the upper state.

As demonstrated in the *Justification* below, Einstein was able to show that the two coefficients of stimulated absorption and emission are equal, and that the coefficient of spontaneous emission is related to them by

$$A = \left(\frac{8\pi h\nu^3}{c^3}\right)B \tag{6.11}$$

Justification 6.2 *The relation between the Einstein coefficients*

At thermal equilibrium, the total rates of emission and absorption are equal, so

$$NB\rho = N'(A + B'\rho)$$

This expression rearranges into

$$\rho = \frac{N'A}{NB - N'B'} = \frac{A/B}{N/N' - B'/B} = \frac{A/B}{e^{h\nu/kT} - B'/B}$$

We have used the Boltzmann expression for the ratio of populations of states of energies E and E' in the last step:

$$\frac{N'}{N} = e^{-h\nu/kT} \qquad h\nu = E' - E$$

This result has the same form as the Planck distribution (eqn 6.7), which describes the radiation density at thermal equilibrium. Indeed, when we compare the two expressions for ρ, we can conclude that $B' = B$ and that A is related to B by eqn 6.11.

The growth of the importance of spontaneous emission with increasing frequency is a very important conclusion, as we shall see when we consider the operation of lasers (Section 7.5). The equality of the coefficients of stimulated emission and absorption implies that, if two states happen to have equal populations, then the rate of stimulated emission is equal to the rate of stimulated absorption, and there is then no net absorption.

Spontaneous emission can be largely ignored at the relatively low frequencies of rotational and vibrational transitions, and the intensities of these transitions can be discussed in terms of stimulated emission and absorption. Then the net rate of absorption is given by

$$W_{\text{net}} = NB\rho - N'B'\rho = (N - N')B\rho \tag{6.12}$$

and is proportional to the population difference of the two states involved in the transition.

(b) Selection rules and transition moments

We met the concept of a 'selection rule' in Sections 3.3 and 5.6 as a statement about whether a transition is forbidden or allowed. Selection rules also apply to molecular spectra, and the form they take depends on the type of transition. The underlying classical idea is that, for the molecule to be able to interact with the electromagnetic field and absorb or create a photon of frequency ν, it must possess, at least transiently, a dipole oscillating at that frequency. We saw in Section 2.10 that this transient dipole is expressed quantum mechanically in terms of the transition dipole moment, $\boldsymbol{\mu}_{\text{fi}}$, between states ψ_{i} and ψ_{f}:

$$\boldsymbol{\mu}_{\text{fi}} = \int \psi_{\text{f}}^{*}\hat{\boldsymbol{\mu}}\psi_{\text{i}}\,d\tau \qquad [6.13]$$

where $\hat{\boldsymbol{\mu}}$ is the electric dipole moment operator. The size of the transition dipole can be regarded as a measure of the charge redistribution that accompanies a transition: a transition will be active (and generate or absorb photons) only if the accompanying charge redistribution is dipolar (Fig. 6.6).

We know from time-dependent perturbation theory (Section 2.10) that the transition rate is proportional to $|\mu_{\text{fi}}|^2$. It follows that the coefficient of stimulated absorption

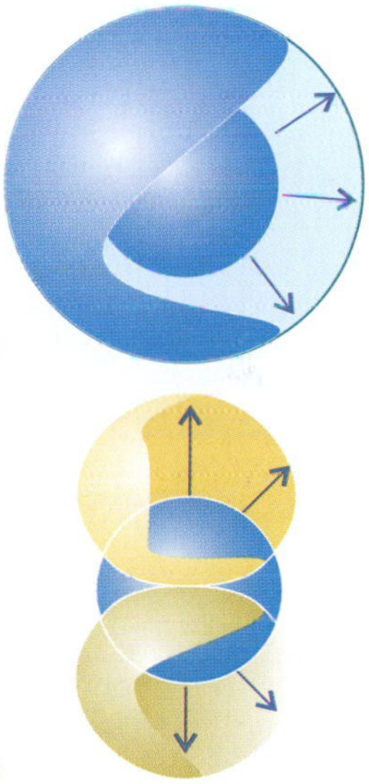
(a)
(b)

Fig. 6.6 (a) When a 1*s* electron becomes a 2*s* electron, there is a spherical migration of charge; there is no dipole moment associated with this migration of charge; this transition is electric-dipole forbidden. (b) In contrast, when a 1*s* electron becomes a 2*p* electron, there is a dipole associated with the charge migration; this transition is allowed. (There are subtle effects arising from the sign of the wavefunction that give the charge migration a dipolar character, which this diagram does not attempt to convey.)

(and emission), and therefore the intensity of the transition, is also proportional to $|\mu_{fi}|^2$. A detailed analysis gives

$$B = \frac{|\mu_{fi}|^2}{6\varepsilon_0\hbar^2} \tag{6.14}$$

Only if the transition moment is nonzero does the transition contribute to the spectrum. It follows that, to identify the selection rules, we must establish the conditions for which $\boldsymbol{\mu}_{fi} \neq 0$.

A **gross selection rule** specifies the general features a molecule must have if it is to have a spectrum of a given kind. For instance, we shall see that a molecule gives a rotational spectrum only if it has a permanent electric dipole moment. This rule, and others like it for other types of transition, will be explained in the relevant sections of the chapter. A detailed study of the transition moment leads to the **specific selection rules** that express the allowed transitions in terms of the changes in quantum numbers. We have already encountered examples of specific selection rules when discussing atomic spectra (Section 3.3), such as the rule $\Delta l = \pm 1$ for the angular momentum quantum number.

6.3 Linewidths

A number of effects contribute to the widths of spectroscopic lines. Some contributions to linewidths can be modified by changing the conditions, and to achieve high resolutions we need to know how to minimize these contributions. Other contributions cannot be changed, and represent an inherent limitation on resolution.

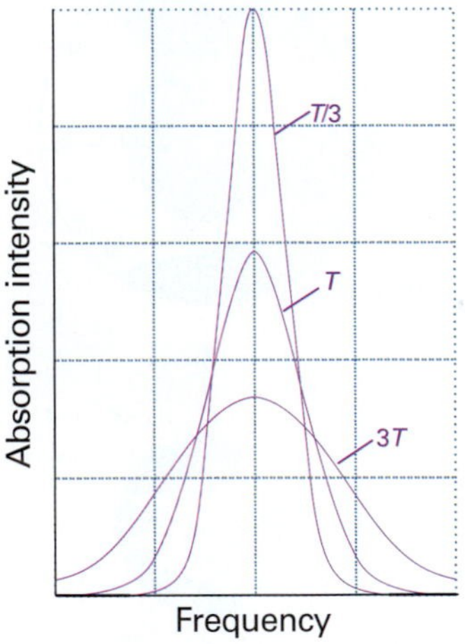

Fig. 6.7 The Gaussian shape of a Doppler-broadened spectral line reflects the Maxwell distribution of speeds in the sample at the temperature of the experiment. Notice that the line broadens as the temperature is increased.

Exploration In a spectrometer that makes use of *phase-sensitive detection* the output signal is proportional to the first derivative of the signal intensity, $dI/d\nu$. Plot the resulting line shape for various temperatures. How is the separation of the peaks related to the temperature?

(a) Doppler broadening

The study of gaseous samples is very important, as it can inform our understanding of atmospheric chemistry. In some cases, meaningful spectroscopic data can be obtained only from gaseous samples. For example, they are essential for rotational spectroscopy, for only in gases can molecules rotate freely.

One important broadening process in gaseous samples is the **Doppler effect**, in which radiation is shifted in frequency when the source is moving towards or away from the observer. When a source emitting electromagnetic radiation of frequency ν moves with a speed s relative to an observer, the observer detects radiation of frequency

$$\nu_{\text{receding}} = \nu\left(\frac{1-s/c}{1+s/c}\right)^{1/2} \qquad \nu_{\text{approaching}} = \nu\left(\frac{1+s/c}{1-s/c}\right)^{1/2} \tag{6.15}$$

where c is the speed of light (see *Further reading* for derivations). For nonrelativistic speeds ($s \ll c$), these expressions simplify to

$$\nu_{\text{receding}} \approx \frac{\nu}{1+s/c} \qquad \nu_{\text{approaching}} \approx \frac{\nu}{1-s/c} \tag{6.16}$$

Molecules reach high speeds in all directions in a gas, and a stationary observer detects the corresponding Doppler-shifted range of frequencies. Some molecules approach the observer, some move away; some move quickly, others slowly. The detected spectral 'line' is the absorption or emission profile arising from all the resulting Doppler shifts. As shown in the following *Justification*, the profile reflects the distribution of molecular velocities parallel to the line of sight, which is a bell-shaped Gaussian curve. The Doppler line shape is therefore also a Gaussian (Fig. 6.7), and we show in the

Justification that, when the temperature is T and the mass of the molecule is m, then the observed width of the line at half-height (in terms of frequency or wavelength) is

$$\delta\nu_{obs}=\frac{2\nu}{c}\left(\frac{2kT\ln 2}{m}\right)^{1/2} \qquad \delta\lambda_{obs}=\frac{2\lambda}{c}\left(\frac{2kT\ln 2}{m}\right)^{1/2} \tag{6.17}$$

For a molecule like N_2 at room temperature ($T \approx 300$ K), $\delta\nu/\nu \approx 2.3 \times 10^{-6}$. For a typical rotational transition wavenumber of 1 cm^{-1} (corresponding to a frequency of 30 GHz), the linewidth is about 70 kHz. Doppler broadening increases with temperature because the molecules acquire a wider range of speeds. Therefore, to obtain spectra of maximum sharpness, it is best to work with cold samples.

Comment 6.3

A Gaussian function of the general form $y(x) = ae^{-(x-b)^2/2\sigma^2}$, where a, b, and σ are constants, has a maximum $y(b) = a$ and a width at half-height $\delta x = 2\sigma(2\ln 2)^{1/2}$.

A note on good practice You will often hear people speak of 'a frequency as so many wavenumbers'. This usage is doubly wrong. First, *frequency* and *wavenumber* are two distinct physical observables with different units, and should be distinguished. Second, 'wavenumber' is not a unit, it is an observable with the dimensions of 1/length and commonly reported in reciprocal centimetres (cm^{-1}).

Justification 6.3 *Doppler broadening*

We know from the Boltzmann distribution that the probability that a gas molecule of mass m and speed s in a sample with temperature T has kinetic energy $E_K = \frac{1}{2}ms^2$ is proportional to $e^{-ms^2/2kT}$. The observed frequencies, ν_{obs}, emitted or absorbed by the molecule are related to its speed by eqn 6.16:

$$\nu_{obs}=\nu\left(\frac{1}{1\pm s/c}\right)$$

where ν is the unshifted frequency. When $s \ll c$, the Doppler shift in the frequency is

$$\nu_{obs}-\nu\approx\pm\nu s/c$$

which implies a symmetrical distribution of observed frequencies with respect to molecular speeds. More specifically, the intensity I of a transition at ν_{obs} is proportional to the probability of finding the molecule that emits or absorbs at ν_{obs}, so it follows from the Boltzmannm distribution and the expression for the Doppler shift that

$$I(\nu_{obs})\propto e^{-mc^2(\nu_{obs}-\nu)^2/2\nu^2kT}$$

which has the form of a Gaussian function. The width at half-height can be calculated directly from the exponent (see *Comment 6.3*) to give eqn 6.17.

(b) Lifetime broadening

It is found that spectroscopic lines from gas-phase samples are not infinitely sharp even when Doppler broadening has been largely eliminated by working at low temperatures. The same is true of the spectra of samples in condensed phases and solution. This residual broadening is due to quantum mechanical effects. Specifically, when the Schrödinger equation is solved for a system that is changing with time, it is found that it is impossible to specify the energy levels exactly. If on average a system survives in a state for a time τ (tau), the **lifetime** of the state, then its energy levels are blurred to an extent of order δE, where

$$\delta E \approx \frac{\hbar}{\tau} \tag{6.18}$$

This expression is reminiscent of the Heisenberg uncertainty principle (eqn 1.40), and consequently this **lifetime broadening** is often called 'uncertainty broadening'. When the energy spread is expressed as a wavenumber through $\delta E = hc\delta\tilde{\nu}$, and the values of the fundamental constants introduced, this relation becomes

$$\delta\tilde{\nu} \approx \frac{5.3\ \text{cm}^{-1}}{\tau/\text{ps}} \tag{6.19}$$

No excited state has an infinite lifetime; therefore, all states are subject to some lifetime broadening and, the shorter the lifetimes of the states involved in a transition, the broader the corresponding spectral lines.

Two processes are responsible for the finite lifetimes of excited states. The dominant one for low frequency transitions is **collisional deactivation**, which arises from collisions between molecules or with the walls of the container. If the **collisional lifetime**, the mean time between collisions, is τ_{col}, the resulting collisional linewidth is $\delta E_{col} \approx \hbar/\tau_{col}$. Because $\tau_{col} = 1/z$, where z is the collision frequency, and from the kinetic model of gases (Volume 1, Section 1.3) we know that z is proportional to the pressure, we see that the collisional linewidth is proportional to the pressure. The collisional linewidth can therefore be minimized by working at low pressures.

The rate of spontaneous emission cannot be changed. Hence it is a natural limit to the lifetime of an excited state, and the resulting lifetime broadening is the **natural linewidth** of the transition. The natural linewidth is an intrinsic property of the transition, and cannot be changed by modifying the conditions. Natural linewidths depend strongly on the transition frequency (they increase with the coefficient of spontaneous emission A and therefore as ν^3), so low frequency transitions (such as the microwave transitions of rotational spectroscopy) have very small natural linewidths, and collisional and Doppler line-broadening processes are dominant. The natural lifetimes of electronic transitions are very much shorter than for vibrational and rotational transitions, so the natural linewidths of electronic transitions are much greater than those of vibrational and rotational transitions. For example, a typical electronic excited state natural lifetime is about 10^{-8} s (10 ns), corresponding to a natural width of about 5×10^{-4} cm^{-1} (15 MHz). A typical rotational state natural lifetime is about 10^3 s, corresponding to a natural linewidth of only 5×10^{-15} cm^{-1} (of the order of 10^{-4} Hz).

IMPACT ON ASTROPHYSICS

I6.1 Rotational and vibrational spectroscopy of interstellar space

Observations by the Cosmic Background Explorer (COBE) satellite support the long-held hypothesis that the distribution of energy in the current Universe can be modelled by a Planck distribution (eqn. 1.5) with $T = 2.726 \pm 0.001$ K, the bulk of the radiation spanning the microwave region of the spectrum. This *cosmic microwave background radiation* is the residue of energy released during the Big Bang, the event that brought the Universe into existence. Very small fluctuations in the background temperature are believed to account for the large-scale structure of the Universe.

The interstellar space in our galaxy is a little warmer than the cosmic background and consists largely of dust grains and gas clouds. The dust grains are carbon-based compounds and silicates of aluminium, magnesium, and iron, in which are embedded trace amounts of methane, water, and ammonia. Interstellar clouds are significant because it is from them that new stars, and consequently new planets, are formed. The hottest clouds are plasmas with temperatures of up to 10^6 K and densities of only about 3×10^3 particles m^{-3}. Colder clouds range from 0.1 to 1000 solar masses (1 solar

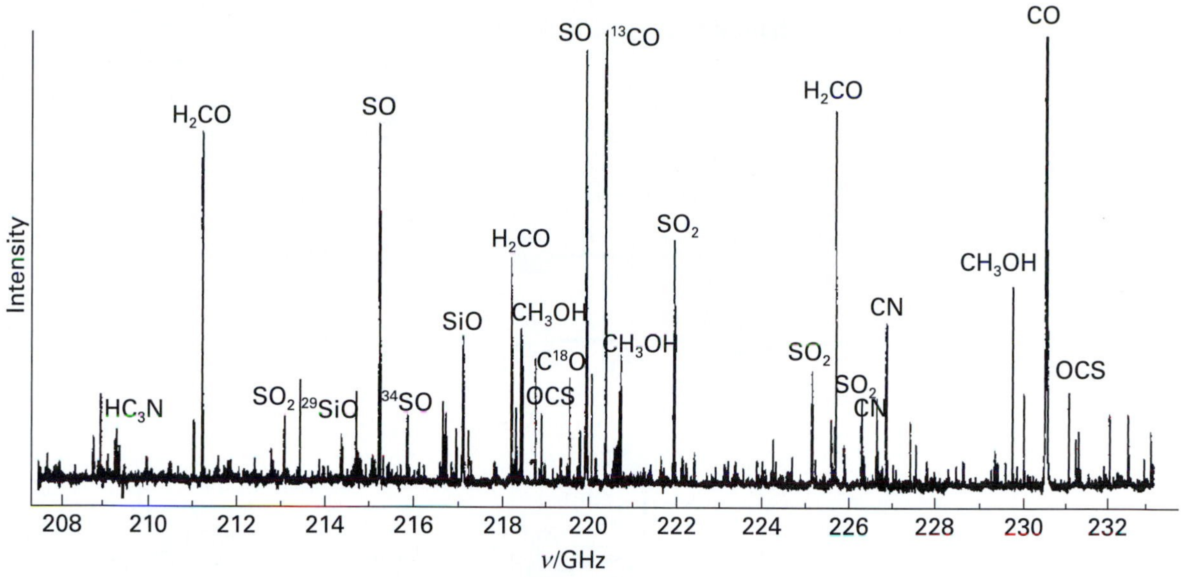

Fig. 6.8 Rotational spectrum of the Orion nebula, showing spectral fingerprints of diatomic and polyatomic molecules present in the interstellar cloud. Adapted from G.A. Blake *et al.*, *Astrophys. J.* **315**, 621 (1987).

mass = 2×10^{30} kg), have a density of about 5×10^5 particles m^{-3}, consist largely of hydrogen atoms, and have a temperature of about 80 K. There are also colder and denser clouds, some with masses greater than 500 000 solar masses, densities greater than 10^9 particles m^{-3}, and temperatures that can be lower than 10 K. They are also called *molecular clouds*, because they are composed primarily of H_2 and CO gas in a proportion of about 10^5 to 1. There are also trace amounts of larger molecules. To place the densities in context, the density of liquid water at 298 K and 1 bar is about 3×10^{28} particles m^{-3}.

It follows from the the Boltzmann distribution and the low temperature of a molecular cloud that the vast majority of a cloud's molecules are in their vibrational and electronic ground states. However, rotational excited states are populated at 10–100 K and decay by spontaneous emission. As a result, the spectrum of the cloud in the radiofrequency and microwave regions consists of sharp lines corresponding to rotational transitions (Fig. 6.8). The emitted light is collected by Earth-bound or space-borne radiotelescopes, telescopes with antennas and detectors optimized for the collection and analysis of radiation in the microwave–radiowave range of the spectrum. Earth-bound radiotelescopes are often located at the tops of high mountains, as atmospheric water vapour can reabsorb microwave radiation from space and hence interfere with the measurement.

Over 100 interstellar molecules have been identified by their rotational spectra, often by comparing radiotelescope data with spectra obtained in the laboratory or calculated by computational methods. The experiments have revealed the presence of trace amounts (with abundances of less than 10^{-8} relative to hydrogen) of neutral molecules, ions, and radicals. Examples of neutral molecules include hydrides, oxides (including water), sulfides, halogenated compounds, nitriles, hydrocarbons, aldehydes, alcohols, ethers, ketones, and amides. The largest molecule detected by rotational spectroscopy is the nitrile $HC_{11}N$.

Interstellar space can also be investigated with vibrational spectroscopy by using a combination of telescopes and infrared detectors. The experiments are conducted

Table 6.1 Moments of inertia*

1. *Diatomic molecules*

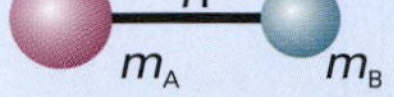

$$I = \mu R^2 \qquad \mu = \frac{m_A m_B}{m}$$

2. *Triatomic linear rotors*

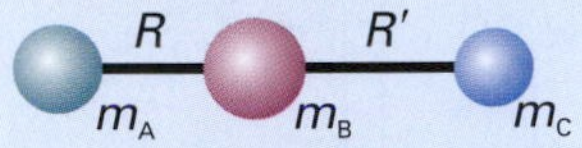

$$I = m_A R^2 + m_C R'^2 - \frac{(m_A R - m_C R')^2}{m}$$

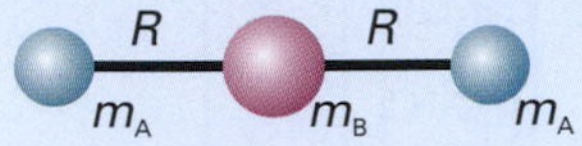

$$I = 2m_A R^2$$

3. *Symmetric rotors*

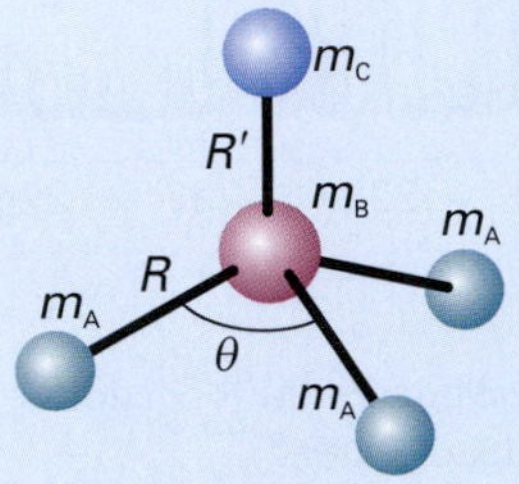

$$I_{\parallel} = 2m_A(1-\cos\theta)R^2$$

$$I_{\perp} = m_A(1-\cos\theta)R^2 + \frac{m_A}{m}(m_B + m_C)(1 + 2\cos\theta)R^2 + \frac{m_C}{m}\{(3m_A + m_B)R' + 6m_A R[\tfrac{1}{3}(1+2\cos\theta)]^{1/2}\}R'$$

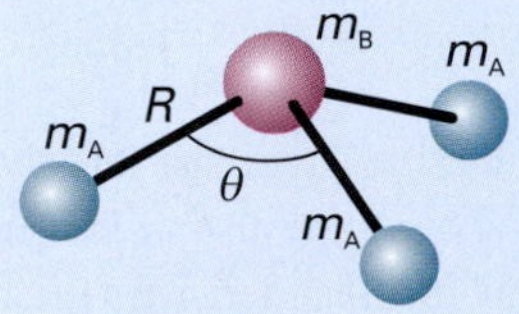

$$I_{\parallel} = 2m_A(1-\cos\theta)R^2$$

$$I_{\perp} = m_A(1-\cos\theta)R^2 + \frac{m_A m_B}{m}(1 + 2\cos\theta)R^2$$

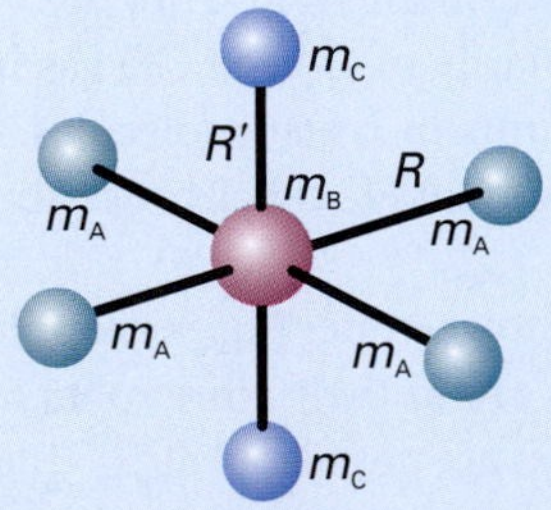

$$I_{\parallel} = 4m_A R^2$$

$$I_{\perp} = 2m_A R^2 + 2m_C R'^2$$

4. *Spherical rotors*

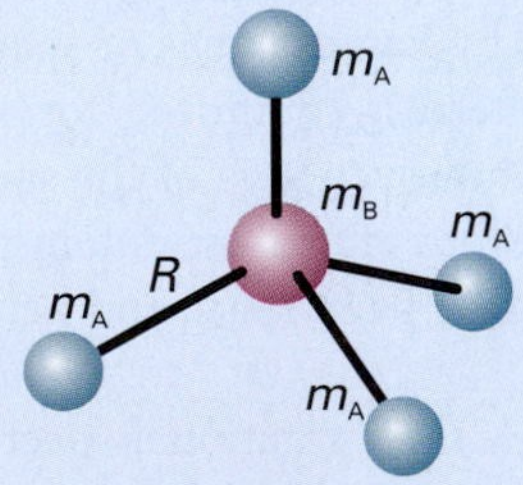

$$I = \tfrac{8}{3}m_A R^2$$

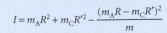

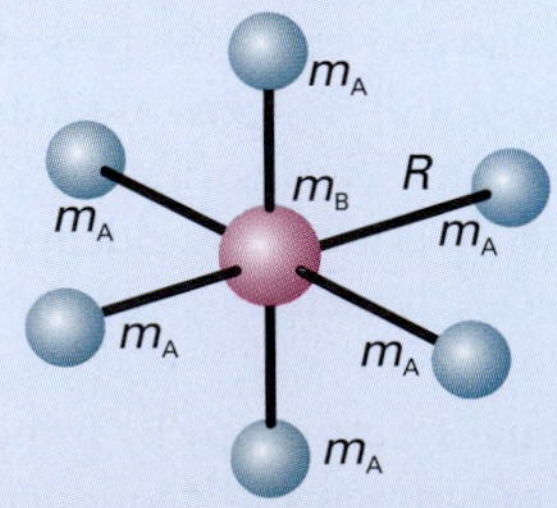

$$I = 4m_A R^2$$

* In each case, m is the total mass of the molecule.

primarily in space-borne telescopes because the Earth's atmosphere absorbs a great deal of infrared radiation (see *Impact I6.2*). In most cases, absorption by an inter-stellar species is detected against the background of infrared radiation emitted by a nearby star. The data can detect the presence of gas and solid water, CO, and CO_2 in molecular clouds. In certain cases, infrared emission can be detected, but these events are rare because interstellar space is too cold and does not provide enough energy to promote a significant number of molecules to vibrational excited states. However, infrared emissions can be observed if molecules are occasionally excited by high-energy photons emitted by hot stars in the vicinity of the cloud. For example, the polycyclic aromatic hydrocarbons hexabenzocoronene ($C_{48}H_{24}$) and circumcoronene ($C_{54}H_{18}$) have been identified from characteristic infrared emissions.

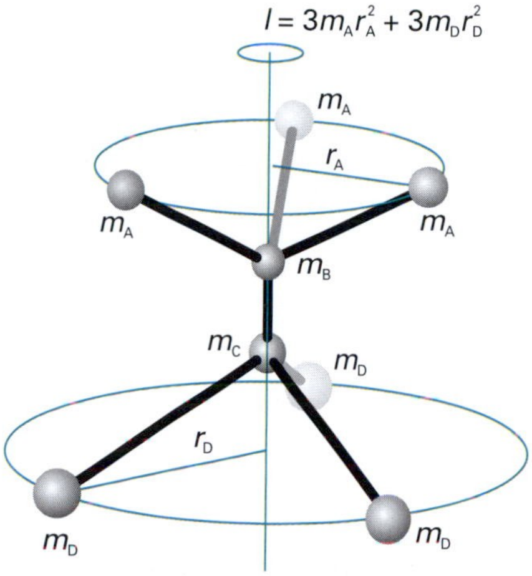

Fig. 6.9 The definition of moment of inertia. In this molecule there are three identical atoms attached to the B atom and three different but mutually identical atoms attached to the C atom. In this example, the centre of mass lies on an axis passing through the B and C atom, and the perpendicular distances are measured from this axis.

Pure rotation spectra

The general strategy we adopt for discussing molecular spectra and the information they contain is to find expressions for the energy levels of molecules and then to calculate the transition frequencies by applying the selection rules. We then predict the appearance of the spectrum by taking into account the transition moments and the populations of the states. In this section we illustrate the strategy by considering the rotational states of molecules.

6.4 Moments of inertia

The key molecular parameter we shall need is the **moment of inertia**, I, of the molecule (Section 2.6). The moment of inertia of a molecule is defined as the mass of each atom multiplied by the square of its distance from the rotational axis through the centre of mass of the molecule (Fig. 6.9):

$$I = \sum_i m_i r_i^2 \qquad [6.20]$$

where r_i is the perpendicular distance of the atom i from the axis of rotation. The moment of inertia depends on the masses of the atoms present and the molecular geometry, so we can suspect (and later shall see explicitly) that rotational spectroscopy will give information about bond lengths and bond angles.

In general, the rotational properties of any molecule can be expressed in terms of the moments of inertia about three perpendicular axes set in the molecule (Fig. 6.10). The convention is to label the moments of inertia I_a, I_b, and I_c, with the axes chosen so that $I_c \geq I_b \geq I_a$. For linear molecules, the moment of inertia around the internuclear axis is zero. The explicit expressions for the moments of inertia of some symmetrical molecules are given in Table 6.1.

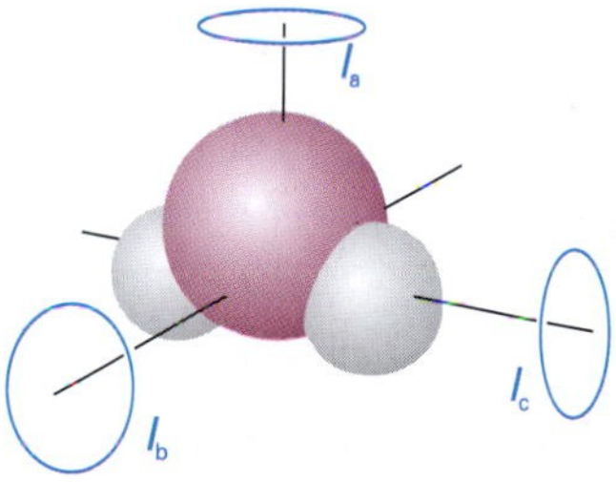

Fig. 6.10 An asymmetric rotor has three different moments of inertia; all three rotation axes coincide at the centre of mass of the molecule.

Example 6.1 *Calculating the moment of inertia of a molecule*

Calculate the moment of inertia of an H_2O molecule around the axis defined by the bisector of the HOH angle (**1**). The HOH bond angle is 104.5° and the bond length is 95.7 pm.

Method According to eqn 6.20, the moment of inertia is the sum of the masses multiplied by the squares of their distances from the axis of rotation. The latter can be expressed by using trigonometry and the bond angle and bond length.

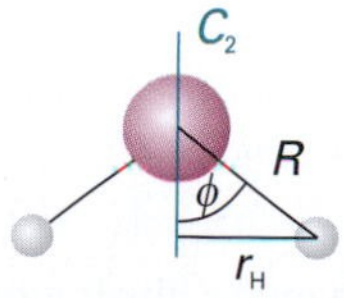

1

Answer From eqn 6.20,

$$I = \sum_i m_i r_i^2 = m_H r_H^2 + 0 + m_H r_H^2 = 2m_H r_H^2$$

If the bond angle of the molecule is denoted 2ϕ and the bond length is R, trigonometry gives $r_H = R \sin \phi$. It follows that

$$I = 2m_H R^2 \sin^2 \phi$$

Substitution of the data gives

$$I = 2 \times (1.67 \times 10^{-27}\ \text{kg}) \times (9.57 \times 10^{-11}\ \text{m})^2 \times \sin^2 52.3° = 1.91 \times 10^{-47}\ \text{kg m}^2$$

Note that the mass of the O atom makes no contribution to the moment of inertia for this mode of rotation as the atom is immobile while the H atoms circulate around it.

A note on good practice The mass to use in the calculation of the moment of inertia is the actual atomic mass, not the element's molar mass; don't forget to convert from atomic mass units (u, formerly amu) to kilograms.

Self-test 6.1 Calculate the moment of inertia of a $CH^{35}Cl_3$ molecule around a rotational axis that contains the C—H bond. The C—Cl bond length is 177 pm and the HCCl angle is 107°; $m(^{35}Cl) = 34.97$ u. [4.99×10^{-45} kg m^2]

We shall suppose initially that molecules are **rigid rotors**, bodies that do not distort under the stress of rotation. Rigid rotors can be classified into four types (Fig. 6.11):

Spherical rotors have three equal moments of inertia (examples: CH_4, SiH_4, and SF_6).

Symmetric rotors have two equal moments of inertia (examples: NH_3, CH_3Cl, and CH_3CN).

Linear rotors have one moment of inertia (the one about the molecular axis) equal to zero (examples: CO_2, HCl, OCS, and HC≡CH).

Asymmetric rotors have three different moments of inertia (examples: H_2O, H_2CO, and CH_3OH).

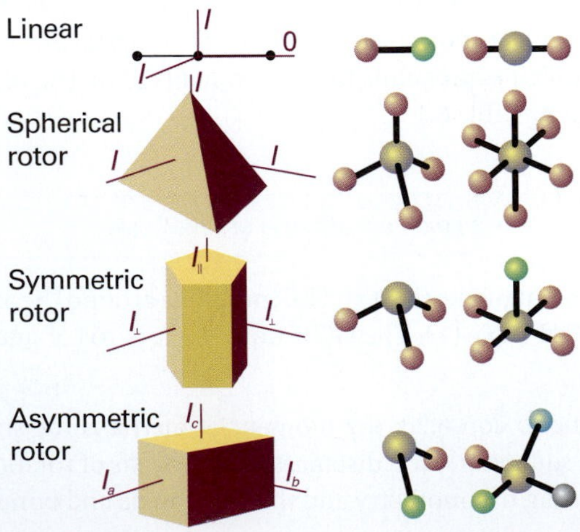

Fig. 6.11 A schematic illustration of the classification of rigid rotors.

6.5 The rotational energy levels

The rotational energy levels of a rigid rotor may be obtained by solving the appropriate Schrödinger equation. Fortunately, however, there is a much less onerous short cut to the exact expressions that depends on noting the classical expression for the energy of a rotating body, expressing it in terms of the angular momentum, and then importing the quantum mechanical properties of angular momentum into the equations.

The classical expression for the energy of a body rotating about an axis a is

$$E_a = \tfrac{1}{2}I_a\omega_a^2 \tag{6.21}$$

where ω_a is the angular velocity (in radians per second, rad s^{-1}) about that axis and I_a is the corresponding moment of inertia. A body free to rotate about three axes has an energy

$$E = \tfrac{1}{2}I_a\omega_a^2 + \tfrac{1}{2}I_b\omega_b^2 + \tfrac{1}{2}I_c\omega_c^2$$

Because the classical angular momentum about the axis a is $J_a = I_a\omega_a$, with similar expressions for the other axes, it follows that

$$E = \frac{J_a^2}{2I_a} + \frac{J_b^2}{2I_b} + \frac{J_c^2}{2I_c} \tag{6.22}$$

This is the key equation. We described the quantum mechanical properties of angular momentum in Section 2.7b, and can now make use of them in conjunction with this equation to obtain the rotational energy levels.

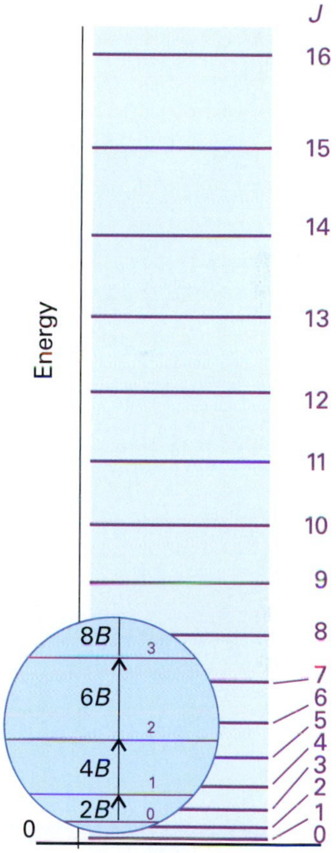

Fig. 6.12 The rotational energy levels of a linear or spherical rotor. Note that the energy separation between neighbouring levels increases as J increases.

(a) Spherical rotors

When all three moments of inertia are equal to some value I, as in CH_4 and SF_6, the classical expression for the energy is

$$E = \frac{J_a^2 + J_b^2 + J_c^2}{2I} = \frac{\mathcal{J}^2}{2I}$$

where $\mathcal{J}^2 = J_a^2 + J_b^2 + J_c^2$ is the square of the magnitude of the angular momentum. We can immediately find the quantum expression by making the replacement

$$\mathcal{J}^2 \rightarrow J(J+1)\hbar^2 \qquad J = 0, 1, 2, \ldots$$

Therefore, the energy of a spherical rotor is confined to the values

$$E_J = J(J+1)\frac{\hbar^2}{2I} \qquad J = 0, 1, 2, \ldots \tag{6.23}$$

The resulting ladder of energy levels is illustrated in Fig. 6.12. The energy is normally expressed in terms of the **rotational constant**, B, of the molecule, where

$$hcB = \frac{\hbar^2}{2I} \qquad \text{so} \qquad B = \frac{\hbar}{4\pi cI} \qquad [6.24]$$

The expression for the energy is then

$$E_J = hcBJ(J+1) \qquad J = 0, 1, 2, \ldots \tag{6.25}$$

The rotational constant as defined by eqn 6.25 is a wavenumber. The energy of a rotational state is normally reported as the **rotational term**, $F(J)$, a wavenumber, by division by hc:

$$F(J) = BJ(J+1) \tag{6.26}$$

Comment 6.4

The definition of B as a wavenumber is convenient when we come to vibration–rotation spectra. However, for pure rotational spectroscopy it is more common to define B as a frequency. Then $B = \hbar/4\pi cI$ and the energy is $E = hBJ(J+1)$.

The separation of adjacent levels is

$$F(J) - F(J-1) = 2BJ \tag{6.27}$$

Because the rotational constant decreases as I increases, we see that large molecules have closely spaced rotational energy levels. We can estimate the magnitude of the separation by considering CCl_4: from the bond lengths and masses of the atoms we find $I = 4.85 \times 10^{-45}$ kg m^2, and hence $B = 0.0577$ cm^{-1}.

(b) Symmetric rotors

In symmetric rotors, two moments of inertia are equal but different from the third (as in CH_3Cl, NH_3, and C_6H_6); the unique axis of the molecule is its **principal axis** (or *figure axis*). We shall write the unique moment of inertia (that about the principal axis) as $I_\parallel$ and the other two as $I_\perp$. If $I_\parallel > I_\perp$, the rotor is classified as **oblate** (like a pancake, and C_6H_6); if $I_\parallel < I_\perp$ it is classified as **prolate** (like a cigar, and CH_3Cl). The classical expression for the energy, eqn 6.22, becomes

$$E = \frac{J_b^2 + J_c^2}{2I_\perp} + \frac{J_a^2}{2I_\parallel}$$

Again, this expression can be written in terms of $\mathcal{J}^2 = J_a^2 + J_b^2 + J_c^2$:

$$E = \frac{\mathcal{J}^2 - J_a^2}{2I_\perp} + \frac{J_a^2}{2I_\parallel} = \frac{\mathcal{J}^2}{2I} + \left(\frac{1}{2I_\parallel} - \frac{1}{2I_\perp}\right)J_a^2 \tag{6.28}$$

Now we generate the quantum expression by replacing $\mathcal{J}^2$ by $J(J+1)\hbar^2$, where J is the angular momentum quantum number. We also know from the quantum theory of angular momentum (Section 2.7b) that the component of angular momentum about any axis is restricted to the values $K\hbar$, with $K = 0, \pm 1, \ldots, \pm J$. ($K$ is the quantum number used to signify a component on the principal axis; M_J is reserved for a component on an externally defined axis.) Therefore, we also replace J_a^2 by $K^2\hbar^2$. It follows that the rotational terms are

$$F(J,K) = BJ(J+1) + (A-B)K^2 \qquad J = 0, 1, 2, \ldots \qquad K = 0, \pm 1, \ldots, \pm J \tag{6.29}$$

with

$$A = \frac{\hbar}{4\pi c I_\parallel} \qquad B = \frac{\hbar}{4\pi c I_\perp} \tag*{[6.30]}$$

Equation 6.29 matches what we should expect for the dependence of the energy levels on the two distinct moments of inertia of the molecule. When $K = 0$, there is no component of angular momentum about the principal axis, and the energy levels depend only on $I_\perp$ (Fig. 6.13). When $K = \pm J$, almost all the angular momentum arises from rotation around the principal axis, and the energy levels are determined largely by $I_\parallel$. The sign of K does not affect the energy because opposite values of K correspond to opposite senses of rotation, and the energy does not depend on the sense of rotation.

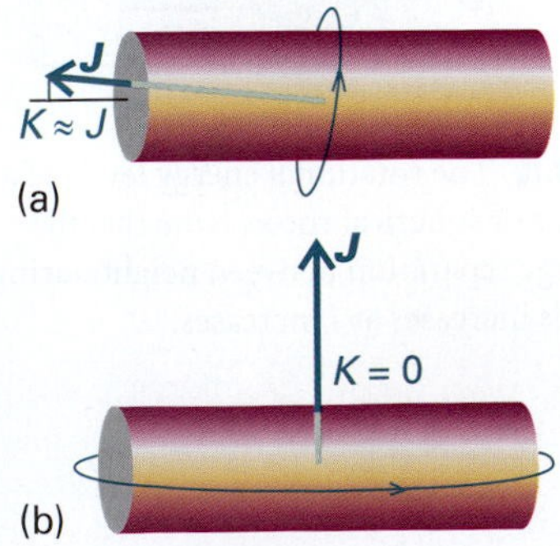

Fig. 6.13 The significance of the quantum number K. (a) When $|K|$ is close to its maximum value, J, most of the molecular rotation is around the figure axis. (b) When $K = 0$ the molecule has no angular momentum about its principal axis: it is undergoing end-over-end rotation.

Example 6.2 *Calculating the rotational energy levels of a molecule*

A $^{14}NH_3$ molecule is a symmetric rotor with bond length 101.2 pm and HNH bond angle 106.7°. Calculate its rotational terms.

A note on good practice To calculate moments of inertia precisely, we need to specify the nuclide.

Method Begin by calculating the rotational constants A and B by using the expressions for moments of inertia given in Table 6.1. Then use eqn 6.29 to find the rotational terms.

Answer Substitution of $m_A = 1.0078$ u, $m_B = 14.0031$ u, $R = 101.2$ pm, and $\theta = 106.7°$ into the second of the symmetric rotor expressions in Table 6.1 gives $I_{\parallel} = 4.4128 \times 10^{-47}$ kg m^2 and $I_{\perp} = 2.8059 \times 10^{-47}$ kg m^2. Hence, $A = 6.344$ cm^{-1} and $B = 9.977$ cm^{-1}. It follows from eqn 6.29 that

$$F(J,K)/\text{cm}^{-1} = 9.977J(J+1) - 3.633K^2$$

Upon multiplication by c, $F(J,K)$ acquires units of frequency:

$$F(J,K)/\text{GHz} = 299.1J(J+1) - 108.9K^2$$

For $J = 1$, the energy needed for the molecule to rotate mainly about its figure axis ($K = \pm J$) is equivalent to 16.32 cm^{-1} (489.3 GHz), but end-over-end rotation ($K = 0$) corresponds to 19.95 cm^{-1} (598.1 GHz).

Self-test 6.2 A $CH_3{}^{35}Cl$ molecule has a C—Cl bond length of 178 pm, a C—H bond length of 111 pm, and an HCH angle of 110.5°. Calculate its rotational energy terms.

[$F(J,K)/\text{cm}^{-1} = 0.444J(J+1) + 4.58K^2$; also $F(J,K)/\text{GHz} = 13.3J(J+1) + 137K^2$]

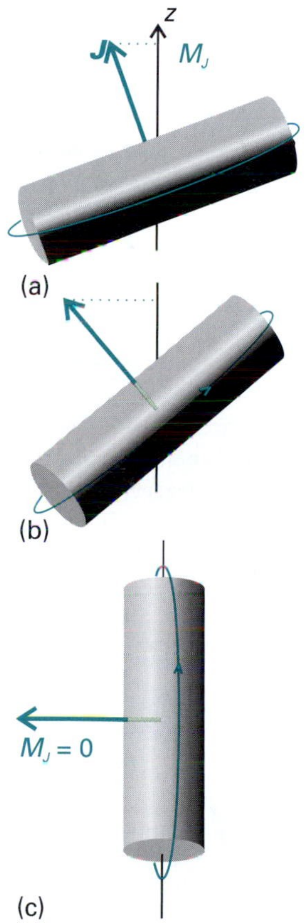

Fig. 6.14 The significance of the quantum number M_J. (a) When M_J is close to its maximum value, J, most of the molecular rotation is around the laboratory z-axis. (b) An intermediate value of M_J. (c) When $M_J = 0$ the molecule has no angular momentum about the z-axis All three diagrams correspond to a state with $K = 0$; there are corresponding diagrams for different values of K, in which the angular momentum makes a different angle to the molecule's principal axis.

(c) Linear rotors

For a linear rotor (such as CO_2, HCl, and C_2H_2), in which the nuclei are regarded as mass points, the rotation occurs only about an axis perpendicular to the line of atoms and there is zero angular momentum around the line. Therefore, the component of angular momentum around the figure axis of a linear rotor is identically zero, and $K \equiv 0$ in eqn 6.29. The rotational terms of a linear molecule are therefore

$$F(J) = BJ(J+1) \qquad J = 0, 1, 2, \ldots \tag{6.31}$$

This expression is the same as eqn 6.26 but we have arrived at it in a significantly different way: here $K \equiv 0$ but for a spherical rotor $A = B$.

(d) Degeneracies and the Stark effect

The energy of a symmetric rotor depends on J and K, and each level except those with $K = 0$ is doubly degenerate: the states with K and $-K$ have the same energy. However, we must not forget that the angular momentum of the molecule has a component on an external, laboratory-fixed axis. This component is quantized, and its permitted values are $M_J\hbar$, with $M_J = 0, \pm1, \ldots, \pm J$, giving $2J + 1$ values in all (Fig. 6.14). The quantum number M_J does not appear in the expression for the energy, but it is necessary for a complete specification of the state of the rotor. Consequently, all $2J + 1$ orientations of the rotating molecule have the same energy. It follows that a symmetric rotor level is $2(2J + 1)$-fold degenerate for $K \neq 0$ and $(2J + 1)$-fold degenerate for $K = 0$. A linear rotor has K fixed at 0, but the angular momentum may still have $2J + 1$ components on the laboratory axis, so its degeneracy is $2J + 1$.

A spherical rotor can be regarded as a version of a symmetric rotor in which A is equal to B: The quantum number K may still take any one of $2J + 1$ values, but the energy is independent of which value it takes. Therefore, as well as having a $(2J + 1)$-fold degeneracy arising from its orientation in space, the rotor also has a $(2J + 1)$-fold degeneracy arising from its orientation with respect to an arbitrary axis in the molecule. The overall degeneracy of a symmetric rotor with quantum number J is therefore $(2J+1)^2$. This degeneracy increases very rapidly: when $J = 10$, for instance, there are 441 states of the same energy.

The degeneracy associated with the quantum number M_J (the orientation of the rotation in space) is partly removed when an electric field is applied to a polar molecule (e.g. HCl or NH_3), as illustrated in Fig. 6.15. The splitting of states by an electric field is called the **Stark effect**. For a linear rotor in an electric field $\mathcal{E}$, the energy of the state with quantum numbers J and M_J is given by

$$E(J,M_J) = hcBJ(J+1) + a(J,M_J)\mu^2\mathcal{E}^2 \quad (6.32a)$$

where (see *Further reading* for a derivation)

$$a(J,M_J) = \frac{J(J+1) - 3M_J^2}{2hcBJ(J+1)(2J-1)(2J+3)} \quad (6.32b)$$

Note that the energy of a state with quantum number M_J depends on the square of the permanent electric dipole moment, μ. The observation of the Stark effect can therefore be used to measure this property, but the technique is limited to molecules that are sufficiently volatile to be studied by rotational spectroscopy. However, as spectra can be recorded for samples at pressures of only about 1 Pa and special techniques (such as using an intense laser beam or an electrical discharge) can be used to vaporize even some quite nonvolatile substances, a wide variety of samples may be studied. Sodium chloride, for example, can be studied as diatomic NaCl molecules at high temperatures.

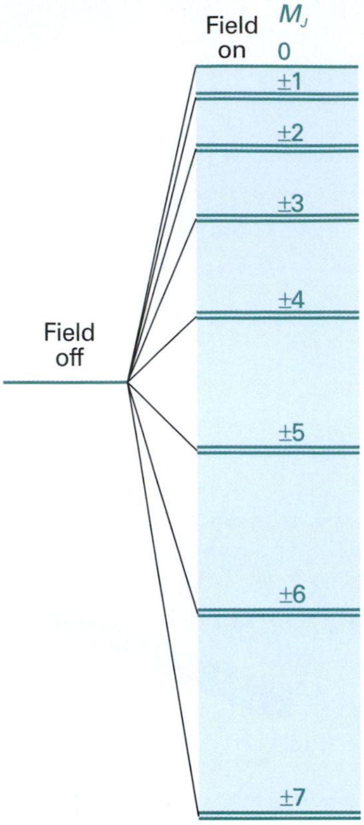

Fig. 6.15 The effect of an electric field on the energy levels of a polar linear rotor. All levels are doubly degenerate except that with $M_J = 0$.

(e) Centrifugal distortion

We have treated molecules as rigid rotors. However, the atoms of rotating molecules are subject to centrifugal forces that tend to distort the molecular geometry and change the moments of inertia (Fig. 6.16). The effect of centrifugal distortion on a diatomic molecule is to stretch the bond and hence to increase the moment of inertia. As a result, centrifugal distortion reduces the rotational constant and consequently the energy levels are slightly closer than the rigid-rotor expressions predict. The effect is usually taken into account largely empirically by subtracting a term from the energy and writing

$$F(J) = BJ(J+1) - D_J J^2(J+1)^2 \quad (6.33)$$

The parameter D_J is the **centrifugal distortion constant**. It is large when the bond is easily stretched. The centrifugal distortion constant of a diatomic molecule is related to the vibrational wavenumber of the bond, $\tilde{\nu}$ (which, as we shall see later, is a measure of its stiffness), through the approximate relation (see Problem 6.22)

$$D_J = \frac{4B^3}{\tilde{\nu}^2} \quad (6.34)$$

Hence the observation of the convergence of the rotational levels as J increases can be interpreted in terms of the rigidity of the bond.

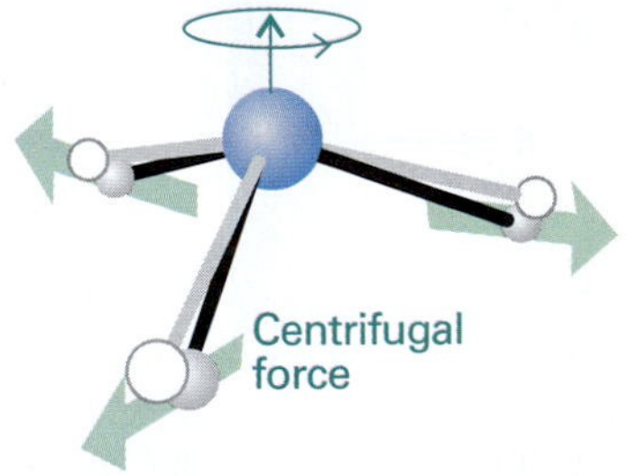

Fig. 6.16 The effect of rotation on a molecule. The centrifugal force arising from rotation distorts the molecule, opening out bond angles and stretching bonds slightly. The effect is to increase the moment of inertia of the molecule and hence to decrease its rotational constant.

6.6 Rotational transitions

Typical values of B for small molecules are in the region of 0.1 to10 cm^{-1} (for example, 0.356 cm^{-1} for NF_3 and 10.59 cm^{-1} for HCl), so rotational transitions lie in the microwave region of the spectrum. The transitions are detected by monitoring the net absorption of microwave radiation. Modulation of the transmitted intensity can be achieved by varying the energy levels with an oscillating electric field. In this **Stark**

modulation, an electric field of about 10^5 V m^{-1} and a frequency of between 10 and 100 kHz is applied to the sample.

(a) Rotational selection rules

We have already remarked (Section 6.2) that the gross selection rule for the observation of a pure rotational spectrum is that a molecule must have a permanent electric dipole moment. That is, *for a molecule to give a pure rotational spectrum, it must be polar*. The classical basis of this rule is that a polar molecule appears to possess a fluctuating dipole when rotating, but a nonpolar molecule does not (Fig. 6.17). The permanent dipole can be regarded as a handle with which the molecule stirs the electromagnetic field into oscillation (and vice versa for absorption). Homonuclear diatomic molecules and symmetrical linear molecules such as CO_2 are rotationally inactive. Spherical rotors cannot have electric dipole moments unless they become distorted by rotation, so they are also inactive except in special cases. An example of a spherical rotor that does become sufficiently distorted for it to acquire a dipole moment is SiH_4, which has a dipole moment of about 8.3 μD by virtue of its rotation when $J \approx 10$ (for comparison, HCl has a permanent dipole moment of 1.1 D. The pure rotational spectrum of SiH_4 has been detected by using long path lengths (10 m) through high-pressure (4 atm) samples.

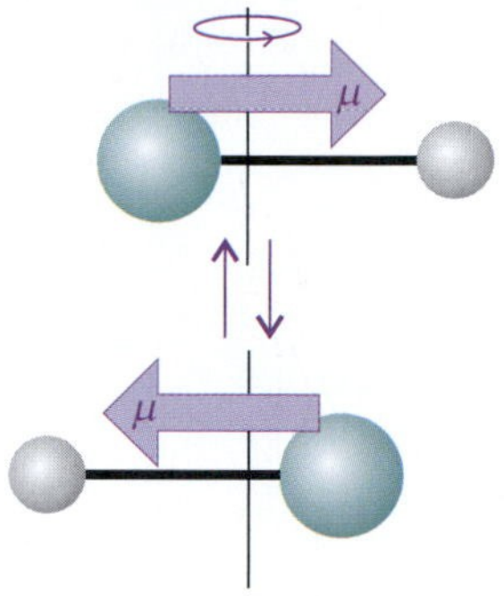

Fig. 6.17 To a stationary observer, a rotating polar molecule looks like an oscillating dipole that can stir the electromagnetic field into oscillation (and vice versa for absorption). This picture is the classical origin of the gross selection rule for rotational transitions.

Illustration 6.2 *Identifying rotationally active molecules*

Of the molecules N_2, CO_2, OCS, H_2O, $CH_2{=}CH_2$, C_6H_6, only OCS and H_2O are polar, so only these two molecules have microwave spectra.

Self-test 6.3 Which of the molecules H_2, NO, N_2O, CH_4 can have a pure rotational spectrum? [NO, N_2O]

The specific rotational selection rules are found by evaluating the transition dipole moment between rotational states. We show in *Further information 6.2* that, for a linear molecule, the transition moment vanishes unless the following conditions are fulfilled:

$$\Delta J = \pm 1 \qquad \Delta M_J = 0, \pm 1 \tag{6.35}$$

The transition $\Delta J = +1$ corresponds to absorption and the transition $\Delta J = -1$ corresponds to emission. The allowed change in J in each case arises from the conservation of angular momentum when a photon, a spin-1 particle, is emitted or absorbed (Fig. 6.18).

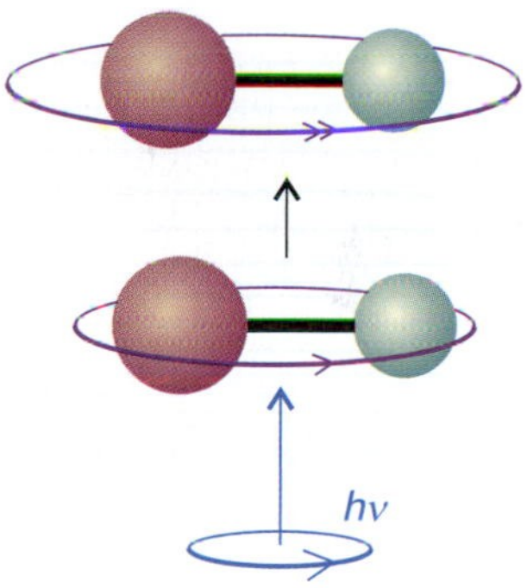

Fig. 6.18 When a photon is absorbed by a molecule, the angular momentum of the combined system is conserved. If the molecule is rotating in the same sense as the spin of the incoming photon, then J increases by 1.

When the transition moment is evaluated for all possible relative orientations of the molecule to the line of flight of the photon, it is found that the total $J+1 \leftrightarrow J$ transition intensity is proportional to

$$|\mu_{J+1,J}|^2 = \left(\frac{J+1}{2J+1}\right)\mu_0^2 \rightarrow \tfrac{1}{2}\mu_0^2 \qquad \text{for } J \gg 1 \tag{6.36}$$

where μ_0 is the permanent electric dipole moment of the molecule. The intensity is proportional to the square of the permanent electric dipole moment, so strongly polar molecules give rise to much more intense rotational lines than less polar molecules.

For symmetric rotors, an additional selection rule states that $\Delta K = 0$. To understand this rule, consider the symmetric rotor NH_3, where the electric dipole moment lies parallel to the figure axis. Such a molecule cannot be accelerated into different states of rotation around the figure axis by the absorption of radiation, so $\Delta K = 0$.

(b) The appearance of rotational spectra

When these selection rules are applied to the expressions for the energy levels of a rigid symmetric or linear rotor, it follows that the wavenumbers of the allowed $J + 1 \leftarrow J$ absorptions are

$$\tilde{\nu}(J+1 \leftarrow J) = 2B(J+1) \qquad J = 0, 1, 2, \ldots \tag{6.37}$$

When centrifugal distortion is taken into account, the corresponding expression is

$$\tilde{\nu}(J+1 \leftarrow J) = 2B(J+1) - 4D_J(J+1)^3 \tag{6.38}$$

However, because the second term is typically very small compared with the first, the appearance of the spectrum closely resembles that predicted from eqn 6.37.

Example 6.3 *Predicting the appearance of a rotational spectrum*

Predict the form of the rotational spectrum of $^{14}NH_3$.

Method We calculated the energy levels in Example 6.2. The $^{14}NH_3$ molecule is a polar symmetric rotor, so the selection rules $\Delta J = \pm 1$ and $\Delta K = 0$ apply. For absorption, $\Delta J = +1$ and we can use eqn 6.37. Because $B = 9.977$ cm^{-1}, we can draw up the following table for the $J + 1 \leftarrow J$ transitions.

J	0	1	2	3	...
$\tilde{\nu}$/cm^{-1}	19.95	39.91	59.86	79.82	...
ν/GHz	598.1	1197	1795	2393	...

The line spacing is 19.95 cm^{-1} (598.1 GHz).

Self-test 6.4 Repeat the problem for $C^{35}ClH_3$ (see Self-test 6.2 for details).

[Lines of separation 0.888 cm^{-1} (26.6 GHz)]

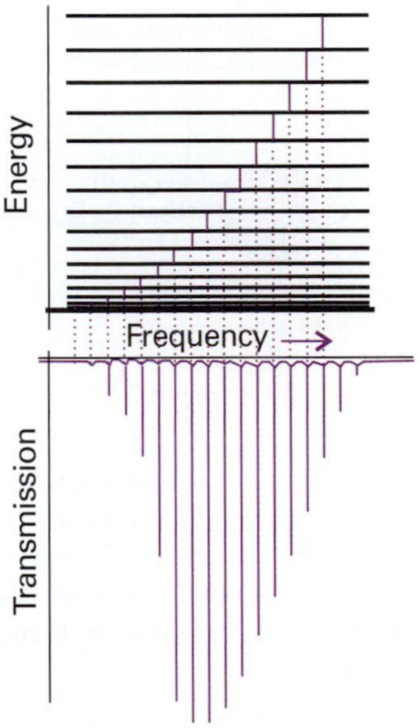

Fig. 6.19 The rotational energy levels of a linear rotor, the transitions allowed by the selection rule $\Delta J = \pm 1$, and a typical pure rotational absorption spectrum (displayed here in terms of the radiation transmitted through the sample). The intensities reflect the populations of the initial level in each case and the strengths of the transition dipole moments.

The form of the spectrum predicted by eqn 6.37 is shown in Fig. 6.19. The most significant feature is that it consists of a series of lines with wavenumbers $2B$, $4B$, $6B$, ... and of separation $2B$. The measurement of the line spacing gives B, and hence the moment of inertia perpendicular to the principal axis of the molecule. Because the masses of the atoms are known, it is a simple matter to deduce the bond length of a diatomic molecule. However, in the case of a polyatomic molecule such as OCS or NH_3, the analysis gives only a single quantity, $I_\perp$, and it is not possible to infer both bond lengths (in OCS) or the bond length and bond angle (in NH_3). This difficulty can be overcome by using isotopically substituted molecules, such as ABC and A′BC; then, by assuming that R(A—B) = R(A′—B), both A—B and B—C bond lengths can be extracted from the two moments of inertia. A famous example of this procedure is the study of OCS; the actual calculation is worked through in Problem 6.10. The assumption that bond lengths are unchanged by isotopic substitution is only an approximation, but it is a good approximation in most cases.

The intensities of spectral lines increase with increasing J and pass through a maximum before tailing off as J becomes large. The most important reason for the maximum in intensity is the existence of a maximum in the population of rotational levels. The Boltzmann distribution implies that the population of each state decays exponentially with increasing J, but the degeneracy of the levels increases. Specifically, the population of a rotational energy level J is given by the Boltzmann expression

$$N_J \propto N g_J e^{-E_J/kT}$$

where N is the total number of molecules and g_J is the degeneracy of the level J. The value of J corresponding to a maximum of this expression is found by treating J as a continuous variable, differentiating with respect to J, and then setting the result equal to zero. The result is (see Problem 6.24)

$$J_{\max} \approx \left(\frac{kT}{2hcB}\right)^{1/2} - \tfrac{1}{2} \qquad (6.39)$$

For a typical molecule (for example, OCS, with $B = 0.2\ \text{cm}^{-1}$) at room temperature, $kT \approx 1000hcB$, so $J_{\max} \approx 30$. However, it must be recalled that the intensity of each transition also depends on the value of J (eqn 6.36) and on the population difference between the two states involved in the transition (see Section 6.2). Hence the value of J corresponding to the most intense line is not quite the same as the value of J for the most highly populated level.

6.7 Rotational Raman spectra

The gross selection rule for rotational Raman transitions is *that the molecule must be anisotropically polarizable*. We begin by explaining what this means. A formal derivation of this rule is given in *Further information 6.2*.

The distortion of a molecule in an electric field is determined by its polarizability, α. More precisely, if the strength of the field is $\mathcal{E}$, then the molecule acquires an induced dipole moment of magnitude

$$\mu = \alpha\mathcal{E} \qquad (6.40)$$

in addition to any permanent dipole moment it may have. An atom is isotropically polarizable. That is, the same distortion is induced whatever the direction of the applied field. The polarizability of a spherical rotor is also isotropic. However, non-spherical rotors have polarizabilities that do depend on the direction of the field relative to the molecule, so these molecules are anisotropically polarizable (Fig. 6.20). The electron distribution in H_2, for example, is more distorted when the field is applied parallel to the bond than when it is applied perpendicular to it, and we write $\alpha_{\parallel} > \alpha_{\perp}$.

All linear molecules and diatomics (whether homonuclear or heteronuclear) have anisotropic polarizabilities, and so are rotationally Raman active. This activity is one reason for the importance of rotational Raman spectroscopy, for the technique can be used to study many of the molecules that are inaccessible to microwave spectroscopy. Spherical rotors such as CH_4 and SF_6, however, are rotationally Raman inactive as well as microwave inactive. This inactivity does not mean that such molecules are never found in rotationally excited states. Molecular collisions do not have to obey such restrictive selection rules, and hence collisions between molecules can result in the population of any rotational state.

We show in *Further information 6.2* that the specific rotational Raman selection rules are

$$\text{Linear rotors:} \quad \Delta J = 0, \pm 2 \qquad (6.41)$$

$$\text{Symmetric rotors:} \quad \Delta J = 0, \pm 1, \pm 2; \qquad \Delta K = 0$$

The $\Delta J = 0$ transitions do not lead to a shift of the scattered photon's frequency in pure rotational Raman spectroscopy, and contribute to the unshifted Rayleigh radiation.

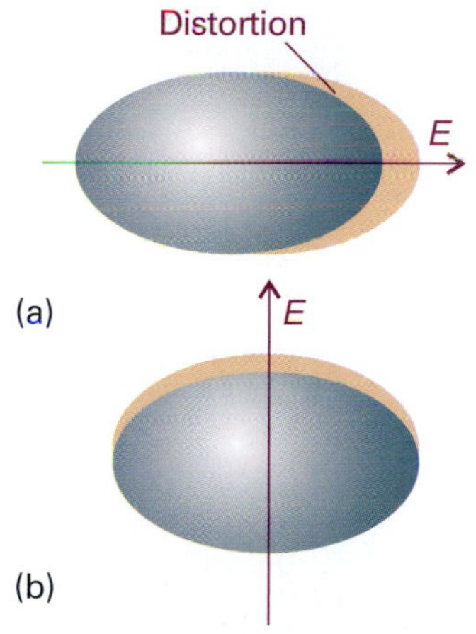

Fig. 6.20 An electric field applied to a molecule results in its distortion, and the distorted molecule acquires a contribution to its dipole moment (even if it is nonpolar initially). The polarizability may be different when the field is applied (a) parallel or (b) perpendicular to the molecular axis (or, in general, in different directions relative to the molecule); if that is so, then the molecule has an anisotropic polarizability.

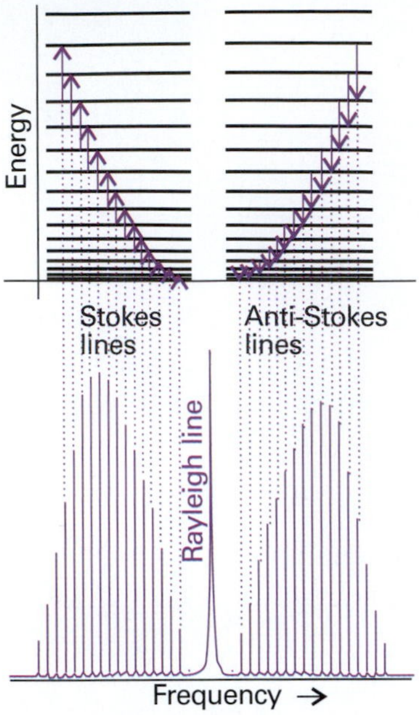

Fig. 6.21 The rotational energy levels of a linear rotor and the transitions allowed by the $\Delta J = \pm 2$ Raman selection rules. The form of a typical rotational Raman spectrum is also shown. The Rayleigh line is much stronger than depicted in the figure; it is shown as a weaker line to improve visualization of the Raman lines.

We can predict the form of the Raman spectrum of a linear rotor by applying the selection rule $\Delta J = \pm 2$ to the rotational energy levels (Fig. 6.21). When the molecule makes a transition with $\Delta J = +2$, the scattered radiation leaves the molecule in a higher rotational state, so the wavenumber of the incident radiation, initially $\tilde{\nu}_i$, is decreased. These transitions account for the Stokes lines in the spectrum:

$$\tilde{\nu}(J+2 \leftarrow J) = \tilde{\nu}_i - \{F(J+2) - F(J)\} = \tilde{\nu}_i - 2B(2J+3) \qquad (6.42a)$$

The Stokes lines appear to low frequency of the incident radiation and at displacements $6B$, $10B$, $14B$, . . . from $\tilde{\nu}_i$ for $J = 0, 1, 2, \ldots$. When the molecule makes a transition with $\Delta J = -2$, the scattered photon emerges with increased energy. These transitions account for the anti-Stokes lines of the spectrum:

$$\tilde{\nu}(J-2 \leftarrow J) = \tilde{\nu}_i + \{F(J) - F(J-2)\} = \tilde{\nu}_i + 2B(2J-1) \qquad (6.42b)$$

The anti-Stokes lines occur at displacements of $6B$, $10B$, $14B$, . . . (for $J = 2, 3, 4, \ldots$; $J = 2$ is the lowest state that can contribute under the selection rule $\Delta J = -2$) to high frequency of the incident radiation. The separation of adjacent lines in both the Stokes and the anti-Stokes regions is $4B$, so from its measurement $I_\perp$ can be determined and then used to find the bond lengths exactly as in the case of microwave spectroscopy.

Example 6.4 *Predicting the form of a Raman spectrum*

Predict the form of the rotational Raman spectrum of $^{14}N_2$, for which $B = 1.99\ \mathrm{cm^{-1}}$, when it is exposed to monochromatic 336.732 nm laser radiation.

Method The molecule is rotationally Raman active because end-over-end rotation modulates its polarizability as viewed by a stationary observer. The Stokes and anti-Stokes lines are given by eqn 6.42.

Answer Because $\lambda_i = 336.732$ nm corresponds to $\tilde{\nu}_i = 29\,697.2\ \mathrm{cm^{-1}}$, eqns 6.42a and 6.42b give the following line positions:

J	0	1	2	3
Stokes lines				
$\tilde{\nu}/\mathrm{cm^{-1}}$	29 685.3	29 677.3	29 669.3	29 661.4
λ/nm	336.868	336.958	337.048	337.139
Anti-Stokes lines				
$\tilde{\nu}/\mathrm{cm^{-1}}$			29 709.1	29 717.1
λ/nm			336.597	336.507

There will be a strong central line at 336.732 nm accompanied on either side by lines of increasing and then decreasing intensity (as a result of transition moment and population effects). The spread of the entire spectrum is very small, so the incident light must be highly monochromatic.

Self-test 6.5 Repeat the calculation for the rotational Raman spectrum of NH_3 ($B = 9.977\ \mathrm{cm^{-1}}$).

[Stokes lines at 29 637.3, 29 597.4, 29 557.5, 29 517.6 $\mathrm{cm^{-1}}$, anti-Stokes lines at 29 757.1, 29 797.0 $\mathrm{cm^{-1}}$.]

6.8 Nuclear statistics and rotational states

If eqn 6.42 is used in conjunction with the rotational Raman spectrum of CO_2, the rotational constant is inconsistent with other measurements of C—O bond lengths.

The results are consistent only if it is supposed that the molecule can exist in states with even values of J, so the Stokes lines are $2 \leftarrow 0, 4 \leftarrow 2, \ldots$ and not $2 \leftarrow 0, 3 \leftarrow 1, \ldots$.

The explanation of the missing lines is the Pauli principle and the fact that O nuclei are spin-0 bosons: just as the Pauli principle excludes certain electronic states, so too does it exclude certain molecular rotational states. The form of the Pauli principle given in *Justification 3.4* states that, when two identical bosons are exchanged, the overall wavefunction must remain unchanged in every respect, including sign. In particular, when a CO_2 molecule rotates through 180°, two identical O nuclei are interchanged, so the overall wavefunction of the molecule must remain unchanged. However, inspection of the form of the rotational wavefunctions (which have the same form as the *s*, *p*, etc. orbitals of atoms) shows that they change sign by $(-1)^J$ under such a rotation (Fig. 6.22). Therefore, only even values of J are permissible for CO_2, and hence the Raman spectrum shows only alternate lines.

The selective occupation of rotational states that stems from the Pauli principle is termed **nuclear statistics**. Nuclear statistics must be taken into account whenever a rotation interchanges equivalent nuclei. However, the consequences are not always as simple as for CO_2 because there are complicating features when the nuclei have nonzero spin: there may be several different relative nuclear spin orientations consistent with even values of J and a different number of spin orientations consistent with odd values of J. For molecular hydrogen and fluorine, for instance, with their two identical spin-$\frac{1}{2}$ nuclei, we show in the *Justification* below that there are three times as many ways of achieving a state with odd J than with even J, and there is a corresponding 3:1 alternation in intensity in their rotational Raman spectra (Fig. 6.23). In general, for a homonuclear diatomic molecule with nuclei of spin I, the numbers of ways of achieving states of odd and even J are in the ratio

$$\frac{\text{Number of ways of achieving odd } J}{\text{Number of ways of achieving even } J} = \begin{cases} (I+1)/I & \text{for half-integral spin nuclei} \\ I/(I+1) & \text{for integral spin nuclei} \end{cases} \tag{6.43}$$

For hydrogen, $I = \frac{1}{2}$, and the ratio is 3:1. For N_2, with $I = 1$, the ratio is 1:2.

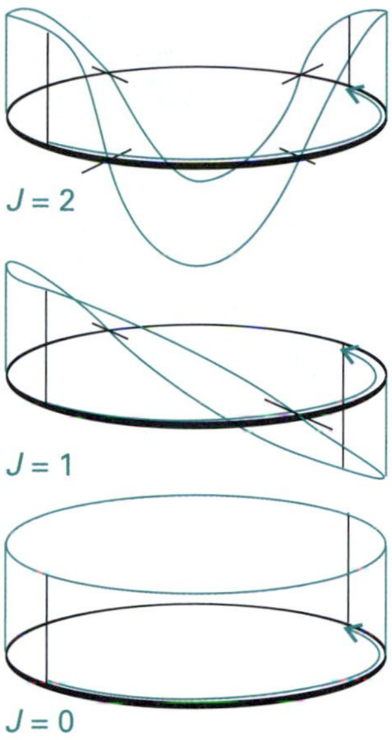

Fig. 6.22 The symmetries of rotational wavefunctions (shown here, for simplicity as a two-dimensional rotor) under a rotation through 180°. Wavefunctions with J even do not change sign; those with J odd do change sign.

Justification 6.4 *The effect of nuclear statistics on rotational spectra*

Hydrogen nuclei are fermions, so the Pauli principle requires the overall wavefunction to change sign under particle interchange. However, the rotation of an H_2 molecule through 180° has a more complicated effect than merely relabelling the nuclei, because it interchanges their spin states too if the nuclear spins are paired (↑↓) but not if they are parallel (↑↑).

For the overall wavefunction of the molecule to change sign when the spins are parallel, the rotational wavefunction must change sign. Hence, only odd values of J are allowed. In contrast, if the nuclear spins are paired, their wavefunction is $\alpha(A)\beta(B) - \alpha(B)\beta(A)$, which changes sign when α and β are exchanged in order to bring about a simple $A \leftrightarrow B$ interchange overall (Fig. 6.24). Therefore, for the overall wavefunction to change sign in this case requires the rotational wavefunction *not* to change sign. Hence, only even values of J are allowed if the nuclear spins are paired.

As there are three nuclear spin states with parallel spins (just like the triplet state of two parallel electrons, as in Fig. 3.24), but only one state with paired spins (the analogue of the singlet state of two electrons, see Fig. 3.18), it follows that the populations of the odd J and even J states should be in the ratio of 3:1, and hence the intensities of transitions originating in these levels will be in the same ratio.

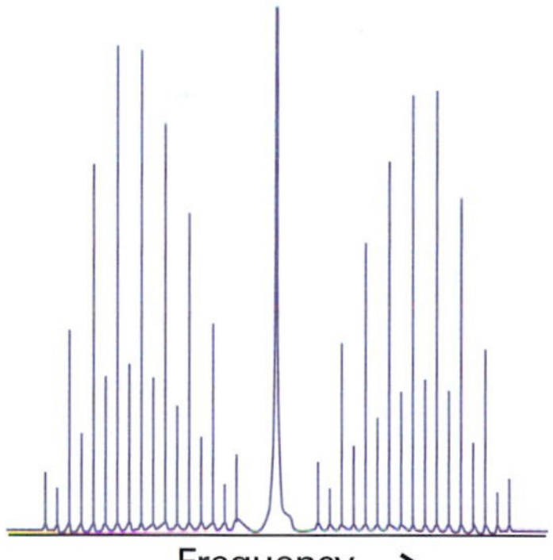

Fig. 6.23 The rotational Raman spectrum of a diatomic molecule with two identical spin-$\frac{1}{2}$ nuclei shows an alternation in intensity as a result of nuclear statistics. The Rayleigh line is much stronger than depicted in the figure; it is shown as a weaker line to improve visualization of the Raman lines.

Different relative nuclear spin orientations change into one another only very slowly, so an H_2 molecule with parallel nuclear spins remains distinct from one with

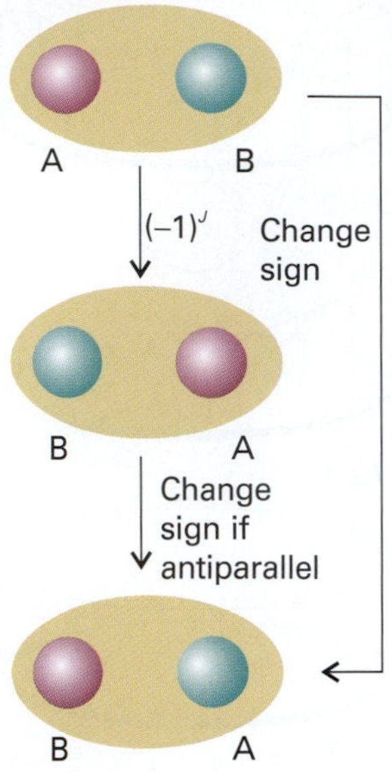

Fig. 6.24 The interchange of two identical fermion nuclei results in the change in sign of the overall wavefunction. The relabelling can be thought of as occurring in two steps: the first is a rotation of the molecule; the second is the interchange of unlike spins (represented by the different colours of the nuclei). The wavefunction changes sign in the second step if the nuclei have antiparallel spins.

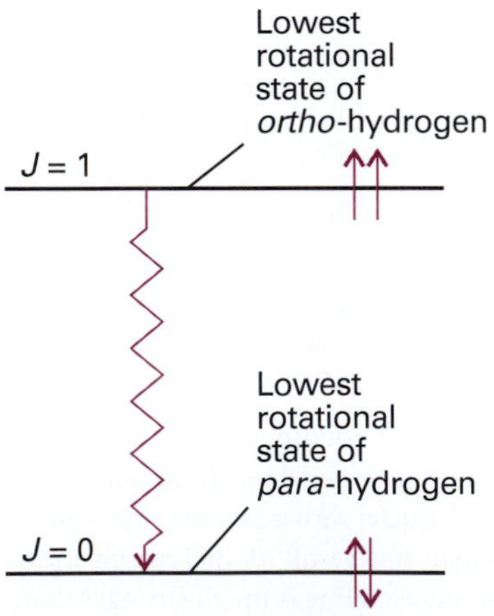

Fig. 6.25 When hydrogen is cooled, the molecules with parallel nuclear spins accumulate in their lowest available rotational state, the one with $J = 0$. They can enter the lowest rotational state ($J = 0$) only if the spins change their relative orientation and become antiparallel. This is a slow process under normal circumstances, so energy is slowly released.

paired nuclear spins for long periods. The two forms of hydrogen can be separated by physical techniques, and stored. The form with parallel nuclear spins is called ***ortho*-hydrogen** and the form with paired nuclear spins is called ***para*-hydrogen**. Because *ortho*-hydrogen cannot exist in a state with $J = 0$, it continues to rotate at very low temperatures and has an effective rotational zero-point energy (Fig. 6.25). This energy is of some concern to manufacturers of liquid hydrogen, for the slow conversion of *ortho*-hydrogen into *para*-hydrogen (which can exist with $J = 0$) as nuclear spins slowly realign releases rotational energy, which vaporizes the liquid. Techniques are used to accelerate the conversion of *ortho*-hydrogen to *para*-hydrogen to avoid this problem. One such technique is to pass hydrogen over a metal surface: the molecules adsorb on the surface as atoms, which then recombine in the lower energy *para*-hydrogen form.

The vibrations of diatomic molecules

In this section, we adopt the same strategy of finding expressions for the energy levels, establishing the selection rules, and then discussing the form of the spectrum. We shall also see how the simultaneous excitation of rotation modifies the appearance of a vibrational spectrum.

6.9 Molecular vibrations

We base our discussion on Fig. 6.26, which shows a typical potential energy curve (as in Fig. 4.1) of a diatomic molecule. In regions close to R_e (at the minimum of the curve) the potential energy can be approximated by a parabola, so we can write

$$V = \tfrac{1}{2}kx^2 \qquad x = R - R_e \tag{6.44}$$

where k is the **force constant** of the bond. The steeper the walls of the potential (the stiffer the bond), the greater the force constant.

To see the connection between the shape of the molecular potential energy curve and the value of k, note that we can expand the potential energy around its minimum by using a Taylor series:

$$V(x) = V(0) + \left(\frac{dV}{dx}\right)_0 x + \tfrac{1}{2}\left(\frac{d^2V}{dx^2}\right)_0 x^2 + \cdots \tag{6.45}$$

The term $V(0)$ can be set arbitrarily to zero. The first derivative of V is 0 at the minimum. Therefore, the first surviving term is proportional to the square of the displacement. For small displacements we can ignore all the higher terms, and so write

$$V(x) \approx \tfrac{1}{2}\left(\frac{d^2V}{dx^2}\right)_0 x^2 \tag{6.46}$$

Therefore, the first approximation to a molecular potential energy curve is a parabolic potential, and we can identify the force constant as

$$k = \left(\frac{d^2V}{dx^2}\right)_0 \qquad [6.47]$$

We see that if the potential energy curve is sharply curved close to its minimum, then k will be large. Conversely, if the potential energy curve is wide and shallow, then k will be small (Fig. 6.27).

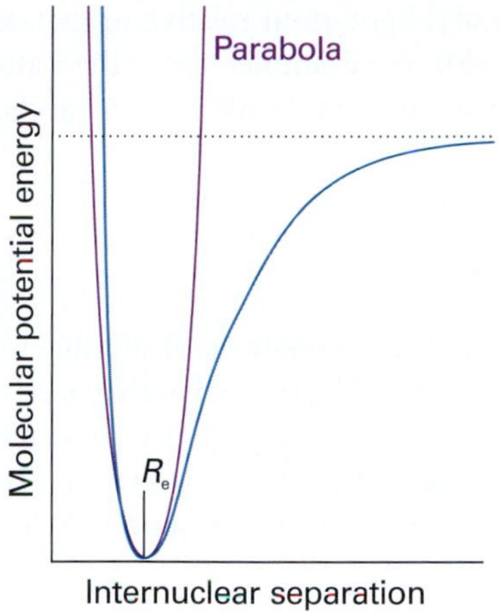

Fig. 6.26 A molecular potential energy curve can be approximated by a parabola near the bottom of the well. The parabolic potential leads to harmonic oscillations.
At high excitation energies the parabolic approximation is poor (the true potential is less confining), and it is totally wrong near the dissociation limit.

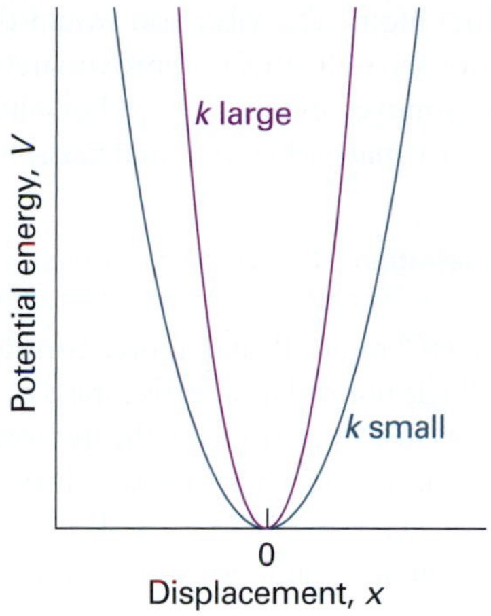

Fig. 6.27 The force constant is a measure of the curvature of the potential energy close to the equilibrium extension of the bond. A strongly confining well (one with steep sides, a stiff bond) corresponds to high values of k.

Comment 6.5

It is often useful to express a function $f(x)$ in the vicinity of $x = a$ as an infinite Taylor series of the form:

$$f(x) = f(a) + \left(\frac{df}{dx}\right)_a (x-a) + \frac{1}{2!}\left(\frac{d^2f}{dx^2}\right)_a (x-a)^2 + \cdots + \frac{1}{n!}\left(\frac{d^nf}{dx^n}\right)_a (x-a)^n + \cdots$$

where $n = 0, 1, 2, \ldots$.

The Schrödinger equation for the relative motion of two atoms of masses m_1 and m_2 with a parabolic potential energy is

$$-\frac{\hbar^2}{2m_{\text{eff}}}\frac{d^2\psi}{dx^2} + \tfrac{1}{2}kx^2\psi = E\psi \qquad (6.48)$$

where m_{eff} is the **effective mass**:

$$m_{\text{eff}} = \frac{m_1 m_2}{m_1 + m_2} \qquad [6.49]$$

These equations are derived in the same way as in *Further information 3.1*, but here the separation of variables procedure is used to separate the relative motion of the atoms from the motion of the molecule as a whole. In that context, the effective mass is called the 'reduced mass', and the name is widely used in this context too.

The Schrödinger equation in eqn 6.48 is the same as eqn 2.24 for a particle of mass m undergoing harmonic motion. Therefore, we can use the results of Section 2.4 to write down the permitted vibrational energy levels:

$$E_v = (v + \tfrac{1}{2})\hbar\omega \qquad \omega = \left(\frac{k}{m_{\text{eff}}}\right)^{1/2} \qquad v = 0, 1, 2, \ldots \qquad (6.50)$$

The **vibrational terms** of a molecule, the energies of its vibrational states expressed in wavenumbers, are denoted $G(v)$, with $E_v = hcG(v)$, so

$$G(v) = (v + \tfrac{1}{2})\tilde{\nu} \qquad \tilde{\nu} = \frac{1}{2\pi c}\left(\frac{k}{m_{\text{eff}}}\right)^{1/2} \qquad (6.51)$$

The vibrational wavefunctions are the same as those discussed in Section 2.5.

It is important to note that the vibrational terms depend on the effective mass of the molecule, not directly on its total mass. This dependence is physically reasonable, for if atom 1 were as heavy as a brick wall, then we would find $m_{\text{eff}} \approx m_2$, the mass of the lighter atom. The vibration would then be that of a light atom relative to that of a stationary wall (this is approximately the case in HI, for example, where the I atom barely moves and $m_{eff} \approx m_H$). For a homonuclear diatomic molecule $m_1 = m_2$, and the effective mass is half the total mass: $m_{eff} = \frac{1}{2}m$.

Illustration 6.3 *Calculating a vibrational wavenumber*

An HCl molecule has a force constant of 516 N m^{-1}, a reasonably typical value for a single bond. The effective mass of $^1H^{35}Cl$ is 1.63×10^{-27} kg (note that this mass is very close to the mass of the hydrogen atom, 1.67×10^{-27} kg, so the Cl atom is acting like a brick wall). These values imply $\omega = 5.63 \times 10^{14}$ s^{-1}, $\nu = 89.5$ THz (1 THz $= 10^{12}$ Hz), $\tilde{\nu} = 2990$ cm^{-1}, $\lambda = 3.35$ μm. These characteristics correspond to electromagnetic radiation in the infrared region.

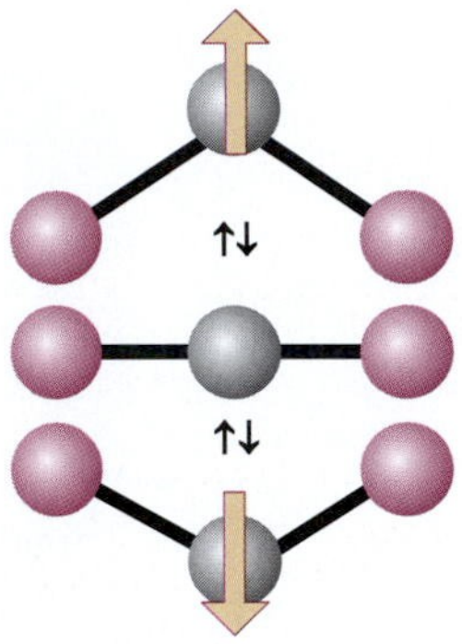

Fig. 6.28 The oscillation of a molecule, even if it is nonpolar, may result in an oscillating dipole that can interact with the electromagnetic field.

6.10 Selection rules

The gross selection rule for a change in vibrational state brought about by absorption or emission of radiation is that *the electric dipole moment of the molecule must change when the atoms are displaced relative to one another*. Such vibrations are said to be **infrared active**. The classical basis of this rule is that the molecule can shake the electromagnetic field into oscillation if its dipole changes as it vibrates, and vice versa (Fig. 6.28); its formal basis is given in *Further information 6.2*. Note that the molecule need not have a permanent dipole: the rule requires only a change in dipole moment, possibly from zero. Some vibrations do not affect the molecule's dipole moment (e.g. the stretching motion of a homonuclear diatomic molecule), so they neither absorb nor generate radiation: such vibrations are said to be **infrared inactive**. Homonuclear diatomic molecules are infrared inactive because their dipole moments remain zero however long the bond; heteronuclear diatomic molecules are infrared active.

Illustration 6.4 *Identifying infrared active molecules*

Of the molecules N_2, CO_2, OCS, H_2O, $CH_2{=}CH_2$, and C_6H_6, all except N_2 possess at least one vibrational mode that results in a change of dipole moment, so all except N_2 can show a vibrational absorption spectrum. Not all the modes of complex molecules are vibrationally active. For example, the symmetric stretch of CO_2, in which the O—C—O bonds stretch and contract symmetrically is inactive because it leaves the dipole moment unchanged (at zero).

Self-test 6.6 Which of the molecules H_2, NO, N_2O, and CH_4 have infrared active vibrations? [NO, N_2O, CH_4]

The specific selection rule, which is obtained from an analysis of the expression for the transition moment and the properties of integrals over harmonic oscillator wavefunctions (as shown in *Further information 6.2*), is

$$\Delta v = \pm 1 \tag{6.52}$$

Transitions for which $\Delta v = +1$ correspond to absorption and those with $\Delta v = -1$ correspond to emission.

It follows from the specific selection rules that the wavenumbers of allowed vibrational transitions, which are denoted $\Delta G_{v+\frac{1}{2}}$ for the transition $v+1 \leftarrow v$, are

$$\Delta G_{v+\frac{1}{2}} = G(v+1) - G(v) = \tilde{\nu} \tag{6.53}$$

As we have seen, $\tilde{\nu}$ lies in the infrared region of the electromagnetic spectrum, so vibrational transitions absorb and generate infrared radiation.

At room temperature $kT/hc \approx 200\ \text{cm}^{-1}$, and most vibrational wavenumbers are significantly greater than $200\ \text{cm}^{-1}$. It follows from the Boltzmann distribution that almost all the molecules will be in their vibrational ground states initially. Hence, the dominant spectral transition will be the **fundamental transition**, $1 \leftarrow 0$. As a result, the spectrum is expected to consist of a single absorption line. If the molecules are formed in a vibrationally excited state, such as when vibrationally excited HF molecules are formed in the reaction $H_2 + F_2 \rightarrow 2\,HF^*$, the transitions $5 \rightarrow 4$, $4 \rightarrow 3$, . . . may also appear (in emission). In the harmonic approximation, all these lines lie at the same frequency, and the spectrum is also a single line. However, as we shall now show, the breakdown of the harmonic approximation causes the transitions to lie at slightly different frequencies, so several lines are observed.

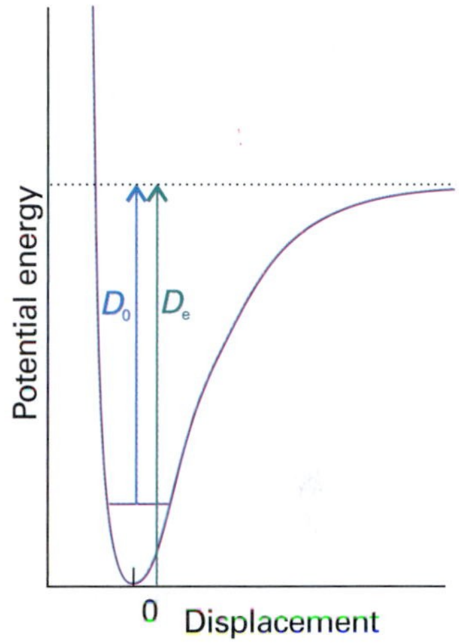

Fig. 6.29 The dissociation energy of a molecule, D_0, differs from the depth of the potential well, D_e, on account of the zero-point energy of the vibrations of the bond.

6.11 Anharmonicity

The vibrational terms in eqn 6.53 are only approximate because they are based on a parabolic approximation to the actual potential energy curve. A parabola cannot be correct at all extensions because it does not allow a bond to dissociate. At high vibrational excitations the swing of the atoms (more precisely, the spread of the vibrational wavefunction) allows the molecule to explore regions of the potential energy curve where the parabolic approximation is poor and additional terms in the Taylor expansion of V (eqn 6.45) must be retained. The motion then becomes **anharmonic**, in the sense that the restoring force is no longer proportional to the displacement. Because the actual curve is less confining than a parabola, we can anticipate that the energy levels become less widely spaced at high excitations.

(a) The convergence of energy levels

One approach to the calculation of the energy levels in the presence of anharmonicity is to use a function that resembles the true potential energy more closely. The **Morse potential energy** is

$$V = hcD_e\{1 - e^{-a(R-R_e)}\}^2 \qquad a = \left(\frac{m_{\text{eff}}\omega^2}{2hcD_e}\right)^{1/2} \tag{6.54}$$

where D_e is the depth of the potential minimum (Fig. 6.29). Near the well minimum the variation of V with displacement resembles a parabola (as can be checked by expanding the exponential as far as the first term) but, unlike a parabola, eqn 6.54 allows for dissociation at large displacements. The Schrödinger equation can be solved for the Morse potential and the permitted energy levels are

$$G(v) = (v+\tfrac{1}{2})\tilde{\nu} - (v+\tfrac{1}{2})^2 x_e\tilde{\nu} \qquad x_e = \frac{a^2\hbar}{2m_{\text{eff}}\omega} = \frac{\tilde{\nu}}{4D_e} \tag{6.55}$$

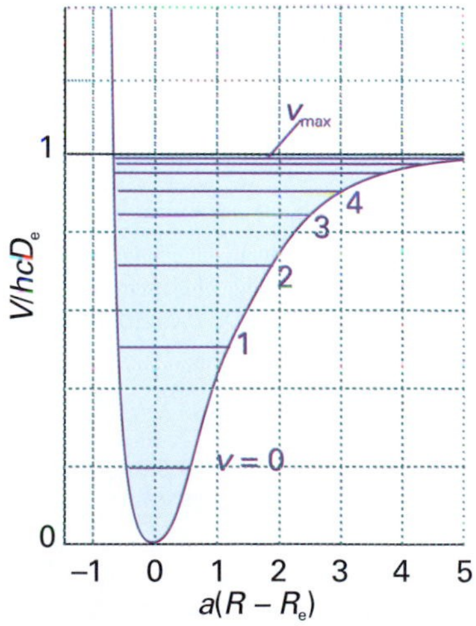

Fig. 6.30 The Morse potential energy curve reproduces the general shape of a molecular potential energy curve. The corresponding Schrödinger equation can be solved, and the values of the energies obtained. The number of bound levels is finite.

The parameter x_e is called the **anharmonicity constant**. The number of vibrational levels of a Morse oscillator is finite, and $v = 0, 1, 2, \ldots, v_{\text{max}}$, as shown in Fig. 6.30 (see also Problem 6.26). The second term in the expression for G subtracts from the first with increasing effect as v increases, and hence gives rise to the convergence of the levels at high quantum numbers.

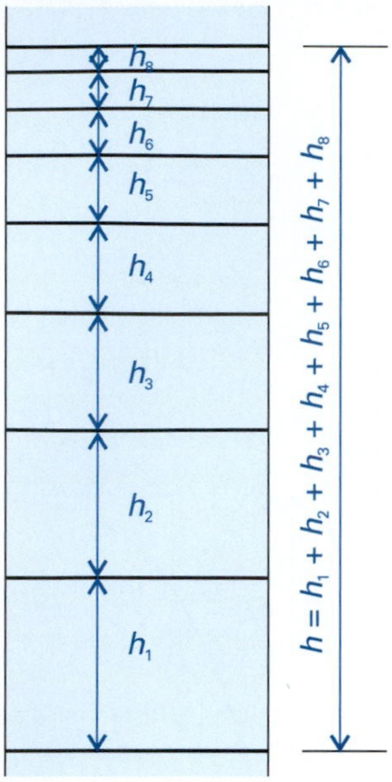

Fig. 6.31 The dissociation energy is the sum of the separations of the vibrational energy levels up to the dissociation limit just as the length of a ladder is the sum of the separations of its rungs.

Although the Morse oscillator is quite useful theoretically, in practice the more general expression

$$G(v) = (v + \tfrac{1}{2})\tilde{\nu} - (v + \tfrac{1}{2})^2 x_e \tilde{\nu} + (v + \tfrac{1}{2})^3 y_e \tilde{\nu} + \cdots \tag{6.56}$$

where x_e, y_e, . . . are empirical dimensionless constants characteristic of the molecule, is used to fit the experimental data and to find the dissociation energy of the molecule. When anharmonicities are present, the wavenumbers of transitions with $\Delta v = +1$ are

$$\Delta G_{v+\frac{1}{2}} = \tilde{\nu} - 2(v+1)x_e\tilde{\nu} + \cdots \tag{6.57}$$

Equation 6.57 shows that, when $x_e > 0$, the transitions move to lower wavenumbers as v increases.

Anharmonicity also accounts for the appearance of additional weak absorption lines corresponding to the transitions $2 \leftarrow 0$, $3 \leftarrow 0$, . . . , even though these first, second, . . . **overtones** are forbidden by the selection rule $\Delta v = \pm 1$. The first overtone, for example, gives rise to an absorption at

$$G(v+2) - G(v) = 2\tilde{\nu} - 2(2v+3)x_e\tilde{\nu} + \cdots \tag{6.58}$$

The reason for the appearance of overtones is that the selection rule is derived from the properties of harmonic oscillator wavefunctions, which are only approximately valid when anharmonicity is present. Therefore, the selection rule is also only an approximation. For an anharmonic oscillator, all values of Δv are allowed, but transitions with $\Delta v > 1$ are allowed only weakly if the anharmonicity is slight.

(b) The Birge–Sponer plot

When several vibrational transitions are detectable, a graphical technique called a **Birge–Sponer plot** may be used to determine the dissociation energy, D_0, of the bond. The basis of the Birge–Sponer plot is that the sum of successive intervals $\Delta G_{v+\frac{1}{2}}$ from the zero-point level to the dissociation limit is the dissociation energy:

$$D_0 = \Delta G_{1/2} + \Delta G_{3/2} + \cdots = \sum_v \Delta G_{v+\frac{1}{2}} \tag{6.59}$$

just as the height of the ladder is the sum of the separations of its rungs (Fig. 6.31). The construction in Fig. 6.32 shows that the area under the plot of $\Delta G_{v+\frac{1}{2}}$ against $v + \frac{1}{2}$ is equal to the sum, and therefore to D_0. The successive terms decrease linearly when only the x_e anharmonicity constant is taken into account and the inaccessible part of the spectrum can be estimated by linear extrapolation. Most actual plots differ from the linear plot as shown in Fig. 6.32, so the value of D_0 obtained in this way is usually an overestimate of the true value.

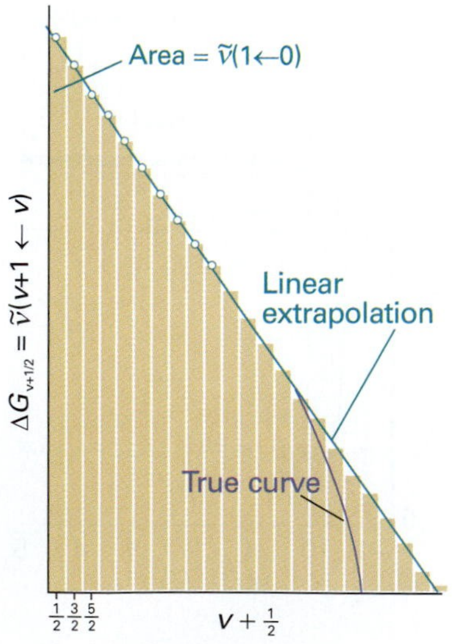

Fig. 6.32 The area under a plot of transition wavenumber against vibrational quantum number is equal to the dissociation energy of the molecule. The assumption that the differences approach zero linearly is the basis of the Birge–Sponer extrapolation.

Example 6.5 *Using a Birge–Sponer plot*

The observed vibrational intervals of H_2^+ lie at the following values for $1 \leftarrow 0$, $2 \leftarrow 1$, . . . respectively (in cm^{-1}): 2191, 2064, 1941, 1821, 1705, 1591, 1479, 1368, 1257, 1145, 1033, 918, 800, 677, 548, 411. Determine the dissociation energy of the molecule.

Method Plot the separations against $v + \frac{1}{2}$, extrapolate linearly to the point cutting the horizontal axis, and then measure the area under the curve.

Answer The points are plotted in Fig. 6.33, and a linear extrapolation is shown as a dotted line. The area under the curve (use the formula for the area of a triangle or count the squares) is 214. Each square corresponds to 100 cm^{-1} (refer to the scale of the vertical axis); hence the dissociation energy is 21 400 cm^{-1} (corresponding to 256 kJ mol^{-1}).

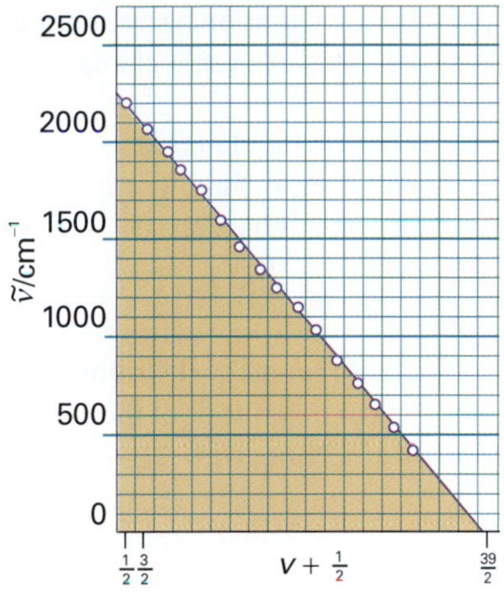

Fig. 6.33 The Birge–Sponer plot used in Example 6.5. The area is obtained simply by counting the squares beneath the line or using the formula for the area of a right triangle.

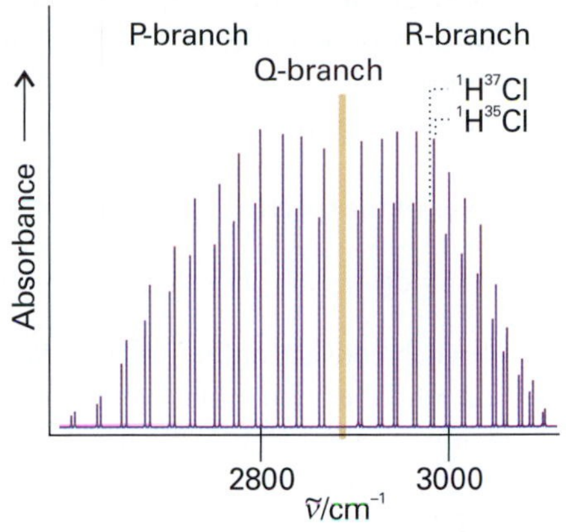

Fig. 6.34 A high-resolution vibration–rotation spectrum of HCl. The lines appear in pairs because $H^{35}Cl$ and $H^{37}Cl$ both contribute (their abundance ratio is 3:1). There is no Q branch, because $\Delta J = 0$ is forbidden for this molecule.

Self-test 6.7 The vibrational levels of HgH converge rapidly, and successive intervals are 1203.7 (which corresponds to the $1 \leftarrow 0$ transition), 965.6, 632.4, and 172 cm^{-1}. Estimate the dissociation energy. [35.6 kJ mol^{-1}]

6.12 Vibration–rotation spectra

Each line of the high resolution vibrational spectrum of a gas-phase heteronuclear diatomic molecule is found to consist of a large number of closely spaced components (Fig. 6.34). Hence, molecular spectra are often called **band spectra**. The separation between the components is less than 10 cm^{-1}, which suggests that the structure is due to rotational transitions accompanying the vibrational transition. A rotational change should be expected because classically we can think of the transition as leading to a sudden increase or decrease in the instantaneous bond length. Just as ice-skaters rotate more rapidly when they bring their arms in, and more slowly when they throw them out, so the molecular rotation is either accelerated or retarded by a vibrational transition.

(a) Spectral branches

A detailed analysis of the quantum mechanics of simultaneous vibrational and rotational changes shows that the rotational quantum number J changes by ± 1 during the vibrational transition of a diatomic molecule. If the molecule also possesses angular momentum about its axis, as in the case of the electronic orbital angular momentum of the paramagnetic molecule NO, then the selection rules also allow $\Delta J = 0$.

The appearance of the vibration–rotation spectrum of a diatomic molecule can be discussed in terms of the combined vibration–rotation terms, S:

$$S(v,J) = G(v) + F(J) \tag{6.60}$$

If we ignore anharmonicity and centrifugal distortion,

$$S(v,J) = (v + \tfrac{1}{2})\tilde{\nu} + BJ(J+1) \tag{6.61}$$

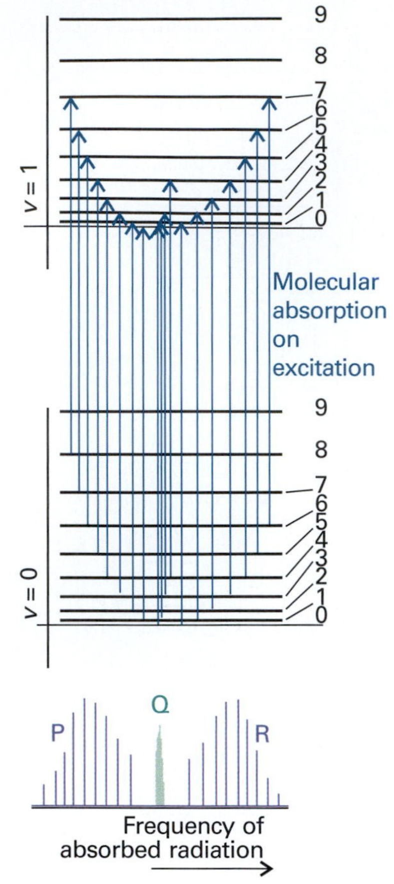

Fig. 6.35 The formation of P, Q, and R branches in a vibration–rotation spectrum. The intensities reflect the populations of the initial rotational levels.

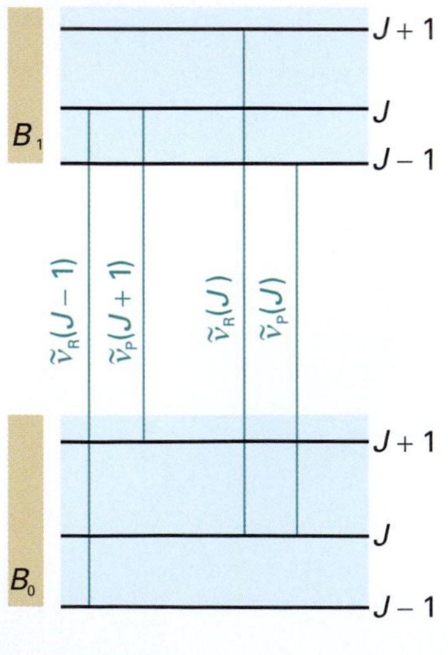

Fig. 6.36 The method of combination differences makes use of the fact that some transitions share a common level.

In a more detailed treatment, B is allowed to depend on the vibrational state because, as v increases, the molecule swells slightly and the moment of inertia changes. We shall continue with the simple expression initially.

When the vibrational transition $v+1 \leftarrow v$ occurs, J changes by ± 1 and in some cases by 0 (when $\Delta J = 0$ is allowed). The absorptions then fall into three groups called **branches** of the spectrum. The **P branch** consists of all transitions with $\Delta J = -1$:

$$\tilde{\nu}_P(J) = S(v+1, J-1) - S(v,J) = \tilde{\nu} - 2BJ \tag{6.62a}$$

This branch consists of lines at $\tilde{\nu} - 2B$, $\tilde{\nu} - 4B, \ldots$ with an intensity distribution reflecting both the populations of the rotational levels and the magnitude of the $J-1 \leftarrow J$ transition moment (Fig. 6.35). The **Q branch** consists of all lines with $\Delta J = 0$, and its wavenumbers are all

$$\tilde{\nu}_Q(J) = S(v+1, J) - S(v,J) = \tilde{\nu} \tag{6.62b}$$

for all values of J. This branch, when it is allowed (as in NO), appears at the vibrational transition wavenumber. In Fig. 6.35 there is a gap at the expected location of the Q branch because it is forbidden in HCl. The **R branch** consists of lines with $\Delta J = +1$:

$$\tilde{\nu}_R(J) = S(v+1, J+1) - S(v,J) = \tilde{\nu} + 2B(J+1) \tag{6.62c}$$

This branch consists of lines displaced from $\tilde{\nu}$ to high wavenumber by $2B, 4B, \ldots$.

The separation between the lines in the P and R branches of a vibrational transition gives the value of B. Therefore, the bond length can be deduced without needing to take a pure rotational microwave spectrum. However, the latter is more precise.

(b) Combination differences

The rotational constant of the vibrationally excited state, B_1 (in general, B_v), is in fact slightly smaller than that of the ground vibrational state, B_0, because the anharmonicity of the vibration results in a slightly extended bond in the upper state. As a result, the Q branch (if it exists) consists of a series of closely spaced lines. The lines of the R branch converge slightly as J increases; and those of the P branch diverge:

$$\begin{aligned}
\tilde{\nu}_P(J) &= \tilde{\nu} - (B_1 + B_0)J + (B_1 - B_0)J^2 \\
\tilde{\nu}_Q(J) &= \tilde{\nu} + (B_1 - B_0)J(J+1) \\
\tilde{\nu}_R(J) &= \tilde{\nu} + (B_1 + B_0)(J+1) + (B_1 - B_0)(J+1)^2
\end{aligned} \tag{6.63}$$

To determine the two rotational constants individually, we use the method of **combination differences**. This procedure is used widely in spectroscopy to extract information about a particular state. It involves setting up expressions for the difference in the wavenumbers of transitions to a common state; the resulting expression then depends solely on properties of the other state.

As can be seen from Fig. 6.36, the transitions $\tilde{\nu}_R(J-1)$ and $\tilde{\nu}_P(J+1)$ have a common upper state, and hence can be anticipated to depend on B_0. Indeed, it is easy to show from eqn 6.63 that

$$\tilde{\nu}_R(J-1) - \tilde{\nu}_P(J+1) = 4B_0(J+\tfrac{1}{2}) \tag{6.64a}$$

Therefore, a plot of the combination difference against $J+\frac{1}{2}$ should be a straight line of slope $4B_0$, so the rotational constant of the molecule in the state $v = 0$ can be determined. (Any deviation from a straight line is a consequence of centrifugal distortion, so that effect can be investigated too.) Similarly, $\tilde{\nu}_R(J)$ and $\tilde{\nu}_P(J)$ have a common lower state, and hence their combination difference gives information about the upper state:

$$\tilde{\nu}_R(J) - \tilde{\nu}_P(J) = 4B_1(J + \tfrac{1}{2}) \qquad (6.64b)$$

The two rotational constants of $^1H^{35}Cl$ found in this way are $B_0 = 10.440\ cm^{-1}$ and $B_1 = 10.136\ cm^{-1}$.

6.13 Vibrational Raman spectra of diatomic molecules

The gross selection rule for vibrational Raman transitions is that *the polarizability should change as the molecule vibrates.* As homonuclear and heteronuclear diatomic molecules swell and contract during a vibration, the control of the nuclei over the electrons varies, and hence the molecular polarizability changes. Both types of diatomic molecule are therefore vibrationally Raman active. The specific selection rule for vibrational Raman transitions in the harmonic approximation is $\Delta v = \pm 1$. The formal basis for the gross and specific selection rules is given in *Further information 6.2.*

The lines to high frequency of the incident radiation, the anti-Stokes lines, are those for which $\Delta v = -1$. The lines to low frequency, the Stokes lines, correspond to $\Delta v = +1$. The intensities of the anti-Stokes and Stokes lines are governed largely by the Boltzmann populations of the vibrational states involved in the transition. It follows that anti-Stokes lines are usually weak because very few molecules are in an excited vibrational state initially.

In gas-phase spectra, the Stokes and anti-Stokes lines have a branch structure arising from the simultaneous rotational transitions that accompany the vibrational excitation (Fig. 6.37). The selection rules are $\Delta J = 0, \pm 2$ (as in pure rotational Raman spectroscopy), and give rise to the **O branch** ($\Delta J = -2$), the **Q branch** ($\Delta J = 0$), and the **S branch** ($\Delta J = +2$):

$$\begin{aligned} \tilde{\nu}_O(J) &= \tilde{\nu}_i - \tilde{\nu} - 2B + 4BJ \\ \tilde{\nu}_Q(J) &= \tilde{\nu}_i - \tilde{\nu} \\ \tilde{\nu}_S(J) &= \tilde{\nu}_i - \tilde{\nu} - 6B - 4BJ \end{aligned} \qquad (6.65)$$

Note that, unlike in infrared spectroscopy, a Q branch is obtained for all linear molecules. The spectrum of CO, for instance, is shown in Fig. 6.38: the structure of the Q branch arises from the differences in rotational constants of the upper and lower vibrational states.

The information available from vibrational Raman spectra adds to that from infrared spectroscopy because homonuclear diatomics can also be studied. The spectra can be interpreted in terms of the force constants, dissociation energies, and bond lengths, and some of the information obtained is included in Table 6.2.

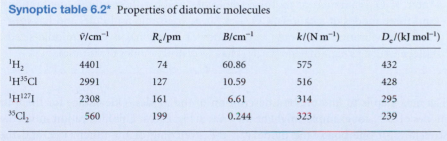

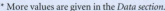

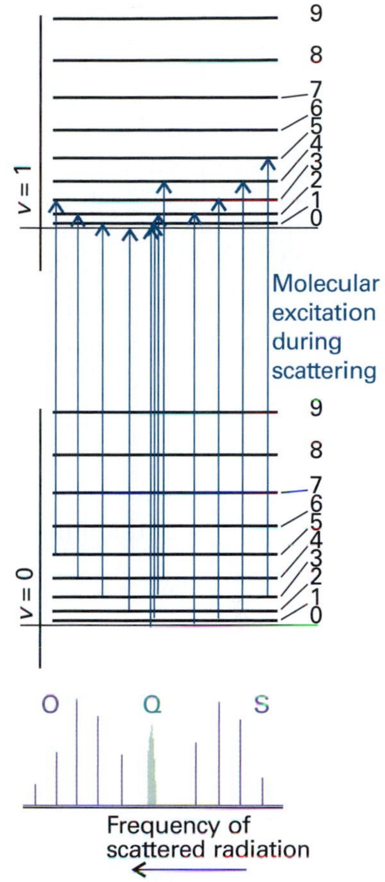

Fig. 6.37 The formation of O, Q, and S branches in a vibration–rotation Raman spectrum of a linear rotor. Note that the frequency scale runs in the opposite direction to that in Fig. 6.35, because the higher energy transitions (on the right) extract more energy from the incident beam and leave it at lower frequency.

Synoptic table 6.2* Properties of diatomic molecules

	$\tilde{\nu}/cm^{-1}$	R_e/pm	B/cm^{-1}	$k/(N\ m^{-1})$	$D_e/(kJ\ mol^{-1})$
$^1H_2$	4401	74	60.86	575	432
$^1H^{35}Cl$	2991	127	10.59	516	428
$^1H^{127}I$	2308	161	6.61	314	295
$^{35}Cl_2$	560	199	0.244	323	239

* More values are given in the *Data section.*

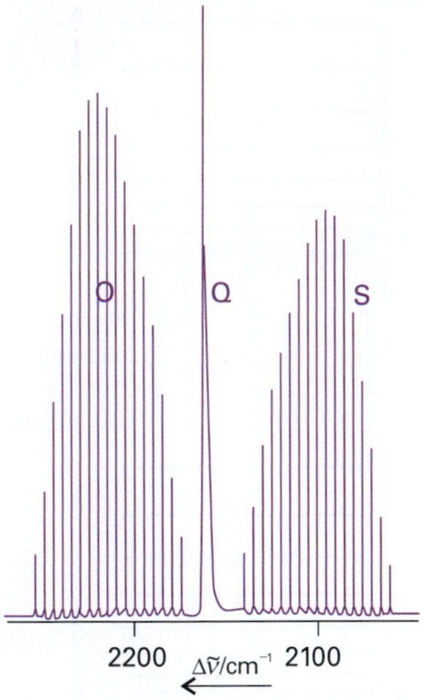

Fig. 6.38 The structure of a vibrational line in the vibrational Raman spectrum of carbon monoxide, showing the O, Q, and S branches.

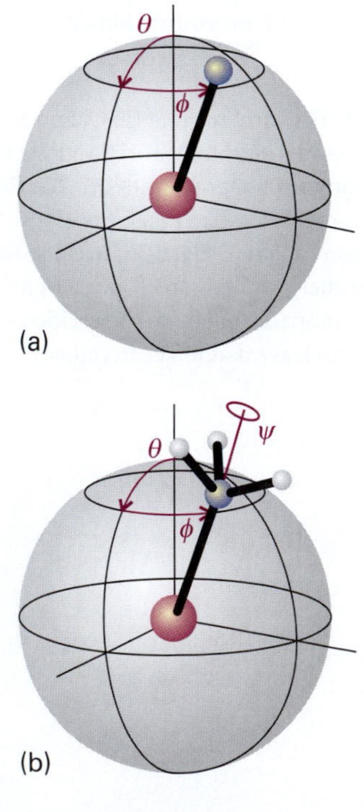

Fig. 6.39 (a) The orientation of a linear molecule requires the specification of two angles. (b) The orientation of a nonlinear molecule requires the specification of three angles.

The vibrations of polyatomic molecules

There is only one mode of vibration for a diatomic molecule, the bond stretch. In polyatomic molecules there are several modes of vibration because all the bond lengths and angles may change and the vibrational spectra are very complex. Nonetheless, we shall see that infrared and Raman spectroscopy can be used to obtain information about the structure of systems as large as animal and plant tissues (see *Impact I6.3*).

6.14 Normal modes

We begin by calculating the total number of vibrational modes of a polyatomic molecule. We then see that we can choose combinations of these atomic displacements that give the simplest description of the vibrations.

As shown in the *Justification* below, for a nonlinear molecule that consists of N atoms, there are $3N - 6$ independent modes of vibration. If the molecule is linear, there are $3N - 5$ independent vibrational modes.

Justification 6.5 *The number of vibrational modes*

The total number of coordinates needed to specify the locations of N atoms is $3N$. Each atom may change its location by varying one of its three coordinates (x, y, and z), so the total number of displacements available is $3N$. These displacements can be grouped together in a physically sensible way. For example, three coordinates are needed to specify the location of the centre of mass of the molecule, so three of the $3N$ displacements correspond to the translational motion of the molecule as a whole. The remaining $3N - 3$ are non-translational 'internal' modes of the molecule.

Two angles are needed to specify the orientation of a linear molecule in space: in effect, we need to give only the latitude and longitude of the direction in which the molecular axis is pointing (Fig. 6.39a). However, three angles are needed for a nonlinear molecule because we also need to specify the orientation of the molecule around the direction defined by the latitude and longitude (Fig. 6.39b). Therefore, two (linear) or three (nonlinear) of the $3N - 3$ internal displacements are rotational. This leaves $3N - 5$ (linear) or $3N - 6$ (nonlinear) displacements of the atoms relative to one another: these are the vibrational modes. It follows that the number of modes of vibration N_{vib} is $3N - 5$ for linear molecules and $3N - 6$ for nonlinear molecules.

Illustration 6.5 *Determining the number of vibrational modes*

Water, H_2O, is a nonlinear triatomic molecule, and has three modes of vibration (and three modes of rotation); CO_2 is a linear triatomic molecule, and has four modes of vibration (and only two modes of rotation). Even a middle-sized molecule such as naphthalene ($C_{10}H_8$) has 48 distinct modes of vibration.

The next step is to find the best description of the modes. One choice for the four modes of CO_2, for example, might be the ones in Fig. 6.40a. This illustration shows the stretching of one bond (the mode ν_L), the stretching of the other (ν_R), and the two perpendicular bending modes (ν_2). The description, while permissible, has a disadvantage: when one CO bond vibration is excited, the motion of the C atom sets the other CO bond in motion, so energy flows backwards and forwards between ν_L and ν_R. Moreover, the position of the centre of mass of the molecule varies in the course of either vibration.

The description of the vibrational motion is much simpler if linear combinations of ν_L and ν_R are taken. For example, one combination is ν_1 in Fig. 6.40b: this mode is the **symmetric stretch**. In this mode, the C atom is buffeted simultaneously from each side and the motion continues indefinitely. Another mode is ν_3, the **antisymmetric stretch**, in which the two O atoms always move in the same direction and opposite to that of the C atom. Both modes are independent in the sense that, if one is excited, then it does not excite the other. They are two of the 'normal modes' of the molecule, its independent, collective vibrational displacements. The two other normal modes are the bending modes ν_2. In general, a **normal mode** is an independent, synchronous motion of atoms or groups of atoms that may be excited without leading to the excitation of any other normal mode and without involving translation or rotation of the molecule as a whole.

The four normal modes of CO_2, and the N_{vib} normal modes of polyatomics in general, are the key to the description of molecular vibrations. Each normal mode, q, behaves like an independent harmonic oscillator (if anharmonicities are neglected), so each has a series of terms

$$G_q(v) = (v + \tfrac{1}{2})\tilde{\nu}_q \qquad \tilde{\nu}_q = \frac{1}{2\pi c}\left(\frac{k_q}{m_q}\right)^{1/2} \tag{6.66}$$

where $\tilde{\nu}_q$ is the wavenumber of mode q and depends on the force constant k_q for the mode and on the effective mass m_q of the mode. The effective mass of the mode is a measure of the mass that is swung about by the vibration and in general is a complicated function of the masses of the atoms. For example, in the symmetric stretch of CO_2, the C atom is stationary, and the effective mass depends on the masses of only the O atoms. In the antisymmetric stretch and in the bends, all three atoms move, so all contribute to the effective mass. The three normal modes of H_2O are shown in Fig. 6.41: note that the predominantly bending mode (ν_2) has a lower frequency than the others, which are predominantly stretching modes. It is generally the case that the frequencies of bending motions are lower than those of stretching modes. One point that must be appreciated is that only in special cases (such as the CO_2 molecule) are the normal modes purely stretches or purely bends. In general, a normal mode is a composite motion of simultaneous stretching and bending of bonds. Another point in this connection is that heavy atoms generally move less than light atoms in normal modes.

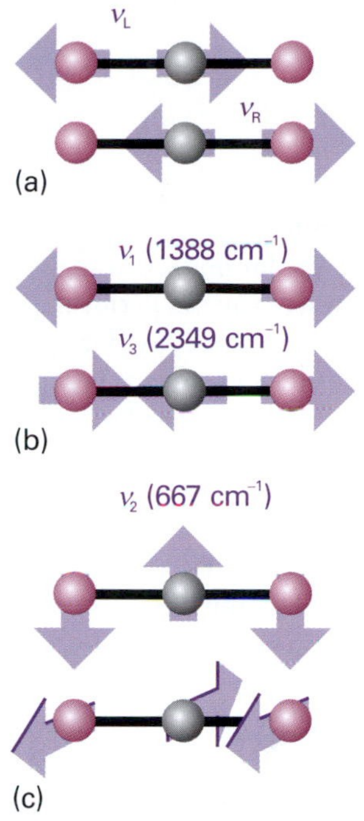

Fig. 6.40 Alternative descriptions of the vibrations of CO_2. (a) The stretching modes are not independent, and if one C—O group is excited the other begins to vibrate. They are not normal modes of vibration of the molecule. (b) The symmetric and antisymmetric stretches are independent, and one can be excited without affecting the other: they are normal modes. (c) The two perpendicular bending motions are also normal modes.

6.15 Infrared absorption spectra of polyatomic molecules

The gross selection rule for infrared activity is that *the motion corresponding to a normal mode should be accompanied by a change of dipole moment.* Deciding whether this is so can sometimes be done by inspection. For example, the symmetric stretch of CO_2 leaves the dipole moment unchanged (at zero, see Fig. 6.40), so this mode is infrared inactive. The antisymmetric stretch, however, changes the dipole moment because the molecule becomes unsymmetrical as it vibrates, so this mode is infrared active. Because the dipole moment change is parallel to the principal axis, the transitions arising from this mode are classified as **parallel bands** in the spectrum. Both bending modes are infrared active: they are accompanied by a changing dipole perpendicular to the principal axis, so transitions involving them lead to a **perpendicular band** in the spectrum. The latter bands eliminate the linearity of the molecule, and as a result a Q branch is observed; a parallel band does not have a Q branch.

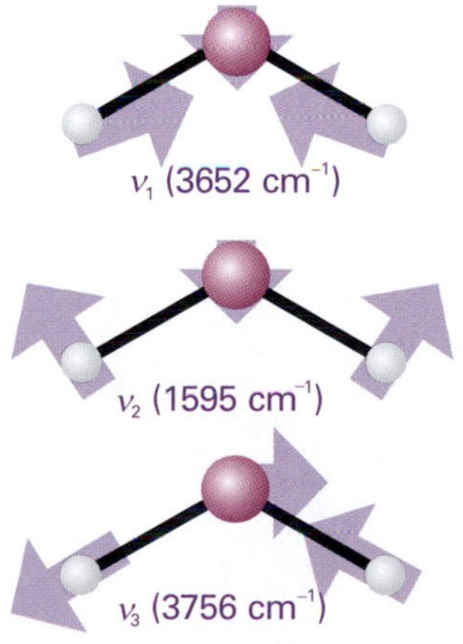

Fig. 6.41 The three normal modes of H_2O. The mode ν_2 is predominantly bending, and occurs at lower wavenumber than the other two.

The active modes are subject to the specific selection rule $\Delta v_q = \pm 1$ in the harmonic approximation, so the wavenumber of the fundamental transition (the 'first harmonic') of each active mode is $\tilde{\nu}_q$. From the analysis of the spectrum, a picture may be constructed of the stiffness of various parts of the molecule, that is, we can establish its **force field**, the set of force constants corresponding to all the displacements of the atoms. The force field may also be estimated by using the semi-empirical, *ab initio*, and DFT computational techniques described in Section 4.7. Superimposed on the simple force field scheme are the complications arising from anharmonicities and the effects of molecular rotation. Very often the sample is a liquid or a solid, and the molecules are unable to rotate freely. In a liquid, for example, a molecule may be able to rotate through only a few degrees before it is struck by another, so it changes its rotational state frequently. This random changing of orientation is called **tumbling**.

Comment 6.6
The web site for this text contains links to sites where you can perform quantum mechanical calculations of frequencies and atomic displacements of normal modes of simple molecules.

Comment 6.7
The web site for this text contains links to databases of infrared spectra.

The lifetimes of rotational states in liquids are very short, so in most cases the rotational energies are ill-defined. Collisions occur at a rate of about 10^{13} s^{-1} and, even allowing for only a 10 per cent success rate in knocking the molecule into another rotational state, a lifetime broadening (eqn 6.19) of more than 1 cm^{-1} can easily result. The rotational structure of the vibrational spectrum is blurred by this effect, so the infrared spectra of molecules in condensed phases usually consist of broad lines spanning the entire range of the resolved gas-phase spectrum, and showing no branch structure.

Synoptic table 6.3* Typical vibrational wavenumbers

Vibration type	$\tilde{\nu}$/cm^{-1}
C—H stretch	2850–2960
C—H bend	1340–1465
C—C stretch, bend	700–1250
C=C stretch	1620–1680

* More values are given in the *Data section*.

One very important application of infrared spectroscopy to condensed phase samples, and for which the blurring of the rotational structure by random collisions is a welcome simplification, is to chemical analysis. The vibrational spectra of different groups in a molecule give rise to absorptions at characteristic frequencies because a normal mode of even a very large molecule is often dominated by the motion of a small group of atoms. The intensities of the vibrational bands that can be identified with the motions of small groups are also transferable between molecules. Consequently, the molecules in a sample can often be identified by examining its infrared spectrum and referring to a table of characteristic frequencies and intensities (Table 6.3).

IMPACT ON ENVIRONMENTAL SCIENCE

I6.2 Global warming[1]

Solar energy strikes the top of the Earth's atmosphere at a rate of 343 W m^{-2}. About 30 per cent of this energy is reflected back into space by the Earth or the atmosphere. The Earth–atmosphere system absorbs the remaining energy and re-emits it into space as black-body radiation, with most of the intensity being carried by infrared radiation in the range 200–2500 cm^{-1} (4–50 μm). The Earth's average temperature is maintained by an energy balance between solar radiation absorbed by the Earth and black-body radiation emitted by the Earth.

The trapping of infrared radiation by certain gases in the atmosphere is known as the *greenhouse effect*, so called because it warms the Earth as if the planet were enclosed in a huge greenhouse. The result is that the natural greenhouse effect raises the average surface temperature well above the freezing point of water and creates an environment in which life is possible. The major constituents to the Earth's atmosphere, O_2 and N_2, do not contribute to the greenhouse effect because homonuclear diatomic molecules cannot absorb infrared radiation. However, the minor atmospheric gases, water vapour and CO_2, do absorb infrared radiation and hence are responsible for the

[1] This section is based on a similar contribution initially prepared by Loretta Jones and appearing in *Chemical principles*, Peter Atkins and Loretta Jones, W.H. Freeman and Co., New York (2005).

greenhouse effect (Fig. 6.42). Water vapour absorbs strongly in the ranges 1300–1900 cm^{-1} (5.3–7.7 μm) and 3550–3900 cm^{-1} (2.6–2.8 μm), whereas CO_2 shows strong absorption in the ranges 500–725 cm^{-1} (14–20 μm) and 2250–2400 cm^{-1} (4.2–4.4 μm).

Increases in the levels of greenhouse gases, which also include methane, dinitrogen oxide, ozone, and certain chlorofluorocarbons, as a result of human activity have the potential to enhance the natural greenhouse effect, leading to significant warming of the planet. This problem is referred to as *global warming*, which we now explore in some detail.

The concentration of water vapour in the atmosphere has remained steady over time, but concentrations of some other greenhouse gases are rising. From about the year 1000 until about 1750, the CO_2 concentration remained fairly stable, but, since then, it has increased by 28 per cent. The concentration of methane, CH_4, has more than doubled during this time and is now at its highest level for 160 000 years (160 ka; a is the SI unit denoting 1 year). Studies of air pockets in ice cores taken from Antarctica show that increases in the concentration of both atmospheric CO_2 and CH_4 over the past 160 ka correlate well with increases in the global surface temperature.

Human activities are primarily responsible for the rising concentrations of atmospheric CO_2 and CH_4. Most of the atmospheric CO_2 comes from the burning of hydrocarbon fuels, which began on a large scale with the Industrial Revolution in the middle of the nineteenth century. The additional methane comes mainly from the petroleum industry and from agriculture.

The temperature of the surface of the Earth has increased by about 0.5 K since the late nineteenth century (Fig. 6.43). If we continue to rely on hydrocarbon fuels and current trends in population growth and energy are not reversed, then by the middle of the twenty-first century, the concentration of CO_2 in the atmosphere will be about twice its value prior to the Industrial Revolution. The Intergovernmental Panel on Climate Change (IPCC) estimated in 1995 that, by the year 2100, the Earth will undergo an increase in temperature of 3 K. Furthermore, the rate of temperature change is likely to be greater than at any time in the last 10 ka. To place a temperature rise of 3 K in perspective, it is useful to consider that the average temperature of the Earth during the last ice age was only 6 K colder than at present. Just as cooling the planet (for example, during an ice age) can lead to detrimental effects on ecosystems, so too can a dramatic warming of the globe. One example of a significant change in the

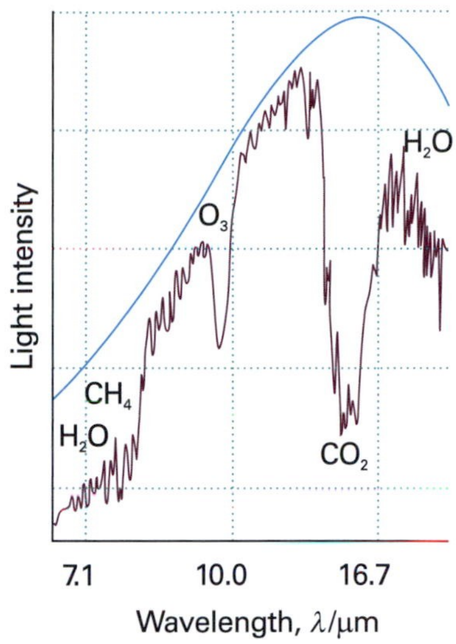

Fig. 6.42 The intensity of infrared radiation that would be lost from Earth in the absence of greenhouse gases is shown by the blue line. The purple line is the intensity of the radiation actually emitted. The maximum wavelength of radiation absorbed by each greenhouse gas is indicated.

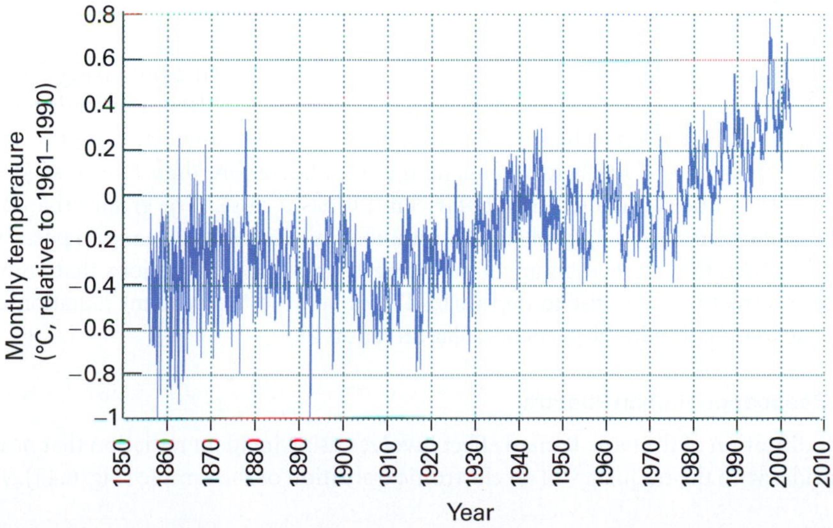

Fig. 6.43 The average change in surface temperature of the Earth from 1855 to 2002.

environment caused by a temperature increase of 3 K is a rise in sea level by about 0.5 m, which is sufficient to alter weather patterns and submerge currently coastal ecosystems.

Computer projections for the next 200 years predict further increases in atmospheric CO_2 levels and suggest that, to maintain CO_2 at its current concentration, we would have to reduce hydrocarbon fuel consumption immediately by about 50 per cent. Clearly, in order to reverse global warming trends, we need to develop alternatives to fossil fuels, such as hydrogen (which can be used in fuel cells) and solar energy technologies.

6.16 Vibrational Raman spectra of polyatomic molecules

The normal modes of vibration of molecules are Raman active if they are accompanied by a changing polarizability. It is sometimes quite difficult to judge by inspection when this is so. The symmetric stretch of CO_2, for example, alternately swells and contracts the molecule: this motion changes the polarizability of the molecule, so the mode is Raman active. The other modes of CO_2 leave the polarizability unchanged, so they are Raman inactive.

A more exact treatment of infrared and Raman activity of normal modes leads to the **exclusion rule:**

> If the molecule has a centre of symmetry, then no modes can be both infrared and Raman active.

(A mode may be inactive in both.) Because it is often possible to judge intuitively if a mode changes the molecular dipole moment, we can use this rule to identify modes that are not Raman active. The rule applies to CO_2 but to neither H_2O nor CH_4 because they have no centre of symmetry. In general, it is necessary to use group theory to predict whether a mode is infrared or Raman active (Section 6.17).

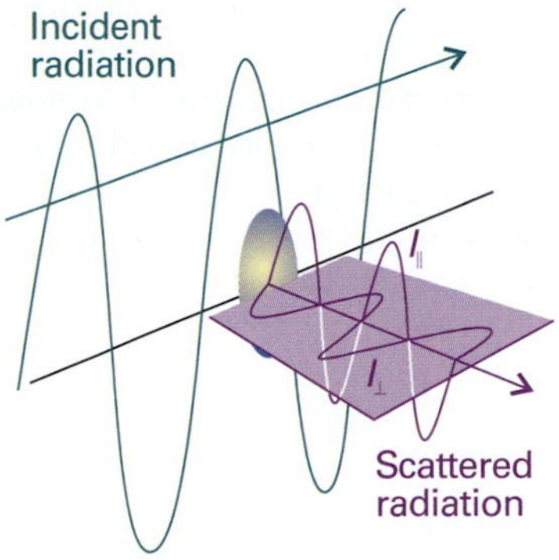

Fig. 6.44 The definition of the planes used for the specification of the depolarization ratio, ρ, in Raman scattering.

(a) Depolarization

The assignment of Raman lines to particular vibrational modes is aided by noting the state of polarization of the scattered light. The **depolarization ratio**, ρ, of a line is the ratio of the intensities, I, of the scattered light with polarizations perpendicular and parallel to the plane of polarization of the incident radiation:

$$\rho = \frac{I_{\perp}}{I_{\parallel}} \qquad [6.67]$$

To measure ρ, the intensity of a Raman line is measured with a polarizing filter (a 'half-wave plate') first parallel and then perpendicular to the polarization of the incident beam. If the emergent light is not polarized, then both intensities are the same and ρ is close to 1; if the light retains its initial polarization, then $I_{\perp} = 0$, so $\rho = 0$ (Fig. 6.44). A line is classified as **depolarized** if it has ρ close to or greater than 0.75 and as **polarized** if $\rho < 0.75$. Only totally symmetrical vibrations give rise to polarized lines in which the incident polarization is largely preserved. Vibrations that are not totally symmetrical give rise to depolarized lines because the incident radiation can give rise to radiation in the perpendicular direction too.

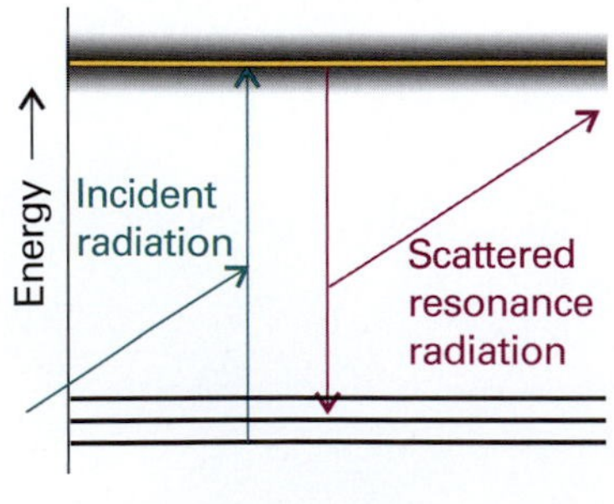

Fig. 6.45 In the *resonance Raman effect*, the incident radiation has a frequency corresponding to an actual electronic excitation of the molecule. A photon is emitted when the excited state returns to a state close to the ground state.

(b) Resonance Raman spectra

A modification of the basic Raman effect involves using incident radiation that nearly coincides with the frequency of an electronic transition of the sample (Fig. 6.45). The technique is then called **resonance Raman spectroscopy**. It is characterized by a much greater intensity in the scattered radiation. Furthermore, because it is often the case that only a few vibrational modes contribute to the more intense scattering, the spectrum is greatly simplified.

Resonance Raman spectroscopy is used to study biological molecules that absorb strongly in the ultraviolet and visible regions of the spectrum. Examples include the pigments β-carotene and chlorophyll, which capture solar energy during plant photosynthesis. The resonance Raman spectra of Fig. 6.46 show vibrational transitions from only the few pigment molecules that are bound to very large proteins dissolved in an aqueous buffer solution. This selectivity arises from the fact that water (the solvent), amino acid residues, and the peptide group do not have electronic transitions at the laser wavelengths used in the experiment, so their *conventional* Raman spectra are weak compared to the enhanced spectra of the pigments. Comparison of the spectra in Figs. 6.46a and 6.46b also shows that, with proper choice of excitation wavelength, it is possible to examine individual classes of pigments bound to the same protein: excitation at 488 nm, where β-carotene absorbs strongly, shows vibrational bands from β-carotene only, whereas excitation at 442 nm, where chlorophyll *a* and β-carotene absorb, reveals features from both types of pigments.

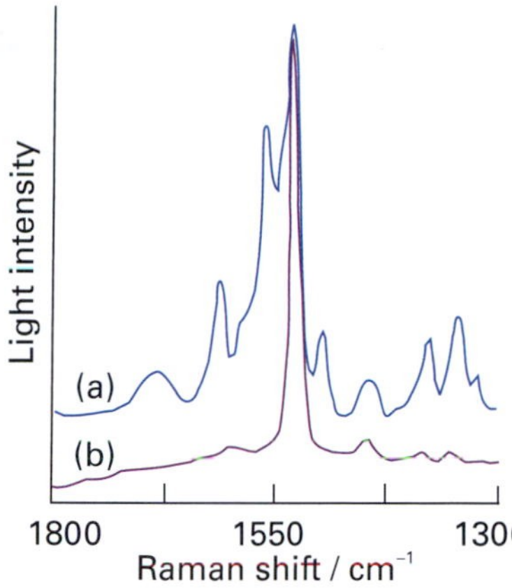

Fig. 6.46 The resonance Raman spectra of a protein complex that is responsible for some of the initial electron transfer events in plant photosynthesis. (a) Laser excitation of the sample at 407 nm shows Raman bands due to both chlorophyll *a* and β-carotene bound to the protein because both pigments absorb light at this wavelength. (b) Laser excitation at 488 nm shows Raman bands from β-carotene only because chlorophyll *a* does not absorb light very strongly at this wavelength. (Adapted from D.F. Ghanotakis *et al.*, *Biochim. Biophys. Acta* **974**, 44 (1989).)

(c) Coherent anti-Stokes Raman spectroscopy

The intensity of Raman transitions may be enhanced by **coherent anti-Stokes Raman spectroscopy** (CARS, Fig. 6.47). The technique relies on the fact that, if two laser beams of frequencies ν_1 and ν_2 pass through a sample, then they may mix together and give rise to coherent radiation of several different frequencies, one of which is

$$\nu' = 2\nu_1 - \nu_2 \qquad (6.68)$$

Suppose that ν_2 is varied until it matches any Stokes line from the sample, such as the one with frequency $\nu_1 - \Delta\nu$; then the coherent emission will have frequency

$$\nu' = 2\nu_1 - (\nu_1 - \Delta\nu) = \nu_1 + \Delta\nu \qquad (6.69)$$

which is the frequency of the corresponding anti-Stokes line. This coherent radiation forms a narrow beam of high intensity.

An advantage of CARS is that it can be used to study Raman transitions in the presence of competing incoherent background radiation, and so can be used to observe the Raman spectra of species in flames. One example is the vibration–rotation CARS spectrum of N_2 gas in a methane–air flame shown in Fig 6.48.

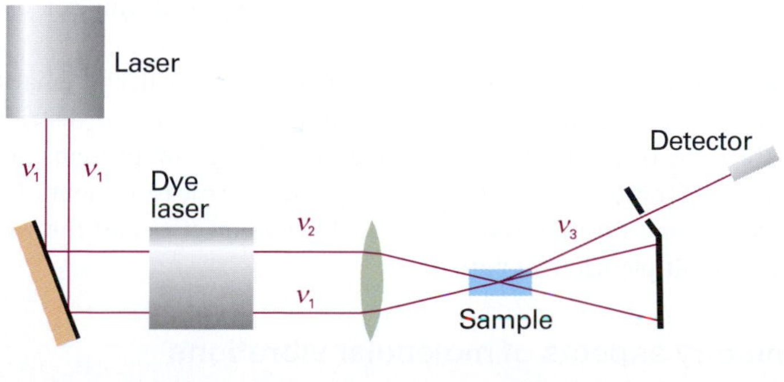

Fig. 6.47 The experimental arrangement for the CARS experiment.

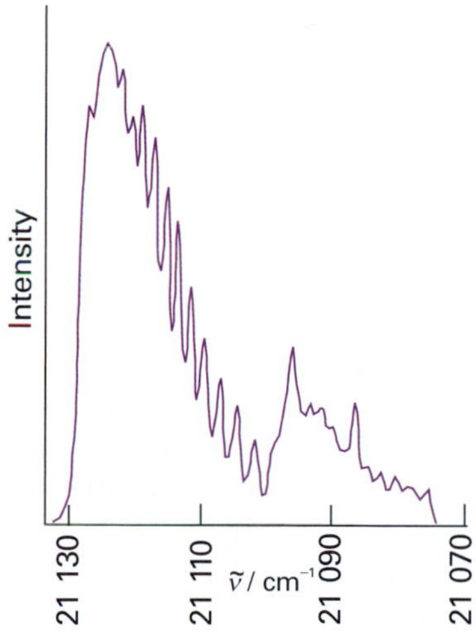

Fig. 6.48 CARS spectrum of a methane–air flame at 2104 K. The peaks correspond to the Q branch of the vibration–rotation spectrum of N_2 gas. (Adapted from J.F. Verdieck *et al.*, *J. Chem. Educ.* **59**, 495 (1982).)

IMPACT ON BIOCHEMISTRY

I6.3 Vibrational microscopy

Optical microscopes can now be combined with infrared and Raman spectrometers and the vibrational spectra of specimens as small as single biological cells obtained. The techniques of *vibrational microscopy* provide details of cellular events that cannot be observed with traditional light or electron microscopy.

The principles behind the operation of infrared and Raman microscopes are simple: radiation illuminates a small area of the sample, and the transmitted, reflected, or scattered light is first collected by a microscope and then analysed by a spectrometer. The sample is then moved by very small increments along a plane perpendicular to the direction of illumination and the process is repeated until vibrational spectra for all sections the sample are obtained. The size of a sample that can be studied by vibrational microscopy depends on a number of factors, such as the area of illumination and the excitance and wavelength of the incident radiation. Up to a point, the smaller the area that is illuminated, the smaller the area from which a spectrum can be obtained. High excitance is required to increase the rate of arrival of photons at the detector from small illuminated areas. For this reason, lasers and synchrotron radiation (see *Further information 6.1*) are the preferred radiation sources.

In a conventional light microscope, an image is constructed from a pattern of diffracted light waves that emanate from the illuminated object. As a result, some information about the specimen is lost by destructive interference of scattered light waves. Ultimately, this *diffraction limit* prevents the study of samples that are much smaller than the wavelength of light used as a probe. In practice, two objects will appear as distinct images under a microscope if the distance between their centres is greater than the *Airy radius*, $r_{\text{Airy}} = 0.61\lambda/a$, where λ is the wavelength of the incident beam of radiation and a is the numerical aperture of the *objective lens*, the lens that collects light scattered by the object. The numerical aperture of the objective lens is defined as $a = n_r \sin \alpha$, where n_r is the refractive index of the lens material (the greater the refractive index, the greater the bending of a ray of light by the lens) and the angle α is the half-angle of the widest cone of scattered light that can collected by the lens (so the lens collects light beams sweeping a cone with angle 2α). Use of the best equipment makes it possible to probe areas as small as 9 μm^2 by vibrational microscopy.

Figure 6.49 shows the infrared spectra of a single mouse cell, living and dying. Both spectra have features at 1545 cm^{-1} and 1650 cm^{-1} that are due to the peptide carbonyl groups of proteins and a feature at 1240 cm^{-1} that is due to the phosphodiester (PO_2^-) groups of lipids. The dying cell shows an additional absorption at 1730 cm^{-1}, which is due to the ester carbonyl group from an unidentified compound. From a plot of the intensities of individual absorption features as a function of position in the cell, it has been possible to map the distribution of proteins and lipids during cell division and cell death.

Vibrational microscopy has also been used in biomedical and pharmaceutical laboratories. Examples include the determination of the size and distribution of a drug in a tablet, the observation of conformational changes in proteins of cancerous cells upon administration of anti-tumour drugs, and the measurement of differences between diseased and normal tissue, such as diseased arteries and the white matter from brains of multiple sclerosis patients.

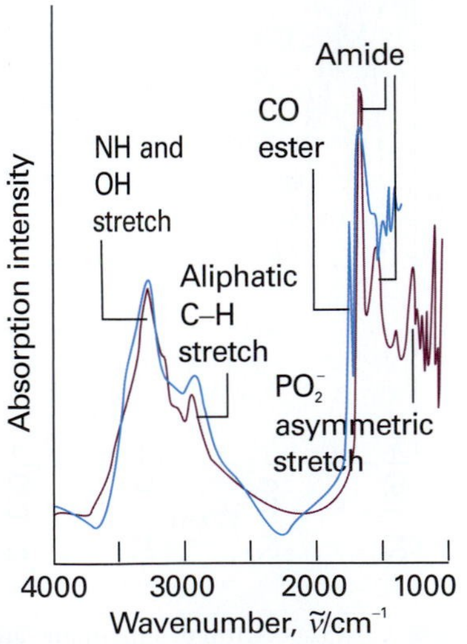

Fig. 6.49 Infrared absorption spectra of a single mouse cell: (purple) living cell, (blue) dying cell. Adapted from N. Jamin *et al.*, *Proc. Natl. Acad. Sci. USA* **95**, 4837 (1998).

6.17 Symmetry aspects of molecular vibrations

One of the most powerful ways of dealing with normal modes, especially of complex molecules, is to classify them according to their symmetries. Each normal mode must belong to one of the symmetry species of the molecular point group, as discussed in Chapter 5.

Example 6.6 *Identifying the symmetry species of a normal mode*

Establish the symmetry species of the normal mode vibrations of CH_4, which belongs to the group T_d.

Method The first step in the procedure is to identify the symmetry species of the irreducible representations spanned by all the 3N displacements of the atoms, using the characters of the molecular point group. Find these characters by counting 1 if the displacement is unchanged under a symmetry operation, −1 if it changes sign, and 0 if it is changed into some other displacement. Next, subtract the symmetry species of the translations. Translational displacements span the same symmetry species as x, y, and z, so they can be obtained from the right-most column of the character table. Finally, subtract the symmetry species of the rotations, which are also given in the character table (and denoted there by R_x, R_y, or R_z).

Answer There are $3\times5=15$ degrees of freedom, of which $(3\times5)-6=9$ are vibrations. Refer to Fig. 6.50. Under E, no displacement coordinates are changed, so the character is 15. Under C_3, no displacements are left unchanged, so the character is 0. Under the C_2 indicated, the z-displacement of the central atom is left unchanged, whereas its x- and y-components both change sign. Therefore $\chi(C_2)=1-1-1+0+0+\cdots=-1$. Under the S_4 indicated, the z-displacement of the central atom is reversed, so $\chi(S_4)=-1$. Under σ_d, the x- and z-displacements of C, H_3, and H_4 are left unchanged and the y-displacements are reversed; hence $\chi(\sigma_d)=3+3-3=3$. The characters are therefore 15, 0, −1, −1, 3. By decomposing the direct product (Section 5.5a), we find that this representation spans $A_1+E+T_1+3T_2$. The translations span T_2; the rotations span T_1. Hence, the nine vibrations span A_1+E+2T_2.

The modes themselves are shown in Fig. 6.51. We shall see that symmetry analysis gives a quick way of deciding which modes are active.

Self-test 6.8 Establish the symmetry species of the normal modes of H_2O.

$[2A_1+B_2]$

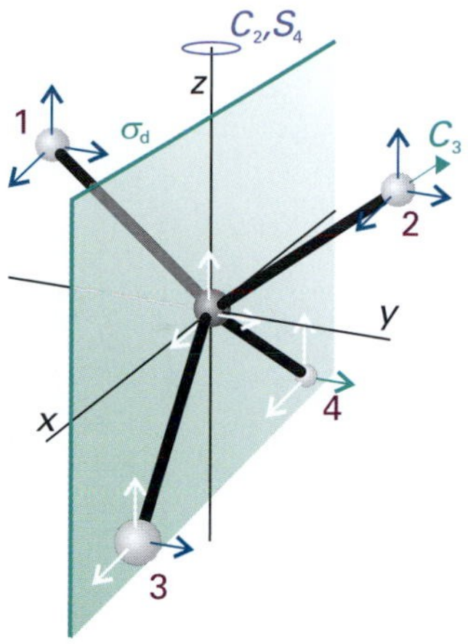

Fig. 6.50 The atomic displacements of CH_4 and the symmetry elements used to calculate the characters.

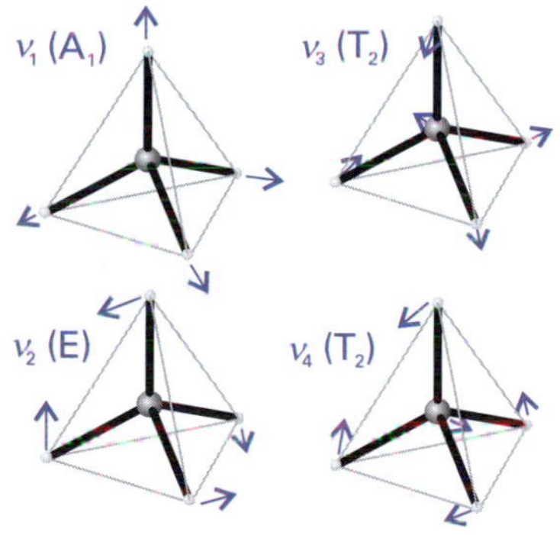

Fig. 6.51 Typical normal modes of vibration of a tetrahedral molecule. There are in fact two modes of symmetry species E and three modes of each T_2 symmetry species.

(a) Infrared activity of normal modes

It is best to use group theory to judge the activities of more complex modes of vibration. This is easily done by checking the character table of the molecular point group for the symmetry species of the irreducible representations spanned by x, y, and z, for their species are also the symmetry species of the components of the electric dipole moment. Then apply the following rule:

If the symmetry species of a normal mode is the same as any of the symmetry species of x, y, or z, then the mode is infrared active.

Justification 6.6 *Using group theory to identify infrared active normal modes*

The rule hinges on the form of the transition dipole moment between the ground-state vibrational wavefunction, ψ_0, and that of the first excited state, ψ_1. The x-component is

$$\mu_{x,10} = -e\int \psi_1^* x \psi_0 \, d\tau \tag{6.70}$$

with similar expressions for the two other components of the transition moment. The ground-state vibrational wavefunction is a Gaussian function of the form e^{-x^2}, so it is symmetrical in x. The wavefunction for the first excited state gives a non-vanishing integral only if it is proportional to x, for then the integrand is proportional to x^2 rather than to xy or xz. Consequently, the excited state wavefunction must have the same symmetry as the displacement x.

Example 6.7 *Identifying infrared active modes*

Which modes of CH_4 are infrared active?

Method Refer to the T_d character table to establish the symmetry species of x, y, and z for this molecule, and then use the rule given above.

Answer The functions x, y, and z span T_2. We found in Example 6.6 that the symmetry species of the normal modes are $A_1 + E + 2T_2$. Therefore, only the T_2 modes are infrared active. The distortions accompanying these modes lead to a changing dipole moment. The A_1 mode, which is inactive, is the symmetrical 'breathing' mode of the molecule.

Self-test 6.9 Which of the normal modes of H_2O are infrared active? [All three]

(b) Raman activity of normal modes

Group theory provides an explicit recipe for judging the Raman activity of a normal mode. In this case, the symmetry species of the quadratic forms (x^2, xy, etc.) listed in the character table are noted (they transform in the same way as the polarizability), and then we use the following rule:

> If the symmetry species of a normal mode is the same as the symmetry species of a quadratic form, then the mode is Raman active.

Illustration 6.6 *Identifying Raman active modes*

To decide which of the vibrations of CH_4 are Raman active, refer to the T_d character table. It was established in Example 6.6 that the symmetry species of the normal modes are $A_1 + E + 2T_2$. Because the quadratic forms span $A_1 + E + T_2$, all the normal modes are Raman active. By combining this information with that in Example 6.6, we see how the infrared and Raman spectra of CH_4 are assigned. The assignment of spectral features to the T_2 modes is straightforward because these are the only modes that are both infrared and Raman active. This leaves the A_1 and E modes to be assigned in the Raman spectrum. Measurement of the depolarization ratio distinguishes between these modes because the A_1 mode, being totally symmetric, is polarized and the E mode is depolarized.

Self-test 6.10 Which of the vibrational modes of H_2O are Raman active? [All three]

Checklist of key ideas

☐ 1. Emission spectroscopy is based on the detection of a transition from a state of high energy to a state of lower energy; absorption spectroscopy is based on the detection of the net absorption of nearly monochromatic incident radiation as the radiation is swept over a range of frequencies.

☐ 2. In Raman spectroscopy molecular energy levels are explored by examining the frequencies present in scattered radiation. Stokes and anti-Stokes radiation are scattered radiation at a lower and higher frequency, respectively, than the incident radiation. Rayleigh radiation is the component of radiation scattered into the forward direction without change of frequency.

☐ 3. The Beer–Lambert law is $I(\tilde{\nu}) = I_0(\tilde{\nu})10^{-\varepsilon(\tilde{\nu})[J]l}$, where $I(\tilde{\nu})$ is the transmitted intensity, $I_0(\tilde{\nu})$ is the incident intensity, and $\varepsilon(\tilde{\nu})$ is the molar absorption coefficient.

☐ 4. The transmittance, $T = I/I_0$, and the absorbance, A, of a sample at a given wavenumber are related by $A = -\log T$.

☐ 5. The integrated absorption coefficient, $\mathcal{A}$, is the sum of the absorption coefficients over the entire band, $\mathcal{A} = \int_{\text{band}} \varepsilon(\tilde{\nu})\mathrm{d}\tilde{\nu}$.

☐ 6. Stimulated absorption is the radiation-driven transition from a low energy state to one of higher energy. Stimulated emission is the radiation-driven transition from a high energy state to one of lower energy. Spontaneous emission is radiative emission independent of the intensity of the radiation (of any frequency) that is already present.

☐ 7. The natural linewidth of a spectral line is due to spontaneous emission. Spectral lines are affected by Doppler broadening, lifetime broadening, and collisional deactivation of excited states.

☐ 8. A rigid rotor is a body that does not distort under the stress of rotation. A spherical rotor is a rigid rotor with three equal moments of inertia. A symmetric rotor is a rigid rotor with two equal moments of inertia. A linear rotor is a rigid rotor with one moment of inertia equal to zero. An asymmetric rotor is a rigid rotor with three different moments of inertia.

☐ 9. The rotational terms of a spherical rotor are $F(J) = BJ(J+1)$ with $B = \hbar/4\pi cI$, $J = 0, 1, 2, \ldots$, and are $(2J+1)^2$-fold degenerate.

☐ 10. The principal axis (figure axis) is the unique axis of a symmetric top. In an oblate top, $I_\parallel > I_\perp$. In a prolate top, $I_\parallel < I_\perp$.

☐ 11. The rotational terms of a symmetric rotor are $F(J,K) = BJ(J+1) + (A-B)K^2$, $J = 0, 1, 2, \ldots$, $K = 0, \pm 1, \ldots, \pm J$.

☐ 12. The rotational terms of a linear rotor are $F(J) = BJ(J+1)$, $J = 0, 1, 2, \ldots$ and are $(2J+1)$-fold degenerate.

☐ 13. The centrifugal distortion constant, D_J, is the empirical constant in the expression $F(J) = BJ(J+1) - D_J J^2(J+1)^2$ that takes into account centrifugal distortion, $D_J \approx 4B^3/\tilde{\nu}^2$.

☐ 14. The gross rotational selection rule for microwave spectra is: for a molecule to give a pure rotational spectrum, it must be polar. The specific rotational selection rule is: $\Delta J = \pm 1$, $\Delta M_J = 0, \pm 1$, $\Delta K = 0$. The rotational wavenumbers in the absence and presence of centrifugal distortion are given by eqns 6.37 and 6.38, respectively.

☐ 15. The gross selection rule for rotational Raman spectra is: the molecule must be anisotropically polarizable. The specific selection rules are: (i) linear rotors, $\Delta J = 0, \pm 2$; (ii) symmetric rotors, $\Delta J = 0, \pm 1, \pm 2$; $\Delta K = 0$.

☐ 16. The appearance of rotational spectra is affected by nuclear statistics, the selective occupation of rotational states that stems from the Pauli principle.

☐ 17. The vibrational energy levels of a diatomic molecule modelled as a harmonic oscillator are $E_v = (v + \tfrac{1}{2})\hbar\omega$, $\omega = (k/m_{\text{eff}})^{1/2}$; the vibrational terms are $G(v) = (v + \tfrac{1}{2})\tilde{\nu}$, $\tilde{\nu} = (1/2\pi c)(k/m_{\text{eff}})^{1/2}$.

☐ 18. The gross selection rule for infrared spectra is: the electric dipole moment of the molecule must change when the atoms are displaced relative to one another. The specific selection rule is: $\Delta v = \pm 1$.

☐ 19. Morse potential energy, eqn 6.54, describes anharmonic motion, oscillatory motion in which the restoring force is not proportional to the displacement; the vibrational terms of a Morse oscillator are given by eqn 6.55.

☐ 20. A Birge–Sponer plot is a graphical procedure for determining the dissociation energy of a bond.

☐ 21. The P branch consists of vibration–rotation infrared transitions with $\Delta J = -1$; the Q branch has transitions with $\Delta J = 0$; the R branch has transitions with $\Delta J = +1$.

☐ 22. The gross selection rule for vibrational Raman spectra is: the polarizability must change as the molecule vibrates. The specific selection rule is: $\Delta v = \pm 1$.

☐ 23. A normal mode is an independent, synchronous motion of atoms or groups of atoms that may be excited without leading to the excitation of any other normal mode. The number of normal modes is $3N - 6$ (for nonlinear molecules) or $3N - 5$ (linear molecules).

☐ 24. A symmetric stretch is a symmetrically stretching vibrational mode. An antisymmetric stretch is a stretching mode, one half of which is the mirror image of the other half.

☐ 25. The exclusion rule states that, if the molecule has a centre of symmetry, then no modes can be both infrared and Raman active.

☐ 26. The depolarization ratio, ρ, the ratio of the intensities, I, of the scattered light with polarizations perpendicular and parallel to the plane of polarization of the incident radiation, $\rho = I_\perp/I_\parallel$. A depolarized line is a line with ρ close to or greater than 0.75. A polarized line is a line with $\rho < 0.75$.

☐ 27. Resonance Raman spectroscopy is a Raman technique in which the frequency of the incident radiation nearly coincides with the frequency of an electronic transition of the sample. Coherent anti-Stokes Raman spectroscopy (CARS) is a

Raman technique that relies on the use of two incident beams of radiation.

☐ 28. A normal mode is infrared active if its symmetry species is the same as any of the symmetry species of *x*, *y*, or *z*, then the mode is infrared active. A normal mode is Raman active if its symmetry species is the same as the symmetry species of a quadratic form.

Further reading

Articles and texts

L. Glasser, Fourier transforms for chemists. Part I. Introduction to the Fourier transform. *J. Chem. Educ.* **64**, A228 (1987). Part II. Fourier transforms in chemistry and spectroscopy. *J. Chem. Educ.* **64**, A260 (1987).

H.-U. Gremlich and B. Yan, *Infrared and Raman spectroscopy of biological materials.* Marcel Dekker, New York (2001).

G. Herzberg, *Molecular spectra and molecular structure I. Spectra of diatomic molecules.* Krieger, Malabar (1989).

G. Herzberg, *Molecular spectra and molecular structure II. Infrared and Raman spectra of polyatomic molecules.* Van Nostrand–Reinhold, New York (1945).

J.C. Lindon, G.E. Tranter, and J.L. Holmes (ed.), *Encyclopedia of spectroscopy and spectrometry*. Academic Press, San Diego (2000).

D.P. Strommen, Specific values of the depolarization ratio in Raman spectroscopy: Their origins and significance. *J. Chem. Educ.* **69**, 803 (1992).

E.B. Wilson, J.C. Decius, and P.C. Cross, *Molecular vibrations.* Dover, New York (1980).

J.M. Brown and A. Carrington, *Rotational spectroscopy of diatomic molecules.* Cambridge University Press (2003).

Sources of data and information

M.E. Jacox, *Vibrational and electronic energy levels of polyatomic transient molecules.* Journal of Physical and Chemical Reference Data, Monograph No. 3 (1994).

K.P. Huber and G. Herzberg, *Molecular spectra and molecular structure IV. Constants of diatomic molecules.* Van Nostrand-Reinhold, New York (1979).

B. Schrader, *Raman/IR atlas of organic compounds.* VCH, New York (1989).

G. Socrates, *Infrared and Raman characteristic group frequencies: tables and charts.* Wiley, New York (2000).

Further information

Further information 6.1 *Spectrometers*

Here we provide additional detail on the principles of operation of spectrometers, describing radiation sources, dispersing elements, detectors, and Fourier transform techniques.

Sources of radiation

Sources of radiation are either *monochromatic*, those spanning a very narrow range of frequencies around a central value, or *polychromatic*, those spanning a wide range of frequencies. Monochromatic sources that can be tuned over a range of frequencies include the *klystron* and the *Gunn diode*, which operate in the microwave range, and lasers, which are discussed in Chapter 7.

Polychromatic sources that take advantage of black-body radiation from hot materials. For far infrared radiation with 35 cm^{-1} $< \tilde{\nu} <$ 200 cm^{-1}, a typical source is a mercury arc inside a quartz envelope, most of the radiation being generated by the hot quartz. Either a *Nernst filament* or a *globar* is used as a source of mid-infrared radiation with 200 cm^{-1} $< \tilde{\nu} <$ 4000 cm^{-1}. The Nernst filament consists of a ceramic filament of lanthanoid oxides that is heated to temperatures ranging from 1200 to 2000 K. The globar consists of a rod of silicon carbide, which is heated electrically to about 1500 K.

A *quartz–tungsten–halogen lamp* consists of a tungsten filament that, when heated to about 3000 K, emits light in the range 320 nm $< \lambda <$ 2500 nm. Near the surface of the lamp's quartz envelope, iodine atoms and tungsten atoms ejected from the filament combine to make a variety of tungsten–iodine compounds that decompose at the hot filament, replenishing it with tungsten atoms.

A *gas discharge lamp* is a common source of ultraviolet and visible radiation. In a *xenon discharge lamp*, an electrical discharge excites xenon atoms to excited states, which then emit ultraviolet radiation. At pressures exceeding 1 kPa, the output consists of sharp lines on a broad, intense background due to emission from a mixture of ions formed by the electrical discharge. These high-pressure xenon lamps have emission profiles similar to that of a black body heated to 6000 K. In a *deuterium lamp*, excited D_2 molecules dissociate into electronically excited D atoms, which emit intense radiation between 200–400 nm.

For certain applications, synchrotron radiation is generated in a *synchrotron storage ring*, which consists of an electron beam travelling in a circular path with circumferences of up to several hunderd metres. As electrons travelling in a circle are constantly accelerated by the forces that constrain them to their path, they generate radiation (Fig. 6.52). Synchrotron radiation spans a wide range of frequencies,

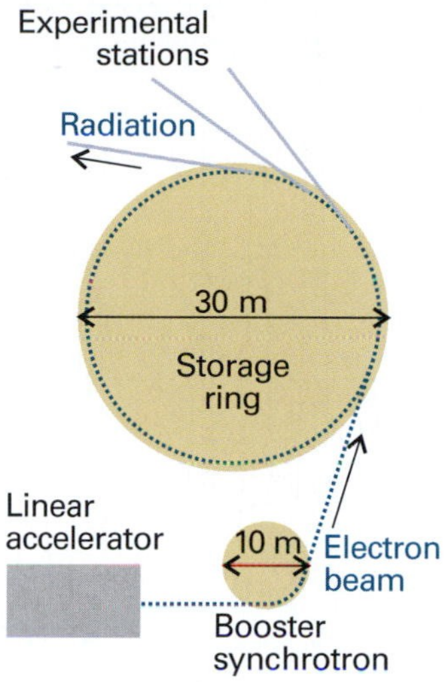

Fig. 6.52 A synchrotron storage ring. The electrons injected into the ring from the linear accelerator and booster synchrotron are accelerated to high speed in the main ring. An electron in a curved path is subject to constant acceleration, and an accelerated charge radiates electromagnetic energy.

including the infrared and X-rays. Except in the microwave region, synchrotron radiation is much more intense than can be obtained by most conventional sources.

The dispersing element

The dispersing element in most absorption spectrometers operating in the ultraviolet to near-infrared region of the spectrum is a **diffraction grating**, which consists of a glass or ceramic plate into which fine grooves have been cut and covered with a reflective aluminium coating. The grating causes interference between waves reflected from its surface, and constructive interference occurs when

$$n\lambda = d(\sin\theta - \sin\phi) \quad (6.71)$$

where $n = 1, 2, \ldots$ is the *diffraction order*, λ is the wavelength of the diffracted radiation, d is the distance between grooves, θ is the angle of incidence of the beam, and ϕ is the angle of emergence of the beam (Fig. 6.53). For given values of n and θ, larger differences in ϕ are

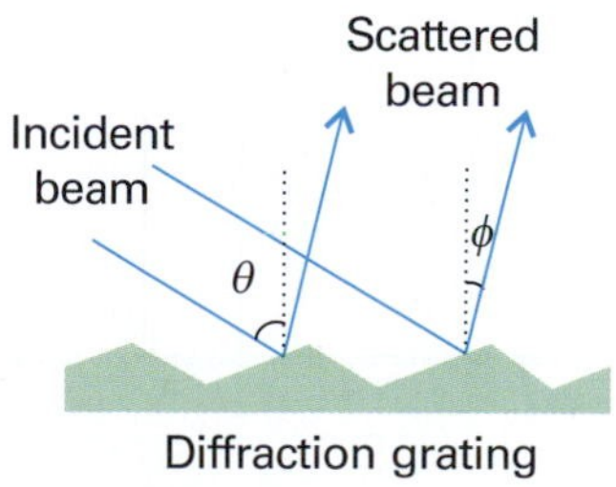

Fig. 6.53 One common dispersing element is a diffraction grating, which separates wavelengths spatially as a result of the scattering of light by fine grooves cut into a coated piece of glass. When a polychromatic light beam strikes the surface at an angle θ, several light beams of different wavelengths emerge at different angles ϕ (eqn 6.71).

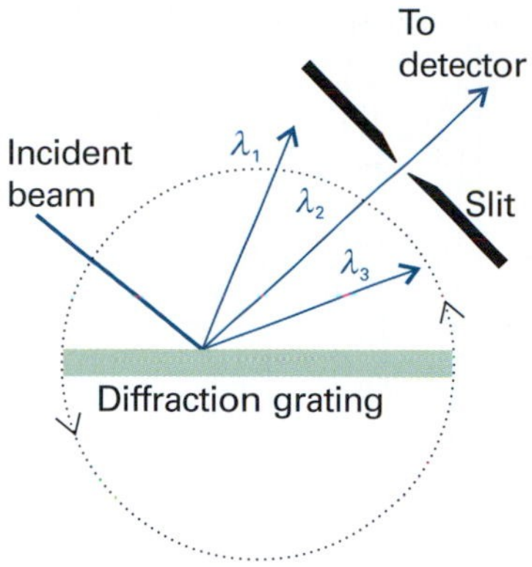

Fig. 6.54 A polychromatic beam is dispersed by a diffraction grating into three component wavelengths λ_1, λ_2, and λ_3. In the configuration shown, only radiation with λ_2 passes through a narrow slit and reaches the detector. Rotating the diffraction grating as shown by the double arrows allows λ_1 or λ_3 to reach the detector.

observed for different wavelengths when d is similar to the wavelength of radiation being analysed. Wide angular separation results in wide spatial separation between wavelengths some distance away from the grating, where a detector is placed.

In a **monochromator**, a narrow exit slit allows only a narrow range of wavelengths to reach the detector (Fig. 6.54). Turning the grating around an axis perpendicular to the incident and diffracted beams allows different wavelengths to be analysed; in this way, the absorption spectrum is built up one narrow wavelength range at a time. Typically, the grating is swept through an angle that investigates only the first order of diffraction ($n = 1$). In a **polychromator**, there is no slit and a broad range of wavelengths can be analysed simultaneously by *array detectors*, such as those discussed below.

Fourier transform techniques

In a Fourier transform instrument, the diffraction grating is replaced by a Michelson interferometer, which works by splitting the beam from the sample into two and introducing a varying path difference, p, into one of them (Fig. 6.55). When the two components

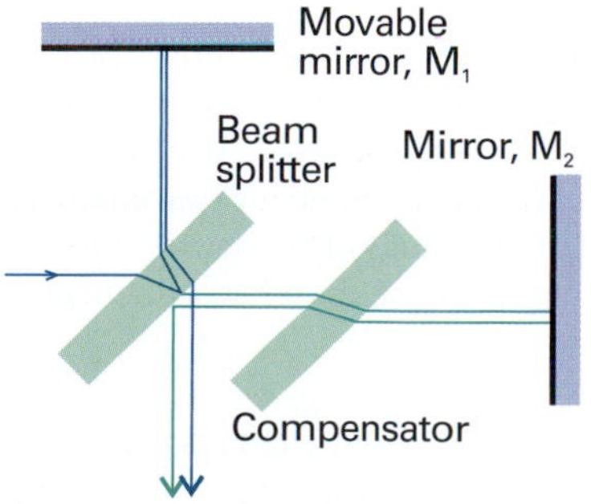

Fig. 6.55 A Michelson interferometer. The beam-splitting element divides the incident beam into two beams with a path difference that depends on the location of the mirror M_1. The compensator ensures that both beams pass through the same thickness of material.

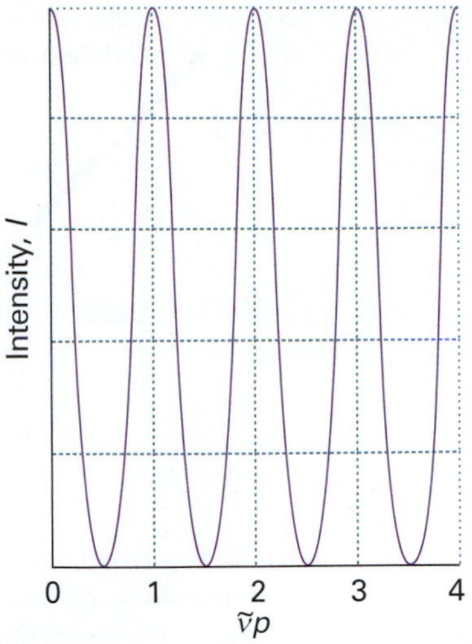

Fig. 6.56 An interferogram produced as the path length p is changed in the interferometer shown in Fig. 6.55. Only a single frequency component is present in the signal, so the graph is a plot of the function $I(p) = I_0(1 + \cos 2\pi\tilde{\nu}p)$, where I_0 is the intensity of the radiation.

Exploration Referring to Fig. 6.55, the mirror M_1 moves in finite distance increments, so the path difference p is also incremented in finite steps. Explore the effect of increasing the step size on the shape of the interferogram for a monochromatic beam of wavenumber $\tilde{\nu}$ and intensity I_0. That is, draw plots of $I(p)/I_0$ against $\tilde{\nu}p$, each with a different number of data points spanning the same total distance path taken by the movable mirror M_1.

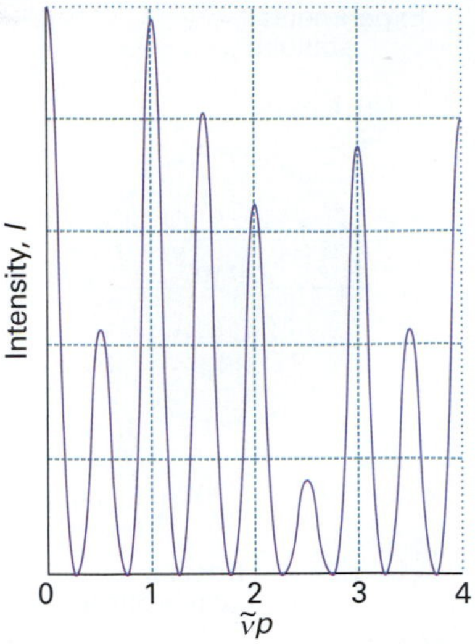

Fig. 6.57 An interferogram obtained when several (in this case, three) frequencies are present in the radiation.

Exploration For a signal consisting of only a few monochromatic beams, the integral in eqn 6.73 can be replaced by a sum over the finite number of wavenumbers. Use this information to draw your own version of Fig. 6.57. Then, go on to explore the effect of varying the wavenumbers and intensities of the three components of the radiation on the shape of the interferogram.

recombine, there is a phase difference between them, and they interfere either constructively or destructively depending on the difference in path lengths. The detected signal oscillates as the two components alternately come into and out of phase as the path difference is changed (Fig. 6.56). If the radiation has wavenumber $\tilde{\nu}$, the intensity of the detected signal due to radiation in the range of wavenumbers $\tilde{\nu}$ to $\tilde{\nu} + d\tilde{\nu}$, which we denote $I(p,\tilde{\nu})d\tilde{\nu}$, varies with p as

$$I(p,\tilde{\nu})d\tilde{\nu} = I(\tilde{\nu})(1 + \cos 2\pi\tilde{\nu}p)d\tilde{\nu} \qquad (6.72)$$

Hence, the interferometer converts the presence of a particular wavenumber component in the signal into a variation in intensity of the radiation reaching the detector. An actual signal consists of radiation spanning a large number of wavenumbers, and the total intensity at the detector, which we write $I(p)$, is the sum of contributions from all the wavenumbers present in the signal (Fig. 6.57):

$$I(p) = \int_0^\infty I(p,\tilde{\nu})d\tilde{\nu} = \int_0^\infty I(\tilde{\nu})(1 + \cos 2\pi\tilde{\nu}p)d\tilde{\nu} \qquad (6.73)$$

The problem is to find $I(\tilde{\nu})$, the variation of intensity with wavenumber, which is the spectrum we require, from the record of values of $I(p)$. This step is a standard technique of mathematics, and is the 'Fourier transformation' step from which this form of spectroscopy takes its name. Specifically:

$$I(\tilde{\nu}) = 4\int_0^\infty \{I(p) - \tfrac{1}{2}I(0)\} \cos 2\pi\tilde{\nu}p \, dp \qquad (6.74)$$

where $I(0)$ is given by eqn 6.73 with $p = 0$. This integration is carried out numerically in a computer connected to the spectrometer, and the output, $I(\tilde{\nu})$, is the transmission spectrum of the sample (Fig. 6.58).

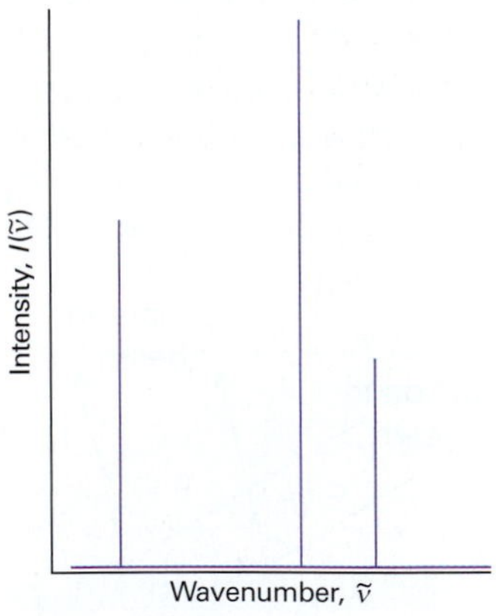

Fig. 6.58 The three frequency components and their intensities that account for the appearance of the interferogram in Fig. 6.57. This spectrum is the Fourier transform of the interferogram, and is a depiction of the contributing frequencies.

Exploration Calculate the Fourier transforms of the functions you generated in the previous *Exploration*.

A major advantage of the Fourier transform procedure is that all the radiation emitted by the source is monitored continuously. This is in contrast to a spectrometer in which a monochromator discards most of the generated radiation. As a result, Fourier transform spectrometers have a higher sensitivity than conventional spectrometers.

Detectors

A **detector** is a device that converts radiation into an electric current or voltage for appropriate signal processing and display. Detectors may consist of a single radiation sensing element or of several small elements arranged in one or two-dimensional arrays.

A microwave detector is typically a *crystal diode* consisting of a tungsten tip in contact with a semiconductor. The most common detectors found in commercial infrared spectrometers are sensitive in the mid-infrared region. In a *photovoltaic device* the potential difference changes upon exposure to infrared radiation. In a *pyroelectric device* the capacitance is sensitive to temperature and hence the presence of infrared radiation.

A common detector for work in the ultraviolet and visible ranges is the *photomultiplier tube* (PMT), in which the photoelectric effect (Section 1.2a) is used to generate an electrical signal proportional to the intensity of light that strikes the detector. In a PMT, photons first strike the *photocathode*, usually a metallic surface to which a large negative potential is applied. Each electron ejected from the photocathode is accelerated by a potential difference to another metallic surface, called the *dynode*, from which more electrons are ejected. After travelling through a chain of dynodes, with each dynode at a more positive potential than the preceding dynode in the chain, all the ejected electrons are collected at a final metallic surface called the *anode*. Depending on how the detector is constructed, a PMT can produce up to 10^8 electrons per photon that strikes the photocathode. This sensitivity is important for the detection of light from weak sources, but can pose problems as well. At room temperature, a small number of electrons on the surfaces of the photocathode and dynodes have sufficient energy to be ejected even in the dark. When amplified through the dynode chain, these electrons give rise to a *dark current*, which interferes with measurements on the sample of interest. To minimize the dark current, it is common to lower the temperature of the PMT detector.

A common, but less sensitive, alternative to the PMT is the *photodiode*, a solid-state device that conducts electricity when struck by photons because light-induced electron transfer reactions in the detector material create mobile charge carriers (negatively charged electrons and positively charged 'holes'). In an *avalanche photodiode*, the photo-generated electrons are accelerated through a very large electrical potential difference. The high-energy electrons then collide with other atoms in the solid and ionize them, thus creating an avalanche of secondary charge carriers and increasing the sensitivity of the device toward photons.

The *charge-coupled device* (CCD) is a two-dimensional array of several million small photodiode detectors. With a CCD, a wide range of wavelengths that emerge from a polychromator are detected simultaneously, thus eliminating the need to measure light intensity one narrow wavelength range at a time. CCD detectors are the imaging devices in digital cameras, but are also used widely in spectroscopy to measure absorption, emission, and Raman scattering. In Raman microscopy, a CCD detector can be used in a variation of the technique known as *Raman imaging*: a special optical filter allows only one Stokes line to reach the two-dimensional detector, which then contains a map of the distribution of the intensity of that line in the illuminated area.

Resolution

A number of factors determine a spectrometer's resolution, the smallest observable separation between two closely spaced spectral bands. We have already seen that the ability of a diffraction grating to disperse light depends on the distance between the grating's grooves and on the wavelength of the incident radiation. Furthermore, the distance between the grating and the slit placed in front of a detector must be long enough and the slit's width must be narrow enough so full advantage can be taken of the grating's dispersing ability (Fig. 6.54). It follows that a bad choice of grating, slit width, and detector placement may result in the failure to distinguish between closely spaced lines or to measure the actual linewidth of any one band in the spectrum.

The resolution of Fourier transform spectrometers is determined by the maximum path length difference, p_{max}, of the interferometer:

$$\Delta\tilde{\nu} = \frac{1}{2p_{max}} \tag{6.75}$$

To achieve a resolution of 0.1 cm^{-1} requires a maximum path length difference of 5 cm.

Assuming that all instrumental factors have been optimized, the highest resolution is obtained when the sample is gaseous and at such low pressure that collisions between the molecules are infrequent (see Section 6.3b). In liquids and solids, the actual linewidths can be so broad that the sample itself can limit the resolution.

Further information 6.2 *Selection rules for rotational and vibrational spectroscopy*

Here we derive the gross and specific selection rules for microwave, infrared, and rotational and vibrational Raman spectroscopy. The starting point for our discussion is the total wavefunction for a molecule, which can be written as

$$\psi_{total} = \psi_{c.m.}\psi$$

where $\psi_{c.m.}$ describes the motion of the centre of mass and ψ describes the internal motion of the molecule. If we neglect the effect of electron spin, the Born–Oppenheimer approximation allows us to write ψ as the product of an electronic part, $|\varepsilon\rangle$, a vibrational part, $|v\rangle$, and a rotational part, which for a diatomic molecule can be represented by the spherical harmonics $Y_{J,M_J}(\theta,\phi)$ (Section 2.7). To simplify the form of the integrals that will soon follow, we are using the Dirac bracket notation introduced in *Further information 2.1*. The transition dipole moment for a spectroscopic transition can now be written as:

$$\boldsymbol{\mu}_{fi} = \langle \varepsilon_f v_f Y_{J,f,M_J,f} | \hat{\boldsymbol{\mu}} | \varepsilon_i v_i Y_{J,i,M_J,i} \rangle \tag{6.76}$$

and our task is to explore conditions for which this integral vanishes or has a non-zero value.

Microwave spectra

During a pure rotational transition the molecule does not change electronic or vibrational states, so that $\langle \varepsilon_f v_f| = \langle \varepsilon_i v_i| = \langle \varepsilon v|$ and we identify $\boldsymbol{\mu}_{\varepsilon v} = \langle \varepsilon v|\hat{\boldsymbol{\mu}}|\varepsilon v\rangle$ with the *permanent* electric dipole moment of the molecule in the state εv. Equation 6.76 becomes

$$\boldsymbol{\mu}_{fi} = \langle Y_{J,f,M_{J,f}}|\hat{\boldsymbol{\mu}}_{\varepsilon v}|Y_{J,i,M_{J,i}}\rangle$$

The electric dipole moment has components $\mu_{\varepsilon v,x}$, $\mu_{\varepsilon v,y}$, and $\mu_{\varepsilon v,z}$, which, in spherical polar coordinates, are written in terms of μ_0, the magnitude of the vector, and the angles θ and ϕ as

$$\mu_{\varepsilon v,x} = \mu_0 \sin\theta\cos\phi \qquad \mu_{\varepsilon v,y} = \mu_0 \sin\theta \sin\phi \qquad \mu_{\varepsilon v,z} = \mu_0 \cos\theta$$

Here, we have taken the z-axis to be coincident with the figure axis. The transition dipole moment has three components, given by:

$$\mu_{fi,x} = \mu_0\langle Y_{J,f,M_{J,f}}|\sin\theta\cos\phi|Y_{J,i,M_{J,i}}\rangle$$
$$\mu_{fi,y} = \mu_0\langle Y_{J,f,M_{J,f}}|\sin\theta\sin\phi|Y_{J,i,M_{J,i}}\rangle$$
$$\mu_{fi,z} = \mu_0\langle Y_{J,f,M_{J,f}}|\cos\theta|Y_{J,i,M_{J,i}}\rangle$$

We see immediately that the molecule must have a permanent dipole moment in order to have a microwave spectrum. This is the gross selection rule for microwave spectroscopy.

For the specific selection rules we need to examine the conditions for which the integrals do not vanish, and we must consider each component. For the z-component, we simplify the integral by using $\cos\theta \propto Y_{1,0}$ (Table 2.3). It follows that

$$\mu_{fi,z} \propto \langle Y_{J,f,M_{J,f}}|Y_{1,0}|Y_{J,i,M_{J,i}}\rangle$$

According to the properties of the spherical harmonics (*Comment* 6.8), this integral vanishes unless $J_f - J_i = \pm 1$ and $M_{J,f} - M_{J,i} = 0$. These are two of the selection rules stated in eqn 13.35.

Comment 6.8

An important 'triple integral' involving the spherical harmonics is

$$\int_0^{\pi}\int_0^{2\pi} Y_{l'',m_l''}(\theta,\phi)^* Y_{l',m_l'}(\theta,\phi) Y_{l,m_l}(\theta,\phi)\sin\theta\,d\theta\,d\phi = 0$$

unless $m_l'' = m_l' + m_l$ and lines of length l'', l', and l can form a triangle.

For the x- and y-components, we use $\cos\phi = \frac{1}{2}(e^{i\phi} + e^{-i\phi})$ and $\sin\phi = -\frac{1}{2}i(e^{i\phi} - e^{-i\phi})$ to write $\sin\theta\cos\phi \propto Y_{1,1} + Y_{1,-1}$ and $\sin\theta\sin\phi \propto Y_{1,1} - Y_{1,-1}$. It follows that

$$\mu_{fi,x} \propto \langle Y_{J,f,M_{J,f}}|(Y_{1,1} + Y_{1,-1})|Y_{J,i,M_{J,i}}\rangle$$
$$\mu_{fi,y} \propto \langle Y_{J,f,M_{J,f}}|(Y_{1,1} - Y_{1,-1})|Y_{J,i,M_{J,i}}\rangle$$

According to the properties of the spherical harmonics, these integrals vanish unless $J_f - J_i = \pm 1$ and $M_{J,f} - M_{J,i} = \pm 1$. This completes the selection rules of eqn 6.35.

Rotational Raman spectra

We understand the origin of the gross and specific selection rules for rotational Raman spectroscopy by using a diatomic molecule as an example. The incident electric field of a wave of electromagnetic radiation of frequency ω_i induces a molecular dipole moment that is given by

$$\mu_{ind} = \alpha\mathcal{E}(t) = \alpha\mathcal{E}\cos\omega_i t$$

If the molecule is rotating at a circular frequency ω_R, to an external observer its polarizability is also time dependent (if it is anisotropic), and we can write

$$\alpha = \alpha_0 + \Delta\alpha\cos 2\omega_R t$$

where $\Delta\alpha = \alpha_{\parallel} - \alpha_{\perp}$ and α ranges from $\alpha_0 + \Delta\alpha$ to $\alpha_0 - \Delta\alpha$ as the molecule rotates. The 2 appears because the polarizability returns to its initial value twice each revolution (Fig. 6.59). Substituting this expression into the expression for the induced dipole moment gives

$$\begin{aligned}\mu_{ind} &= (\alpha_0 + \Delta\alpha\cos 2\omega_R t)\times(\mathcal{E}\cos\omega_i t)\\ &= \alpha_0\mathcal{E}\cos\omega_i t + \mathcal{E}\Delta\alpha\cos 2\omega_R t\cos\omega_i t\\ &= \alpha_0\mathcal{E}\cos\omega_i t + \tfrac{1}{2}\mathcal{E}\Delta\alpha\{\cos(\omega_i + 2\omega_R)t + \cos(\omega_i - 2\omega_R)t\}\end{aligned}$$

This calculation shows that the induced dipole has a component oscillating at the incident frequency (which generates Rayleigh radiation), and that it also has two components at $\omega_i \pm 2\omega_R$, which give rise to the shifted Raman lines. These lines appear only if $\Delta\alpha \neq 0$; hence the polarizability must be anisotropic for there to be Raman lines. This is the gross selection rule for rotational Raman spectroscopy. We also see that the distortion induced in the molecule by the incident electric field returns to its initial value after a rotation of 180° (that is, twice a revolution). This is the origin of the specific selection rule $\Delta J = \pm 2$.

We now use a quantum mechanical formalism to understand the selection rules. First, we write the x-, y-, and z-components of the induced dipole moment as

$$\mu_{ind,x} = \mu_x\sin\theta\cos\phi \qquad \mu_{ind,y} = \mu_y\sin\theta\sin\phi \qquad \mu_{ind,z} = \mu_z\cos\theta$$

where μ_x, μ_y, and μ_z are the components of the electric dipole moment of the molecule and the z-axis is coincident with the molecular figure axis. The incident electric field also has components along the x-, y-, and z-axes:

$$\mathcal{E}_x = \mathcal{E}\sin\theta\cos\phi \qquad \mathcal{E}_y = \mathcal{E}\sin\theta\sin\phi \qquad \mathcal{E}_z = \mathcal{E}\cos\theta$$

Using eqn 6.40 and the preceding equations, it follows that

$$\begin{aligned}\mu_{ind} &= \alpha_{\perp}\mathcal{E}_x\sin\theta\cos\phi + \alpha_{\perp}\mathcal{E}_y\sin\theta\sin\phi + \alpha_{\parallel}\mathcal{E}\cos\theta\\ &= \alpha_{\perp}\mathcal{E}\sin^2\theta + \alpha_{\parallel}\mathcal{E}\cos^2\theta\end{aligned}$$

By using the spherical harmonic $Y_{2,0}(\theta,\phi) = (5/16\pi)^{1/2}(3\cos^2\theta - 1)$ and the relation $\sin^2\theta = 1 - \cos^2\theta$, it follows that:

$$\mu_{ind} = \left\{\tfrac{1}{3}\alpha_{\parallel} + \tfrac{2}{3}\alpha_{\perp} + \tfrac{4}{3}\left(\frac{\pi}{5}\right)^{1/2}\Delta\alpha Y_{2,0}(\theta,\phi)\right\}\mathcal{E}$$

For a transition between two rotational states, we calculate the integral $\langle Y_{J,f,M_{J,f}}|\mu_{ind}|Y_{J,i,M_{J,i}}\rangle$, which has two components:

$$(\tfrac{1}{3}\alpha_{\parallel} + \tfrac{2}{3}\alpha_{\perp})\langle Y_{J_f,M_{J,f}}|(Y_{J_i,M_{J,i}}\rangle \quad\text{and}\quad \mathcal{E}\Delta\alpha\langle Y_{J_f,M_{J,f}}|Y_{2,0}Y_{J_i,M_{J,i}}\rangle$$

According to the properties of the spherical harmonics (Table 2.3), the first integral vanishes unless $J_f - J_i = 0$ and the second integral vanishes unless $J_f - J_i = \pm 2$ and $\Delta\alpha \neq 0$. These are the gross and specific selection rules for linear rotors.

Infrared spectra

The gross selection rule for infrared spectroscopy is based on an analysis of the transition dipole moment $\langle v_f|\hat{\boldsymbol{\mu}}|v_i\rangle$, which arises from eqn 6.76 when the molecule does not change electronic or rotational

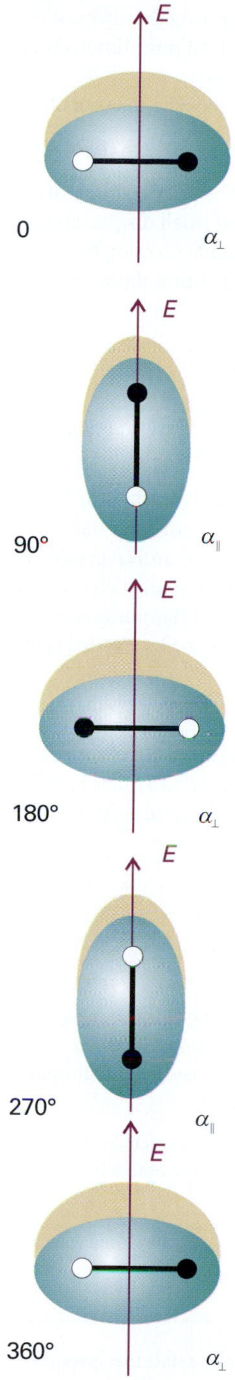

Fig. 6.59 The distortion induced in a molecule by an applied electric field returns to its initial value after a rotation of only 180° (that is, twice a revolution). This is the origin of the $\Delta J = \pm 2$ selection rule in rotational Raman spectroscopy.

states. For simplicity, we shall consider a one-dimensional oscillator (like a diatomic molecule). The electric dipole moment operator depends on the location of all the electrons and all the nuclei in the molecule, so it varies as the internuclear separation changes

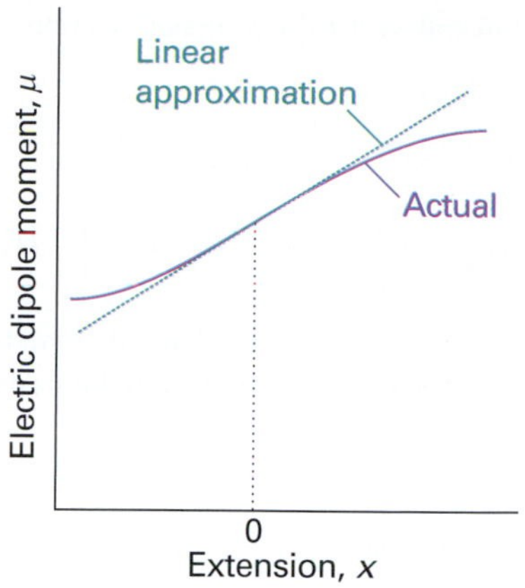

Fig. 6.60 The electric dipole moment of a heteronuclear diatomic molecule varies as shown by the purple curve. For small displacements the change in dipole moment is proportional to the displacement.

(Fig. 6.60). If we think of the dipole moment as arising from two partial charges $\pm\delta q$ separated by a distance $R = R_e + x$, we can write its variation with displacement from the equilibrium separation, x, as

$$\mu = R\delta q = R_e\delta q + x\delta q = \mu_0 + x\delta q$$

where μ_0 is the electric dipole moment operator when the nuclei have their equilibrium separation. It then follows that, with $\mathrm{f} \neq \mathrm{i}$,

$$\langle v_\mathrm{f}|\hat{\mu}|v_\mathrm{i}\rangle = \mu_0\langle v_\mathrm{f}|v_\mathrm{i}\rangle + \delta q\langle v_\mathrm{f}|x|v_\mathrm{i}\rangle$$

The term proportional to μ_0 is zero because the states with different values of v are orthogonal. It follows that the transition dipole moment is

$$\langle v_\mathrm{f}|\hat{\mu}|v_\mathrm{i}\rangle = \langle v_\mathrm{f}|x|v_\mathrm{i}\rangle\delta q$$

Because

$$\delta q = \frac{\mathrm{d}\mu}{\mathrm{d}x}$$

we can write the transition dipole moment more generally as

$$\langle v_\mathrm{f}|\hat{\mu}|v_\mathrm{i}\rangle = \langle v_\mathrm{f}|x|v_\mathrm{i}\rangle\left(\frac{\mathrm{d}\mu}{\mathrm{d}x}\right)$$

and we see that the right-hand side is zero unless the dipole moment varies with displacement. This is the gross selection rule for infrared spectroscopy.

The specific selection rule is determined by considering the value of $\langle v_\mathrm{f}|x|v_\mathrm{i}\rangle$. We need to write out the wavefunctions in terms of the Hermite polynomials given in Section 2.5 and then to use their properties (Example 2.4 should be reviewed, for it gives further details of the calculation). We note that $x = \alpha y$ with $\alpha = (\hbar^2/m_{\mathrm{eff}}k)^{1/4}$ (eqn 2.28; note that in this context α is not the polarizability). Then we write

$$\langle v_\mathrm{f}|x|v_\mathrm{i}\rangle = N_{v_\mathrm{f}}N_{v_\mathrm{i}}\int_{-\infty}^{\infty} H_{v_\mathrm{f}}xH_{v_\mathrm{i}}\mathrm{e}^{-y^2}\mathrm{d}x = \alpha^2 N_{v_\mathrm{f}}N_{v_\mathrm{i}}\int_{-\infty}^{\infty} H_{v_\mathrm{f}}yH_{v_\mathrm{i}}\mathrm{e}^{-y^2}\mathrm{d}y$$

To evaluate the integral we use the recursion relation

$$yH_v = vH_{v-1} + \tfrac{1}{2}H_{v+1}$$

which turns the matrix element into

$$\langle v_f|x|v_i\rangle = \alpha^2 N_{v_f}N_{v_i}\left\{v_i\int_{-\infty}^{\infty} H_{v_f}H_{v_i-1}e^{-y^2}dy + \tfrac{1}{2}\int_{-\infty}^{\infty} H_{v_f}H_{v_i+1}e^{-y^2}dy\right\}$$

The first integral is zero unless $v_f = v_i - 1$ and that the second is zero unless $v_f = v_i + 1$. It follows that the transition dipole moment is zero unless $\Delta v = \pm 1$.

Comment 6.9

An important integral involving Hermite polynomials is

$$\int_{-\infty}^{\infty} H_{v'}H_v e^{-y^2}dy = \begin{cases} 0 & \text{if } v' \neq v \\ \pi^{1/2}2^v v! & \text{if } v' = v \end{cases}$$

Vibrational Raman spectra

The gross selection rule for vibrational Raman spectroscopy is based on an analysis of the transition dipole moment $\langle \varepsilon v_f|\hat{\mu}|\varepsilon v_i\rangle$, which is written from eqn 6.76 by using the Born–Oppenheimer approximation and neglecting the effect of rotation and electron spin. For simplicity, we consider a one-dimensional harmonic oscillator (like a diatomic molecule).

First, we use eqn 6.40 to write the transition dipole moment as

$$\mu_{fi} = \langle \varepsilon v_f|\hat{\mu}|\varepsilon v_i\rangle = \langle \varepsilon v_f|\alpha|\varepsilon v_i\rangle\mathcal{E} = \langle v_f|\alpha(x)|v_i\rangle\mathcal{E}$$

where $\alpha(x) = \langle \varepsilon|\alpha|\varepsilon\rangle$ is the polarizability of the molecule, which we expect to be a function of small displacements x from the equilibrium bond length of the molecule (Section 6.13). Next, we expand $\alpha(x)$ as a Taylor series, so the transition dipole moment becomes

$$\mu_{fi} = \left\langle v_f \middle| \alpha(0) + \left(\frac{d\alpha}{dx}\right)_0 x + \cdots \middle| v_i\right\rangle\mathcal{E}$$

$$= \langle v_f|v_i\rangle\alpha(0)\mathcal{E} + \left(\frac{d\alpha}{dx}\right)_0 \langle v_f|x|v_i\rangle\mathcal{E} + \cdots$$

The term containing $\langle v_f|v_i\rangle$ vanishes for $f \neq i$ because the harmonic oscillator wavefunctions are orthogonal. Therefore, the vibration is Raman active if $(d\alpha/dx)_0 \neq 0$ and $\langle v_f|x|v_i\rangle \neq 0$. Therefore, the polarizability of the molecule must change during the vibration; this is the gross selection rule of Raman spectroscopy. Also, we already know that $\langle v_f|x|v_i\rangle \neq 0$ if $v_f - v_i = \pm 1$; this is the specific selection rule of Raman spectroscopy.

Each end-of-chapter discussion question, exercise, and problem for which a solution is available in the Student Solution Manual has a corresponding bracketed number referring to the Student Solutions Manual number for this problem, for example [SSM 13.1]. Each end-of-chapter exercise and problem for which a solution is available in the Instructors' Solution Manual has a corresponding bracketed number referring to the Instructor Solutions Manual number for this problem, for example [ISM 13.2].

Discussion questions

6.1 Describe the physical origins of linewidths in the absorption and emission spectra of gases, liquids, and solids. [SSM **13.1**]

6.2 Discuss the physical origins of the gross selection rules for microwave and infrared spectroscopy. [ISM **13.2**]

6.3 Discuss the physical origins of the gross selection rules for rotational and vibrational Raman spectroscopy. [SSM **13.3**]

6.4 Consider a diatomic molecule that is highly susceptible to centrifugal distortion in its ground vibrational state. Do you expect excitation to high rotational energy levels to change the equilibrium bond length of this molecule? Justify your answer. [ISM **13.4**]

6.5 Suppose that you wish to characterize the normal modes of benzene in the gas phase. Why is it important to obtain both infrared absorption and Raman spectra of your sample? [SSM **13.5**]

Exercises

6.1a Calculate the ratio of the Einstein coefficients of spontaneous and stimulated emission, A and B, for transitions with the following characteristics: (a) 70.8 pm X-rays, (b) 500 nm visible light, (c) 3000 cm^{-1} infrared radiation. [ISM **13.1a**]

6.1b Calculate the ratio of the Einstein coefficients of spontaneous and stimulated emission, A and B, for transitions with the following characteristics: (a) 500 MHz radiofrequency radiation, (e) 3.0 cm microwave radiation. [SSM **13.1b**]

6.2a What is the Doppler-shifted wavelength of a red (660 nm) traffic light approached at 80 km h^{-1}? [ISM **13.2a**]

6.2b At what speed of approach would a red (660 nm) traffic light appear green (520 nm)? [SSM **13.2b**]

6.3a Estimate the lifetime of a state that gives rise to a line of width (a) 0.10 cm^{-1}, (b) 1.0 cm^{-1}. [ISM **13.3a**]

6.3b Estimate the lifetime of a state that gives rise to a line of width (a) 100 MHZ, (b) 2.14 cm^{-1}. [SSM **13.3b**]

6.4a A molecule in a liquid undergoes about 1.0×10^{13} collisions in each second. Suppose that (a) every collision is effective in deactivating the molecule vibrationally and (b) that one collision in 100 is effective. Calculate the width (in cm^{-1}) of vibrational transitions in the molecule. [ISM **13.4a**]

6.4b A molecule in a gas undergoes about 1.0×10^9 collisions in each second. Suppose that (a) every collision is effective in deactivating the molecule rotationally and (b) that one collision in 10 is effective. Calculate the width (in hertz) of rotational transitions in the molecule. [SSM **13.4b**]

6.5a Calculate the frequency of the $J = 4 \leftarrow 3$ transition in the pure rotational spectrum of $^{14}N^{16}O$. The equilibrium bond length is 115 pm. [ISM **13.5a**]

6.5b Calculate the frequency of the $J = 3 \leftarrow 2$ transition in the pure rotational spectrum of $^{12}C^{16}O$. The equilibrium bond length is 112.81 pm. [SSM **13.5b**]

6.6a If the wavenumber of the $J = 3 \leftarrow 2$ rotational transition of $^1H^{35}Cl$ considered as a rigid rotator is 63.56 cm^{-1}, what is (a) the moment of inertia of the molecule, (b) the bond length? [ISM **13.6a**]

6.6b If the wavenumber of the $J = 1 \leftarrow 0$ rotational transition of $^1H^{81}Br$ considered as a rigid rotator is 16.93 cm^{-1}, what is (a) the moment of inertia of the molecule, (b) the bond length? [SSM **13.6b**]

6.7a Given that the spacing of lines in the microwave spectrum of $^{27}Al^1H$ is constant at 12.604 cm^{-1}, calculate the moment of inertia and bond length of the molecule ($m(^{27}Al) = 26.9815$ u). [ISM **13.7a**]

6.7b Given that the spacing of lines in the microwave spectrum of $^{35}Cl^{19}F$ is constant at 1.033 cm^{-1}, calculate the moment of inertia and bond length of the molecule ($m(^{35}Cl) = 34.9688$ u, $m(^{19}F) = 18.9984$ u). [SSM **13.7b**]

6.8a The rotational constant of $^{127}I^{35}Cl$ is 0.1142 cm^{-1}. Calculate the ICl bond length ($m(^{35}Cl) = 34.9688$ u, $m(^{127}I) = 126.9045$ u). [ISM **13.8a**]

6.8b The rotational constant of $^{12}C^{16}O_2$ is 0.39021 cm^{-1}. Calculate the bond length of the molecule ($m(^{12}C) = 12$ u exactly, $m(^{16}O) = 15.9949$ u). [SSM **13.8b**]

6.9a Determine the HC and CN bond lengths in HCN from the rotational constants $B(^1H^{12}C^{14}N) = 44.316$ GHz and $B(^2H^{12}C^{14}N) = 36.208$ GHz. [ISM **13.9a**]

6.9b Determine the CO and CS bond lengths in OCS from the rotational constants $B(^{16}O^{12}C^{32}S) = 6081.5$ MHz, $B(^{16}O^{12}C^{34}S) = 5932.8$ MHz. [SSM **13.9b**]

6.10a The wavenumber of the incident radiation in a Raman spectrometer is 20 487 cm^{-1}. What is the wavenumber of the scattered Stokes radiation for the $J = 2 \leftarrow 0$ transition of $^{14}N_2$? [ISM **13.10a**]

6.10b The wavenumber of the incident radiation in a Raman spectrometer is 20 623 cm^{-1}. What is the wavenumber of the scattered Stokes radiation for the $J = 4 \leftarrow 2$ transition of $^{16}O_2$? [SSM **13.10b**]

6.11a The rotational Raman spectrum of $^{35}Cl_2$ ($m(^{35}Cl) = 34.9688$ u) shows a series of Stokes lines separated by 0.9752 cm^{-1} and a similar series of anti-Stokes lines. Calculate the bond length of the molecule. [ISM **13.11a**]

6.11b The rotational Raman spectrum of $^{19}F_2$ ($m(^{19}F) = 18.9984$ u) shows a series of Stokes lines separated by 3.5312 cm^{-1} and a similar series of anti-Stokes lines. Calculate the bond length of the molecule. [SSM **13.11b**]

6.12a Which of the following molecules may show a pure rotational microwave absorption spectrum: (a) H_2, (b) HCl, (c) CH_4, (d) CH_3Cl, (e) CH_2Cl_2? [ISM **13.12a**]

6.12b Which of the following molecules may show a pure rotational microwave absorption spectrum: (a) H_2O, (b) H_2O_2, (c) NH_3, (d) N_2O? [SSM **13.12b**]

6.13a Which of the following molecules may show a pure rotational Raman spectrum: (a) H_2, (b) HCl, (c) CH_4, (d) CH_3Cl? [ISM **13.13a**]

6.13b Which of the following molecules may show a pure rotational Raman spectrum: (a) CH_2Cl_2, (b) CH_3CH_3, (c) SF_6, (d) N_2O? [SSM **13.13b**]

6.14a An object of mass 1.0 kg suspended from the end of a rubber band has a vibrational frequency of 2.0 Hz. Calculate the force constant of the rubber band. [ISM **13.14a**]

6.14b An object of mass 2.0 g suspended from the end of a spring has a vibrational frequency of 3.0 Hz. Calculate the force constant of the spring. [SSM **13.14b**]

6.15a Calculate the percentage difference in the fundamental vibration wavenumber of $^{23}Na^{35}Cl$ and $^{23}Na^{37}Cl$ on the assumption that their force constants are the same. [ISM **13.15a**]

6.15b Calculate the percentage difference in the fundamental vibration wavenumber of $^1H^{35}Cl$ and $^2H^{37}Cl$ on the assumption that their force constants are the same. [SSM **13.15b**]

6.16a The wavenumber of the fundamental vibrational transition of $^{35}Cl_2$ is 564.9 cm^{-1}. Calculate the force constant of the bond ($m(^{35}Cl) = 34.9688$ u). [ISM **13.16a**]

6.16b The wavenumber of the fundamental vibrational transition of $^{79}Br^{81}Br$ is 323.2 cm^{-1}. Calculate the force constant of the bond ($m(^{79}Br) = 78.9183$ u, $m(^{81}Br) = 80.9163$ u). [SSM **13.16b**]

6.17a Calculate the relative numbers of Cl_2 molecules ($\tilde{\nu} = 559.7$ cm^{-1}) in the ground and first excited vibrational states at (a) 298 K, (b) 500 K. [ISM **13.17a**]

6.17b Calculate the relative numbers of Br_2 molecules ($\tilde{\nu} = 321$ cm^{-1}) in the second and first excited vibrational states at (a) 298 K, (b) 800 K. [SSM **13.17b**]

6.18a The hydrogen halides have the following fundamental vibrational wavenumbers: 4141.3 cm^{-1} (HF); 2988.9 cm^{-1} ($H^{35}Cl$); 2649.7 cm^{-1} ($H^{81}Br$); 2309.5 cm^{-1} ($H^{127}I$). Calculate the force constants of the hydrogen–halogen bonds. [ISM **13.18a**]

6.18b From the data in Exercise 6.18a, predict the fundamental vibrational wavenumbers of the deuterium halides. [SSM **13.18b**]

6.19a For $^{16}O_2$, ΔG values for the transitions $v = 1 \leftarrow 0$, $2 \leftarrow 0$, and $3 \leftarrow 0$ are, respectively, 1556.22, 3088.28, and 4596.21 cm^{-1}. Calculate $\tilde{\nu}$ and x_e. Assume y_e to be zero. [ISM **13.19a**]

6.19b For $^{14}N_2$, ΔG values for the transitions $v = 1 \leftarrow 0$, $2 \leftarrow 0$, and $3 \leftarrow 0$ are, respectively, 2345.15, 4661.40, and 6983.73 cm^{-1}. Calculate $\tilde{\nu}$ and x_e. Assume y_e to be zero. [SSM **13.19b**]

6.20a The first five vibrational energy levels of HCl are at 1481.86, 4367.50, 7149.04, 9826.48, and 12 399.8 cm^{-1}. Calculate the dissociation energy of the molecule in reciprocal centimetres and electronvolts. [ISM **13.20a**]

6.20b The first five vibrational energy levels of HI are at 1144.83, 3374.90, 5525.51, 7596.66, and 9588.35 cm^{-1}. Calculate the dissociation energy of the molecule in reciprocal centimetres and electronvolts. [SSM **13.20b**]

6.21a Infrared absorption by $^1H^{81}Br$ gives rise to an R branch from $v = 0$. What is the wavenumber of the line originating from the rotational state with $J = 2$? Use the information in Table 6.2. [ISM **13.21a**]

6.21b Infrared absorption by $^1H^{127}I$ gives rise to an R branch from $v = 0$. What is the wavenumber of the line originating from the rotational state with $J = 2$? Use the information in Table 6.2. [SSM **13.21b**]

6.22a Which of the following molecules may show infrared absorption spectra: (a) H_2, (b) HCl, (c) CO_2, (d) H_2O? [ISM **13.22a**]

6.22b Which of the following molecules may show infrared absorption spectra: (a) CH_3CH_3, (b) CH_4, (c) CH_3Cl, (d) N_2? [SSM **13.22b**]

6.23a How many normal modes of vibration are there for the following molecules: (a) H_2O, (b) H_2O_2, (c) C_2H_4? [ISM **13.23a**]

6.23b How many normal modes of vibration are there for the following molecules: (a) C_6H_6, (b) $C_6H_6CH_3$, (c) HC≡C–C≡CH. [SSM **13.23b**]

6.24a Which of the three vibrations of an AB_2 molecule are infrared or Raman active when it is (a) angular, (b) linear? [ISM 13.24a]

6.24b Which of the vibrations of an AB_3 molecule are infrared or Raman active when it is (a) trigonal planar, (b) trigonal pyramidal? [SSM 13.24b]

6.25a Consider the vibrational mode that corresponds to the uniform expansion of the benzene ring. Is it (a) Raman, (b) infrared active? [ISM 13.25a]

6.25b Consider the vibrational mode that corresponds to the boat-like bending of a benzene ring. Is it (a) Raman, (b) infrared active? [SSM 13.25b]

6.26a The molecule CH_2Cl_2 belongs to the point group C_{2v}. The displacements of the atoms span $5A_1 + 2A_2 + 4B_1 + 4B_2$. What are the symmetries of the normal modes of vibration? [ISM 13.26a]

6.26b A carbon disulfide molecule belongs to the point group $D_{\infty h}$. The nine displacements of the three atoms span $A_{1g} + A_{1u} + A_{2g} + 2E_{1u} + E_{1g}$. What are the symmetries of the normal modes of vibration? [SSM 13.26b]

Problems*

Numerical problems

6.1 Use mathematical software to evaluate the Planck distribution at any temperature and wavelength or frequency, and evaluate integrals for the energy density of the radiation between any two wavelengths. Calculate the total energy density in the visible region (700 nm to 400 nm) for a black body at (a) 1500 K, a typical operating temperature for globars, (b) 2500 K, a typical operating temperature for tungsten filament lamps, (c) 5800 K, the surface temperature of the Sun. What are the classical values at these temperatures? [SSM 13.1]

6.2 Calculate the Doppler width (as a fraction of the transition wavelength) for any kind of transition in (a) HCl, (b) ICl at 25°C. What would be the widths of the rotational and vibrational transitions in these molecules (in MHz and cm^{-1}, respectively), given $B(ICl) = 0.1142\ cm^{-1}$ and $\tilde{\nu}(ICl) = 384\ cm^{-1}$ and additional information in Table 6.2. [ISM 13.2]

6.3 The collision frequency z of a molecule of mass m in a gas at a pressure p is $z = 4\sigma(kT/\pi m)^{1/2}p/kT$, where σ is the collision cross-section. Find an expression for the collision-limited lifetime of an excited state assuming that every collision is effective. Estimate the width of rotational transition in HCl ($\sigma = 0.30\ nm^2$) at 25°C and 1.0 atm. To what value must the pressure of the gas be reduced in order to ensure that collision broadening is less important than Doppler broadening? [SSM 13.3]

6.4 The rotational constant of NH_3 is equivalent to 298 GHz. Compute the separation of the pure rotational spectrum lines in GHz, cm^{-1}, and mm, and show that the value of B is consistent with an N—H bond length of 101.4 pm and a bond angle of 106.78°. [ISM 13.4]

6.5 The rotational constant for CO is $1.9314\ cm^{-1}$ and $1.6116\ cm^{-1}$ in the ground and first excited vibrational states, respectively. By how much does the internuclear distance change as a result of this transition? [SSM 13.5]

6.6 Pure rotational Raman spectra of gaseous C_6H_6 and C_6D_6 yield the following rotational constants: $B(C_6H_6) = 0.189\ 60\ cm^{-1}$, $B(C_6D_6) = 0.156\ 81\ cm^{-1}$. The moments of inertia of the molecules about any axis perpendicular to the C_6 axis were calculated from these data as $I(C_6H_6) = 1.4759 \times 10^{-45}\ kg\ m^2$, $I(C_6D_6) = 1.7845 \times 10^{-45}\ kg\ m^2$. Calculate the CC, CH, and CD bond lengths. [ISM 13.6]

6.7 Rotational absorption lines from $^1H^{35}Cl$ gas were found at the following wavenumbers (R.L. Hausler and R.A. Oetjen, *J. Chem. Phys.* **21**, 1340 (1953)): 83.32, 104.13, 124.73, 145.37, 165.89, 186.23, 206.60, 226.86 cm^{-1}. Calculate the moment of inertia and the bond length of the molecule. Predict the positions of the corresponding lines in $^2H^{35}Cl$. [SSM 13.7]

6.8 Is the bond length in HCl the same as that in DCl? The wavenumbers of the $J = 1 \leftarrow 0$ rotational transitions for $H^{35}Cl$ and $^2H^{35}Cl$ are 20.8784 and 10.7840 cm^{-1}, respectively. Accurate atomic masses are 1.007825 u and 2.0140 u for ^1H and ^2H, respectively. The mass of ^{35}Cl is 34.96885 u. Based on this information alone, can you conclude that the bond lengths are the same or different in the two molecules? [ISM 13.8]

6.9 Thermodynamic considerations suggest that the copper monohalides CuX should exist mainly as polymers in the gas phase, and indeed it proved difficult to obtain the monomers in sufficient abundance to detect spectroscopically. This problem was overcome by flowing the halogen gas over copper heated to 1100 K (E.L. Manson, F.C. de Lucia, and W. Gordy, *J. Chem. Phys.* **63**, 2724 (1975)). For CuBr the J = 13–14, 14–15, and 15–16 transitions occurred at 84 421.34, 90 449.25, and 96 476.72 MHz, respectively. Calculate the rotational constant and bond length of CuBr. [SSM 13.9]

6.10 The microwave spectrum of $^{16}O^{12}CS$ (C.H. Townes, A.N. Holden, and F.R. Merritt, *Phys. Rev.* **74**, 1113 (1948)) gave absorption lines (in GHz) as follows:

J	1	2	3	4
^{32}S	24.325 92	36.488 82	48.651 64	60.814 08
^{34}S	23.732 33		47.462 40	

Use the expressions for moments of inertia in Table 6.1 and assume that the bond lengths are unchanged by substitution; calculate the CO and CS bond lengths in OCS. [ISM 13.10]

6.11‡ In a study of the rotational spectrum of the linear FeCO radical, K. Tanaka, M. Shirasaka, and T. Tanaka (*J. Chem. Phys.* **106**, 6820 (1997)) report the following $J + 1 \leftarrow J$ transitions:

J	24	25	26	27	28	29
$\tilde{\nu}/m^{-1}$	214 777.7	223 379.0	231 981.2	240 584.4	249 188.5	257 793.5

Evaluate the rotational constant of the molecule. Also, estimate the value of J for the most highly populated rotational energy level at 298 K and at 100 K. [SSM 13.11]

6.12 The vibrational energy levels of NaI lie at the wavenumbers 142.81, 427.31, 710.31, and 991.81 cm^{-1}. Show that they fit the expression $(v + \frac{1}{2})\tilde{\nu} - (v + \frac{1}{2})^2 x\tilde{\nu}$, and deduce the force constant, zero-point energy, and dissociation energy of the molecule. [ISM 13.12]

6.13 Predict the shape of the nitronium ion, NO_2^+, from its Lewis structure and the VSEPR model. It has one Raman active vibrational mode at 1400 cm^{-1}, two strong IR active modes at 2360 and 540 cm^{-1}, and one weak IR mode at 3735 cm^{-1}. Are these data consistent with the predicted shape of the molecule? Assign the vibrational wavenumbers to the modes from which they arise. [SSM 13.13]

6.14 At low resolution, the strongest absorption band in the infrared absorption spectrum of $^{12}C^{16}O$ is centred at 2150 cm^{-1}. Upon closer examination at higher resolution, this band is observed to be split into two sets

* Problems denoted with the symbol ‡ were supplied by Charles Trapp, Carmen Giunta, and Marshall Cady.

of closely spaced peaks, one on each side of the centre of the spectrum at 2143.26 cm^{-1}. The separation between the peaks immediately to the right and left of the centre is 7.655 cm^{-1}. Make the harmonic oscillator and rigid rotor approximations and calculate from these data: (a) the vibrational wavenumber of a CO molecule, (b) its molar zero-point vibrational energy, (c) the force constant of the CO bond, (d) the rotational constant B, and (e) the bond length of CO. [ISM **13.14**]

6.15 The HCl molecule is quite well described by the Morse potential with $D_e = 5.33$ eV, $\tilde{\nu} = 2989.7$ cm^{-1}, and $x\tilde{\nu} = 52.05$ cm^{-1}. Assuming that the potential is unchanged on deuteration, predict the dissociation energies (D_0) of (a) HCl, (b) DCl. [SSM **13.15**]

6.16 The Morse potential (eqn 6.54) is very useful as a simple representation of the actual molecular potential energy. When RbH was studied, it was found that $\tilde{\nu} = 936.8$ cm^{-1} and $x\tilde{\nu} = 14.15$ cm^{-1}. Plot the potential energy curve from 50 pm to 800 pm around $R_e = 236.7$ pm. Then go on to explore how the rotation of a molecule may weaken its bond by allowing for the kinetic energy of rotation of a molecule and plotting $V^* = V + hcBJ(J+1)$ with $B = \hbar/4\pi c\mu R^2$. Plot these curves on the same diagram for $J = 40$, 80, and 100, and observe how the dissociation energy is affected by the rotation. (Taking $B = 3.020$ cm^{-1} at the equilibrium bond length will greatly simplify the calculation.) [ISM **13.16**]

6.17‡ F. Luo, G.C. McBane, G. Kim, C.F. Giese, and W.R. Gentry (*J. Chem. Phys.* **98**, 3564 (1993)) reported experimental observation of the He_2 complex, a species that had escaped detection for a long time. The fact that the observation required temperatures in the neighbourhood of 1 mK is consistent with computational studies that suggest that hcD_e for He_2 is about 1.51×10^{-23} J, hcD_0 about 2×10^{-26} J, and R_e about 297 pm. (a) Estimate the fundamental vibrational wavenumber, force constant, moment of inertia, and rotational constant based on the harmonic oscillator and rigid-rotor approximations. (b) Such a weakly bound complex is hardly likely to be rigid. Estimate the vibrational wavenumber and anharmonicity constant based on the Morse potential. [SSM **13.17**]

6.18 As mentioned in Section 6.15, the semi-empirical, *ab initio*, and DFT methods discussed in Chapter 4 can be used to estimate the force field of a molecule. The molecule's vibrational spectrum can be simulated, and it is then possible to determine the correspondence between a vibrational frequency and the atomic displacements that give rise to a normal mode. (a) Using molecular modelling software[3] and the computational method of your choice (semi-empirical, *ab initio*, or DFT methods), calculate the fundamental vibrational wavenumbers and visualize the vibrational normal modes of SO_2 in the gas phase. (b) The experimental values of the fundamental vibrational wavenumbers of SO_2 in the gas phase are 525 cm^{-1}, 1151 cm^{-1}, and 1336 cm^{-1}. Compare the calculated and experimental values. Even if agreement is poor, is it possible to establish a correlation between an experimental value of the vibrational wavenumber with a specific vibrational normal mode? [ISM **13.18**]

6.19 Consider the molecule CH_3Cl. (a) To what point group does the molecule belong? (b) How many normal modes of vibration does the molecule have? (c) What are the symmetries of the normal modes of vibration for this molecule? (d) Which of the vibrational modes of this molecule are infrared active? (e) Which of the vibrational modes of this molecule are Raman active? [SSM **13.19**]

6.20 Suppose that three conformations are proposed for the nonlinear molecule H_2O_2 (**2**, **3**, and **4**). The infrared absorption spectrum of gaseous H_2O_2 has bands at 870, 1370, 2869, and 3417 cm^{-1}. The Raman spectrum of the same sample has bands at 877, 1408, 1435, and 3407 cm^{-1}. All bands correspond to fundamental vibrational wavenumbers and you may assume that: (i) the 870 and 877 cm^{-1} bands arise from the same normal mode, and (ii) the 3417 and 3407 cm^{-1} bands arise from the same normal mode. (a) If H_2O_2 were linear, how many normal modes of vibration would it have? (b) Give the symmetry point group of each of the three proposed conformations of nonlinear H_2O_2. (c) Determine which of the proposed conformations is inconsistent with the spectroscopic data. Explain your reasoning. [ISM **13.20**]

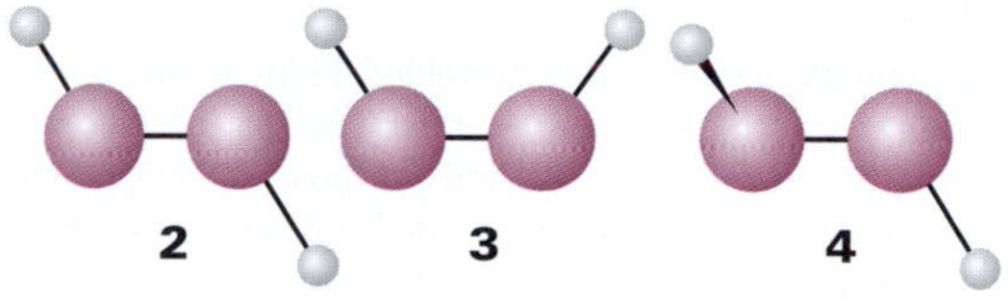

2 **3** **4**

Theoretical problems

6.21 Show that the moment of inertia of a diatomic molecule composed of atoms of masses m_A and m_B and bond length R is equal to $m_{eff}R^2$, where $m_{eff} = m_A m_B/(m_A + m_B)$. [SSM **13.21**]

6.22 Derive eqn 6.34 for the centrifugal distortion constant D_J of a diatomic molecule of effective mass m_{eff}. Treat the bond as an elastic spring with force constant k and equilibrium length r_e that is subjected to a centrifugal distortion to a new length r_c. Begin the derivation by letting the particles experience a restoring force of magnitude $k(r_c - r_e)$ that is countered perfectly by a centrifugal force $m_{eff}\omega^2 r_c$, where ω is the angular velocity of the rotating molecule. Then introduce quantum mechanical effects by writing the angular momentum as $\{J(J+1)\}^{1/2}\hbar$. Finally, write an expression for the energy of the rotating molecule, compare it with eqn 6.33, and write an expression for D_J. For help with the classical aspects of this derivation, see *Appendix 3*. [ISM **13.22**]

6.23 In the group theoretical language developed in Chapter 5, a spherical rotor is a molecule that belongs to a cubic or icosahedral point group, a symmetric rotor is a molecule with at least a threefold axis of symmetry, and an asymmetric rotor is a molecule without a threefold (or higher) axis. Linear molecules are linear rotors. Classify each of the following molecules as a spherical, symmetric, linear, or asymmetric rotor and justify your answers with group theoretical arguments: (a) CH_4, (b) CH_3CN, (c) CO_2, (d) CH_3OH, (e) benzene, (f) pyridine. [SSM **13.23**]

6.24 Derive an expression for the value of J corresponding to the most highly populated rotational energy level of a diatomic rotor at a temperature T remembering that the degeneracy of each level is $2J + 1$. Evaluate the expression for ICl (for which $B = 0.1142$ cm^{-1}) at 25°C. Repeat the problem for the most highly populated level of a spherical rotor, taking note of the fact that each level is $(2J+1)^2$-fold degenerate. Evaluate the expression for CH_4 (for which $B = 5.24$ cm^{-1}) at 25°C. [ISM **13.24**]

6.25 The moments of inertia of the linear mercury(II) halides are very large, so the O and S branches of their vibrational Raman spectra show little rotational structure. Nevertheless, the peaks of both branches can be identified and have been used to measure the rotational constants of the molecules (R.J.H. Clark and D.M. Rippon, *J. Chem. Soc. Faraday Soc. II*, **69**, 1496 (1973)). Show, from a knowledge of the value of J corresponding to the intensity maximum, that the separation of the peaks of the O and S branches is given by the Placzek–Teller relation $\delta\tilde{\nu} = (32BkT/hc)^{1/2}$. The following widths were obtained at the temperatures stated:

	$HgCl_2$	$HgBr_2$	HgI_2
θ/°C	282	292	292
$\delta\tilde{\nu}$/cm^{-1}	23.8	15.2	11.4

Calculate the bond lengths in the three molecules. [SSM **13.25**]

[3] The web site contains links to molecular modelling freeware and to other sites where you may perform molecular orbital calculations directly from your web browser.

6.26 Confirm that a Morse oscillator has a finite number of bound states, the states with $V < hcD_e$. Determine the value of v_{max} for the highest bound state. [ISM **13.26**]

Applications: to biology, environmental science, and astrophysics

6.27 The protein haemerythrin is responsible for binding and carrying O_2 in some invertebrates. Each protein molecule has two Fe^{2+} ions that are in very close proximity and work together to bind one molecule of O_2. The Fe_2O_2 group of oxygenated haemerythrin is coloured and has an electronic absorption band at 500 nm. The resonance Raman spectrum of oxygenated haemerythrin obtained with laser excitation at 500 nm has a band at 844 cm^{-1} that has been attributed to the O—O stretching mode of bound $^{16}O_2$. (a) Why is resonance Raman spectroscopy and not infrared spectroscopy the method of choice for the study of the binding of O_2 to haemerythrin? (b) Proof that the 844 cm^{-1} band arises from a bound O_2 species may be obtained by conducting experiments on samples of haemerythrin that have been mixed with $^{18}O_2$, instead of $^{16}O_2$. Predict the fundamental vibrational wavenumber of the $^{18}O—^{18}O$ stretching mode in a sample of haemerythrin that has been treated with $^{18}O_2$. (c) The fundamental vibrational wavenumbers for the O—O stretching modes of O_2, O_2^- (superoxide anion), and O_2^{2-} (peroxide anion) are 1555, 1107, and 878 cm^{-1}, respectively. Explain this trend in terms of the electronic structures of O_2, O_2^-, and O_2^{2-}. *Hint:* Review Section 4.4. What are the bond orders of O_2, O_2^-, and O_2^{2-}? (d) Based on the data given above, which of the following species best describes the Fe_2O_2 group of haemerythrin: $Fe_2^{2+}O_2$, $Fe^{2+}Fe^{3+}O_2^-$, or $Fe_2^{3+}O_2^{2-}$? Explain your reasoning. (e) The resonance Raman spectrum of haemerythrin mixed with $^{16}O^{18}O$ has two bands that can be attributed to the O—O stretching mode of bound oxygen. Discuss how this observation may be used to exclude one or more of the four proposed schemes (**5–8**) for binding of O_2 to the Fe_2 site of haemerythrin. [SSM **13.27**]

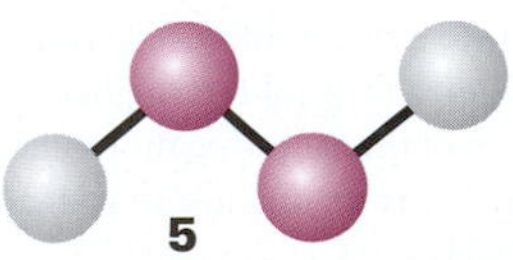

5

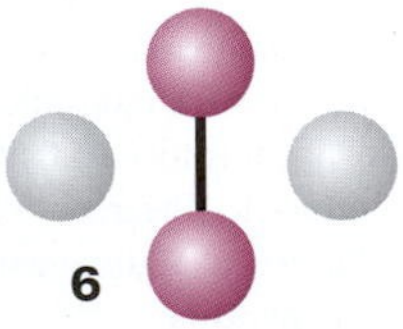

6

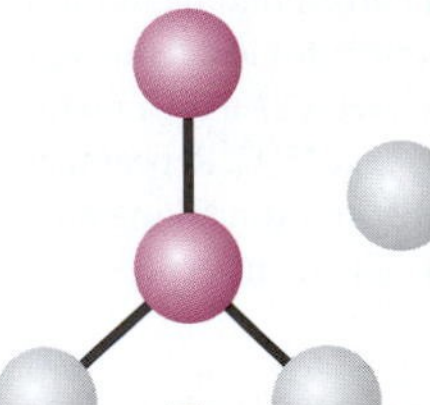

7

8

6.28‡ A mixture of carbon dioxide (2.1 per cent) and helium, at 1.00 bar and 298 K in a gas cell of length 10 cm has an infrared absorption band centred at 2349 cm^{-1} with absorbances, $A(\tilde{\nu})$, described by:

$$A(\tilde{\nu}) = \frac{a_1}{1 + a_2(\tilde{\nu} - a_3)^2} + \frac{a_4}{1 + a_5(\tilde{\nu} - a_6)^2}$$

where the coefficients are $a_1 = 0.932$, $a_2 = 0.005050\ cm^2$, $a_3 = 2333\ cm^{-1}$, $a_4 = 1.504$, $a_5 = 0.01521\ cm^2$, $a_6 = 2362\ cm^{-1}$. (a) Draw graphs of $A(\tilde{\nu})$ and $\varepsilon(\tilde{\nu})$. What is the origin of both the band and the band width? What are the allowed and forbidden transitions of this band? (b) Calculate the transition wavenumbers and absorbances of the band with a simple harmonic oscillator-rigid rotor model and compare the result with the experimental spectra. The CO bond length is 116.2 pm. (c) Within what height, h, is basically all the infrared emission from the Earth in this band absorbed by atmospheric carbon dioxide? The mole fraction of CO_2 in the atmosphere is 3.3×10^{-4} and $T/K = 288 - 0.0065(h/m)$ below 10 km. Draw a surface plot of the atmospheric transmittance of the band as a function of both height and wavenumber. [ISM **13.28**]

6.29 In Problem 3.27, we saw that Doppler shifts of atomic spectral lines are used to estimate the speed of recession or approach of a star. From the discussion in Section 6.3a, it is easy to see that Doppler broadening of an atomic spectral line depends on the temperature of the star that emits the radiation. A spectral line of $^{48}Ti^{8+}$ (of mass 47.95 u) in a distant star was found to be shifted from 654.2 nm to 706.5 nm and to be broadened to 61.8 pm. What is the speed of recession and the surface temperature of the star? [SSM **13.29**]

6.30 A. Dalgarno, in Chemistry in the interstellar medium, *Frontiers of Astrophysics*, E.H. Avrett (ed.), Harvard University Press, Cambridge (1976), notes that although both CH and CN spectra show up strongly in the interstellar medium in the constellation Ophiuchus, the CN spectrum has become the standard for the determination of the temperature of the cosmic microwave background radiation. Demonstrate through a calculation why CH would not be as useful for this purpose as CN. The rotational constant B_0 for CH is 14.190 cm^{-1}. [ISM **13.30**]

6.31‡ There is a gaseous interstellar cloud in the constellation Ophiuchus that is illuminated from behind by the star ζ-Ophiuci. Analysis of the electronic–vibrational–rotational absorption lines obtained by H.S. Uhler and R.A. Patterson (*Astrophys. J.* **42**, 434 (1915)) shows the presence of CN molecules in the interstellar medium. A strong absorption line in the ultraviolet region at $\lambda = 387.5$ nm was observed corresponding to the transition $J = 0–1$. Unexpectedly, a second strong absorption line with 25 per cent of the intensity of the first was found at a slightly longer wavelength ($\Delta\lambda = 0.061$ nm) corresponding to the transition $J = 1–1$ (here allowed). Calculate the temperature of the CN molecules. Gerhard Herzberg, who was later to receive the Nobel Prize for his contributions to spectroscopy, calculated the temperature as 2.3 K. Although puzzled by this result, he did not realize its full significance. If he had, his prize might have been for the discovery of the cosmic microwave background radiation. [SSM **13.31**]

6.32‡ The H_3^+ ion has recently been found in the interstellar medium and in the atmospheres of Jupiter, Saturn, and Uranus. The rotational energy levels of H_3^+, an oblate symmetric rotor, are given by eqn 6.29, with C replacing A, when centrifugal distortion and other complications are ignored. Experimental values for vibrational–rotational constants are $\tilde{\nu}(E') = 2521.6\ cm^{-1}$, $B = 43.55\ cm^{-1}$, and $C = 20.71\ cm^{-1}$. (a) Show that, for a nonlinear planar molecule (such as H_3^+), $I_C = 2I_B$. The rather large discrepancy with the experimental values is due to factors ignored in eqn 6.29. (b) Calculate an approximate value of the H—H bond length in H_3^+. (c) The value of R_e obtained from the best quantum mechanical calculations by J.B. Anderson (*J. Chem. Phys.* **96**, 3702 (1991)) is 87.32 pm. Use this result to calculate the values of the rotational constants B and C. (d) Assuming that the geometry and force constants are the same in D_3^+ and H_3^+, calculate the spectroscopic constants of D_3^+. The molecular ion D_3^+ was first produced by J.T. Shy, J.W. Farley, W.E. Lamb Jr, and W.H. Wing (*Phys. Rev. Lett* **45**, 535 (1980)) who observed the $\nu_2(E')$ band in the infrared. [ISM **13.32**]

6.33 The space immediately surrounding stars, also called the *circumstellar space*, is significantly warmer because stars are very intense black-body emitters with temperatures of several thousand kelvin. Discuss how such factors as cloud temperature, particle density, and particle velocity may affect the rotational spectrum of CO in an interstellar cloud. What new features in the spectrum of CO can be observed in gas ejected from and still near a star with temperatures of about 1000 K, relative to gas in a cloud with temperature of about 10 K? Explain how these features may be used to distinguish between circumstellar and interstellar material on the basis of the rotational spectrum of CO. [SSM **13.33**]

Molecular spectroscopy 2: electronic transitions

7

Simple analytical expressions for the electronic energy levels of molecules cannot be given, so this chapter concentrates on the qualitative features of electronic transitions. A common theme throughout the chapter is that electronic transitions occur within a stationary nuclear framework. We pay particular attention to spontaneous radiative decay processes, which include fluorescence and phosphorescence. A specially important example of stimulated radiative decay is that responsible for the action of lasers, and we see how this stimulated emission may be achieved and employed.

The energies needed to change the electron distributions of molecules are of the order of several electronvolts (1 eV is equivalent to about 8000 cm^{-1} or 100 kJ mol^{-1}). Consequently, the photons emitted or absorbed when such changes occur lie in the visible and ultraviolet regions of the spectrum (Table 7.1).

One of the revolutions that has occurred in physical chemistry in recent years is the application of lasers to spectroscopy and kinetics. Lasers have brought unprecedented precision to spectroscopy, made Raman spectroscopy a widely useful technique, and have made it possible to study chemical reactions on a femtosecond time scale. We shall see the principles of their action in this chapter and their applications throughout the rest of the book.

The characteristics of electronic transitions

In the ground state of a molecule the nuclei are at equilibrium in the sense that they experience no net force from the electrons and other nuclei in the molecule. Immediately after an electronic transition they are subjected to different forces and

Synoptic table 7.1* Colour, frequency, and energy of light

Colour	λ/nm	$\nu/(10^{14}$ Hz)	E/(kJ mol^{-1})
Infrared	>1000	<3.0	<120
Red	700	4.3	170
Yellow	580	5.2	210
Blue	470	6.4	250
Ultraviolet	<300	>10	>400

* More values are given in the *Data section*.

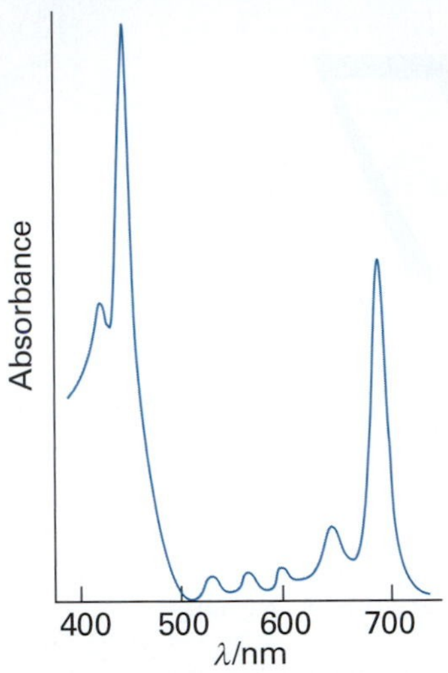

Fig. 7.1 The absorption spectrum of chlorophyll in the visible region. Note that it absorbs in the red and blue regions, and that green light is not absorbed.

the molecule may respond by starting to vibrate. The resulting vibrational structure of electronic transitions can be resolved for gaseous samples, but in a liquid or solid the lines usually merge together and result in a broad, almost featureless band (Fig. 7.1). Superimposed on the vibrational transitions that accompany the electronic transition of a molecule in the gas phase is an additional branch structure that arises from rotational transitions. The electronic spectra of gaseous samples are therefore very complicated but rich in information.

7.1 The electronic spectra of diatomic molecules

We examine some general features of electronic transitions by using diatomic molecules as examples. We begin by assigning term symbols to ground and excited electronic states. Then we use the symmetry designations to formulate selection rules. Finally, we examine the origin of vibrational structure in electronic spectra.

(a) Term symbols

The term symbols of linear molecules (the analogues of the symbols ^{2}P, etc. for atoms) are constructed in a similar way to those for atoms, but now we must pay attention to the component of total orbital angular momentum about the internuclear axis, $\Lambda\hbar$. The value of $|\Lambda|$ is denoted by the symbols Σ, Π, Δ, . . . for $|\Lambda| = 0, 1, 2\ldots$, respectively. These labels are the analogues of S, P, D, . . . for atoms. The value of Λ is the sum of the values of λ, the quantum number for the component $\lambda\hbar$ of orbital angular momentum of an individual electron around the internuclear axis. A single electron in a σ orbital has $\lambda = 0$: the orbital is cylindrically symmetrical and has no angular nodes when viewed along the internuclear axis. Therefore, if that is the only electron present, $\Lambda = 0$. The term symbol for H_2^+ is therefore Σ.

Comment 7.1

It is important to distinguish between the (upright) term symbol Σ and the (sloping) quantum number $\mathit{\Sigma}$.

As in atoms, we use a superscript with the value of $2S+1$ to denote the multiplicity of the term. The component of total spin angular momentum about the internuclear axis is denoted $\mathit{\Sigma}$, where $\mathit{\Sigma} = S, S-1, S-2, \ldots, -S$. For H_2^+, because there is only one electron, $S = s = \frac{1}{2}$ $(\mathit{\Sigma} = \pm\frac{1}{2})$ and the term symbol is $^2\Sigma$, a doublet term. The overall parity of the term is added as a right subscript. For H_2^+, the parity of the only occupied orbital is g (Section 4.3c), so the term itself is also g, and in full dress is $^2\Sigma_g$. If there are several electrons, the overall parity is calculated by using

$$\mathrm{g}\times\mathrm{g}=\mathrm{g} \qquad \mathrm{u}\times\mathrm{u}=\mathrm{g} \qquad \mathrm{u}\times\mathrm{g}=\mathrm{u} \tag{7.1}$$

These rules are generated by interpreting g as +1 and u as −1. The term symbol for the ground state of any closed-shell homonuclear diatomic molecule is $^1\Sigma_g$ because the spin is zero (a singlet term in which all electrons paired), there is no orbital angular momentum from a closed shell, and the overall parity is g.

A π electron in a diatomic molecule has one unit of orbital angular momentum about the internuclear axis ($\lambda = \pm 1$), and if it is the only electron outside a closed shell, gives rise to a Π term. If there are two π electrons (as in the ground state of O_2, with configuration $1\pi_u^4 1\pi_g^2$) then the term symbol may be either Σ (if the electrons are travelling in opposite directions, which is the case if they occupy different π orbitals, one with $\lambda = +1$ and the other with $\lambda = -1$) or Δ (if they are travelling in the same direction, which is the case if they occupy the same π orbital, both $\lambda = +1$, for instance). For O_2, the two π electrons occupy different orbitals with parallel spins (a triplet term), so the ground term is $^3\Sigma$. The overall parity of the molecule is

$$(\text{closed shell})\times\mathrm{g}\times\mathrm{g}=\mathrm{g}$$

The term symbol is therefore $^3\Sigma_g$.

Table 7.2 Properties of O_2 in its lower electronic states*

Configuration†	Term	Relative energy/cm^{-1}	$\tilde{\nu}$/cm^{-1}	R_e/pm
$\pi_u^2\pi_u^2\pi_g^1\pi_g^1$	$^3\Sigma_g^-$	0	1580	120.74
$\pi_u^2\pi_u^2\pi_g^2\pi_g^0$	$^1\Delta_g$	7 882.39	1509	121.55
$\pi_u^2\pi_u^2\pi_g^1\pi_g^1$	$^1\Sigma_g^+$	13 120.9	1433	122.68
$\pi_u^2\pi_u^1\pi_g^2\pi_g^1$	$^3\Sigma_u^+$	35 713	819	142
$\pi_u^2\pi_u^1\pi_g^2\pi_g^1$	$^3\Sigma_u^-$	49 363	700	160

* Adapted from G. Herzberg, *Spectra of diatomic molecules*, Van Nostrand, New York (1950) and D.C. Harris and M.D. Bertolucci, *Symmetry and spectroscopy: an introduction to vibrational and electronic spectroscopy*, Dover, New York (1989).

† The configuration $\pi_u^2\pi_u^1\pi_g^2\pi_g^1$ should also give rise to a $^3\Delta_u$ term, but electronic transitions to or from this state have not been observed.

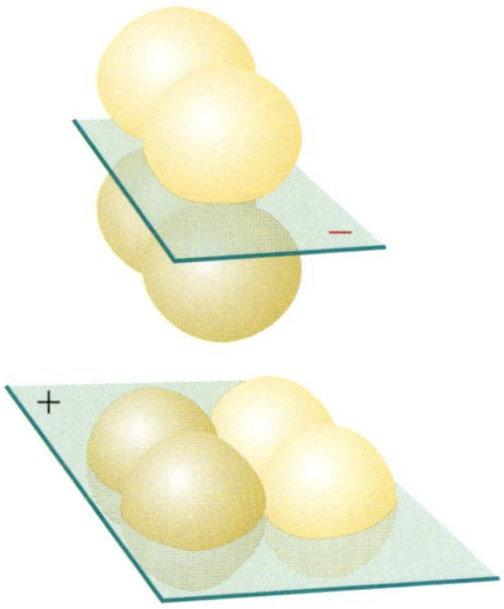

Fig. 7.2 The + or – on a term symbol refers to the overall symmetry of a configuration under reflection in a plane containing the two nuclei.

For Σ terms, a ± superscript denotes the behaviour of the molecular wavefunction under reflection in a plane containing the nuclei (Fig. 7.2). If, for convenience, we think of O_2 as having one electron in $1\pi_{g,x}$, which changes sign under reflection in the *yz*-plane, and the other electron in $1\pi_{g,y}$, which does not change sign under reflection in the same plane, then the overall reflection symmetry is

$$(\text{closed shell}) \times (+) \times (-) = (-)$$

and the full term symbol of the ground electronic state of O_2 is $^3\Sigma_g^-$.

The term symbols of excited electronic states are constructed in a similar way. For example, the term symbol for the excited state of O_2 formed by placing two electrons in a $1\pi_{g,x}$ (or in a $1\pi_{g,y}$) orbital is $^1\Delta_g$ because $|\Lambda| = 2$ (two electrons in the same π orbital), the spin is zero (all electrons are paired), and the overall parity is (closed shell) × g × g = g. Table 7.2 and Fig. 7.3 summarize the configurations, term symbols, and energies of the ground and some excited states of O_2.

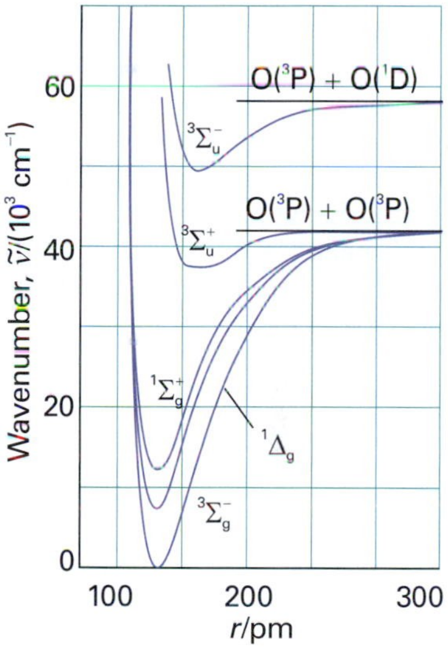

Fig. 7.3 The electronic states of dioxygen.

(b) Selection rules

A number of selection rules govern which transitions will be observed in the electronic spectrum of a molecule. The selection rules concerned with changes in angular momentum are

$$\Delta\Lambda = 0, \pm1 \qquad \Delta S = 0 \qquad \Delta\Sigma = 0 \qquad \Delta\Omega = 0, \pm1$$

where $\Omega = \Lambda + \Sigma$ is the quantum number for the component of total angular momentum (orbital and spin) around the internuclear axis (Fig. 7.4). As in atoms (Section 3.9), the origins of these rules are conservation of angular momentum during a transition and the fact that a photon has a spin of 1.

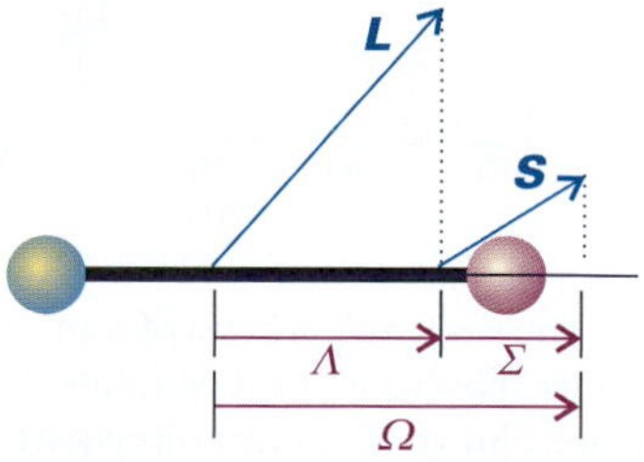

Fig. 7.4 The coupling of spin and orbital angular momenta in a linear molecule: only the components along the internuclear axis are conserved.

There are two selection rules concerned with changes in symmetry. First, for Σ terms, only $\Sigma^+ \leftrightarrow \Sigma^+$ and $\Sigma^- \leftrightarrow \Sigma^-$ transitions are allowed. Second, the **Laporte selection rule** for centrosymmetric molecules (those with a centre of inversion) and atoms states that:

> The only allowed transitions are transitions that are accompanied by a change of parity.

That is, u → g and g → u transitions are allowed, but g → g and u → u transitions are forbidden.

Justification 7.1 *The Laporte selection rule*

The last two selection rules result from the fact that the electric-dipole transition moment

$$\boldsymbol{\mu}_{\rm fi} = \int \psi_{\rm f}^{*}\hat{\mu}\psi_{\rm i}\,{\rm d}\tau$$

vanishes unless the integrand is invariant under all symmetry operations of the molecule. The three components of the dipole moment operator transform like x, y, and z, and are all u. Therefore, for a g $\rightarrow$ g transition, the overall parity of the transition dipole moment is g $\times$ u $\times$ g = u, so it must be zero. Likewise, for a u $\rightarrow$ u transition, the overall parity is u $\times$ u $\times$ u = u, so the transition dipole moment must also vanish. Hence, transitions without a change of parity are forbidden. The z-component of the dipole moment operator, the only component of $\boldsymbol{\mu}$ responsible for $\Sigma \leftrightarrow \Sigma$ transitions, has (+) symmetry. Therefore, for a (+) $\leftrightarrow$ (−) transition, the overall symmetry of the transition dipole moment is $(+)\times(+)\times(-) = (-)$, so it must be zero. Therefore, for Σ terms, $\Sigma^+ \leftrightarrow \Sigma^-$ transitions are not allowed.

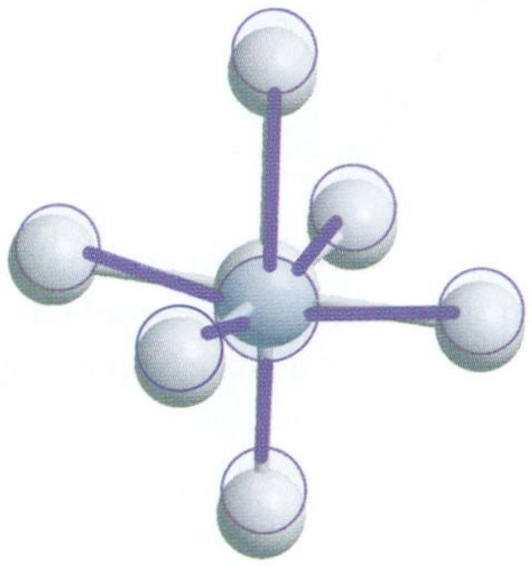

Fig. 7.5 A d–d transition is parity-forbidden because it corresponds to a g–g transition. However, a vibration of the molecule can destroy the inversion symmetry of the molecule and the g,u classification no longer applies. The removal of the centre of symmetry gives rise to a vibronically allowed transition.

A forbidden g $\rightarrow$ g transition can become allowed if the centre of symmetry is eliminated by an asymmetrical vibration, such as the one shown in Fig. 7.5. When the centre of symmetry is lost, g $\rightarrow$ g and u $\rightarrow$ u transitions are no longer parity-forbidden and become weakly allowed. A transition that derives its intensity from an asymmetrical vibration of a molecule is called a **vibronic transition**.

Self-test 7.1 Which of the following electronic transitions are allowed in O_2: ${}^3\Sigma_g^- \leftrightarrow {}^1\Delta_g$, ${}^3\Sigma_g^- \leftrightarrow {}^1\Sigma_g^+$, ${}^3\Sigma_g^- \leftrightarrow {}^3\Delta_u$, ${}^3\Sigma_g^- \leftrightarrow {}^3\Sigma_u^+$, ${}^3\Sigma_g^- \leftrightarrow {}^3\Sigma_u^-$? [${}^3\Sigma_g^- \leftrightarrow {}^3\Sigma_u^-$]

(c) Vibrational structure

To account for the vibrational structure in electronic spectra of molecules (Fig. 7.6), we apply the **Franck–Condon principle**:

> Because the nuclei are so much more massive than the electrons, an electronic transition takes place very much faster than the nuclei can respond.

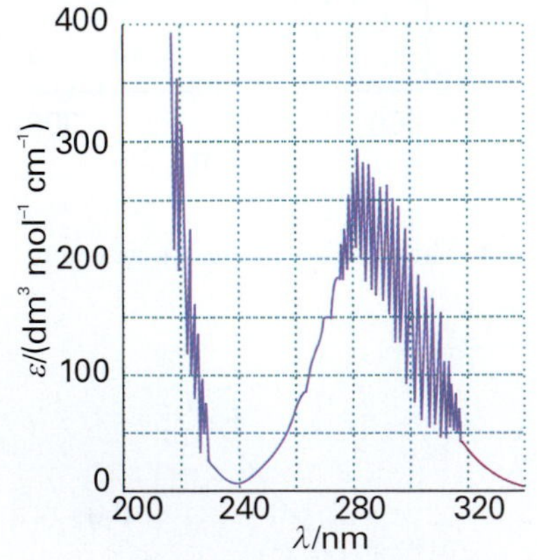

Fig. 7.6 The electronic spectra of some molecules show significant vibrational structure. Shown here is the ultraviolet spectrum of gaseous SO_2 at 298 K. As explained in the text, the sharp lines in this spectrum are due to transitions from a lower electronic state to different vibrational levels of a higher electronic state.

As a result of the transition, electron density is rapidly built up in new regions of the molecule and removed from others. The initially stationary nuclei suddenly experience a new force field, to which they respond by beginning to vibrate and (in classical terms) swing backwards and forwards from their original separation (which was maintained during the rapid electronic excitation). The stationary equilibrium separation of the nuclei in the initial electronic state therefore becomes a stationary turning point in the final electronic state (Fig. 7.7).

The quantum mechanical version of the Franck–Condon principle refines this picture. Before the absorption, the molecule is in the lowest vibrational state of its lowest electronic state (Fig. 7.8); the most probable location of the nuclei is at their equilibrium separation, R_e. The electronic transition is most likely to take place when the nuclei have this separation. When the transition occurs, the molecule is excited to the state represented by the upper curve. According to the Franck–Condon principle, the nuclear framework remains constant during this excitation, so we may imagine the transition as being up the vertical line in Fig. 7.7. The vertical line is the origin of the expression **vertical transition**, which is used to denote an electronic transition that occurs without change of nuclear geometry.

The vertical transition cuts through several vibrational levels of the upper electronic state. The level marked * is the one in which the nuclei are most probably at the same initial separation R_e (because the vibrational wavefunction has maximum amplitude there), so this vibrational state is the most probable state for the termination of the transition. However, it is not the only accessible vibrational state because several nearby states have an appreciable probability of the nuclei being at the separation R_e. Therefore, transitions occur to all the vibrational states in this region, but most intensely to the state with a vibrational wavefunction that peaks most strongly near R_e.

The vibrational structure of the spectrum depends on the relative horizontal position of the two potential energy curves, and a long **vibrational progression**, a lot of vibrational structure, is stimulated if the upper potential energy curve is appreciably displaced horizontally from the lower. The upper curve is usually displaced to greater equilibrium bond lengths because electronically excited states usually have more antibonding character than electronic ground states.

The separation of the vibrational lines of an electronic absorption spectrum depends on the vibrational energies of the *upper* electronic state. Hence, electronic absorption spectra may be used to assess the force fields and dissociation energies of electronically excited molecules (for example, by using a Birge–Sponer plot, as in Problem 7.2).

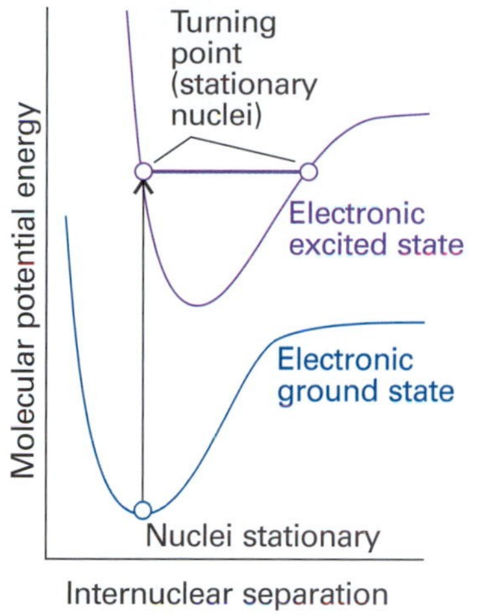

Fig. 7.7 According to the Franck–Condon principle, the most intense vibronic transition is from the ground vibrational state to the vibrational state lying vertically above it. Transitions to other vibrational levels also occur, but with lower intensity.

(d) Franck–Condon factors

The quantitative form of the Franck–Condon principle is derived from the expression for the transition dipole moment, $\boldsymbol{\mu}_{\rm fi} = \langle {\rm f}|\boldsymbol{\mu}|{\rm i}\rangle$. The dipole moment operator is a sum over all nuclei and electrons in the molecule:

$$\hat{\mu} = -e\sum_i \boldsymbol{r}_i + e\sum_I Z_I \boldsymbol{R}_I \qquad (7.2)$$

where the vectors are the distances from the centre of charge of the molecule. The intensity of the transition is proportional to the square modulus, $|\boldsymbol{\mu}_{\rm fi}|^2$, of the magnitude of the transition dipole moment (eqn 2.70), and we show in the *Justification* below that this intensity is proportional to the square modulus of the overlap integral, $S(v_{\rm f},v_{\rm i})$, between the vibrational states of the initial and final electronic states. This overlap integral is a measure of the match between the vibrational wavefunctions in the upper and lower electronic states: $S = 1$ for a perfect match and $S = 0$ when there is no similarity.

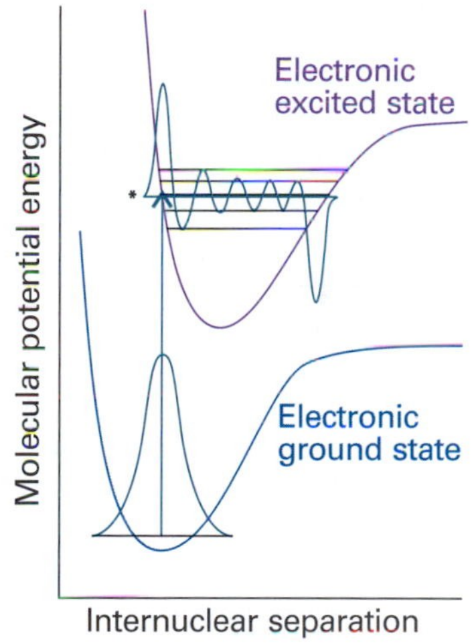

Fig. 7.8 In the quantum mechanical version of the Franck–Condon principle, the molecule undergoes a transition to the upper vibrational state that most closely resembles the vibrational wavefunction of the vibrational ground state of the lower electronic state. The two wavefunctions shown here have the greatest overlap integral of all the vibrational states of the upper electronic state and hence are most closely similar.

Justification 7.2 *The Franck–Condon approximation*

The overall state of the molecule consists of an electronic part, $|\varepsilon\rangle$, and a vibrational part, $|v\rangle$. Therefore, within the Born–Oppenheimer approximation, the transition dipole moment factorizes as follows:

$$\boldsymbol{\mu}_{\rm fi} = \langle \varepsilon_{\rm f} v_{\rm f}|\{-e\sum_i \boldsymbol{r}_i + e\sum_I Z_I \boldsymbol{R}_I\}|\varepsilon_{\rm i} v_{\rm i}\rangle$$

$$= -e\sum_i \langle \varepsilon_{\rm f}|\boldsymbol{r}_i|\varepsilon_{\rm i}\rangle\langle v_{\rm f}|v_{\rm i}\rangle + e\sum_I Z_I\langle \varepsilon_{\rm f}|\varepsilon_{\rm i}\rangle\langle v_{\rm f}|\boldsymbol{R}_I|v_{\rm i}\rangle$$

The second term on the right of the second row is zero, because $\langle \varepsilon_{\rm f}|\varepsilon_{\rm i}\rangle = 0$ for two different electronic states (they are orthogonal). Therefore,

$$\boldsymbol{\mu}_{\rm fi} = -e\sum_i \langle \varepsilon_{\rm f}|\boldsymbol{r}_i|\varepsilon_{\rm i}\rangle\langle v_{\rm f}|v_{\rm i}\rangle = \boldsymbol{\mu}_{\varepsilon_{\rm f},\varepsilon_{\rm i}} S(v_{\rm f},v_{\rm i}) \qquad (7.3)$$

where

$$\boldsymbol{\mu}_{\varepsilon_{\rm f},\varepsilon_{\rm i}} = -e\sum_i \langle \varepsilon_{\rm f}|\boldsymbol{r}_i|\varepsilon_{\rm i}\rangle \qquad S(v_{\rm f},v_{\rm i}) = \langle v_{\rm f}|v_{\rm i}\rangle \qquad (7.4)$$

The matrix element $\boldsymbol{\mu}_{\varepsilon_{\rm f},\varepsilon_{\rm i}}$ is the electric-dipole transition moment arising from the redistribution of electrons (and a measure of the 'kick' this redistribution gives to the electromagnetic field, and vice versa for absorption). The factor $S(v_{\rm f},v_{\rm i})$, is the overlap integral between the vibrational state $|v_{\rm i}\rangle$ in the initial electronic state of the molecule, and the vibrational state $|v_{\rm f}\rangle$ in the final electronic state of the molecule.

Because the transition intensity is proportional to the square of the magnitude of the transition dipole moment, the intensity of an absorption is proportional to $|S(v_f,v_i)|^2$, which is known as the **Franck–Condon factor** for the transition. It follows that, the greater the overlap of the vibrational state wavefunction in the upper electronic state with the vibrational wavefunction in the lower electronic state, the greater the absorption intensity of that particular simultaneous electronic and vibrational transition. This conclusion is the basis of the illustration in Fig. 7.8, where we see that the vibrational wavefunction of the ground state has the greatest overlap with the vibrational states that have peaks at similar bond lengths in the upper electronic state.

Example 7.1 *Calculating a Franck–Condon factor*

Consider the transition from one electronic state to another, their bond lengths being R_e and R_e' and their force constants equal. Calculate the Franck–Condon factor for the 0–0 transition and show that the transition is most intense when the bond lengths are equal.

Method We need to calculate $S(0,0)$, the overlap integral of the two ground-state vibrational wavefunctions, and then take its square. The difference between harmonic and anharmonic vibrational wavefunctions is negligible for $v = 0$, so harmonic oscillator wavefunctions can be used (Table 2.1).

Answer We use the (real) wavefunctions

$$\psi_0 = \left(\frac{1}{\alpha\pi^{1/2}}\right)^{1/2} e^{-x^2/2\alpha^2} \qquad \psi_0' = \left(\frac{1}{\alpha\pi^{1/2}}\right)^{1/2} e^{-x'^2/2\alpha^2}$$

where $x = R - R_e$ and $x' = R - R_e'$, with $\alpha = (\hbar^2/mk)^{1/4}$ (Section 2.5a). The overlap integral is

$$S(0,0) = \langle 0|0\rangle = \int_{-\infty}^{\infty} \psi_0'\psi_0 dR = \frac{1}{\alpha\pi^{1/2}}\int_{-\infty}^{\infty} e^{-(x^2+x'^2)/2\alpha^2} dx$$

We now write $\alpha z = R - \frac{1}{2}(R_e + R_e')$, and manipulate this expression into

$$S(0,0) = \frac{1}{\pi^{1/2}} e^{-(R_e-R_e')^2/4\alpha^2} \int_{-\infty}^{\infty} e^{-z^2} dz$$

The value of the integral is $\pi^{1/2}$. Therefore, the overlap integral is

$$S(0,0) = e^{-(R_e-R_e')^2/4\alpha^2}$$

and the Franck–Condon factor is

$$S(0,0)^2 = e^{-(R_e-R_e')^2/2\alpha^2}$$

This factor is equal to 1 when $R_e' = R_e$ and decreases as the equilibrium bond lengths diverge from each other (Fig. 7.9).

For Br_2, $R_e = 228$ pm and there is an upper state with $R_e' = 266$ pm. Taking the vibrational wavenumber as 250 cm^{-1} gives $S(0,0)^2 = 5.1 \times 10^{-10}$, so the intensity of the 0–0 transition is only 5.1×10^{-10} what it would have been if the potential curves had been directly above each other.

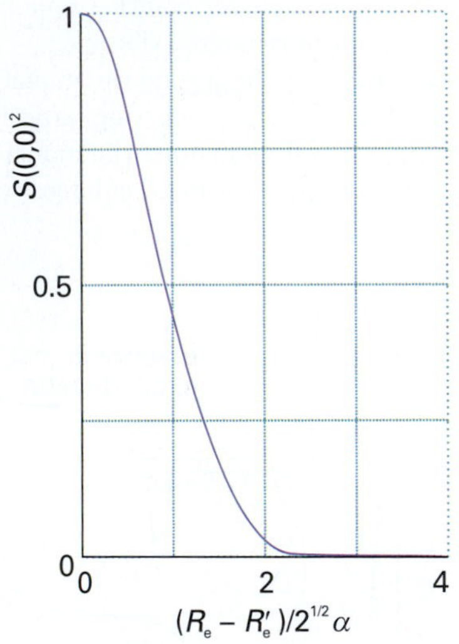

Fig. 7.9 The Franck–Condon factor for the arrangement discussed in Example 7.1.

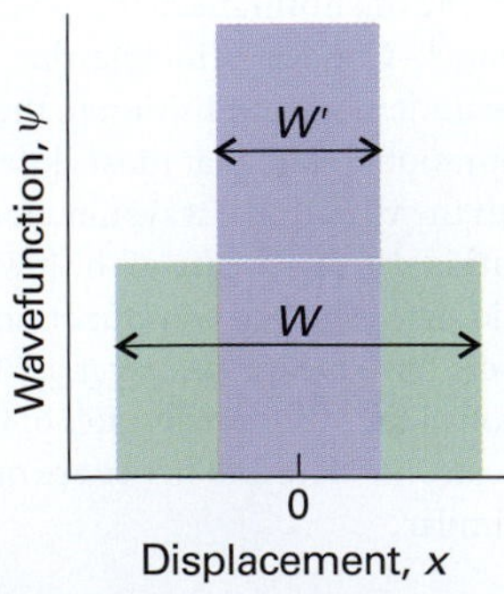

Fig. 7.10 The model wavefunctions used in Self-test 7.2.

Self-test 7.2 Suppose the vibrational wavefunctions can be approximated by rectangular functions of width W and W', centred on the equilibrium bond lengths (Fig. 7.10). Find the corresponding Franck–Condon factors when the centres are coincident and $W' < W$. [$S^2 = W'/W$]

(e) Rotational structure

Just as in vibrational spectroscopy, where a vibrational transition is accompanied by rotational excitation, so rotational transitions accompany the excitation of the vibrational excitation that accompanies electronic excitation. We therefore see P, Q, and R branches for each vibrational transition, and the electronic transition has a very rich structure. However, the principal difference is that electronic excitation can result in much larger changes in bond length than vibrational excitation causes alone, and the rotational branches have a more complex structure than in vibration–rotation spectra.

We suppose that the rotational constants of the electronic ground and excited states are B and B', respectively. The rotational energy levels of the initial and final states are

$$E(J) = hcBJ(J+1) \qquad E(J') = hcB'J'(J'+1)$$

and the rotational transitions occur at the following positions relative to the vibrational transition of wavenumber $\tilde{\nu}$ that they accompany:

$$\text{P branch } (\Delta J = -1): \quad \tilde{\nu}_P(J) = \tilde{\nu} - (B'+B)J + (B'-B)J^2 \qquad (7.5a)$$

$$\text{Q branch } (\Delta J = 0): \quad \tilde{\nu}_Q(J) = \tilde{\nu} + (B'-B)J(J+1) \qquad (7.5b)$$

$$\text{R branch } (\Delta J = +1): \quad \tilde{\nu}_R(J) = \tilde{\nu} + (B'+B)(J+1) + (B'-B)(J+1)^2 \qquad (7.5c)$$

(These are the analogues of eqn 6.63.) First, suppose that the bond length in the electronically excited state is greater than that in the ground state; then $B' < B$ and $B' - B$ is negative. In this case the lines of the R branch converge with increasing J and when J is such that $|B' - B|(J+1) > B' + B$ the lines start to appear at successively decreasing wavenumbers. That is, the R branch has a **band head** (Fig. 7.11a). When the bond is shorter in the excited state than in the ground state, $B' > B$ and $B' - B$ is positive. In this case, the lines of the P branch begin to converge and go through a head when J is such that $(B' - B)J > B' + B$ (Fig. 7.11b).

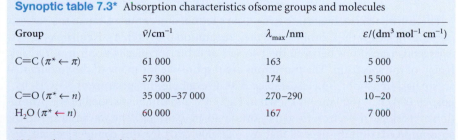

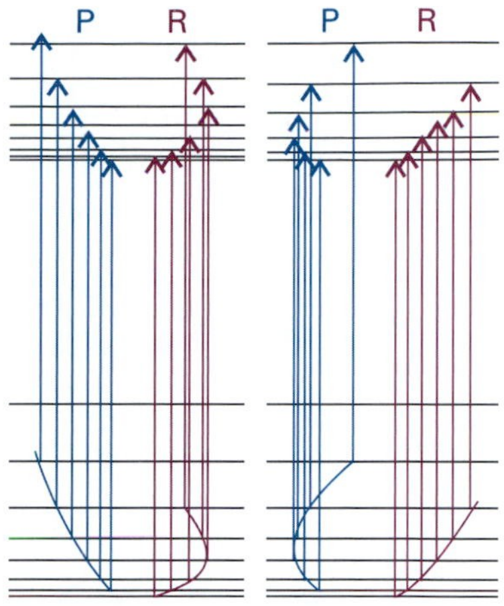

Fig. 7.11 When the rotational constants of a diatomic molecule differ significantly in the initial and final states of an electronic transition, the P and R branches show a head. (a) The formation of a head in the R branch when $B' < B$; (b) the formation of a head in the P branch when $B' > B$.

7.2 The electronic spectra of polyatomic molecules

The absorption of a photon can often be traced to the excitation of specific types of electrons or to electrons that belong to a small group of atoms in a polyatomic molecule. For example, when a carbonyl group (> C=O) is present, an absorption at about 290 nm is normally observed, although its precise location depends on the nature of the rest of the molecule. Groups with characteristic optical absorptions are called **chromophores** (from the Greek for 'colour bringer'), and their presence often accounts for the colours of substances (Table 7.3).

Comment 7.2
The web site for this text contains links to databases of electronic spectra.

(a) *d–d* transitions

In a free atom, all five *d* orbitals of a given shell are degenerate. In a *d*-metal complex, where the immediate environment of the atom is no longer spherical, the *d* orbitals

Synoptic table 7.3* Absorption characteristics ofsome groups and molecules

Group	$\tilde{\nu}$/cm^{-1}	λ_{max}/nm	ε/(dm^3 mol^{-1} cm^{-1})
C=C ($\pi^* \leftarrow \pi$)	61 000	163	5 000
	57 300	174	15 500
C=O ($\pi^* \leftarrow n$)	35 000–37 000	270–290	10–20
H_2O ($\pi^* \leftarrow n$)	60 000	167	7 000

* More values are given in the *Data section.*

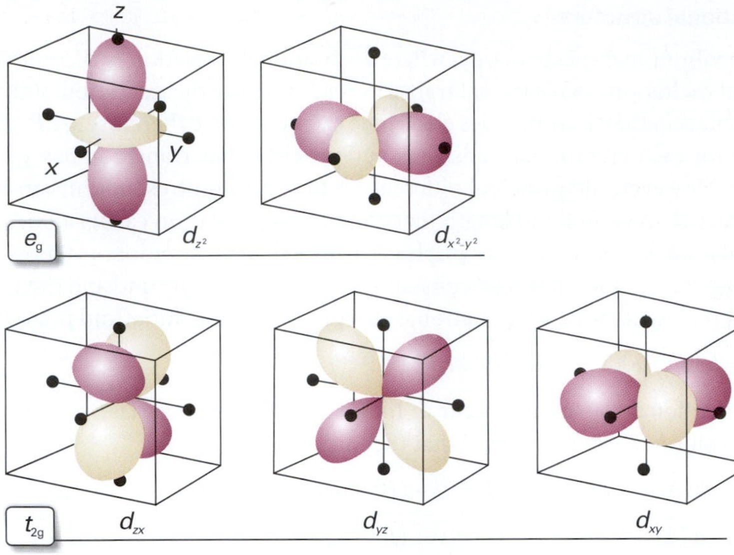

Fig. 7.12 The classification of *d*-orbitals in an octahedral environment.

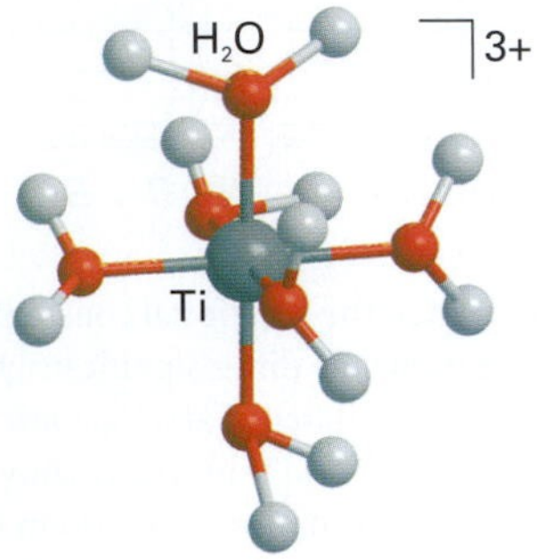

1 $[Ti(OH_2)_6]^{3+}$

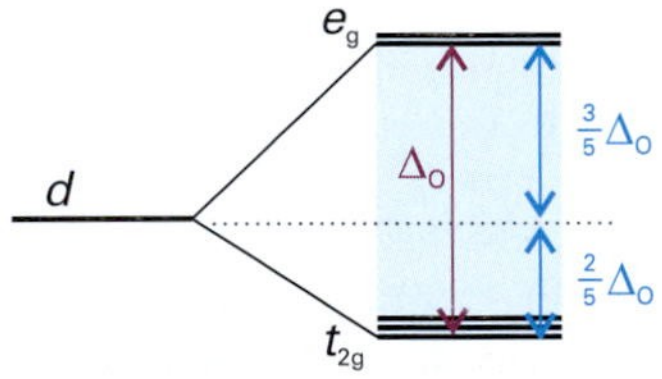

2

are not all degenerate, and electrons can absorb energy by making transitions between them. We show in the following *Justification* that in an octahedral complex, such as $[Ti(OH_2)_6]^{3+}$ (**1**), the five *d* orbitals of the central atom are split into two sets (**2**), a triply degenerate set labelled t_{2g} and a doubly degenerate set labelled e_g. The three t_{2g} orbitals lie below the two e_g orbitals; the difference in energy is denoted Δ_O and called the **ligand-field splitting parameter** (the O denoting octahedral symmetry).

Justification 7.3 *The splitting of* d*-orbitals in an octahedral* d*-metal complex*

In an octahedral *d*-metal complex, six identical ions or molecules, the *ligands*, are at the vertices of a regular octahedron, with the metal ion at its centre. The ligands can be regarded as point negative charges that are repelled by the *d*-electrons of the central ion. Figure 7.12 shows the consequence of this arrangement: the five *d*-orbitals fall into two groups, with $d_{x^2-y^2}$ and d_{z^2} pointing directly towards the ligand positions, and d_{xy}, d_{yz}, and d_{zx} pointing between them. An electron occupying an orbital of the former group has a less favourable potential energy than when it occupies any of the three orbitals of the other group, and so the *d*-orbitals split into the two sets shown in (**2**) with an energy difference Δ_O: a triply degenerate set comprising the d_{xy}, d_{yz}, and d_{zx} orbitals and labelled t_{2g}, and a doubly degenerate set comprising the $d_{x^2-y^2}$ and d_{z^2} orbitals and labelled e_g.

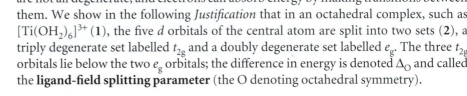

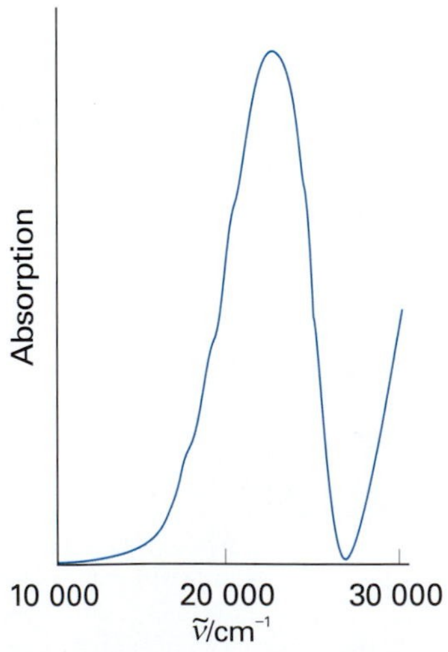

Fig. 7.13 The electronic absorption spectrum of $[Ti(OH_2)_6]^{3+}$ in aqueous solution.

The *d*-orbitals also divide into two sets in a tetrahedral complex, but in this case the *e* orbitals lie below the t_2 orbitals and their separation is written Δ_T. Neither Δ_O nor Δ_T is large, so transitions between the two sets of orbitals typically occur in the visible region of the spectrum. The transitions are responsible for many of the colours that are so characteristic of *d*-metal complexes. As an example, the spectrum of $[Ti(OH_2)_6]^{3+}$ near 20 000 cm^{-1} (500 nm) is shown in Fig. 7.13, and can be ascribed to the promotion of its single *d* electron from a t_{2g} orbital to an e_g orbital. The wavenumber of the absorption maximum suggests that $\Delta_O \approx 20\ 000$ cm^{-1} for this complex, which corresponds to about 2.5 eV.

According to the Laporte rule (Section 7.1b), *d–d* transitions are parity-forbidden in octahedral complexes because they are g → g transitions (more specifically $e_g \leftarrow t_{2g}$ transitions). However, *d–d* transitions become weakly allowed as vibronic transitions as a result of coupling to asymmetrical vibrations such as that shown in Fig. 7.5.

(b) Charge-transfer transitions

A complex may absorb radiation as a result of the transfer of an electron from the ligands into the *d*-orbitals of the central atom, or vice versa. In such **charge-transfer transitions** the electron moves through a considerable distance, which means that the transition dipole moment may be large and the absorption is correspondingly intense. This mode of chromophore activity is shown by the permanganate ion, MnO_4^-, and accounts for its intense violet colour (which arises from strong absorption within the range 420–700 nm). In this oxoanion, the electron migrates from an orbital that is largely confined to the O atom ligands to an orbital that is largely confined to the Mn atom. It is therefore an example of a **ligand-to-metal charge-transfer transition** (LMCT). The reverse migration, a **metal-to-ligand charge-transfer transition** (MLCT), can also occur. An example is the transfer of a *d* electron into the antibonding π orbitals of an aromatic ligand. The resulting excited state may have a very long lifetime if the electron is extensively delocalized over several aromatic rings, and such species can participate in photochemically induced redox reactions.

The intensities of charge-transfer transitions are proportional to the square of the transition dipole moment, in the usual way. We can think of the transition moment as a measure of the distance moved by the electron as it migrates from metal to ligand or vice versa, with a large distance of migration corresponding to a large transition dipole moment and therefore a high intensity of absorption. However, because the integrand in the transition dipole is proportional to the product of the initial and final wavefunctions, it is zero unless the two wavefunctions have nonzero values in the same region of space. Therefore, although large distances of migration favour high intensities, the diminished overlap of the initial and final wavefunctions for large separations of metal and ligands favours low intensities (see Problem 7.17). We encounter similar considerations when we examine electron transfer reactions (Volume 1, Chapter 11), which can be regarded as a special type of charge-transfer transition.

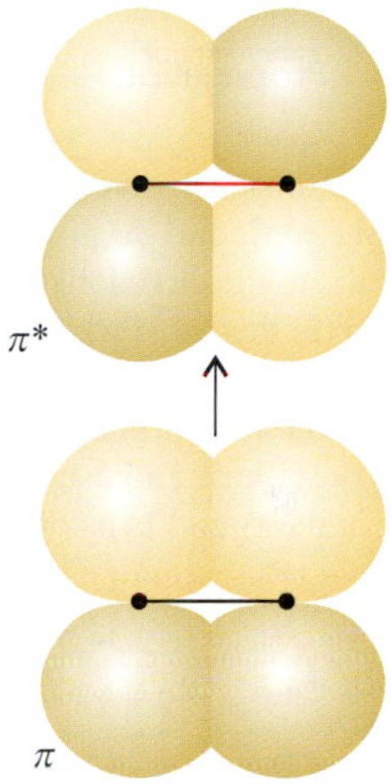

Fig. 7.14 A C=C double bond acts as a chromophore. One of its important transitions is the $\pi^* \leftarrow \pi$ transition illustrated here, in which an electron is promoted from a π orbital to the corresponding antibonding orbital.

(c) $\pi^* \leftarrow \pi$ and $\pi^* \leftarrow n$ transitions

Absorption by a C=C double bond results in the excitation of a π electron into an antibonding π^* orbital (Fig. 7.14). The chromophore activity is therefore due to a $\pi^* \leftarrow \pi$ transition (which is normally read 'π to π-star transition'). Its energy is about 7 eV for an unconjugated double bond, which corresponds to an absorption at 180 nm (in the ultraviolet). When the double bond is part of a conjugated chain, the energies of the molecular orbitals lie closer together and the $\pi^* \leftarrow \pi$ transition moves to longer wavelength; it may even lie in the visible region if the conjugated system is long enough. An important example of an $\pi^* \leftarrow \pi$ transition is provided by the photochemical mechanism of vision (*Impact I7.1*).

The transition responsible for absorption in carbonyl compounds can be traced to the lone pairs of electrons on the O atom. The Lewis concept of a 'lone pair' of electrons is represented in molecular orbital theory by a pair of electrons in an orbital confined largely to one atom and not appreciably involved in bond formation. One of these electrons may be excited into an empty π^* orbital of the carbonyl group (Fig. 7.15), which gives rise to a $\pi^* \leftarrow n$ transition (an '*n* to π-star transition'). Typical absorption energies are about 4 eV (290 nm). Because $\pi^* \leftarrow n$ transitions in carbonyls are symmetry forbidden, the absorptions are weak.

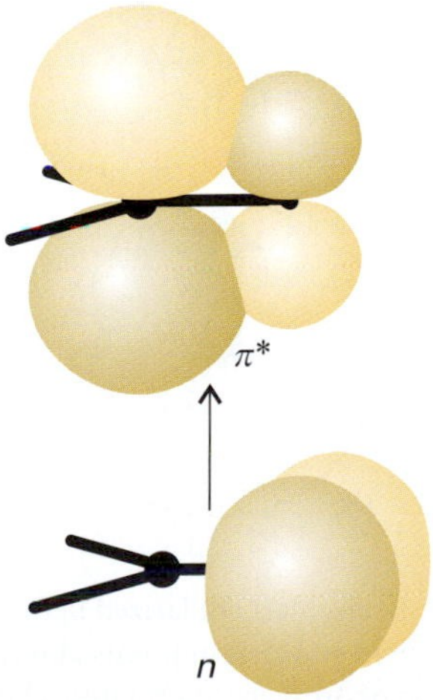

Fig. 7.15 A carbonyl group (C=O) acts as a chromophore primarily on account of the excitation of a nonbonding O lone-pair electron to an antibonding CO π orbital.

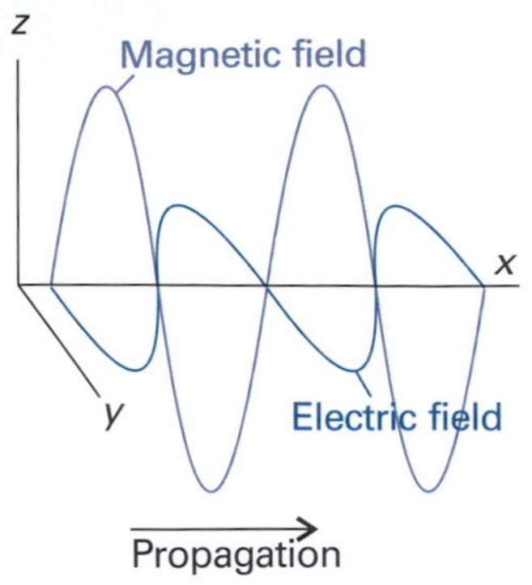

Fig. 7.16 Electromagnetic radiation consists of a wave of electric and magnetic fields perpendicular to the direction of propagation (in this case the *x*-direction), and mutually perpendicular to each other. This illustration shows a plane-polarized wave, with the electric and magnetic fields oscillating in the *xy* and *xz* planes, respectively.

(d) Circular dichroism spectroscopy

Electronic spectra can reveal additional details of molecular structure when experiments are conducted with **polarized light**, electromagnetic radiation with electric and magnetic fields that oscillate only in certain directions. Light is **plane polarized** when the electric and magnetic fields each oscillate in a single plane (Fig. 7.16). The plane of polarization may be oriented in any direction around the direction of propagation (the *x*-direction in Fig. 7.16), with the electric and magnetic fields perpendicular to that direction (and perpendicular to each other). An alternative mode of polarization is **circular polarization**, in which the electric and magnetic fields rotate around the direction of propagation in either a clockwise or a counter-clockwise sense but remain perpendicular to it and each other.

When plane-polarized radiation passes through samples of certain kinds of matter, the plane of polarization is rotated around the direction of propagation. This rotation is the familiar phenomenon of optical activity, observed when the molecules in the sample are chiral (Section 5.3b). Chiral molecules have a second characteristic: they absorb left and right circularly polarized light to different extents. In a circularly polarized ray of light, the electric field describes a helical path as the wave travels through space (Fig. 7.17), and the rotation may be either clockwise or counterclockwise. The differential absorption of left- and right-circularly polarized light is called **circular dichroism.** In terms of the absorbances for the two components, A_L and A_R, the circular dichroism of a sample of molar concentration [J] is reported as

$$\Delta\varepsilon = \varepsilon_L - \varepsilon_R = \frac{A_L - A_R}{[J]l} \qquad (7.6)$$

where l is the path length of the sample.

Circular dichroism is a useful adjunct to visible and UV spectroscopy. For example, the CD spectra of the enantiomeric pairs of chiral *d*-metal complexes are distinctly different, whereas there is little difference between their absorption spectra (Fig. 7.18). Moreover, CD spectra can be used to assign the absolute configuration of complexes by comparing the observed spectrum with the CD spectrum of a similar complex of known handedness. The CD spectra of biological polymers, such as proteins and nucleic acids, give similar structural information. In these cases the spectrum of the polymer chain arises from the chirality of individual monomer units and, in addition, a contribution from the three-dimensional structure of the polymer itself.

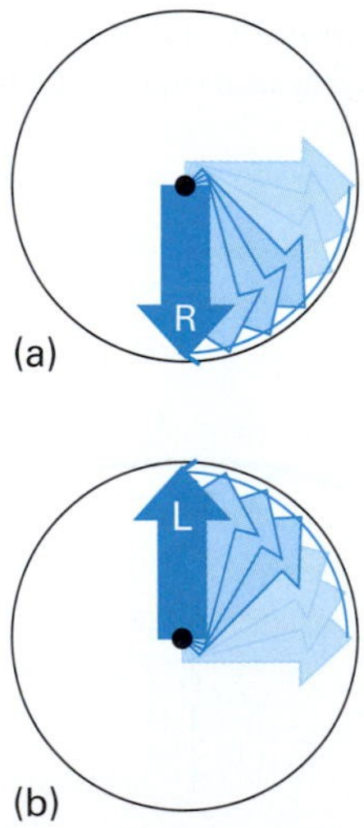

Fig. 7.17 In circularly polarized light, the electric field at different points along the direction of propagation rotates. The arrays of arrows in these illustrations show the view of the electric field when looking toward the oncoming ray: (a) right-circularly polarized, (b) left-circularly polarized light.

IMPACT ON BIOCHEMISTRY

I7.1 Vision

The eye is an exquisite photochemical organ that acts as a transducer, converting radiant energy into electrical signals that travel along neurons. Here we concentrate on the events taking place in the human eye, but similar processes occur in all animals. Indeed, a single type of protein, rhodopsin, is the primary receptor for light throughout the animal kingdom, which indicates that vision emerged very early in evolutionary history, no doubt because of its enormous value for survival.

Photons enter the eye through the cornea, pass through the ocular fluid that fills the eye, and fall on the retina. The ocular fluid is principally water, and passage of light through this medium is largely responsible for the *chromatic aberration* of the eye, the blurring of the image as a result of different frequencies being brought to slightly different focuses. The chromatic aberration is reduced to some extent by the tinted region called the *macular pigment* that covers part of the retina. The pigments in this region are the carotene-like xanthophylls (**3**), which absorb some of the blue light and hence help to sharpen the image. They also protect the photoreceptor molecules from

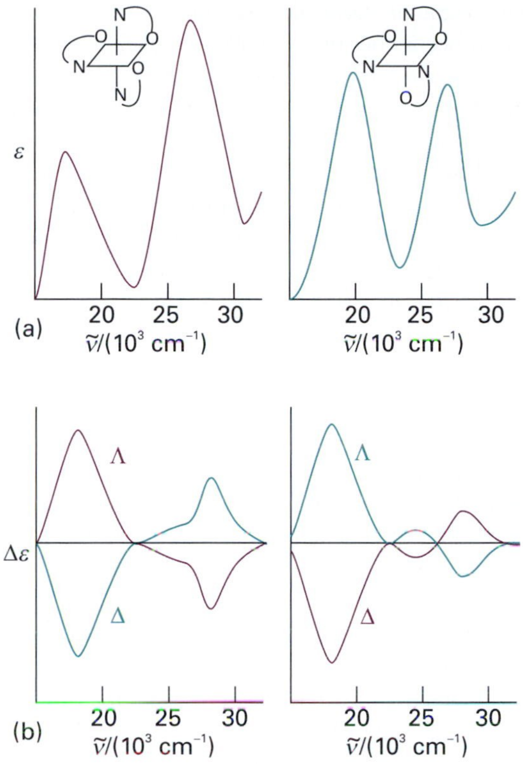

Fig. 7.18 (a) The absorption spectra of two isomers, denoted mer and fac, of $[Co(ala)_3]$, where ala is the conjugate base of alanine, and (b) the corresponding CD spectra. The left- and right-handed forms of these isomers give identical absorption spectra. However, the CD spectra are distinctly different, and the absolute configurations (denoted Λ and Δ) have been assigned by comparison with the CD spectra of a complex of known absolute configuration.

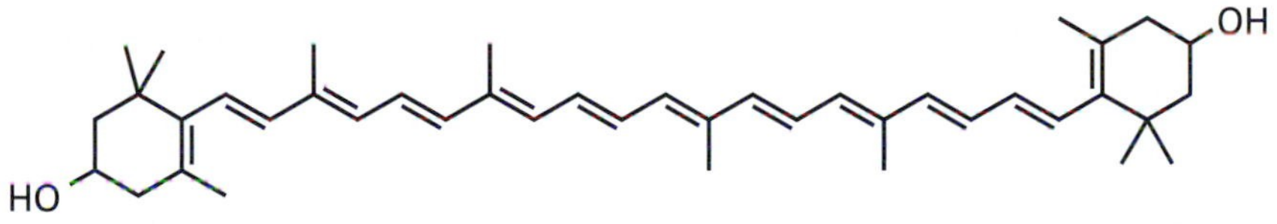

3 A xanthophyll

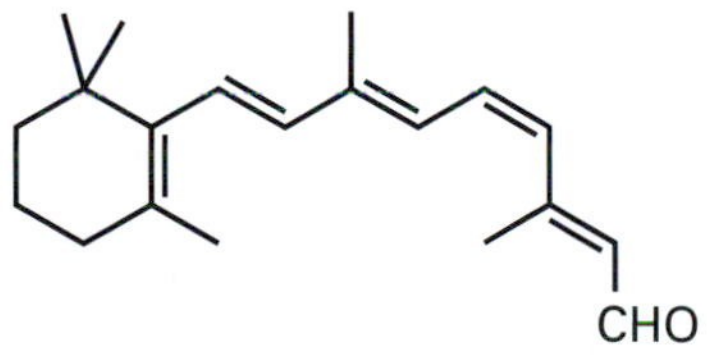

4 11-*cis*-retinal

too great a flux of potentially dangerous high energy photons. The xanthophylls have delocalized electrons that spread along the chain of conjugated double bonds, and the $\pi^* \leftarrow \pi$ transition lies in the visible.

About 57 per cent of the photons that enter the eye reach the retina; the rest are scattered or absorbed by the ocular fluid. Here the primary act of vision takes place, in which the chromophore of a rhodopsin molecule absorbs a photon in another $\pi^* \leftarrow \pi$ transition. A rhodopsin molecule consists of an opsin protein molecule to which is attached a 11-*cis*-retinal molecule (**4**). The latter resembles half a carotene molecule, showing Nature's economy in its use of available materials. The attachment is by the formation of a protonated Schiff's base, utilizing the —CHO group of the chromophore and the terminal NH_2 group of the sidechain, a lysine residue from opsin. The free 11-*cis*-retinal molecule absorbs in the ultraviolet, but attachment to the opsin protein molecule shifts the absorption into the visible region. The rhodopsin molecules are situated in the membranes of special cells (the 'rods' and the 'cones') that cover the retina. The opsin molecule is anchored into the cell membrane by two hydrophobic groups and largely surrounds the chromophore (Fig. 7.19).

Immediately after the absorption of a photon, the 11-*cis*-retinal molecule undergoes photoisomerization into all-*trans*-retinal (**5**). Photoisomerization takes about

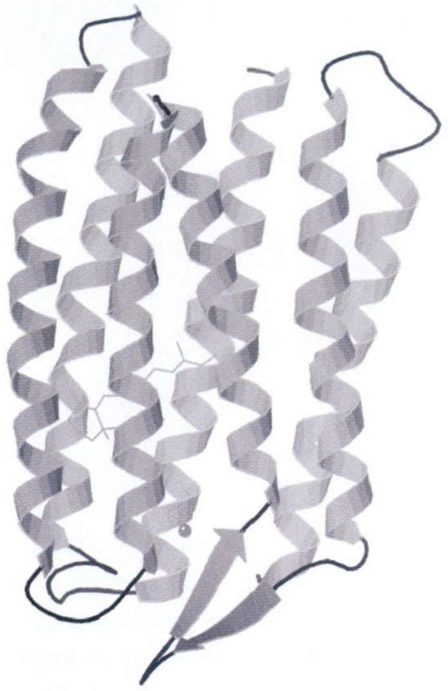

Fig. 7.19 The structure of the rhodopsin molecule, consisting of an opsin protein to which is attached an 11-*cis*-retinal molecule embedded in the space surrounded by the helical regions.

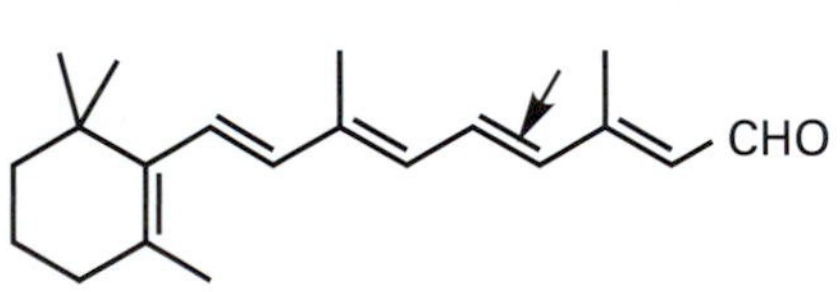

5 All-*trans*-retinal

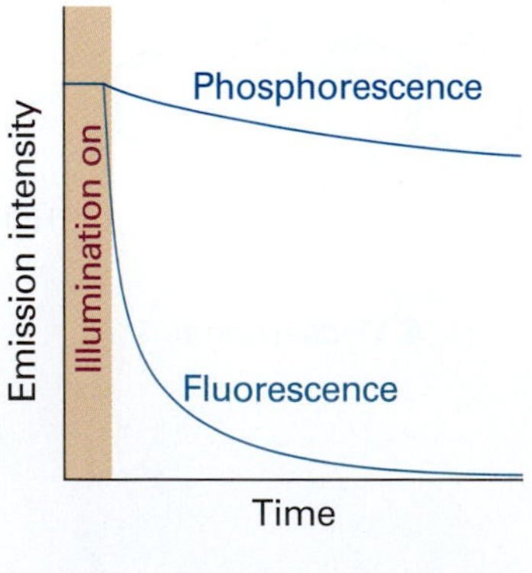

Fig. 7.20 The empirical (observation-based) distinction between fluorescence and phosphorescence is that the former is extinguished very quickly after the exciting source is removed, whereas the latter continues with relatively slowly diminishing intensity.

200 fs and about 67 pigment molecules isomerize for every 100 photons that are absorbed. The process occurs because the $\pi^* \leftarrow \pi$ excitation of an electron loosens one of the π-bonds (the one indicated by the arrow in **5**), its torsional rigidity is lost, and one part of the molecule swings round into its new position. At that point, the molecule returns to its ground state, but is now trapped in its new conformation. The straightened tail of all-*trans*-retinal results in the molecule taking up more space than 11-*cis*-retinal did, so the molecule presses against the coils of the opsin molecule that surrounds it. In about 0.25–0.50 ms from the initial absorption event, the rhodopsin molecule is activated both by the isomerization of retinal and deprotonation of its Schiff's base tether to opsin, forming an intermediate known as *metarhodopsin II*.

In a sequence of biochemical events known as the *biochemical cascade*, metarhodopsin II activates the protein transducin, which in turn activates a phosphodiesterase enzyme that hydrolyses cyclic guanine monophosphate (cGMP) to GMP. The reduction in the concentration of cGMP causes ion channels, proteins that mediate the movement of ions across biological membranes, to close and the result is a sizable change in the transmembrane potential. The pulse of electric potential travels through the optical nerve and into the optical cortex, where it is interpreted as a signal and incorporated into the web of events we call 'vision'.

The resting state of the rhodopsin molecule is restored by a series of nonradiative chemical events powered by ATP. The process involves the escape of all-*trans*-retinal as all-*trans*-retinol (in which —CHO has been reduced to $-CH_2OH$) from the opsin molecule by a process catalysed by the enzyme rhodopsin kinase and the attachment of another protein molecule, arrestin. The free all-*trans*-retinol molecule now undergoes enzyme-catalysed isomerization into 11-*cis*-retinol followed by dehydrogenation to form 11-*cis*-retinal, which is then delivered back into an opsin molecule. At this point, the cycle of excitation, photoisomerization, and regeneration is ready to begin again.

The fates of electronically excited states

A **radiative decay process** is a process in which a molecule discards its excitation energy as a photon. A more common fate is **nonradiative decay**, in which the excess energy is transferred into the vibration, rotation, and translation of the surrounding molecules. This thermal degradation converts the excitation energy completely into thermal motion of the environment (that is, to 'heat'). An excited molecule may also take part in a chemical reaction.

7.3 Fluorescence and phosphorescence

In **fluorescence**, spontaneous emission of radiation occurs within a few nanoseconds after the exciting radiation is extinguished (Fig. 7.20). In **phosphorescence**, the spontaneous emission may persist for long periods (even hours, but characteristically seconds or fractions of seconds). The difference suggests that fluorescence is a fast conversion of absorbed radiation into re-emitted energy, and that phosphorescence involves the storage of energy in a reservoir from which it slowly leaks.

(a) Fluorescence

Figure 7.21 shows the sequence of steps involved in fluorescence. The initial absorption takes the molecule to an excited electronic state, and if the absorption spectrum were monitored it would look like the one shown in Fig. 7.22a. The excited molecule

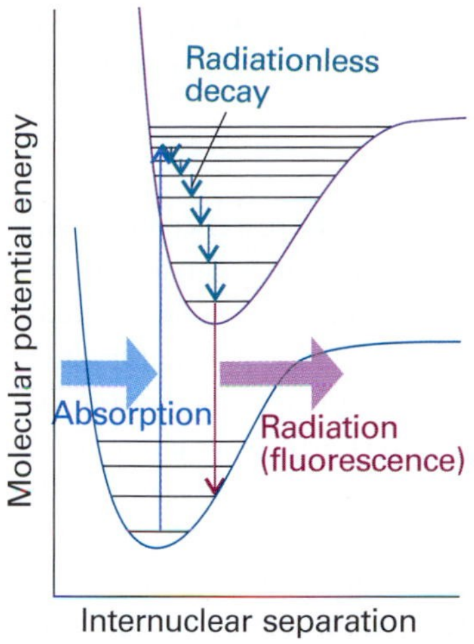

Fig. 7.21 The sequence of steps leading to fluorescence. After the initial absorption, the upper vibrational states undergo radiationless decay by giving up energy to the surroundings. A radiative transition then occurs from the vibrational ground state of the upper electronic state.

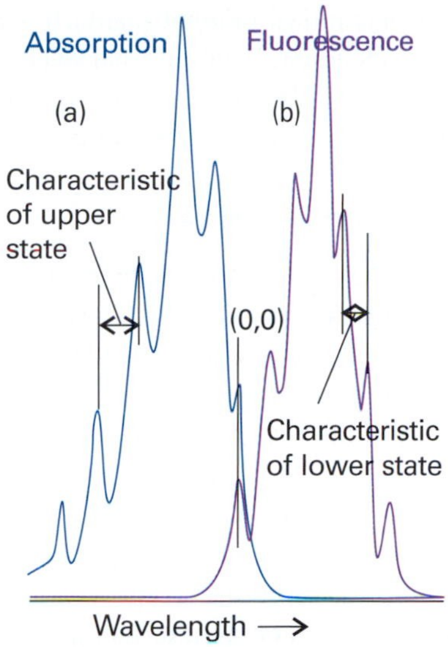

Fig. 7.22 An absorption spectrum (a) shows a vibrational structure characteristic of the upper state. A fluorescence spectrum (b) shows a structure characteristic of the lower state; it is also displaced to lower frequencies (but the 0–0 transitions are coincident) and resembles a mirror image of the absorption.

is subjected to collisions with the surrounding molecules, and as it gives up energy nonradiatively it steps down the ladder of vibrational levels to the lowest vibrational level of the electronically excited molecular state. The surrounding molecules, however, might now be unable to accept the larger energy difference needed to lower the molecule to the ground electronic state. It might therefore survive long enough to undergo spontaneous emission, and emit the remaining excess energy as radiation. The downward electronic transition is vertical (in accord with the Franck–Condon principle) and the fluorescence spectrum has a vibrational structure characteristic of the *lower* electronic state (Fig. 7.22b).

Provided they can be seen, the 0–0 absorption and fluorescence transitions can be expected to be coincident. The absorption spectrum arises from 1–0, 2–0, . . . transitions that occur at progressively higher wavenumber and with intensities governed by the Franck–Condon principle. The fluorescence spectrum arises from 0–0, 0–1, . . . *downward* transitions that hence occur with decreasing wavenumbers. The 0–0 absorption and fluorescence peaks are not always exactly coincident, however, because the solvent may interact differently with the solute in the ground and excited states (for instance, the hydrogen bonding pattern might differ). Because the solvent molecules do not have time to rearrange during the transition, the absorption occurs in an environment characteristic of the solvated ground state; however, the fluorescence occurs in an environment characteristic of the solvated excited state (Fig. 7.23).

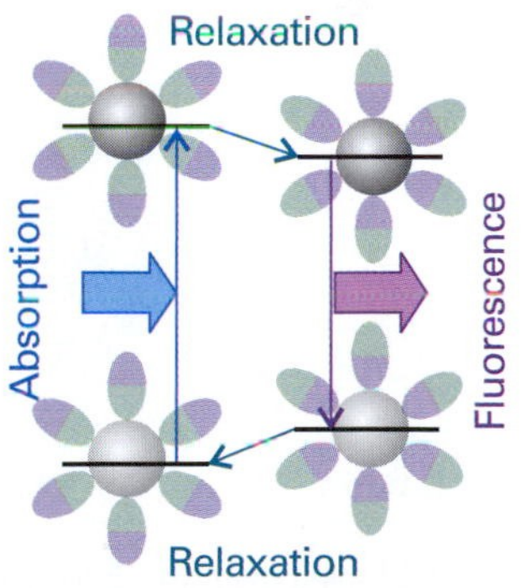

Fig. 7.23 The solvent can shift the fluorescence spectrum relative to the absorption spectrum. On the left we see that the absorption occurs with the solvent (the ellipses) in the arrangement characteristic of the ground electronic state of the molecule (the sphere). However, before fluorescence occurs, the solvent molecules relax into a new arrangement, and that arrangement is preserved during the subsequent radiative transition.

Fluorescence occurs at lower frequencies (longer wavelengths) than the incident radiation because the emissive transition occurs after some vibrational energy has been discarded into the surroundings. The vivid oranges and greens of fluorescent dyes are an everyday manifestation of this effect: they absorb in the ultraviolet and blue, and fluoresce in the visible. The mechanism also suggests that the intensity of the fluorescence ought to depend on the ability of the solvent molecules to accept the

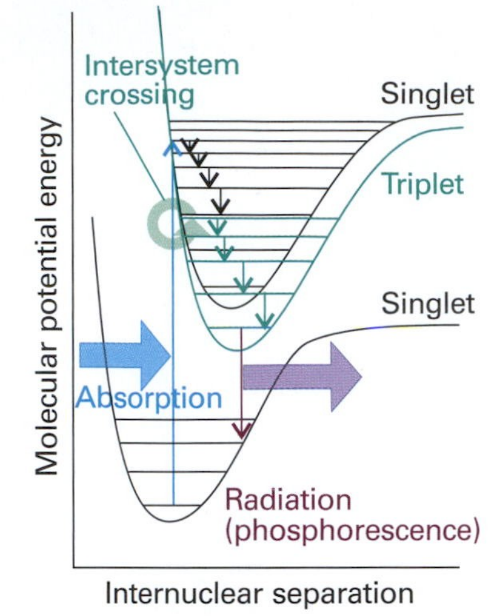

Fig. 7.24 The sequence of steps leading to phosphorescence. The important step is the intersystem crossing, the switch from a singlet state to a triplet state brought about by spin–orbit coupling. The triplet state acts as a slowly radiating reservoir because the return to the ground state is spin-forbidden.

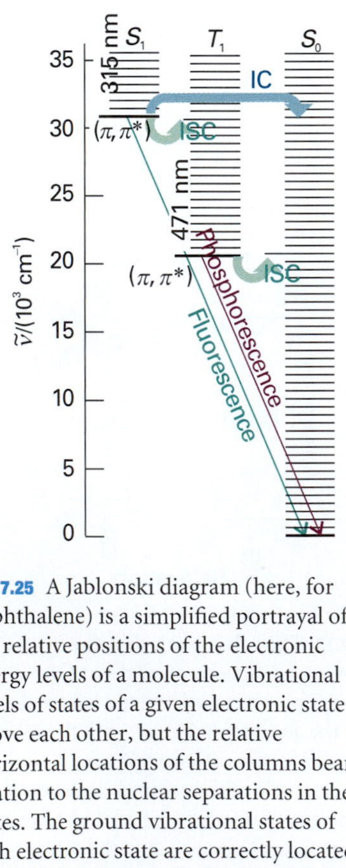

Fig. 7.25 A Jablonski diagram (here, for naphthalene) is a simplified portrayal of the relative positions of the electronic energy levels of a molecule. Vibrational levels of states of a given electronic state lie above each other, but the relative horizontal locations of the columns bear no relation to the nuclear separations in the states. The ground vibrational states of each electronic state are correctly located vertically but the other vibrational states are shown only schematically. (IC: internal conversion; ISC: intersystem crossing.)

electronic and vibrational quanta. It is indeed found that a solvent composed of molecules with widely spaced vibrational levels (such as water) can in some cases accept the large quantum of electronic energy and so extinguish, or 'quench', the fluorescence. The mechanism of fluorescence quenching is discussed in Volume 1, Chapter 10.

(b) Phosphorescence

Figure 7.24 shows the sequence of events leading to phosphorescence for a molecule with a singlet ground state. The first steps are the same as in fluorescence, but the presence of a triplet excited state plays a decisive role. The singlet and triplet excited states share a common geometry at the point where their potential energy curves intersect. Hence, if there is a mechanism for unpairing two electron spins (and achieving the conversion of ↑↓ to ↑↑), the molecule may undergo **intersystem crossing**, a nonradiative transition between states of different multiplicity, and become a triplet state. We saw in the discussion of atomic spectra (Section 3.9d) that singlet–triplet transitions may occur in the presence of spin–orbit coupling, and the same is true in molecules. We can expect intersystem crossing to be important when a molecule contains a moderately heavy atom (such as S), because then the spin–orbit coupling is large.

If an excited molecule crosses into a triplet state, it continues to deposit energy into the surroundings. However, it is now stepping down the triplet's vibrational ladder, and at the lowest energy level it is trapped because the triplet state is at a lower energy than the corresponding singlet (recall Hund's rule, Section 6.7). The solvent cannot absorb the final, large quantum of electronic excitation energy, and the molecule cannot radiate its energy because return to the ground state is spin-forbidden (Section 7.1). The radiative transition, however, is not totally forbidden because the spin–orbit coupling that was responsible for the intersystem crossing also breaks the selection rule. The molecules are therefore able to emit weakly, and the emission may continue long after the original excited state was formed.

The mechanism accounts for the observation that the excitation energy seems to get trapped in a slowly leaking reservoir. It also suggests (as is confirmed experimentally) that phosphorescence should be most intense from solid samples: energy transfer is then less efficient and intersystem crossing has time to occur as the singlet excited state steps slowly past the intersection point. The mechanism also suggests that the phosphorescence efficiency should depend on the presence of a moderately heavy atom (with strong spin–orbit coupling), which is in fact the case. The confirmation of the mechanism is the experimental observation (using the sensitive magnetic resonance techniques described in Chapter 8) that the sample is paramagnetic while the reservoir state, with its unpaired electron spins, is populated.

The various types of nonradiative and radiative transitions that can occur in molecules are often represented on a schematic **Jablonski diagram** of the type shown in Fig. 7.25.

IMPACT ON BIOCHEMISTRY
I7.2 Fluorescence microscopy

Apart from a small number of co-factors, such as the chlorophylls and flavins, the majority of the building blocks of proteins and nucleic acids do not fluoresce strongly. Four notable exceptions are the amino acids tryptophan ($\lambda_{\text{abs}} \approx 280$ nm and $\lambda_{\text{fluor}} \approx 348$ nm in water), tyrosine ($\lambda_{\text{abs}} \approx 274$ nm and $\lambda_{\text{fluor}} \approx 303$ nm in water), and phenylalanine ($\lambda_{\text{abs}} \approx 257$ nm and $\lambda_{\text{fluor}} \approx 282$ nm in water), and the oxidized form of the sequence serine–tyrosine–glycine (**6**) found in the green fluorescent protein (GFP) of certain jellyfish. The wild type of GFP from *Aequora victoria* absorbs strongly at 395 nm and emits maximally at 509 nm.

In **fluorescence microscopy**, images of biological cells at work are obtained by attaching a large number of fluorescent molecules to proteins, nucleic acids, and membranes and then measuring the distribution of fluorescence intensity within the illuminated area. A common fluorescent label is GFP. With proper filtering to remove light due to Rayleigh scattering of the incident beam, it is possible to collect light from the sample that contains only fluorescence from the label. However, great care is required to eliminate fluorescent impurities from the sample.

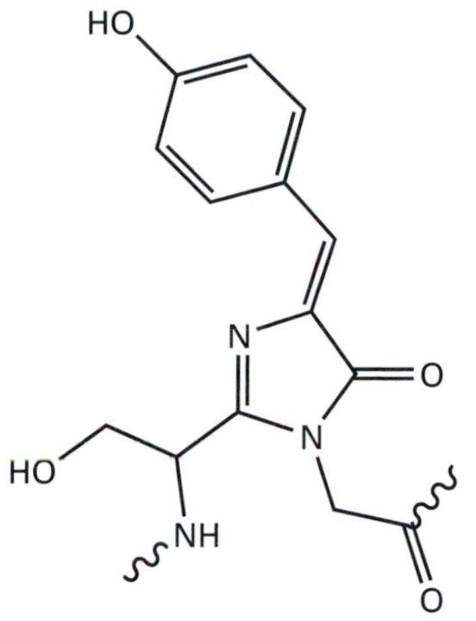

6 The chromophore of GFP

7.4 Dissociation and predissociation

Another fate for an electronically excited molecule is **dissociation**, the breaking of bonds (Fig. 7.26). The onset of dissociation can be detected in an absorption spectrum by seeing that the vibrational structure of a band terminates at a certain energy. Absorption occurs in a continuous band above this **dissociation limit** because the final state is an unquantized translational motion of the fragments. Locating the dissociation limit is a valuable way of determining the bond dissociation energy.

In some cases, the vibrational structure disappears but resumes at higher photon energies. This **predissociation** can be interpreted in terms of the molecular potential energy curves shown in Fig. 7.27. When a molecule is excited to a vibrational level, its electrons may undergo a redistribution that results in it undergoing an **internal conversion**, a radiationless conversion to another state of the same multiplicity. An internal conversion occurs most readily at the point of intersection of the two molecular potential energy curves, because there the nuclear geometries of the two states are the same. The state into which the molecule converts may be dissociative, so the states near the intersection have a finite lifetime, and hence their energies are imprecisely defined. As a result, the absorption spectrum is blurred in the vicinity of the intersection. When the incoming photon brings enough energy to excite the molecule

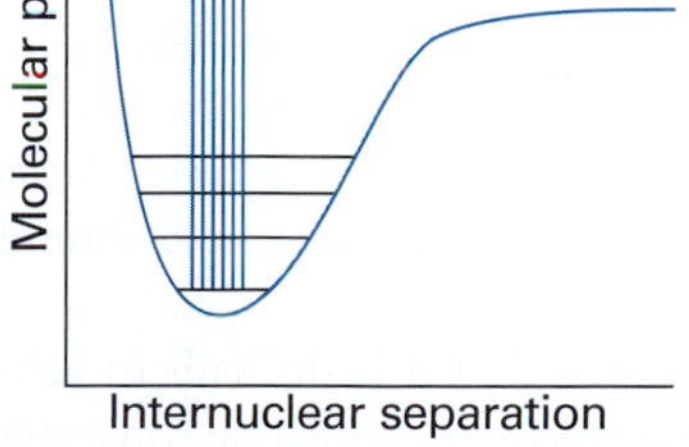

Fig. 7.26 When absorption occurs to unbound states of the upper electronic state, the molecule dissociates and the absorption is a continuum. Below the dissociation limit the electronic spectrum shows a normal vibrational structure.

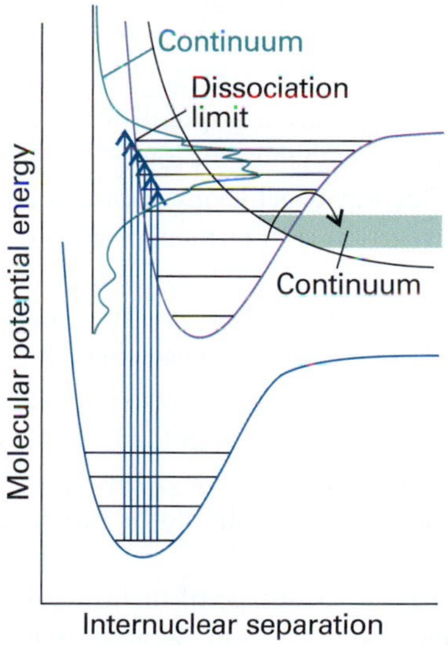

Fig. 7.27 When a dissociative state crosses a bound state, as in the upper part of the illustration, molecules excited to levels near the crossing may dissociate. This process is called predissociation, and is detected in the spectrum as a loss of vibrational structure that resumes at higher frequencies.

to a vibrational level high above the intersection, the internal conversion does not occur (the nuclei are unlikely to have the same geometry). Consequently, the levels resume their well-defined, vibrational character with correspondingly well-defined energies, and the line structure resumes on the high-frequency side of the blurred region.

Lasers

Lasers have transformed chemistry as much as they have transformed the everyday world. They lie very much on the frontier of physics and chemistry, for their operation depends on details of optics and, in some cases, of solid-state processes. In this section, we discuss the mechanisms of laser action, and then explore their applications in chemistry. In *Further information 7.1*, we discuss the modes of operation of a number of commonly available laser systems.

7.5 General principles of laser action

The word laser is an acronym formed from **l**ight **a**mplification by **s**timulated **e**mission of **r**adiation. In stimulated emission, an excited state is stimulated to emit a photon by radiation of the same frequency; the more photons that are present, the greater the probability of the emission. The essential feature of laser action is positive-feedback: the more photons present of the appropriate frequency, the more photons of that frequency that will be stimulated to form.

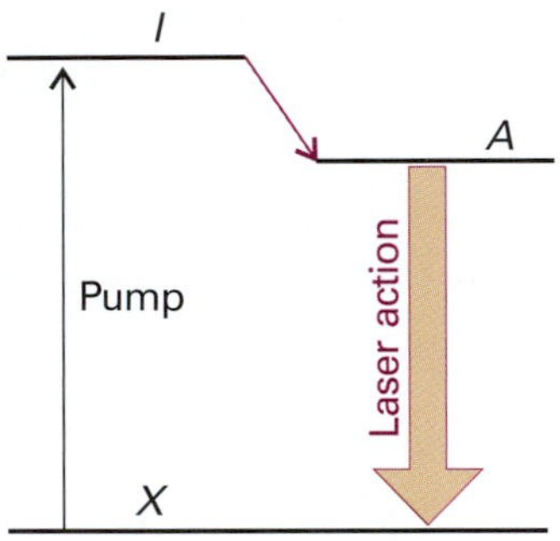

Fig. 7.28 The transitions involved in one kind of three-level laser. The pumping pulse populates the intermediate state *I*, which in turn populates the laser state *A*. The laser transition is the stimulated emission $A \rightarrow X$.

(a) Population inversion

One requirement of laser action is the existence of a **metastable excited state**, an excited state with a long enough lifetime for it to participate in stimulated emission. Another requirement is the existence of a greater population in the metastable state than in the lower state where the transition terminates, for then there will be a net emission of radiation. Because at thermal equilibrium the opposite is true, it is necessary to achieve a **population inversion** in which there are more molecules in the upper state than in the lower.

One way of achieving population inversion is illustrated in Fig. 7.28. The molecule is excited to an intermediate state *I*, which then gives up some of its energy nonradiatively and changes into a lower state *A*; the laser transition is the return of *A* to the ground state *X*. Because three energy levels are involved overall, this arrangement leads to a **three-level laser**. In practice, *I* consists of many states, all of which can convert to the upper of the two laser states *A*. The $I \leftarrow X$ transition is stimulated with an intense flash of light in the process called **pumping**. The pumping is often achieved with an electric discharge through xenon or with the light of another laser. The conversion of *I* to *A* should be rapid, and the laser transitions from *A* to *X* should be relatively slow.

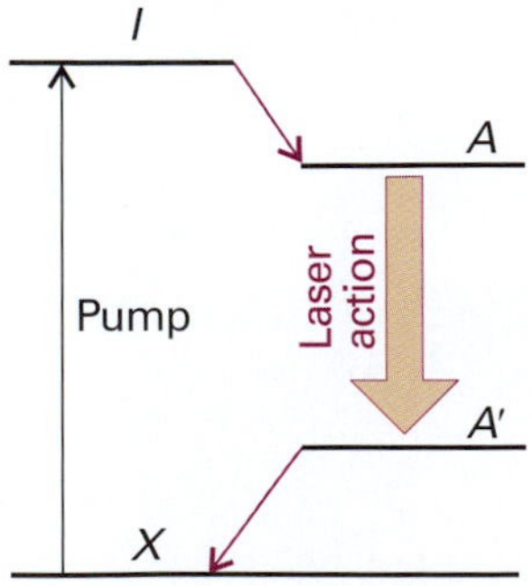

Fig. 7.29 The transitions involved in a four-level laser. Because the laser transition terminates in an excited state (A'), the population inversion between *A* and A' is much easier to achieve.

The disadvantage of this three-level arrangement is that it is difficult to achieve population inversion, because so many ground-state molecules must be converted to the excited state by the pumping action. The arrangement adopted in a **four-level laser** simplifies this task by having the laser transition terminate in a state A' other than the ground state (Fig. 7.29). Because A' is unpopulated initially, any population in *A* corresponds to a population inversion, and we can expect laser action if *A* is sufficiently metastable. Moreover, this population inversion can be maintained if the $A' \leftarrow X$ transitions are rapid, for these transitions will deplete any population in A' that stems from the laser transition, and keep the state A' relatively empty.

(b) Cavity and mode characteristics

The laser medium is confined to a cavity that ensures that only certain photons of a particular frequency, direction of travel, and state of polarization are generated abundantly. The cavity is essentially a region between two mirrors, which reflect the light back and forth. This arrangement can be regarded as a version of the particle in a box, with the particle now being a photon. As in the treatment of a particle in a box (Section 2.1), the only wavelengths that can be sustained satisfy

$$n \times \tfrac{1}{2}\lambda = L \tag{7.7}$$

where n is an integer and L is the length of the cavity. That is, only an integral number of half-wavelengths fit into the cavity; all other waves undergo destructive interference with themselves. In addition, not all wavelengths that can be sustained by the cavity are amplified by the laser medium (many fall outside the range of frequencies of the laser transitions), so only a few contribute to the laser radiation. These wavelengths are the **resonant modes** of the laser.

Photons with the correct wavelength for the resonant modes of the cavity and the correct frequency to stimulate the laser transition are highly amplified. One photon might be generated spontaneously, and travel through the medium. It stimulates the emission of another photon, which in turn stimulates more (Fig. 7.30). The cascade of energy builds up rapidly, and soon the cavity is an intense reservoir of radiation at all the resonant modes it can sustain. Some of this radiation can be withdrawn if one of the mirrors is partially transmitting.

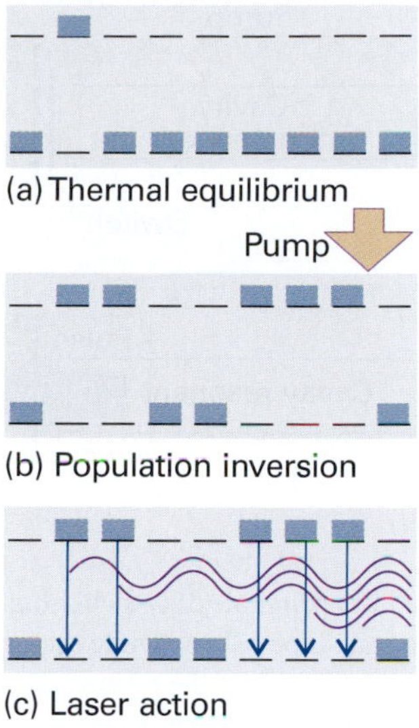

Fig. 7.30 A schematic illustration of the steps leading to laser action. (a) The Boltzmann population of states, with more atoms in the ground state. (b) When the initial state absorbs, the populations are inverted (the atoms are pumped to the excited state). (c) A cascade of radiation then occurs, as one emitted photon stimulates another atom to emit, and so on. The radiation is coherent (phases in step).

The resonant modes of the cavity have various natural characteristics, and to some extent may be selected. Only photons that are travelling strictly parallel to the axis of the cavity undergo more than a couple of reflections, so only they are amplified, all others simply vanishing into the surroundings. Hence, laser light generally forms a beam with very low divergence. It may also be polarized, with its electric vector in a particular plane (or in some other state of polarization), by including a polarizing filter into the cavity or by making use of polarized transitions in a solid medium.

Laser radiation is **coherent** in the sense that the electromagnetic waves are all in step. In **spatial coherence** the waves are in step across the cross-section of the beam emerging from the cavity. In **temporal coherence** the waves remain in step along the beam. The latter is normally expressed in terms of a **coherence length**, l_C, the distance over which the waves remain coherent, and is related to the range of wavelengths, $\Delta\lambda$ present in the beam:

$$l_C = \frac{\lambda^2}{2\Delta\lambda} \tag{7.8}$$

If the beam were perfectly monochromatic, with strictly one wavelength present, $\Delta\lambda$ would be zero and the waves would remain in step for an infinite distance. When many wavelengths are present, the waves get out of step in a short distance and the coherence length is small. A typical light bulb gives out light with a coherence length of only about 400 nm; a He–Ne laser with $\Delta\lambda \approx 2$ pm has a coherence length of about 10 cm.

(c) *Q*-switching

A laser can generate radiation for as long as the population inversion is maintained. A laser can operate continuously when heat is easily dissipated, for then the population of the upper level can be replenished by pumping. When overheating is a problem, the laser can be operated only in pulses, perhaps of microsecond or millisecond duration, so that the medium has a chance to cool or the lower state discard its population.

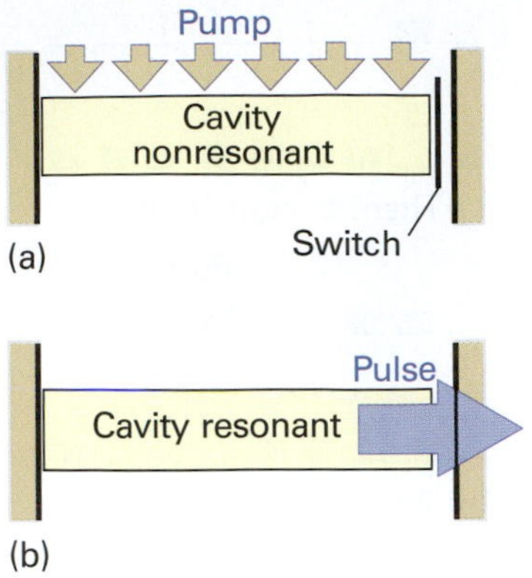

Fig. 7.31 The principle of *Q*-switching. The excited state is populated while the cavity is nonresonant. Then the resonance characteristics are suddenly restored, and the stimulated emission emerges in a giant pulse.

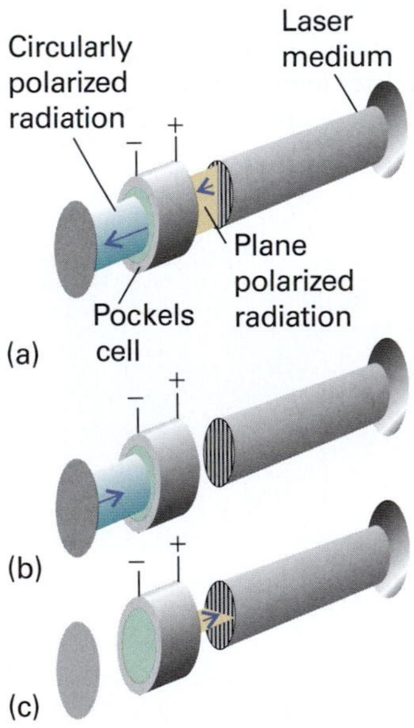

Fig. 7.32 The principle of a Pockels cell. When light passes through a cell that is 'on', its plane of polarization is rotated and so the laser cavity is non-resonant (its *Q*-factor is reduced). In this sequence, (a) the plane polarized ray becomes circularly polarized, (b) is reflected, and (c) emerges from the Pockels cell with perpendicular plane polarization. When the cell is turned off, no change of polarization occurs, and the cavity becomes resonant.

However, it is sometimes desirable to have pulses of radiation rather than a continuous output, with a lot of power concentrated into a brief pulse. One way of achieving pulses is by **Q-switching**, the modification of the resonance characteristics of the laser cavity. The name comes from the '*Q*-factor' used as a measure of the quality of a resonance cavity in microwave engineering.

Example 7.2 *Relating the power and energy of a laser*

A laser rated at 0.10 J can generate radiation in 3.0 ns pulses at a pulse repetition rate of 10 Hz. Assuming that the pulses are rectangular, calculate the peak power output and the average power output of this laser.

Method The power output is the energy released in an interval divided by the duration of the interval, and is expressed in watts ($1\ \text{W} = 1\ \text{J s}^{-1}$). To calculate the peak power output, P_{peak}, we divide the energy released during the pulse divided by the duration of the pulse. The average power output, P_{average}, is the total energy released by a large number of pulses divided by the duration of the time interval over which the total energy was measured. So, the average power is simply the energy released by one pulse multiplied by the pulse repetition rate.

Answer From the data,

$$P_{\text{peak}} = \frac{0.10\ \text{J}}{3.0 \times 10^{-9}\ \text{s}} = 3.3 \times 10^{7}\ \text{J s}^{-1}$$

That is, the peak power output is 33 MW. The pulse repetition rate is 10 Hz, so ten pulses are emitted by the laser for every second of operation. It follows that the average power output is

$$P_{\text{average}} = 0.10\ \text{J} \times 10\ \text{s}^{-1} = 1.0\ \text{J s}^{-1} = 1.0\ \text{W}$$

The peak power is much higher than the average power because this laser emits light for only 30 ns during each second of operation.

Self-test 7.3 Calculate the peak power and average power output of a laser with a pulse energy of 2.0 mJ, a pulse duration of 30 ps, and a pulse repetition rate of 38 MHz. [$P_{\text{peak}} = 67$ MW, $P_{\text{average}} = 76$ kW]

The aim of *Q*-switching is to achieve a healthy population inversion in the absence of the resonant cavity, then to plunge the population-inverted medium into a cavity, and hence to obtain a sudden pulse of radiation. The switching may be achieved by impairing the resonance characteristics of the cavity in some way while the pumping pulse is active, and then suddenly to improve them (Fig. 7.31). One technique is to use a **Pockels cell**, which is an electro-optical device based on the ability of some crystals, such as those of potassium dihydrogenphosphate (KH_2PO_4), to convert plane-polarized light to circularly polarized light when an electrical potential difference is applied. If a Pockels cell is made part of a laser cavity, then its action and the change in polarization that occurs when light is reflected from a mirror convert light polarized in one plane into reflected light polarized in the perpendicular plane (Fig. 7.32). As a result, the reflected light does not stimulate more emission. However, if the cell is suddenly turned off, the polarization effect is extinguished and all the energy stored in the cavity can emerge as an intense pulse of stimulated emission. An alternative technique is to use a **saturable absorber**, typically a solution of a dye that loses its ability to absorb when many of its molecules have been excited by intense radiation. The dye then suddenly becomes transparent and the cavity becomes resonant. In practice, Q-switching can give pulses of about 5 ns duration.

(d) Mode locking

The technique of **mode locking** can produce pulses of picosecond duration and less. A laser radiates at a number of different frequencies, depending on the precise details of the resonance characteristics of the cavity and in particular on the number of half-wavelengths of radiation that can be trapped between the mirrors (the cavity modes). The resonant modes differ in frequency by multiples of $c/2L$ (as can be inferred from eqn 7.8 with $\nu = c/\lambda$). Normally, these modes have random phases relative to each other. However, it is possible to lock their phases together. Then interference occurs to give a series of sharp peaks, and the energy of the laser is obtained in short bursts (Fig. 7.33). The sharpness of the peaks depends on the range of modes superimposed, and the wider the range, the narrower the pulses. In a laser with a cavity of length 30 cm, the peaks are separated by 2 ns. If 1000 modes contribute, the width of the pulses is 4 ps.

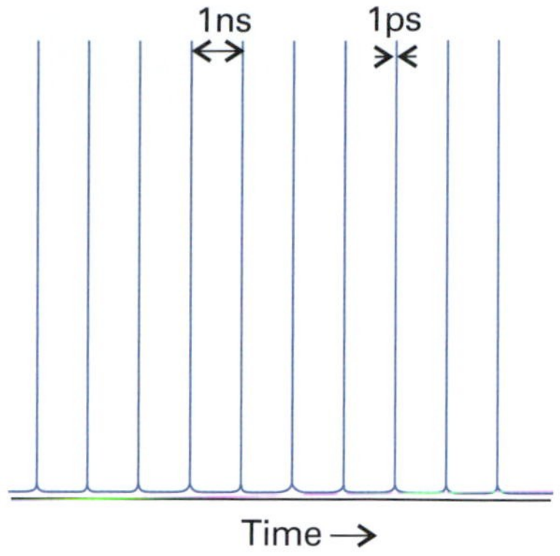

Fig. 7.33 The output of a mode-locked laser consists of a stream of very narrow pulses separated by an interval equal to the time it takes for light to make a round trip inside the cavity.

Justification 7.4 *The origin of mode locking*

The general expression for a (complex) wave of amplitude $\mathcal{E}_0$ and frequency ω is $\mathcal{E}_0 e^{i\omega t}$. Therefore, each wave that can be supported by a cavity of length L has the form

$$\mathcal{E}_n(t) = \mathcal{E}_0 e^{2\pi(\nu + nc/2L)t}$$

where ν is the lowest frequency. A wave formed by superimposing N modes with $n = 0, 1, \ldots, N-1$ has the form

$$\mathcal{E}(t) = \sum_{n=0}^{N-1} \mathcal{E}_n(t) = \mathcal{E}_0 e^{2\pi i\nu t} \sum_{n=0}^{N-1} e^{i\pi nct/L}$$

The sum is a geometrical progression:

$$\sum_{n=0}^{N-1} e^{i\pi nct/L} = 1 + e^{i\pi ct/L} + e^{2i\pi ct/L} + \cdots$$

$$= \frac{\sin(N\pi ct/2L)}{\sin(\pi ct/2L)} \times e^{(N-1)i\pi ct/2L}$$

The intensity, I, of the radiation is proportional to the square modulus of the total amplitude, so

$$I \propto \mathcal{E}^*\mathcal{E} = \mathcal{E}_0^2 \frac{\sin^2(N\pi ct/2L)}{\sin^2(\pi ct/2L)}$$

This function is shown in Fig. 7.34. We see that it is a series of peaks with maxima separated by $t = 2L/c$, the round-trip transit time of the light in the cavity, and that the peaks become sharper as N is increased.

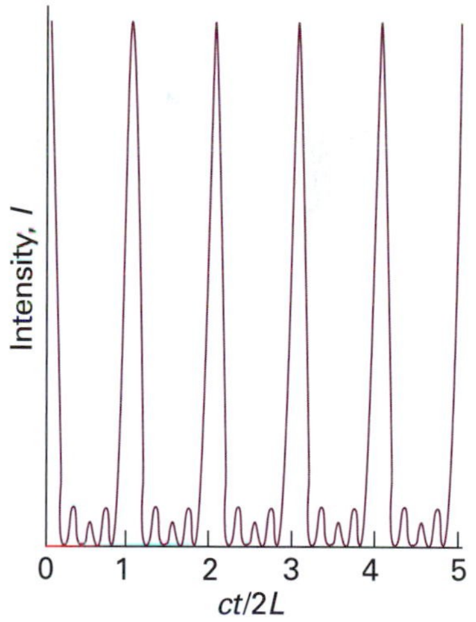

Fig. 7.34 The function derived in *Justification 7.4* showing in more detail the structure of the pulses generated by a mode-locked laser.

Mode locking is achieved by varying the Q-factor of the cavity periodically at the frequency $c/2L$. The modulation can be pictured as the opening of a shutter in synchrony with the round-trip travel time of the photons in the cavity, so only photons making the journey in that time are amplified. The modulation can be achieved by linking a prism in the cavity to a transducer driven by a radiofrequency source at a frequency $c/2L$. The transducer sets up standing-wave vibrations in the prism and modulates the

Table 7.4 Characteristics of laser radiation and their chemical applications

Characteristic	Advantage	Application
High power	Multiphoton process	Nonlinear spectroscopy
		Saturation spectroscopy
	Low detector noise	Improved sensitivity
	High scattering intensity	Raman spectroscopy
Monochromatic	High resolution	Spectroscopy
	State selection	Isotope separation
		Photochemically precise
		State-to-state reaction dynamics
Collimated beam	Long path lengths	Sensitivity
	Forward-scattering observable	Nonlinear Raman spectroscopy
Coherent	Interference between separate beams	CARS
Pulsed	Precise timing of excitation	Fast reactions
		Relaxation
		Energy transfer

loss it introduces into the cavity. The optical properties of some materials can be exploited to bring about mode-locking.

7.6 Applications of lasers in chemistry

Laser radiation has five striking characteristics (Table 7.4). Each of them (sometimes in combination with the others) opens up interesting opportunities in spectroscopy, giving rise to 'laser spectroscopy' and, in photochemistry, giving rise to 'laser photochemistry'. What follows is only an initial listing of applications of lasers to chemistry.

(a) Multiphoton spectroscopy

The large number of photons in an incident beam generated by a laser gives rise to a qualitatively different branch of spectroscopy, for the photon density is so high that more than one photon may be absorbed by a single molecule and give rise to **multiphoton processes**. One application of multiphoton processes is that states inaccessible by conventional one-photon spectroscopy become observable because the overall transition occurs with no change of parity. For example, in one-photon spectroscopy, only $g \leftrightarrow u$ transitions are observable; in two-photon spectroscopy, however, the overall outcome of absorbing two photons is a $g \rightarrow g$ or a $u \rightarrow u$ transition.

(b) Raman spectroscopy

Raman spectroscopy was revitalized by the introduction of lasers. An intense excitation beam increases the intensity of scattered radiation, so the use of laser sources increases the sensitivity of Raman spectroscopy. A well-defined beam also implies that the detector can be designed to collect only the radiation that has passed through a sample, and can be screened much more effectively against stray scattered light, which can obscure the Raman signal. The monochromaticity of laser radiation is also a great advantage, for it makes possible the observation of scattered light that differs by only fractions of reciprocal centimetres from the incident radiation. Such high resolution is particularly useful for observing the rotational structure of Raman lines because rotational transitions are of the order of a few reciprocal centimetres. Monochromaticity also allows observations to be made very close to absorption frequencies, giving rise to the techniques of Fourier-transform Raman spectroscopy (Section 6.1) and resonance Raman spectroscopy (Section 6.16b).

The availability of nondivergent beams makes possible a qualitatively different kind of spectroscopy. The beam is so well-defined that it is possible to observe Raman transitions very close to the direction of propagation of the incident beam. This configuration is employed in the technique called **stimulated Raman spectroscopy**. In this form of spectroscopy, the Stokes and anti-Stokes radiation in the forward direction are powerful enough to undergo more scattering and hence give up or acquire more quanta of energy from the molecules in the sample. This multiple scattering results in lines of frequency $\nu_i \pm 2\nu_M$, $\nu_i \pm 3\nu_M$, and so on, where ν_i is the frequency of the incident radiation and ν_M the frequency of a molecular excitation.

(c) Precision-specified transitions

The monochromatic character of laser radiation is a very powerful characteristic because it allows us to excite specific states with very high precision. One consequence of state-specificity for photochemistry is that the illumination of a sample may be photochemically precise and hence efficient in stimulating a reaction, because its frequency can be tuned exactly to an absorption. The specific excitation of a particular excited state of a molecule may greatly enhance the rate of a reaction even at low temperatures. The rate of a reaction is generally increased by raising the temperature because the energies of the various modes of motion of the molecule are enhanced. However, this enhancement increases the energy of all the modes, even those that do not contribute appreciably to the reaction rate. With a laser we can excite the kinetically significant mode, so rate enhancement is achieved most efficiently. An example is the reaction

$$BCl_3 + C_6H_6 \rightarrow C_6H_5\text{—}BCl_2 + HCl$$

which normally proceeds only above 600°C in the presence of a catalyst; exposure to 10.6 μm CO_2 laser radiation results in the formation of products at room temperature without a catalyst. The commercial potential of this procedure is considerable (provided laser photons can be produced sufficiently cheaply), because heat-sensitive compounds, such as pharmaceuticals, may perhaps be made at lower temperatures than in conventional reactions.

A related application is the study of **state-to-state reaction dynamics**, in which a specific state of a reactant molecule is excited and we monitor not only the rate at which it forms products but also the states in which they are produced. Studies such as these give highly detailed information about the deployment of energy in chemical reactions (Volume 1, Chapter 11).

(d) Isotope separation

The precision state-selectivity of lasers is also of considerable potential for laser isotope separation. Isotope separation is possible because two **isotopomers**, or species that differ only in their isotopic composition, have slightly different energy levels and hence slightly different absorption frequencies.

One approach is to use **photoionization**, the ejection of an electron by the absorption of electromagnetic radiation. Direct photoionization by the absorption of a single photon does not distinguish between isotopomers because the upper level belongs to a continuum; to distinguish isotopomers it is necessary to deal with discrete states. At least two absorption processes are required. In the first step, a photon excites an atom to a higher state; in the second step, a photon achieves photoionization from that state (Fig. 7.35). The energy separation between the two states involved in the first step

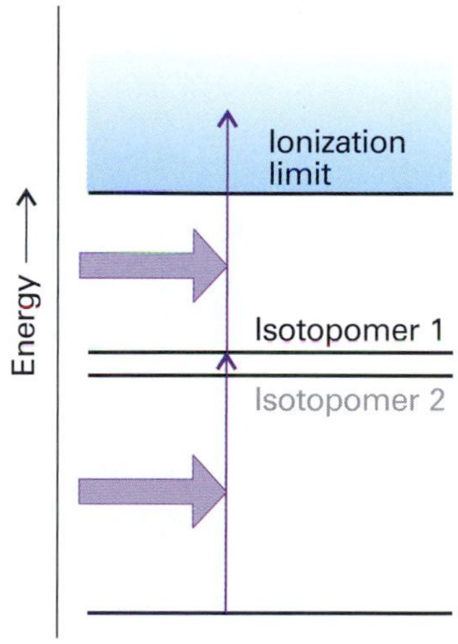

Fig. 7.35 In one method of isotope separation, one photon excites an isotopomer to an excited state, and then a second photon achieves photoionization. The success of the first step depends on the nuclear mass.

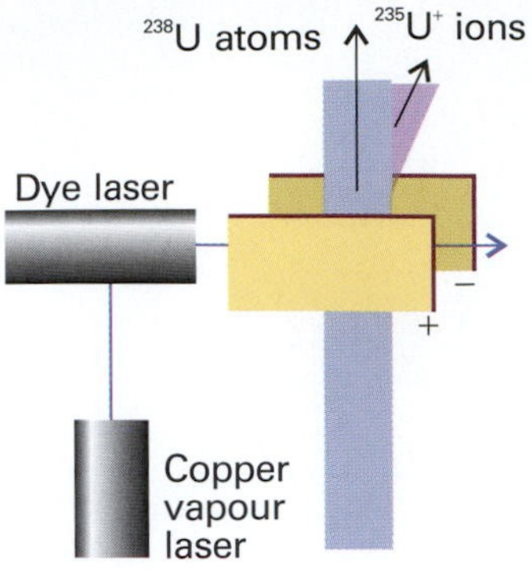

Fig. 7.36 An experimental arrangement for isotope separation. The dye laser, which is pumped by a copper-vapour laser, photoionizes the U atoms selectively according to their mass, and the ions are deflected by the electric field applied between the plates.

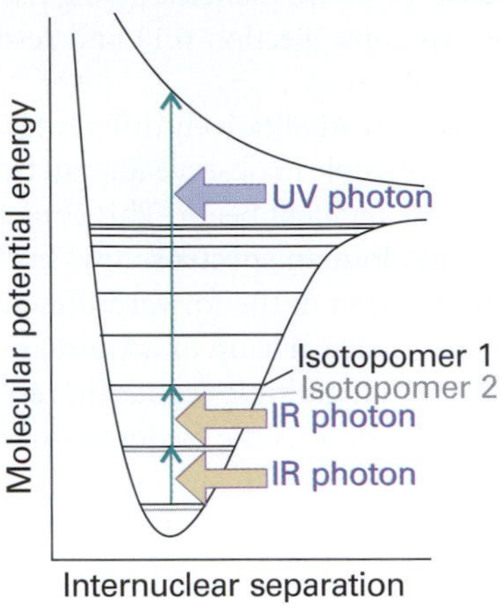

Fig. 7.37 Isotopomers may be separated by making use of their selective absorption of infrared photons followed by photodissociation with an ultraviolet photon.

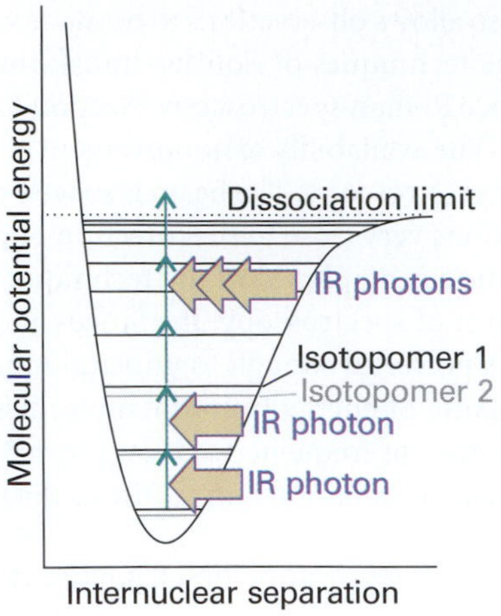

Fig. 7.38 In an alternative scheme for separating isotopomers, multiphoton absorption of infrared photons is used to reach the dissociation limit of a ground electronic state.

depends on the nuclear mass. Therefore, if the laser radiation is tuned to the appropriate frequency, only one of the isotopomers will undergo excitation and hence be available for photoionization in the second step. An example of this procedure is the photoionization of uranium vapour, in which the incident laser is tuned to excite ^{235}U but not ^{238}U. The ^{235}U atoms in the atomic beam are ionized in the two-step process; they are then attracted to a negatively charged electrode, and may be collected (Fig. 7.36). This procedure is being used in the latest generation of uranium separation plants.

Molecular isotopomers are used in techniques based on **photodissociation**, the fragmentation of a molecule following absorption of electromagnetic radiation. The key problem is to achieve both mass selectivity (which requires excitation to take place between discrete states) and dissociation (which requires excitation to continuum states). In one approach, two lasers are used: an infrared photon excites one isotopomer selectively to a higher vibrational level, and then an ultraviolet photon completes the process of photodissociation (Fig. 7.37). An alternative procedure is to make use of multiphoton absorption within the ground electronic state (Fig. 7.38); the efficiency of absorption of the first few photons depends on the match of their frequency to the energy level separations, so it is sensitive to nuclear mass. The absorbed photons open the door to a subsequent influx of enough photons to complete the dissociation process. The isotopomers $^{32}SF_6$ and $^{34}SF_6$ have been separated in this way.

In a third approach, a selectively vibrationally excited species may react with another species and give rise to products that can be separated chemically. This procedure has been employed successfully to separate isotopes of B, N, O, and, most efficiently, H. A variation on this procedure is to achieve selective **photoisomerization**, the conversion of a species to one of its isomers (particularly a geometrical isomer) on absorption of electromagnetic radiation. Once again, the initial absorption, which is isotope selective, opens the way to subsequent further absorption and the formation of a geometrical isomer that can be separated chemically. The approach has been used with the photoisomerization of CH_3NC to CH_3CN.

A different, more physical approach, that of **photodeflection**, is based on the recoil that occurs when a photon is absorbed by an atom and the linear momentum of the photon (which is equal to h/λ) is transferred to the atom. The atom is deflected from its original path only if the absorption actually occurs, and the incident radiation can be tuned to a particular isotope. The deflection is very small, so an atom must absorb dozens of photons before its path is changed sufficiently to allow collection. For instance, if a Ba atom absorbs about 50 photons of 550 nm light, it will be deflected by only about 1 mm after a flight of 1 m.

(e) Time-resolved spectroscopy

The ability of lasers to produce pulses of very short duration is particularly useful in chemistry when we want to monitor processes in time. Q-switched lasers produce nanosecond pulses, which are generally fast enough to study reactions with rates controlled by the speed with which reactants can move through a fluid medium. However, when we want to study the rates at which energy is converted from one mode to another within a molecule, we need femtosecond and picosecond pulses. These timescales are available from mode-locked lasers.

In **time-resolved spectroscopy**, laser pulses are used to obtain the absorption, emission, or Raman spectrum of reactants, intermediates, products, and even transition states of reactions. It is also possible to study energy transfer, molecular rotations, vibrations, and conversion from one mode of motion to another. Some of the information obtained from time-resolved spectroscopy is described in Volume 1, Chapters 9 to 11. Here, we describe some of the experimental techniques that employ pulsed lasers.

The arrangement shown in Fig. 7.39 is often used to study ultrafast chemical reactions that can be initiated by light, such as the initial events of vision (*Impact I7.1*). A strong and short laser pulse, the *pump*, promotes a molecule A to an excited electronic state A^* that can either emit a photon (as fluorescence or phosphorescence) or react with another species B to yield a product C:

$$\begin{array}{ll} A + h\nu \rightarrow A^* & \text{(absorption)} \\ A^* \rightarrow A & \text{(emission)} \\ A^* + B \rightarrow [AB] \rightarrow C & \text{(reaction)} \end{array}$$

Here [AB] denotes either an intermediate or an activated complex. The rates of appearance and disappearance of the various species are determined by observing time-dependent changes in the absorption spectrum of the sample during the course of the reaction. This monitoring is done by passing a weak pulse of white light, the *probe*, through the sample at different times after the laser pulse. Pulsed 'white' light can be generated directly from the laser pulse by the phenomenon of **continuum generation**, in which focusing an ultrafast laser pulse on a vessel containing a liquid such as water, carbon tetrachloride, CaF, or sapphire results in an outgoing beam with a wide distribution of frequencies. A time delay between the strong laser pulse and the 'white' light pulse can be introduced by allowing one of the beams to travel a longer distance before reaching the sample. For example, a difference in travel distance of $\Delta d = 3$ mm corresponds to a time delay $\Delta t = \Delta d/c \approx 10$ ps between two beams, where c is the speed of light. The relative distances travelled by the two beams in Fig. 7.39 are controlled by directing the 'white' light beam to a motorized stage carrying a pair of mirrors.

Variations of the arrangement in Fig. 7.39 allow for the observation of fluorescence decay kinetics of A^* and time-resolved Raman spectra during the course of the reaction. The fluorescence lifetime of A^* can be determined by exciting A as before and measuring the decay of the fluorescence intensity after the pulse with a fast photodetector system. In this case, continuum generation is not necessary. Time-resolved resonance Raman spectra of A, A^*, B, [AB], or C can be obtained by initiating the reaction with a strong laser pulse of a certain wavelength and then, some time later,

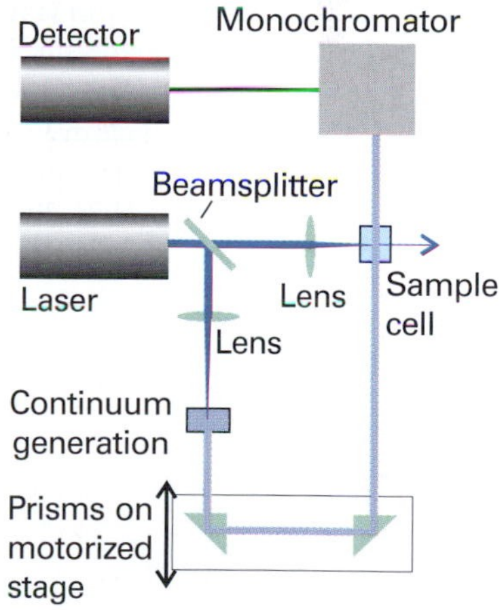

Fig. 7.39 A configuration used for time-resolved absorption spectroscopy, in which the same pulsed laser is used to generate a monochromatic pump pulse and, after continuum generation in a suitable liquid, a 'white' light probe pulse. The time delay between the pump and probe pulses may be varied by moving the motorized stage in the direction shown by the double arrow.

irradiating the sample with another laser pulse that can excite the resonance Raman spectrum of the desired species. Also in this case continuum generation is not necessary. Instead, the Raman excitation beam may be generated in a dye laser (see *Further information 7.1*) or by stimulated Raman scattering of the laser pulse in a medium such as $H_2(g)$ or $CH_4(g)$.

(f) Spectroscopy of single molecules

There is great interest in the development of new experimental probes of very small specimens. On the one hand, our understanding of biochemical processes, such as enzymatic catalysis, protein folding, and the insertion of DNA into the cell's nucleus, will be enhanced if it is possible to visualize individual biopolymers at work. On the other hand, techniques that can probe the structure, dynamics, and reactivity of single molecules will be needed to advance research on nanometre-sized materials.

We saw in *Impact I6.3* that it is possible to obtain the vibrational spectrum of samples with areas of more than 10 μm^2. Fluorescence microscopy (*Impact I7.2*) has also been used for many years to image biological cells, but the diffraction limit prevents the visualization of samples that are smaller than the wavelength of light used as a probe (*Impact I6.3*). Most molecules—including biological polymers—have dimensions that are much smaller than visible wavelengths, so special techniques had to be developed to make single-molecule spectroscopy possible.

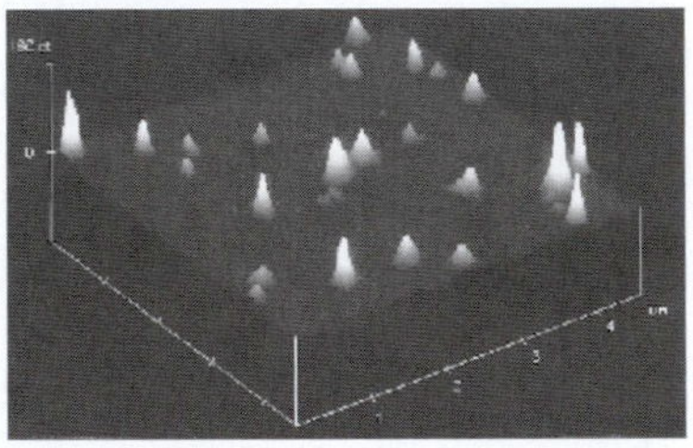

Fig. 7.40 Image of a 4.5 μm × 4.5 μm sample of oxazine-720 dye molecules embedded in a polymer film and obtained with NSOM. Each peak corresponds to a single dye molecule. Reproduced with permission from X.S. Xie. *Acc. Chem. Res.* 1996, **29**, 598.

The bulk of the work done in the field of single-molecule spectroscopy is based on fluorescence microscopy with laser excitation. The laser is the radiation source of choice because it provides the high excitance required to increase the rate of arrival of photons on to the detector from small illuminated areas. Two techniques are commonly used to circumvent the diffraction limit. First, the concentration of the sample is kept so low that, on average, only one fluorescent molecule is in the illuminated area. Second, special strategies are used to illuminate very small volumes. In **near-field optical microscopy** (NSOM), a very thin metal-coated optical fibre is used to deliver light to a small area. It is possible to construct fibres with tip diameters in the range of 50 to 100 nm, which are indeed smaller than visible wavelengths. The fibre tip is placed very close to the sample, in a region known as the *near field*, where, according to classical physics, photons do not diffract. Figure 7.40 shows the image of a 4.5 μm × 4.5 μm sample of oxazine 720 dye molecules embedded in a polymer film and obtained with NSOM by measuring the fluorescence intensity as the tip is scanned over the film surface. Each peak corresponds to a single dye molecule.

In **far-field confocal microscopy**, laser light focused by an objective lens is used to illuminate about 1 μm^3 of a very dilute sample placed beyond the near field. This illumination scheme is limited by diffraction and, as a result, data from far-field microscopy have less structural detail than data from NSOM. However, far-field microscopes are very easy to construct and the technique can be used to probe single molecules as long as there is one molecule, on average, in the illuminated area.

In the **wide-field epifluorescence method**, a two-dimensional array detector (*Further information 6.1*) detects fluorescence excited by a laser and scattered scattered back from the sample (Fig. 7.41a). If the fluorescing molecules are well separated in the specimen, then it is possible to obtain a map of the distribution of fluorescent molecules in the illuminated area. For example, Fig. 7.41b shows how epifluorescence microscopy can be used to observe single molecules of the major histocompatibility (MHC) protein on the surface of a cell.

Though still a relatively new technique, single-molecule spectroscopy has already been used to address important problems in chemistry and biology. Nearly all the techniques discussed in this text measure the average value of a property in a large ensemble of molecules. Single-molecule methods allow a chemist to study the nature

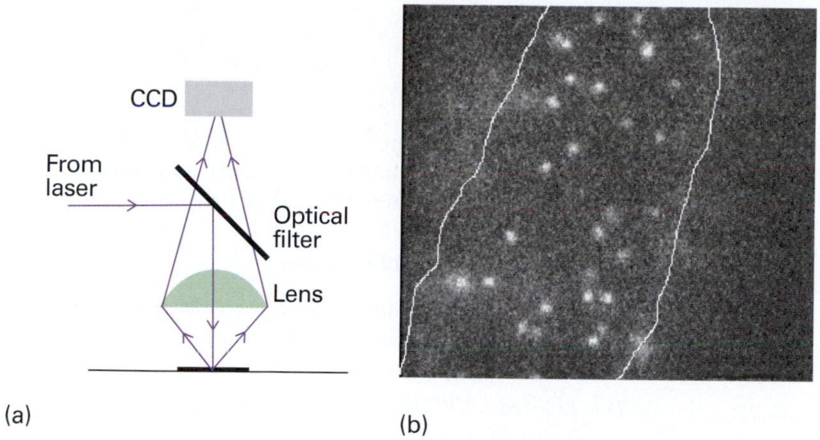

Fig. 7.41 (a) Layout of an epifluorescence microscope. Laser radiation is diverted to a sample by a special optical filter that reflects radiation with a specified wavelength (in this case the laser excitation wavelength) but transmits radiation with other wavelengths (in this case, wavelengths at which the fluorescent label emits). A CCD detector (see *Further information 6.1*) analyses the spatial distribution of the fluorescence signal from the illuminated area. (b) Observation of fluorescence from single MHC proteins that have been labelled with a fluorescent marker and are bound to the surface of a cell (the area shown has dimensions of 12 μm × 12 μm). Image provided by Professor W.E. Moerner, Stanford University, USA.

of distributions of physical and chemical properties in an ensemble of molecules. For example, it is possible to measure the fluorescence lifetime of a molecule by moving the laser focus to a location on the sample that contains a molecule and then measuring the decay of fluorescence intensity after excitation with a pulsed laser. Such studies have shown that not every molecule in a sample has the same fluorescence lifetime, probably because each molecule interacts with its immediate environment in a slightly different way. These details are not apparent from conventional measurements of fluorescence lifetimes, in which many molecules are excited electronically and only an average lifetime for the ensemble can be measured.

Checklist of key ideas

☐ 1. The selection rules for electronic transitions that are concerned with changes in angular momentum are: $\Delta\Lambda = 0, \pm 1$, $\Delta S = 0$, $\Delta\Sigma = 0$, $\Delta\Omega = 0, \pm 1$.

☐ 2. The Laporte selection rule (for centrosymmetric molecules) states that the only allowed transitions are transitions that are accompanied by a change of parity.

☐ 3. The Franck–Condon principle states that, because the nuclei are so much more massive than the electrons, an electronic transition takes place very much faster than the nuclei can respond.

☐ 4. The intensity of an electronic transition is proportional to the Franck–Condon factor, the quantity $|S(v_f,v_i)|^2$, with $S(v_f,v_i) = \langle v_f|v_i\rangle$.

☐ 5. Examples of electronic transitions include *d–d* transitions in *d*-metal complexes, charge-transfer transitions (a transition in which an electron moves from metal to ligand or from ligand to metal in a complex), $\pi^* \leftarrow \pi$, and $\pi^* \leftarrow n$ transitions.

☐ 6. A Jablonski diagram is a schematic diagram of the various types of nonradiative and radiative transitions that can occur in molecules.

☐ 7. Fluorescence is the spontaneous emission of radiation arising from a transition between states of the same multiplicity.

☐ 8. Phosphorescence is the spontaneous emission of radiation arising from a transition between states of different multiplicity.

☐ 9. Intersystem crossing is a nonradiative transition between states of different multiplicity.

☐ 10. Internal conversion is a nonradiative transition between states of the same multiplicity.

☐ 11. Laser action depends on the achievement of population inversion, an arrangement in which there are more molecules in an upper state than in a lower state, and the stimulated emission of radiation.

☐ 12. Resonant modes are the wavelengths that can be sustained by an optical cavity and contribute to the laser action. *Q*-switching is the modification of the resonance characteristics of the laser cavity and, consequently, of the laser output.

☐ 13. Mode locking is a technique for producing pulses of picosecond duration and less by matching the phases of many resonant cavity modes.

☐ 14. Applications of lasers in chemistry include multiphoton spectroscopy, Raman spectroscopy, precision-specified transitions, isotope separation, time-resolved spectroscopy, and single-molecule spectroscopy.

Further reading

Articles and texts

G. Herzberg, *Molecular spectra and molecular structure I. Spectra of diatomic molecules*. Krieger, Malabar (1989).

G. Herzberg, *Molecular spectra and molecular structure III. Electronic spectra and electronic structure of polyatomic molecules*. Van Nostrand–Reinhold, New York (1966).

J.R. Lakowicz, *Principles of fluorescence spectroscopy*. Kluwer/Plenum, New York (1999).

J.C. Lindon, G.E. Tranter, and J.L. Holmes (ed.), *Encyclopedia of spectroscopy and spectrometry*. Academic Press, San Diego (2000).

G.R. van Hecke and K.K. Karukstis, *A guide to lasers in chemistry*. Jones and Bartlett, Boston (1998).

G. Steinmeyer, D.H. Sutter, L. Gallmann, N. Matuschek, and U. Keller, Frontiers in ultrashort pulse generation: pushing the limits in linear and nonlinear optics. *Science* 1999, **286**, 1507.

Sources of data and information

M.E. Jacox, *Vibrational and electronic energy levels of polyatomic transient molecules*. Journal of Physical and Chemical Reference Data , Monograph No. 3 (1994).

D.R. Lide (ed.), *CRC Handbook of Chemistry and Physics*, Sections 9 and 10, CRC Press, Boca Raton (2000).

Further information

Further information 7.1 *Examples of practical lasers*

Figure 7.42 summarizes the requirements for an efficient laser. In practice, the requirements can be satisfied by using a variety of different systems, and this section reviews some that are commonly available. We also include some lasers that operate by using other than electronic transitions.

Comment 7.3

The web site for this text contains links to databases on the optical properties of laser materials.

Gas lasers

Because gas lasers can be cooled by a rapid flow of the gas through the cavity, they can be used to generate high powers. The pumping is normally achieved using a gas that is different from the gas responsible for the laser emission itself.

In the **helium–neon laser** the active medium is a mixture of helium and neon in a mole ratio of about 5:1 (Fig. 7.43). The initial step is the excitation of an He atom to the metastable $1s^12s^1$ configuration by using an electric discharge (the collisions of electrons and ions cause transitions that are not restricted by electric-dipole selection rules). The excitation energy of this transition happens to match an excitation energy of neon, and during an He–Ne collision efficient transfer of energy may occur, leading to the

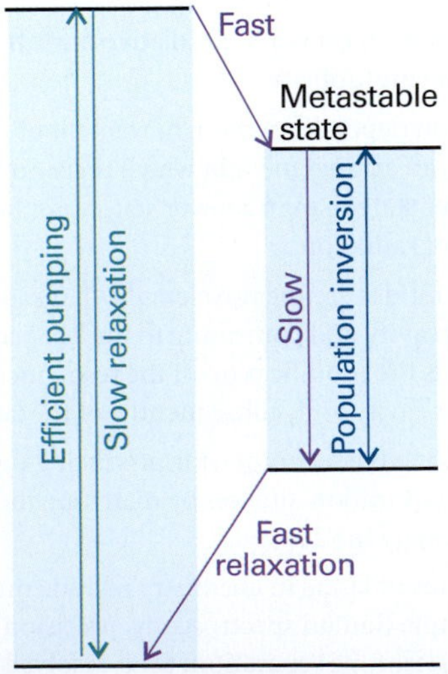

Fig. 7.42 A summary of the features needed for efficient laser action.

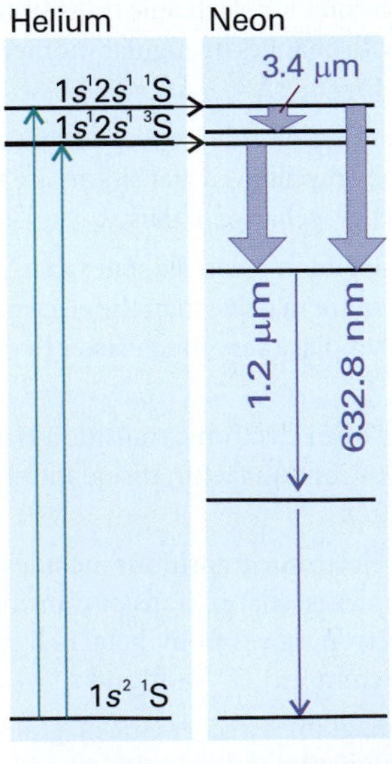

Fig. 7.43 The transitions involved in a helium–neon laser. The pumping (of the neon) depends on a coincidental matching of the helium and neon energy separations, so excited He atoms can transfer their excess energy to Ne atoms during a collision.

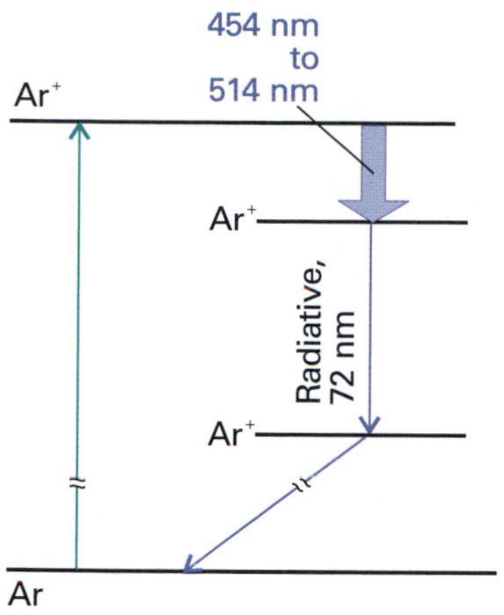

Fig. 7.44 The transitions involved in an argon-ion laser.

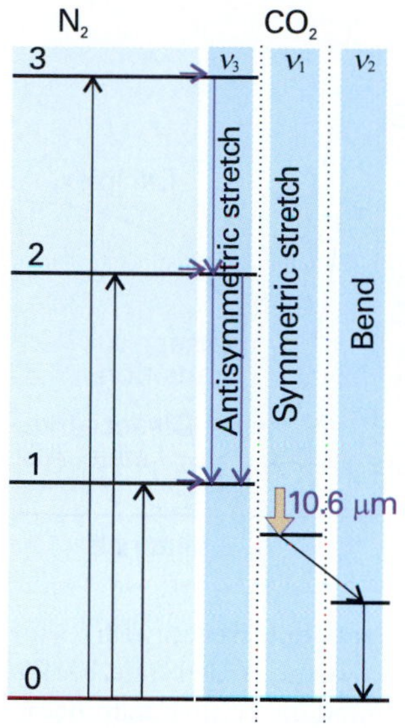

Fig. 7.45 The transitions involved in a carbon dioxide laser. The pumping also depends on the coincidental matching of energy separations; in this case the vibrationally excited N_2 molecules have excess energies that correspond to a vibrational excitation of the antisymmetric stretch of CO_2. The laser transition is from $\nu_3 = 1$ to $\nu_1 = 1$.

production of highly excited, metastable Ne atoms with unpopulated intermediate states. Laser action generating 633 nm radiation (among about 100 other lines) then occurs.

The **argon-ion laser** (Fig. 7.44), one of a number of 'ion lasers', consists of argon at about 1 Torr, through which is passed an electric discharge. The discharge results in the formation of Ar^+ and Ar^{2+} ions in excited states, which undergo a laser transition to a lower state. These ions then revert to their ground states by emitting hard ultraviolet radiation (at 72 nm), and are then neutralized by a series of electrodes in the laser cavity. One of the design problems is to find materials that can withstand this damaging residual radiation. There are many lines in the laser transition because the excited ions may make transitions to many lower states, but two strong emissions from Ar^+ are at 488 nm (blue) and 514 nm (green); other transitions occur elsewhere in the visible region, in the infrared, and in the ultraviolet. The **krypton-ion laser** works similarly. It is less efficient, but gives a wider range of wavelengths, the most intense being at 647 nm (red), but it can also generate yellow, green, and violet lines. Both lasers are widely used in laser light shows (for this application argon and krypton are often used simultaneously in the same cavity) as well as laboratory sources of high-power radiation.

The **carbon dioxide laser** works on a slightly different principle (Fig. 7.45), for its radiation (between 9.2 μm and 10.8 μm, with the strongest emission at 10.6 μm, in the infrared) arises from vibrational transitions. Most of the working gas is nitrogen, which becomes vibrationally excited by electronic and ionic collisions in an electric discharge. The vibrational levels happen to coincide with the ladder of antisymmetric stretch (ν_3, see Fig. 6.40) energy levels of CO_2, which pick up the energy during a collision. Laser action then occurs from the lowest excited level of ν_3 to the lowest excited level of the symmetric stretch (ν_1), which has remained unpopulated during the collisions. This transition is allowed by anharmonicities in the molecular potential energy. Some helium is included in the gas to help remove energy from this state and maintain the population inversion.

In a **nitrogen laser**, the efficiency of the stimulated transition (at 337 nm, in the ultraviolet, the transition $C^3\Pi_u \rightarrow B^3\Pi_g$) is so great that a single passage of a pulse of radiation is enough to generate laser radiation and mirrors are unnecessary: such lasers are said to be **superradiant**.

Chemical and exciplex lasers

Chemical reactions may also be used to generate molecules with nonequilibrium, inverted populations. For example, the photolysis of Cl_2 leads to the formation of Cl atoms which attack H_2 molecules in the mixture and produce HCl and H. The latter then attacks Cl_2 to produce vibrationally excited ('hot') HCl molecules. Because the newly formed HCl molecules have nonequilibrium vibrational populations, laser action can result as they return to lower states. Such processes are remarkable examples of the direct conversion of chemical energy into coherent electromagnetic radiation.

The population inversion needed for laser action is achieved in a more underhand way in **exciplex lasers**, for in these (as we shall see) the lower state does not effectively exist. This odd situation is achieved by forming an **exciplex**, a combination of two atoms that survives only in an excited state and which dissociates as soon as the excitation energy has been discarded. An exciplex can be formed in a mixture of xenon, chlorine, and neon (which acts as a buffer gas). An electric discharge through the mixture produces excited Cl atoms, which attach to the Xe atoms to give the exciplex XeCl*. The exciplex survives for about 10 ns, which is time for it to participate in laser action at 308 nm (in the ultraviolet). As soon as XeCl* has discarded a photon, the atoms separate because the molecular potential energy curve of the ground state is dissociative, and the ground state of the exciplex cannot become populated (Fig. 7.46). The KrF* exciplex laser is another example: it produces radiation at 249 nm.

Comment 7.4

The term 'excimer laser' is also widely encountered and used loosely when 'exciplex laser' is more appropriate. An exciplex has the form AB* whereas an excimer, an excited dimer, is AA*.

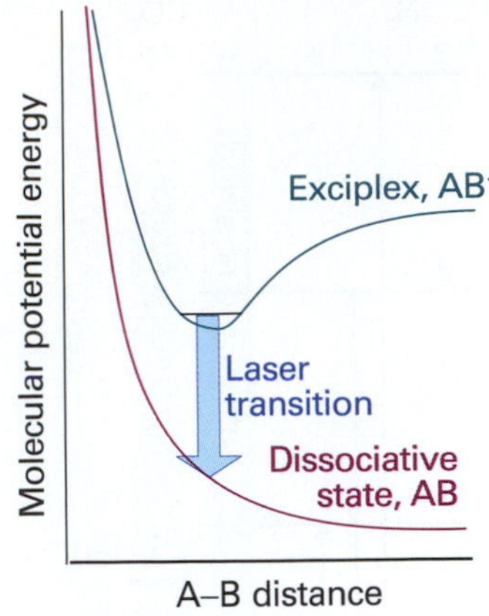

Fig. 7.46 The molecular potential energy curves for an exciplex. The species can survive only as an excited state, because on discarding its energy it enters the lower, dissociative state. Because only the upper state can exist, there is never any population in the lower state.

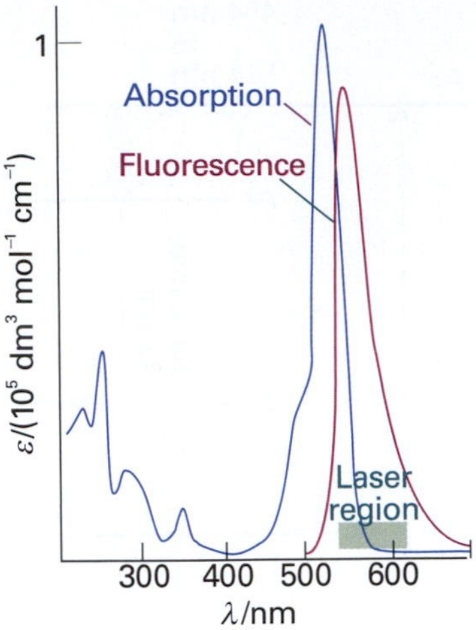

Fig. 7.47 The optical absorption spectrum of the dye Rhodamine 6G and the region used for laser action.

Dye lasers

Gas lasers and most solid state lasers operate at discrete frequencies and, although the frequency required may be selected by suitable optics, the laser cannot be tuned continuously. The tuning problem is overcome by using a titanium sapphire laser (see above) or a **dye laser**, which has broad spectral characteristics because the solvent broadens the vibrational structure of the transitions into bands. Hence, it is possible to scan the wavelength continuously (by rotating the diffraction grating in the cavity) and achieve laser action at any chosen wavelength. A commonly used dye is rhodamine 6G in methanol (Fig. 7.47). As the gain is very high, only a short length of the optical path need be through the dye. The excited states of the active medium, the dye, are sustained by another laser or a flash lamp, and the dye solution is flowed through the laser cavity to avoid thermal degradation (Fig. 7.48).

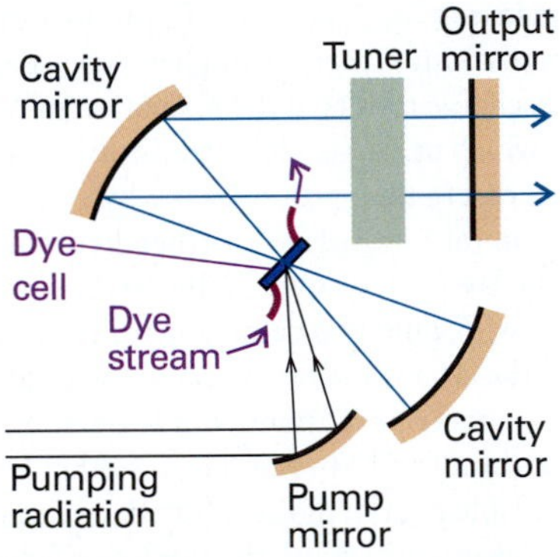

Fig. 7.48 The configuration used for a dye laser. The dye is flowed through the cell inside the laser cavity. The flow helps to keep it cool and prevents degradation.

Each end-of-chapter discussion question, exercise, and problem for which a solution is available in the Student Solution Manual has a corresponding bracketed number referring to the Student Solutions Manual number for this problem, for example [SSM 14.1]. Each end-of-chapter exercise and problem for which a solution is available in the Instructors' Solution Manual has a corresponding bracketed number referring to the Instructor Solutions Manual number for this problem, for example [ISM 14.2].

Discussion questions

7.1 Explain the origin of the term symbol $^3\Sigma_g^-$ for the ground state of dioxygen. [SSM **14.1**]

7.2 Explain the basis of the Franck–Condon principle and how it leads to the formation of a vibrational progression. [ISM **14.2**]

7.3 How do the band heads in P and R branches arise? Could the Q branch show a head? [SSM **14.3**]

7.4 Explain how colour can arise from molecules. [ISM **14.4**]

7.5 Describe the mechanism of fluorescence. To what extent is a fluorescence spectrum not the exact mirror image of the corresponding absorption spectrum? [SSM **14.5**]

7.6 What is the evidence for the correctness of the mechanism of fluorescence? [ISM **14.6**]

7.7 Describe the principles of laser action, with actual examples. [SSM **14.7**]

7.8 What features of laser radiation are applied in chemistry? Discuss two applications of lasers in chemistry. [ISM **14.8**]

Exercises

7.1a The term symbol for the ground state of N_2^+ is $^2\Sigma_g^+$. What is the total spin and total orbital angular momentum of the molecule? Show that the term symbol agrees with the electron configuration that would be predicted using the building-up principle. [ISM **14.1a**]

7.1b One of the excited states of the C_2 molecule has the valence electron configuration $1\sigma_g^2 1\sigma_u^2 1\pi_u^3 1\pi_g^1$. Give the multiplicity and parity of the term. [SSM **14.1b**]

7.2a The molar absorption coefficient of a substance dissolved in hexane is known to be 855 dm^3 mol^{-1} cm^{-1} at 270 nm. Calculate the percentage reduction in intensity when light of that wavelength passes through 2.5 mm of a solution of concentration 3.25 mmol dm^{-3}. [ISM **14.2a**]

7.2b The molar absorption coefficient of a substance dissolved in hexane is known to be 327 dm^3 mol^{-1} cm^{-1} at 300 nm. Calculate the percentage reduction in intensity when light of that wavelength passes through 1.50 mm of a solution of concentration 2.22 mmol dm^{-3}. [SSM **14.2b**]

7.3a A solution of an unknown component of a biological sample when placed in an absorption cell of path length 1.00 cm transmits 20.1 per cent of light of 340 nm incident upon it. If the concentration of the component is 0.111 mmol dm^{-3}, what is the molar absorption coefficient? [ISM **14.3a**]

7.3b When light of wavelength 400 nm passes through 3.5 mm of a solution of an absorbing substance at a concentration 0.667 mmol dm^{-3}, the transmission is 65.5 per cent. Calculate the molar absorption coefficient of the solute at this wavelength and express the answer in cm^2 mol^{-1}. [SSM **14.3b**]

7.4a The molar absorption coefficient of a solute at 540 nm is 286 dm^3 mol^{-1} cm^{-1}. When light of that wavelength passes through a 6.5 mm cell containing a solution of the solute, 46.5 per cent of the light is absorbed. What is the concentration of the solution? [ISM **14.4a**]

7.4b The molar absorption coefficient of a solute at 440 nm is 323 dm^3 mol^{-1} cm^{-1}. When light of that wavelength passes through a 7.50 mm cell containing a solution of the solute, 52.3 per cent of the light is absorbed. What is the concentration of the solution? [SSM **14.4b**]

7.5a The absorption associated with a particular transition begins at 230 nm, peaks sharply at 260 nm, and ends at 290 nm. The maximum value of the molar absorption coefficient is 1.21×10^4 dm^3 mol^{-1} cm^{-1}. Estimate the integrated absorption coefficient of the transition assuming a triangular lineshape (see eqn 6.5). [ISM **14.5a**]

7.5b The absorption associated with a certain transition begins at 199 nm, peaks sharply at 220 nm, and ends at 275 nm. The maximum value of the molar absorption coefficient is 2.25×10^4 dm^3 mol^{-1} cm^{-1}. Estimate the integrated absorption coefficient of the transition assuming an inverted parabolic lineshape (Fig. 7.49; use eqn 6.5). [SSM **14.5b**]

7.6a The two compounds, 2,3-dimethyl-2-butene and 2,5-dimethyl-2,4-hexadiene, are to be distinguished by their ultraviolet absorption spectra. The maximum absorption in one compound occurs at 192 nm and in the other at 243 nm. Match the maxima to the compounds and justify the assignment. [ISM **14.6a**]

7.6b 1,3,5-hexatriene (a kind of 'linear' benzene) was converted into benzene itself. On the basis of a free-electron molecular orbital model (in which hexatriene is treated as a linear box and benzene as a ring), would you expect the lowest energy absorption to rise or fall in energy? [SSM **14.6b**]

7.7a The following data were obtained for the absorption by Br_2 in carbon tetrachloride using a 2.0 mm cell. Calculate the molar absorption coefficient of bromine at the wavelength employed:

$[Br_2]$/(mol dm^{-3})	0.0010	0.0050	0.0100	0.0500
T/(per cent)	81.4	35.6	12.7	3.0×10^{-3}

[ISM **14.7a**]

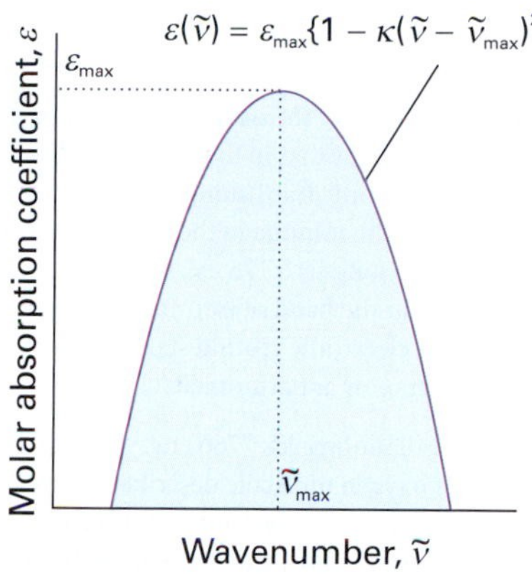

Fig. 7.49

7.7b The following data were obtained for the absorption by a dye dissolved in methylbenzene using a 2.50 mm cell. Calculate the molar absorption coefficient of the dye at the wavelength employed:

[dye]/(mol dm^{-3})	0.0010	0.0050	0.0100	0.0500
T/(per cent)	73	21	4.2	1.33×10^{-5}

[SSM **14.7b**]

7.8a A 2.0-mm cell was filled with a solution of benzene in a non-absorbing solvent. The concentration of the benzene was 0.010 mol dm^{-3} and the wavelength of the radiation was 256 nm (where there is a maximum in the absorption). Calculate the molar absorption coefficient of benzene at this wavelength given that the transmission was 48 per cent. What will the transmittance be in a 4.0-mm cell at the same wavelength? [ISM **14.8a**]

7.8b A 2.50-mm cell was filled with a solution of a dye. The concentration of the dye was 15.5 mmol dm^{-3}. Calculate the molar absorption coefficient of benzene at this wavelength given that the transmission was 32 per cent. What will the transmittance be in a 4.50-mm cell at the same wavelength? [SSM **14.8b**]

7.9a A swimmer enters a gloomier world (in one sense) on diving to greater depths. Given that the mean molar absorption coefficient of sea water in the visible region is 6.2×10^{-3} dm^3 mol^{-1} cm^{-1}, calculate the depth at which a diver will experience (a) half the surface intensity of light, (b) one tenth the surface intensity. [ISM **14.9a**]

7.9b Given that the maximum molar absorption coefficient of a molecule containing a carbonyl group is 30 dm^3 mol^{-1} cm^{-1} near 280 nm, calculate the thickness of a sample that will result in (a) half the initial intensity of radiation, (b) one-tenth the initial intensity. [SSM **14.9b**]

7.10a The electronic absorption bands of many molecules in solution have half-widths at half-height of about 5000 cm^{-1}. Estimate the integrated absorption coefficients of bands for which (a) $\varepsilon_{max} \approx 1 \times 10^4$ dm^3 mol^{-1} cm^{-1}, (b) $\varepsilon_{max} \approx 5 \times 10^2$ dm^3 mol^{-1} cm^{-1}. [ISM **14.10a**]

7.10b The electronic absorption band of a compound in solution had a Gaussian lineshape and a half-width at half-height of 4233 cm^{-1} and $\varepsilon_{max} = 1.54 \times 10^4$ dm^3 mol^{-1} cm^{-1}. Estimate the integrated absorption coefficient. [SSM **14.10b**]

7.11a The photoionization of H_2 by 21 eV photons produces H_2^+. Explain why the intensity of the $v = 2 \leftarrow 0$ transition is stronger than that of the $0 \leftarrow 0$ transition. [ISM **14.11a**]

7.11b The photoionization of F_2 by 21 eV photons produces F_2^+. Would you expect the $2 \leftarrow 0$ transition to be weaker or stronger than the $0 \leftarrow 0$ transition? Justify your answer. [SSM **14.11b**]

Problems*

Numerical problems

7.1 The vibrational wavenumber of the oxygen molecule in its electronic ground state is 1580 cm^{-1}, whereas that in the first excited state ($B\ ^3\Sigma_u^-$), to which there is an allowed electronic transition, is 700 cm^{-1}. Given that the separation in energy between the minima in their respective potential energy curves of these two electronic states is 6.175 eV, what is the wavenumber of the lowest energy transition in the band of transitions originating from the $v = 0$ vibrational state of the electronic ground state to this excited state? Ignore any rotational structure or anharmonicity. [SSM **14.1**]

7.2 A Birge–Sponer extrapolation yields 7760 cm^{-1} as the area under the curve for the B state of the oxygen molecule described in Problem 7.1. Given that the B state dissociates to ground-state atoms (at zero energy, ^3P) and 15 870 cm^{-1} (^1D) and the lowest vibrational state of the B state is 49 363 cm^{-1} above the lowest vibrational state of the ground electronic state, calculate the dissociation energy of the molecular ground state to the ground-state atoms. [ISM **14.2**]

7.3 The electronic spectrum of the IBr molecule shows two low-lying, well defined convergence limits at 14 660 and 18 345 cm^{-1}. Energy levels for the iodine and bromine atoms occur at 0, 7598; and 0, 3685 cm^{-1}, respectively. Other atomic levels are at much higher energies. What possibilities exist for the numerical value of the dissociation energy of IBr? Decide which is the correct possibility by calculating this quantity from $\Delta_f H^\ominus(\mathrm{IBr, g}) = +40.79$ kJ mol^{-1} and the dissociation energies of I_2(g) and Br_2(g) which are 146 and 190 kJ mol^{-1}, respectively. [SSM **14.3**]

7.4 In many cases it is possible to assume that an absorption band has a Gaussian lineshape (one proportional to e^{-x^2}) centred on the band maximum. Assume such a line shape, and show that $A \approx 1.0645\varepsilon_{max}\Delta\tilde{\nu}_{1/2}$, where $\Delta\tilde{\nu}_{1/2}$ is the width at half-height. The absorption spectrum of azoethane ($CH_3CH_2N_2$) between 24 000 cm^{-1} and 34 000 cm^{-1} is shown in Fig. 7.50. First, estimate A for the band by assuming that it is Gaussian. Then integrate the absorption band graphically. The latter can be done either by ruling and counting squares, or by tracing the lineshape on to paper and weighing. A more sophisticated procedure would be to use mathematical software to fit a polynomial to the absorption band (or a Gaussian), and then to integrate the result analytically. [ISM **14.4**]

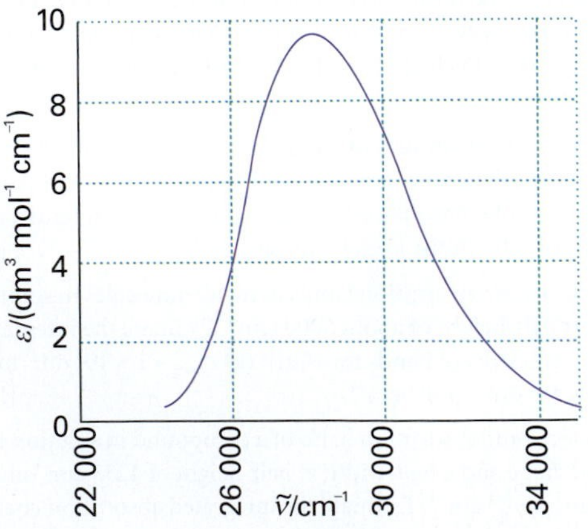

Fig. 7.50

7.5 A lot of information about the energy levels and wavefunctions of small inorganic molecules can be obtained from their ultraviolet spectra. An example of a spectrum with considerable vibrational structure, that of gaseous SO_2 at 25°C, is shown in Fig. 7.6. Estimate the integrated absorption coefficient for the transition. transition. What electronic states are accessible from the A_1 ground state of this C_v molecule by electric dipole transitions? [SSM **14.5**]

7.6‡ J.G. Dojahn, E.C.M. Chen, and W.E. Wentworth (*J. Phys. Chem.* **100**, 9649 (1996)) characterized the potential energy curves of the ground and electronic states of homonuclear diatomic halogen anions. These anions have a $^2\Sigma_u^+$ ground state and $^2\Pi_g$, $^2\Pi_u$, and $^2\Sigma_g^+$ excited states. To which of the excited states are transitions by absorption of photons allowed? Explain. [ISM **14.6**]

7.7 A transition of particular importance in O_2 gives rise to the 'Schumann–Runge band' in the ultraviolet region. The wavenumbers (in cm^{-1}) of transitions from the ground state to the vibrational levels of the first excited state ($^3\Sigma_u^-$) are 50 062.6, 50 725.4, 51 369.0, 51 988.6, 52 579.0, 53 143.4, 53 679.6, 54 177.0, 54 641.8, 55 078.2, 55 460.0, 55 803.1, 56 107.3, 56 360.3, 56 570.6. What is the dissociation energy of the upper electronic state? (Use a Birge–Sponer plot.) The same excited state is known to dissociate into one ground-state O atom and one excited-state atom with an energy 190 kJ mol^{-1} above the ground state. (This excited atom is responsible for a great deal of photochemical mischief in the atmosphere.) Ground-state O_2 dissociates into two ground-state atoms. Use this information to calculate the dissociation energy of ground-state O_2 from the Schumann–Runge data. [SSM **14.7**]

7.8 The compound $CH_3CH{=}CHCHO$ has a strong absorption in the ultraviolet at 46 950 cm^{-1} and a weak absorption at 30 000 cm^{-1}. Justify these features in terms of the structure of the molecule. [ISM **14.8**]

7.9 Aromatic hydrocarbons and I_2 form complexes from which charge-transfer electronic transitions are observed. The hydrocarbon acts as an electron donor and I_2 as an electron acceptor. The energies $h\nu_{max}$ of the charge-transfer transitions for a number of hydrocarbon–I_2 complexes are given below:

Hydrocarbon	benzene	biphenyl	naphthalene	phenanthrene	pyrene	anthracene
$h\nu_{max}$/eV	4.184	3.654	3.452	3.288	2.989	2.890

Investigate the hypothesis that there is a correlation between the energy of the HOMO of the hydrocarbon (from which the electron comes in the charge–transfer transition) and $h\nu_{max}$. Use one of the molecular electronic structure methods discussed in Chapter 4 to determine the energy of the HOMO of each hydrocarbon in the data set.[1] [SSM **14.9**]

7.10 A certain molecule fluoresces at a wavelength of 400 nm with a half-life of 1.0 ns. It phosphoresces at 500 nm. If the ratio of the transition probabilities for stimulated emission for the $S^* \rightarrow S$ to the $T \rightarrow S$ transitions is 1.0×10^5, what is the half-life of the phosphorescent state? [ISM **14.10**]

7.11 Consider some of the precautions that must be taken when conducting single-molecule spectroscopy experiments. (a) What is the molar concentration of a solution in which there is, on average, one solute molecule in 1.0 μm³ (1.0 fL) of solution? (b) It is important to use pure solvents in single-molecule

* Problems denoted with the symbol ‡ were supplied by Charles Trapp and Carmen Giunta.

[1] The web site contains links to molecular modelling freeware and to other sites where you may perform molecular orbital calculations directly from your web browser.

spectroscopy because optical signals from fluorescent impurities in the solvent may mask optical signals from the solute. Suppose that water containing a fluorescent impurity of molar mass 100 g mol^{-1} is used as solvent and that analysis indicates the presence of 0.10 mg of impurity per 1.0 kg of solvent. On average, how many impurity molecules will be present in 1.0 μm^3 of solution? You may take the density of water as 1.0 g cm^{-3}. Comment on the suitability of this solvent for single-molecule spectroscopy experiments. [SSM **14.11**]

7.12 Light-induced degradation of molecules, also called *photobleaching*, is a serious problem in single-molecule spectroscopy. A molecule of a fluorescent dye commonly used to label biopolymers can withstand about 10^6 excitations by photons before light-induced reactions destroy its π system and the molecule no longer fluoresces. For how long will a single dye molecule fluoresce while being excited by 1.0 mW of 488 nm radiation from a continuous-wave argon ion laser? You may assume that the dye has an absorption spectrum that peaks at 488 nm and that every photon delivered by the laser is absorbed by the molecule. [ISM **14.12**]

Theoretical problems

7.13 Assume that the electronic states of the π electrons of a conjugated molecule can be approximated by the wavefunctions of a particle in a one-dimensional box, and that the dipole moment can be related to the displacement along this length by $\mu = -ex$. Show that the transition probability for the transition $n=1 \rightarrow n=2$ is nonzero, whereas that for $n=1 \rightarrow n=3$ is zero. *Hint.* The following relations will be useful:

$$\sin x \sin y = \tfrac{1}{2}\cos(x-y) - \tfrac{1}{2}\cos(x+y)$$

$$\int x \cos ax \, dx = \frac{1}{a^2}\cos ax + \frac{x}{a}\sin ax$$

[SSM **14.13**]

7.14 Use a group theoretical argument to decide which of the following transitions are electric-dipole allowed: (a) the $\pi^* \leftarrow \pi$ transition in ethene, (b) the $\pi^* \leftarrow n$ transition in a carbonyl group in a C_{2v} environment. [ISM **14.14**]

7.15 Suppose that you are a colour chemist and had been asked to intensify the colour of a dye without changing the type of compound, and that the dye in question was a polyene. Would you choose to lengthen or to shorten the chain? Would the modification to the length shift the apparent colour of the dye towards the red or the blue? [SSM **14.15**]

7.16 One measure of the intensity of a transition of frequency ν is the *oscillator strength*, *f*, which is defined as

$$f = \frac{8\pi^2 m_e \nu |\mu_{fi}|^2}{3he^2}$$

Consider an electron in an atom to be oscillating harmonically in one dimension (the three-dimensional version of this model was used in early attempts to describe atomic structure). The wavefunctions for such an electron are those in Table 2.1. Show that the oscillator strength for the transition of this electron from its ground state is exactly $\tfrac{1}{3}$. [ISM **14.16**]

7.17 Estimate the oscillator strength (see Problem 7.16) of a charge-transfer transition modelled as the migration of an electron from an H1*s* orbital on one atom to another H1*s* orbital on an atom a distance *R* away. Approximate the transition moment by $-eRS$ where *S* is the overlap integral of the two orbitals. Sketch the oscillator strength as a function of *R* using the curve for *S* given in Fig. 4.29. Why does the intensity fall to zero as *R* approaches 0 and infinity? [SSM **14.17**]

7.18 The line marked A in Fig. 7.51 is the fluorescence spectrum of benzophenone in solid solution in ethanol at low temperatures observed when the sample is illuminated with 360 nm light. What can be said about the vibrational energy levels of the carbonyl group in (a) its ground electronic state and (b) its excited electronic state? When naphthalene is illuminated with 360 nm light it does not absorb, but the line marked B in the illustration is the phosphorescence spectrum of a solid solution of a mixture of naphthalene and benzophenone in ethanol. Now a component of fluorescence from naphthalene can be detected. Account for this observation. [ISM **14.18**]

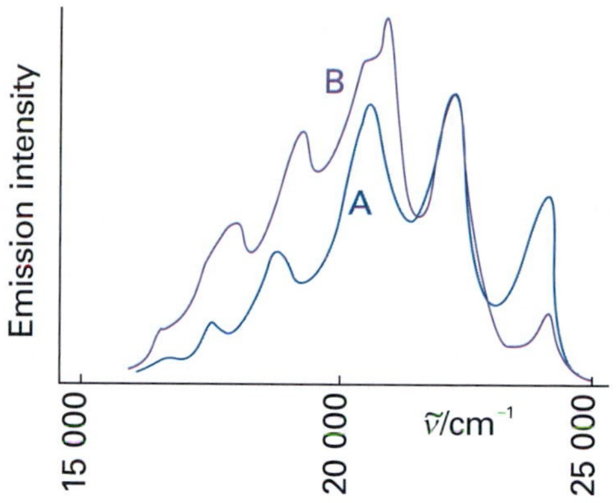

Fig. 7.51

7.19 The fluorescence spectrum of anthracene vapour shows a series of peaks of increasing intensity with individual maxima at 440 nm, 410 nm, 390 nm, and 370 nm followed by a sharp cut-off at shorter wavelengths. The absorption spectrum rises sharply from zero to a maximum at 360 nm with a trail of peaks of lessening intensity at 345 nm, 330 nm, and 305 nm. Account for these observations. [SSM **14.19**]

7.20 The Beer–Lambert law states that the absorbance of a sample at a wavenumber $\tilde{\nu}$ is proportional to the molar concentration [J] of the absorbing species J and to the length *l* of the sample (eqn 6.4). In this problem you will show that the intensity of fluorescence emission from a sample of J is also proportional to [J] and *l*. Consider a sample of J that is illuminated with a beam of intensity $I_0(\tilde{\nu})$ at the wavenumber $\tilde{\nu}$. Before fluorescence can occur, a fraction of $I_0(\tilde{\nu})$ must be absorbed and an intensity $I(\tilde{\nu})$ will be transmitted. However, not all of the absorbed intensity is emitted and the intensity of fluorescence depends on the fluorescence quantum yield, ϕ_f, the efficiency of photon emission. The fluorescence quantum yield ranges from 0 to 1 and is proportional to the ratio of the integral of the fluorescence spectrum over the integrated absorption coefficient. Because of a Stokes shift of magnitude $\Delta\tilde{\nu}_{Stokes}$, fluorescence occurs at a wavenumber $\tilde{\nu}_f$, with $\tilde{\nu}_f + \Delta\tilde{\nu}_{Stokes} = \tilde{\nu}$. It follows that the fluorescence intensity at $\tilde{\nu}_f$, $I_f(\tilde{\nu}_f)$, is proportional to ϕ_f and to the intensity of exciting radiation that is absorbed by J, $I_{abs}(\tilde{\nu}) = I_0(\tilde{\nu}) - I(\tilde{\nu})$. (a) Use the Beer–Lambert law to express $I_{abs}(\tilde{\nu})$ in terms of $I_0(\tilde{\nu})$, [J], *l*, and $\varepsilon(\tilde{\nu})$, the molar absorption coefficient of J at $\tilde{\nu}$. (b) Use your result from part (a) to show that $I_f(\tilde{\nu}_f) \propto I_0(\tilde{\nu})\varepsilon(\tilde{\nu})\phi_f[J]l$. [ISM **14.20**]

7.21 Spin angular momentum is conserved when a molecule dissociates into atoms. What atom multiplicities are permitted when (a) an O_2 molecule, (b) an N_2 molecule dissociates into atoms? [SSM **14.21**]

Applications: to biochemistry, environmental science, and astrophysics

7.22 The protein haemerythrin (Her) is responsible for binding and carrying O_2 in some invertebrates. Each protein molecule has two Fe^{2+} ions that are in very close proximity and work together to bind one molecule of O_2. The Fe_2O_2 group of oxygenated haemerythrin is coloured and has an electronic absorption band at 500 nm. Figure 7.52 shows the UV-visible absorption spectrum of a derivative of haemerythrin in the presence of different concentrations of CNS^- ions. What may be inferred from the spectrum? [ISM **14.22**]

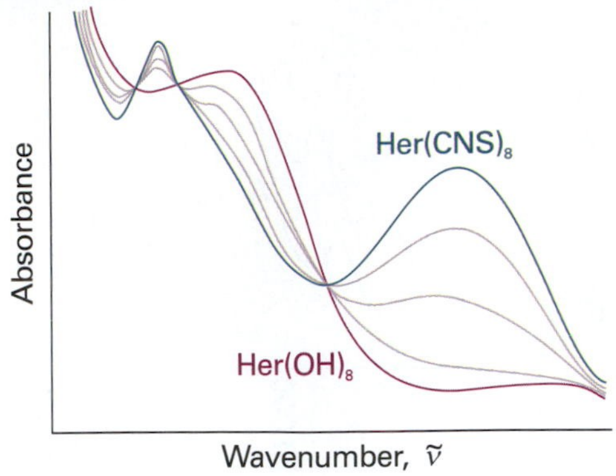

Fig. 7.52

7.23 The flux of visible photons reaching Earth from the North Star is about 4×10^3 mm^{-2} s^{-1}. Of these photons, 30 per cent are absorbed or scattered by the atmosphere and 25 per cent of the surviving photons are scattered by the surface of the cornea of the eye. A further 9 per cent are absorbed inside the cornea. The area of the pupil at night is about 40 mm^2 and the response time of the eye is about 0.1 s. Of the photons passing through the pupil, about 43 per cent are absorbed in the ocular medium. How many photons from the North Star are focused on to the retina in 0.1 s? For a continuation of this story, see R.W. Rodieck, *The first steps in seeing*, Sinauer, Sunderland (1998). [SSM **14.23**]

7.24 Use molecule (**7**) as a model of the *trans* conformation of the chromophore found in rhodopsin. In this model, the methyl group bound to the nitrogen atom of the protonated Schiff's base replaces the protein. (a) Using molecular modelling software and the computational method of your instructor's choice, calculate the energy separation between the HOMO and LUMO of (**7**). (b) Repeat the calculation for the 11-*cis* form of (**7**). (c) Based on your results from parts (a) and (b), do you expect the experimental frequency for the $\pi^* \leftarrow \pi$ visible absorption of the *trans* form of (**7**) to be higher or lower than that for the 11-*cis* form of (**7**)? [ISM **14.24**]

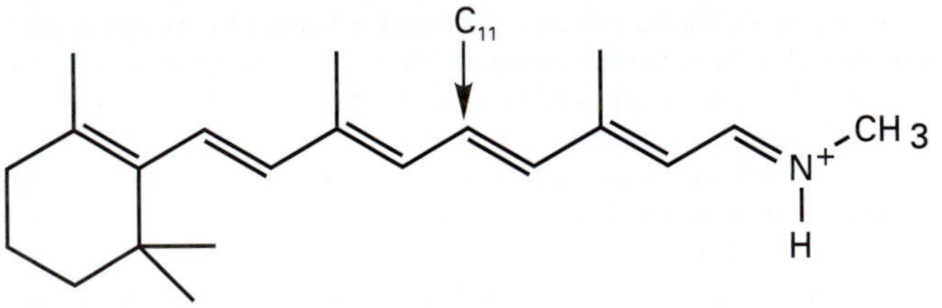

7

7.25‡ Ozone absorbs ultraviolet radiation in a part of the electromagnetic spectrum energetic enough to disrupt DNA in biological organisms and that is absorbed by no other abundant atmospheric constituent. This spectral range, denoted UV-B, spans the wavelengths of about 290 nm to 320 nm. The molar extinction coefficient of ozone over this range is given in the table below (W.B. DeMore, S.P. Sander, D.M. Golden, R.F. Hampson, M.J. Kurylo, C.J. Howard, A.R. Ravishankara, C.E. Kolb, and M.J. Molina, *Chemical kinetics and photochemical data for use in stratospheric modeling: Evaluation Number 11*, JPL Publication 94–26 (1994)).

λ/nm	292.0	296.3	300.8	305.4	310.1	315.0	320.0
ε/(dm^3 mol^{-1} cm^{-1})	1512	865	477	257	135.9	69.5	34.5

Compute the integrated absorption coefficient of ozone over the wavelength range 290–320 nm. (*Hint.* $\varepsilon(\tilde{\nu})$ can be fitted to an exponential function quite well.) [SSM **14.25**]

7.26‡ The abundance of ozone is typically inferred from measurements of UV absorption and is often expressed in terms of *Dobson units* (DU): 1 DU is equivalent to a layer of pure ozone 10^{-3} cm thick at 1 atm and 0°C. Compute the absorbance of UV radiation at 300 nm expected for an ozone abundance of 300 DU (a typical value) and 100 DU (a value reached during seasonal Antarctic ozone depletions) given a molar absorption coefficient of 476 dm^3 mol^{-1} cm^{-1}. [ISM **14.26**]

7.27‡ G.C.G. Wachewsky, R. Horansky, and V. Vaida (*J. Phys. Chem.* **100**, 11559 (1996)) examined the UV absorption spectrum of CH_3I, a species of interest in connection with stratospheric ozone chemistry. They found the integrated absorption coefficient to be dependent on temperature and pressure to an extent inconsistent with internal structural changes in isolated CH_3I molecules; they explained the changes as due to dimerization of a substantial fraction of the CH_3I, a process that would naturally be pressure and temperature dependent. (a) Compute the integrated absorption coefficient over a triangular lineshape in the range 31 250 to 34 483 cm^{-1} and a maximal molar absorption coefficient of 150 dm^3 mol^{-1} cm^{-1} at 31 250 cm^{-1}. (b) Suppose 1 per cent of the CH_3I units in a sample at 2.4 Torr and 373 K exists as dimers. Compute the absorbance expected at 31 250 cm^{-1} in a sample cell of length 12.0 cm. (c) Suppose 18 per cent of the CH_3I units in a sample at 100 Torr and 373 K exists as dimers. Compute the absorbance expected at 31 250 cm^{-1} in a sample cell of length 12.0 cm; compute the molar absorption coefficient that would be inferred from this absorbance if dimerization was not considered. [SSM **14.27**]

7.28‡ The molecule Cl_2O_2 is believed to participate in the seasonal depletion of ozone over Antarctica. M. Schwell, H.-W. Jochims, B. Wassermann, U. Rockland, R. Flesch, and E. Rühl (*J. Phys. Chem.* **100**, 10070 (1996)) measured the ionization energies of Cl_2O_2 by photoelectron spectroscopy in which the ionized fragments were detected using a mass spectrometer. From their data, we can infer that the ionization enthalpy of Cl_2O_2 is 11.05 eV and the enthalpy of the dissociative ionization $Cl_2O_2 \rightarrow Cl + OClO^+ + e^-$ is 10.95 eV. They used this information to make some inferences about the structure of Cl_2O_2. Computational studies had suggested that the lowest energy isomer is ClOOCl, but that $ClClO_2$ (C_{2v}) and ClOClO are not very much higher in energy. The Cl_2O_2 in the photoionization step is the lowest energy isomer, whatever its structure may be, and its enthalpy of formation had previously been reported as +133 kJ mol^{-1}. The Cl_2O_2 in the dissociative ionization step is unlikely to be ClOOCl, for the product can be derived from it only with substantial rearrangement. Given $\Delta_f H^{\ominus}(OClO^+) = +1096$ kJ mol^{-1} and $\Delta_f H^{\ominus}(e^-) = 0$, determine whether the Cl_2O_2 in the dissociative ionization is the same as that in the photoionization. If different, how much greater is its $\Delta_f H^{\ominus}$? Are these results consistent with or contradictory to the computational studies? [ISM **14.28**]

7.29‡ One of the principal methods for obtaining the electronic spectra of unstable radicals is to study the spectra of comets, which are almost entirely due to radicals. Many radical spectra have been found in comets, including that due to CN. These radicals are produced in comets by the absorption of far ultraviolet solar radiation by their parent compounds. Subsequently, their fluorescence is excited by sunlight of longer wavelength. The spectra of comet Hale–Bopp (C/1995 O1) have been the subject of many recent studies. One such study is that of the fluorescence spectrum of CN in the comet at large heliocentric distances by R.M. Wagner and D.G. Schleicher (*Science* **275**, 1918 (1997)), in which the authors determine the spatial distribution and rate of production of CN in the coma. The (0–0) vibrational band is centred on 387.6 nm and the weaker (1–1) band with relative intensity 0.1 is centred on 386.4 nm. The band heads for (0–0) and (0–1) are known to be 388.3 and 421.6 nm, respectively. From these data, calculate the energy of the excited S_1 state relative to the ground S_0 state, the vibrational wavenumbers and the difference in the vibrational wavenumbers of the two states, and the relative populations of the $v = 0$ and $v = 1$ vibrational levels of the S_1 state. Also estimate the effective temperature of the molecule in the excited S_1 state. Only eight rotational levels of the S_1 state are thought to be populated. Is that observation consistent with the effective temperature of the S_1 state? [SSM **14.29**]

Molecular spectroscopy 3: magnetic resonance

8

One of the most widely used spectroscopic procedures in chemistry makes use of the classical concept of resonance. The chapter begins with an account of conventional nuclear magnetic resonance, which shows how the resonance frequency of a magnetic nucleus is affected by its electronic environment and the presence of magnetic nuclei in its vicinity. Then we turn to the modern versions of NMR, which are based on the use of pulses of electromagnetic radiation and the processing of the resulting signal by Fourier transform techniques. The experimental techniques for electron paramagnetic resonance resemble those used in the early days of NMR. The information obtained is very useful for the determination of the properties of species with unpaired electrons.

When two pendulums share a slightly flexible support and one is set in motion, the other is forced into oscillation by the motion of the common axle. As a result, energy flows between the two pendulums. The energy transfer occurs most efficiently when the frequencies of the two pendulums are identical. The condition of strong effective coupling when the frequencies of two oscillators are identical is called **resonance**. Resonance is the basis of a number of everyday phenomena, including the response of radios to the weak oscillations of the electromagnetic field generated by a distant transmitter. In this chapter we explore some spectroscopic applications that, as originally developed (and in some cases still), depend on matching a set of energy levels to a source of monochromatic radiation and observing the strong absorption that occurs at resonance.

The effect of magnetic fields on electrons and nuclei

The Stern–Gerlach experiment (Section 2.8) provided evidence for electron spin. It turns out that many nuclei also possess spin angular momentum. Orbital and spin angular momenta give rise to magnetic moments, and to say that electrons and nuclei have magnetic moments means that, to some extent, they behave like small bar magnets. First, we establish how the energies of electrons and nuclei depend on the strength of an external field. Then we see how to use this dependence to study the structure and dynamics of complex molecules.

8.1 The energies of electrons in magnetic fields

Classically, the energy of a magnetic moment $\boldsymbol{\mu}$ in a magnetic field $\boldsymbol{\mathcal{B}}$ is equal to the scalar product

Comment 8.1
More formally, $\mathcal{B}$ is the magnetic induction and is measured in tesla, T; 1 T = 1 kg s^{-2} A^{-1}. The (non-SI) unit gauss, G, is also occasionally used: 1 T = 10^4 G.

Comment 8.2
The scalar product (or 'dot product') of two vectors $\boldsymbol{a}$ and $\boldsymbol{b}$ is given by $\boldsymbol{a}\cdot\boldsymbol{b} = ab\cos\theta$, where a and b are the magnitudes of $\boldsymbol{a}$ and $\boldsymbol{b}$, respectively, and θ is the angle between them.

$$E = -\boldsymbol{\mu}\cdot\boldsymbol{\mathcal{B}} \tag{8.1}$$

Quantum mechanically, we write the hamiltonian as

$$\hat{H} = -\hat{\boldsymbol{\mu}}\cdot\boldsymbol{\mathcal{B}} \tag{8.2}$$

To write an expression for $\hat{\boldsymbol{\mu}}$, we recall from Section 2.8 that the magnetic moment is proportional to the angular momentum. For an electron possessing orbital angular momentum we write

$$\hat{\boldsymbol{\mu}} = \gamma_e \hat{\boldsymbol{l}} \qquad \text{and} \qquad \hat{H} = -\gamma_e \boldsymbol{\mathcal{B}}\cdot\hat{\boldsymbol{l}} \tag{8.3}$$

where $\hat{\boldsymbol{l}}$ is the orbital angular momentum operator and

$$\gamma_e = -\frac{e}{2m_e} \tag{8.4}$$

γ_e is called the **magnetogyric ratio** of the electron: The negative sign (arising from the sign of the electron's charge) shows that the orbital moment is antiparallel to its orbital angular momentum (as was depicted in Fig 3.26).

For a magnetic field $\mathcal{B}_0$ along the z-direction, eqn 8.3 becomes

$$\hat{\mu}_z = \gamma_e \hat{l}_z \qquad \text{and} \qquad \hat{H} = -\gamma_e \mathcal{B}_0 \hat{l}_z = -\hat{\mu}_z \mathcal{B}_0 \tag{8.5a}$$

Because the eigenvalues of the operator $\hat{\boldsymbol{l}}_z$ are $m_l\hbar$, the z-component of the orbital magnetic moment and the energy of interaction are, respectively,

$$\mu_z = \gamma_e m_l \hbar \qquad \text{and} \qquad E = -\gamma_e m_l \hbar \mathcal{B}_0 = m_l \mu_B \mathcal{B}_0 \tag{8.5b}$$

where the **Bohr magneton**, μ_B, is

$$\mu_B = -\gamma_e \hbar = \frac{e\hbar}{2m_e} = 9.724 \times 10^{-24}\ \mathrm{J\,T^{-1}} \qquad [8.6]$$

The Bohr magneton is often regarded as the fundamental quantum of magnetic moment.

The spin magnetic moment of an electron, which has a spin quantum number $s = \frac{1}{2}$ (Section 2.8), is also proportional to its spin angular momentum. However, instead of eqn 8.3, the spin magnetic moment and hamiltonian operators are, respectively,

$$\hat{\boldsymbol{\mu}} = g_e \gamma_e \hat{\boldsymbol{s}} \qquad \text{and} \qquad \hat{H} = -g_e \gamma_e \boldsymbol{\mathcal{B}}\cdot\hat{\boldsymbol{s}} \tag{8.7}$$

where $\hat{\boldsymbol{s}}$ is the spin angular momentum operator and the extra factor g_e is called the **g-value** of the electron: $g_e = 2.002\,319\ldots$. The g-value arises from relativistic effects and from interactions of the electron with the electromagnetic fluctuations of the vacuum that surrounds the electron. For a magnetic field $\mathcal{B}_0$ in the z-direction,

$$\hat{\mu}_z = g_e \gamma_e \hat{s}_z \qquad \text{and} \qquad \hat{H} = -g_e \gamma_e \mathcal{B}_0 \hat{s}_z \tag{8.8a}$$

Because the eigenvalues of the operator $\hat{s}_z$ are $m_s\hbar$ with $m_s = +\frac{1}{2}(\alpha)$ and $m_s = -\frac{1}{2}(\beta)$, it follows that the energies of an electron spin in a magnetic field are

$$\mu_z = g_e \gamma_e m_s \hbar \qquad \text{and} \qquad E_{m_s} = -g_e \gamma_e m_s \hbar \mathcal{B}_0 = g_e \mu_B m_s \mathcal{B}_0 \tag{8.8b}$$

with $m_s = \pm\frac{1}{2}$.

In the absence of a magnetic field, the states with different values of m_l and m_s are degenerate. When a field is present, the degeneracy is removed: the state with $m_s = +\frac{1}{2}$ moves up in energy by $\frac{1}{2}g_e\mu_B\mathcal{B}_0$ and the state with $m_s = -\frac{1}{2}$ moves down by $\frac{1}{2}g_e\mu_B\mathcal{B}_0$. The different energies arising from an interaction with an external field are sometimes represented on the vector model by picturing the vectors as **precessing**, or sweeping

Table 8.1 Nuclear constitution and the nuclear spin quantum number*

Number of protons	Number of neutrons	I
even	even	0
odd	odd	integer (1, 2, 3, . . .)
even	odd	half-integer ($\frac{1}{2}, \frac{3}{2}, \frac{5}{2}, \ldots$)
odd	even	half-integer ($\frac{1}{2}, \frac{3}{2}, \frac{5}{2}, \ldots$)

* The spin of a nucleus may be different if it is in an excited state; throughout this chapter we deal only with the ground state of nuclei.

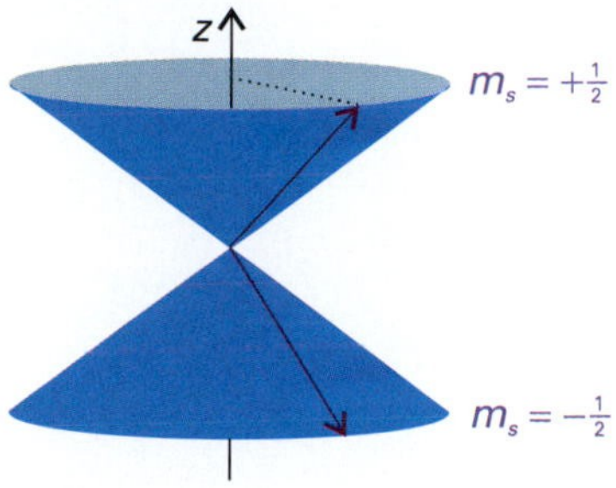

Fig. 8.1 The interactions between the m_s states of an electron and an external magnetic field may be visualized as the precession of the vectors representing the angular momentum.

round their cones (Fig. 8.1), with the rate of precession equal to the **Larmor frequency**, ν_L:

$$\nu_L = \frac{\gamma_e \mathcal{B}_0}{2\pi} \tag{8.9}$$

Equation 8.9 shows that the Larmor frequency increases with the strength of the magnetic field. For a field of 1 T, the Larmor frequency is 30 GHz.

8.2 The energies of nuclei in magnetic fields

The spin quantum number, I, of a nucleus is a fixed characteristic property of a nucleus and is either an integer or a half-integer (Table 8.1). A nucleus with spin quantum number I has the following properties:

1. An angular momentum of magnitude $\{I(I+1)\}^{1/2}\hbar$.
2. A component of angular momentum $m_I\hbar$ on a specified axis ('the z-axis'), where $m_I = I, I-1, \ldots, -I$.
3. If $I > 0$, a magnetic moment with a constant magnitude and an orientation that is determined by the value of m_I.

According to the second property, the spin, and hence the magnetic moment, of the nucleus may lie in $2I+1$ different orientations relative to an axis. A proton has $I = \frac{1}{2}$ and its spin may adopt either of two orientations; a ^{14}N nucleus has $I = 1$ and its spin may adopt any of three orientations; both ^{12}C and ^{16}O have $I = 0$ and hence zero magnetic moment.

The energy of interaction between a nucleus with a magnetic moment μ and an external magnetic field $\mathcal{B}$ may be calculated by using operators analogous to those of eqn 8.3:

$$\hat{\mu} = \gamma \hat{I} \qquad \text{and} \qquad H = -\gamma \mathcal{B} \cdot \hat{I} \tag{8.10a}$$

where γ is the magnetogyric ratio of the specified nucleus, an empirically determined characteristic arising from the internal structure of the nucleus (Table 8.2). The corresponding energies are

$$E_{m_I} = -\mu_z \mathcal{B}_0 = -\gamma \hbar \mathcal{B}_0 m_I \tag{8.10b}$$

As for electrons, the nuclear spin may be pictured as precessing around the direction of the applied field at a rate proportional to the applied field. For protons, a field of 1 T corresponds to a Larmor frequency (eqn 8.9, with γ_e replaced by γ) of about 40 MHz.

Synoptic table 8.2* Nuclear spin properties

Nuclide	Natural abundance/%	Spin I	g-factor, g_I	Magnetogyric ratio, $\gamma/(10^7\ \mathrm{T^{-1}\ s^{-1}})$	NMR frequency at 1 T, ν/MHz
^{1}n		$\frac{1}{2}$	−3.826	−18.32	29.165
^{1}H	99.98	$\frac{1}{2}$	5.586	26.75	42.577
^{2}H	0.02	1	0.857	4.10	6.536
^{13}C	1.11	$\frac{1}{2}$	1.405	6.73	10.705
^{14}N	99.64	1	0.404	1.93	3.076

* More values are given in the *Data section*.

The magnetic moment of a nucleus is sometimes expressed in terms of the **nuclear g-factor**, g_I, a characteristic of the nucleus, and the **nuclear magneton**, μ_N, a quantity independent of the nucleus, by using

$$\gamma\hbar = g_I\mu_N \qquad \mu_N = \frac{e\hbar}{2m_p} = 5.051\times10^{-27}\ \mathrm{J\ T^{-1}} \tag{8.11}$$

where m_p is the mass of the proton. Nuclear g-factors vary between −6 and +6 (see Table 8.2): positive values of g_I and γ denote a magnetic moment that is parallel to the spin; negative values indicate that the magnetic moment and spin are antiparallel. For the remainder of this chapter we shall assume that γ is positive, as is the case for the majority of nuclei. In such cases, the states with negative values of m_I lie above states with positive values of m_I. The nuclear magneton is about 2000 times smaller than the Bohr magneton, so nuclear magnetic moments—and consequently the energies of interaction with magnetic fields—are about 2000 times weaker than the electron spin magnetic moment.

8.3 Magnetic resonance spectroscopy

In its original form, the magnetic resonance experiment is the resonant absorption of radiation by nuclei or unpaired electrons in a magnetic field. From eqn 8.8b, the separation between the $m_s = -\frac{1}{2}$ and $m_s = +\frac{1}{2}$ levels of an electron spin in a magnetic field $\mathcal{B}_0$ is

$$\Delta E = E_\alpha - E_\beta = g_e\mu_B\mathcal{B}_0 \tag{8.12a}$$

If the sample is exposed to radiation of frequency ν, the energy separations come into resonance with the radiation when the frequency satisfies the **resonance condition** (Fig. 8.2):

$$h\nu = g_e\mu_B\mathcal{B}_0 \tag{8.12b}$$

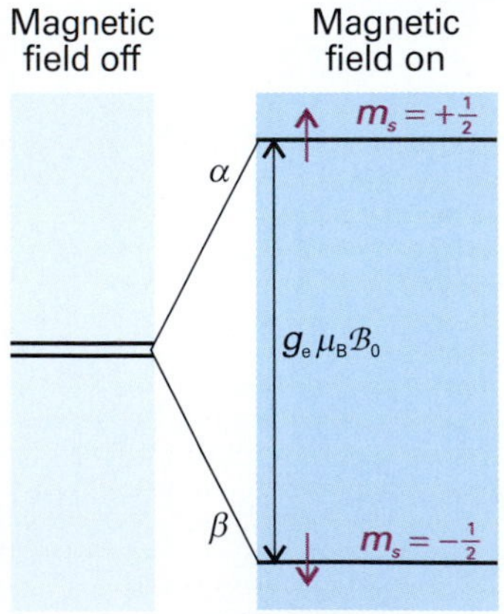

Fig. 8.2 Electron spin levels in a magnetic field. Note that the β state is lower in energy than the α state (because the magnetogyric ratio of an electron is negative). Resonance is achieved when the frequency of the incident radiation matches the frequency corresponding to the energy separation.

At resonance there is strong coupling between the electron spins and the radiation, and strong absorption occurs as the spins make the transition $\beta \rightarrow \alpha$. **Electron paramagnetic resonance** (EPR), or **electron spin resonance** (ESR), is the study of molecules and ions containing unpaired electrons by observing the magnetic fields at which they come into resonance with monochromatic radiation. Magnetic fields of about 0.3 T (the value used in most commercial EPR spectrometers) correspond to resonance with an electromagnetic field of frequency 10 GHz (10^{10} Hz) and wavelength 3 cm. Because 3 cm radiation falls in the microwave region of the electromagnetic spectrum, EPR is a microwave technique.

The energy separation between the $m_I = +\frac{1}{2}$ ($\uparrow$ or α) and the $m_I = -\frac{1}{2}$ ($\downarrow$ or β) states of **spin-$\frac{1}{2}$ nuclei**, which are nuclei with $I = \frac{1}{2}$, is

$$\Delta E = E_\beta - E_\alpha = \tfrac{1}{2}\gamma\hbar\mathcal{B}_0 - (-\tfrac{1}{2}\gamma\hbar\mathcal{B}_0) = \gamma\hbar\mathcal{B}_0 \qquad (8.13a)$$

and resonant absorption occurs when the resonance condition (Fig. 8.3)

$$h\nu = \gamma\hbar\mathcal{B}_0 \qquad (8.13b)$$

is fulfilled. Because $\gamma\hbar\mathcal{B}_0/h$ is the Larmor frequency of the nucleus, this resonance occurs when the frequency of the electromagnetic field matches the Larmor frequency ($\nu = \nu_L$). In its simplest form, **nuclear magnetic resonance** (NMR) is the study of the properties of molecules containing magnetic nuclei by applying a magnetic field and observing the frequency of the resonant electromagnetic field. Larmor frequencies of nuclei at the fields normally employed (about 12 T) typically lie in the radiofrequency region of the electromagnetic spectrum (close to 500 MHz), so NMR is a radiofrequency technique.

For much of this chapter we consider spin-$\frac{1}{2}$ nuclei, but NMR is applicable to nuclei with any non-zero spin. As well as protons, which are the most common nuclei studied by NMR, spin-$\frac{1}{2}$ nuclei include ^{13}C, ^{19}F, and ^{31}P.

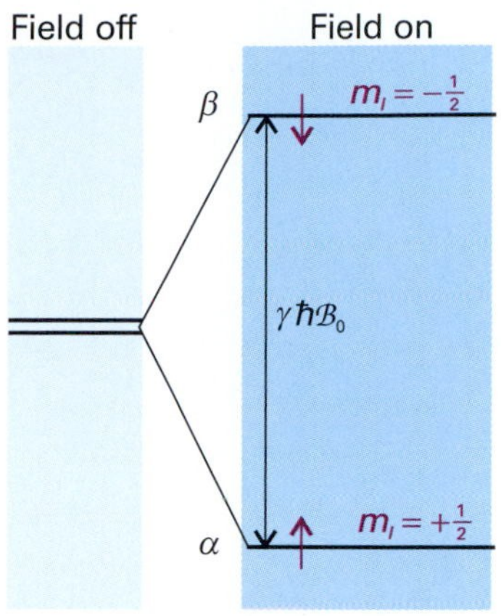

Fig. 8.3 The nuclear spin energy levels of a spin-$\frac{1}{2}$ nucleus with positive magnetogyric ratio (for example, ^{1}H or ^{13}C) in a magnetic field. Resonance occurs when the energy separation of the levels matches the energy of the photons in the electromagnetic field.

Nuclear magnetic resonance

Although the NMR technique is simple in concept, NMR spectra can be highly complex. However, they have proved invaluable in chemistry, for they reveal so much structural information. A magnetic nucleus is a very sensitive, non-invasive probe of the surrounding electronic structure.

8.4 The NMR spectrometer

An NMR spectrometer consists of the appropriate sources of radiofrequency electromagnetic radiation and a magnet that can produce a uniform, intense field. In simple instruments, the magnetic field is provided by a permanent magnet. For serious work, a **superconducting magnet** capable of producing fields of the order of 10 T and more is used (Fig. 8.4). The sample is placed in the cylindrically wound magnet. In some cases the sample is rotated rapidly to average out magnetic inhomogeneities. However, sample spinning can lead to irreproducible results, and is often avoided. Although a superconducting magnet operates at the temperature of liquid helium (4 K), the sample itself is normally at room temperature.

The intensity of an NMR transition depends on a number of factors. We show in the following *Justification* that

$$Intensity \propto (N_\alpha - N_\beta)\mathcal{B}_0 \qquad (8.14a)$$

where

$$N_\alpha - N_\beta \approx \frac{N\gamma\hbar\mathcal{B}_0}{2kT} \qquad (8.14b)$$

with N the total number of spins ($N = N_\alpha + N_\beta$). It follows that decreasing the temperature increases the intensity by increasing the population difference. By combining these two equations we see that the intensity is proportional to $\mathcal{B}_0^2$, so NMR transitions can be enhanced significantly by increasing the strength of the applied magnetic field. We shall also see (Section 8.6) that the use of high magnetic fields simplifies the

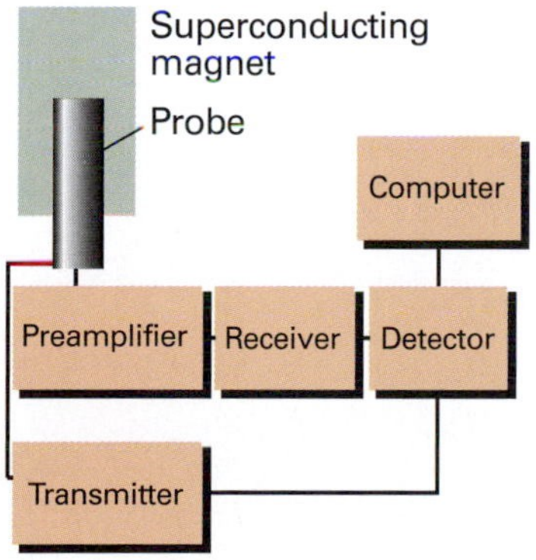

Fig. 8.4 The layout of a typical NMR spectrometer. The link from the transmitter to the detector indicates that the high frequency of the transmitter is subtracted from the high frequency received signal to give a low frequency signal for processing.

appearance of spectra and so allows them to be interpreted more readily. We also conclude that absorptions of nuclei with large magnetogyric ratios (^{1}H, for instance) are more intense than those with small magnetogyric ratios (^{13}C, for instance)

Justification 8.1 *Intensities in NMR spectra*

From the general considerations of transition intensities in Section 6.2, we know that the rate of absorption of electromagnetic radiation is proportional to the population of the lower energy state (N_α in the case of a proton NMR transition) and the rate of stimulated emission is proportional to the population of the upper state (N_β). At the low frequencies typical of magnetic resonance, we can neglect spontaneous emission as it is very slow. Therefore, the net rate of absorption is proportional to the difference in populations, and we can write

$$\text{Rate of absorption} \propto N_\alpha - N_\beta$$

The intensity of absorption, the rate at which energy is absorbed, is proportional to the product of the rate of absorption (the rate at which photons are absorbed) and the energy of each photon, and the latter is proportional to the frequency ν of the incident radiation (through $E = h\nu$). At resonance, this frequency is proportional to the applied magnetic field (through $\nu = \nu_L = \gamma\mathcal{B}_0/2\pi$), so we can write

$$\text{Intensity of absorption} \propto (N_\alpha - N_\beta)\mathcal{B}_0$$

as in eqn 8.14a. To write an expression for the population difference, we use the Boltzmann distribution (Section 9.1) to write the ratio of populations as

$$\frac{N_\beta}{N_\alpha} = e^{-\Delta E/kT} \approx 1 - \frac{\Delta E}{kT} = 1 - \frac{\gamma\mathcal{B}_0}{kT}$$

where $\Delta E = E_\beta - E_\alpha$. The expansion of the exponential term is appropriate for $\Delta E \ll kT$, a condition usually met for nuclear spins. It follows after rearrangement that

$$\frac{N_\alpha - N_\beta}{N_\alpha + N_\beta} = \frac{N_\alpha(1 - N_\beta/N_\alpha)}{N_\alpha(1 + N_\beta/N_\alpha)} = \frac{1 - N_\beta/N_\alpha}{1 + N_\beta/N_\alpha}$$

$$\approx \frac{1 - (1 - \gamma\hbar\mathcal{B}_0/kT)}{1 + (1 - \gamma\hbar\mathcal{B}_0/kT)} \approx \frac{\gamma\hbar\mathcal{B}_0/kT}{2}$$

Then, with $N_\alpha + N_\beta = N$, the total number of spins, we obtain eqn 8.14b.

Comment 8.3
The expansion of an exponential function used here is $e^{-x} = 1 - x + \frac{1}{2}x^2 - \ldots$. If $x \ll 1$, then $e^{-x} \approx 1 - x$.

8.5 The chemical shift

Nuclear magnetic moments interact with the *local* magnetic field. The local field may differ from the applied field because the latter induces electronic orbital angular momentum (that is, the circulation of electronic currents) which gives rise to a small additional magnetic field $\delta\mathcal{B}$ at the nuclei. This additional field is proportional to the applied field, and it is conventional to write

$$\delta\mathcal{B} = -\sigma\mathcal{B}_0 \qquad [8.15]$$

where the dimensionless quantity σ is called the **shielding constant** of the nucleus (σ is usually positive but may be negative). The ability of the applied field to induce an electronic current in the molecule, and hence affect the strength of the resulting local magnetic field experienced by the nucleus, depends on the details of the electronic structure near the magnetic nucleus of interest, so nuclei in different chemical groups have different shielding constants. The calculation of reliable values of the shielding constant is very difficult, but trends in it are quite well understood and we concentrate on them.

(a) The δ scale of chemical shifts

Because the total local field is

$$\mathcal{B}_{\text{loc}} = \mathcal{B}_0 + \delta\mathcal{B} = (1-\sigma)\mathcal{B}_0 \tag{8.16}$$

the nuclear Larmor frequency is

$$\nu_{\text{L}} = \frac{\gamma\mathcal{B}_{\text{loc}}}{2\pi} = (1-\sigma)\frac{\gamma\mathcal{B}_0}{2\pi} \tag{8.17}$$

This frequency is different for nuclei in different environments. Hence, different nuclei, even of the same element, come into resonance at different frequencies.

It is conventional to express the resonance frequencies in terms of an empirical quantity called the **chemical shift**, which is related to the difference between the resonance frequency, ν, of the nucleus in question and that of a reference standard, ν°:

$$\delta = \frac{\nu - \nu^\circ}{\nu^\circ} \times 10^6 \qquad [8.18]$$

The standard for protons is the proton resonance in tetramethylsilane ($Si(CH_3)_4$, commonly referred to as TMS), which bristles with protons and dissolves without reaction in many liquids. Other references are used for other nuclei. For ^{13}C, the reference frequency is the ^{13}C resonance in TMS; for ^{31}P it is the ^{31}P resonance in 85 per cent H_3PO_4(aq). The advantage of the δ-scale is that shifts reported on it are independent of the applied field (because both numerator and denominator are proportional to the applied field).

Illustration 8.1 *Using the chemical shift*

From eqn 8.18,

$$\nu - \nu^\circ = \nu^\circ\delta \times 10^{-6}$$

A nucleus with $\delta = 1.00$ in a spectrometer operating at 500 MHz will have a shift relative to the reference equal to

$$\nu - \nu^\circ = (500\text{ MHz}) \times 1.00 \times 10^{-6} = 500\text{ Hz}$$

In a spectrometer operating at 100 MHz, the shift relative to the reference would be only 100 Hz.

A note on good practice In much of the literature, chemical shifts are reported in 'parts per million', ppm, in recognition of the factor of 10^6 in the definition. This practice is unnecessary.

The relation between δ and σ is obtained by substituting eqn 8.17 into eqn 8.18:

$$\delta = \frac{(1-\sigma)\mathcal{B}_0 - (1-\sigma^\circ)\mathcal{B}_0}{(1-\sigma^\circ)\mathcal{B}_0} \times 10^6 = \frac{\sigma^\circ - \sigma}{1-\sigma^\circ} \times 10^6 \approx (\sigma^\circ - \sigma) \times 10^6 \tag{8.19}$$

As the shielding, σ, gets smaller, δ *increases*. Therefore, we speak of nuclei with large chemical shift as being strongly **deshielded**. Some typical chemical shifts are given in Fig. 8.5. As can be seen from the illustration, the nuclei of different elements have

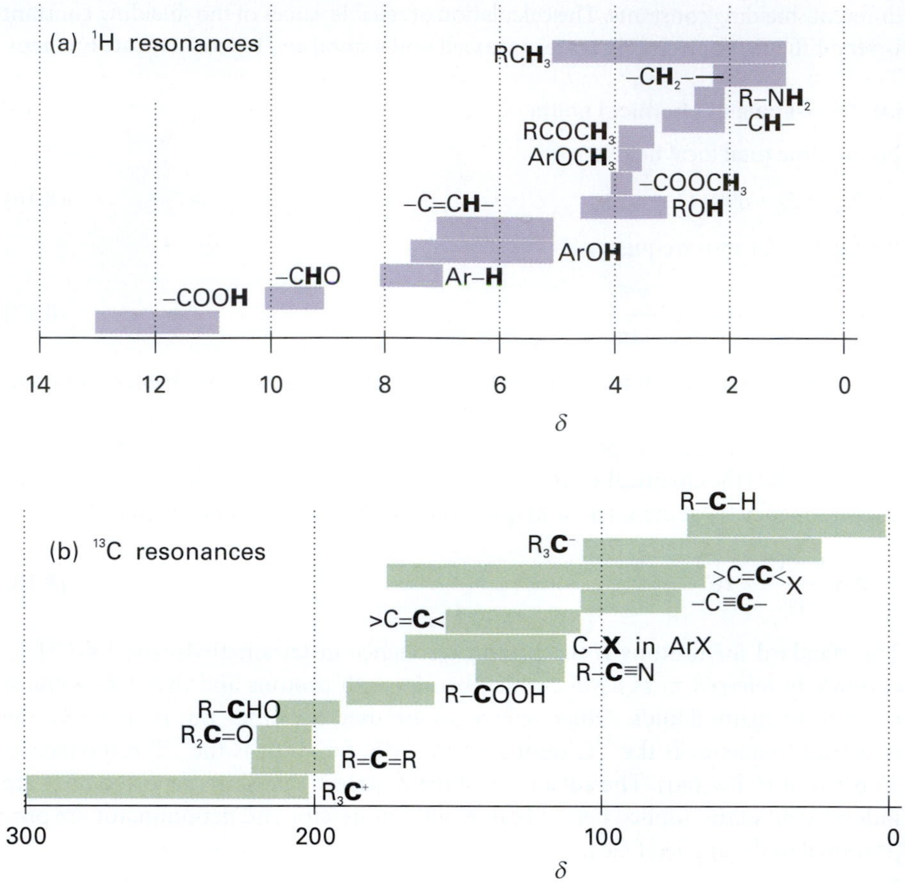

Fig. 8.5 The range of typical chemical shifts for (a) ^{1}H resonances and (b) ^{13}C resonances.

very different ranges of chemical shifts. The ranges exhibit the variety of electronic environments of the nuclei in molecules: the heavier the element, the greater the number of electrons around the nucleus and hence the greater the range of shieldings.

By convention, NMR spectra are plotted with δ increasing from right to left. Consequently, in a given applied magnetic field the Larmor frequency also increases from right to left. In the original continuous wave (CW) spectrometers, in which the radiofrequency was held constant and the magnetic field varied (a 'field sweep experiment'), the spectrum was displayed with the applied magnetic field increasing from left to right: a nucleus with a small chemical shift experiences a relatively low local magnetic field, so it needs a higher applied magnetic field to bring it into resonance with the radiofrequency field. Consequently, the right-hand (low chemical shift) end of the spectrum became known as the 'high field end' of the spectrum.

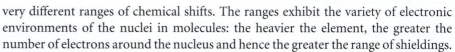

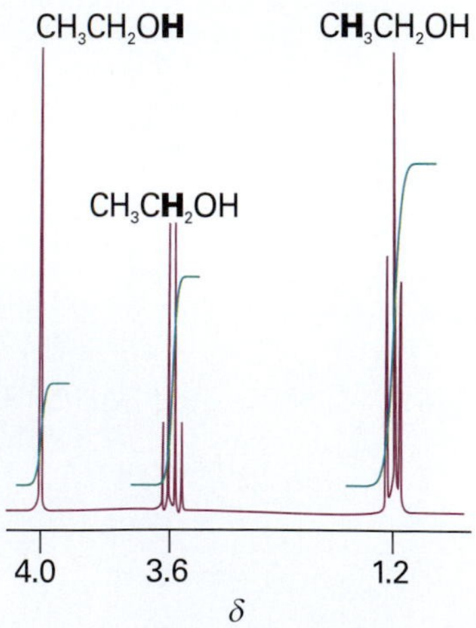

Fig. 8.6 The ^{1}H-NMR spectrum of ethanol. The bold letters denote the protons giving rise to the resonance peak, and the step-like curve is the integrated signal.

(b) Resonance of different groups of nuclei

The existence of a chemical shift explains the general features of the spectrum of ethanol shown in Fig. 8.6. The CH_3 protons form one group of nuclei with $\delta \approx 1$. The two CH_2 protons are in a different part of the molecule, experience a different local magnetic field, and resonate at $\delta \approx 3$. Finally, the OH proton is in another environment, and has a chemical shift of $\delta \approx 4$. The increasing value of δ (that is, the decrease in shielding) is consistent with the electron-withdrawing power of the O atom: it reduces the electron density of the OH proton most, and that proton is strongly deshielded. It reduces the electron density of the distant methyl protons least, and those nuclei are least deshielded.

The relative intensities of the signals (the areas under the absorption lines) can be used to help distinguish which group of lines corresponds to which chemical group. The determination of the area under an absorption line is referred to as the **integration** of the signal (just as any area under a curve may be determined by mathematical integration). Spectrometers can integrate the absorption automatically (as indicated in Fig. 8.6). In ethanol the group intensities are in the ratio 3:2:1 because there are three CH_3 protons, two CH_2 protons, and one OH proton in each molecule. Counting the number of magnetic nuclei as well as noting their chemical shifts helps to identify a compound present in a sample.

(c) The origin of shielding constants

The calculation of shielding constants is difficult, even for small molecules, for it requires detailed information about the distribution of electron density in the ground and excited states and the excitation energies of the molecule. Nevertheless, considerable success has been achieved with the calculation for diatomic molecules and small molecules such as H_2O and CH_4 and even large molecules, such as proteins, are within the scope of some types of calculation. Nevertheless, it is easier to understand the different contributions to chemical shifts by studying the large body of empirical information now available for large molecules.

The empirical approach supposes that the observed shielding constant is the sum of three contributions:

$$\sigma = \sigma(\text{local}) + \sigma(\text{neighbour}) + \sigma(\text{solvent}) \tag{8.20}$$

The **local contribution**, σ(local), is essentially the contribution of the electrons of the atom that contains the nucleus in question. The **neighbouring group contribution**, σ(neighbour), is the contribution from the groups of atoms that form the rest of the molecule. The **solvent contribution**, σ(solvent), is the contribution from the solvent molecules.

(d) The local contribution

It is convenient to regard the local contribution to the shielding constant as the sum of a **diamagnetic contribution**, σ_d, and a **paramagnetic contribution**, σ_p:

$$\sigma(\text{local}) = \sigma_d + \sigma_p \tag{8.21}$$

A diamagnetic contribution to σ(local) opposes the applied magnetic field and shields the nucleus in question. A paramagnetic contribution to σ(local) reinforces the applied magnetic field and deshields the nucleus in question. Therefore, $\sigma_d > 0$ and $\sigma_p < 0$. The total local contribution is positive if the diamagnetic contribution dominates, and is negative if the paramagnetic contribution dominates.

The diamagnetic contribution arises from the ability of the applied field to generate a circulation of charge in the ground-state electron distribution of the atom. The circulation generates a magnetic field that opposes the applied field and hence shields the nucleus. The magnitude of σ_d depends on the electron density close to the nucleus and can be calculated from the **Lamb formula** (see *Further reading* for a derivation):

$$\sigma_d = \frac{e^2\mu_0}{12\pi m_e}\left\langle \frac{1}{r} \right\rangle \tag{8.22}$$

where μ_0 is the vacuum permeability (a fundamental constant, see inside the front cover) and r is the electron–nucleus distance.

Illustration 8.2 *Calculating the diamagnetic contribution to the chemical shift of a proton*

To calculate σ_d for the proton in a free H atom, we need to calculate the expectation value of $1/r$ for a hydrogen 1*s* orbital. Wavefunctions are given in Table 3.1, and a useful integral is given in Example 1.7. Because $d\tau = r^2 dr \sin\theta \, d\theta d\phi$, we can write

$$\left\langle \frac{1}{r} \right\rangle = \int \frac{\psi^*\psi}{r} d\tau = \frac{1}{\pi a_0^3} \int_0^{2\pi} d\phi \int_0^{\pi} \sin\theta d\theta \int_0^{\infty} re^{-2r/a_0} dr = \frac{4}{a_0^3} \int_0^{\infty} re^{-2r/a_0} dr = \frac{1}{a_0}$$

Therefore,

$$\sigma_d = \frac{e^2 \mu_0}{12\pi m_e a_0}$$

With the values of the fundamental constants inside the front cover, this expression evaluates to 1.78×10^{-5}.

The diamagnetic contribution is the only contribution in closed-shell free atoms. It is also the only contribution to the local shielding for electron distributions that have spherical or cylindrical symmetry. Thus, it is the only contribution to the local shielding from inner cores of atoms, for cores remain spherical even though the atom may be a component of a molecule and its valence electron distribution highly distorted. The diamagnetic contribution is broadly proportional to the electron density of the atom containing the nucleus of interest. It follows that the shielding is decreased if the electron density on the atom is reduced by the influence of an electronegative atom nearby. That reduction in shielding translates into an increase in deshielding, and hence to an increase in the chemical shift δ as the electronegativity of a neighbouring atom increases (Fig. 8.7). That is, as the electronegativity increases, δ decreases.

The local paramagnetic contribution, σ_p, arises from the ability of the applied field to force electrons to circulate through the molecule by making use of orbitals that are unoccupied in the ground state. It is zero in free atoms and around the axes of linear molecules (such as ethyne, HC≡CH) where the electrons can circulate freely and a field applied along the internuclear axis is unable to force them into other orbitals. We can expect large paramagnetic contributions from small atoms in molecules with low-lying excited states. In fact, the paramagnetic contribution is the dominant local contribution for atoms other than hydrogen.

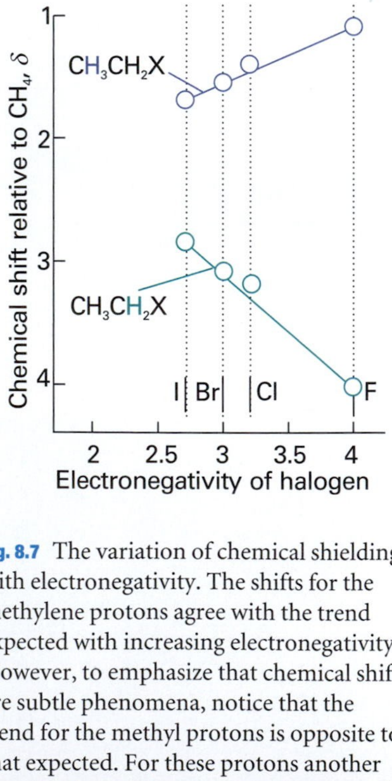

Fig. 8.7 The variation of chemical shielding with electronegativity. The shifts for the methylene protons agree with the trend expected with increasing electronegativity. However, to emphasize that chemical shifts are subtle phenomena, notice that the trend for the methyl protons is opposite to that expected. For these protons another contribution (the magnetic anisotropy of C—H and C—X bonds) is dominant.

(e) Neighbouring group contributions

The neighbouring group contribution arises from the currents induced in nearby groups of atoms. Consider the influence of the neighbouring group X on the proton H in a molecule such as H—X. The applied field generates currents in the electron distribution of X and gives rise to an induced magnetic moment proportional to the applied field; the constant of proportionality is the magnetic susceptibility, χ (chi), of the group X. The proton H is affected by this induced magnetic moment in two ways. First, the strength of the additional magnetic field the proton experiences is inversely proportional to the cube of the distance r between H and X. Second, the field at H depends on the anisotropy of the magnetic susceptibility of X, the variation of χ with the angle that X makes to the applied field. We assume that the magnetic susceptibility of X has two components, $\chi_\parallel$ and $\chi_\perp$, which are parallel and perpendicular to the axis of symmetry of X, respectively. The axis of symmetry of X makes an angle θ to the

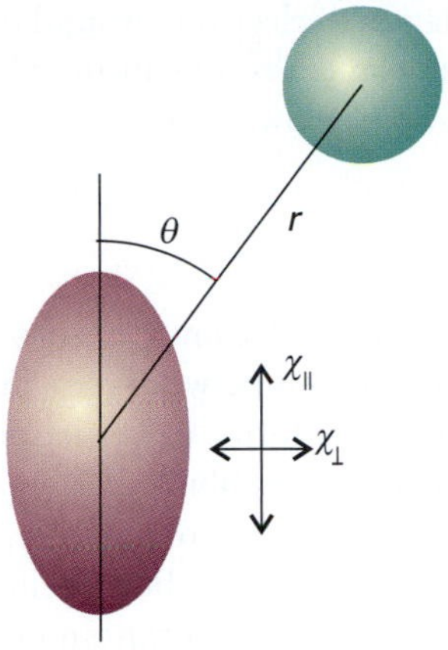

1

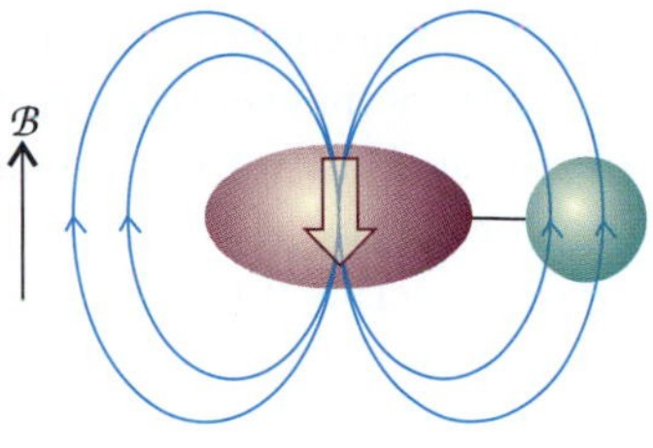

2

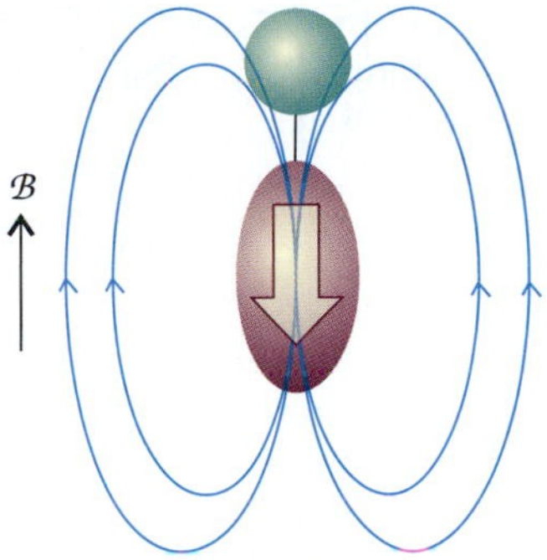

3

vector connecting X to H (**1**, where X is represented by the ellipse and H is represented by the circle).

To examine the effect of anisotropy of the magnetic susceptibility of X on the shielding constant, consider the case $\theta = 0$ for a molecule H—X that is free to tumble (**2** and **3**). Some of the time the H—X axis will be perpendicular to the applied field and then only $\chi_{\perp}$ will contribute to the induced magnetic moment that shields X from the applied field. The result is deshielding of the proton H, or σ(neighbour) < 0 (**2**). When the applied field is parallel to the H—X axis, only $\chi_{\parallel}$ contributes to the induced magnetic moment at X. The result is shielding of the proton H (**3**). We conclude that, as the molecule tumbles and the H—X axis takes all possible angles with respect to the applied field, the effects of anisotropic magnetic susceptibility do not average to zero because $\chi_{\parallel} \neq \chi_{\perp}$.

Self-test 8.1 For a tumbling H—X molecule, show that when $\theta = 90°$: (a) contributions from the $\chi_{\perp}$ component lead to shielding of H, or σ(neighbour) > 0, and (b) contributions from the $\chi_{\parallel}$ component lead to deshielding of H, or σ(neighbour) < 0. Comparison between the $\theta = 0$ and $\theta = 90°$ cases shows that the patterns of shielding and deshielding by neighbouring groups depend not only on differences between $\chi_{\parallel}$ and $\chi_{\perp}$, but also the angle θ.

[Draw diagrams similar to **2** and **3** where the $\chi_{\perp}$ component is parallel to the H—X axis and then analyse the problem as above.]

To a good approximation, the shielding constant σ(neighbour) depends on the distance r, the difference $\chi_{\parallel} - \chi_{\perp}$, as (see *Further reading* for a derivation)

$$\sigma(\text{neighbour}) \propto (\chi_{\parallel} - \chi_{\perp})\left(\frac{1 - 3\cos^2\theta}{r^3}\right) \qquad (8.23)$$

where $\chi_{\parallel}$ and $\chi_{\perp}$ are both negative for a diamagnetic group X. Equation 8.23 shows that the neighbouring group contribution may be positive or negative according to the relative magnitudes of the two magnetic susceptibilities and the relative orientation of the nucleus with respect to X. The latter effect is easy to anticipate: if $54.7° < \theta < 125.3°$, then $1 - 3\cos^2\theta$ is positive, but it is negative otherwise (Fig. 8.8).

A special case of a neighbouring group effect is found in aromatic compounds. The strong anisotropy of the magnetic susceptibility of the benzene ring is ascribed to the

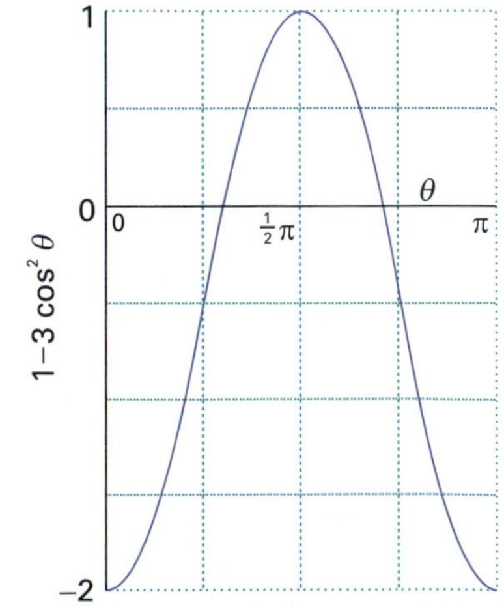

Fig. 8.8 The variation of the function $1 - 3\cos^2\theta$ with the angle θ.

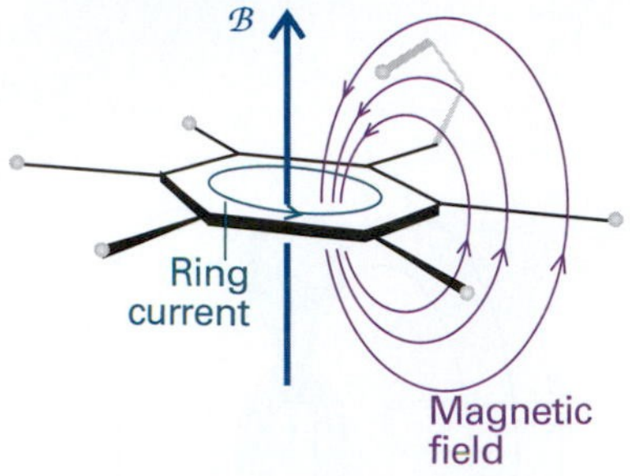

Fig. 8.9 The shielding and deshielding effects of the ring current induced in the benzene ring by the applied field. Protons attached to the ring are deshielded but a proton attached to a substituent that projects above the ring is shielded.

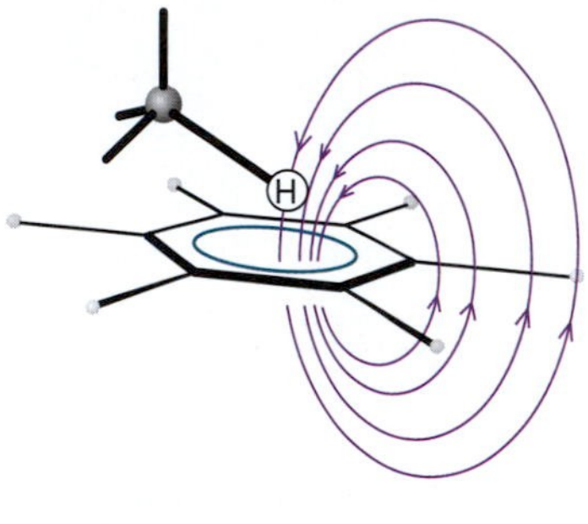

Fig. 8.10 An aromatic solvent (benzene here) can give rise to local currents that shield or deshield a proton in a solvent molecule. In this relative orientation of the solvent and solute, the proton on the solute molecule is shielded.

ability of the field to induce a **ring current**, a circulation of electrons around the ring, when it is applied perpendicular to the molecular plane. Protons in the plane are deshielded (Fig. 8.9), but any that happen to lie above or below the plane (as members of substituents of the ring) are shielded.

(f) The solvent contribution

A solvent can influence the local magnetic field experienced by a nucleus in a variety of ways. Some of these effects arise from specific interactions between the solute and the solvent (such as hydrogen-bond formation and other forms of Lewis acid–base complex formation). The anisotropy of the magnetic susceptibility of the solvent molecules, especially if they are aromatic, can also be the source of a local magnetic field. Moreover, if there are steric interactions that result in a loose but specific interaction between a solute molecule and a solvent molecule, then protons in the solute molecule may experience shielding or deshielding effects according to their location relative to the solvent molecule (Fig. 8.10). We shall see that the NMR spectra of species that contain protons with widely different chemical shifts are easier to interpret than those in which the shifts are similar, so the appropriate choice of solvent may help to simplify the appearance and interpretation of a spectrum.

8.6 The fine structure

The splitting of resonances into individual lines in Fig. 8.6 is called the **fine structure** of the spectrum. It arises because each magnetic nucleus may contribute to the local field experienced by the other nuclei and so modify their resonance frequencies. The strength of the interaction is expressed in terms of the **scalar coupling constant**, J, and reported in hertz (Hz). The scalar coupling constant is so called because the energy of interaction it describes is proportional to the scalar product of the two interacting spins: $E \propto \boldsymbol{I}_1 \cdot \boldsymbol{I}_2$. The constant of proportionality in this expression is $hJ/\hbar^2$, because each angular momentum is proportional to $\hbar$.

Spin coupling constants are independent of the strength of the applied field because they do not depend on the latter for their ability to generate local fields. If the resonance line of a particular nucleus is split by a certain amount by a second nucleus, then the resonance line of the second nucleus is split by the first to the same extent.

(a) The energy levels of coupled systems

It will be useful for later discussions to consider an NMR spectrum in terms of the energy levels of the nuclei and the transitions between them. In NMR, letters far apart in the alphabet (typically A and X) are used to indicate nuclei with very different chemical shifts; letters close together (such as A and B) are used for nuclei with similar chemical shifts. We shall consider first an AX system, a molecule that contains two spin-$\frac{1}{2}$ nuclei A and X with very different chemical shifts in the sense that the difference in chemical shift corresponds to a frequency that is large compared to their spin–spin coupling.

The energy level diagram for a single spin-$\frac{1}{2}$ nucleus and its single transition were shown in Fig. 8.3, and nothing more needs to be said. For a spin-$\frac{1}{2}$ AX system there are four spin states:

$$\alpha_A\alpha_X \qquad \alpha_A\beta_X \qquad \beta_A\alpha_X \quad \beta_A\beta_X$$

The energy depends on the orientation of the spins in the external magnetic field, and if spin–spin coupling is neglected

$$E = -\gamma\hbar(1-\sigma_A)\mathcal{B}m_A - \gamma\hbar(1-\sigma_X)\mathcal{B}m_X = -h\nu_A m_A - h\nu_X m_X \qquad (8.24)$$

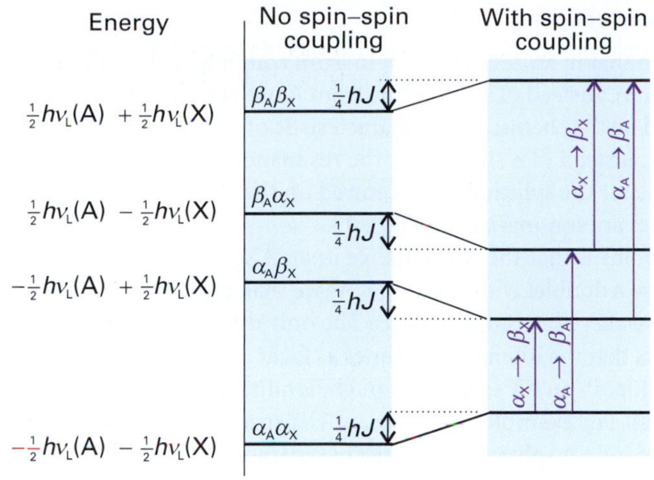

Fig. 8.11 The energy levels of an AX system. The four levels on the left are those of the two spins in the absence of spin–spin coupling. The four levels on the right show how a positive spin–spin coupling constant affects the energies. The transitions shown are for $\beta \leftarrow \alpha$ of A or X, the other nucleus (X or A, respectively) remaining unchanged. We have exaggerated the effect for clarity; in practice, the splitting caused by spin–spin coupling is much smaller than that caused by the applied field.

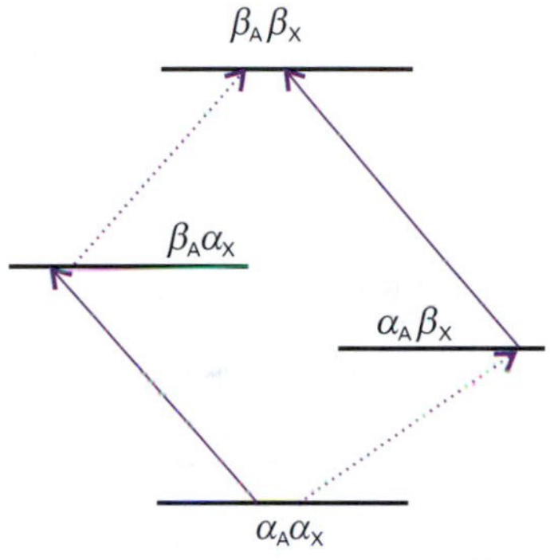

Fig. 8.12 An alternative depiction of the energy levels and transitions shown in Fig. 8.11. Once again, we have exaggerated the effect of spin–spin coupling.

where ν_A and ν_X are the Larmor frequencies of A and X and m_A and m_X are their quantum numbers. This expression gives the four lines on the left of Fig. 8.11. The spin–spin coupling depends on the relative orientation of the two nuclear spins, so it is proportional to the product $m_A m_X$. Therefore, the energy including spin–spin coupling is

$$E = -h\nu_A m_A - h\nu_X m_X + hJm_A m_X \tag{8.25}$$

If $J > 0$, a lower energy is obtained when $m_A m_X < 0$, which is the case if one spin is α and the other is β. A higher energy is obtained if both spins are α or both spins are β. The opposite is true if $J < 0$. The resulting energy level diagram (for $J > 0$) is shown on the right of Fig. 8.11. We see that the $\alpha\alpha$ and $\beta\beta$ states are both raised by $\frac{1}{4}hJ$ and that the $\alpha\beta$ and $\beta\alpha$ states are both lowered by $\frac{1}{4}hJ$.

When a transition of nucleus A occurs, nucleus X remains unchanged. Therefore, the A resonance is a transition for which $\Delta m_A = +1$ and $\Delta m_X = 0$. There are two such transitions, one in which $\beta_A \leftarrow \alpha_A$ occurs when the X nucleus is α_X, and the other in which $\beta_A \leftarrow \alpha_A$ occurs when the X nucleus is β_X. They are shown in Fig. 8.11 and in a slightly different form in Fig. 8.12. The energies of the transitions are

$$\Delta E = h\nu_A \pm \tfrac{1}{2}hJ \tag{8.26a}$$

Therefore, the A resonance consists of a doublet of separation J centred on the chemical shift of A (Fig. 8.13). Similar remarks apply to the X resonance, which consists of two transitions according to whether the A nucleus is α or β (as shown in Fig. 8.12). The transition energies are

$$\Delta E = h\nu_X \pm \tfrac{1}{2}hJ \tag{8.26b}$$

It follows that the X resonance also consists of two lines of separation J, but they are centred on the chemical shift of X (as shown in Fig. 8.13).

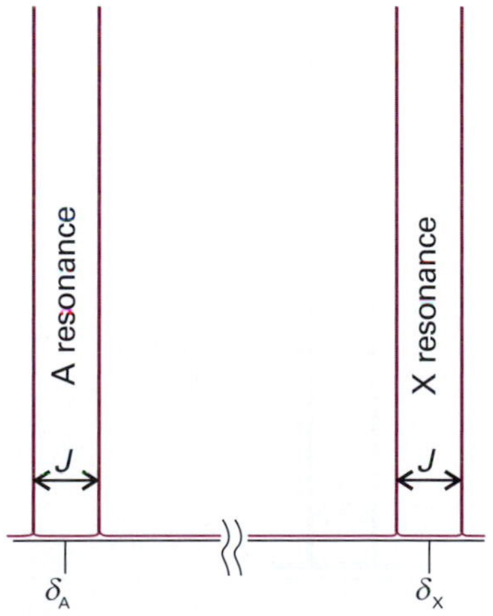

Fig. 8.13 The effect of spin–spin coupling on an AX spectrum. Each resonance is split into two lines separated by J. The pairs of resonances are centred on the chemical shifts of the protons in the absence of spin–spin coupling.

(b) Patterns of coupling

We have seen that in an AX system, spin–spin coupling will result in four lines in the NMR spectrum. Instead of a single line from A, we get a doublet of lines separated by J and centred on the chemical shift characteristic of A. The same splitting occurs in the X resonance: instead of a single line, the resonance is a doublet with splitting J (the same value as for the splitting of A) centred on the chemical shift characteristic of X. These features are summarized in Fig. 8.13.

A subtle point is that the X resonance in an AX_n species (such as an AX_2 or AX_3 species) is also a doublet with splitting J. As we shall explain below, *a group of equivalent nuclei resonates like a single nucleus*. The only difference for the X resonance of an AX_n species is that the intensity is n times as great as that of an AX species (Fig. 8.14). The A resonance in an AX_n species, though, is quite different from the A resonance in an AX species. For example, consider an AX_2 species with two equivalent X nuclei. The resonance of A is split into a doublet of separation J by one X, and each line of that doublet is split again by the same amount by the second X (Fig. 8.15). This splitting results in three lines in the intensity ratio 1:2:1 (because the central frequency can be obtained in two ways). The A resonance of an A_nX_2 species would also be a 1:2:1 triplet of splitting J, the only difference being that the intensity of the A resonance would be n times as great as that of AX_2.

Three equivalent X nuclei (an AX_3 species) split the resonance of A into four lines of intensity ratio 1:3:3:1 and separation J (Fig. 8.16). The X resonance, though, is still a doublet of separation J. In general, n equivalent spin-$\frac{1}{2}$ nuclei split the resonance of a nearby spin or group of equivalent spins into $n+1$ lines with an intensity distribution given by 'Pascal's triangle' in which each entry is the sum of the two entries immediately above (**4**). The easiest way of constructing the pattern of fine structure is to draw a diagram in which successive rows show the splitting of a subsequent proton. The procedure is illustrated in Fig. 8.17 and was used in Figs. 8.15 and 8.16. It is easily extended to molecules containing nuclei with $I > \frac{1}{2}$ (Fig. 8.18).

1
1 1
1 2 1
1 3 3 1
1 4 6 4 1
1 5 10 10 5 1

4

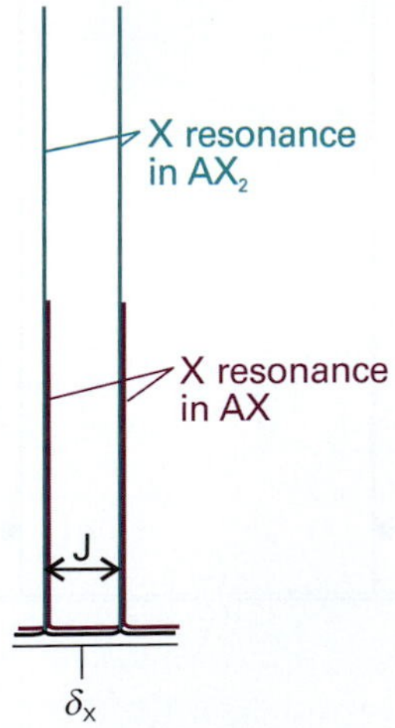

Fig. 8.14 The X resonance of an AX_2 species is also a doublet, because the two equivalent X nuclei behave like a single nucleus; however, the overall absorption is twice as intense as that of an AX species.

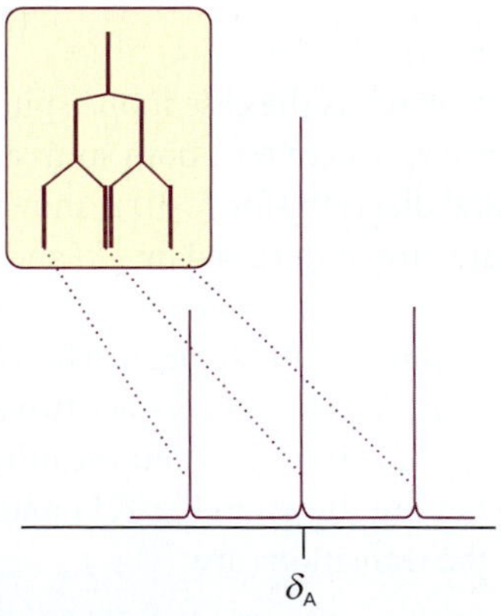

Fig. 8.15 The origin of the 1:2:1 triplet in the A resonance of an AX_2 species. The resonance of A is split into two by coupling with one X nucleus (as shown in the inset), and then each of those two lines is split into two by coupling to the second X nucleus. Because each X nucleus causes the same splitting, the two central transitions are coincident and give rise to an absorption line of double the intensity of the outer lines.

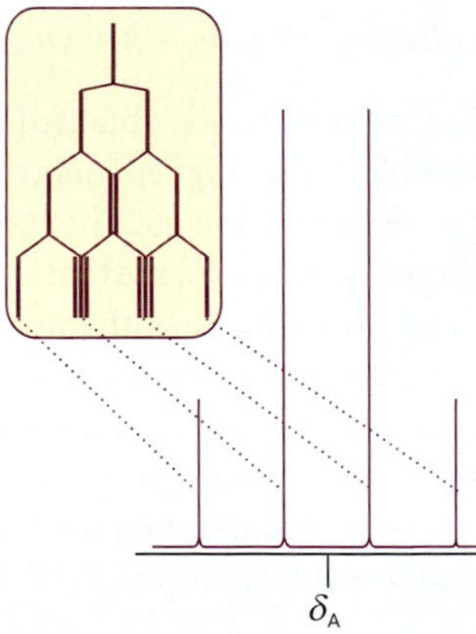

Fig. 8.16 The origin of the 1:3:3:1 quartet in the A resonance of an AX_3 species. The third X nucleus splits each of the lines shown in Fig. 8.15 for an AX_2 species into a doublet, and the intensity distribution reflects the number of transitions that have the same energy.

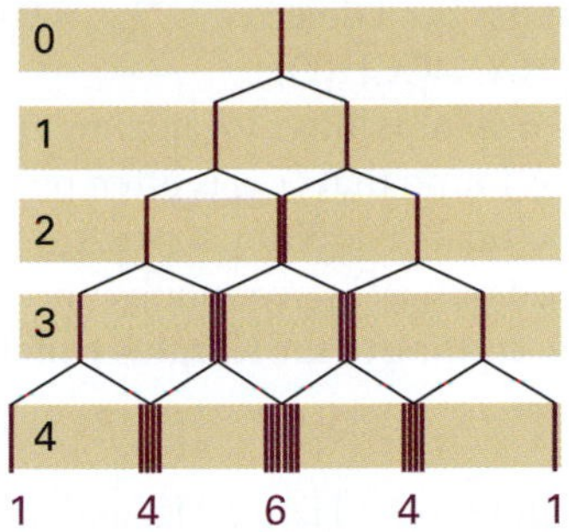

Fig. 8.17 The intensity distribution of the A resonance of an AX_n resonance can be constructed by considering the splitting caused by 1, 2, . . . *n* protons, as in Figs. 8.15 and 8.16. The resulting intensity distribution has a binomial distribution and is given by the integers in the corresponding row of Pascal's triangle. Note that, although the lines have been drawn side-by-side for clarity, the members of each group are coincident. Four protons, in AX_4, split the A resonance into a 1:4:6:4:1 quintet.

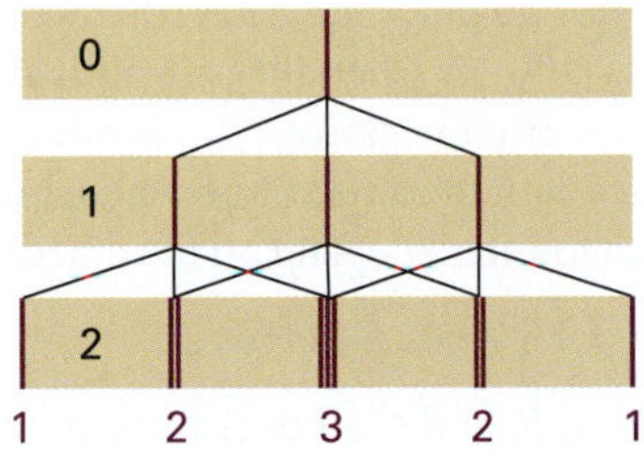

Fig. 8.18 The intensity distribution arising from spin–spin interaction with nuclei with $I = 1$ can be constructed similarly, but each successive nucleus splits the lines into three equal intensity components. Two equivalent spin-1 nuclei give rise to a 1:2:3:2:1 quintet.

Example 8.1 *Accounting for the fine structure in a spectrum*

Account for the fine structure in the NMR spectrum of the C—H protons of ethanol.

Method Consider how each group of equivalent protons (for instance, three methyl protons) split the resonances of the other groups of protons. There is no splitting within groups of equivalent protons. Each splitting pattern can be decided by referring to Pascal's triangle.

Answer The three protons of the CH_3 group split the resonance of the CH_2 protons into a 1:3:3:1 quartet with a splitting J. Likewise, the two protons of the CH_2 group split the resonance of the CH_3 protons into a 1:2:1 triplet with the same splitting J. The CH_2 and CH_3 protons all interact with the OH proton, but these couplings do not cause any splitting because the OH protons migrate rapidly from molecule to molecule and their effect averages to zero.

Self-test 8.2 What fine-structure can be expected for the protons in $^{14}NH_4^+$? The spin quantum number of nitrogen is 1. [1:1:1 triplet from N]

(c) The magnitudes of coupling constants

The scalar coupling constant of two nuclei separated by N bonds is denoted ^NJ, with subscripts for the types of nuclei involved. Thus, $^1J_{CH}$ is the coupling constant for a proton joined directly to a ^{13}C atom, and $^2J_{CH}$ is the coupling constant when the same two nuclei are separated by two bonds (as in $^{13}C—C—H$). A typical value of $^1J_{CH}$ is in the range 120 to 250 Hz; $^2J_{CH}$ is between −10 and +20 Hz. Both ^3J and ^4J can give detectable effects in a spectrum, but couplings over larger numbers of bonds can generally

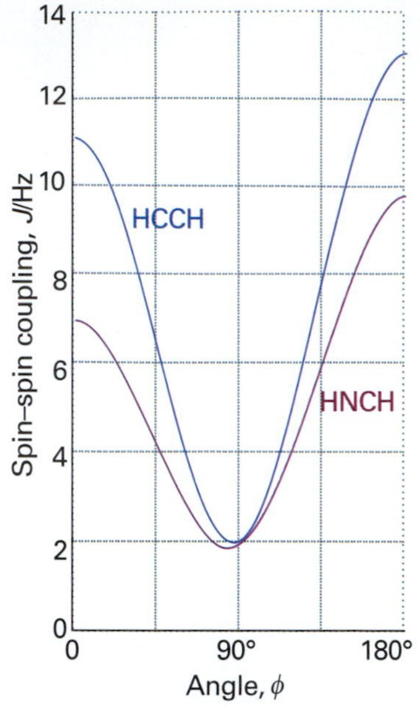

Fig. 8.19 The variation of the spin–spin coupling constant with angle predicted by the Karplus equation for an HCCH group and an HNCH group.

Exploration Draw a family of curves showing the variation of $^3J_{HH}$ with ϕ for which $A = +7.0$ Hz, $B = -1.0$ Hz, and C varies slightly from a typical value of +5.0 Hz. What is the effect of changing the value of the parameter C on the shape of the curve? In a similar fashion, explore the effect of the values of A and B on the shape of the curve.

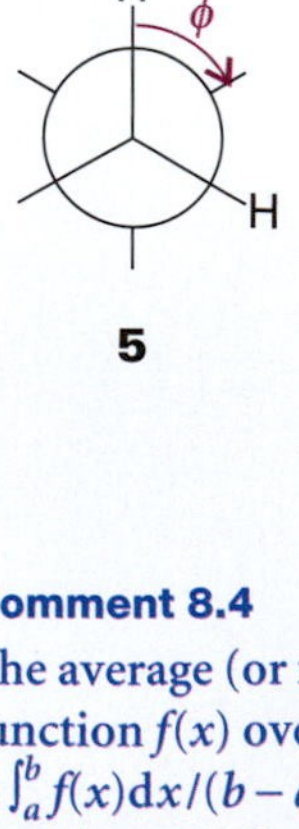

5

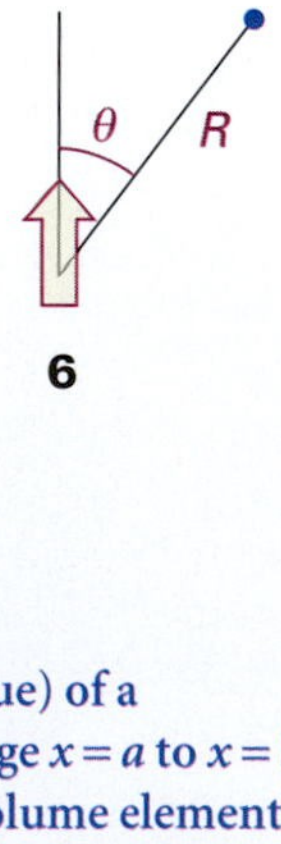

6

Comment 8.4

The average (or mean value) of a function $f(x)$ over the range $x = a$ to $x = b$ is $\int_a^b f(x)\mathrm{d}x/(b-a)$. The volume element in polar coordinates is proportional to $\sin\theta\,\mathrm{d}\theta$, and θ ranges from 0 to π. Therefore the average value of $(1 - 3\cos^2\theta)$ is $\int_0^\pi (1-3\cos^2\theta)\sin\theta\,\mathrm{d}\theta/\pi = 0$.

be ignored. One of the longest range couplings that has been detected is $^9J_{HH} = 0.4$ Hz between the CH_3 and CH_2 protons in $CH_3C{\equiv}CC{\equiv}CC{\equiv}CCH_2OH$.

The sign of J_{XY} indicates whether the energy of two spins is lower when they are parallel ($J < 0$) or when they are antiparallel ($J > 0$). It is found that $^1J_{CH}$ is often positive, $^2J_{HH}$ is often negative, $^3J_{HH}$ is often positive, and so on. An additional point is that J varies with the angle between the bonds (Fig. 8.19). Thus, a $^3J_{HH}$ coupling constant is often found to depend on the dihedral angle ϕ (**5**) according to the **Karplus equation**:

$$J = A + B\cos\phi + C\cos 2\phi \tag{8.27}$$

with A, B, and C empirical constants with values close to +7 Hz, −1 Hz, and +5 Hz, respectively, for an HCCH fragment. It follows that the measurement of $^3J_{HH}$ in a series of related compounds can be used to determine their conformations. The coupling constant $^1J_{CH}$ also depends on the hybridization of the C atom, as the following values indicate:

	sp	sp^2	sp^3
$^1J_{CH}$/Hz:	250	160	125

(d) The origin of spin–spin coupling

Spin–spin coupling is a very subtle phenomenon, and it is better to treat J as an empirical parameter than to use calculated values. However, we can get some insight into its origins, if not its precise magnitude—or always reliably its sign—by considering the magnetic interactions within molecules.

A nucleus with spin projection m_I gives rise to a magnetic field with z-component $\mathcal{B}_{nuc}$ at a distance R, where, to a good approximation,

$$\mathcal{B}_{nuc} = -\frac{\gamma\hbar\mu_0}{4\pi R^3}(1 - 3\cos^2\theta)m_I \tag{8.28}$$

The angle θ is defined in (**6**). The magnitude of this field is about 0.1 mT when R = 0.3 nm, corresponding to a splitting of resonance signal of about 10^4 Hz, and is of the order of magnitude of the splitting observed in solid samples (see Section 8.3a).

In a liquid, the angle θ sweeps over all values as the molecule tumbles, and $1–3\cos^2\theta$ averages to zero. Hence the direct dipolar interaction between spins cannot account for the fine structure of the spectra of rapidly tumbling molecules. The direct interaction does make an important contribution to the spectra of solid samples and is a very useful indirect source of structure information through its involvement in spin relaxation (Section 8.11).

Spin–spin coupling in molecules in solution can be explained in terms of the **polarization mechanism**, in which the interaction is transmitted through the bonds. The simplest case to consider is that of $^1J_{XY}$ where X and Y are spin-$\frac{1}{2}$ nuclei joined by an electron-pair bond (Fig. 8.20). The coupling mechanism depends on the fact that in some atoms it is favourable for the nucleus and a nearby electron spin to be parallel (both α or both β), but in others it is favourable for them to be antiparallel (one α and the other β). The electron–nucleus coupling is magnetic in origin, and may be either a dipolar interaction between the magnetic moments of the electron and nuclear spins or a **Fermi contact interaction**. A pictorial description of the Fermi contact interaction is as follows. First, we regard the magnetic moment of the nucleus as arising from the circulation of a current in a tiny loop with a radius similar to that of the nucleus (Fig. 8.21). Far from the nucleus the field generated by this loop is indistinguishable from the field generated by a point magnetic dipole. Close to the loop, however, the field differs from that of a point dipole. The magnetic interaction between this non-dipolar field and the electron's magnetic moment is the contact

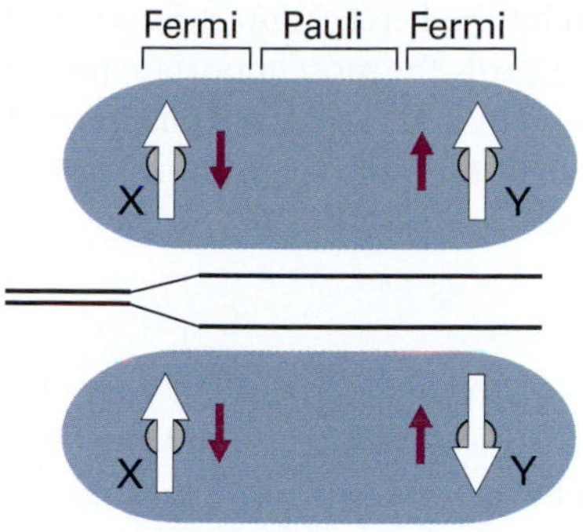

Fig. 8.20 The polarization mechanism for spin–spin coupling ($^1J_{HH}$). The two arrangements have slightly different energies. In this case, J is positive, corresponding to a lower energy when the nuclear spins are antiparallel.

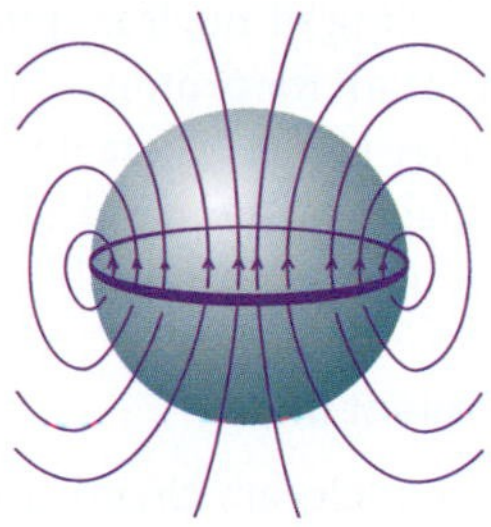

Fig. 8.21 The origin of the Fermi contact interaction. From far away, the magnetic field pattern arising from a ring of current (representing the rotating charge of the nucleus, the pale green sphere) is that of a point dipole. However, if an electron can sample the field close to the region indicated by the sphere, the field distribution differs significantly from that of a point dipole. For example, if the electron can penetrate the sphere, then the spherical average of the field it experiences is not zero.

interaction. The lines of force depicted in Fig. 8.21 correspond to those for a proton with α spin. The lower energy state of an electron spin in such a field is the β state. In conclusion, the contact interaction depends on the very close approach of an electron to the nucleus and hence can occur only if the electron occupies an *s* orbital (which is the reason why $^1J_{CH}$ depends on the hybridization ratio). We shall suppose that it is energetically favourable for an electron spin and a nuclear spin to be antiparallel (as is the case for a proton and an electron in a hydrogen atom).

If the X nucleus is α, a β electron of the bonding pair will tend to be found nearby (because that is energetically favourable for it). The second electron in the bond, which must have α spin if the other is β, will be found mainly at the far end of the bond (because electrons tend to stay apart to reduce their mutual repulsion). Because it is energetically favourable for the spin of Y to be antiparallel to an electron spin, a Y nucleus with β spin has a lower energy, and hence a lower Larmor frequency, than a Y nucleus with α spin. The opposite is true when X is β, for now the α spin of Y has the lower energy. In other words, the antiparallel arrangement of nuclear spins lies lower in energy than the parallel arrangement as a result of their magnetic coupling with the bond electrons. That is, $^1J_{HH}$ is positive.

To account for the value of $^2J_{XY}$, as in H—C—H, we need a mechanism that can transmit the spin alignments through the central C atom (which may be ^{12}C, with no nuclear spin of its own). In this case (Fig. 8.22), an X nucleus with α spin polarizes the electrons in its bond, and the α electron is likely to be found closer to the C nucleus. The more favourable arrangement of two electrons on the same atom is with their spins parallel (Hund's rule, Section 3.4d), so the more favourable arrangement is for the α electron of the neighbouring bond to be close to the C nucleus. Consequently, the β electron of that bond is more likely to be found close to the Y nucleus, and therefore that nucleus will have a lower energy if it is α. Hence, according to this mechanism, the lower Larmor frequency of Y will be obtained if its spin is parallel to that of X. That is, $^2J_{HH}$ is negative.

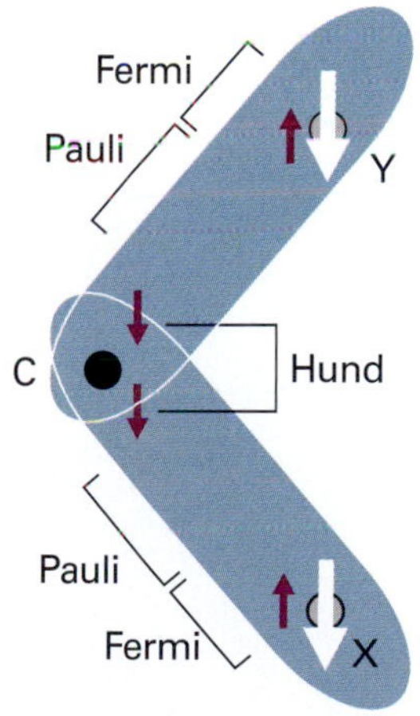

Fig. 8.22 The polarization mechanism for $^2J_{HH}$ spin–spin coupling. The spin information is transmitted from one bond to the next by a version of the mechanism that accounts for the lower energy of electrons with parallel spins in different atomic orbitals (Hund's rule of maximum multiplicity). In this case, $J < 0$, corresponding to a lower energy when the nuclear spins are parallel.

The coupling of nuclear spin to electron spin by the Fermi contact interaction is most important for proton spins, but it is not necessarily the most important mechanism for other nuclei. These nuclei may also interact by a dipolar mechanism with the electron magnetic moments and with their orbital motion, and there is no simple way of specifying whether J will be positive or negative.

(e) Equivalent nuclei

A group of nuclei are **chemically equivalent** if they are related by a symmetry operation of the molecule and have the same chemical shifts. Chemically equivalent nuclei are nuclei that would be regarded as 'equivalent' according to ordinary chemical criteria. Nuclei are **magnetically equivalent** if, as well as being chemically equivalent, they also have identical spin–spin interactions with any other magnetic nuclei in the molecule.

The difference between chemical and magnetic equivalence is illustrated by CH_2F_2 and $H_2C{=}CF_2$. In each of these molecules the protons are chemically equivalent: they are related by symmetry and undergo the same chemical reactions. However, although the protons in CH_2F_2 are magnetically equivalent, those in $CH_2{=}CF_2$ are not. One proton in the latter has a *cis* spin-coupling interaction with a given F nucleus whereas the other proton has a *trans* interaction with it. In constrast, in CH_2F_2 both protons are connected to a given F nucleus by identical bonds, so there is no distinction between them. Strictly speaking, the CH_3 protons in ethanol (and other compounds) are magnetically inequivalent on account of their different interactions with the CH_2 protons in the next group. However, they are in practice made magnetically equivalent by the rapid rotation of the CH_3 group, which averages out any differences. Magnetically inequivalent species can give very complicated spectra (for instance, the proton and ^{19}F spectra of $H_2C{=}CF_2$ each consist of 12 lines), and we shall not consider them further.

An important feature of chemically equivalent magnetic nuclei is that, although they do couple together, the coupling has no effect on the appearance of the spectrum. The reason for the invisibility of the coupling is set out in the following *Justification*, but qualitatively it is that all allowed nuclear spin transitions are *collective* reorientations of groups of equivalent nuclear spins that do not change the relative orientations of the spins within the group (Fig. 8.23). Then, because the relative orientations of nuclear spins are not changed in any transition, the magnitude of the coupling between them is undetectable. Hence, an isolated CH_3 group gives a single, unsplit line because all the allowed transitions of the group of three protons occur without change of their relative orientations.

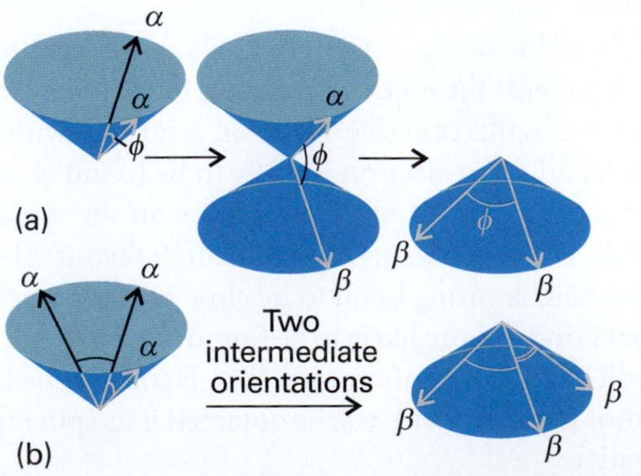

Fig. 8.23 (a) A group of two equivalent nuclei realigns as a group, without change of angle between the spins, when a resonant absorption occurs. Hence it behaves like a single nucleus and the spin–spin coupling between the individual spins of the group is undetectable. (b) Three equivalent nuclei also realign as a group without change of their relative orientations.

Justification 8.2 *The energy levels of an A_2 system*

Consider an A_2 system of two chemically equivalent spin-$\frac{1}{2}$ nuclei. First, consider the energy levels in the absence of spin–spin coupling. There are four spin states that (just as for two electrons) can be classified according to their total spin I (the analogue of S for two electrons) and their total projection M_I on the z-axis. The states are analogous to those we developed for two electrons in singlet and triplet states:

Spins parallel, $I = 1$:	$M_I = +1$	$\alpha\alpha$
	$M_I = 0$	$(1/2^{1/2})\{\alpha\beta + \beta\alpha\}$
	$M_I = -1$	$\beta\beta$
Spins paired, $I = 0$:	$M_I = 0$	$(1/2^{1/2})\{\alpha\beta - \beta\alpha\}$

The effect of a magnetic field on these four states is shown in Fig. 8.24: the energies of the two states with $M_I = 0$ are unchanged by the field because they are composed of equal proportions of α and β spins.

As remarked in Section 8.6a, the spin–spin coupling energy is proportional to the scalar product of the vectors representing the spins, $E = (hJ/\hbar^2)\boldsymbol{I}_1 \cdot \boldsymbol{I}_2$. The scalar product can be expressed in terms of the total nuclear spin by noting that

$$I^2 = (\boldsymbol{I}_1 + \boldsymbol{I}_2)\cdot(\boldsymbol{I}_1 + \boldsymbol{I}_2) = I_1^2 + I_2^2 + 2\boldsymbol{I}_1 \cdot \boldsymbol{I}_2$$

rearranging this expression to

$$\boldsymbol{I}_1 \cdot \boldsymbol{I}_2 = \tfrac{1}{2}\{I^2 - I_1^2 - I_2^2\}$$

and replacing the magnitudes by their quantum mechanical values:

$$\boldsymbol{I}_1 \cdot \boldsymbol{I}_2 = \tfrac{1}{2}\{I(I+1) - I_1(I_1+1) - I_2(I_2+1)\}\hbar^2$$

Then, because $I_1 = I_2 = \frac{1}{2}$, it follows that

$$E = \tfrac{1}{2}hJ\{I(I+1) - \tfrac{3}{2}\}$$

For parallel spins, $I = 1$ and $E = +\frac{1}{4}hJ$; for antiparallel spins $I = 0$ and $E = -\frac{3}{4}hJ$, as in Fig. 8.24. We see that three of the states move in energy in one direction and the fourth (the one with antiparallel spins) moves three times as much in the opposite direction. The resulting energy levels are shown on the right in Fig. 8.24.

Comment 8.5

As in Section 3.7, the states we have selected in *Justification 8.2* are those with a definite resultant, and hence a well defined value of I. The + sign in $\alpha\beta + \beta\alpha$ signifies an in-phase alignment of spins and $I = 1$; the − sign in $\alpha\beta - \beta\alpha$ signifies an alignment out of phase by π, and hence $I = 0$. See Fig. 3.24.

The NMR spectrum of the A_2 species arises from transitions between the levels. However, the radiofrequency field affects the two equivalent protons equally, so it cannot change the orientation of one proton relative to the other; therefore, the transitions take place within the set of states that correspond to parallel spin (those labelled $I = 1$), and no spin-parallel state can change to a spin-antiparallel state (the state with $I = 0$). Put another way, the allowed transitions are subject to the selection rule $\Delta I = 0$. This selection rule is in addition to the rule $\Delta M_I = \pm 1$ that arises from the conservation of angular momentum and the unit spin of the photon. The allowed transitions are shown in Fig. 8.24: we see that there are only two transitions, and that they occur at the same resonance frequency that the nuclei would have in the absence of spin–spin coupling. Hence, the spin–spin coupling interaction does not affect the appearance of the spectrum.

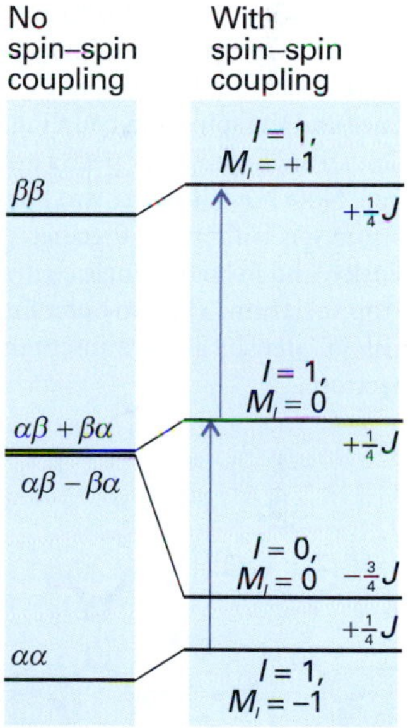

Fig. 8.24 The energy levels of an A_2 system in the absence of spin–spin coupling are shown on the left. When spin–spin coupling is taken into account, the energy levels on the right are obtained. Note that the three states with total nuclear spin $I = 1$ correspond to parallel spins and give rise to the same increase in energy (J is positive); the one state with $I = 0$ (antiparallel nuclear spins) has a lower energy in the presence of spin–spin coupling. The only allowed transitions are those that preserve the angle between the spins, and so take place between the three states with $I = 1$. They occur at the same resonance frequency as they would have in the absence of spin–spin coupling.

(f) Strongly coupled nuclei

NMR spectra are usually much more complex than the foregoing simple analysis suggests. We have described the extreme case in which the differences in chemical shifts are much greater than the spin–spin coupling constants. In such cases it is simple to identify groups of magnetically equivalent nuclei and to think of the groups of nuclear spins as reorientating relative to each other. The spectra that result are called **first-order spectra**.

Transitions cannot be allocated to definite groups when the differences in their chemical shifts are comparable to their spin–spin coupling interactions. The complicated spectra that are then obtained are called **strongly coupled spectra** (or 'second-order spectra') and are much more difficult to analyse (Fig. 8.25). Because the difference in resonance frequencies increases with field, but spin–spin coupling constants are independent of it, a second-order spectrum may become simpler (and first-order) at high fields because individual groups of nuclei become identifiable again.

A clue to the type of analysis that is appropriate is given by the notation for the types of spins involved. Thus, an AX spin system (which consists of two nuclei with a large chemical shift difference) has a first-order spectrum. An AB system, on the other hand (with two nuclei of similar chemical shifts), gives a spectrum typical of a strongly coupled system. An AX system may have widely different Larmor frequencies because A and X are nuclei of different elements (such as ^{13}C and ^{1}H), in which case they form a **heteronuclear spin system**. AX may also denote a **homonuclear spin system** in which the nuclei are of the same element but in markedly different environments.

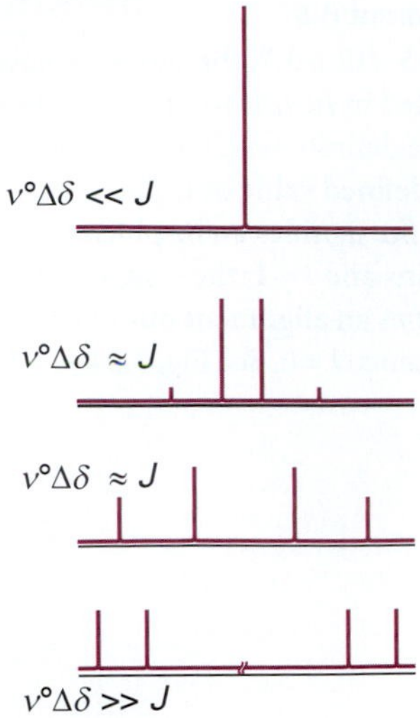

Fig. 8.25 The NMR spectra of an A_2 system (top) and an AX system (bottom) are simple 'first-order' spectra. At intermediate relative values of the chemical shift difference and the spin–spin coupling, complex 'strongly coupled' spectra are obtained. Note how the inner two lines of the bottom spectrum move together, grow in intensity, and form the single central line of the top spectrum. The two outer lines diminish in intensity and are absent in the top spectrum.

8.7 Conformational conversion and exchange processes

The appearance of an NMR spectrum is changed if magnetic nuclei can jump rapidly between different environments. Consider a molecule, such as *N,N*-dimethylformamide, that can jump between conformations; in its case, the methyl shifts depend on whether they are *cis* or *trans* to the carbonyl group (Fig. 8.26). When the jumping rate is low, the spectrum shows two sets of lines, one each from molecules in each conformation. When the interconversion is fast, the spectrum shows a single line at the mean of the two chemical shifts. At intermediate inversion rates, the line is very broad. This maximum broadening occurs when the lifetime, τ, of a conformation gives rise to a linewidth that is comparable to the difference of resonance frequencies, $\delta\nu$, and both broadened lines blend together into a very broad line. Coalescence of the two lines occurs when

$$\tau = \frac{\sqrt{2}}{\pi\delta\nu} \tag{8.29}$$

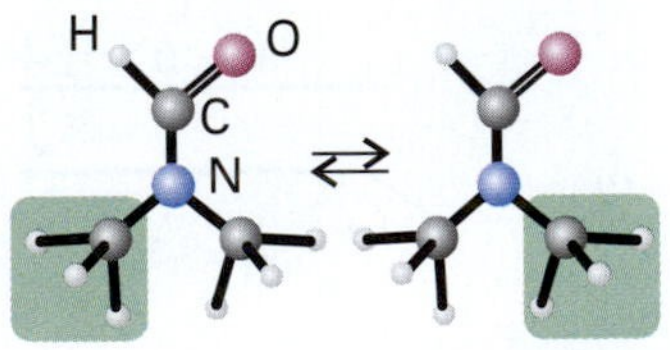

Fig. 8.26 When a molecule changes from one conformation to another, the positions of its protons are interchanged and jump between magnetically distinct environments.

Example 8.2 *Interpreting line broadening*

The NO group in *N,N*-dimethylnitrosamine, $(CH_3)_2N—NO$, rotates about the N—N bond and, as a result, the magnetic environments of the two CH_3 groups are interchanged. The two CH_3 resonances are separated by 390 Hz in a 600 MHz spectrometer. At what rate of interconversion will the resonance collapse to a single line?

Method Use eqn 8.29 for the average lifetimes of the conformations. The rate of interconversion is the inverse of their lifetime.

Answer With $\delta\nu = 390$ Hz,

$$\tau = \frac{\sqrt{2}}{\pi \times (390\ \text{s}^{-1})} = 1.2\ \text{ms}$$

It follows that the signal will collapse to a single line when the interconversion rate exceeds about 830 s^{-1}. The dependence of the rate of exchange on the temperature is used to determine the energy barrier to interconversion.

Self-test 8.3 What would you deduce from the observation of a single line from the same molecule in a 300 MHz spectrometer?

[Conformation lifetime less than 2.3 ms]

A similar explanation accounts for the loss of fine structure in solvents able to exchange protons with the sample. For example, hydroxyl protons are able to exchange with water protons. When this **chemical exchange** occurs, a molecule ROH with an α-spin proton (we write this ROH_α) rapidly converts to ROH_β and then perhaps to ROH_α again because the protons provided by the solvent molecules in successive exchanges have random spin orientations. Therefore, instead of seeing a spectrum composed of contributions from both ROH_α and ROH_β molecules (that is, a spectrum showing a doublet structure due to the OH proton) we see a spectrum that shows no splitting caused by coupling of the OH proton (as in Fig. 8.6). The effect is observed when the lifetime of a molecule due to this chemical exchange is so short that the lifetime broadening is greater than the doublet splitting. Because this splitting is often very small (a few hertz), a proton must remain attached to the same molecule for longer than about 0.1 s for the splitting to be observable. In water, the exchange rate is much faster than that, so alcohols show no splitting from the OH protons. In dry dimethylsulfoxide (DMSO), the exchange rate may be slow enough for the splitting to be detected.

Pulse techniques in NMR

Modern methods of detecting the energy separation between nuclear spin states are more sophisticated than simply looking for the frequency at which resonance occurs. One of the best analogies that has been suggested to illustrate the difference between the old and new ways of observing an NMR spectrum is that of detecting the spectrum of vibrations of a bell. We could stimulate the bell with a gentle vibration at a gradually increasing frequency, and note the frequencies at which it resonated with the stimulation. A lot of time would be spent getting zero response when the stimulating frequency was between the bell's vibrational modes. However, if we were simply to hit the bell with a hammer, we would immediately obtain a clang composed of all the frequencies that the bell can produce. The equivalent in NMR is to monitor the radiation nuclear spins emit as they return to equilibrium after the appropriate stimulation. The resulting **Fourier-transform NMR** gives greatly increased sensitivity, so opening up the entire periodic table to the technique. Moreover, multiple-pulse FT-NMR gives chemists unparalleled control over the information content and display of spectra. We need to understand how the equivalent of the hammer blow is delivered and how the signal is monitored and interpreted. These features are generally expressed in terms of the vector model of angular momentum introduced in Section 2.7d.

8.8 The magnetization vector

Consider a sample composed of many identical spin-$\frac{1}{2}$ nuclei. As we saw in Section 2.7d, an angular momentum can be represented by a vector of length $\{I(I+1)\}^{1/2}$ units

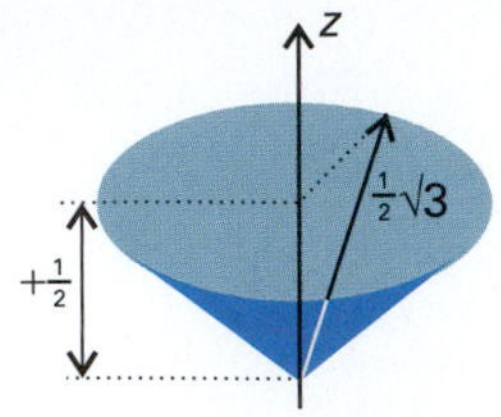

Fig. 8.27 The vector model of angular momentum for a single spin-$\frac{1}{2}$ nucleus. The angle around the z-axis is indeterminate.

with a component of length m_I units along the z-axis. As the uncertainty principle does not allow us to specify the x- and y-components of the angular momentum, all we know is that the vector lies somewhere on a cone around the z-axis. For $I = \frac{1}{2}$, the length of the vector is $\frac{1}{2}\sqrt{3}$ and it makes an angle of 55° to the z-axis (Fig. 8.27).

In the absence of a magnetic field, the sample consists of equal numbers of α and β nuclear spins with their vectors lying at random angles on the cones. These angles are unpredictable, and at this stage we picture the spin vectors as stationary. The **magnetization, M**, of the sample, its net nuclear magnetic moment, is zero (Fig. 8.28a).

(a) The effect of the static field

Two changes occur in the magnetization when a magnetic field is present. First, the energies of the two orientations change, the α spins moving to low energy and the β spins to high energy (provided $\gamma > 0$). At 10 T, the Larmor frequency for protons is 427 MHz, and in the vector model the individual vectors are pictured as precessing at this rate. This motion is a pictorial representation of the difference in energy of the spin states (it is not an actual representation of reality). As the field is increased, the Larmor frequency increases and the precession becomes faster. Secondly, the populations of the two spin states (the numbers of α and β spins) at thermal equilibrium change, and there will be more α spins than β spins. Because $h\nu_L/kT \approx 7 \times 10^{-5}$ for protons at 300 K and 10 T, it follows from the Boltzmann distribution that $N_\beta/N_\alpha = e^{-h\nu_L/kT}$ is only slightly less than 1. That is, there is only a tiny imbalance of populations, and it is even smaller for other nuclei with their smaller magnetogyric ratios. However, despite its smallness, the imbalance means that there is a net magnetization that we can represent by a vector M pointing in the z-direction and with a length proportional to the population difference (Fig. 8.28b).

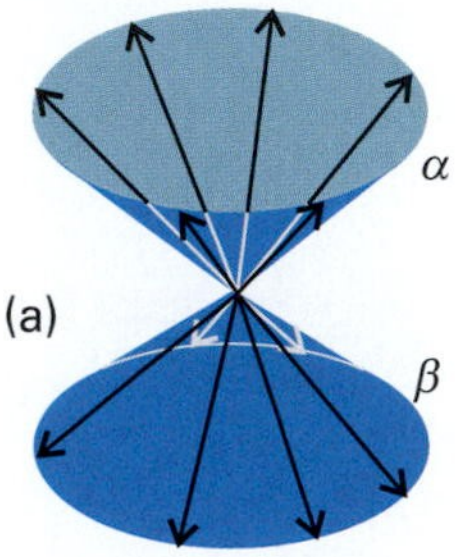

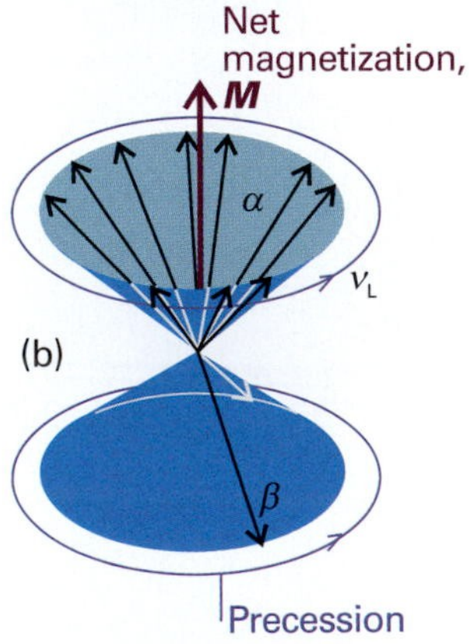

Fig. 8.28 The magnetization of a sample of spin-$\frac{1}{2}$ nuclei is the resultant of all their magnetic moments. (a) In the absence of an externally applied field, there are equal numbers of α and β spins at random angles around the z-axis (the field direction) and the magnetization is zero. (b) In the presence of a field, the spins precess around their cones (that is, there is an energy difference between the α and β states) and there are slightly more α spins than β spins. As a result, there is a net magnetization along the z-axis.

(b) The effect of the radiofrequency field

We now consider the effect of a radiofrequency field circularly polarized in the xy-plane, so that the magnetic component of the electromagnetic field (the only component we need to consider) is rotating around the z-direction, the direction of the applied field $\mathcal{B}_0$, in the same sense as the Larmor precession. The strength of the rotating magnetic field is $\mathcal{B}_1$. Suppose we choose the frequency of this field to be equal to the Larmor frequency of the spins, $\nu_L = (\gamma/2\pi)\mathcal{B}_0$; this choice is equivalent to selecting the resonance condition in the conventional experiment. The nuclei now experience a steady $\mathcal{B}_1$ field because the rotating magnetic field is in step with the precessing spins (Fig. 8.29a). Just as the spins precess about the strong static field $\mathcal{B}_0$ at a frequency $\gamma\mathcal{B}_0/2\pi$, so under the influence of the field $\mathcal{B}_1$ they precess about $\mathcal{B}_1$ at a frequency $\gamma\mathcal{B}_1/2\pi$.

To interpret the effects of radiofrequency pulses on the magnetization, it is often useful to look at the spin system from a different perspective. If we were to imagine stepping on to a platform, a so-called **rotating frame**, that rotates around the direction of the applied field at the radiofrequency, then the nuclear magnetization appears stationary if the radiofrequency is the same as the Larmor frequency (Fig. 8.29b). If the $\mathcal{B}_1$ field is applied in a pulse of duration $\pi/2\gamma\mathcal{B}_1$, the magnetization tips through 90° in the rotating frame and we say that we have applied a **90° pulse**, or a 'π/2 pulse' (Fig. 8.30a). The duration of the pulse depends on the strength of the $\mathcal{B}_1$ field, but is typically of the order of microseconds.

Now imagine stepping out of the rotating frame. To an external observer (the role played by a radiofrequency coil) in this stationary frame, the magnetization vector is now rotating at the Larmor frequency in the xy-plane (Fig. 8.30b). The rotating magnetization induces in the coil a signal that oscillates at the Larmor frequency and that

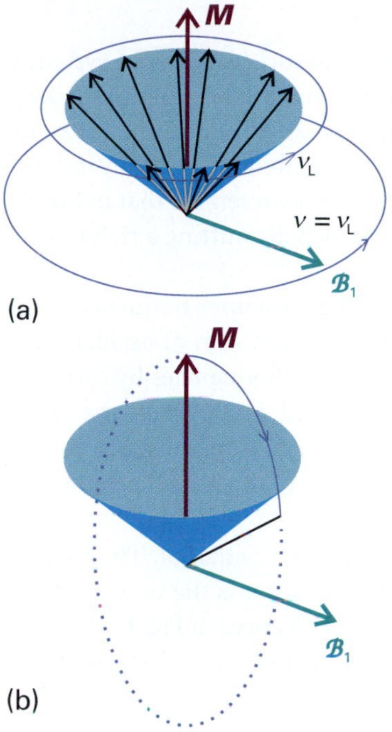

Fig. 8.29 (a) In a resonance experiment, a circularly polarized radiofrequency magnetic field $\mathcal{B}_1$ is applied in the *xy*-plane (the magnetization vector lies along the *z*-axis). (b) If we step into a frame rotating at the radiofrequency, $\mathcal{B}_1$ appears to be stationary, as does the magnetization ***M*** if the Larmor frequency is equal to the radiofrequency. When the two frequencies coincide, the magnetization vector of the sample rotates around the direction of the $\mathcal{B}_1$ field.

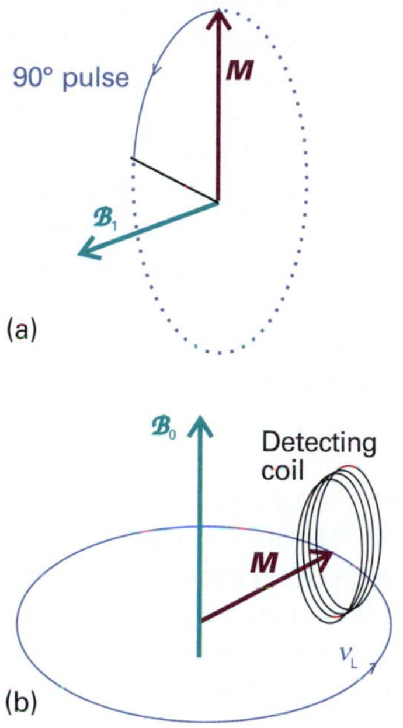

Fig. 8.30 (a) If the radiofrequency field is applied for a certain time, the magnetization vector is rotated into the *xy*-plane. (b) To an external stationary observer (the coil), the magnetization vector is rotating at the Larmor frequency, and can induce a signal in the coil.

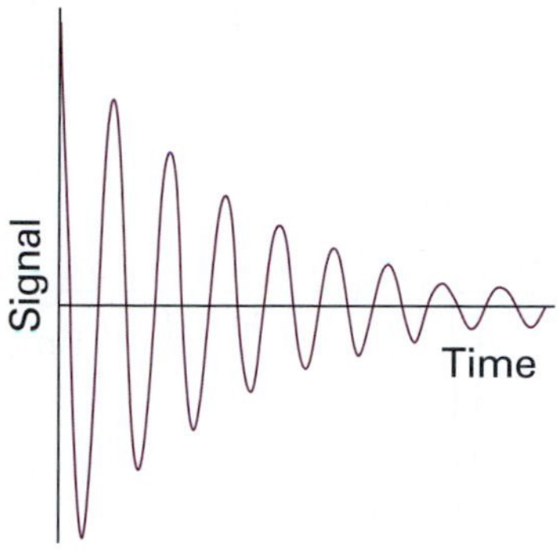

Fig. 8.31 A simple free-induction decay of a sample of spins with a single resonance frequency.

can be amplified and processed. In practice, the processing takes place after subtraction of a constant high frequency component (the radiofrequency used for $\mathcal{B}_1$), so that all the signal manipulation takes place at frequencies of a few kilohertz.

As time passes, the individual spins move out of step (partly because they are precessing at slightly different rates, as we shall explain later), so the magnetization vector shrinks exponentially with a time constant T_2 and induces an ever weaker signal in the detector coil. The form of the signal that we can expect is therefore the oscillating-decaying **free-induction decay** (FID) shown in Fig. 8.31. The *y*-component of the magnetization varies as

$$M_y(t) = M_0 \cos(2\pi\nu_L t)\, e^{-t/T_2} \qquad (8.30)$$

We have considered the effect of a pulse applied at exactly the Larmor frequency. However, virtually the same effect is obtained off resonance, provided that the pulse is applied close to ν_L. If the difference in frequency is small compared to the inverse of the duration of the 90° pulse, the magnetization will end up in the *xy*-plane. Note that we do not need to know the Larmor frequency beforehand: the short pulse is the analogue of the hammer blow on the bell, exciting a range of frequencies. The detected signal shows that a particular resonant frequency is present.

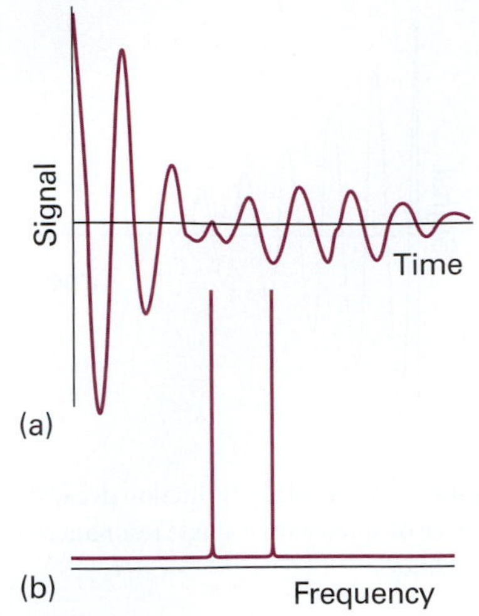

Fig. 8.32 (a) A free induction decay signal of a sample of AX species and (b) its analysis into its frequency components.

Exploration The *Living graphs* section of the text's web site has an applet that allows you to calculate and display the FID curve from an AX system. Explore the effect on the shape of the FID curve of changing the chemical shifts (and therefore the Larmor frequencies) of the A and X nuclei.

Comment 8.6

The web site for this text contains links to databases of NMR spectra and to sites that allow for interactive simulation of NMR spectra.

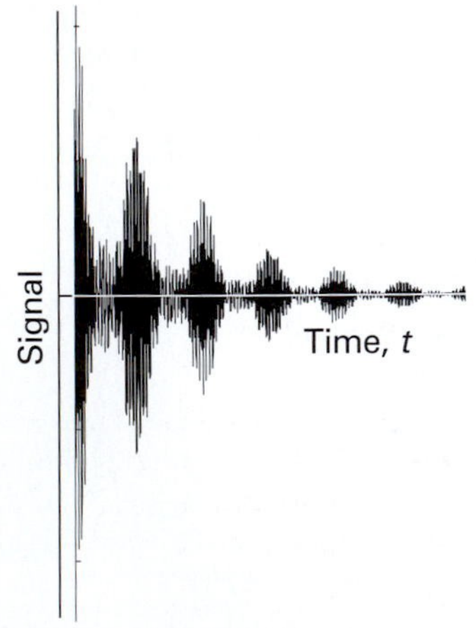

Fig. 8.33 A free induction decay signal of a sample of ethanol. Its Fourier transform is the frequency-domain spectrum shown in Fig. 8.6. The total length of the image corresponds to about 1 s.

(c) Time- and frequency-domain signals

We can think of the magnetization vector of a homonuclear AX spin system with $J = 0$ as consisting of two parts, one formed by the A spins and the other by the X spins. When the 90° pulse is applied, both magnetization vectors are rotated into the *xy*-plane. However, because the A and X nuclei precess at different frequencies, they induce two signals in the detector coils, and the overall FID curve may resemble that in Fig. 8.32a. The composite FID curve is the analogue of the struck bell emitting a rich tone composed of all the frequencies at which it can vibrate.

The problem we must address is how to recover the resonance frequencies present in a free-induction decay. We know that the FID curve is a sum of oscillating functions, so the problem is to analyse it into its component frequencies by carrying out a Fourier transformation (*Further information* 6.2 and 8.1). When the signal in Fig. 8.32a is transformed in this way, we get the frequency-domain spectrum shown in Fig. 8.32b. One line represents the Larmor frequency of the A nuclei and the other that of the X nuclei.

The FID curve in Fig. 8.33 is obtained from a sample of ethanol. The frequency-main spectrum obtained from it by Fourier transformation is the one that we have already discussed (Fig. 8.6). We can now see why the FID curve in Fig. 8.33 is so complex: it arises from the precession of a magnetization vector that is composed of eight components, each with a characteristic frequency.

8.9 Spin relaxation

There are two reasons why the component of the magnetization vector in the *xy*-plane shrinks. Both reflect the fact that the nuclear spins are not in thermal equilibrium with their surroundings (for then $\boldsymbol{M}$ lies parallel to z). The return to equilibrium is the process called **spin relaxation**.

(a) Longitudinal and transverse relaxation

At thermal equilibrium the spins have a Boltzmann distribution, with more α spins than β spins; however, a magnetization vector in the *xy*-plane immediately after a 90° pulse has equal numbers of α and β spins.

Now consider the effect of a **180° pulse**, which may be visualized in the rotating frame as a flip of the net magnetization vector from one direction along the z-axis to the opposite direction. That is, the 180° pulse leads to population inversion of the spin system, which now has more β spins than α spins. After the pulse, the populations revert to their thermal equilibrium values exponentially. As they do so, the z-component of magnetization reverts to its equilibrium value M_0 with a time constant called the **longitudinal relaxation time**, T_1 (Fig. 8.34):

$$M_z(t) - M_0 \propto e^{-t/T_1} \tag{8.31}$$

Because this relaxation process involves giving up energy to the surroundings (the 'lattice') as β spins revert to α spins, the time constant T_1 is also called the **spin–lattice relaxation time.** Spin–lattice relaxation is caused by local magnetic fields that fluctuate at a frequency close to the resonance frequency of the $\alpha \rightarrow \beta$ transition. Such fields can arise from the tumbling motion of molecules in a fluid sample. If molecular tumbling is too slow or too fast compared to the resonance frequency, it will give rise to a fluctuating magnetic field with a frequency that is either too low or too high to stimulate a spin change from β to α, so T_1 will be long. Only if the molecule tumbles at about the resonance frequency will the fluctuating magnetic field be able to induce spin changes effectively, and only then will T_1 be short. The rate of molecular tumbling

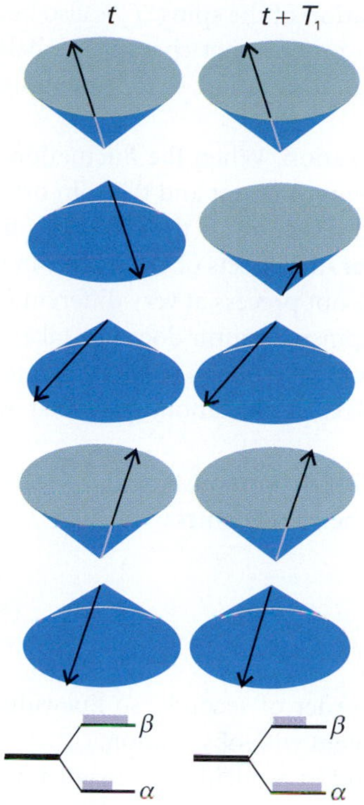

Fig. 8.34 In longitudinal relaxation the spins relax back towards their thermal equilibrium populations. On the left we see the precessional cones representing spin-$\frac{1}{2}$ angular momenta, and they do not have their thermal equilibrium populations (there are more α-spins than β-spins). On the right, which represents the sample a long time after a time T_1 has elapsed, the populations are those characteristic of a Boltzmann distribution. In actuality, T_1 is the time constant for relaxation to the arrangement on the right and $T_1 \ln 2$ is the half-life of the arrangement on the left.

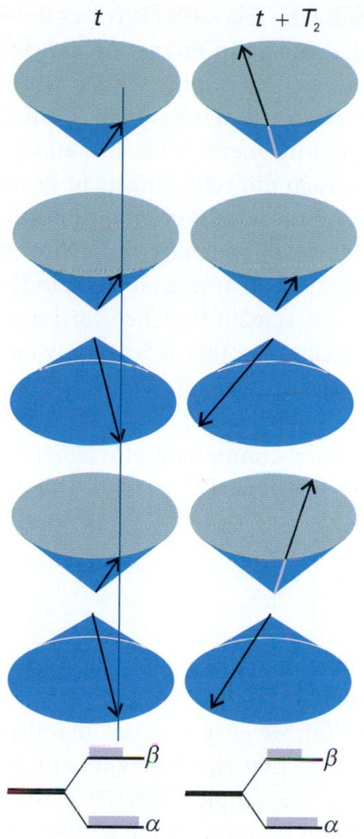

Fig. 8.36 The transverse relaxation time, T_2, is the time constant for the phases of the spins to become randomized (another condition for equilibrium) and to change from the orderly arrangement shown on the left to the disorderly arrangement on the right (long after a time T_2 has elapsed). Note that the populations of the states remain the same; only the relative phase of the spins relaxes. In actuality, T_2 is the time constant for relaxation to the arrangement on the right and $T_2 \ln 2$ is the half-life of the arrangement on the left.

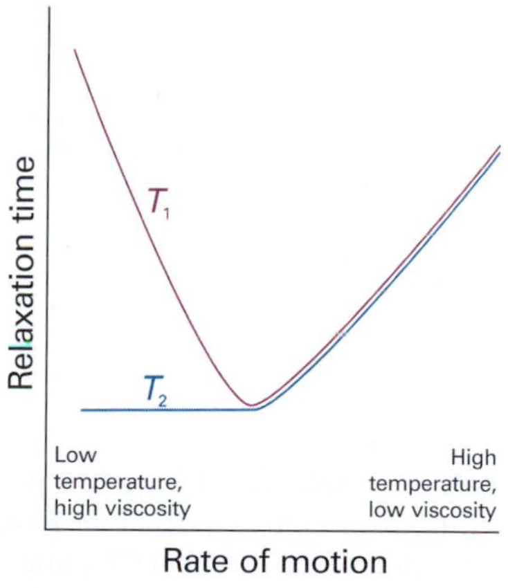

Fig. 8.35 The variation of the two relaxation times with the rate at which the molecules move (either by tumbling or migrating through the solution). The horizontal axis can be interpreted as representing temperature or viscosity. Note that, at rapid rates of motion, the two relaxation times coincide.

increases with temperature and with reducing viscosity of the solvent, so we can expect a dependence like that shown in Fig. 8.35.

A second aspect of spin relaxation is the fanning-out of the spins in the *xy*-plane if they precess at different rates (Fig. 8.36). The magnetization vector is large when all the spins are bunched together immediately after a 90° pulse. However, this orderly bunching of spins is not at equilibrium and, even if there were no spin–lattice relaxation, we would expect the individual spins to spread out until they were uniformly distributed with all possible angles around the *z*-axis. At that stage, the component of magnetization vector in the plane would be zero. The randomization of the spin directions occurs exponentially with a time constant called the **transverse relaxation time**, T_2:

$$M_y(t) \propto e^{-t/T_2} \qquad (8.32)$$

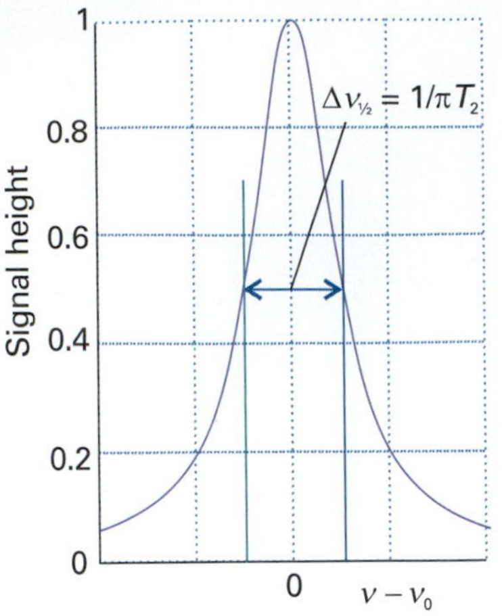

Fig. 8.37 A Lorentzian absorption line. The width at half-height is inversely proportional to the parameter T_2 and the longer the transverse relaxation time, the narrower the line.

Exploration The *Living graphs* section of the text's web site has an applet that allows you to calculate and display Lorenztian absorption lines. Explore the effect of the parameter T_2 on the width and the maximal intensity of a Lorentzian line. Rationalize your observations.

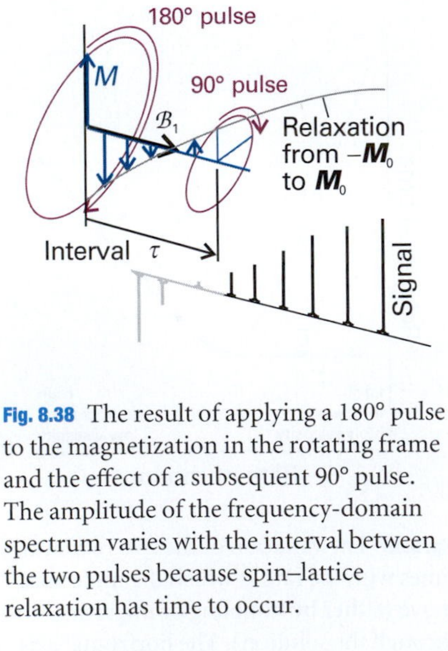

Fig. 8.38 The result of applying a 180° pulse to the magnetization in the rotating frame and the effect of a subsequent 90° pulse. The amplitude of the frequency-domain spectrum varies with the interval between the two pulses because spin–lattice relaxation has time to occur.

Because the relaxation involves the relative orientation of the spins, T_2 is also known as the **spin–spin relaxation time**. Any relaxation process that changes the balance between α and β spins will also contribute to this randomization, so the time constant T_2 is almost always less than or equal to T_1.

Local magnetic fields also affect spin–spin relaxation. When the fluctuations are slow, each molecule lingers in its local magnetic environment and the spin orientations randomize quickly around the applied field direction. If the molecules move rapidly from one magnetic environment to another, the effects of differences in local magnetic field average to zero: individual spins do not precess at very different rates, they can remain bunched for longer, and spin–spin relaxation does not take place as quickly. In other words, slow molecular motion corresponds to short T_2 and fast motion corresponds to long T_2 (as shown in Fig. 8.35). Calculations show that, when the motion is fast, $T_2 \approx T_1$.

If the y-component of magnetization decays with a time constant T_2, the spectral line is broadened (Fig. 8.37), and its width at half-height becomes

$$\Delta\nu_{1/2} = \frac{1}{\pi T_2} \qquad (8.33)$$

Typical values of T_2 in proton NMR are of the order of seconds, so linewidths of around 0.1 Hz can be anticipated, in broad agreement with observation.

So far, we have assumed that the equipment, and in particular the magnet, is perfect, and that the differences in Larmor frequencies arise solely from interactions within the sample. In practice, the magnet is not perfect, and the field is different at different locations in the sample. The inhomogeneity broadens the resonance, and in most cases this **inhomogeneous broadening** dominates the broadening we have discussed so far. It is common to express the extent of inhomogeneous broadening in terms of an **effective transverse relaxation time**, T_2^*, by using a relation like eqn 8.33, but writing

$$T_2^* = \frac{1}{\pi\Delta\nu_{1/2}} \qquad [8.34]$$

where $\Delta\nu_{1/2}$ is the observed width at half-height of a line with a Lorenztian shape of the form $I \propto 1/(1 + \nu^2)$. As an example, consider a line in a spectrum with a width of 10 Hz. It follows from eqn 8.34 that the effective transverse relaxation time is

$$T_2^* = \frac{1}{\pi \times (10\ \mathrm{s}^{-1})} = 32\ \mathrm{ms}$$

(b) The measurement of T_1

The longitudinal relaxation time T_1 can be measured by the **inversion recovery technique**. The first step is to apply a 180° pulse to the sample. A 180° pulse is achieved by applying the $\mathcal{B}_1$ field for twice as long as for a 90° pulse, so the magnetization vector precesses through 180° and points in the $-z$-direction (Fig. 8.38). No signal can be seen at this stage because there is no component of magnetization in the xy-plane (where the coil can detect it). The β spins begin to relax back into α spins, and the magnetization vector first shrinks exponentially, falling through zero to its thermal equilibrium value, M_z. After an interval τ, a 90° pulse is applied that rotates the magnetization into the xy-plane, where it generates an FID signal. The frequency-domain spectrum is then obtained by Fourier transformation.

The intensity of the spectrum obtained in this way depends on the length of the magnetization vector that is rotated into the *xy*-plane. The length of that vector changes exponentially as the interval between the two pulses is increased, so the intensity of the spectrum also changes exponentially with increasing τ. We can therefore measure T_1 by fitting an exponential curve to the series of spectra obtained with different values of τ.

(c) Spin echoes

The measurement of T_2 (as distinct from T_2^*) depends on being able to eliminate the effects of inhomogeneous broadening. The cunning required is at the root of some of the most important advances that have been made in NMR since its introduction.

A **spin echo** is the magnetic analogue of an audible echo: transverse magnetization is created by a radiofrequency pulse, decays away, is reflected by a second pulse, and grows back to form an echo. The sequence of events is shown in Fig. 8.39. We can consider the overall magnetization as being made up of a number of different magnetizations, each of which arises from a **spin packet** of nuclei with very similar precession frequencies. The spread in these frequencies arises because the applied field $\mathcal{B}_0$ is inhomogeneous, so different parts of the sample experience different fields. The precession frequencies also differ if there is more than one chemical shift present. As will be seen, the importance of a spin echo is that it can suppress the effects of both field inhomogeneities and chemical shifts.

First, a 90° pulse is applied to the sample. We follow events by using the rotating frame, in which $\mathcal{B}_1$ is stationary along the *x*-axis and causes the magnetization to be into the *xy*-plane. The spin packets now begin to fan out because they have different Larmor frequencies, with some above the radiofrequency and some below. The detected signal depends on the resultant of the spin-packet magnetization vectors, and decays with a time-constant T_2^* because of the combined effects of field inhomogeneity and spin–spin relaxation.

After an evolution period τ, a 180° pulse is applied to the sample; this time, about the *y*-axis of the rotating frame (the axis of the pulse is changed from *x* to *y* by a 90° phase shift of the radiofrequency radiation). The pulse rotates the magnetization vectors of the faster spin packets into the positions previously occupied by the slower spin packets, and vice versa. Thus, as the vectors continue to precess, the fast vectors are now behind the slow; the fan begins to close up again, and the resultant signal begins to grow back into an echo. At time 2τ, all the vectors will once more be aligned along the *y*-axis, and the fanning out caused by the field inhomogeneity is said to have been **refocused**: the spin echo has reached its maximum. Because the effects of field inhomogeneities have been suppressed by the refocusing, the echo signal will have been attenuated by the factor $e^{-2\tau/T_2}$ caused by spin–spin relaxation alone. After the time 2τ, the magnetization will continue to precess, fanning out once again, giving a resultant that decays with time constant T_2^*.

The important feature of the technique is that the size of the echo is independent of any local fields that remain constant during the two τ intervals. If a spin packet is 'fast' because it happens to be composed of spins in a region of the sample that experiences higher than average fields, then it remains fast throughout both intervals, and what it gains on the first interval it loses on the second interval. Hence, the size of the echo is independent of inhomogeneities in the magnetic field, for these remain constant. The true transverse relaxation arises from fields that vary on a molecular distance scale, and there is no guarantee that an individual 'fast' spin will remain 'fast' in the refocusing phase: the spins within the packets therefore spread with a time constant T_2. Hence, the effects of the true relaxation are not refocused, and the size of the echo decays with the time constant T_2 (Fig. 8.40).

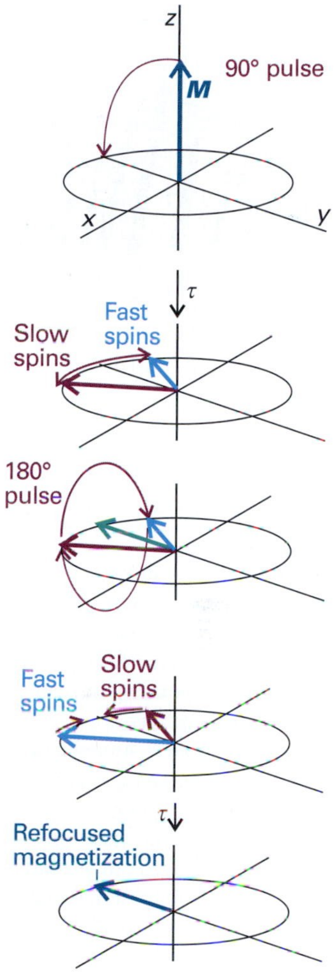

Fig. 8.39 The sequence of pulses leading to the observation of a spin echo.

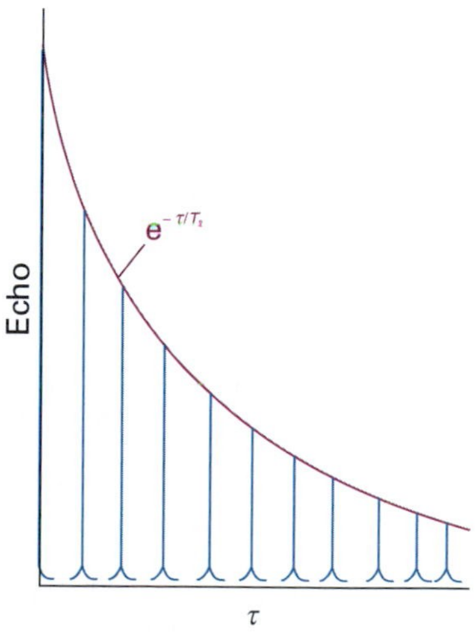

Fig. 8.40 The exponential decay of spin echoes can be used to determine the transverse relaxation time.

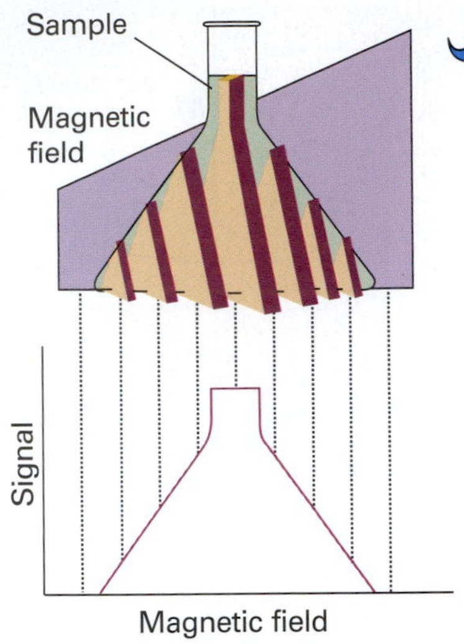

Fig. 8.41 In a magnetic field that varies linearly over a sample, all the protons within a given slice (that is, at a given field value) come into resonance and give a signal of the corresponding intensity. The resulting intensity pattern is a map of the numbers in all the slices, and portrays the shape of the sample. Changing the orientation of the field shows the shape along the corresponding direction, and computer manipulation can be used to build up the three-dimensional shape of the sample.

IMPACT ON MEDICINE

I8.1 Magnetic resonance imaging

One of the most striking applications of nuclear magnetic resonance is in medicine. *Magnetic resonance imaging* (MRI) is a portrayal of the concentrations of protons in a solid object. The technique relies on the application of specific pulse sequences to an object in an inhomogeneous magnetic field.

If an object containing hydrogen nuclei (a tube of water or a human body) is placed in an NMR spectrometer and exposed to a *homogeneous* magnetic field, then a single resonance signal will be detected. Now consider a flask of water in a magnetic field that varies linearly in the z-direction according to $\mathcal{B}_0 + G_z z$, where G_z is the field gradient along the z-direction (Fig. 8.41). Then the water protons will be resonant at the frequencies

$$\nu_L(z) = \frac{\gamma}{2\pi}(\mathcal{B}_0 + G_z z) \qquad (8.35)$$

(Similar equations may be written for gradients along the x- and y-directions.) Application of a 90° radiofrequency pulse with $\nu = \nu_L(z)$ will result in a signal with an intensity that is proportional to the numbers of protons at the position z. This is an example of *slice selection*, the application of a selective 90° pulse that excites nuclei in a specific region, or slice, of the sample. It follows that the intensity of the NMR signal will be a projection of the numbers of protons on a line parallel to the field gradient. The image of a three-dimensional object such as a flask of water can be obtained if the slice selection technique is applied at different orientations (see Fig. 8.41). In *projection reconstruction*, the projections can be analysed on a computer to reconstruct the three-dimensional distribution of protons in the object.

In practice, the NMR signal is not obtained by direct analysis of the FID curve after application of a single 90° pulse. Instead, spin echoes are often detected with several variations of the 90°–τ–180° pulse sequence (Section 8.9c). In *phase encoding*, field gradients are applied during the evolution period and the detection period of a spin-echo pulse sequence. The first step consists of a 90° pulse that results in slice selection along the z-direction. The second step consists of application of a *phase gradient*, a field gradient along the y-direction, during the evolution period. At each position along the gradient, a spin packet will precess at a different Larmor frequency due to chemical shift effects and the field inhomogeneity, so each packet will dephase to a different extent by the end of the evolution period. We can control the extent of dephasing by changing the duration of the evolution period, so Fourier transformation on τ gives information about the location of a proton along the y-direction.[1] For each value of τ, the next steps are application of the 180° pulse and then of a *read gradient*, a field gradient along the x-direction, during detection of the echo. Protons at different positions along x experience different fields and will resonate at different frequencies. Therefore Fourier transformation of the FID gives different signals for protons at different positions along x.

A common problem with the techniques described above is image contrast, which must be optimized in order to show spatial variations in water content in the sample. One strategy for solving this problem takes advantage of the fact that the relaxation times of water protons are shorter for water in biological tissues than for the pure liquid. Furthermore, relaxation times from water protons are also different in healthy and diseased tissues. A T_1-*weighted image* is obtained by repeating the spin echo sequence

[1] For technical reasons, it is more common to vary the magnitude of the phase gradient. See *Further reading* for details.

before spin–lattice relaxation can return the spins in the sample to equilibrium. Under these conditions, differences in signal intensities are directly related to differences in T_1. A *T_2-weighted image* is obtained by using an evolution period τ that is relatively long. Each point on the image is an echo signal that behaves in the manner shown in Fig. 8.40, so signal intensities are strongly dependent on variations in T_2. However, allowing so much of the decay to occur leads to weak signals even for those protons with long spin–spin relaxation times. Another strategy involves the use of *contrast agents*, paramagnetic compounds that shorten the relaxation times of nearby protons. The technique is particularly useful in enhancing image contrast and in diagnosing disease if the contrast agent is distributed differently in healthy and diseased tissues.

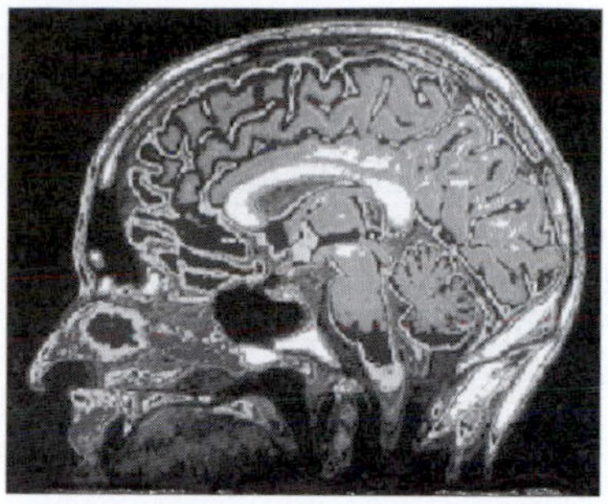

Fig. 8.42 The great advantage of MRI is that it can display soft tissue, such as in this cross-section through a patient's head. (Courtesy of the University of Manitoba.)

The MRI technique is used widely to detect physiological abnormalities and to observe metabolic processes. With *functional MRI*, blood flow in different regions of the brain can be studied and related to the mental activities of the subject. The technique is based on differences in the magnetic properties of deoxygenated and oxygenated haemoglobin, the iron-containing protein that transports O_2 in red blood cells. The more paramagnetic deoxygenated haemoglobin affects the proton resonances of tissue differently from the oxygenated protein. Because there is greater blood flow in active regions of the brain than in inactive regions, changes in the intensities of proton resonances due to changes in levels of oxygenated haemoglobin can be related to brain activity.

The special advantage of MRI is that it can image *soft* tissues (Fig. 8.42), whereas X-rays are largely used for imaging hard, bony structures and abnormally dense regions, such as tumours. In fact, the invisibility of hard structures in MRI is an advantage, as it allows the imaging of structures encased by bone, such as the brain and the spinal cord. X-rays are known to be dangerous on account of the ionization they cause; the high magnetic fields used in MRI may also be dangerous but, apart from anecdotes about the extraction of loose fillings from teeth, there is no convincing evidence of their harmfulness, and the technique is considered safe.

8.10 Spin decoupling

Carbon-13 is a **dilute-spin species** in the sense that it is unlikely that more than one ^{13}C nucleus will be found in any given small molecule (provided the sample has not been enriched with that isotope; the natural abundance of ^{13}C is only 1.1 per cent). Even in large molecules, although more than one ^{13}C nucleus may be present, it is unlikely that they will be close enough to give an observable splitting. Hence, it is not normally necessary to take into account ^{13}C—^{13}C spin–spin coupling within a molecule.

Protons are **abundant-spin species** in the sense that a molecule is likely to contain many of them. If we were observing a ^{13}C-NMR spectrum, we would obtain a very complex spectrum on account of the coupling of the one ^{13}C nucleus with many of the protons that are present. To avoid this difficulty, ^{13}C-NMR spectra are normally observed using the technique of **proton decoupling**. Thus, if the CH_3 protons of ethanol are irradiated with a second, strong, resonant radiofrequency pulse, they undergo rapid spin reorientations and the ^{13}C nucleus senses an average orientation. As a result, its resonance is a single line and not a 1:3:3:1 quartet. Proton decoupling has the additional advantage of enhancing sensitivity, because the intensity is concentrated into a single transition frequency instead of being spread over several transition frequencies (see Section 8.11). If care is taken to ensure that the other parameters on which the strength of the signal depends are kept constant, the intensities of proton-decoupled spectra are proportional to the number of ^{13}C nuclei present. The technique is widely used to characterize synthetic polymers.

Comment 8.7

In a dipole–dipole interaction between two nuclei, one nucleus influences the behaviour of another nucleus in much the same way that the orientation of a bar magnet is influenced by the presence of another bar magnet nearby.

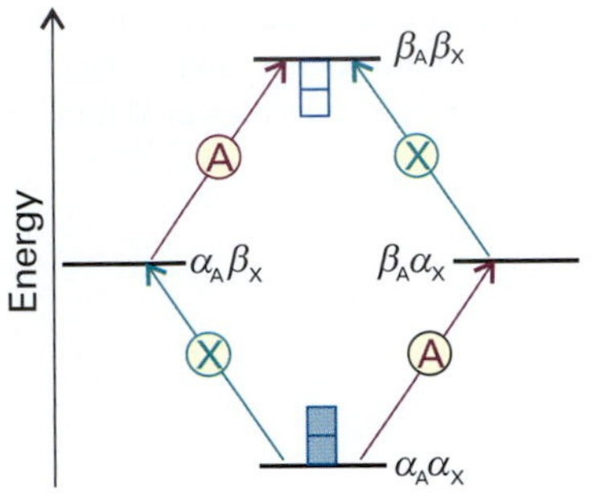

Fig. 8.43 The energy levels of an AX system and an indication of their relative populations. Each grey square above the line represents an excess population and each white square below the line represents a population deficit. The transitions of A and X are marked.

8.11 The nuclear Overhauser effect

We have seen already that one advantage of protons in NMR is their high magnetogyric ratio, which results in relatively large Boltzmann population differences and hence greater resonance intensities than for most other nuclei. In the steady-state **nuclear Overhauser effect** (NOE), spin relaxation processes involving internuclear dipole–dipole interactions are used to transfer this population advantage to another nucleus (such as ^{13}C or another proton), so that the latter's resonances are modified.

To understand the effect, we consider the populations of the four levels of a homonuclear (for instance, proton) AX system; these were shown in Fig. 8.12. At thermal equilibrium, the population of the $\alpha_A\alpha_X$ level is the greatest, and that of the $\beta_A\beta_X$ level is the least; the other two levels have the same energy and an intermediate population. The thermal equilibrium absorption intensities reflect these populations as shown in Fig. 8.43. Now consider the combined effect of spin relaxation and keeping the X spins saturated. When we saturate the X transition, the populations of the X levels are equalized ($N_{\alpha X} = N_{\beta X}$) and all transitions involving $\alpha_X \leftrightarrow \beta_X$ spin flips are no longer observed. At this stage there is no change in the populations of the A levels. If that were all there were to happen, all we would see would be the loss of the X resonance and no effect on the A resonance.

Now consider the effect of spin relaxation. Relaxation can occur in a variety of ways if there is a dipolar interaction between the A and X spins. One possibility is for the magnetic field acting between the two spins to cause them both to flop from β to α, so the $\alpha_A\alpha_X$ and $\beta_A\beta_X$ states regain their thermal equilibrium populations. However, the populations of the $\alpha_A\beta_X$ and $\beta_A\alpha_X$ levels remain unchanged at the values characteristic of saturation. As we see from Fig. 8.44, the population difference between the states joined by transitions of A is now greater than at equilibrium, so the resonance absorption is enhanced. Another possibility is for the dipolar interaction between the two spins to cause α to flip to β and β to flop to α. This transition equilibrates the populations of $\alpha_A\beta_X$ and $\beta_A\alpha_X$ but leaves the $\alpha_A\alpha_X$ and $\beta_A\beta_X$ populations unchanged. Now we see from the illustration that the population differences in the states involved in the A transitions are decreased, so the resonance absorption is diminished.

Which effect wins? Does the NOE enhance the A absorption or does it diminish it? As in the discussion of relaxation times in Section 8.9, the efficiency of the intensity-enhancing $\beta_A\beta_X \leftrightarrow \alpha_A\alpha_X$ relaxation is high if the dipole field oscillates at a frequency close to the transition frequency, which in this case is about 2ν; likewise, the efficiency

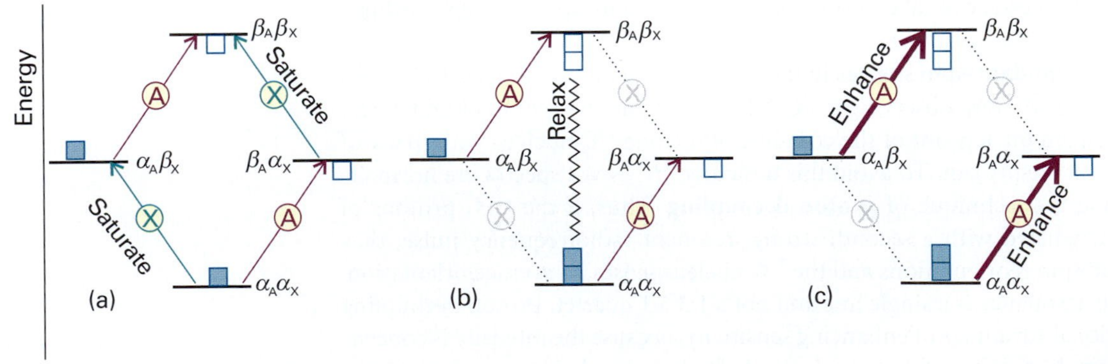

Fig. 8.44 (a) When the X transition is saturated, the populations of its two states are equalized and the population excess and deficit become as shown (using the same symbols as in Fig. 8.43). (b) Dipole–dipole relaxation relaxes the populations of the highest and lowest states, and they regain their original populations. (c) The A transitions reflect the difference in populations resulting from the preceding changes, and are enhanced compared with those shown in Fig. 8.43.

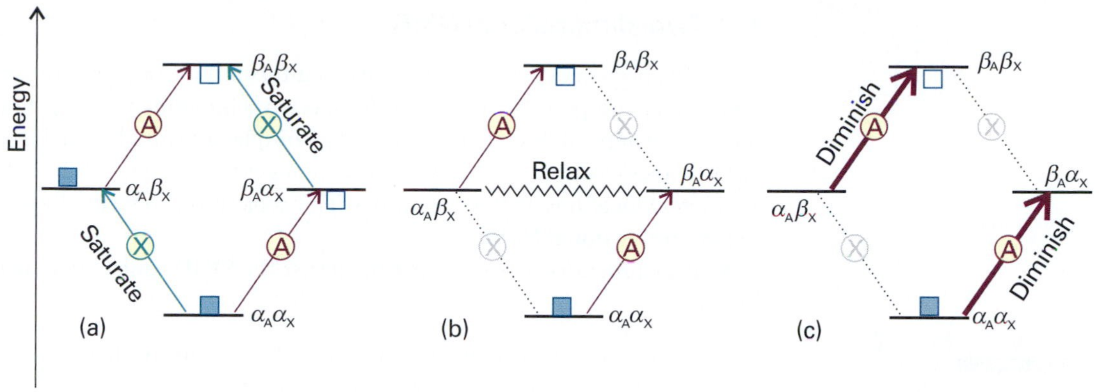

Fig. 8.45 (a) When the X transition is saturated, just as in Fig. 8.44 the populations of its two states are equalized and the population excess and deficit become as shown. (b) Dipole–dipole relaxation relaxes the populations of the two intermediate states, and they regain their original populations. (c) The A transitions reflect the difference in populations resulting from the preceding changes, and are diminished compared with those shown in Fig. 8.41.

of the intensity-diminishing $\alpha_A\beta_X \leftrightarrow \beta_A\alpha_X$ relaxation is high if the dipole field is stationary (as there is no frequency difference between the initial and final states). A large molecule rotates so slowly that there is very little motion at 2ν, so we expect an intensity decrease (Fig. 8.45). A small molecule rotating rapidly can be expected to have substantial motion at 2ν, and a consequent enhancement of the signal. In practice, the enhancement lies somewhere between the two extremes and is reported in terms of the parameter η (eta), where

$$\eta = \frac{I_A - I_A^\circ}{I_A^\circ} \qquad [8.36]$$

Here I_A° and I_A are the intensities of the NMR signals due to nucleus A before and after application of the long ($> T_1$) radiofrequency pulse that saturates transitions due to the X nucleus. When A and X are nuclei of the same species, such as protons, η lies between −1 (diminution) and $+\frac{1}{2}$ (enhancement). However, η also depends on the values of the magnetogyric ratios of A and X. In the case of maximal enhancement it is possible to show that

$$\eta = \frac{\gamma_X}{2\gamma_A} \qquad (8.37)$$

where γ_A and γ_X are the magnetogyric ratios of nuclei A and X, respectively. For ^{13}C close to a saturated proton, the ratio evaluates to 1.99, which shows that an enhancement of about a factor of 2 can be achieved.

The NOE is also used to determine interproton distances. The Overhauser enhancement of a proton A generated by saturating a spin X depends on the fraction of A's spin–lattice relaxation that is caused by its dipolar interaction with X. Because the dipolar field is proportional to r^{-3}, where r is the internuclear distance, and the relaxation effect is proportional to the square of the field, and therefore to r^{-6}, the NOE may be used to determine the geometries of molecules in solution. The determination of the structure of a small protein in solution involves the use of several hundred NOE measurements, effectively casting a net over the protons present. The enormous importance of this procedure is that we can determine the conformation of biological macromolecules in an aqueous environment and do not need to try to make the single crystals that are essential for an X-ray diffraction investigation.

8.12 Two-dimensional NMR

An NMR spectrum contains a great deal of information and, if many protons are present, is very complex. Even a first-order spectrum is complex, for the fine structure of different groups of lines can overlap. The complexity would be reduced if we could use two axes to display the data, with resonances belonging to different groups lying at different locations on the second axis. This separation is essentially what is achieved in **two-dimensional NMR**.

All two-dimensional NMR experiments use the **PEMD pulse structure**, which consists of:

P: a *preparation period*, in which the spins first return to thermal equilibrium and then are excited by one or more radiofrequency pulses

E: an *evolution period* of duration t_1, during which the spins precess under the influence of their chemical shifts and spin–spin couplings

M: a *mixing period*, in which pulses may be used to transfer information between spins

D: a *detection period* of duration t_2, during which the FID is recorded.

Now we shall see how the PEMD pulse structure can be used to devise experiments that reveal spin–spin couplings and internuclear distances in small and large molecules.

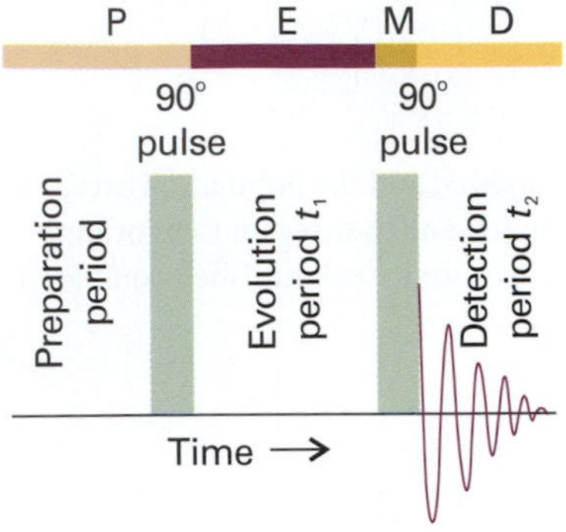

Fig. 8.46 The pulse sequence used in correlation spectroscopy (COSY). The preparation period is much longer than either T_1 or T_2, so the spins have time to relax before the next cycle of pulses begins. Acquisitions of free-induction decays are taken during t_2 for a set of different evolution times t_1. Fourier transformation on both variables t_1 and t_2 results in a two-dimensional spectrum, such as that shown in Fig 8.52.

(a) Correlation spectroscopy

Much modern NMR work makes use of techniques such as **correlation spectroscopy** (COSY) in which a clever choice of pulses and Fourier transformation techniques makes it possible to determine all spin–spin couplings in a molecule. The basic COSY experiment uses the simplest of all two-dimensional pulse sequences, consisting of two consecutive 90° pulses (Fig. 8.46).

To see how we can obtain a two-dimensional spectrum from a COSY experiment, we consider a trivial but illustrative example: the spectrum of a compound containing one proton, such as trichloromethane (chloroform, $CHCl_3$). Figure 8.47 shows the effect of the pulse sequence on the magnetization of the sample, which is aligned initially along the z-axis with a magnitude M_0. A 90° pulse applied in the x-direction (in the stationary frame) tilts the magnetization vector toward the y-axis. Then, during the evolution period, the magnetization vector rotates in the xy-plane with a frequency ν. At a time t_1 the vector will have swept through an angle $2\pi\nu t_1$ and the magnitude of the magnetization will have decayed by spin–spin relaxation to $M = M_0 e^{-t_1/T_2}$. By trigonometry, the magnitudes of the components of the magnetization vector are:

$$M_x = M \sin 2\pi\nu t_1 \qquad M_y = M \cos 2\pi\nu t_1 \qquad M_z = 0 \tag{8.38a}$$

Application of the second 90° pulse parallel to the x-axis tilts the magnetization again and the resulting vector has components with magnitudes (once again, in the stationary frame)

$$M_x = M \sin 2\pi\nu t_1 \qquad M_y = 0 \qquad M_z = M \cos 2\pi\nu t_1 \tag{8.38b}$$

The FID is detected over a period t_2 and Fourier transformation yields a signal over a frequency range ν_2 with a peak at ν, the resonance frequency of the proton. The signal intensity is related to M_x, the magnitude of the magnetization that is rotating around the xy-plane at the time of application of the detection pulse, so it follows that the signal strength varies sinusoidally with the duration of the evolution period. That is, if we were to acquire a series of spectra at different evolution times t_1, then we would obtain data as shown in Fig. 8.48a.

Comment 8.8

A vector, v, of length v, in the xy-plane and its two components, v_x and v_y, can be thought of as forming a right-angled triangle, with v the length of the hypotenuse (see the illustration). If θ is the angle that v_y makes with v, then it follows that $v_x = v \sin\theta$ and $v_y = v\cos\theta$.

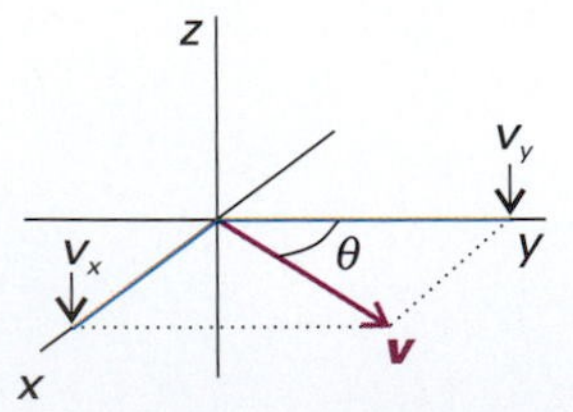

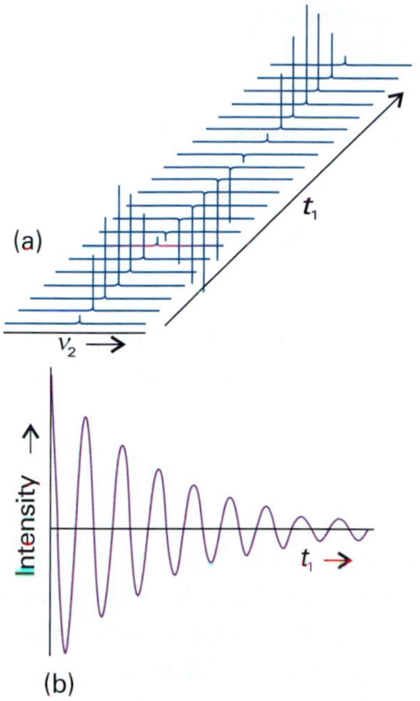

Fig. 8.48 (a) Spectra acquired for different evolution times t_1 between two 90° pulses. (b) A plot of the maximum intensity of each absorption line against t_1. Fourier transformation of this plot leads to a spectrum centred at ν, the resonance frequency of the protons in the sample.

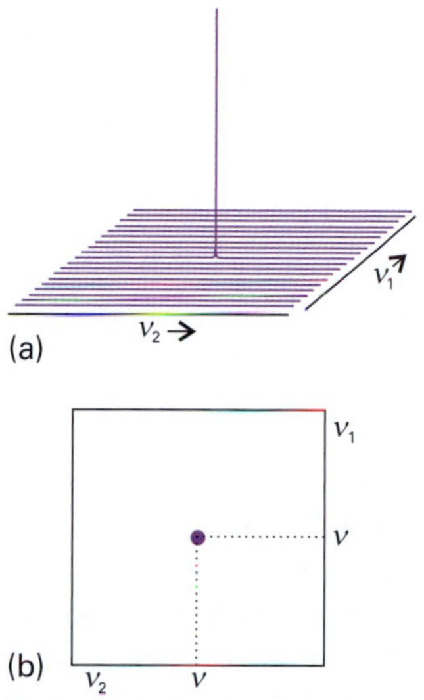

Fig. 8.49 (a) The two-dimensional NMR spectrum of the sample discussed in Figs. 8.47 and 8.48. See the text for an explanation of how the spectrum is obtained from a series of Fourier transformations of the data. (b) The contour plot of the spectrum in (a).

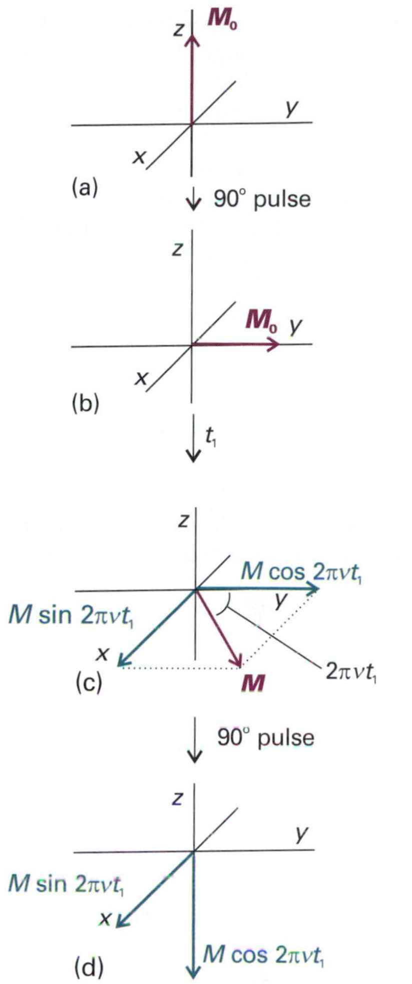

Fig. 8.47 (a) The effect of the pulse sequence shown in Fig. 8.46 on the magnetization $\boldsymbol{M}_0$ of a sample of a compound with only one proton. (b) A 90° pulse applied in the x-direction tilts the magnetization vector toward the y-axis. (c) After a time t_1 has elapsed, the vector will have swept through an angle $2\pi\nu t_1$ and the magnitude of the magnetization will have decayed to $\boldsymbol{M}$. The magnitudes of the components of $\boldsymbol{M}$ are $M_x = M \sin 2\pi\nu t_1$, $M_y = M \cos 2\pi\nu t_1$, and $M_z = 0$. (d) Application of the second 90° pulse parallel to the x-axis tilts the magnetization again and the resulting vector has components with magnitude $M_x = M \sin 2\pi\nu t_1$, $M_y = 0$, and $M_z = M \cos 2\pi\nu t_1$. The FID is detected at this stage of the experiment.

A plot of the maximum intensity of each absorption band in Fig. 8.48a against t_1 has the form shown in Fig. 8.48b. The plot resembles an FID curve with the oscillating component having a frequency ν, so Fourier transformation yields a signal over a frequency range ν_1 with a peak at ν. If we continue the process by first plotting signal intensity against t_1 for several frequencies along the ν_2 axis and then carrying out Fourier transformations, we generate a family of curves that can be pooled together into a three-dimensional plot of $I(\nu_1,\nu_2)$, the signal intensity as a function of the frequencies ν_1 and ν_2 (Fig. 8.49a). This plot is referred to as a *two-dimensional NMR spectrum* because Fourier transformations were performed in two variables. The most common representation of the data is as a contour plot, such as the one shown in Fig. 8.49b.

The experiment described above is not necessary for as simple a system as chloroform because the information contained in the two-dimensional spectrum could have been obtained much more quickly through the conventional, one-dimensional approach. However, when the one-dimensional spectrum is complex, the COSY experiment shows which spins are related by spin–spin coupling. To justify this statement, we now examine a spin-coupled AX system.

From our discussion so far, we know that the key to the COSY technique is the effect of the second 90° pulse. In this more complex example we consider its role for the four energy levels of an AX system (as shown in Fig. 8.12). At thermal equilibrium, the population of the $\alpha_A\alpha_X$ level is the greatest, and that of the $\beta_A\beta_X$ level is the least; the other two levels have the same energy and an intermediate population. After the first 90° pulse, the spins are no longer at thermal equilibrium. If a second 90° pulse

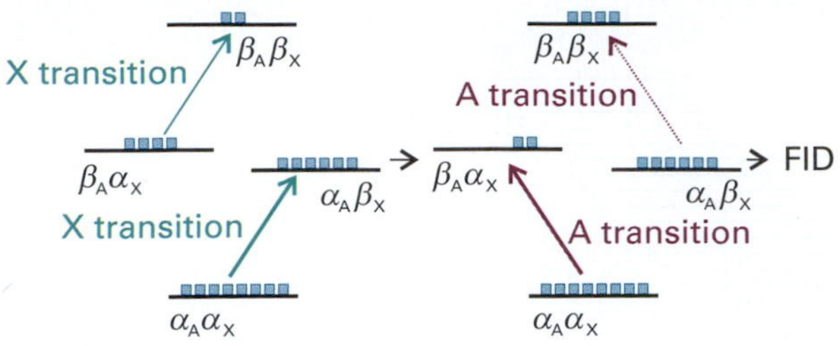

Fig. 8.50 An example of the change in the population of energy levels of an AX spin system that results from the second 90° pulse of a COSY experiment. Each square represents the same large number of spins. In this example, we imagine that the pulse affects the X spins first, and then the A spins. Excitation of the X spins inverts the populations of the $\beta_A\beta_X$ and $\beta_A\alpha_X$ levels and does not affect the populations of the $\alpha_A\alpha_X$ and $\alpha_A\beta_X$ levels. As a result, excitation of the A spins by the pulse generates an FID in which one of the two A transitions has increased in intensity and the other has decreased. That is, magnetization has been transferred from the X spins to the A spins. Similar schemes can be written to show that magnetization can be transferred from the A spins to the X spins.

is applied at a time t_1 that is short compared to the spin–lattice relaxation time T_1, the extra input of energy causes further changes in the populations of the four states. The changes in populations of the four states of the AX system will depend on how far the individual magnetizations have precessed during the evolution period. It is difficult to visualize these changes because the A spins are affecting the X spins and vice-versa.

For simplicity, we imagine that the second pulse induces X and A transitions sequentially. Depending on the evolution time t_1, the 90° pulse may leave the population differences across each of the two X transitions unchanged, inverted, or somewhere in between. Consider the extreme case in which one population difference is inverted and the other unchanged (Fig. 8.50). Excitation of the A transitions will now generate an FID in which one of the two A transitions has increased in intensity (because the population difference is now greater), and the other has decreased (because the population difference is now smaller). The overall effect is that precession of the X spins during the evolution period determines the amplitudes of the signals from the A spins obtained during the detection period. As the evolution time t_1 is increased, the intensities of the signals from A spins oscillate with frequencies determined by the frequencies of the two X transitions. Of course, it is just as easy to turn our scenario around and to conclude that the intensities of signals from X spins oscillate with frequencies determined by the frequencies of the A transitions.

This transfer of information between spins is at the heart of two-dimensional NMR spectroscopy: it leads to the *correlation* between different signals in a spectrum. In this case, information transfer tells us that there is spin–spin coupling between A and X. So, just as before, if we conduct a series of experiments in which t_1 is incremented, Fourier transformation of the FIDs on t_2 yields a set of spectra $I(t_1,F_2)$ in which the signal amplitudes oscillate as a function of t_1. A second Fourier transformation, now on t_1, converts these oscillations into a two-dimensional spectrum $I(F_1,F_2)$. The signals are spread out in F_1 according to their precession frequencies during the detection period. Thus, if we apply the COSY pulse sequence (Fig. 8.46) to the AX spin system, the result is a two-dimensional spectrum that contains four groups of signals in F_1 and F_2 centred on the two chemical shifts (Fig. 8.51). Each group consists of a block of four signals separated by J. The **diagonal peaks** are signals centred on (δ_A,δ_A) and (δ_X,δ_X) and lie along the diagonal $F_1 = F_2$. That is, the spectrum along the diagonal is equivalent to the one-dimensional spectrum obtained with the conventional NMR technique (Fig. 8.13). The **cross-peaks** (or *off-diagonal peaks*) are signals centred on (δ_A,δ_X) and (δ_X,δ_A) and owe their existence to the coupling between A and X.

Although information from two-dimensional NMR spectroscopy is trivial in an AX system, it can be of enormous help in the interpretation of more complex spectra, leading to a map of the couplings between spins and to the determination of the bonding network in complex molecules. Indeed, the spectrum of a synthetic or biological polymer that would be impossible to interpret in one-dimensional NMR but can often be interpreted reasonably rapidly by two-dimensional NMR. Below we illustrate the procedure by assigning the resonances in the COSY spectrum of an amino acid.

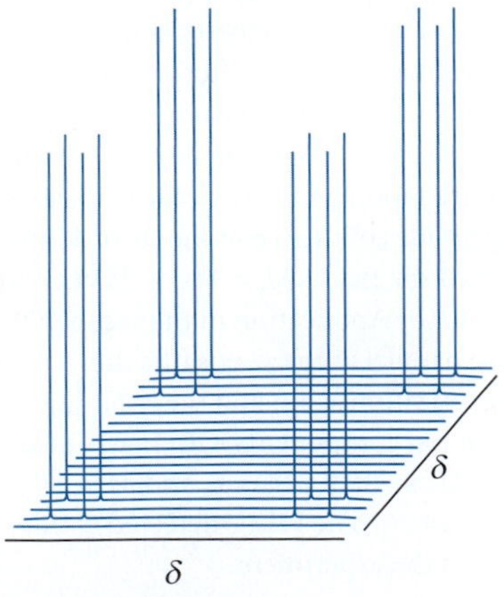

Fig. 8.51 A representation of the two-dimensional NMR spectrum obtained by application of the COSY pulse sequence to an AX spin system.

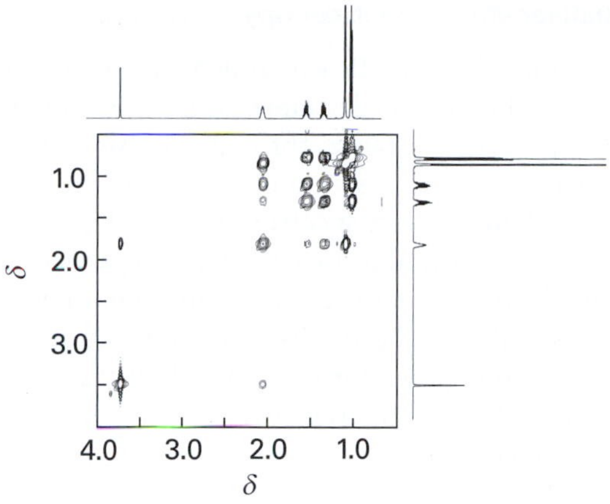

Fig. 8.52 Proton COSY spectrum of isoleucine. (Adapted from K.E. van Holde, W.C. Johnson, and P.S. Ho, *Principles of physical biochemistry*, p. 508, Prentice Hall, Upper Saddle River (1998).)

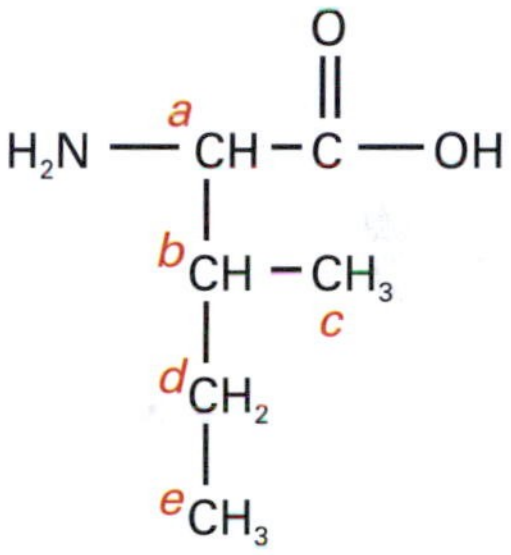

7 Isoleucine

Illustration 8.3 *The COSY spectrum of isoleucine*

Figure 8.52 is a portion of the COSY spectrum of the amino acid isoleucine (7), showing the resonances associated with the protons bound to the carbon atoms. We begin the assignment process by considering which protons should be interacting by spin–spin coupling. From the known molecular structure, we conclude that:

1. The C_a—H proton is coupled only to the C_b—H proton.
2. The C_b—H protons are coupled to the C_a—H, C_c—H, and C_d—H protons.
3. The inequivalent C_d—H protons are coupled to the C_b—H and C_e—H protons.

We now note that:

• The resonance with $\delta = 3.6$ shares a cross-peak with only one other resonance at $\delta = 1.9$, which in turn shares cross-peaks with resonances at $\delta = 1.4$, 1.2, and 0.9. This identification is consistent with the resonances at $\delta = 3.6$ and 1.9 corresponding to the C_a—H and C_b—H protons, respectively.

• The proton with resonance at $\delta = 0.8$ is not coupled to the C_b—H protons, so we assign the resonance at $\delta = 0.8$ to the C_e—H protons.

• The resonances at $\delta = 1.4$ and 1.2 do not share cross-peaks with the resonance at $\delta = 0.9$.

In the light of the expected couplings, we assign the resonance at $\delta = 0.9$ to the C_c—H protons and the resonances at $\delta = 1.4$ and 1.2 to the inequivalent C_d—H protons.

Our simplified description of the COSY experiment does not reveal some important details. For example, the second 90° pulse actually mixes the spin state transitions caused by the first 90° pulse (hence the term 'mixing period'). Each of the four transitions (two for A and two for X) generated by the first pulse can be converted into any of the other three, or into formally forbidden **multiple quantum transitions**, which have $|\Delta m| > 1$. The latter transitions cannot generate any signal in the receiver coil of the spectrometer, but their existence can be demonstrated by applying a third pulse to mix them back into the four observable single quantum transitions. Many modern NMR experiments exploit multiple quantum transitions to filter out unwanted signals and to simplify spectra for interpretation.

(b) Nuclear Overhauser effect spectroscopy

Many different two-dimensional NMR experiments are based on the PEMD pulse structure. We have seen that the steady-state nuclear Overhauser effect can provide information about internuclear distances through analysis of enhancement patterns in the NMR spectrum before and after saturation of selected resonances. In its dynamic analogue, **nuclear Overhauser effect spectroscopy** (NOESY), the second 90° pulse of the COSY experiment is replaced by two 90° pulses separated by a time delay during which dipole–dipole interactions cause magnetization to be exchanged between neighbouring spins (Fig. 8.50). The results of double Fourier transformation is a spectrum in which the cross-peaks form a map of all the NOE interactions in a molecule. Because the nuclear Overhauser effect depends on the inverse sixth power of the separation between nuclei (see Section 8.11), NOESY data reveal internuclear distances up to about 0.5 nm.

8.13 Solid-state NMR

The principal difficulty with the application of NMR to solids is the low resolution characteristic of solid samples. Nevertheless, there are good reasons for seeking to overcome these difficulties. They include the possibility that a compound of interest is unstable in solution or that it is insoluble, so conventional solution NMR cannot be employed. Moreover, many species are intrinsically interesting as solids, and it is important to determine their structures and dynamics. Synthetic polymers are particularly interesting in this regard, and information can be obtained about the arrangement of molecules, their conformations, and the motion of different parts of the chain. This kind of information is crucial to an interpretation of the bulk properties of the polymer in terms of its molecular characteristics. Similarly, inorganic substances, such as the zeolites that are used as molecular sieves and shape-selective catalysts, can be studied using solid-state NMR, and structural problems can be resolved that cannot be tackled by X-ray diffraction.

Problems of resolution and linewidth are not the only features that plague NMR studies of solids, but the rewards are so great that considerable efforts have been made to overcome them and have achieved notable success. Because molecular rotation has almost ceased (except in special cases, including 'plastic crystals' in which the molecules continue to tumble), spin–lattice relaxation times are very long but spin–spin relaxation times are very short. Hence, in a pulse experiment, there is a need for lengthy delays—of several seconds—between successive pulses so that the spin system has time to revert to equilibrium. Even gathering the murky information may therefore be a lengthy process. Moreover, because lines are so broad, very high powers of radiofrequency radiation may be required to achieve saturation. Whereas solution pulse NMR uses transmitters of a few tens of watts, solid-state NMR may require transmitters rated at several hundreds of watts.

(a) The origins of linewidths in solids

There are two principal contributions to the linewidths of solids. One is the direct magnetic dipolar interaction between nuclear spins. As we saw in the discussion of spin–spin coupling, a nuclear magnetic moment will give rise to a local magnetic field, which points in different directions at different locations around the nucleus. If we are interested only in the component parallel to the direction of the applied magnetic field (because only this component has a significant effect), then we can use a classical expression to write the magnitude of the local magnetic field as

$$\mathcal{B}_{\text{loc}} = -\frac{\gamma\hbar\mu_0 m_I}{4\pi R^3}(1-3\cos^2\theta) \quad (8.39)$$

Unlike in solution, this field is not motionally averaged to zero. Many nuclei may contribute to the total local field experienced by a nucleus of interest, and different nuclei in a sample may experience a wide range of fields. Typical dipole fields are of the order of 10^{-3} T, which corresponds to splittings and linewidths of the order of 10^4 Hz.

A second source of linewidth is the anisotropy of the chemical shift. We have seen that chemical shifts arise from the ability of the applied field to generate electron currents in molecules. In general, this ability depends on the orientation of the molecule relative to the applied field. In solution, when the molecule is tumbling rapidly, only the average value of the chemical shift is relevant. However, the anisotropy is not averaged to zero for stationary molecules in a solid, and molecules in different orientations have resonances at different frequencies. The chemical shift anisotropy also varies with the angle between the applied field and the principal axis of the molecule as $1-3\cos^2\theta$.

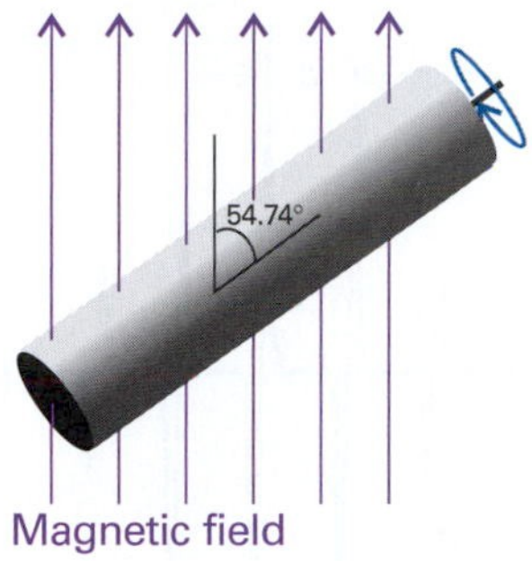

Fig. 8.53 In magic angle spinning, the sample spins at 54.74° (that is, arccos $(\frac{1}{3})^{1/2}$) to the applied magnetic field. Rapid motion at this angle averages dipole–dipole interactions and chemical shift anisotropies to zero.

(b) The reduction of linewidths

Fortunately, there are techniques available for reducing the linewidths of solid samples. One technique, **magic-angle spinning** (MAS), takes note of the $1-3\cos^2\theta$ dependence of both the dipole–dipole interaction and the chemical shift anisotropy. The 'magic angle' is the angle at which $1-3\cos^2\theta=0$, and corresponds to 54.74°. In the technique, the sample is spun at high speed at the magic angle to the applied field (Fig. 8.53). All the dipolar interactions and the anisotropies average to the value they would have at the magic angle, but at that angle they are zero. The difficulty with MAS is that the spinning frequency must not be less than the width of the spectrum, which is of the order of kilohertz. However, gas-driven sample spinners that can be rotated at up to 25 kHz are now routinely available, and a considerable body of work has been done.

Pulsed techniques similar to those described in the previous section may also be used to reduce linewidths. The dipolar field of protons, for instance, may be reduced by a decoupling procedure. However, because the range of coupling strengths is so large, radiofrequency power of the order of 1 kW is required. Elaborate pulse sequences have also been devised that reduce linewidths by averaging procedures that make use of twisting the magnetization vector through an elaborate series of angles.

Electron paramagnetic resonance

Electron paramagnetic resonance (EPR) is less widely applicable than NMR because it cannot be detected in normal, spin-paired molecules and the sample must possess unpaired electron spins. It is used to study radicals formed during chemical reactions or by radiation, radicals that act as probes of biological structure, many *d*-metal complexes, and molecules in triplet states (such as those involved in phosphorescence, Section 7.3b). The sample may be a gas, a liquid, or a solid, but the free rotation of molecules in the gas phase gives rise to complications.

8.14 The EPR spectrometer

Both Fourier-transform (FT) and continuous wave (CW) EPR spectrometers are available. The FT-EPR instrument is based on the concepts developed in Section 8.8, except that pulses of microwaves are used to excite electron spins in the sample. The layout of the more common CW-EPR spectrometer is shown in Fig. 8.54. It consists

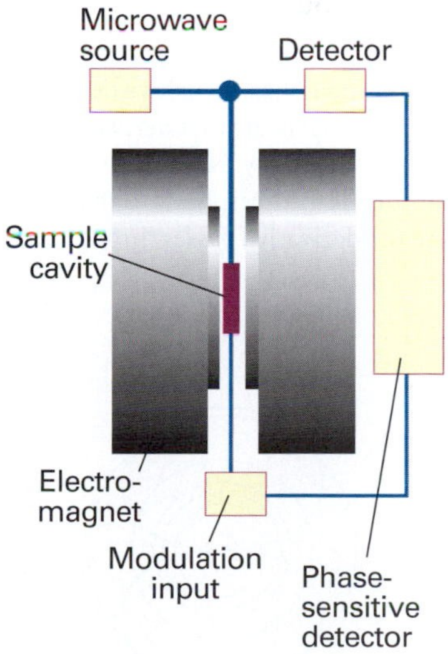

Fig. 8.54 The layout of a continuous-wave EPR spectrometer. A typical magnetic field is 0.3 T, which requires 9 GHz (3 cm) microwaves for resonance.

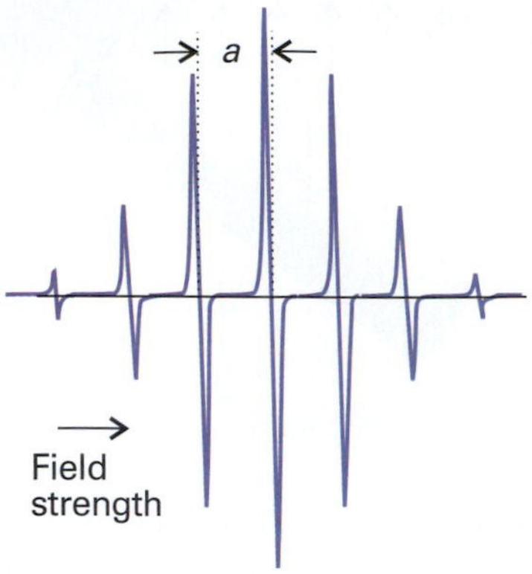

Fig. 8.55 The EPR spectrum of the benzene radical anion, $C_6H_6^-$, in fluid solution. *a* is the hyperfine splitting of the spectrum; the centre of the spectrum is determined by the *g*-value of the radical.

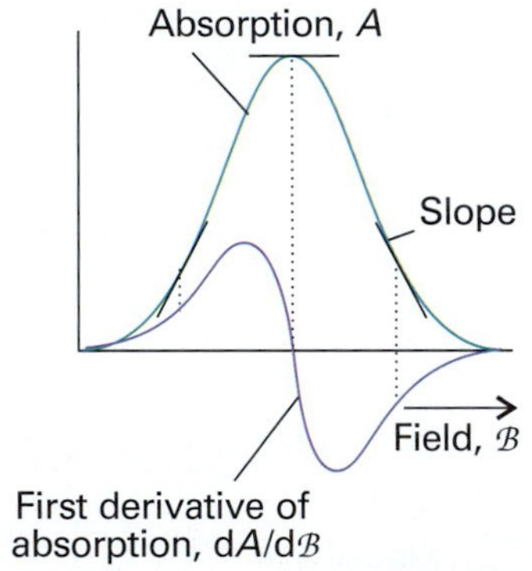

Fig. 8.56 When phase-sensitive detection is used, the signal is the first derivative of the absorption intensity. Note that the peak of the absorption corresponds to the point where the derivative passes through zero.

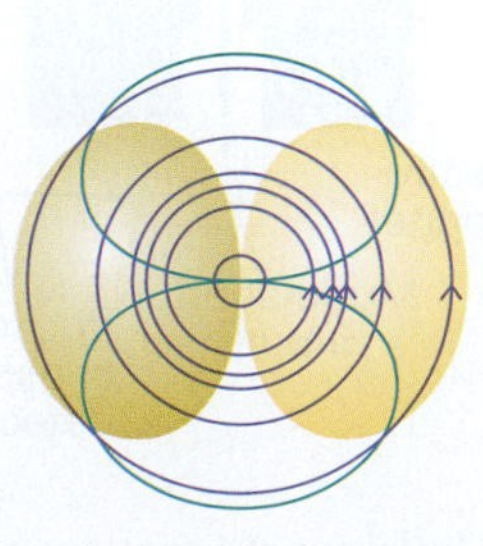

Fig. 8.57 An applied magnetic field can induce circulation of electrons that makes use of excited state orbitals (shown as a green outline).

of a microwave source (a klystron or a Gunn oscillator), a cavity in which the sample is inserted in a glass or quartz container, a microwave detector, and an electromagnet with a field that can be varied in the region of 0.3 T. The EPR spectrum is obtained by monitoring the microwave absorption as the field is changed, and a typical spectrum (of the benzene radical anion, $C_6H_6^-$) is shown in Fig. 8.55. The peculiar appearance of the spectrum, which is in fact the first-derivative of the absorption, arises from the detection technique, which is sensitive to the slope of the absorption curve (Fig. 8.56).

8.15 The *g*-value

Equation 8.13b gives the resonance frequency for a transition between the $m_s = -\frac{1}{2}$ and the $m_s = +\frac{1}{2}$ levels of a 'free' electron in terms of the *g*-value $g_e \approx 2.0023$. The magnetic moment of an unpaired electron in a radical also interacts with an external field, but the *g*-value is different from that for a free electron because of local magnetic fields induced by the molecular framework of the radical. Consequently, the resonance condition is normally written as

$$h\nu = g\mu_B \mathcal{B}_0 \tag{8.40}$$

where *g* is the **g-value** of the radical.

Illustration 8.4 *Calculating the g-value of an organic radical*

The centre of the EPR spectrum of the methyl radical occurred at 329.40 mT in a spectrometer operating at 9.2330 GHz (radiation belonging to the X band of the microwave region). Its *g*-value is therefore

$$g = \frac{h\nu}{\mu_B \mathcal{B}} = \frac{(6.626\,08 \times 10^{-34}\ \mathrm{J\ s}) \times (9.2330 \times 10^{9}\ \mathrm{s^{-1}})}{(9.2740 \times 10^{-24}\ \mathrm{J\ T^{-1}}) \times (0.329\,40\ \mathrm{T})} = 2.0027$$

Self-test 8.4 At what magnetic field would the methyl radical come into resonance in a spectrometer operating at 34.000 GHz (radiation belonging to the Q band of the microwave region)? [1.213 T]

The *g*-value in a molecular environment (a radical or a *d*-metal complex) is related to the ease with which the applied field can stir up currents through the molecular framework and the strength of the magnetic field the currents generate. Therefore, the *g*-value gives some information about electronic structure and plays a similar role in EPR to that played by shielding constants in NMR.

Electrons can migrate through the molecular framework by making use of excited states (Fig. 8.57). This additional path for circulation of electrons gives rise to a local magnetic field that adds to the applied field. Therefore, we expect that the ease of stirring up currents to be inversely proportional to the separation of energy levels, ΔE, in the molecule. As we saw in Section 3.8, the strength of the field generated by electronic currents in atoms (and analogously in molecules) is related to the extent of coupling between spin and orbital angular momenta. That is, the local field strength is proportional to the molecular spin–orbit coupling constant, ξ.

We can conclude from the discussion above that the *g*-value of a radical or *d*-metal complex differs from g_e, the 'free-electron' *g*-value, by an amount that is proportional to $\xi/\Delta E$. This proportionality is widely observed. Many organic radicals have *g*-values close to 2.0027 and inorganic radicals have *g*-values typically in the range 1.9 to 2.1. The *g*-values of paramagnetic *d*-metal complexes often differ considerably from g_e, varying from 0 to 6, because in them ΔE is small (on account of the splitting of *d*-orbitals brought about by interactions with ligands, as we saw in Section 7.2).

Just as in the case of the chemical shift in NMR spectroscopy, the *g*-value is anisotropic: that is, its magnitude depends on the orientation of the radical with respect to the applied field. In solution, when the molecule is tumbling rapidly, only the average value of the *g*-value is observed. Therefore, anisotropy of the *g*-value is observed only for radicals trapped in solids.

8.16 Hyperfine structure

The most important feature of EPR spectra is their **hyperfine structure**, the splitting of individual resonance lines into components. In general in spectroscopy, the term 'hyperfine structure' means the structure of a spectrum that can be traced to interactions of the electrons with nuclei other than as a result of the latter's point electric charge. The source of the hyperfine structure in EPR is the magnetic interaction between the electron spin and the magnetic dipole moments of the nuclei present in the radical.

(a) The effects of nuclear spin

Consider the effect on the EPR spectrum of a single H nucleus located somewhere in a radical. The proton spin is a source of magnetic field, and depending on the orientation of the nuclear spin, the field it generates adds to or subtracts from the applied field. The total local field is therefore

$$\mathcal{B}_{\text{loc}} = \mathcal{B} + am_I \qquad m_I = \pm\tfrac{1}{2} \tag{8.41}$$

where a is the **hyperfine coupling constant**. Half the radicals in a sample have $m_I = +\frac{1}{2}$, so half resonate when the applied field satisfies the condition

$$h\nu = g\mu_B(\mathcal{B} + \tfrac{1}{2}a), \qquad \text{or} \qquad \mathcal{B} = \frac{h\nu}{g\mu_B} - \tfrac{1}{2}a \tag{8.42a}$$

The other half (which have $m_I = -\frac{1}{2}$) resonate when

$$h\nu = g\mu_B(\mathcal{B} - \tfrac{1}{2}a), \qquad \text{or} \qquad \mathcal{B} = \frac{h\nu}{g\mu_B} + \tfrac{1}{2}a \tag{8.42b}$$

Therefore, instead of a single line, the spectrum shows two lines of half the original intensity separated by a and centred on the field determined by g (Fig. 8.58).

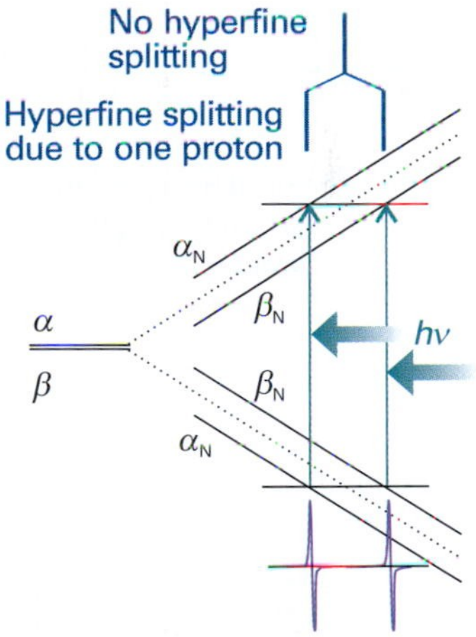

Fig. 8.58 The hyperfine interaction between an electron and a spin-$\frac{1}{2}$ nucleus results in four energy levels in place of the original two. As a result, the spectrum consists of two lines (of equal intensity) instead of one. The intensity distribution can be summarized by a simple stick diagram. The diagonal lines show the energies of the states as the applied field is increased, and resonance occurs when the separation of states matches the fixed energy of the microwave photon.

If the radical contains an ^{14}N atom ($I = 1$), its EPR spectrum consists of three lines of equal intensity, because the ^{14}N nucleus has three possible spin orientations, and each spin orientation is possessed by one-third of all the radicals in the sample. In general, a spin-I nucleus splits the spectrum into $2I + 1$ hyperfine lines of equal intensity.

When there are several magnetic nuclei present in the radical, each one contributes to the hyperfine structure. In the case of equivalent protons (for example, the two CH_2 protons in the radical CH_3CH_2) some of the hyperfine lines are coincident. It is not hard to show that, if the radical contains N equivalent protons, then there are $N + 1$ hyperfine lines with a binomial intensity distribution (the intensity distribution given by Pascal's triangle). The spectrum of the benzene radical anion in Fig. 8.55, which has seven lines with intensity ratio 1:6:15:20:15:6:1, is consistent with a radical containing six equivalent protons. More generally, if the radical contains N equivalent nuclei with spin quantum number I, then there are $2NI + 1$ hyperfine lines with an intensity distribution based on a modified version of Pascal's triangle as shown in the following *Example*.

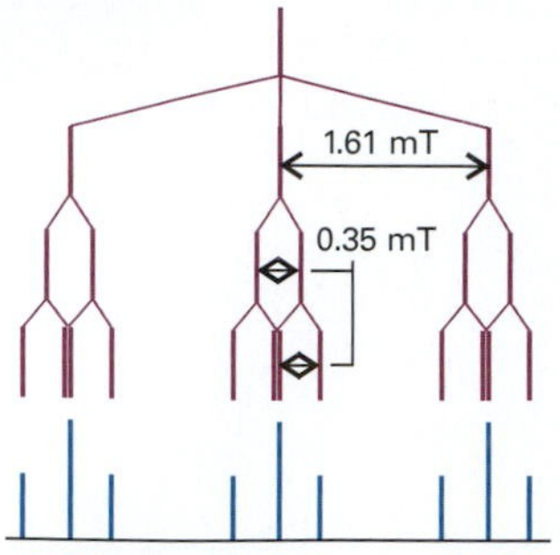

Fig. 8.59 The analysis of the hyperfine structure of radicals containing one ^{14}N nucleus ($I = 1$) and two equivalent protons.

Example 8.3 *Predicting the hyperfine structure of an EPR spectrum*

A radical contains one ^{14}N nucleus ($I = 1$) with hyperfine constant 1.61 mT and two equivalent protons ($I = \frac{1}{2}$) with hyperfine constant 0.35 mT. Predict the form of the EPR spectrum.

Method We should consider the hyperfine structure that arises from each type of nucleus or group of equivalent nuclei in succession. So, split a line with one nucleus, then each of those lines is split by a second nucleus (or group of nuclei), and so on. It is best to start with the nucleus with the largest hyperfine splitting; however, any choice could be made, and the order in which nuclei are considered does not affect the conclusion.

Answer The ^{14}N nucleus gives three hyperfine lines of equal intensity separated by 1.61 mT. Each line is split into doublets of spacing 0.35 mT by the first proton, and each line of these doublets is split into doublets with the same 0.35 mT splitting (Fig. 8.59). The central lines of each split doublet coincide, so the proton splitting gives 1:2:1 triplets of internal splitting 0.35 mT. Therefore, the spectrum consists of three equivalent 1:2:1 triplets.

Self-test 8.5 Predict the form of the EPR spectrum of a radical containing three equivalent ^{14}N nuclei. [Fig. 8.60]

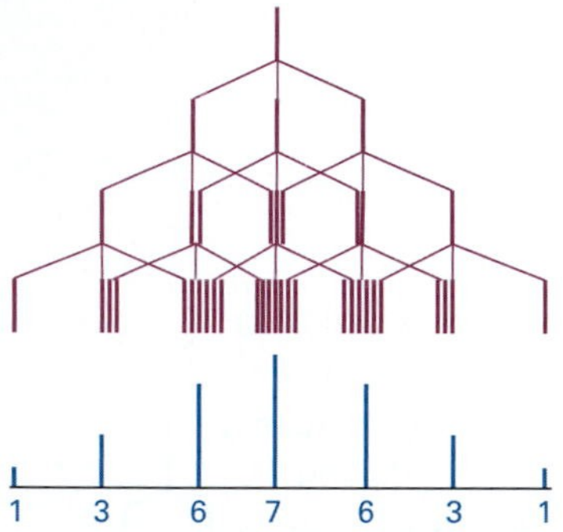

Fig. 8.60 The analysis of the hyperfine structure of radicals containing three equivalent ^{14}N nuclei.

The hyperfine structure of an EPR spectrum is a kind of fingerprint that helps to identify the radicals present in a sample. Moreover, because the magnitude of the splitting depends on the distribution of the unpaired electron near the magnetic nuclei present, the spectrum can be used to map the molecular orbital occupied by the unpaired electron. For example, because the hyperfine splitting in $C_6H_6^-$ is 0.375 mT, and one proton is close to a C atom with one-sixth the unpaired electron spin density (because the electron is spread uniformly around the ring), the hyperfine splitting caused by a proton in the electron spin entirely confined to a single adjacent C atom should be $6 \times 0.375\ \text{mT} = 2.25\ \text{mT}$. If in another aromatic radical we find a hyperfine splitting constant a, then the **spin density**, ρ, the probability that an unpaired electron is on the atom, can be calculated from the **McConnell equation:**

$$a = Q\rho \tag{8.43}$$

with $Q = 2.25$ mT. In this equation, ρ is the spin density on a C atom and a is the hyperfine splitting observed for the H atom to which it is attached.

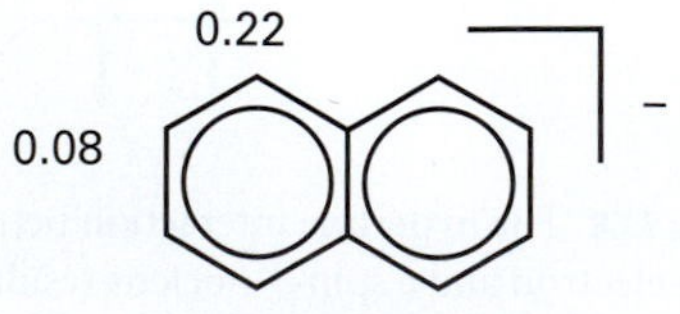

8

Illustration 8.5 *Using the McConnell equation*

The hyperfine structure of the EPR spectrum of the radical anion (naphthalene)$^-$ can be interpreted as arising from two groups of four equivalent protons. Those at the 1, 4, 5, and 8 positions in the ring have $a = 0.490$ mT and for those in the 2, 3, 6, and 7 positions have $a = 0.183$ mT. The densities obtained by using the McConnell equation are 0.22 and 0.08, respectively (**8**).

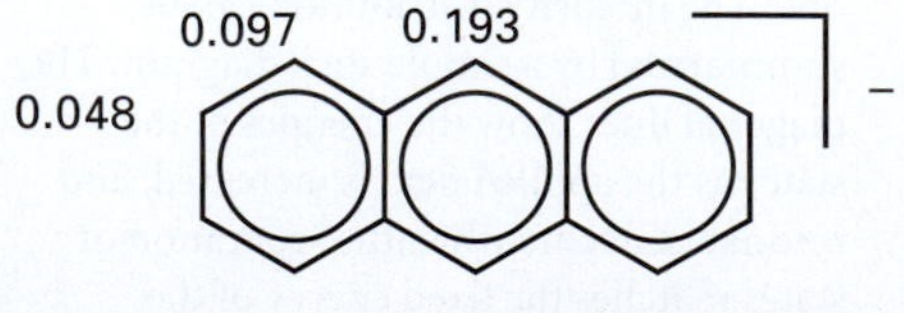

9

Self-test 8.6 The spin density in (anthracene)$^-$ is shown in (**9**). Predict the form of its EPR spectrum.

[A 1:2:1 triplet of splitting 0.43 mT split into a 1:4:6:4:1 quintet of splitting 0.22 mT, split into a 1:4:6:4:1 quintet of plitting 0.11 mT, $3 \times 5 \times 5 = 75$ lines in all]

(b) The origin of the hyperfine interaction

The hyperfine interaction is an interaction between the magnetic moments of the unpaired electron and the nuclei. There are two contributions to the interaction.

An electron in a *p* orbital does not approach the nucleus very closely, so it experiences a field that appears to arise from a point magnetic dipole. The resulting interaction is called the **dipole–dipole interaction**. The contribution of a magnetic nucleus to the local field experienced by the unpaired electron is given by an expression like that in eqn 8.28. A characteristic of this type of interaction is that it is anisotropic. Furthermore, just as in the case of NMR, the dipole–dipole interaction averages to zero when the radical is free to tumble. Therefore, hyperfine structure due to the dipole–dipole interaction is observed only for radicals trapped in solids.

An *s* electron is spherically distributed around a nucleus and so has zero average dipole–dipole interaction with the nucleus even in a solid sample. However, because an *s* electron has a nonzero probability of being at the nucleus, it is incorrect to treat the interaction as one between two point dipoles. An *s* electron has a Fermi contact interaction with the nucleus, which as we saw in Section 8.6d is a magnetic interaction that occurs when the point dipole approximation fails. The contact interaction is isotropic (that is, independent of the radical's orientation), and consequently is shown even by rapidly tumbling molecules in fluids (provided the spin density has some *s* character).

The dipole–dipole interactions of *p* electrons and the Fermi contact interaction of *s* electrons can be quite large. For example, a 2*p* electron in a nitrogen atom experiences an average field of about 3.4 mT from the ^{14}N nucleus. A 1*s* electron in a hydrogen atom experiences a field of about 50 mT as a result of its Fermi contact interaction with the central proton. More values are listed in Table 8.3. The magnitudes of the contact interactions in radicals can be interpreted in terms of the *s* orbital character of the molecular orbital occupied by the unpaired electron, and the dipole–dipole interaction can be interpreted in terms of the *p* character. The analysis of hyperfine structure therefore gives information about the composition of the orbital, and especially the hybridization of the atomic orbitals (see Problem 8.11).

We still have the source of the hyperfine structure of the $C_6H_6^-$ anion and other aromatic radical anions to explain. The sample is fluid, and as the radicals are tumbling the hyperfine structure cannot be due to the dipole–dipole interaction. Moreover, the protons lie in the nodal plane of the π orbital occupied by the unpaired electron, so the structure cannot be due to a Fermi contact interaction. The explanation lies in a **polarization mechanism** similar to the one responsible for spin–spin coupling in NMR. There is a magnetic interaction between a proton and the α electrons ($m_s = +\frac{1}{2}$) which results in one of the electrons tending to be found with a greater probability nearby (Fig. 8.61). The electron with opposite spin is therefore more likely to be close to the C atom at the other end of the bond. The unpaired electron on the C atom has a lower energy if it is parallel to that electron (Hund's rule favours parallel electrons on atoms), so the unpaired electron can detect the spin of the proton indirectly. Calculation using this model leads to a hyperfine interaction in agreement with the observed value of 2.25 mT.

Synoptic table 8.3* Hyperfine coupling constants for atoms, a/mT

Nuclide	Isotropic coupling	Anisotropic coupling
^{1}H	50.8 (1*s*)	
^{2}H	7.8 (1*s*)	
^{14}N	55.2 (2*s*)	3.4 (2*p*)
^{19}F	1720 (2*s*)	108.4 (2*p*)

* More values are given in the *Data section*.

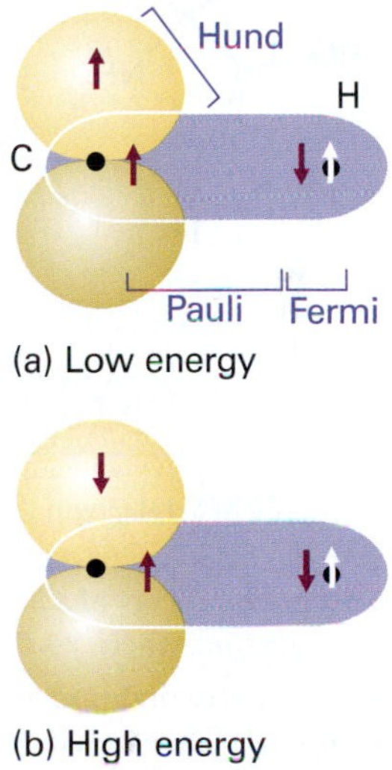

Fig. 8.61 The polarization mechanism for the hyperfine interaction in π-electron radicals. The arrangement in (a) is lower in energy than that in (b), so there is an effective coupling between the unpaired electron and the proton.

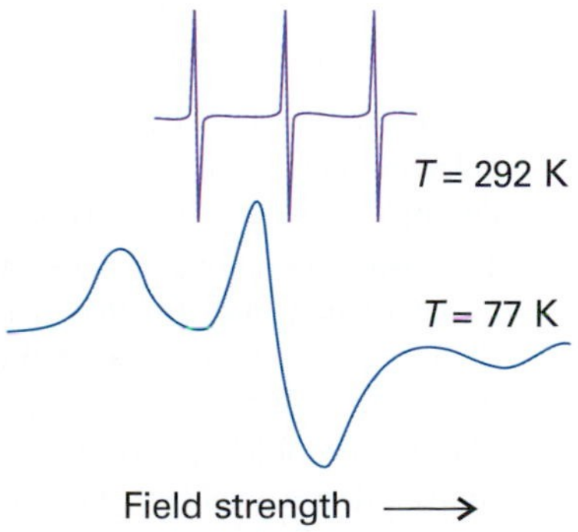

Fig. 8.62 ESR spectra of the di-*tert*-butyl nitroxide radical at 292 K (top) and 77 K (bottom). Adapted from J.R. Bolton, in *Biological applications of electron spin resonance*, H.M. Swartz, J.R. Bolton, and D.C. Borg (ed.), Wiley, New York (1972).

IMPACT ON BIOCHEMISTRY
I8.2 Spin probes

We saw in Sections 8.14 and 8.15 that anisotropy of the *g*-value and of the nuclear hyperfine interactions can be observed when a radical is immobilized in a solid. Figure 8.62 shows the variation of the lineshape of the EPR spectrum of the di-*tert*-butyl nitroxide radical (**10**) with temperature. At 292 K, the radical tumbles freely and isotropic hyperfine coupling to the ^{14}N nucleus gives rise to three sharp peaks. At 77 K,

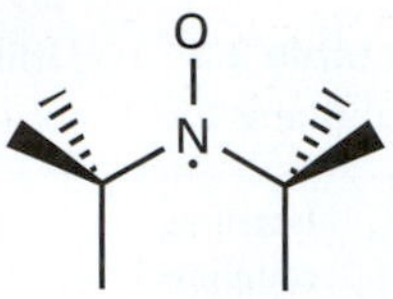

10

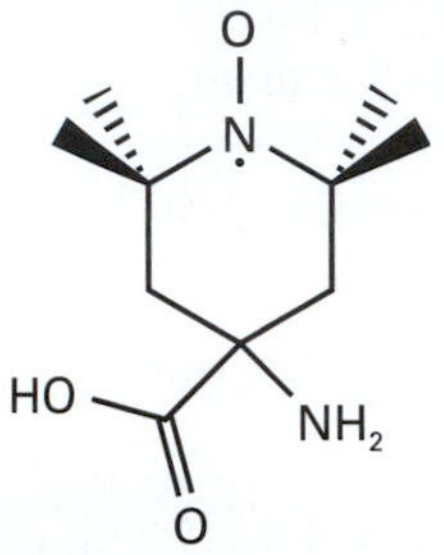

11

motion of the radical is restricted. Both isotropic and anisotropic hyperfine couplings determine the appearance of the spectrum, which now consists of three broad peaks.

A *spin probe* (or *spin label*) is a radical that interacts with a biopolymer and with an EPR spectrum that reports on the dynamical properties of the biopolymer. The ideal spin probe is one with a spectrum that broadens significantly as its motion is restricted to a relatively small extent. Nitroxide spin probes have been used to show that the hydrophobic interiors of biological membranes, once thought to be rigid, are in fact very fluid and individual lipid molecules move laterally through the sheet-like structure of the membrane.

Just as chemical exchange can broaden proton NMR spectra (Section 8.7), electron exchange between two radicals can broaden EPR spectra. Therefore, the distance between two spin probe molecules may be measured from the linewidths of their EPR spectra. The effect can be used in a number of biochemical studies. For example, the kinetics of association of two polypeptides labelled with the synthetic amino acid 2,2,6,6,-tetramethylpiperidine-1-oxyl-4-amino-4-carboxylic acid (**11**) may be studied by measuring the change in linewidth of the label with time. Alternatively, the thermodynamics of association may be studied by examining the temperature dependence of the linewidth.

Checklist of key ideas

☐ 1. The energy of an electron in a magnetic field $\mathcal{B}_0$ is $E_{m_s} = -g_e\gamma_e\hbar\mathcal{B}_0 m_s$, where γ_e is the magnetogyric ratio of the electron. The energy of a nucleus in a magnetic field $\mathcal{B}_0$ is $E_{m_I} = -\gamma\hbar\mathcal{B}_0 m_I$, where γ is the nuclear magnetogyric ratio.

☐ 2. The resonance condition for an electron in a magnetic field is $h\nu = g_e\mu_B\mathcal{B}_0$. The resonance condition for a nucleus in a magnetic field is $h\nu = \gamma\hbar\mathcal{B}_0$.

☐ 3. Nuclear magnetic resonance (NMR) is the observation of the frequency at which magnetic nuclei in molecules come into resonance with an electromagnetic field when the molecule is exposed to a strong magnetic field; NMR is a radiofrequency technique.

☐ 4. Electron paramagnetic resonance (EPR) is the observation of the frequency at which an electron spin comes into resonance with an electromagnetic field when the molecule is exposed to a strong magnetic field; EPR is a microwave technique.

☐ 5. The intensity of an NMR or EPR transition increases with the difference in population of α and β states and the strength of the applied magnetic field (as $\mathcal{B}_0^2$).

☐ 6. The chemical shift of a nucleus is the difference between its resonance frequency and that of a reference standard; chemical shifts are reported on the δ scale, in which $\delta = (\nu - \nu^\circ) \times 10^6/\nu^\circ$.

☐ 7. The observed shielding constant is the sum of a local contribution, a neighbouring group contribution, and a solvent contribution.

☐ 8. The fine structure of an NMR spectrum is the splitting of the groups of resonances into individual lines; the strength of the interaction is expressed in terms of the spin–spin coupling constant, J.

☐ 9. N equivalent spin-$\frac{1}{2}$ nuclei split the resonance of a nearby spin or group of equivalent spins into $N+1$ lines with an intensity distribution given by Pascal's triangle.

☐ 10. Spin–spin coupling in molecules in solution can be explained in terms of the polarization mechanism, in which the interaction is transmitted through the bonds.

☐ 11. The Fermi contact interaction is a magnetic interaction that depends on the very close approach of an electron to the nucleus and can occur only if the electron occupies an *s* orbital.

☐ 12. Coalescence of the two lines occurs in conformational interchange or chemical exchange when the lifetime, τ, of the states is related to their resonance frequency difference, $\delta\nu$, by $\tau = 2^{1/2}/\pi\delta\nu$.

☐ 13. In Fourier-transform NMR, the spectrum is obtained by mathematical analysis of the free-induction decay of magnetization, the response of nuclear spins in a sample to the application of one or more pulses of radiofrequency radiation.

☐ 14. Spin relaxation is the nonradiative return of a spin system to an equilibrium distribution of populations in which the transverse spin orientations are random; the system returns exponentially to the equilibrium population distribution with a time constant called the spin–lattice relaxation time, T_1.

☐ 15. The spin–spin relaxation time, T_2, is the time constant for the exponential return of the system to random transverse spin orientations.

☐ 16. In proton decoupling of ^{13}C-NMR spectra, protons are made to undergo rapid spin reorientations and the ^{13}C nucleus senses an average orientation. As a result, its resonance is a single line and not a group of lines.

☐ 17. The nuclear Overhauser effect (NOE) is the modification of one resonance by the saturation of another.

☐ 18. In two-dimensional NMR, spectra are displayed in two axes, with resonances belonging to different groups lying at different locations on the second axis. An example of a two-dimensional NMR technique is correlation spectroscopy (COSY), in which all spin–spin couplings in a molecule are determined. Another example is nuclear Overhauser effect spectroscopy (NOESY), in which internuclear distances up to about 0.5 nm are determined.

☐ 19. Magic-angle spinning (MAS) is technique in which the NMR linewidths in a solid sample are reduced by spinning at an angle of 54.74° to the applied magnetic field.

☐ 20. The EPR resonance condition is written $h\nu = g\mu_B\mathcal{B}$, where g is the g-value of the radical; the deviation of g from $g_e = 2.0023$ depends on the ability of the applied field to induce local electron currents in the radical.

☐ 21. The hyperfine structure of an EPR spectrum is its splitting of individual resonance lines into components by the magnetic interaction between the electron and nuclei with spin.

☐ 22. If a radical contains N equivalent nuclei with spin quantum number I, then there are $2NI+1$ hyperfine lines with an intensity distribution given by a modified version of Pascal's triangle.

☐ 23. The hyperfine structure due to a hydrogen attached to an aromatic ring is converted to spin density, ρ, on the neighbouring carbon atom by using the McConnell equation: $a = Q\rho$ with $Q = 2.25$ mT.

Further reading

Articles and texts

N.M. Atherton, *Principles of electron spin resonance*. Ellis Horwood/Prentice Hall, Hemel Hempstead (1993).

E.D. Becker, *High resolution NMR: theory and chemical applications*. Academic Press, San Diego (2000).

R. Freeman, *Spin choreography: basic steps in high resolution NMR*. Oxford University Press (1998).

J.C. Lindon, G.E. Tranter, and J.L. Holmes (ed.), *Encyclopedia of spectroscopy and spectrometry*. Academic Press, San Diego (2000).

R.W. King and K.R. Williams. The Fourier transform in chemistry. Part 1. Nuclear magnetic resonance: Introduction. *J. Chem. Educ.* **66**, A213 (1989); Part 2. Nuclear magnetic resonance: The single pulse experiment. *Ibid.* A243; Part 3. Multiple-pulse experiments. *J. Chem. Educ.* **67**, A93 (1990); Part 4. NMR: Two-dimensional methods. *Ibid.* A125; A glossary of NMR terms. *Ibid.* A100.

M.T. Vlaardingerbroek and J.A. de Boer, *Magnetic resonance imaging: theory and practice*. Springer, Berlin (1999).

M.H. Levitt, *Spin dynamics*. Wiley (2001).

Sources of data and information

D.R. Lide (ed.), *CRC Handbook of Chemistry and Physics*, Section 9, CRC Press, Boca Raton (2000).

C.P. Poole, Jr. and H.A. Farach (ed.), *Handbook of electron spin resonance: data sources, computer technology, relaxation, and ENDOR*. Vols. 1–2. Springer, Berlin (1997, 1999).

Further information

Further information 8.1 *Fourier transformation of the FID curve*

The analysis of the FID curve is achieved by the standard mathematical technique of Fourier transformation, which we explored in the context of FT infrared spectroscopy (*Further information* 6.2). We start by noting that the signal $S(t)$ in the time domain, the total FID curve, is the sum (more precisely, the integral) over all the contributing frequencies

$$S(t) = \int_{-\infty}^{\infty} I(\nu)e^{-2\pi i\nu t}d\nu \qquad (8.44)$$

Because $e^{2\pi i\nu t} = \cos(2\pi\nu t) + i\sin(2\pi\nu t)$, the expression above is a sum over harmonically oscillating functions, with each one weighted by the intensity $I(\nu)$.

We need $I(\nu)$, the spectrum in the frequency domain; it is obtained by evaluating the integral

$$I(\nu) = 2\,\mathrm{Re}\int_0^{\infty} S(t)e^{2\pi i\nu t}dt \qquad (8.45)$$

where Re means take the real part of the following expression. This integral is very much like an overlap integral: it gives a nonzero value if $S(t)$ contains a component that matches the oscillating function $e^{2i\pi\nu t}$. The integration is carried out at a series of frequencies ν on a computer that is built into the spectrometer.

Each end-of-chapter discussion question, exercise, and problem for which a solution is available in the Student Solution Manual has a corresponding bracketed number referring to the Student Solutions Manual number for this problem, for example [SSM 15.1]. Each end-of-chapter exercise and problem for which a solution is available in the Instructors' Solution Manual has a corresponding bracketed number referring to the Instructor Solutions Manual number for this problem, for example [ISM 15.2].

Discussion questions

8.1 Discuss in detail the origins of the local, neighbouring group, and solvent contributions to the shielding constant. [SSM **15.1**]

8.2 Discuss in detail the effects of a 90° pulse and of a 180° pulse on a system of spin-$\frac{1}{2}$ nuclei in a static magnetic field. [ISM **15.2**]

8.3 Suggest a reason why the relaxation times of ^{13}C nuclei are typically much longer than those of ^{1}H nuclei. [SSM **15.3**]

8.4 Discuss the origins of diagonal and cross peaks in the COSY spectrum of an AX system. [ISM **15.4**]

8.5 Discuss how the Fermi contact interaction and the polarization mechanism contribute to spin–spin couplings in NMR and hyperfine interactions in EPR. [SSM **15.5**]

8.6 Suggest how spin probes could be used to estimate the depth of a crevice in a biopolymer, such as the active site of an enzyme. [ISM **15.6**]

Exercises

8.1a What is the resonance frequency of a proton in a magnetic field of 14.1 T? [ISM **15.1a**]

8.1b What is the resonance frequency of a ^{19}F nucleus in a magnetic field of 16.2 T? [SSM **15.1b**]

8.2a ^{33}S has a nuclear spin of $\frac{3}{2}$ and a nuclear g-factor of 0.4289. Calculate the energies of the nuclear spin states in a magnetic field of 7.500 T. [ISM **15.2a**]

8.2b ^{14}N has a nuclear spin of 1 and a nuclear g-factor of 0.404. Calculate the energies of the nuclear spin states in a magnetic field of 11.50 T. [SSM **15.2b**]

8.3a Calculate the frequency separation of the nuclear spin levels of a ^{13}C nucleus in a magnetic field of 14.4 T given that the magnetogyric ratio is $6.73 \times 10^{7}\ T^{-1}\ s^{-1}$. [ISM **15.3a**]

8.3b Calculate the frequency separation of the nuclear spin levels of a ^{14}N nucleus in a magnetic field of 15.4 T given that the magnetogyric ratio is $1.93 \times 10^{7}\ T^{-1}\ s^{-1}$. [SSM **15.3b**]

8.4a In which of the following systems is the energy level separation the largest? (a) A proton in a 600 MHz NMR spectrometer, (b) a deuteron in the same spectrometer. [ISM **15.4a**]

8.4b In which of the following systems is the energy level separation the largest? (a) A ^{14}N nucleus in (for protons) a 600 MHz NMR spectrometer, (b) an electron in a radical in a field of 0.300 T. [SSM **15.4b**]

8.5a Calculate the energy difference between the lowest and highest nuclear spin states of a ^{14}N nucleus in a 15.00 T magnetic field. [ISM **15.5a**]

8.5b Calculate the magnetic field needed to satisfy the resonance condition for unshielded protons in a 150.0 MHz radiofrequency field. [SSM **15.5b**]

8.6a Use Table 8.2 to predict the magnetic fields at which (a) ^{1}H, (b) ^{2}H, (c) ^{13}C come into resonance at (i) 250 MHz, (ii) 500 MHz. [ISM **15.6a**]

8.6b Use Table 8.2 to predict the magnetic fields at which (a) ^{14}N, (b) ^{19}F, and (c) ^{31}P come into resonance at (i) 300 MHz, (ii) 750 MHz. [SSM **15.6b**]

8.7a Calculate the relative population differences ($\delta N/N$) for protons in fields of (a) 0.30 T, (b) 1.5 T, and (c) 10 T at 25°C. [ISM **15.7a**]

8.7b Calculate the relative population differences ($\delta N/N$) for ^{13}C nuclei in fields of (a) 0.50 T, (b) 2.5 T, and (c) 15.5 T at 25°C. [SSM **15.7b**]

8.8a The first generally available NMR spectrometers operated at a frequency of 60 MHz; today it is not uncommon to use a spectrometer that operates at 800 MHz. What are the relative population differences of ^{13}C spin states in these two spectrometers at 25°C? [ISM **15.8a**]

8.8b What are the relative values of the chemical shifts observed for nuclei in the spectrometers mentioned in Exercise 8.8a in terms of (a) δ values, (b) frequencies? [SSM **15.8b**]

8.9a The chemical shift of the CH_3 protons in acetaldehyde (ethanal) is $\delta = 2.20$ and that of the CHO proton is 9.80. What is the difference in local magnetic field between the two regions of the molecule when the applied field is (a) 1.5 T, (b) 15 T? [ISM **15.9a**]

8.9b The chemical shift of the CH_3 protons in diethyl ether is $\delta = 1.16$ and that of the CH_2 protons is 3.36. What is the difference in local magnetic field between the two regions of the molecule when the applied field is (a) 1.9 T, (b) 16.5 T? [SSM **15.9b**]

8.10a Sketch the appearance of the ^{1}H-NMR spectrum of acetaldehyde (ethanal) using $J = 2.90$ Hz and the data in Exercise 8.9a in a spectrometer operating at (a) 250 MHz, (b) 500 MHz. [ISM **15.10a**]

8.10b Sketch the appearance of the ^{1}H-NMR spectrum of diethyl ether using $J = 6.97$ Hz and the data in Exercise 8.9b in a spectrometer operating at (a) 350 MHz, (b) 650 MHz. [SSM **15.10b**]

8.11a Two groups of protons are made equivalent by the isomerization of a fluxional molecule. At low temperatures, where the interconversion is slow, one group has $\delta = 4.0$ and the other has $\delta = 5.2$. At what rate of interconversion will the two signals merge in a spectrometer operating at 250 MHz? [ISM **15.11a**]

8.11b Two groups of protons are made equivalent by the isomerization of a fluxional molecule. At low temperatures, where the interconversion is slow, one group has $\delta = 5.5$ and the other has $\delta = 6.8$. At what rate of interconversion will the two signals merge in a spectrometer operating at 350 MHz? [SSM **15.11b**]

8.12a Sketch the form of the ^{19}F-NMR spectra of a natural sample of tetrafluoroborate ions, BF_4^-, allowing for the relative abundances of $^{10}BF_4^-$ and $^{11}BF_4^-$. [ISM **15.12a**]

8.12b From the data in Table 8.2, predict the frequency needed for ^{31}P-NMR in an NMR spectrometer designed to observe proton resonance at 500 MHz. Sketch the proton and ^{31}P resonances in the NMR spectrum of PH_4^+. [SSM **15.12b**]

8.13a Sketch the form of an $A_3M_2X_4$ spectrum, where A, M, and X are protons with distinctly different chemical shifts and $J_{AM} > J_{AX} > J_{MX}$. [ISM **15.13a**]

8.13b Sketch the form of an $A_2M_2X_5$ spectrum, where A, M, and X are protons with distinctly different chemical shifts and $J_{AM} > J_{AX} > J_{MX}$. [SSM **15.13b**]

8.14a Which of the following molecules have sets of nuclei that are chemically but not magnetically equivalent? (a) CH_3CH_3, (b) $CH_2{=}CH_2$. [ISM **15.14a**]

8.14b Which of the following molecules have sets of nuclei that are chemically but not magnetically equivalent? (a) $CH_2{=}C{=}CF_2$, (b) *cis*- and *trans*-$[Mo(CO)_4(PH_3)_2]$. [SSM **15.14b**]

8.15a The duration of a 90° or 180° pulse depends on the strength of the $\mathcal{B}_1$ field. If a 90° pulse requires 10 μs, what is the strength of the $\mathcal{B}_1$ field? How long would the corresponding 180° pulse require? [ISM **15.15a**]

8.15b The duration of a 90° or 180° pulse depends on the strength of the $\mathcal{B}_1$ field. If a 180° pulse requires 12.5 μs, what is the strength of the $\mathcal{B}_1$ field? How long would the corresponding 90° pulse require? [SSM **15.15b**]

8.16a What magnetic field would be required in order to use an EPR X–band spectrometer (9 GHz) to observe ^{1}H-NMR and a 300 MHz spectrometer to observe EPR? [ISM **15.16a**]

8.16b Some commercial EPR spectrometers use 8 mm microwave radiation (the Q band). What magnetic field is needed to satisfy the resonance condition? [SSM **15.16b**]

8.17a The centre of the EPR spectrum of atomic hydrogen lies at 329.12 mT in a spectrometer operating at 9.2231 GHz. What is the g-value of the electron in the atom? [ISM **15.17a**]

8.17b The centre of the EPR spectrum of atomic deuterium lies at 330.02 mT in a spectrometer operating at 9.2482 GHz. What is the g-value of the electron in the atom? [SSM **15.17b**]

8.18a A radical containing two equivalent protons shows a three-line spectrum with an intensity distribution 1:2:1. The lines occur at 330.2 mT, 332.5 mT, and 334.8 mT. What is the hyperfine coupling constant for each proton? What is the g-value of the radical given that the spectrometer is operating at 9.319 GHz? [ISM **15.18a**]

8.18b A radical containing three equivalent protons shows a four–line spectrum with an intensity distribution 1:3:3:1. The lines occur at 331.4 mT, 333.6 mT, 335.8 mT, and 338.0 mT. What is the hyperfine coupling constant for each proton? What is the g-value of the radical given that the spectrometer is operating at 9.332 GHz? [SSM **15.18b**]

8.19a A radical containing two inequivalent protons with hyperfine constants 2.0 mT and 2.6 mT gives a spectrum centred on 332.5 mT. At what fields do the hyperfine lines occur and what are their relative intensities? [ISM **15.19a**]

8.19b A radical containing three inequivalent protons with hyperfine constants 2.11 mT, 2.87 mT, and 2.89 mT gives a spectrum centred on 332.8 mT. At what fields do the hyperfine lines occur and what are their relative intensities? [SSM **15.19b**]

8.20a Predict the intensity distribution in the hyperfine lines of the EPR spectra of (a) $\cdot CH_3$, (b) $\cdot CD_3$. [ISM **15.20a**]

8.20b Predict the intensity distribution in the hyperfine lines of the EPR spectra of (a) $\cdot CH_2H_3$, (b) $\cdot CD_2CD_3$. [SSM **15.20b**]

8.21a The benzene radical anion has $g = 2.0025$. At what field should you search for resonance in a spectrometer operating at (a) 9.302 GHz, (b) 33.67 GHz? [ISM **15.21a**]

8.21b The naphthalene radical anion has $g = 2.0024$. At what field should you search for resonance in a spectrometer operating at (a) 9.312 GHz, (b) 33.88 GHz? [SSM **15.21b**]

8.22a The EPR spectrum of a radical with a single magnetic nucleus is split into four lines of equal intensity. What is the nuclear spin of the nucleus? [ISM **15.22a**]

8.22b The EPR spectrum of a radical with two equivalent nuclei of a particular kind is split into five lines of intensity ratio 1:2:3:2:1. What is the spin of the nuclei? [SSM **15.22b**]

8.23a Sketch the form of the hyperfine structures of radicals XH_2 and XD_2, where the nucleus X has $I = \frac{5}{2}$. [ISM **15.23a**]

8.23b Sketch the form of the hyperfine structures of radicals XH_3 and XD_3, where the nucleus X has $I = \frac{3}{2}$. [SSM **15.23b**]

Problems*

Numerical problems

8.1 A scientist investigates the possibility of neutron spin resonance, and has available a commercial NMR spectrometer operating at 300 MHz. What field is required for resonance? What is the relative population difference at room temperature? Which is the lower energy spin state of the neutron? [SSM **15.1**]

8.2 Two groups of protons have $\delta = 4.0$ and $\delta = 5.2$ and are interconverted by a conformational change of a fluxional molecule. In a 60 MHz spectrometer the spectrum collapsed into a single line at 280 K but at 300 MHz the collapse did not occur until the temperature had been raised to 300 K. What is the activation energy of the interconversion? [ISM **15.2**]

8.3‡ Suppose that the FID in Fig. 8.31 was recorded in a 300 MHz spectrometer, and that the interval between maxima in the oscillations in the FID is 0.10 s. What is the Larmor frequency of the nuclei and the spin–spin relaxation time? [SSM **15.3**]

8.4‡ In a classic study of the application of NMR to the measurement of rotational barriers in molecules, P.M. Nair and J.D. Roberts (*J. Am. Chem. Soc.* **79**, 4565 (1957)) obtained the 40 MHz ^{19}F-NMR spectrum of $F_2BrCCBrCl_2$. Their spectra are reproduced in Fig. 15.63. At 193 K the spectrum shows five resonance peaks. Peaks I and III are separated by 160 Hz, as are IV and V. The ratio of the integrated intensities of peak II to peaks I, III, IV, and V is approximately 10 to 1. At 273 K, the five peaks have collapsed into

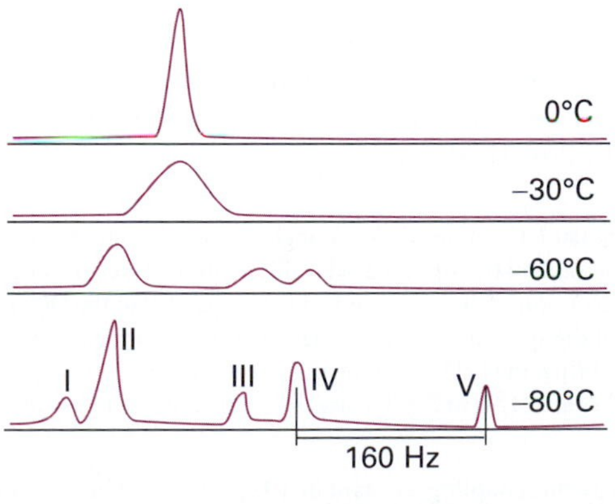

Fig. 8.63

* Problems denoted with the symbol ‡ were supplied by Charles Trapp and Carmen Giunta.

one. Explain the spectrum and its change with temperature. At what rate of interconversion will the spectrum collapse to a single line? Calculate the rotational energy barrier between the rotational isomers on the assumption that it is related to the rate of interconversion between the isomers. [ISM 15.4]

8.5‡ Various versions of the Karplus equation (eqn 8.27) have been used to correlate data on vicinal proton coupling constants in systems of the type $R_1R_2CHCHR_3R_4$. The original version, (M. Karplus, *J. Am. Chem. Soc.* **85**, 2870 (1963)), is $^3J_{HH} = A\cos^2\phi_{HH} + B$. When $R_3 = R_4 = H$, $^3J_{HH} = 7.3$ Hz; when $R_3 = CH_3$ and $R_4 = H$, $^3J_{HH} = 8.0$ Hz; when $R_3 = R_4 = CH_3$, $^3J_{HH} = 11.2$ Hz. Assume that only staggered conformations are important and determine which version of the Karplus equation fits the data better. [SSM 15.5]

8.6‡ It might be unexpected that the Karplus equation, which was first derived for $^3J_{HH}$ coupling constants, should also apply to vicinal coupling between the nuclei of metals such as tin. T.N. Mitchell and B. Kowall (*Magn. Reson. Chem.* **33**, 325 (1995)) have studied the relation between $^3J_{HH}$ and $^3J_{SnSn}$ in compounds of the type $Me_3SnCH_2CHRSnMe_3$ and find that $^3J_{SnSn} = 78.86\,{}^3J_{HH} + 27.84$ Hz. (a) Does this result support a Karplus type equation for tin? Explain your reasoning. (b) Obtain the Karplus equation for $^3J_{SnSn}$ and plot it as a function of the dihedral angle. (c) Draw the preferred conformation. [ISM 15.6]

8.7 Figure 8.64 shows the proton COSY spectrum of 1-nitropropane. Account for the appearance of off-diagonal peaks in the spectrum. [SSM 15.7]

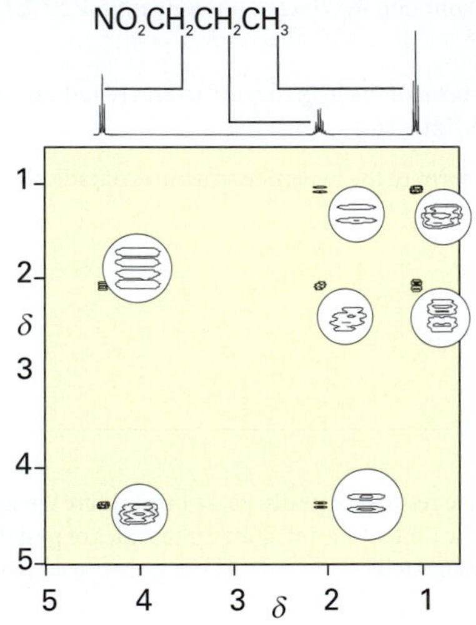

Fig. 8.64 The COSY spectrum of 1-nitropropane ($NO_2CH_2CH_2CH_3$). The circles show enhanced views of the spectral features. (Spectrum provided by Prof. G. Morris.)

8.8 The angular NO_2 molecule has a single unpaired electron and can be trapped in a solid matrix or prepared inside a nitrite crystal by radiation damage of NO_2^- ions. When the applied field is parallel to the OO direction the centre of the spectrum lies at 333.64 mT in a spectrometer operating at 9.302 GHz. When the field lies along the bisector of the ONO angle, the resonance lies at 331.94 mT. What are the *g*-values in the two orientations? [ISM 15.8]

8.9 The hyperfine coupling constant in $\cdot CH_3$ is 2.3 mT. Use the information in Table 8.3 to predict the splitting between the hyperfine lines of the spectrum of $\cdot CD_3$. What are the overall widths of the hyperfine spectra in each case? [SSM 15.9]

8.10 The *p*-dinitrobenzene radical anion can be prepared by reduction of *p*-dinitrobenzene. The radical anion has two equivalent N nuclei ($I = 1$) and four equivalent protons. Predict the form of the EPR spectrum using $a(N) = 0.148$ mT and $a(H) = 0.112$ mT. [ISM 15.10]

8.11 When an electron occupies a 2*s* orbital on an N atom it has a hyperfine interaction of 55.2 mT with the nucleus. The spectrum of NO_2 shows an isotropic hyperfine interaction of 5.7 mT. For what proportion of its time is the unpaired electron of NO_2 occupying a 2*s* orbital? The hyperfine coupling constant for an electron in a 2*p* orbital of an N atom is 3.4 mT. In NO_2 the anisotropic part of the hyperfine coupling is 1.3 mT. What proportion of its time does the unpaired electron spend in the 2*p* orbital of the N atom in NO_2? What is the total probability that the electron will be found on (a) the N atoms, (b) the O atoms? What is the hybridization ratio of the N atom? Does the hybridization support the view that NO_2 is angular? [SSM 15.11]

8.12 The hyperfine coupling constants observed in the radical anions (**12**), (**13**), and (**14**) are shown (in millitesla, mT). Use the value for the benzene radical anion to map the probability of finding the unpaired electron in the π orbital on each C atom. [ISM 15.12]

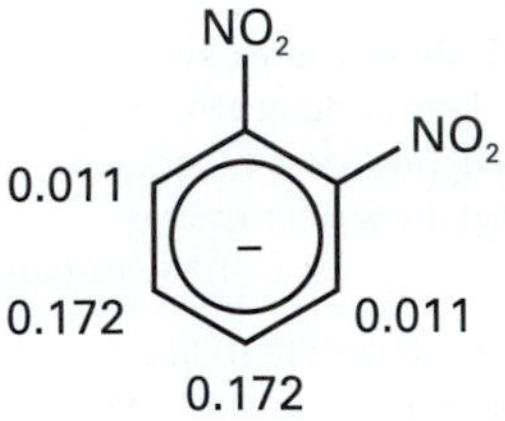

12

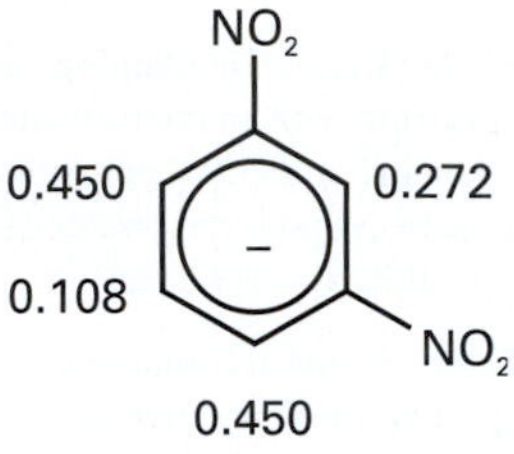

13

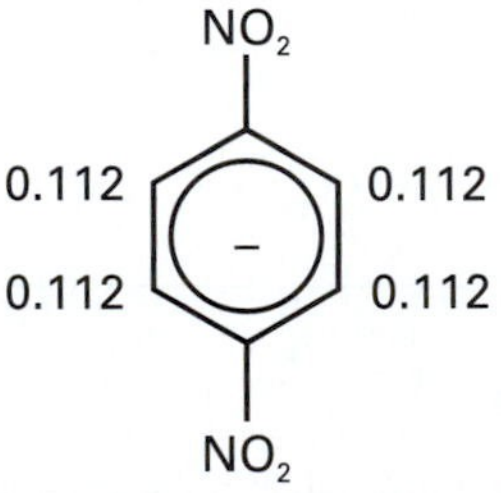

14

Theoretical problems

8.13 Calculate σ_d for a hydrogenic atom with atomic number Z. [SSM 15.13]

8.14 In this problem you will use the molecular electronic structure methods described in Chapter 4 to investigate the hypothesis that the magnitude of the ^{13}C chemical shift correlates with the net charge on a ^{13}C atom. (a) Using molecular modelling software[3] and the computational method of your choice, calculate the net charge at the C atom *para* to the substituents in this series of molecules: benzene, phenol, toluene, trifluorotoluene, benzonitrile, and nitrobenzene. (b) The ^{13}C chemical shifts of the *para* C atoms in each of the molecules that you examined in part (a) are given below:

Substituent	OH	CH_3	H	CF_3	CN	NO_2
δ	130.1	128.4	128.5	128.9	129.1	129.4

Is there a linear correlation between net charge and ^{13}C chemical shift of the *para* C atom in this series of molecules? (c) If you did find a correlation in part

[3] The web site contains links to molecular modelling freeware and to other sites where you may perform molecular orbital calculations directly from your web browser.

(b), use the concepts developed in this chapter to explain the physical origins of the correlation. [ISM 15.14]

8.15 The z-component of the magnetic field at a distance R from a magnetic moment parallel to the z-axis is given by eqn 8.28. In a solid, a proton at a distance R from another can experience such a field and the measurement of the splitting it causes in the spectrum can be used to calculate R. In gypsum, for instance, the splitting in the H_2O resonance can be interpreted in terms of a magnetic field of 0.715 mT generated by one proton and experienced by the other. What is the separation of the protons in the H_2O molecule? [SSM 15.15]

8.16 In a liquid, the dipolar magnetic field averages to zero: show this result by evaluating the average of the field given in eqn 8.28. *Hint.* The volume element is $\sin\theta\,d\theta d\phi$ in polar coordinates. [ISM 15.16]

8.17 The shape of a spectral line, $I(\omega)$, is related to the free induction decay signal $G(t)$ by

$$I(\omega) = a\,\mathrm{Re}\int_0^\infty G(t)e^{i\omega t}dt$$

where a is a constant and 'Re' means take the real part of what follows. Calculate the lineshape corresponding to an oscillating, decaying function $G(t) = \cos\omega_0 t\, e^{-t/\tau}$. [SSM 15.17]

8.18 In the language of Problem 8.17, show that, if $G(t) = (a\cos\omega_1 t + b\cos\omega_2 t)e^{-t/\tau}$, then the spectrum consists of two lines with intensities proportional to a and b and located at $\omega = \omega_1$ and ω_2, respectively. [ISM 15.18]

8.19 EPR spectra are commonly discussed in terms of the parameters that occur in the *spin-hamiltonian*, a hamiltonian operator that incorporates various effects involving spatial operators (like the orbital angular momentum) into operators that depend on the spin alone. Show that, if you use $H = -g_e\gamma_e\mathcal{B}_0 s_z - \gamma_e\mathcal{B}_0 l_z$ as the true hamiltonian, then from second-order perturbation theory (and specifically eqn 2.65), the eigenvalues of the spin are the same as those of the spin-hamiltonian $H_{spin} = -g\gamma_e\mathcal{B}_0 s_z$ (note the g in place of g_e) and find an expression for g. [SSM 15.19]

Applications: to biochemistry and medicine

8.20 Interpret the following features of the NMR spectra of hen lysozyme: (a) saturation of a proton resonance assigned to the side chain of methionine-105 changes the intensities of proton resonances assigned to the side chains of tryptophan-28 and tyrosine-23; (b) saturation of proton resonances assigned to tryptophan-28 did not affect the spectrum of tyrosine-23. [ISM 15.20]

8.21 When interacting with a large biopolymer or even larger organelle, a small molecule might not rotate freely in all directions and the dipolar interaction might not average to zero. Suppose a molecule is bound so that, although the vector separating two protons may rotate freely around the z-axis, the colatitude may vary only between 0 and θ'. Average the dipolar field over this restricted range of orientations and confirm that the average vanishes when $\theta' = \pi$ (corresponding to rotation over an entire sphere). What is the average value of the local dipolar field for the H_2O molecule in Problem 8.15 if it is bound to a biopolymer that enables it to rotate up to $\theta' = 30°$? [SSM 15.21]

8.22 Suggest a reason why the spin–lattice relaxation time of benzene (a small molecule) in a mobile, deuterated hydrocarbon solvent increases with temperature whereas that of an oligonucleotide (a large molecule) decreases. [ISM 15.22]

8.23 NMR spectroscopy may be used to determine the equilibrium constant for dissociation of a complex between a small molecule, such as an enzyme inhibitor I, and a protein, such as an enzyme E:

$$\mathrm{EI} \rightleftharpoons \mathrm{E} + \mathrm{I} \qquad K_I = [\mathrm{E}][\mathrm{I}]/[\mathrm{EI}]$$

In the limit of slow chemical exchange, the NMR spectrum of a proton in I would consist of two resonances: one at ν_I for free I and another at ν_{EI} for bound I. When chemical exchange is fast, the NMR spectrum of the same proton in I consists of a single peak with a resonance frequency ν given by $\nu = f_I\nu_I + f_{EI}\nu_{EI}$, where $f_I = [\mathrm{I}]/([\mathrm{I}] + [\mathrm{EI}])$ and $f_{EI} = [\mathrm{EI}]/([\mathrm{I}] + [\mathrm{EI}])$ are, respectively, the fractions of free I and bound I. For the purposes of analysing the data, it is also useful to define the frequency differences $\delta\nu = \nu - \nu_I$ and $\Delta\nu = \nu_{EI} - \nu_I$. Show that, when the initial concentration of I, $[\mathrm{I}]_0$, is much greater than the initial concentration of E, $[\mathrm{E}]_0$, a plot of $[\mathrm{I}]_0$ against $\delta\nu^{-1}$ is a straight line with slope $[\mathrm{E}]_0\Delta\nu$ and y-intercept $-K_I$. [SSM 15.23]

8.24 The molecular electronic structure methods described in Chapter 4 may be used to predict the spin density distribution in a radical. Recent EPR studies have shown that the amino acid tyrosine participates in a number of biological electron transfer reactions, including the processes of water oxidation to O_2 in plant photosystem II and of O_2 reduction to water in cytochrome c oxidase (*Impact I10.2*). During the course of these electron transfer reactions, a tyrosine radical forms, with spin density delocalized over the side chain of the amino acid. (a) The phenoxy radical shown in (**15**) is a suitable model of the tyrosine radical. Using molecular modelling software and the computational method of your choice (semi-empirical or *ab initio* methods), calculate the spin densities at the O atom and at all of the C atoms in (**15**). (b) Predict the form of the EPR spectrum of (**15**). [ISM 15.24]

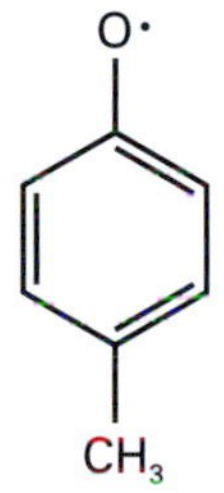

15

8.25 Sketch the EPR spectra of the di-*tert*-butyl nitroxide radical (**10**) at 292 K in the limits of very low concentration (at which electron exchange is negligible), moderate concentration (at which electron exchange effects begin to be observed), and high concentration (at which electron exchange effects predominate). Discuss how the observation of electron exchange between nitroxide spin probes can inform the study of lateral mobility of lipids in a biological membrane. [SSM 15.25]

8.26 You are designing an MRI spectrometer. What field gradient (in microtesla per metre, $\mu T\,m^{-1}$) is required to produce a separation of 100 Hz between two protons separated by the long diameter of a human kidney (taken as 8 cm) given that they are in environments with $\delta = 3.4$? The radiofrequency field of the spectrometer is at 400 MHz and the applied field is 9.4 T. [ISM 15.26]

8.27 Suppose a uniform disk-shaped organ is in a linear field gradient, and that the MRI signal is proportional to the number of protons in a slice of width δx at each horizontal distance x from the centre of the disk. Sketch the shape of the absorption intensity for the MRI image of the disk before any computer manipulation has been carried out. [SSM 15.27]

9 Statistical thermodynamics 1: the concepts

Statistical thermodynamics provides the link between the microscopic properties of matter and its bulk properties. Two key ideas are introduced in this chapter. The first is the Boltzmann distribution, which is used to predict the populations of states in systems at thermal equilibrium. In this chapter we see its derivation in terms of the distribution of particles over available states. The derivation leads naturally to the introduction of the partition function, which is the central mathematical concept of this and the next chapter. We see how to interpret the partition function and how to calculate it in a number of simple cases. We then see how to extract thermodynamic information from the partition function. In the final part of the chapter, we generalize the discussion to include systems that are composed of assemblies of interacting particles. Very similar equations are developed to those in the first part of the chapter, but they are much more widely applicable.

The preceding chapters of this part of the text have shown how the energy levels of molecules can be calculated, determined spectroscopically, and related to their structures. The next major step is to see how a knowledge of these energy levels can be used to account for the properties of matter in bulk. To do so, we now introduce the concepts of **statistical thermodynamics**, the link between individual molecular properties and bulk thermodynamic properties.

The crucial step in going from the quantum mechanics of individual molecules to the thermodynamics of bulk samples is to recognize that the latter deals with the *average* behaviour of large numbers of molecules. For example, the pressure of a gas depends on the average force exerted by its molecules, and there is no need to specify which molecules happen to be striking the wall at any instant. Nor is it necessary to consider the fluctuations in the pressure as different numbers of molecules collide with the wall at different moments. The fluctuations in pressure are very small compared with the steady pressure: it is highly improbable that there will be a sudden lull in the number of collisions, or a sudden surge. Fluctuations in other thermodynamic properties also occur, but for large numbers of particles they are negligible compared to the mean values.

This chapter introduces statistical thermodynamics in two stages. The first, the derivation of the Boltzmann distribution for individual particles, is of restricted applicability, but it has the advantage of taking us directly to a result of central importance in a straightforward and elementary way. We can *use* statistical thermodynamics once we have deduced the Boltzmann distribution. Then (in Section 9.5) we extend the arguments to systems composed of interacting particles.

The distribution of molecular states

We consider a closed system composed of N molecules. Although the total energy is constant at E, it is not possible to be definite about how that energy is shared between the molecules. Collisions result in the ceaseless redistribution of energy not only between the molecules but also among their different modes of motion. The closest we can come to a description of the distribution of energy is to report the **population** of a state, the average number of molecules that occupy it, and to say that on average there are n_i molecules in a state of energy ε_i. The populations of the states remain almost constant, but the precise identities of the molecules in each state may change at every collision.

The problem we address in this section is the calculation of the populations of states for any type of molecule in any mode of motion at any temperature. The only restriction is that the molecules should be independent, in the sense that the total energy of the system is a sum of their individual energies. We are discounting (at this stage) the possibility that in a real system a contribution to the total energy may arise from interactions between molecules. We also adopt the **principle of equal *a priori* probabilities**, the assumption that all possibilities for the distribution of energy are equally probable. *A priori* means in this context loosely 'as far as one knows'. We have no reason to presume otherwise than that, for a collection of molecules at thermal equilibrium, vibrational states of a certain energy, for instance, are as likely to be populated as rotational states of the same energy.

One very important conclusion that will emerge from the following analysis is that the populations of states depend on a single parameter, the 'temperature'. That is, statistical thermodynamics provides a molecular justification for the concept of temperature and some insight into this crucially important quantity.

9.1 Configurations and weights

Any individual molecule may exist in states with energies ε_0, ε_1, ... We shall always take ε_0, the lowest state, as the zero of energy ($\varepsilon_0 = 0$), and measure all other energies relative to that state. To obtain the actual internal energy, U, we may have to add a constant to the calculated energy of the system. For example, if we are considering the vibrational contribution to the internal energy, then we must add the total zero-point energy of any oscillators in the sample.

(a) Instantaneous configurations

At any instant there will be n_0 molecules in the state with energy ε_0, n_1 with ε_1, and so on. The specification of the set of populations n_0, n_1, . . . in the form $\{n_0, n_1, \ldots\}$ is a statement of the instantaneous **configuration** of the system. The instantaneous configuration fluctuates with time because the populations change. We can picture a large number of different instantaneous configurations. One, for example, might be $\{N,0,0,\ldots\}$, corresponding to every molecule being in its ground state. Another might be $\{N-2,2,0,0,\ldots\}$, in which two molecules are in the first excited state. The latter configuration is intrinsically more likely to be found than the former because it can be achieved in more ways: $\{N,0,0,\ldots\}$ can be achieved in only one way, but $\{N-2,2,0,\ldots\}$ can be achieved in $\frac{1}{2}N(N-1)$ different ways (Fig. 9.1; see *Justification 9.1*). At this stage in the argument, we are ignoring the requirement that the total energy of the system should be constant (the second configuration has a higher energy than the first). The constraint of total energy is imposed later in this section.

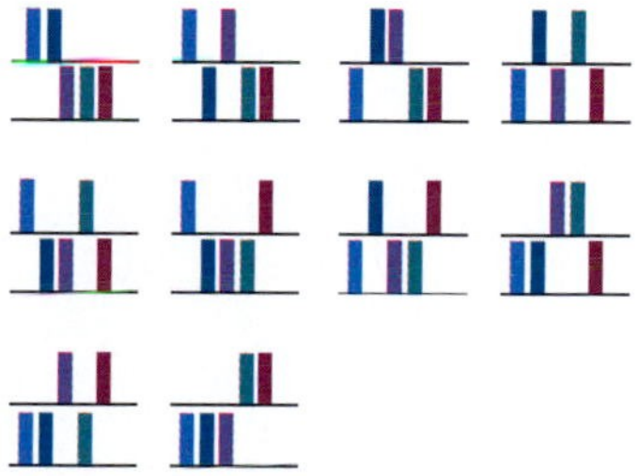

Fig. 9.1 Whereas a configuration $\{5,0,0,\ldots\}$ can be achieved in only one way, a configuration $\{3,2,0,\ldots\}$ can be achieved in the ten different ways shown here, where the tinted blocks represent different molecules.

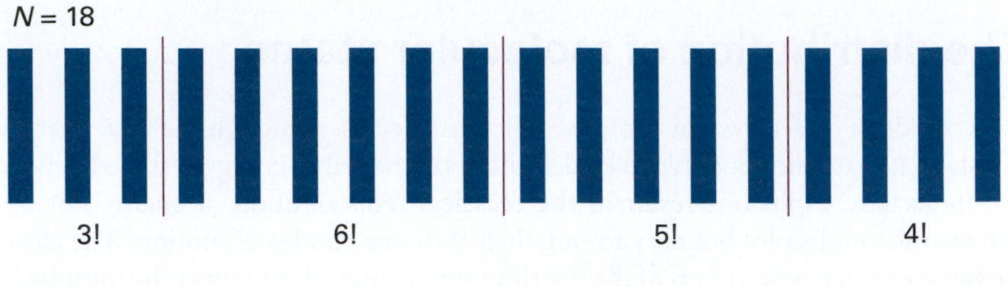

Fig. 9.2 The 18 molecules shown here can be distributed into four receptacles (distinguished by the three vertical lines) in 18! different ways. However, 3! of the selections that put three molecules in the first receptacle are equivalent, 6! that put six molecules into the second receptacle are equivalent, and so on. Hence the number of distinguishable arrangements is 18!/3!6!5!4!.

If, as a result of collisions, the system were to fluctuate between the configurations $\{N,0,0,\ldots\}$ and $\{N-2,2,0,\ldots\}$, it would almost always be found in the second, more likely state (especially if N were large). In other words, a system free to switch between the two configurations would show properties characteristic almost exclusively of the second configuration. A general configuration $\{n_0,n_1,\ldots\}$ can be achieved in W different ways, where W is called the **weight** of the configuration. The weight of the configuration $\{n_0,n_1,\ldots\}$ is given by the expression

$$W = \frac{N!}{n_0!n_1!n_2!\ldots} \tag{9.1}$$

Equation 9.1 is a generalization of the formula $W = \frac{1}{2}N(N-1)$, and reduces to it for the configuration $\{N-2,2,0,\ldots\}$.

Comment 9.1

More formally, W is called the *multinomial coefficient* (see *Appendix* 2). In eqn 9.1, $x!$, x factorial, denotes $x(x-1)(x-2)\ldots 1$, and by definition $0! = 1$.

Justification 9.1 *The weight of a configuration*

First, consider the weight of the configuration $\{N-2,2,0,0,\ldots\}$. One candidate for promotion to an upper state can be selected in N ways. There are $N-1$ candidates for the second choice, so the total number of choices is $N(N-1)$. However, we should not distinguish the choice (Jack, Jill) from the choice (Jill, Jack) because they lead to the same configurations. Therefore, only half the choices lead to distinguishable configurations, and the total number of distinguishable choices is $\frac{1}{2}N(N-1)$.

Now we generalize this remark. Consider the number of ways of distributing N balls into bins. The first ball can be selected in N different ways, the next ball in $N-1$ different ways for the balls remaining, and so on. Therefore, there are $N(N-1)\ldots 1 = N!$ ways of selecting the balls for distribution over the bins. However, if there are n_0 balls in the bin labelled ε_0, there would be $n_0!$ different ways in which the same balls could have been chosen (Fig. 9.2). Similarly, there are $n_1!$ ways in which the n_1 balls in the bin labelled ε_1 can be chosen, and so on. Therefore, the total number of distinguishable ways of distributing the balls so that there are n_0 in bin ε_0, n_1 in bin ε_1, etc. regardless of the order in which the balls were chosen is $N!/n_0!n_1!\ldots$, which is the content of eqn 9.1.

Illustration 9.1 *Calculating the weight of a distribution*

To calculate the number of ways of distributing 20 identical objects with the arrangement 1, 0, 3, 5, 10, 1, we note that the configuration is $\{1,0,3,5,10,1\}$ with $N = 20$; therefore the weight is

$$W = \frac{20!}{1!0!3!5!10!1!} = 9.31 \times 10^8$$

Self-test 9.1 Calculate the weight of the configuration in which 20 objects are distributed in the arrangement 0, 1, 5, 0, 8, 0, 3, 2, 0, 1. $[4.19 \times 10^{10}]$

It will turn out to be more convenient to deal with the natural logarithm of the weight, ln W, rather than with the weight itself. We shall therefore need the expression

$$\ln W = \ln \frac{N!}{n_0!n_1!n_2!\ldots} = \ln N! - \ln(n_0!n_1!n_2!\cdots)$$
$$= \ln N! - (\ln n_0! + \ln n_1! + \ln n_2! + \cdots)$$
$$= \ln N! - \sum_i \ln n_i!$$

where in the first line we have used $\ln(x/y) = \ln x - \ln y$ and in the second $\ln xy = \ln x + \ln y$. One reason for introducing ln W is that it is easier to make approximations. In particular, we can simplify the factorials by using **Stirling's approximation** in the form

$$\ln x! \approx x \ln x - x \tag{9.2}$$

Then the approximate expression for the weight is

$$\ln W = (N \ln N - N) - \sum_i (n_i \ln n_i - n_i) = N \ln N - \sum_i n_i \ln n_i \tag{9.3}$$

The final form of eqn 9.3 is derived by noting that the sum of n_i is equal to N, so the second and fourth terms in the second expression cancel.

Comment 9.2

A more accurate form of Stirling's approximation is

$$x! \approx (2\pi)^{1/2} x^{x+\frac{1}{2}} e^{-x}$$

and is in error by less than 1 per cent when x is greater than about 10. We deal with far larger values of x, and the simplified version in eqn 9.2 is adequate.

(b) The Boltzmann distribution

We have seen that the configuration $\{N-2,2,0,\ldots\}$ dominates $\{N,0,0,\ldots\}$, and it should be easy to believe that there may be other configurations that have a much greater weight than both. We shall see, in fact, that there is a configuration with so great a weight that it overwhelms all the rest in importance to such an extent that the system will almost always be found in it. The properties of the system will therefore be characteristic of that particular dominating configuration. This dominating configuration can be found by looking for the values of n_i that lead to a maximum value of W. Because W is a function of all the n_i, we can do this search by varying the n_i and looking for the values that correspond to $dW = 0$ (just as in the search for the maximum of any function), or equivalently a maximum value of ln W. However, there are two difficulties with this procedure.

The first difficulty is that the only permitted configurations are those corresponding to the specified, constant, total energy of the system. This requirement rules out many configurations; $\{N,0,0,\ldots\}$ and $\{N-2,2,0,\ldots\}$, for instance, have different energies, so both cannot occur in the same isolated system. It follows that, in looking for the configuration with the greatest weight, we must ensure that the configuration also satisfies the condition

$$\text{Constant total energy:} \quad \sum_i n_i \varepsilon_i = E \tag{9.4}$$

where E is the total energy of the system.

The second constraint is that, because the total number of molecules present is also fixed (at N), we cannot arbitrarily vary all the populations simultaneously. Thus, increasing the population of one state by 1 demands that the population of another state must be reduced by 1. Therefore, the search for the maximum value of W is also subject to the condition

$$\text{Constant total number of molecules:} \quad \sum_i n_i = N \tag{9.5}$$

We show in *Further information 9.1* that the populations in the configuration of greatest weight, subject to the two constraints in eqns 9.4 and 9.5, depend on the energy of the state according to the **Boltzmann distribution**:

$$\frac{n_i}{N} = \frac{e^{-\beta\varepsilon_i}}{\sum_i e^{-\beta\varepsilon_i}} \tag{9.6a}$$

where $\varepsilon_0 \le \varepsilon_1 \le \varepsilon_2 \dots$. Equation 9.6a is the justification of the remark that a single parameter, here denoted β, determines the most probable populations of the states of the system. We shall see in Section 9.3b that

$$\beta = \frac{1}{kT} \tag{9.6b}$$

where T is the thermodynamic temperature and k is Boltzmann's constant. In other words, *the thermodynamic temperature is the unique parameter that governs the most probable populations of states of a system at thermal equilibrium.* In *Further information 9.3*, moreover, we see that β is a more natural measure of temperature than T itself.

9.2 The molecular partition function

From now on we write the Boltzmann distribution as

$$p_i = \frac{e^{-\beta\varepsilon_i}}{q} \tag{9.7}$$

where p_i is the fraction of molecules in the state i, $p_i = n_i/N$, and q is the **molecular partition function:**

$$q = \sum_i e^{-\beta\varepsilon_i} \qquad [9.8]$$

The sum in q is sometimes expressed slightly differently. It may happen that several states have the same energy, and so give the same contribution to the sum. If, for example, g_i states have the same energy ε_i (so the level is g_i-fold degenerate), we could write

$$q = \sum_{\text{levels } i} g_i e^{-\beta\varepsilon_i} \tag{9.9}$$

where the sum is now over energy levels (sets of states with the same energy), not individual states.

Example 9.1 *Writing a partition function*

Write an expression for the partition function of a linear molecule (such as HCl) treated as a rigid rotor.

Method To use eqn 9.9 we need to know (a) the energies of the levels, (b) the degeneracies, the number of states that belong to each level. Whenever calculating a partition function, the energies of the levels are expressed relative to 0 for the state of lowest energy. The energy levels of a rigid linear rotor were derived in Section 6.5c.

Answer From eqn 6.31, the energy levels of a linear rotor are $hcBJ(J+1)$, with $J = 0, 1, 2, \dots$. The state of lowest energy has zero energy, so no adjustment need be made to the energies given by this expression. Each level consists of $2J + 1$ degenerate states. Therefore,

$$q = \sum_{J=0}^{\infty} \overbrace{(2J+1)}^{g_J} e^{\overbrace{-\beta hcBJ(J+1)}^{\varepsilon_J}}$$

The sum can be evaluated numerically by supplying the value of B (from spectroscopy or calculation) and the temperature. For reasons explained in Section 10.2b, this expression applies only to unsymmetrical linear rotors (for instance, HCl, not CO_2).

Self-test 9.2 Write the partition function for a two-level system, the lower state (at energy 0) being nondegenerate, and the upper state (at an energy ε) doubly degenerate. $[q = 1 + 2e^{-\beta\varepsilon}]$

(a) An interpretation of the partition function

Some insight into the significance of a partition function can be obtained by considering how q depends on the temperature. When T is close to zero, the parameter $\beta = 1/kT$ is close to infinity. Then every term except one in the sum defining q is zero because each one has the form e^{-x} with $x \to \infty$. The exception is the term with $\varepsilon_0 \equiv 0$ (or the g_0 terms at zero energy if the ground state is g_0-fold degenerate), because then $\varepsilon_0/kT \equiv 0$ whatever the temperature, including zero. As there is only one surviving term when $T = 0$, and its value is g_0, it follows that

$$\lim_{T\to 0} q = g_0 \tag{9.10}$$

That is, at $T = 0$, the partition function is equal to the degeneracy of the ground state.

Now consider the case when T is so high that for each term in the sum $\varepsilon_j/kT \approx 0$. Because $e^{-x} = 1$ when $x = 0$, each term in the sum now contributes 1. It follows that the sum is equal to the number of molecular states, which in general is infinite:

$$\lim_{T\to\infty} q = \infty \tag{9.11}$$

In some idealized cases, the molecule may have only a finite number of states; then the upper limit of q is equal to the number of states. For example, if we were considering only the spin energy levels of a radical in a magnetic field, then there would be only two states ($m_s = \pm\tfrac{1}{2}$). The partition function for such a system can therefore be expected to rise towards 2 as T is increased towards infinity.

We see that *the molecular partition function gives an indication of the number of states that are thermally accessible to a molecule at the temperature of the system.* At $T = 0$, only the ground level is accessible and $q = g_0$. At very high temperatures, virtually all states are accessible, and q is correspondingly large.

Example 9.2 *Evaluating the partition function for a uniform ladder of energy levels*

Evaluate the partition function for a molecule with an infinite number of equally spaced nondegenerate energy levels (Fig. 9.3). These levels can be thought of as the vibrational energy levels of a diatomic molecule in the harmonic approximation.

Method We expect the partition function to increase from 1 at $T = 0$ and approach infinity as T to ∞. To evaluate eqn 9.8 explicitly, note that

$$1 + x + x^2 + \cdots = \frac{1}{1-x}$$

Answer If the separation of neighbouring levels is ε, the partition function is

$$q = 1 + e^{-\beta\varepsilon} + e^{-2\beta\varepsilon} + \cdots = 1 + e^{-\beta\varepsilon} + (e^{-\beta\varepsilon})^2 + \cdots = \frac{1}{1 - e^{-\beta\varepsilon}}$$

This expression is plotted in Fig. 9.4: notice that, as anticipated, q rises from 1 to infinity as the temperature is raised.

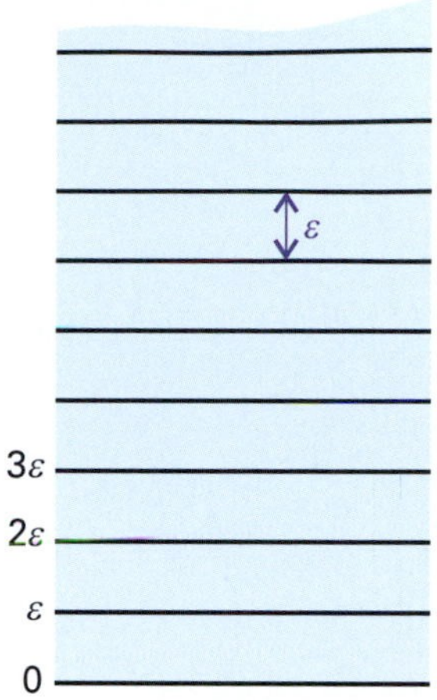

Fig. 9.3 The equally spaced infinite array of energy levels used in the calculation of the partition function. A harmonic oscillator has the same spectrum of levels.

Comment 9.3

The sum of the infinite series $S = 1 + x + x^2 + \cdots$ is obtained by multiplying both sides by x, which gives $xS = x + x^2 + x^3 + \cdots = S - 1$ and hence $S = 1/(1 - x)$.

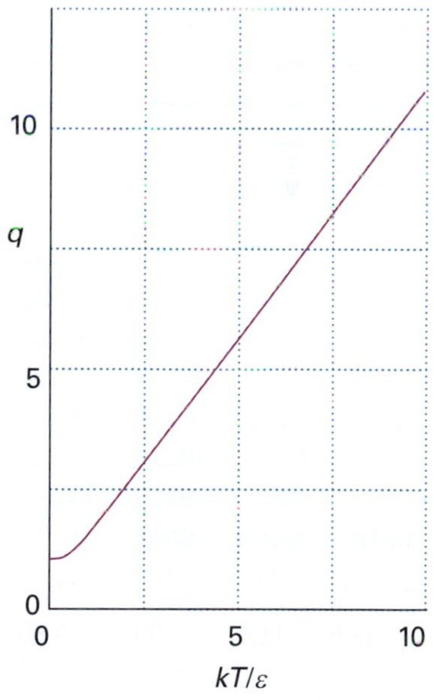

Fig. 9.4 The partition function for the system shown in Fig. 9.3 (a harmonic oscillator) as a function of temperature.

Exploration Plot the partition function of a harmonic oscillator against temperature for several values of the energy separation ε. How does q vary with temperature when T is high, in the sense that $kT \gg \varepsilon$ (or $\beta\varepsilon \ll 1$)?

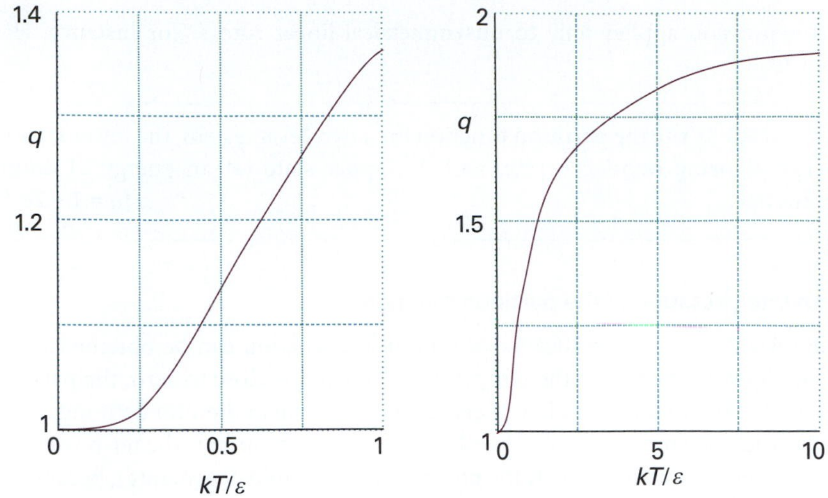

Fig. 9.5 The partition function for a two-level system as a function of temperature. The two graphs differ in the scale of the temperature axis to show the approach to 1 as $T \to 0$ and the slow approach to 2 as $T \to \infty$.

Exploration Consider a three-level system with levels 0, ε, and 2ε. Plot the partition function against kT/ε.

> ***Self-test 9.3*** Find and plot an expression for the partition function of a system with one state at zero energy and another state at the energy ε.
>
> [$q = 1 + e^{-\beta\varepsilon}$, Fig. 9.5]

It follows from eqn 9.8 and the expression for q derived in Example 9.2 for a uniform ladder of states of spacing ε,

$$q = \frac{1}{1 - e^{-\beta\varepsilon}} \tag{9.12}$$

that the fraction of molecules in the state with energy ε_i is

$$p_i = \frac{e^{-\beta\varepsilon_i}}{q} = (1 - e^{-\beta\varepsilon})e^{-\beta\varepsilon_i} \tag{9.13}$$

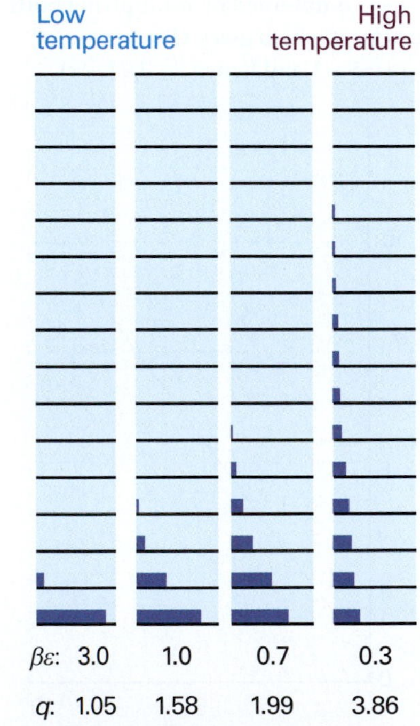

Fig. 9.6 The populations of the energy levels of the system shown in Fig.9.3 at different temperatures, and the corresponding values of the partition function calculated in Example 9.2. Note that $\beta = 1/kT$.

Exploration To visualize the content of Fig. 9.6 in a different way, plot the functions p_0, p_1, p_2, and p_3 against kT/ε.

Figure 9.6 shows how p_i varies with temperature. At very low temperatures, where q is close to 1, only the lowest state is significantly populated. As the temperature is raised, the population breaks out of the lowest state, and the upper states become progressively more highly populated. At the same time, the partition function rises from 1 and its value gives an indication of the range of states populated. The name 'partition function' reflects the sense in which q measures how the total number of molecules is distributed—partitioned—over the available states.

The corresponding expressions for a two-level system derived in Self-test 9.3 are

$$p_0 = \frac{1}{1 + e^{-\beta\varepsilon}} \qquad p_1 = \frac{e^{-\beta\varepsilon}}{1 + e^{-\beta\varepsilon}} \tag{9.14}$$

These functions are plotted in Fig. 9.7. Notice how the populations tend towards equality ($p_0 = \frac{1}{2}$, $p_1 = \frac{1}{2}$) as $T \to \infty$. A common error is to suppose that all the molecules in the system will be found in the upper energy state when $T = \infty$; however, we see

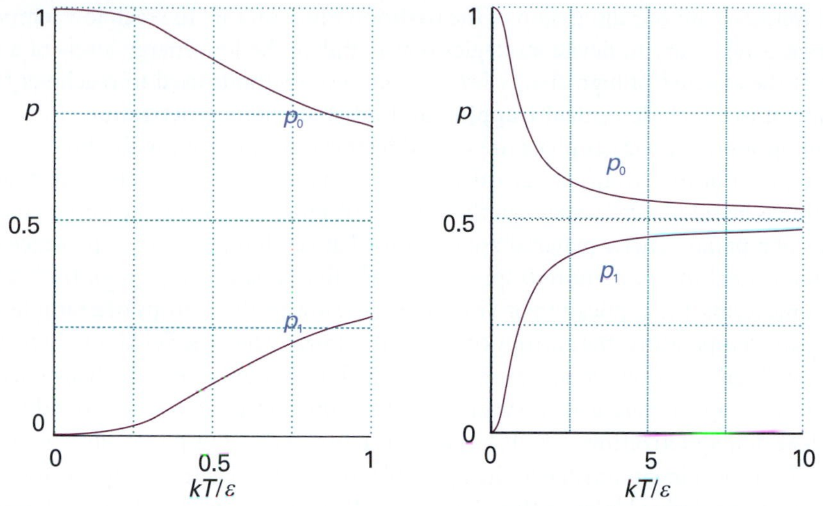

Fig. 9.7 The fraction of populations of the two states of a two-level system as a function of temperature (eqn 9.14). Note that, as the temperature approaches infinity, the populations of the two states become equal (and the fractions both approach 0.5).

Exploration Consider a three-level system with levels 0, ε, and 2ε. Plot the functions p_0, p_1, and p_2 against kT/ε.

from eqn 9.14 that, as $T \to \infty$, the populations of states become equal. The same conclusion is true of multi-level systems too: as $T \to \infty$, all states become equally populated.

Example 9.3 *Using the partition function to calculate a population*

Calculate the proportion of I_2 molecules in their ground, first excited, and second excited vibrational states at 25°C. The vibrational wavenumber is 214.6 cm^{-1}.

Method Vibrational energy levels have a constant separation (in the harmonic approximation, Section 6.9), so the partition function is given by eqn 9.12 and the populations by eqn 9.13. To use the latter equation, we identify the index i with the quantum number v, and calculate p_v for $v = 0$, 1, and 2. At 298.15 K, $kT/hc = 207.226$ cm^{-1}.

Answer First, we note that

$$\beta\varepsilon = \frac{hc\tilde{\nu}}{kT} = \frac{214.6\ \text{cm}^{-1}}{207.226\ \text{cm}^{-1}} = 1.036$$

Then it follows from eqn 9.13 that the populations are

$$p_v = (1 - e^{-\beta\varepsilon})e^{-v\beta\varepsilon} = 0.645e^{-1.036v}$$

Therefore, $p_0 = 0.645$, $p_1 = 0.229$, $p_2 = 0.081$. The I—I bond is not stiff and the atoms are heavy: as a result, the vibrational energy separations are small and at room temperature several vibrational levels are significantly populated. The value of the partition function, $q = 1.55$, reflects this small but significant spread of populations.

Self-test 9.4 At what temperature would the $v = 1$ level of I_2 have (a) half the population of the ground state, (b) the same population as the ground state?

[(a) 445 K, (b) infinite]

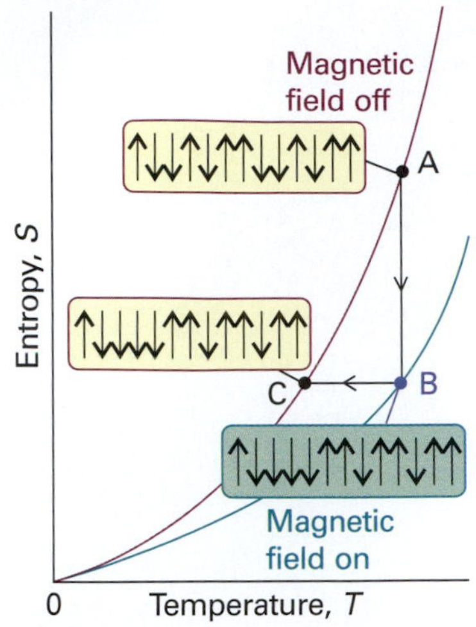

Fig. 9.8 The technique of adiabatic demagnetization is used to attain very low temperatures. The upper curve shows that variation of the entropy of a paramagnetic system in the absence of an applied field. The lower curve shows that variation in entropy when a field is applied and has made the electron magnets more orderly. The isothermal magnetization step is from A to B; the adiabatic demagnetization step (at constant entropy) is from B to C.

It follows from our discussion of the partition function that to reach low temperatures it is necessary to devise strategies that populate the low energy levels of a system at the expense of high energy levels. Common methods used to reach very low temperatures include **optical trapping** and **adiabatic demagnetization**. In optical trapping, atoms in the gas phase are cooled by inelastic collisions with photons from intense laser beams, which act as walls of a very small container. Adiabatic demagnetization is based on the fact that, in the absence of a magnetic field, the unpaired electrons of a paramagnetic material are orientated at random, but in the presence of a magnetic field there are more β spins ($m_s = -\frac{1}{2}$) than α spins ($m_s = +\frac{1}{2}$). In thermodynamic terms, the application of a magnetic field lowers the entropy of a sample and, at a given temperature, the entropy of a sample is lower when the field is on than when it is off. Even lower temperatures can be reached if nuclear spins (which also behave like small magnets) are used instead of electron spins in the technique of **adiabatic nuclear demagnetization**, which has been used to cool a sample of silver to about 280 pK. In certain circumstances it is possible to achieve negative temperatures, and the equations derived later in this chapter can be extended to $T < 0$ with interesting consequences (see *Further information* 9.3).

Illustration 9.2 *Cooling a sample by adiabatic demagnetization*

Consider the situation summarized by Fig. 9.8. A sample of paramagnetic material, such as a *d*- or *f*-metal complex with several unpaired electrons, is cooled to about 1 K by using helium. The sample is then exposed to a strong magnetic field while it is surrounded by helium, which provides thermal contact with the cold reservoir. This magnetization step is isothermal, and energy leaves the system as heat while the electron spins adopt the lower energy state (AB in the illustration). Thermal contact between the sample and the surroundings is now broken by pumping away the helium and the magnetic field is reduced to zero. This step is adiabatic and effectively reversible, so the state of the sample changes from B to C. At the end of this step the sample is the same as it was at A except that it now has a lower entropy. That lower entropy in the absence of a magnetic field corresponds to a lower temperature. That is, adiabatic demagnetization has cooled the sample.

(b) Approximations and factorizations

In general, exact analytical expressions for partition functions cannot be obtained. However, closed approximate expressions can often be found and prove to be very important in a number of chemical and biochemical applications (*Impact* 9.1). For instance, the expression for the partition function for a particle of mass m free to move in a one-dimensional container of length X can be evaluated by making use of the fact that the separation of energy levels is very small and that large numbers of states are accessible at normal temperatures. As shown in the *Justification* below, in this case

$$q_X = \left(\frac{2\pi m}{h^2\beta}\right)^{1/2} X \qquad (9.15)$$

This expression shows that the partition function for translational motion increases with the length of the box and the mass of the particle, for in each case the separation of the energy levels becomes smaller and more levels become thermally accessible. For a given mass and length of the box, the partition function also increases with increasing temperature (decreasing β), because more states become accessible.

Justification 9.2 *The partition function for a particle in a one-dimensional box*

The energy levels of a molecule of mass m in a container of length X are given by eqn 2.4a with $L = X$:

$$E_n = \frac{n^2h^2}{8mX^2} \qquad n = 1, 2, \ldots$$

The lowest level ($n = 1$) has energy $h^2/8mX^2$, so the energies relative to that level are

$$\varepsilon_n = (n^2 - 1)\varepsilon \qquad \varepsilon = h^2/8mX^2$$

The sum to evaluate is therefore

$$q_X = \sum_{n=1}^{\infty} e^{-(n^2-1)\beta\varepsilon}$$

The translational energy levels are very close together in a container the size of a typical laboratory vessel; therefore, the sum can be approximated by an integral:

$$q_X = \int_1^{\infty} e^{-(n^2-1)\beta\varepsilon}dn \approx \int_0^{\infty} e^{-n^2\beta\varepsilon}dn$$

The extension of the lower limit to $n = 0$ and the replacement of $n^2 - 1$ by n^2 introduces negligible error but turns the integral into standard form. We make the substitution $x^2 = n^2\beta\varepsilon$, implying $dn = dx/(\beta\varepsilon)^{1/2}$, and therefore that

$$q_X = \left(\frac{1}{\beta\varepsilon}\right)^{1/2} \overbrace{\int_0^{\infty} e^{-x^2}dx}^{\pi^{1/2}/2} = \left(\frac{1}{\beta\varepsilon}\right)^{1/2}\left(\frac{\pi^{1/2}}{2}\right) = \left(\frac{2\pi m}{h^2\beta}\right)^{1/2} X$$

Another useful feature of partition functions is used to derive expressions when the energy of a molecule arises from several different, independent sources: if the energy is a sum of contributions from independent modes of motion, then the partition function is a product of partition functions for each mode of motion. For instance, suppose the molecule we are considering is free to move in three dimensions. We take the length of the container in the y-direction to be Y and that in the z-direction to be Z. The total energy of a molecule ε is the sum of its translational energies in all three directions:

$$\varepsilon_{n_1n_2n_3} = \varepsilon_{n_1}^{(X)} + \varepsilon_{n_2}^{(Y)} + \varepsilon_{n_3}^{(Z)} \tag{9.16}$$

where n_1, n_2, and n_3 are the quantum numbers for motion in the x-, y-, and z-directions, respectively. Therefore, because $e^{a+b+c} = e^a e^b e^c$, the partition function factorizes as follows:

$$\begin{aligned} q &= \sum_{\text{all } n} e^{-\beta\varepsilon_{n_1}^{(X)}-\beta\varepsilon_{n_2}^{(Y)}-\beta\varepsilon_{n_3}^{(Z)}} = \sum_{\text{all } n} e^{-\beta\varepsilon_{n_1}^{(X)}} e^{-\beta\varepsilon_{n_2}^{(Y)}} e^{-\beta\varepsilon_{n_3}^{(Z)}} \\ &= \left(\sum_{n_1} e^{-\beta\varepsilon_{n_1}^{(X)}}\right)\left(\sum_{n_2} e^{-\beta\varepsilon_{n_2}^{(Y)}}\right)\left(\sum_{n_3} e^{-\beta\varepsilon_{n_3}^{(Z)}}\right) \\ &= q_X q_Y q_Z \end{aligned} \tag{9.17}$$

It is generally true that, if the energy of a molecule can be written as the sum of independent terms, then the partition function is the corresponding product of individual contributions.

Equation 9.15 gives the partition function for translational motion in the x-direction. The only change for the other two directions is to replace the length X by the lengths Y or Z. Hence the partition function for motion in three dimensions is

$$q = \left(\frac{2\pi m}{h^2\beta}\right)^{3/2} XYZ \qquad (9.18)$$

The product of lengths XYZ is the volume, V, of the container, so we can write

$$q = \frac{V}{\Lambda^3} \qquad \Lambda = h\left(\frac{\beta}{2\pi m}\right)^{1/2} = \frac{h}{(2\pi mkT)^{1/2}} \qquad (9.19)$$

The quantity Λ has the dimensions of length and is called the **thermal wavelength** (sometimes the *thermal de Broglie wavelength*) of the molecule. The thermal wavelength decreases with increasing mass and temperature. As in the one-dimensional case, the partition function increases with the mass of the particle (as $m^{3/2}$) and the volume of the container (as V); for a given mass and volume, the partition function increases with temperature (as $T^{3/2}$).

Illustration 9.3 *Calculating the translational partition function*

To calculate the translational partition function of an H_2 molecule confined to a 100 cm^3 vessel at 25°C we use $m = 2.016$ u; then

$$\Lambda = \frac{6.626\times10^{-34}\,\text{J s}}{\{2\pi\times(2.016\times1.6605\times10^{-27}\,\text{kg})\times(1.38\times10^{-23}\,\text{J K}^{-1})\times(298\,\text{K})\}^{1/2}}$$

$$= 7.12\times10^{-11}\,\text{m}$$

where we have used 1 J = 1 kg m^2 s^{-2}. Therefore,

$$q = \frac{1.00\times10^{-4}\,\text{m}^3}{(7.12\times10^{-11}\,\text{m})^3} = 2.77\times10^{26}$$

About 10^{26} quantum states are thermally accessible, even at room temperature and for this light molecule. Many states are occupied if the thermal wavelength (which in this case is 71.2 pm) is small compared with the linear dimensions of the container.

Self-test 9.5 Calculate the translational partition function for a D_2 molecule under the same conditions. [$q = 7.8\times10^{26}$, $2^{3/2}$ times larger]

The validity of the approximations that led to eqn 9.19 can be expressed in terms of the average separation of the particles in the container, d. We do not have to worry about the role of the Pauli principle on the occupation of states if there are many states available for each molecule. Because q is the total number of accessible states, the average number of states per molecule is q/N. For this quantity to be large, we require $V/N\Lambda^3 \gg 1$. However, V/N is the volume occupied by a single particle, and therefore the average separation of the particles is $d = (V/N)^{1/3}$. The condition for there being many states available per molecule is therefore $d^3/\Lambda^3 \gg 1$, and therefore $d \gg \Lambda$. That is, for eqn 9.19 to be valid, *the average separation of the particles must be much greater than their thermal wavelength.* For H_2 molecules at 1 bar and 298 K, the average separation is 3 nm, which is significantly larger than their thermal wavelength (71.2 pm, *Illustration* 9.3).

IMPACT ON BIOCHEMISTRY
I9.1 The helix–coil transition in polypeptides

Proteins are polymers that attain well defined three-dimensional structures both in solution and in biological cells. They are *polypeptides* formed from different amino acids strung together by the *peptide link*, —CONH—. Hydrogen bonds between amino acids of a polypeptide give rise to stable helical or sheet structures, which may collapse into a random coil when certain conditions are changed. The unwinding of a helix into a random coil is a *cooperative transition*, in which the polymer becomes increasingly more susceptible to structural changes once the process has begun. We examine here a model grounded in the principles of statistical thermodynamics that accounts for the cooperativity of the helix–coil transition in polypeptides.

To calculate the fraction of polypeptide molecules present as helix or coil we need to set up the partition function for the various states of the molecule. To illustrate the approach, consider a short polypeptide with four amino acid residues, each labelled h if it contributes to a helical region and c if it contributes to a random coil region. We suppose that conformations *hhhh* and *cccc* contribute terms q_0 and q_4, respectively, to the partition function q. Then we assume that each of the four conformations with one c amino acid (such as *hchh*) contributes q_1. Similarly, each of the six states with two c amino acids contributes a term q_2, and each of the four states with three c amino acids contributes a term q_3. The partition function is then

$$q = q_0 + 4q_1 + 6q_2 + 4q_3 + q_4 = q_0\left(1 + \frac{4q_1}{q_0} + \frac{6q_2}{q_0} + \frac{4q_3}{q_0} + \frac{q_4}{q_0}\right)$$

We shall now suppose that each partition function differs from q_0 only by the energy of each conformation relative to *hhhh*, and write

$$\frac{q_i}{q_0} = e^{-(\varepsilon_i - \varepsilon_0)/kT}$$

Next, we suppose that the conformational transformations are non-cooperative, in the sense that the energy associated with changing one h amino acid into one c amino acid has the same value regardless of how many h or c amino acid residues are in the reactant or product state and regardless of where in the chain the conversion occurs. That is, we suppose that the difference in energy between c^ih^{4-i} and $c^{i+1}h^{3-i}$ has the same value γ for all i. This assumption implies that $\varepsilon_i - \varepsilon_0 = i\gamma$ and therefore that

$$q = q_0(1 + 4s + 6s^2 + 4s^3 + s^4) \qquad s = e^{-\Gamma/RT} \tag{9.20}$$

where $\Gamma = N_A\gamma$ and s is called the *stability parameter*. The term in parentheses has the form of the binomial expansion of $(1 + s)^4$.

$$\frac{q}{q_0} = \sum_{i=0}^{4} C(4,i)s^i \quad \text{with} \quad C(4,i) = \frac{4!}{(4-i)!i!} \tag{9.21}$$

which we interpret as the number of ways in which a state with i c amino acids can be formed.

The extension of this treatment to take into account a longer chain of residues is now straightforward: we simply replace the upper limit of 4 in the sum by n:

$$\frac{q}{q_0} = \sum_{i=0}^{n} C(n,i)s^i \tag{9.22}$$

Comment 9.4

The binomial expansion of $(1 + x)^n$ is

$$(1 + x)^n = \sum_{i=0}^{n} C(n,i)x^i,$$

$$\text{with } C(n,i) = \frac{n!}{(n-i)!i!}$$

A cooperative transformation is more difficult to accommodate, and depends on building a model of how neighbours facilitate each other's conformational change. In the simple *zipper model*, conversion from h to c is allowed only if a residue adjacent to the one undergoing the conversion is already a c residue. Thus, the zipper model allows a transition of the type . . .*hhhch*. . . → . . .*hhhcc*. . . , but not a transition of the type . . .*hhhch*. . . → . . .*hchch*. . . . The only exception to this rule is, of course, the very first conversion from h to c in a fully helical chain. Cooperativity is included in the zipper model by assuming that the first conversion from h to c, called the *nucleation step*, is less favourable than the remaining conversions and replacing s for that step by σs, where $\sigma \ll 1$. Each subsequent step is called a *propagation step* and has a stability parameter s. In Problem 9.24, you are invited to show that the partition function is:

$$q = 1 + \sum_{i=1}^{n} Z(n,i)\sigma s^i \tag{9.23}$$

where $Z(n,i)$ is the number of ways in which a state with a number i of c amino acids can be formed under the strictures of the zipper model. Because $Z(n,i) = n - i + 1$ (see Problem 9.24),

$$q = 1 + \sigma(n+1)\sum_{i=1}^{n} s^i - \sigma\sum_{i=1}^{n} is^i \tag{9.24}$$

After evaluating both geometric series by using the two relations

$$\sum_{i=1}^{n} x^i = \frac{x^{n+1} - x}{x - 1} \qquad \sum_{i=1}^{n} ix^i = \frac{x}{(x-1)^2}[nx^{n+1} - (n+1)x^n + 1]$$

we find

$$q = 1 + \frac{\sigma s[s^{n+1} - (n+1)s^n + 1]}{(s-1)^2}$$

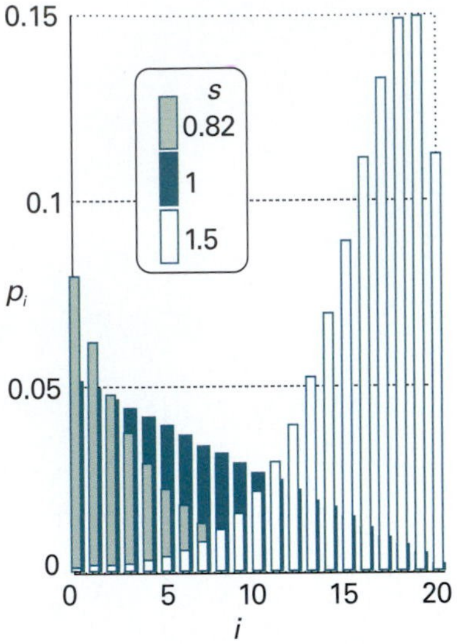

Fig. 9.9 The distribution of p_i, the fraction of molecules that has a number i of c amino acids for $s = 0.8$ ($\langle i \rangle = 1.1$), 1.0 ($\langle i \rangle = 3.8$), and 1.5 ($\langle i \rangle = 15.9$), with $\sigma = 5.0 \times 10^{-3}$.

The fraction $p_i = q_i/q$ of molecules that has a number i of c amino acids is $p_i = [(n-i+1)\sigma s^i]/q$ and the mean value of i is then $\langle i \rangle = \sum_i ip_i$. Figure 9.9 shows the distribution of p_i for various values of s with $\sigma = 5.0 \times 10^{-3}$. We see that most of the polypeptide chains remain largely helical when $s < 1$ and that most of the chains exist largely as random coils when $s > 1$. When $s = 1$, there is a more widespread distribution of length of random coil segments. Because the *degree of conversion*, θ, of a polypeptide with n amino acids to a random coil is defined as $\theta = \langle i \rangle / n$, it is possible to show (see Problem 9.24) that

$$\theta = \frac{1}{n}\frac{\mathrm{d}}{\mathrm{d}(\ln s)}\ln q \tag{9.25}$$

This is a general result that applies to any model of the helix–coil transition in which the partition function q is expressed as a function of the stability parameter s.

A more sophisticated model for the helix–coil transition must allow for helical segments to form in different regions of a long polypeptide chain, with the nascent helices being separated by shrinking coil segments. Calculations based on this more complete *Zimm–Bragg model* give

$$\theta = \tfrac{1}{2}\left(1 + \frac{(s-1) + 2\sigma}{[(s-1)^2 + 4s\sigma]^{1/2}}\right) \tag{9.26}$$

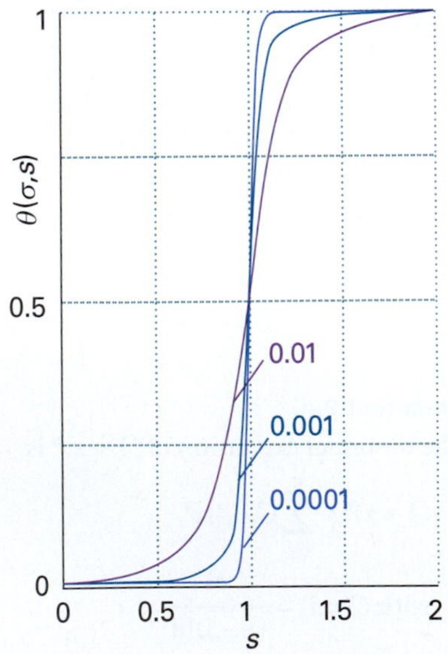

Fig. 9.10 Plots of the degree of conversion θ, against s for several values of σ. The curves show the sigmoidal shape characteristics of cooperative behaviour.

Figure 9.10 shows plots of θ against s for several values of σ. The curves show the sigmoidal shape characteristic of cooperative behaviour. There is a sudden surge of transition to a random coil as s passes through 1 and, the smaller the parameter σ, the greater the sharpness and hence the greater the cooperativity of the transition. That is, the harder it is to get coil formation started, the sharper the transition from helix to coil.

The internal energy and the entropy

The importance of the molecular partition function is that it contains all the information needed to calculate the thermodynamic properties of a system of independent particles. In this respect, q plays a role in statistical thermodynamics very similar to that played by the wavefunction in quantum mechanics: q is a kind of thermal wavefunction.

9.3 The internal energy

We shall begin to unfold the importance of q by showing how to derive an expression for the internal energy of the system.

(a) The relation between *U* and *q*

The total energy of the system relative to the energy of the lowest state is

$$E = \sum_i n_i \varepsilon_i \tag{9.27}$$

Because the most probable configuration is so strongly dominating, we can use the Boltzmann distribution for the populations and write

$$E = \frac{N}{q} \sum_i \varepsilon_i e^{-\beta \varepsilon_i} \tag{9.28}$$

To manipulate this expression into a form involving only q we note that

$$\varepsilon_i e^{-\beta\varepsilon_i} = -\frac{d}{d\beta} e^{-\beta\varepsilon_i}$$

It follows that

$$E = -\frac{N}{q} \sum_i \frac{d}{d\beta} e^{-\beta\varepsilon_i} = -\frac{N}{q}\frac{d}{d\beta} \sum_i e^{-\beta\varepsilon_i} = -\frac{N}{q}\frac{dq}{d\beta} \tag{9.29}$$

Illustration 9.4 *The energy of a two-level system*

From the two-level partition function $q = 1 + e^{-\beta\varepsilon}$, we can deduce that the total energy of N two-level systems is

$$E = -\left(\frac{N}{1+e^{-\beta\varepsilon}}\right)\frac{d}{d\beta}(1+e^{-\beta\varepsilon}) = \frac{N\varepsilon e^{-\beta\varepsilon}}{1+e^{-\beta\varepsilon}} = \frac{N\varepsilon}{1+e^{\beta\varepsilon}}$$

This function is plotted in Fig. 9.11. Notice how the energy is zero at $T=0$, when only the lower state (at the zero of energy) is occupied, and rises to $\frac{1}{2}N\varepsilon$ as $T \to \infty$, when the two levels become equally populated.

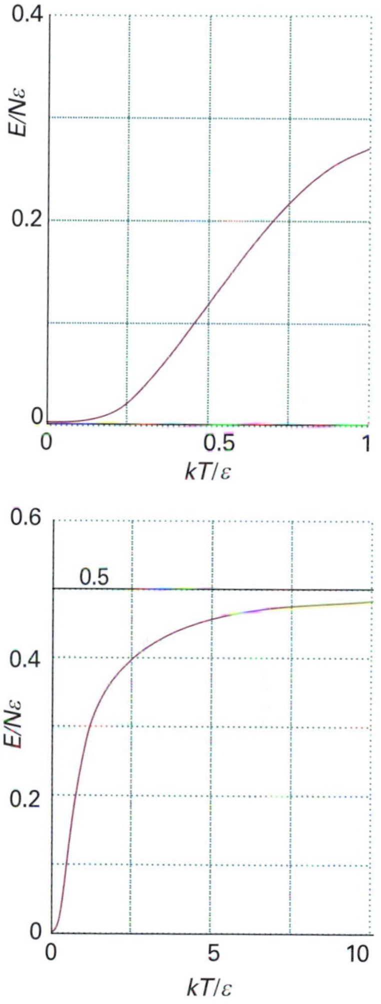

Fig. 9.11 The total energy of a two-level system (expressed as a multiple of $N\varepsilon$) as a function of temperature, on two temperature scales. The graph at the top shows the slow rise away from zero energy at low temperatures; the slope of the graph at $T=0$ is 0 (that is, the heat capacity is zero at $T=0$). The graph below shows the slow rise to 0.5 as $T \to \infty$ as both states become equally populated (see Fig. 9.7).

Exploration Draw graphs similar to those in Fig. 9.11 for a three-level system with levels 0, ε, and 2ε.

There are several points in relation to eqn 9.29 that need to be made. Because $\varepsilon_0 = 0$ (remember that we measure all energies from the lowest available level), E should be interpreted as the value of the internal energy relative to its value at $T=0$, $U(0)$. Therefore, to obtain the conventional internal energy U, we must add the internal energy at $T=0$:

$$U = U(0) + E \tag{16.30}$$

Secondly, because the partition function may depend on variables other than the temperature (for example, the volume), the derivative with respect to β in eqn 9.29 is actually a *partial* derivative with these other variables held constant. The complete expression relating the molecular partition function to the thermodynamic internal energy of a system of independent molecules is therefore

$$U = U(0) - \frac{N}{q}\left(\frac{\partial q}{\partial \beta}\right)_V \qquad (9.31a)$$

An equivalent form is obtained by noting that $\mathrm{d}x/x = \mathrm{d}\ln x$:

$$U = U(0) - N\left(\frac{\partial \ln q}{\partial \beta}\right)_V \qquad (9.31b)$$

These two equations confirm that we need know only the partition function (as a function of temperature) to calculate the internal energy relative to its value at $T = 0$.

(b) The value of β

We now confirm that the parameter β, which we have anticipated is equal to $1/kT$, does indeed have that value. To do so, we compare the equipartition expression for the internal energy of a monatomic perfect gas, which from *Molecular interpretation 2.2* in Volume 1 we know to be

$$U = U(0) + \tfrac{3}{2}nRT \qquad (9.32a)$$

with the value calculated from the translational partition function (see the following *Justification*), which is

$$U = U(0) + \frac{3N}{2\beta} \qquad (9.32b)$$

It follows by comparing these two expressions that

$$\beta = \frac{N}{nRT} = \frac{nN_A}{nN_A kT} = \frac{1}{kT} \qquad (9.33)$$

(We have used $N = nN_A$, where n is the amount of gas molecules, N_A is Avogadro's constant, and $R = N_A k$.) Although we have proved that $\beta = 1/kT$ by examining a very specific example, the translational motion of a perfect monatomic gas, the result is general (see Example 10.1 and *Further reading*).

Justification 9.3 *The internal energy of a perfect gas*

To use eqn 9.31, we introduce the translational partition function from eqn 9.19:

$$\left(\frac{\partial q}{\partial \beta}\right)_V = \left(\frac{\partial}{\partial \beta}\frac{V}{\Lambda^3}\right)_V = V\frac{\mathrm{d}}{\mathrm{d}\beta}\frac{1}{\Lambda^3} = -3\frac{V}{\Lambda^4}\frac{\mathrm{d}\Lambda}{\mathrm{d}\beta}$$

Then we note from the formula for Λ in eqn 9.19 that

$$\frac{\mathrm{d}\Lambda}{\mathrm{d}\beta} = \frac{\mathrm{d}}{\mathrm{d}\beta}\left\{\frac{h\beta^{1/2}}{(2\pi m)^{1/2}}\right\} = \frac{1}{2\beta^{1/2}} \times \frac{h}{(2\pi m)^{1/2}} = \frac{\Lambda}{2\beta}$$

and so obtain

$$\left(\frac{\partial q}{\partial \beta}\right)_V = -\frac{3V}{2\beta\Lambda^3}$$

Then, by eqn 9.31a,

$$U = U(0) - N\left(\frac{\Lambda^3}{V}\right)\left(-\frac{3V}{2\beta\Lambda^3}\right) = U(0) + \frac{3N}{2\beta}$$

as in eqn 9.32b.

9.4 The statistical entropy

If it is true that the partition function contains all thermodynamic information, then it must be possible to use it to calculate the entropy as well as the internal energy. Because entropy is related to the dispersal of energy (Volume 1, Section 3.2) and the partition function is a measure of the number of thermally accessible states, we can be confident that the two are indeed related.

We shall develop the relation between the entropy and the partition function in two stages. In *Further information* 9.2, we justify one of the most celebrated equations in statistical thermodynamics, the **Boltzmann formula** for the entropy:

$$S = k \ln W \qquad [9.34]$$

In this expression, W is the weight of the most probable configuration of the system. In the second stage, we express W in terms of the partition function.

The statistical entropy behaves in exactly the same way as the thermodynamic entropy. Thus, as the temperature is lowered, the value of W, and hence of S, decreases because fewer configurations are compatible with the total energy. In the limit $T \to 0$, $W = 1$, so $\ln W = 0$, because only one configuration (every molecule in the lowest level) is compatible with $E = 0$. It follows that $S \to 0$ as $T \to 0$, which is compatible with the Third Law of thermodynamics, that the entropies of all perfect crystals approach the same value as $T \to 0$.

Now we relate the Boltzmann formula for the entropy to the partition function. To do so, we substitute the expression for $\ln W$ given in eqn 9.3 into eqn 9.34 and, as shown in the *Justification* below, obtain

$$S = \frac{U - U(0)}{T} + Nk \ln q \qquad (9.35)$$

Justification 9.4 *The statistical entropy*

The first stage is to use eqn 9.3 ($\ln W = N \ln N - \sum_i n_i \ln n_i$) and $N = \sum_i n_i$ to write

$$S = k\sum_i (n_i \ln N - n_i \ln n_i) = -k\sum_i n_i \ln \frac{n_i}{N} = -Nk\sum_i p_i \ln p_i$$

where $p_i = n_i/N$, the fraction of molecules in state i. It follows from eqn 9.7 that

$$\ln p_i = -\beta\varepsilon_i - \ln q$$

and therefore that

$$S = -Nk(-\beta\sum_i p_i\varepsilon_i - \sum_i p_i \ln q) = k\beta\{U - U(0)\} + Nk \ln q$$

We have used the fact that the sum over the p_i is equal to 1 and that (from eqns 9.27 and 9.30)

$$N\sum_i p_i\varepsilon_i = \sum_i Np_i\varepsilon_i = \sum_i Np_i\varepsilon_i = \sum_i n_i\varepsilon_i = E = U - U(0)$$

We have already established that $\beta = 1/kT$, so eqn 9.35 immediately follows.

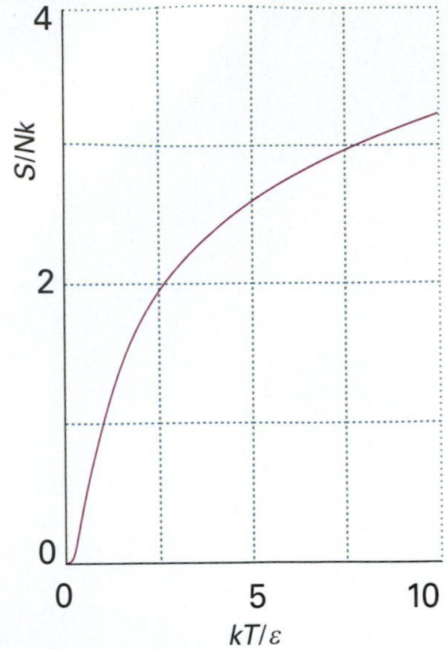

Fig. 9.12 The temperature variation of the entropy of the system shown in Fig. 9.3 (expressed here as a multiple of Nk). The entropy approaches zero as $T \to 0$, and increases without limit as $T \to \infty$.

Exploration Plot the function $\mathrm{d}S/\mathrm{d}T$, the temperature coefficient of the entropy, against kT/ε. Is there a temperature at which this coefficient passes through a maximum? If you find a maximum, explain its physical origins.

Example 9.4 *Calculating the entropy of a collection of oscillators*

Calculate the entropy of a collection of N independent harmonic oscillators, and evaluate it using vibrational data for I_2 vapour at 25°C (Example 9.3).

Method To use eqn 9.35, we use the partition function for a molecule with evenly spaced vibrational energy levels, eqn 9.12. With the partition function available, the internal energy can be found by differentiation (as in eqn 9.31a), and the two expressions then combined to give S.

Answer The molecular partition function as given in eqn 9.12 is

$$q = \frac{1}{1 - e^{-\beta\varepsilon}}$$

The internal energy is obtained by using eqn 9.31a:

$$U - U(0) = -\frac{N}{q}\left(\frac{\partial q}{\partial \beta}\right)_V = \frac{N\varepsilon e^{-\beta\varepsilon}}{1 - e^{-\beta\varepsilon}} = \frac{N\varepsilon}{e^{-\beta\varepsilon} - 1}$$

The entropy is therefore

$$S = Nk\left\{\frac{\beta\varepsilon}{e^{\beta\varepsilon} - 1} - \ln(1 - e^{\beta\varepsilon})\right\}$$

This function is plotted in Fig. 9.12. For I_2 at 25°C, $\beta\varepsilon = 1.036$ (Example 9.3), so $S_m = 8.38\ \mathrm{J\,K^{-1}\,mol^{-1}}$.

Self-test 9.6 Evaluate the molar entropy of N two-level systems and plot the resulting expression. What is the entropy when the two states are equally thermally accessible?

[$S/Nk = \beta\varepsilon/(1 + e^{\beta\varepsilon}) + \ln(1 + e^{-\beta\varepsilon})$; see Fig. 9.13; $S = Nk \ln 2$]

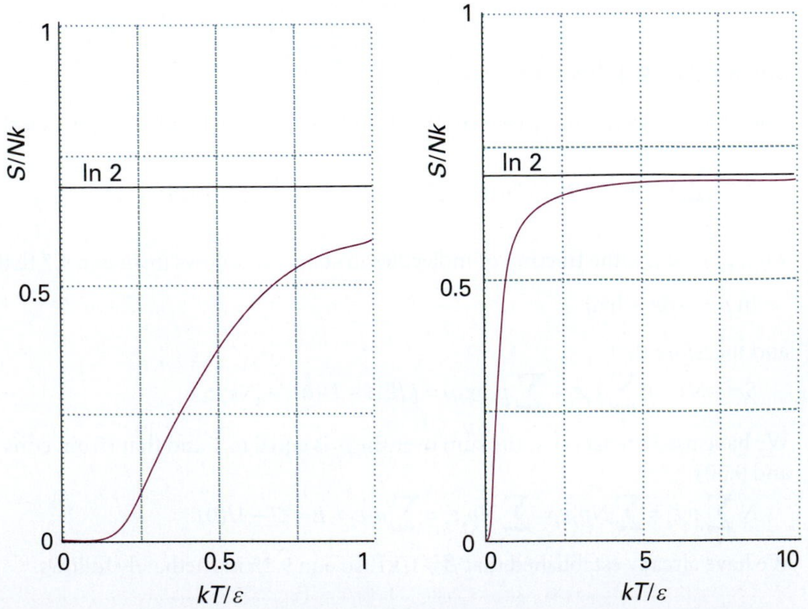

Fig. 9.13 The temperature variation of the entropy of a two-level system (expressed as a multiple of Nk). As $T \to \infty$, the two states become equally populated and S approaches $Nk \ln 2$.

Exploration Draw graphs similar to those in Fig. 9.13 for a three-level system with levels 0, ε, and 2ε.

The canonical partition function

In this section we see how to generalize our conclusions to include systems composed of interacting molecules. We shall also see how to obtain the molecular partition function from the more general form of the partition function developed here.

9.5 The canonical ensemble

The crucial new concept we need when treating systems of interacting particles is the 'ensemble'. Like so many scientific terms, the term has basically its normal meaning of 'collection', but it has been sharpened and refined into a precise significance.

(a) The concept of ensemble

To set up an ensemble, we take a closed system of specified volume, composition, and temperature, and think of it as replicated $\tilde{N}$ times (Fig. 9.14). All the identical closed systems are regarded as being in thermal contact with one another, so they can exchange energy. The total energy of all the systems is $\tilde{E}$ and, because they are in thermal equilibrium with one another, they all have the same temperature, T. This imaginary collection of replications of the actual system with a common temperature is called the **canonical ensemble**. The word 'canon' means 'according to a rule'.

There are two other important ensembles. In the **microcanonical ensemble** the condition of constant temperature is replaced by the requirement that all the systems should have exactly the same energy: each system is individually isolated. In the **grand canonical ensemble** the volume and temperature of each system is the same, but they are open, which means that matter can be imagined as able to pass between the systems; the composition of each one may fluctuate, but now the chemical potential is the same in each system:

Microcanonical ensemble: N, V, E common

Canonical ensemble: N, V, T common

Grand canonical ensemble: μ, V, T common

The important point about an ensemble is that it is a collection of *imaginary* replications of the system, so we are free to let the number of members be as large as we like; when appropriate, we can let $\tilde{N}$ become infinite. The number of members of the ensemble in a state with energy E_i is denoted $\tilde{n}_i$, and we can speak of the configuration of the ensemble (by analogy with the configuration of the system used in Section 9.1) and its weight, $\tilde{W}$. Note that $\tilde{N}$ is unrelated to N, the number of molecules in the actual system; $\tilde{N}$ is the number of imaginary replications of that system.

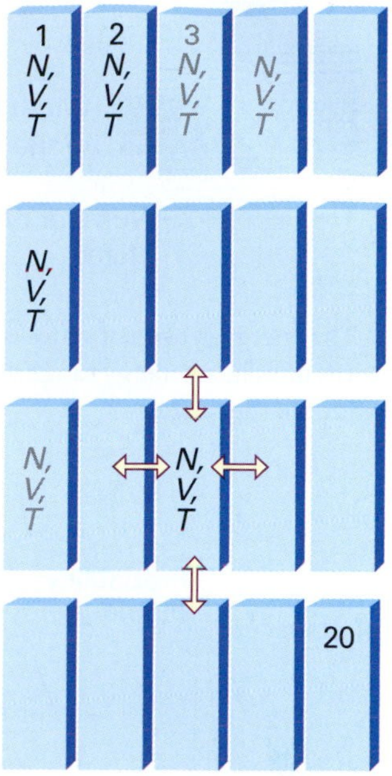

Fig. 9.14 A representation of the canonical ensemble, in this case for $\tilde{N} = 20$. The individual replications of the actual system all have the same composition and volume. They are all in mutual thermal contact, and so all have the same temperature. Energy may be transferred between them as heat, and so they do not all have the same energy. The total energy $\tilde{E}$ of all 20 replications is a constant because the ensemble is isolated overall.

(b) Dominating configurations

Just as in Section 9.1, some of the configurations of the ensemble will be very much more probable than others. For instance, it is very unlikely that the whole of the total energy, $\tilde{E}$, will accumulate in one system. By analogy with the earlier discussion, we can anticipate that there will be a dominating configuration, and that we can evaluate the thermodynamic properties by taking the average over the ensemble using that single, most probable, configuration. In the **thermodynamic limit** of $\tilde{N} \to \infty$, this dominating configuration is overwhelmingly the most probable, and it dominates the properties of the system virtually completely.

The quantitative discussion follows the argument in Section 9.1 with the modification that N and n_i are replaced by $\tilde{N}$ and $\tilde{n}_i$. The weight of a configuration $\{\tilde{n}_0, \tilde{n}_1, \ldots\}$ is

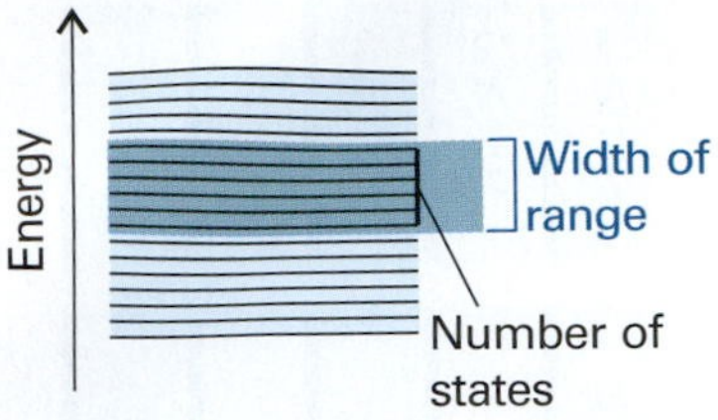

Fig. 9.15 The energy density of states is the number of states in an energy range divided by the width of the range.

$$\tilde{W}=\frac{\tilde{N}!}{\tilde{n}_0!\tilde{n}_1!\ldots} \tag{9.36}$$

The configuration of greatest weight, subject to the constraints that the total energy of the ensemble is constant at $\tilde{E}$ and that the total number of members is fixed at $\tilde{N}$, is given by the **canonical distribution**:

$$\frac{\tilde{n}_i}{\tilde{N}}=\frac{e^{-\beta E_i}}{Q} \qquad Q=\sum_i e^{-\beta E_i} \tag{9.37}$$

The quantity Q, which is a function of the temperature, is called the **canonical partition function.**

(c) Fluctuations from the most probable distribution

The canonical distribution in eqn 9.37 is only apparently an exponentially decreasing function of the energy of the system. We must appreciate that eqn 9.37 gives the probability of occurrence of members in a single state i of the entire system of energy E_i. There may in fact be numerous states with almost identical energies. For example, in a gas the identities of the molecules moving slowly or quickly can change without necessarily affecting the total energy. The density of states, the number of states in an energy range divided by the width of the range (Fig. 9.15), is a very sharply increasing function of energy. It follows that the probability of a member of an ensemble having a specified energy (as distinct from being in a specified state) is given by eqn 9.37, a sharply decreasing function, multiplied by a sharply increasing function (Fig. 9.16). Therefore, the overall distribution is a sharply peaked function. We conclude that most members of the ensemble have an energy very close to the mean value.

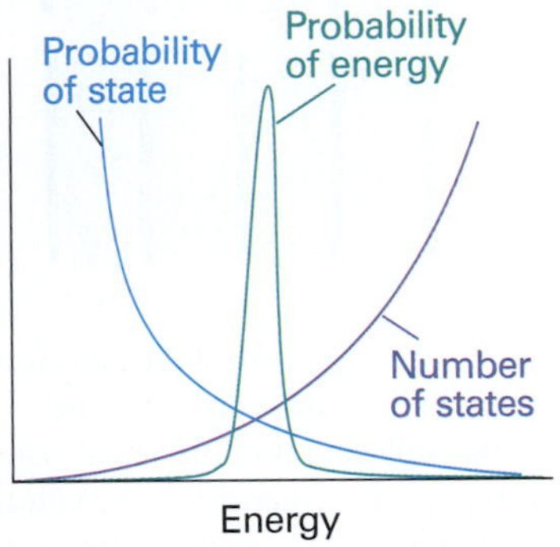

Fig. 9.16 To construct the form of the distribution of members of the canonical ensemble in terms of their energies, we multiply the probability that any one is in a state of given energy, eqn 9.39, by the number of states corresponding to that energy (a steeply rising function). The product is a sharply peaked function at the mean energy, which shows that almost all the members of the ensemble have that energy.

9.6 The thermodynamic information in the partition function

Like the molecular partition function, the canonical partition function carries all the thermodynamic information about a system. However, Q is more general than q because it does not assume that the molecules are independent. We can therefore use Q to discuss the properties of condensed phases and real gases where molecular interactions are important.

(a) The internal energy

If the total energy of the ensemble is $\tilde{E}$, and there are $\tilde{N}$ members, the average energy of a member is $E=\tilde{E}/\tilde{N}$. We use this quantity to calculate the internal energy of the system in the limit of $\tilde{N}$ (and $\tilde{E}$) approaching infinity:

$$U=U(0)+E=U(0)+\tilde{E}/\tilde{N} \qquad \text{as} \qquad \tilde{N}\to\infty \tag{9.38}$$

The fraction, $\tilde{p}_i$, of members of the ensemble in a state i with energy E_i is given by the analogue of eqn 9.7 as

$$\tilde{p}_i=\frac{e^{-\beta E_i}}{Q} \tag{9.39}$$

It follows that the internal energy is given by

$$U=U(0)+\sum_i \tilde{p}_i E_i=U(0)+\frac{1}{Q}\sum_i E_i e^{-\beta E_i} \tag{9.40}$$

By the same argument that led to eqn 9.31,

$$U = U(0) - \frac{1}{Q}\left(\frac{\partial Q}{\partial \beta}\right)_V = U(0) - \left(\frac{\partial \ln Q}{\partial \beta}\right)_V \tag{9.41}$$

(b) The entropy

The total weight, $\tilde{W}$, of a configuration of the ensemble is the product of the average weight W of each member of the ensemble, $\tilde{W} = W^{\tilde{N}}$. Hence, we can calculate S from

$$S = k \ln W = k \ln \tilde{W}^{1/\tilde{N}} = \frac{k}{\tilde{N}} \ln \tilde{W} \tag{9.42}$$

It follows, by the same argument used in Section 9.4, that

$$S = \frac{U - U(0)}{T} + k \ln Q \tag{9.43}$$

9.7 Independent molecules

We shall now see how to recover the molecular partition function from the more general canonical partition function when the molecules are independent. When the molecules are independent and distinguishable (in the sense to be described), the relation between Q and q is

$$Q = q^N \tag{9.44}$$

Justification 9.5 *The relation between Q and q*

The total energy of a collection of N independent molecules is the sum of the energies of the molecules. Therefore, we can write the total energy of a state i of the system as

$$E_i = \varepsilon_i(1) + \varepsilon_i(2) + \cdots + \varepsilon_i(N)$$

In this expression, $\varepsilon_i(1)$ is the energy of molecule 1 when the system is in the state i, $\varepsilon_i(2)$ the energy of molecule 2 when the system is in the same state i, and so on. The canonical partition function is then

$$Q = \sum_i e^{-\beta\varepsilon_i(1) - \beta\varepsilon_i(2) - \cdots - \beta\varepsilon_i(N)}$$

The sum over the states of the system can be reproduced by letting each molecule enter all its own individual states (although we meet an important proviso shortly). Therefore, instead of summing over the states i of the system, we can sum over all the individual states i of molecule 1, all the states i of molecule 2, and so on. This rewriting of the original expression leads to

$$Q = \left(\sum_i e^{-\beta\varepsilon_i}\right)\left(\sum_i e^{-\beta\varepsilon_i}\right)\cdots\left(\sum_i e^{-\beta\varepsilon_i}\right) = \left(\sum_i e^{-\beta\varepsilon_i}\right)^N = q^N$$

(a) Distinguishable and indistinguishable molecules

If all the molecules are identical and free to move through space, we cannot distinguish them and the relation $Q = q^N$ is not valid. Suppose that molecule 1 is in some state a, molecule 2 is in b, and molecule 3 is in c, then one member of the ensemble has an energy $E = \varepsilon_a + \varepsilon_b + \varepsilon_c$. This member, however, is indistinguishable from one formed by putting molecule 1 in state b, molecule 2 in state c, and molecule 3 in state a, or some other permutation. There are six such permutations in all, and $N!$ in general. In the case of indistinguishable molecules, it follows that we have counted too many states in going from the sum over system states to the sum over molecular states, so writing $Q = q^N$ overestimates the value of Q. The detailed argument is quite involved, but at all except very low temperatures it turns out that the correction factor is $1/N!$. Therefore:

- For distinguishable independent molecules: $Q = q^N$ (9.45a)
- For indistinguishable independent molecules: $Q = q^N/N!$ (9.45b)

For molecules to be indistinguishable, they must be of the same kind: an Ar atom is never indistinguishable from a Ne atom. Their identity, however, is not the only criterion. Each identical molecule in a crystal lattice, for instance, can be 'named' with a set of coordinates. Identical molecules in a lattice can therefore be treated as distinguishable because their sites are distinguishable, and we use eqn 9.45a. On the other hand, identical molecules in a gas are free to move to different locations, and there is no way of keeping track of the identity of a given molecule; we therefore use eqn 9.45b.

(b) The entropy of a monatomic gas

An important application of the previous material is the derivation (as shown in the *Justification* below) of the **Sackur–Tetrode equation** for the entropy of a monatomic gas:

$$S = nR \ln\left(\frac{e^{5/2}V}{nN_A\Lambda^3}\right) \qquad \Lambda = \frac{h}{(2\pi mkT)^{1/2}} \tag{9.46a}$$

This equation implies that the molar entropy of a perfect gas of high molar mass is greater than one of low molar mass under the same conditions (because the former has more thermally accessible translational states). Because the gas is perfect, we can use the relation $V = nRT/p$ to express the entropy in terms of the pressure as

$$S = nR \ln\left(\frac{e^{5/2}kT}{p\Lambda^3}\right) \tag{9.46b}$$

Justification 9.6 *The Sackur–Tetrode equation*

For a gas of independent molecules, Q may be replaced by $q^N/N!$, with the result that eqn 9.43 becomes

$$S = \frac{U - U(0)}{T} + Nk \ln q - k \ln N!$$

Because the number of molecules ($N = nN_A$) in a typical sample is large, we can use Stirling's approximation (eqn 9.2) to write

$$S = \frac{U - U(0)}{T} + nR \ln q - nR \ln N + nR$$

The only mode of motion for a gas of atoms is translation, and the partition function is $q = V/\Lambda^3$ (eqn 9.19), where Λ is the thermal wavelength. The internal energy is given by eqn 9.32, so the entropy is

$$S = \tfrac{3}{2}nR + nR\left(\ln \frac{V}{\Lambda^3} - \ln nN_A + 1\right) = nR\left(\ln e^{3/2} + \ln \frac{V}{\Lambda^3} - \ln nN_A + \ln e\right)$$

which rearranges into eqn 9.46.

Example 9.5 *Using the Sackur–Tetrode equation*

Calculate the standard molar entropy of gaseous argon at 25°C.

Method To calculate the molar entropy, S_m, from eqn 9.46b, divide both sides by n. To calculate the standard molar entropy, $S_m^{\ominus}$, set $p = p^{\ominus}$ in the expression for S_m:

$$S_m^{\ominus} = R \ln\left(\frac{e^{5/2}kT}{p^{\ominus}\Lambda^3}\right)$$

Answer The mass of an Ar atom is $m = 39.95$ u. At 25°C, its thermal wavelength is 16.0 pm (by the same kind of calculation as in *Illustration* 9.3). Therefore,

$$S_m^{\ominus} = R \ln\left\{\frac{e^{5/2} \times (4.12 \times 10^{-21}\ \text{J})}{(10^5\ \text{N m}^{-2}) \times (1.60 \times 10^{-11}\ \text{m})^3}\right\} = 18.6R = 155\ \text{J K}^{-1}\ \text{mol}^{-1}$$

We can anticipate, on the basis of the number of accessible states for a lighter molecule, that the standard molar entropy of Ne is likely to be smaller than for Ar; its actual value is 17.60R at 298 K.

Self-test 9.7 Calculate the translational contribution to the standard molar entropy of H_2 at 25°C. [14.2R]

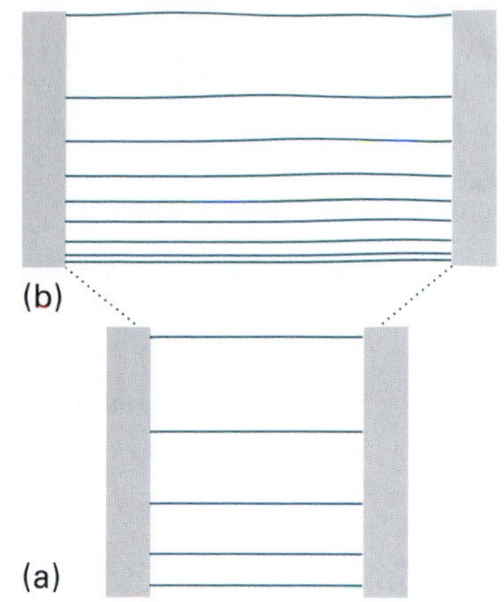

Fig. 9.17 As the width of a container is increased (going from (a) to (b)), the energy levels become closer together (as $1/L^2$), and as a result more are thermally accessible at a given temperature. Consequently, the entropy of the system rises as the container expands.

The Sackur–Tetrode equation implies that, when a monatomic perfect gas expands isothermally from V_i to V_f, its entropy changes by

$$\Delta S = nR \ln(aV_f) - nR \ln(aV_i) = nR \ln \frac{V_f}{V_i} \tag{9.47}$$

where aV is the collection of quantities inside the logarithm of eqn 9.46a. This is exactly the expression we obtained by using classical thermodynamics. Now, though, we see that that classical expression is in fact a consequence of the increase in the number of accessible translational states when the volume of the container is increased (Fig. 9.17).

Checklist of key ideas

☐ 1. The instantaneous configuration of a system of N molecules is the specification of the set of populations $n_0, n_1, \ldots$ of the energy levels $\varepsilon_0, \varepsilon_1, \ldots$. The weight W of a configuration is given by $W = N!/n_0!n_1! \ldots$.

☐ 2. The Boltzmann distribution gives the numbers of molecules in each state of a system at any temperature: $N_i = Ne^{-\beta\varepsilon_i}/q$, $\beta = 1/kT$.

☐ 3. The partition function is defined as $q = \sum_j e^{-\beta\varepsilon_j}$ and is an indication of the number of thermally accessible states at the temperature of interest.

☐ 4. The internal energy is $U(T) = U(0) + E$, with $E = -(N/q)(\partial q/\partial\beta)_V = -N(\partial \ln q/\partial\beta)_V$.

☐ 5. The Boltzmann formula for the entropy is $S = k \ln W$, where W is the number of different ways in which the molecules of a system can be arranged while keeping the same total energy.

☐ 6. The entropy in terms of the partition function is $S = \{U - U(0)\}/T + Nk \ln q$ (distinguishable molecules) or $S = \{U - U(0)\}/T + Nk \ln q - Nk(\ln N - 1)$ (indistinguishable molecules).

☐ 7. The canonical ensemble is an imaginary collection of replications of the actual system with a common temperature.

☐ 8. The canonical distribution is given by $\tilde{n}_i/\tilde{N} = e^{-\beta E_i}/\sum_j e^{-\beta E_j}$. The canonical partition function, $Q = \sum_i e^{-\beta E_i}$.

☐ 9. The internal energy and entropy of an ensemble are, respectively, $U = U(0) - (\partial \ln Q/\partial\beta)_V$ and $S = \{U - U(0)\}/T + k \ln Q$.

☐ 10. For distinguishable independent molecules we write $Q = q^N$. For indistinguishable independent molecules we write $Q = q^N/N!$.

☐ 11. The Sackur–Tetrode equation, eqn 9.46, is an expression for the entropy of a monatomic gas.

Further reading

Articles and texts

D. Chandler, *Introduction to modern statistical mechanics*. Oxford University Press (1987).

D.A. McQuarrie and J.D. Simon, *Molecular thermodynamics*. University Science Books, Sausalito (1999).

K.E. van Holde, W.C. Johnson, and P.S. Ho, *Principles of physical biochemistry*. Prentice Hall, Upper Saddle River (1998).

J. Wisniak, Negative absolute temperatures, a novelty. *J. Chem. Educ.* **77**, 518 (2000).

Further information

Further information 9.1 *The Boltzmann distribution*

We remarked in Section 9.1 that ln W is easier to handle than W. Therefore, to find the form of the Boltzmann distribution, we look for the condition for ln W being a maximum rather than dealing directly with W. Because ln W depends on all the n_i, when a configuration changes and the n_i change to $n_i + dn_i$, the function ln W changes to ln W + d ln W, where

$$\mathrm{d}\ln W = \sum_i \left(\frac{\partial \ln W}{\partial n_i}\right) \mathrm{d}n_i$$

All this expression states is that a change in ln W is the sum of contributions arising from changes in each value of n_i. At a maximum, d ln $W = 0$. However, when the n_i change, they do so subject to the two constraints

$$\sum_i \varepsilon_i \mathrm{d}n_i = 0 \qquad \sum_i \mathrm{d}n_i = 0 \qquad (9.48)$$

The first constraint recognizes that the total energy must not change, and the second recognizes that the total number of molecules must not change. These two constraints prevent us from solving d ln $W = 0$ simply by setting all $(\partial \ln W/\partial n_i) = 0$ because the $\mathrm{d}n_i$ are not all independent.

The way to take constraints into account was devised by the French mathematician Lagrange, and is called the **method of undetermined multipliers**. The technique is described in *Appendix* 2. All we need here is the rule that a constraint should be multiplied by a constant and then added to the main variation equation. The variables are then treated as though they were all independent, and the constants are evaluated at the end of the calculation.

We employ the technique as follows. The two constraints in eqn 16.48 are multiplied by the constants $-\beta$ and α, respectively (the minus sign in $-\beta$ has been included for future convenience), and then added to the expression for d ln W:

$$\mathrm{d}\ln W = \sum_i \left(\frac{\partial \ln W}{\partial n_i}\right) \mathrm{d}n_i + \alpha \sum_i \mathrm{d}n_i - \beta \sum_i \varepsilon_i \mathrm{d}n_i$$

$$= \sum_i \left\{\left(\frac{\partial \ln W}{\partial n_i}\right) + \alpha - \beta\varepsilon_i\right\} \mathrm{d}n_i$$

All the $\mathrm{d}n_i$ are now treated as independent. Hence the only way of satisfying d ln $W = 0$ is to require that, for each i,

$$\frac{\partial \ln W}{\partial n_i} + \alpha - \beta\varepsilon_i = 0 \qquad (9.49)$$

when the n_i have their most probable values.

Differentiation of ln W as given in eqn 9.3 with respect to n_i gives

$$\frac{\partial \ln W}{\partial n_i} = \frac{\partial (N \ln N)}{\partial n_i} - \sum_j \frac{\partial (n_j \ln n_j)}{\partial n_i}$$

The derivative of the first term is obtained as follows:

$$\frac{\partial (N \ln N)}{\partial n_i} = \left(\frac{\partial N}{\partial n_i}\right) \ln N + N\left(\frac{\partial \ln N}{\partial n_i}\right)$$

$$= \ln N + \frac{\partial N}{\partial n_i} = \ln N + 1$$

The ln N in the first term on the right in the second line arises because $N = n_1 + n_2 + \cdots$ and so the derivative of N with respect to any of the n_i is 1: that is, $\partial N/\partial n_i = 1$. The second term on the right in the second line arises because $\partial(\ln N)/\partial n_i = (1/N)\partial N/\partial n_i$. The final 1 is then obtained in the same way as in the preceding remark, by using $\partial N/\partial n_i = 1$.

For the derivative of the second term we first note that

$$\frac{\partial \ln n_j}{\partial n_i} = \frac{1}{n_j}\left(\frac{\partial n_j}{\partial n_i}\right)$$

Morever, if $i \neq j$, n_j is independent of n_i, so $\partial n_j/\partial n_i = 0$. However, if $i = j$,

$$\frac{\partial n_j}{\partial n_i} = \frac{\partial n_j}{\partial n_j} = 1$$

Therefore,

$$\frac{\partial n_j}{\partial n_i} = \delta_{ij}$$

with δ_{ij} the Kronecker delta ($\delta_{ij} = 1$ if $i = j$, $\delta_{ij} = 0$ otherwise). Then

$$\sum_j \frac{\partial(n_j \ln n_j)}{\partial n_i} = \sum_j \left\{ \left(\frac{\partial n_j}{\partial n_i}\right) \ln n_j + n_j \left(\frac{\partial \ln n_j}{\partial n_i}\right) \right\}$$
$$= \sum_j \left\{ \left(\frac{\partial n_j}{\partial n_i}\right) \ln n_j + \left(\frac{\partial n_j}{\partial n_i}\right) \right\}$$
$$= \sum_j \left(\frac{\partial n_j}{\partial n_i}\right)(\ln n_j + 1)$$
$$= \sum_j \delta_{ij}(\ln n_j + 1) = \ln n_i + 1$$

and therefore

$$\frac{\partial \ln W}{\partial n_i} = -(\ln n_i + 1) + (\ln N + 1) = -\ln \frac{n_i}{N}$$

It follows from eqn 9.49 that

$$-\ln \frac{n_i}{N} + \alpha - \beta\varepsilon_i = 0$$

and therefore that

$$\frac{n_i}{N} = \mathrm{e}^{\alpha - \beta\varepsilon_i}$$

At this stage we note that

$$N = \sum_i n_i = \sum_i N\mathrm{e}^{\alpha - \beta\varepsilon_i} = N\mathrm{e}^{\alpha} \sum_i \mathrm{e}^{\beta\varepsilon_i}$$

Because the N cancels on each side of this equality, it follows that

$$\mathrm{e}^{\alpha} = \frac{1}{\sum_j \mathrm{e}^{-\beta\varepsilon_j}} \qquad (9.50)$$

and

$$\frac{n_i}{N} = \mathrm{e}^{\alpha - \beta\varepsilon_i} = \mathrm{e}^{\alpha}\mathrm{e}^{-\beta\varepsilon_i} = \frac{1}{\sum_j \mathrm{e}^{-\beta\varepsilon_j}} \mathrm{e}^{-\beta\varepsilon_i}$$

which is eqn 9.6a.

Further information 9.2 *The Boltzmann formula*

A change in the internal energy

$$U = U(0) + \sum_i n_i\varepsilon_i \qquad (9.51)$$

may arise from either a modification of the energy levels of a system (when ε_i changes to $\varepsilon_i + \mathrm{d}\varepsilon_i$) or from a modification of the populations (when n_i changes to $n_i + \mathrm{d}n_i$). The most general change is therefore

$$\mathrm{d}U = \mathrm{d}U(0) + \sum_i n_i \mathrm{d}\varepsilon_i + \sum_i \varepsilon_i \mathrm{d}n_i \qquad (9.52)$$

Because the energy levels do not change when a system is heated at constant volume (Fig. 9.18), in the absence of all changes other than heating

$$\mathrm{d}U = \sum_i \varepsilon_i \mathrm{d}n_i$$

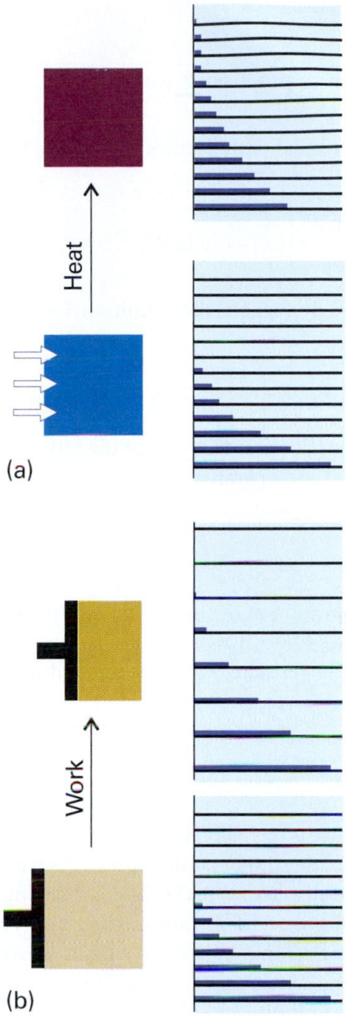

Fig. 9.18 (a) When a system is heated, the energy levels are unchanged but their populations are changed. (b) When work is done on a system, the energy levels themselves are changed. The levels in this case are the one-dimensional particle-in-a-box energy levels of Chapter 2: they depend on the size of the container and move apart as its length is decreased.

We know from thermodynamics (and specifically from *eqn 3.43* of Volume 1) that under the same conditions

$$\mathrm{d}U = \mathrm{d}q_{\mathrm{rev}} = T\mathrm{d}S$$

Therefore,

$$\mathrm{d}S = \frac{\mathrm{d}U}{T} = k\beta \sum_i \varepsilon_i \mathrm{d}n_i \qquad (9.53)$$

For changes in the most probable configuration (the only one we need consider), we rearrange eqn 9.49 to

$$\beta\varepsilon_i = \frac{\partial \ln W}{\partial n_i} + \alpha$$

and find that

$$dS = k\sum_i \left(\frac{\partial \ln W}{\partial n_i}\right) dn_i + k\alpha \sum_i dn_i$$

But because the number of molecules is constant, the sum over the dn_i is zero. Hence

$$dS = k\sum_i \left(\frac{\partial \ln W}{\partial n_i}\right) dn_i = k(d \ln W)$$

This relation strongly suggests the definition $S = k \ln W$, as in eqn 9.34.

Further information 9.3 *Temperatures below zero*

The Boltzmann distribution tells us that the ratio of populations in a two-level system at a temperature T is

$$\frac{N_+}{N_-} = e^{-\varepsilon/kT} \qquad (9.54)$$

where ε is the separation of the upper state N_+ and the lower state N_-. It follows that, if we can contrive the population of the upper state to exceed that of the lower state, then the temperature must have a negative value. Indeed, for a general population,

$$T = \frac{\varepsilon/k}{\ln(N_-/N_+)} \qquad (9.55)$$

and the temperature is formally negative for all $N_+ > N_-$.

All the statistical thermodynamic expressions we have derived apply to $T < 0$ as well as to $T > 0$, the difference being that states with $T < 0$ are not in thermal equilibrium and therefore have to be achieved by techniques that do not rely on the equalization of temperatures of the system and its surroundings. The Third Law of thermodynamics prohibits the achievement of absolute zero in a finite number of steps. However, it is possible to circumvent this restriction in systems that have a finite number of levels or in systems that are effectively finite because they have such weak coupling to their surroundings. The practical realization of such a system is a group of spin-$\frac{1}{2}$ nuclei that have very long relaxation times, such as the ^{19}F nuclei in cold solid LiF. Pulse techniques in NMR can achieve non-equilibrium populations (Section 8.8) as can pumping procedures in laser technologies (Section 7.5). From now on, we shall suppose that these non-equilibrium distributions have been achieved, and will concentrate on the consequences.

The expressions for q, U, and S that we have derived in this chapter are applicable to $T < 0$ as well as to $T > 0$, and are shown in Fig. 9.19. We see that q and U show sharp discontinuities on passing through zero, and $T = +0$ (corresponding to all population in the lower state) is quite distinct from $T = -0$, where all the population is in the upper state. The entropy S is continuous at $T = 0$. But all these functions are continuous if we use $\beta = 1/kT$ as the dependent variable (Fig. 9.20), which shows that β is a more natural, if less familiar, variable than T. Note that $U \rightarrow 0$ as $\beta \rightarrow \infty$ (that is, as $T \rightarrow 0$, when only the lower state is occupied) and $U \rightarrow N\varepsilon$ as $\beta \rightarrow -\infty$ (that is, as $T \rightarrow -0$);

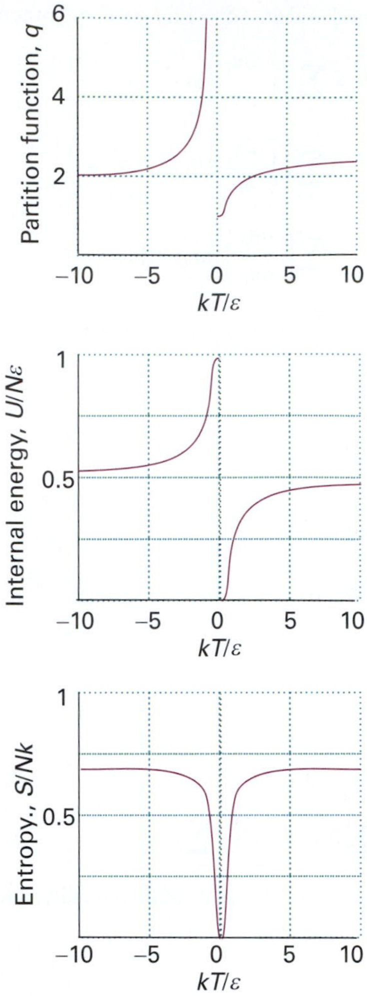

Fig. 9.19 The partition function, internal energy, and entropy of a two-level system extended to negative temperatures.

we see that a state with $T = -0$ is 'hotter' than one with $T = +0$. The entropy of the system is zero on either side of $T = 0$, and rises to $Nk \ln 2$ as $T \rightarrow \pm\infty$. At $T = +0$ only one state is accessible (the lower state), only the upper state is accessible, so the entropy is zero in each case.

We get more insight into the dependence of thermodynamic properties on temperature by noting the thermodynamic result that $T = (\partial S/\partial U)_T$. When S is plotted against U for a two-level system (Fig. 9.21), we see that the entropy rises as energy is supplied to the system (as we would expect) provided that $T > 0$ (the thermal equilibrium regime). However, the entropy decreases as energy is supplied when $T < 0$. This conclusion is consistent with the thermodynamic definition of entropy, $dS = dq_{rev}/T$ (where, of course, q denotes heat and not the partition function). Physically, the increase in entropy for $T > 0$ corresponds to the increasing accessibility of the upper state, and the decrease for $T < 0$ corresponds

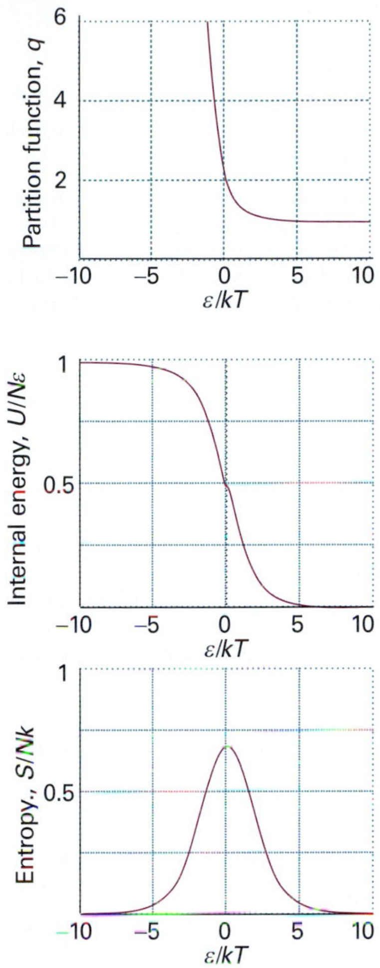

Fig. 9.20 The partition function, internal energy, and entropy of a two-level system extended to negative temperatures but plotted against $\beta = 1/kT$ (modified here to the dimensionless quantity ε/kT).

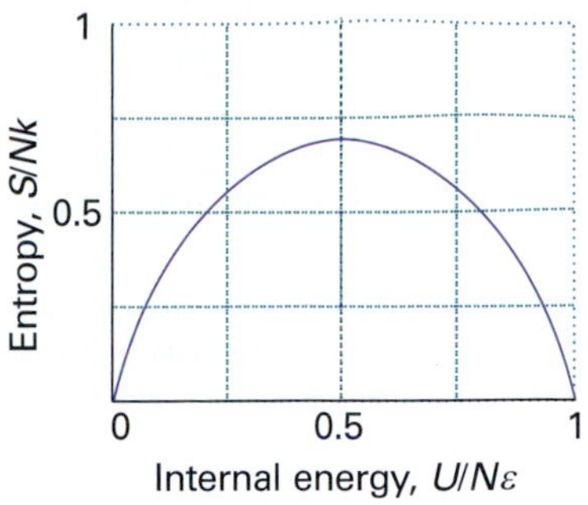

Fig. 9.21 The variation of the entropy with internal energy for a two-level system extended to negative temperatures.

to the shift towards population of the upper state alone as more energy is packed into the system.

The phenomenological laws of thermodynamics survive largely intact at negative temperatures. The First Law (in essence, the conservation of energy) is robust, and independent of how populations are distributed over states. The Second Law survives because the definition of entropy survives (as we have seen above). The efficiency of heat engines (Volume 1, Section 3.2), which is a direct consequence of the Second Law, is still given by $1 - T_{cold}/T_{hot}$. However, if the temperature of the cold reservoir is negative, then the efficiency of the engine may be greater than 1. This condition corresponds to the amplification of signals achieved in lasers. Alternatively, an efficiency greater than 1 implies that heat *can* be converted completely into work provided the heat is withdrawn from a reservoir at $T < 0$. If both reservoirs are at negative temperatures, then the efficiency is less than 1. The Third Law requires a slight amendment on account of the discontinuity of the populations across $T = 0$: it is impossible in a finite number of steps to cool any system down to +0 or to heat any system above –0.

Each end-of-chapter discussion question, exercise, and problem for which a solution is available in the Student Solution Manual has a corresponding bracketed number referring to the Student Solutions Manual number for this problem, for example [SSM 16.1]. Each end-of-chapter exercise and problem for which a solution is available in the Instructors' Solution Manual has a corresponding bracketed number referring to the Instructor Solutions Manual number for this problem, for example [ISM 16.2].

Discussion questions

9.1 Describe the physical significance of the partition function. [SSM **16.1**]

9.2 Explain how the internal energy and entropy of a system composed of two levels vary with temperature. [ISM **16.2**]

9.3 Enumerate the ways by which the parameter β may be identified with $1/kT$. [SSM **16.3**]

9.4 Distinguish between the zipper and Zimm–Bragg models of the helix–coil transition. [ISM **16.4**]

9.5 Explain what is meant by an *ensemble* and why it is useful in statistical thermodynamics. [SSM **16.5**]

9.6 Under what circumstances may identical particles be regarded as distinguishable? [ISM **16.6**]

Exercises

9.1a What are the relative populations of the states of a two-level system when the temperature is infinite? [ISM **16.1a**]

9.1b What is the temperature of a two-level system of energy separation equivalent to 300 cm^{-1} when the population of the upper state is one-half that of the lower state? [SSM **16.1b**]

9.2a Calculate the translational partition function at (a) 300 K and (b) 600 K of a molecule of molar mass 120 g mol^{-1} in a container of volume 2.00 cm^3. [ISM **16.2a**]

9.2b Calculate (a) the thermal wavelength, (b) the translational partition function of an Ar atom in a cubic box of side 1.00 cm at (i) 300 K and (ii) 3000 K. [SSM **16.2b**]

9.3a Calculate the ratio of the translational partition functions of D_2 and H_2 at the same temperature and volume. [ISM **16.3a**]

9.3b Calculate the ratio of the translational partition functions of xenon and helium at the same temperature and volume. [SSM **16.3b**]

9.4a A certain atom has a threefold degenerate ground level, a non-degenerate electronically excited level at 3500 cm^{-1}, and a threefold degenerate level at 4700 cm^{-1}. Calculate the partition function of these electronic states at 1900 K. [ISM **16.4a**]

9.4b A certain atom has a doubly degenerate ground level, a triply degenerate electronically excited level at 1250 cm^{-1}, and a doubly degenerate level at 1300 cm^{-1}. Calculate the partition function of these electronic states at 2000 K. [SSM **16.4b**]

9.5a Calculate the electronic contribution to the molar internal energy at 1900 K for a sample composed of the atoms specified in Exercise 9.4a. [ISM **16.5a**]

9.5b Calculate the electronic contribution to the molar internal energy at 2000 K for a sample composed of the atoms specified in Exercise 9.4b. [SSM **16.5b**]

9.6a A certain molecule has a non-degenerate excited state lying at 540 cm^{-1} above the non-degenerate ground state. At what temperature will 10 per cent of the molecules be in the upper state? [ISM **16.6a**]

9.6b A certain molecule has a doubly degenerate excited state lying at 360 cm^{-1} above the non-degenerate ground state. At what temperature will 15 per cent of the molecules be in the upper state? [SSM **16.6b**]

9.7a An electron spin can adopt either of two orientations in a magnetic field, and its energies are $\pm\mu_B\mathcal{B}$, where μ_B is the Bohr magneton. Deduce an expression for the partition function and mean energy of the electron and sketch the variation of the functions with $\mathcal{B}$. Calculate the relative populations of the spin states at (a) 4.0 K, (b) 298 K when $\mathcal{B}$ = 1.0 T. [ISM **16.7a**]

9.7b A nitrogen nucleus spin can adopt any of three orientations in a magnetic field, and its energies are $0, \pm\gamma_N\hbar\mathcal{B}$, where γ_N is the magnetogyric ratio of the nucleus. Deduce an expression for the partition function and mean energy of the nucleus and sketch the variation of the functions with $\mathcal{B}$. Calculate the relative populations of the spin states at (a) 1.0 K, (b) 298 K when $\mathcal{B}$ = 20.0 T. [SSM **16.7b**]

9.8a Consider a system of distinguishable particles having only two non-degenerate energy levels separated by an energy that is equal to the value of kT at 10 K. Calculate (a) the ratio of populations in the two states at (1) 1.0 K, (2) 10 K, and (3) 100 K, (b) the molecular partition function at 10 K, (c) the molar energy at 10 K, (d) the molar heat capacity at 10 K, (e) the molar entropy at 10 K. [ISM **16.8a**]

9.8b Consider a system of distinguishable particles having only three non-degenerate energy levels separated by an energy which is equal to the value of kT at 25.0 K. Calculate (a) the ratio of populations in the states at (1) 1.00 K, (2) 25.0 K, and (3) 100 K, (b) the molecular partition function at 25.0 K, (c) the molar energy at 25.0 K, (d) the molar heat capacity at 25.0 K, (e) the molar entropy at 25.0 K. [SSM **16.8b**]

9.9a At what temperature would the population of the first excited vibrational state of HCl be 1/e times its population of the ground state? [ISM **16.9a**]

9.9b At what temperature would the population of the first excited rotational level of HCl be 1/e times its population of the ground state? [SSM **16.9b**]

9.10a Calculate the standard molar entropy of neon gas at (a) 200 K, (b) 298.15 K. [ISM **16.10a**]

9.10b Calculate the standard molar entropy of xenon gas at (a) 100 K, (b) 298.15 K. [SSM **16.10b**]

9.11a Calculate the vibrational contribution to the entropy of Cl_2 at 500 K given that the wavenumber of the vibration is 560 cm^{-1}. [ISM **16.11a**]

9.11b Calculate the vibrational contribution to the entropy of Br_2 at 600 K given that the wavenumber of the vibration is 321 cm^{-1}. [SSM **16.11b**]

9.12a Identify the systems for which it is essential to include a factor of $1/N!$ on going from Q to q: (a) a sample of helium gas, (b) a sample of carbon monoxide gas, (c) a solid sample of carbon monoxide, (d) water vapour. [ISM **16.12a**]

9.12b Identify the systems for which it is essential to include a factor of $1/N!$ on going from Q to q: (a) a sample of carbon dioxide gas, (b) a sample of graphite, (c) a sample of diamond, (d) ice. [SSM **16.12b**]

Problems*

Numerical problems

9.1‡ Consider a system A consisting of subsystems A_1 and A_2, for which $W_1 = 1\times10^{20}$ and $W_2 = 2\times10^{20}$. What is the number of configurations available to the combined system? Also, compute the entropies S, S_1, and S_2. What is the significance of this result? [SSM **16.1**]

9.2‡ Consider 1.00×10^{22} ^{4}He atoms in a box of dimensions 1.0 cm × 1.0 cm × 1.0 cm. Calculate the occupancy of the first excited level at 1.0 mK, 2.0 K, and 4.0 K. Do the same for ^{3}He. What conclusions might you draw from the results of your calculations? [ISM **16.2**]

* Problems denoted with the symbol ‡ were supplied by Charles Trapp and Carmen Giunta.

9.3‡ By what factor does the number of available configurations increase when 100 J of energy is added to a system containing 1.00 mol of particles at constant volume at 298 K? [SSM **16.3**]

9.4‡ By what factor does the number of available configurations increase when 20 m^3 of air at 1.00 atm and 300 K is allowed to expand by 0.0010 per cent at constant temperature? [ISM **16.4**]

9.5 Explore the conditions under which the 'integral' approximation for the translational partition function is not valid by considering the translational partition function of an Ar atom in a cubic box of side 1.00 cm. Estimate the temperature at which, according to the integral approximation, $q = 10$ and evaluate the exact partition function at that temperature. [SSM **16.5**]

9.6 A certain atom has a doubly degenerate ground level pair and an upper level of four degenerate states at 450 cm^{-1} above the ground level. In an atomic beam study of the atoms it was observed that 30 per cent of the atoms were in the upper level, and the translational temperature of the beam was 300 K. Are the electronic states of the atoms in thermal equilibrium with the translational states? [ISM **16.6**]

9.7 (a) Calculate the electronic partition function of a tellurium atom at (i) 298 K, (ii) 5000 K by direct summation using the following data:

Term	Degeneracy	Wavenumber/cm^{-1}
Ground	5	0
1	1	4 707
2	3	4 751
3	5	10 559

(b) What proportion of the Te atoms are in the ground term and in the term labelled 2 at the two temperatures? (c) Calculate the electronic contribution to the standard molar entropy of gaseous Te atoms. [SSM **16.7**]

9.8 The four lowest electronic levels of a Ti atom are: $^3F_2$, $^3F_3$, $^3F_4$, and $^5F_1$, at 0, 170, 387, and 6557 cm^{-1}, respectively. There are many other electronic states at higher energies. The boiling point of titanium is 3287°C. What are the relative populations of these levels at the boiling point? *Hint.* The degeneracies of the levels are $2J + 1$. [ISM **16.8**]

9.9 The NO molecule has a doubly degenerate excited electronic level 121.1 cm^{-1} above the doubly degenerate electronic ground term. Calculate and plot the electronic partition function of NO from $T = 0$ to 1000 K. Evaluate (a) the term populations and (b) the electronic contribution to the molar internal energy at 300 K. Calculate the electronic contribution to the molar entropy of the NO molecule at 300 K and 500 K. [SSM **16.9**]

9.10‡ J. Sugar and A. Musgrove (*J. Phys. Chem. Ref. Data* **22**, 1213 (1993)) have published tables of energy levels for germanium atoms and cations from Ge^+ to Ge^{+31}. The lowest-lying energy levels in neutral Ge are as follows:

	$^3P_0$	$^3P_1$	$^3P_2$	$^1D_2$	$^1S_0$
E/cm^{-1}	0	557.1	1410.0	7125.3	16 367.3

Calculate the electronic partition function at 298 K and 1000 K by direct summation. *Hint.* The degeneracy of a level is $2J + 1$. [ISM **16.10**]

9.11 Calculate, by explicit summation, the vibrational partition function and the vibrational contribution to the molar internal energy of I_2 molecules at (a) 100 K, (b) 298 K given that its vibrational energy levels lie at the following wavenumbers above the zero-point energy level: 0, 213.30, 425.39, 636.27, 845.93 cm^{-1}. What proportion of I_2 molecules are in the ground and first two excited levels at the two temperatures? Calculate the vibrational contribution to the molar entropy of I_2 at the two temperatures. [SSM **16.11**]

9.12‡ (a) The standard molar entropy of graphite at 298, 410, and 498 K is 5.69, 9.03, and 11.63 J K^{-1} mol^{-1}, respectively. If 1.00 mol C(graphite) at 298 K is surrounded by thermal insulation and placed next to 1.00 mol C(graphite) at 498 K, also insulated, how many configurations are there altogether for the combined but independent systems? (b) If the same two samples are now placed in thermal contact and brought to thermal equilibrium, the final temperature will be 410 K. (Why might the final temperature not be the average?) How many configurations are there now in the combined system? Neglect any volume changes. (c) Demonstrate that this process is spontaneous. [ISM **16.12**]

Theoretical problems

9.13 A sample consisting of five molecules has a total energy 5ε. Each molecule is able to occupy states of energy $j\varepsilon$, with $j = 0, 1, 2, \ldots$. (a) Calculate the weight of the configuration in which the molecules are distributed evenly over the available states. (b) Draw up a table with columns headed by the energy of the states and write beneath them all configurations that are consistent with the total energy. Calculate the weights of each configuration and identify the most probable configurations. [SSM **16.13**]

9.14 A sample of nine molecules is numerically tractable but on the verge of being thermodynamically significant. Draw up a table of configurations for $N = 9$, total energy 9ε in a system with energy levels $j\varepsilon$ (as in Problem 9.13). Before evaluating the weights of the configurations, guess (by looking for the most 'exponential' distribution of populations) which of the configurations will turn out to be the most probable. Go on to calculate the weights and identify the most probable configuration. [ISM **16.14**]

9.15 The most probable configuration is characterized by a parameter we know as the 'temperature'. The temperatures of the system specified in Problems 9.13 and 9.14 must be such as to give a mean value of ε for the energy of each molecule and a total energy $N\varepsilon$ for the system. (a) Show that the temperature can be obtained by plotting p_j against j, where p_j is the (most probable) fraction of molecules in the state with energy $j\varepsilon$. Apply the procedure to the system in Problem 9.14. What is the temperature of the system when ε corresponds to 50 cm^{-1}? (b) Choose configurations other than the most probable, and show that the same procedure gives a worse straight line, indicating that a temperature is not well-defined for them. [SSM **16.15**]

9.16 A certain molecule can exist in either a non-degenerate singlet state or a triplet state (with degeneracy 3). The energy of the triplet exceeds that of the singlet by ε. Assuming that the molecules are distinguishable (localized) and independent, (a) obtain the expression for the molecular partition function. (b) Find expressions in terms of ε for the molar energy, molar heat capacity, and molar entropy of such molecules and calculate their values at $T = \varepsilon/k$. [ISM **16.16**]

9.17 Consider a system with energy levels $\varepsilon_j = j\varepsilon$ and N molecules. (a) Show that if the mean energy per molecule is $a\varepsilon$, then the temperature is given by

$$\beta = \frac{1}{\varepsilon}\ln\left(1 + \frac{1}{a}\right)$$

Evaluate the temperature for a system in which the mean energy is ε, taking ε equivalent to 50 cm^{-1}. (b) Calculate the molecular partition function q for the system when its mean energy is $a\varepsilon$. (c) Show that the entropy of the system is

$$S/k = (1 + a)\ln(1 + a) - a\ln a$$

and evaluate this expression for a mean energy ε. [SSM **16.17**]

9.18 Consider Stirling's approximation for ln $N!$ in the derivation of the Boltzmann distribution. What difference would it make if (a) a cruder approximation, $N! = N^N$, (b) the better approximation in *Comment* 9.2 were used instead? [ISM **16.18**]

9.19‡ For gases, the canonical partition function, Q, is related to the molecular partition function q by $Q = q^N/N!$. Use the expression for q and general thermodynamic relations to derive the perfect gas law $pV = nRT$. [SSM **16.19**]

Applications: to atmospheric science, astrophysics, and biochemistry

9.20‡ Obtain the barometric formula (Volume 1, Problem 1.27) from the Boltzmann distribution. Recall that the potential energy of a particle at height *h* above the surface of the Earth is *mgh*. Convert the barometric formula from pressure to number density, $\mathcal{N}$. Compare the relative number densities, $\mathcal{N}(h)/\mathcal{N}(0)$, for O_2 and H_2O at $h = 8.0$ km, a typical cruising altitude for commercial aircraft. [ISM **16.20**]

9.21‡ Planets lose their atmospheres over time unless they are replenished. A complete analysis of the overall process is very complicated and depends upon the radius of the planet, temperature, atmospheric composition, and other factors. Prove that the atmosphere of planets cannot be in an equilibrium state by demonstrating that the Boltzmann distribution leads to a uniform finite number density as $r \to \infty$. *Hint.* Recall that in a gravitational field the potential energy is $V(r) = -GMm/r$, where G is the gravitational constant, M is the mass of the planet, and m the mass of the particle. [SSM **16.21**]

9.22‡ Consider the electronic partition function of a perfect atomic hydrogen gas at a density of 1.99×10^{-4} kg m^{-3} and 5780 K. These are the mean conditions within the Sun's photosphere, the surface layer of the Sun that is about 190 km thick. (a) Show that this partition function, which involves a sum over an infinite number of quantum states that are solutions to the Schrödinger equation for an isolated atomic hydrogen atom, is infinite. (b) Develop a theoretical argument for truncating the sum and estimate the maximum number of quantum states that contribute to the sum. (c) Calculate the equilibrium probability that an atomic hydrogen electron is in each quantum state. Are there any general implications concerning electronic states that will be observed for other atoms and molecules? Is it wise to apply these calculations in the study of the Sun's photosphere? [ISM **16.22**]

9.23 Consider a protein P with four distinct sites, with each site capable of binding one ligand L. Show that the possible varieties (configurations) of the species PL_i (with PL_0 denoting P) are given by the binomial coefficients $C(4,i)$. [SSM **16.23**]

9.24 Complete some of the derivations in the discussion of the helix–coil transition in polypeptides (*Impact* I9.1). (a) Show that, within the tenets of the zipper model,

$$q = 1 + \sum_{i=1}^{n} Z(n,i)\sigma s^i$$

and that $Z(n,i) = n - i + 1$ is the number of ways in which an allowed state with a number i of c amino acids can be formed. (b) Using the zipper model, show that $\theta = (1/n)\mathrm{d}(\ln q)/\mathrm{d}(\ln s)$. *Hint.* As a first step, show that $\sum_i i(n-i+1)\sigma s^i = s(\mathrm{d}q/\mathrm{d}s)$. [ISM **16.24**]

9.25 Here you will use the zipper model discussed in *Impact* I16.1 to explore the helix–coil transition in polypeptides.(a) Investigate the effect of the parameter s on the distribution of random coil segments in a polypeptide with $n = 20$ by plotting p_i, the fraction of molecules with a number i of amino acids in a coil region, against i for $s = 0.8$, 1.0, and 1.5, with $\sigma = 5.0 \times 10^{-2}$. Discuss the significance of any effects you discover. (b) The average value of i given by $\langle i \rangle = \sum_i ip_i$. Use the results of the zipper model to calculate $\langle i \rangle$ for all the combinations of s and σ used in Fig. 9.10 and part (a). [SSM **16.25**]

Statistical thermodynamics 2: applications

10

In this chapter we apply the concepts of statistical thermodynamics to the calculation of chemically significant quantities. First, we establish the relations between thermodynamic functions and partition functions. Next, we show that the molecular partition function can be factorized into contributions from each mode of motion and establish the formulas for the partition functions for translational, rotational, and vibrational modes of motion and the contribution of electronic excitation. These contributions can be calculated from spectroscopic data. Finally, we turn to specific applications, which include the mean energies of modes of motion, the heat capacities of substances, and residual entropies. In the final section, we see how to calculate the equilibrium constant of a reaction and through that calculation understand some of the molecular features that determine the magnitudes of equilibrium constants and their variation with temperature.

A partition function is the bridge between thermodynamics, spectroscopy, and quantum mechanics. Once it is known, a partition function can be used to calculate thermodynamic functions, heat capacities, entropies, and equilibrium constants. It also sheds light on the significance of these properties.

Fundamental relations

In this section we see how to obtain any thermodynamic function once we know the partition function. Then we see how to calculate the molecular partition function, and through that the thermodynamic functions, from spectroscopic data.

10.1 The thermodynamic functions

We have already derived (in Chapter 9) the two expressions for calculating the internal energy and the entropy of a system from its canonical partition function, Q:

$$U - U(0) = -\left(\frac{\partial \ln Q}{\partial \beta}\right)_V \qquad S = \frac{U - U(0)}{T} + k \ln Q \tag{10.1}$$

where $\beta = 1/kT$. If the molecules are independent, we can go on to make the substitutions $Q = q^N$ (for distinguishable molecules, as in a solid) or $Q = q^N/N!$ (for indistinguishable molecules, as in a gas). All the thermodynamic functions introduced in Volume 1 are related to U and S, so we have a route to their calculation from Q.

(a) The Helmholtz energy

The Helmholtz energy, A, is defined as $A = U - TS$. This relation implies that $A(0) = U(0)$, so substitution for U and S by using eqn 10.1 leads to the very simple expression

$$A - A(0) = -kT \ln Q \tag{10.2}$$

(b) The pressure

By an argument like that leading to eqn 3.31 of Volume 1, it follows from $A = U - TS$ that $dA = -pdV - SdT$. Therefore, on imposing constant temperature, the pressure and the Helmholtz energy are related by $p = -(\partial A/\partial V)_T$. It then follows from eqn 10.2 that

$$p = kT\left(\frac{\partial \ln Q}{\partial V}\right)_T \tag{10.3}$$

This relation is entirely general, and may be used for any type of substance, including perfect gases, real gases, and liquids. Because Q is in general a function of the volume, temperature, and amount of substance, eqn 10.3 is an equation of state.

Example 10.1 *Deriving an equation of state*

Derive an expression for the pressure of a gas of independent particles.

Method We should suspect that the pressure is that given by the perfect gas law. To proceed systematically, substitute the explicit formula for Q for a gas of independent, indistinguishable molecules (see eqn 9.45 and Table 10.3 at the end of the chapter) into eqn 10.3.

Answer For a gas of independent molecules, $Q = q^N/N!$ with $q = V/\Lambda^3$:

$$p = kT\left(\frac{\partial \ln Q}{\partial V}\right)_T = \frac{kT}{Q}\left(\frac{\partial Q}{\partial V}\right)_T = \frac{NkT}{q}\left(\frac{\partial q}{\partial V}\right)_T$$

$$= \frac{NkT\Lambda^3}{V} \times \frac{1}{\Lambda^3} = \frac{NkT}{V} = \frac{nRT}{V}$$

To derive this relation, we have used

$$\left(\frac{\partial q}{\partial V}\right)_T = \left(\frac{\partial (V/\Lambda^3)}{\partial V}\right)_T = \frac{1}{\Lambda^3}$$

and $NkT = nN_AkT = nRT$. The calculation shows that the equation of state of a gas of independent particles is indeed the perfect gas law.

Self-test 10.1 Derive the equation of state of a sample for which $Q = q^N f/N!$, with $q = V/\Lambda^3$, where f depends on the volume. [$p = nRT/V + kT(\partial \ln f/\partial V)_T$]

(c) The enthalpy

At this stage we can use the expressions for U and p in the definition $H = U + pV$ to obtain an expression for the enthalpy, H, of any substance:

$$H - H(0) = -\left(\frac{\partial \ln Q}{\partial \beta}\right)_V + kTV\left(\frac{\partial \ln Q}{\partial V}\right)_T \tag{10.4}$$

We have already seen that $U - U(0) = \frac{3}{2}nRT$ for a gas of independent particles (eqn 9.32a), and have just shown that $pV = nRT$. Therefore, for such a gas,

$$H - H(0) = \tfrac{5}{2}nRT \qquad (10.5)^{\circ}$$

(d) The Gibbs energy

One of the most important thermodynamic functions for chemistry is the Gibbs energy, $G = H - TS = A + pV$. We can now express this function in terms of the partition function by combining the expressions for A and p:

$$G - G(0) = -kT \ln Q + kTV\left(\frac{\partial \ln Q}{\partial V}\right)_T \qquad (10.6)$$

This expression takes a simple form for a gas of independent molecules because pV in the expression $G = A + pV$ can be replaced by nRT:

$$G - G(0) = -kT \ln Q + nRT \qquad (10.7)^{\circ}$$

Furthermore, because $Q = q^N/N!$, and therefore $\ln Q = N \ln q - \ln N!$, it follows by using Stirling's approximation ($\ln N! \approx N \ln N - N$) that we can write

$$\begin{aligned} G - G(0) &= -NkT \ln q + kT \ln N! + nRT \\ &= -nRT \ln q + kT(N \ln N - N) + nRT \\ &= -nRT \ln \frac{q}{N} \end{aligned} \qquad (10.8)^{\circ}$$

with $N = nN_A$. Now we see another interpretation of the Gibbs energy: it is proportional to the logarithm of the average number of thermally accessible states per molecule.

It will turn out to be convenient to define the **molar partition function**, $q_m = q/n$ (with units mol^{-1}), for then

$$G - G(0) = -nRT \ln \frac{q_m}{N_A} \qquad (10.9)^{\circ}$$

10.2 The molecular partition function

The energy of a molecule is the sum of contributions from its different modes of motion:

$$\varepsilon_i = \varepsilon_i^T + \varepsilon_i^R + \varepsilon_i^V + \varepsilon_i^E \qquad (10.10)$$

where T denotes translation, R rotation, V vibration, and E the electronic contribution. The electronic contribution is not actually a 'mode of motion', but it is convenient to include it here. The separation of terms in eqn 10.10 is only approximate (except for translation) because the modes are not completely independent, but in most cases it is satisfactory. The separation of the electronic and vibrational motions is justified provided only the ground electronic state is occupied (for otherwise the vibrational characteristics depend on the electronic state) and, for the electronic ground state, that the Born–Oppenheimer approximation is valid (Chapter 4). The separation of the vibrational and rotational modes is justified to the extent that the rotational constant is independent of the vibrational state.

Given that the energy is a sum of independent contributions, the partition function factorizes into a product of contributions (recall Section 9.2b):

$$q = \sum_i e^{-\beta\varepsilon_i} = \sum_{i\,(\text{all states})} e^{-\beta\varepsilon_i^{T}-\beta\varepsilon_i^{R}-\beta\varepsilon_i^{V}-\beta\varepsilon_i^{E}}$$

$$= \sum_{i\,(\text{translational})}\ \sum_{i\,(\text{rotational})}\ \sum_{i\,(\text{vibrational})}\ \sum_{i\,(\text{electronic})} e^{-\beta\varepsilon_i^{T}-\beta\varepsilon_i^{R}-\beta\varepsilon_i^{V}-\beta\varepsilon_i^{E}} \tag{10.11}$$

$$= \left(\sum_{i\,(\text{translational})} e^{-\beta\varepsilon_i^{T}}\right)\left(\sum_{i\,(\text{rotational})} e^{-\beta\varepsilon_i^{R}}\right)\left(\sum_{i\,(\text{vibrational})} e^{-\beta\varepsilon_i^{V}}\right)\left(\sum_{i\,(\text{electronic})} e^{-\beta\varepsilon_i^{E}}\right)$$

$$= q^{T}q^{R}q^{V}q^{E}$$

This factorization means that we can investigate each contribution separately.

(a) The translational contribution

The translational partition function of a molecule of mass m in a container of volume V was derived in Section 9.2:

$$q^{T} = \frac{V}{\Lambda^3} \qquad \Lambda = h\left(\frac{\beta}{2\pi m}\right)^{1/2} = \frac{h}{(2\pi mkT)^{1/2}} \tag{10.12}$$

Notice that $q^{T} \to \infty$ as $T \to \infty$ because an infinite number of states becomes accessible as the temperature is raised. Even at room temperature $q^{T} \approx 2 \times 10^{28}$ for an O_2 molecule in a vessel of volume 100 cm^3.

The thermal wavelength, Λ, lets us judge whether the approximations that led to the expression for q^{T} are valid. The approximations are valid if many states are occupied, which requires V/Λ^3 to be large. That will be so if Λ is small compared with the linear dimensions of the container. For H_2 at 25°C, $\Lambda = 71$ pm, which is far smaller than any conventional container is likely to be (but comparable to pores in zeolites or cavities in clathrates). For O_2, a heavier molecule, $\Lambda = 18$ pm. We saw in Section 9.2 that an equivalent criterion of validity is that Λ should be much less than the average separation of the molecules in the sample.

(b) The rotational contribution

As demonstrated in Example 9.1, the partition function of a nonsymmetrical (AB) linear rotor is

$$q^{R} = \sum_J (2J+1)e^{-\beta hcBJ(J+1)} \tag{10.13}$$

The direct method of calculating q^{R} is to substitute the experimental values of the rotational energy levels into this expression and to sum the series numerically.

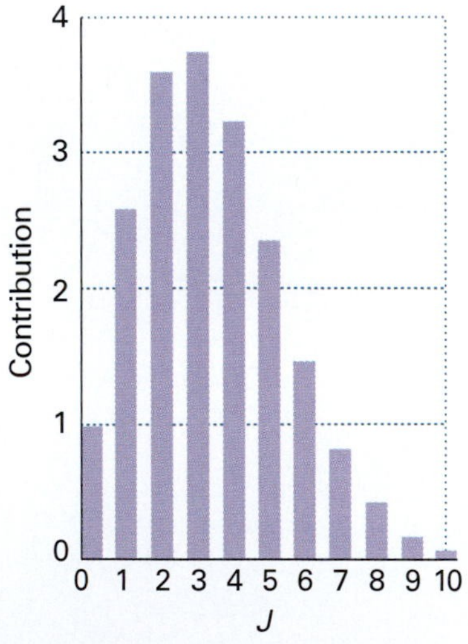

Fig. 10.1 The contributions to the rotational partition function of an HCl molecule at 25°C. The vertical axis is the value of $(2J+1)e^{-\beta hcBJ(J+1)}$. Successive terms (which are proportional to the populations of the levels) pass through a maximum because the population of individual states decreases exponentially, but the degeneracy of the levels increases with J.

Example 10.2 *Evaluating the rotational partition function explicitly*

Evaluate the rotational partition function of $^1H^{35}Cl$ at 25°C, given that B = 10.591 cm^{-1}.

Method We use eqn 10.13 and evaluate it term by term. A useful relation is kT/hc = 207.22 cm^{-1} at 298.15 K. The sum is readily evaluated by using mathematical software.

Answer To show how successive terms contribute, we draw up the following table by using $kT/hcB = 0.051\,11$ (Fig. 10.1):

J	0	1	2	3	4	...	10
$(2J+1)e^{-0.0511J(J+1)}$	1	2.71	3.68	3.79	3.24	...	0.08

The sum required by eqn 10.13 (the sum of the numbers in the second row of the table) is 19.9, hence $q^R = 19.9$ at this temperature. Taking J up to 50 gives $q^R = 19.902$. Notice that about ten J-levels are significantly populated but the number of populated *states* is larger on account of the $(2J+1)$-fold degeneracy of each level. We shall shortly encounter the approximation that $q^R \approx kT/hcB$, which in the present case gives $q^R = 19.6$, in good agreement with the exact value and with much less work.

Self-test 10.2 Evaluate the rotational partition function for HCl at 0°C. [18.26]

At room temperature $kT/hc \approx 200\ \text{cm}^{-1}$. The rotational constants of many molecules are close to $1\ \text{cm}^{-1}$ (Table 6.2) and often smaller (though the very light H_2 molecule, for which $B = 60.9\ \text{cm}^{-1}$, is one exception). It follows that many rotational levels are populated at normal temperatures. When this is the case, the partition function may be approximated by

Linear rotors: $$q^R = \frac{kT}{hcB} \tag{10.14a}$$

Nonlinear rotors: $$q^R = \left(\frac{kT}{hc}\right)^{3/2}\left(\frac{\pi}{ABC}\right)^{1/2} \tag{10.14b}$$

where A, B, and C are the rotational constants of the molecule. However, before using these expressions, read on (to eqns 10.15 and 10.16).

Justification 10.1 *The rotational contribution to the molecular partition function*

When many rotational states are occupied and kT is much larger than the separation between neighbouring states, the sum in the partition function can be approximated by an integral, much as we did for translational motion in *Justification 9.2*:

$$q^R = \int_0^\infty (2J+1)e^{-\beta hcBJ(J+1)}dJ$$

Although this integral looks complicated, it can be evaluated without much effort by noticing that because

$$\frac{d}{dJ}e^{aJ(J+1)} = \left\{\frac{d}{dJ}aJ(J+1)\right\}e^{aJ(J+1)} = a(2J+1)e^{aJ(J+1)}$$

it can also be written as

$$q^R = \frac{1}{\beta hcB}\int_0^\infty \left(\frac{d}{dJ}e^{-\beta hcBJ(J+1)}\right)dJ$$

Then, because the integral of a derivative of a function is the function itself, we obtain

$$q^R = -\frac{1}{\beta hcB}e^{-\beta hcBJ(J+1)}\Big|_0^\infty = \frac{1}{\beta hcB}$$

which (because $\beta = 1/kT$) is eqn 10.14a.

The calculation for a nonlinear molecule is along the same lines, but slightly trickier. First, we note that the energies of a symmetric rotor are

$$E_{J,K,M_J} = hcBJ(J+1) + hc(A-B)K^2$$

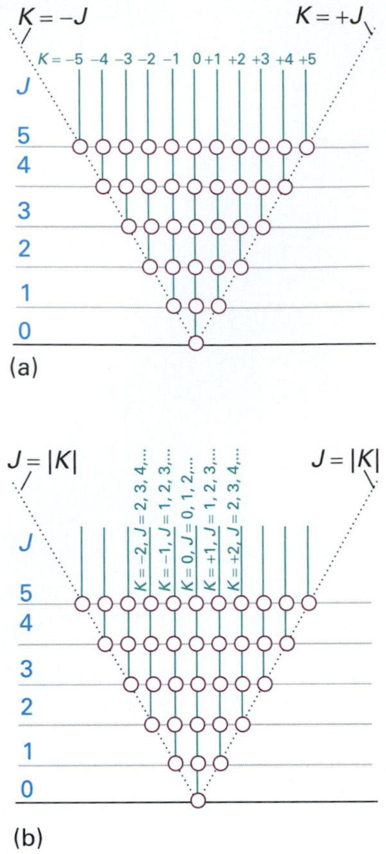

Fig. 10.2 (a) The sum over $J = 0, 1, 2, \ldots$ and $K = J, J-1, \ldots, -J$ (depicted by the circles) can be covered (b) by allowing K to range from $-\infty$ to ∞, with J confined to $|K|$, $|K|+1, \ldots, \infty$ for each value of K.

with $J = 0, 1, 2, \ldots, K = J, J-1, \ldots, -J$, and $M_J = J, J-1, \ldots, -J$. Instead of considering these ranges, we can cover the same values by allowing K to range from $-\infty$ to ∞, with J confined to $|K|, |K|+1, \ldots, \infty$ for each value of K (Fig. 10.2). Because the energy is independent of M_J, and there are $2J+1$ values of M_J for each value of J, each value of J is $2J+1$-fold degenerate. It follows that the partition function

$$q = \sum_{J=0}^{\infty} \sum_{K=-J}^{J} \sum_{M_J=-J}^{J} e^{-E_{JKM_J}/kT}$$

can be written equivalently as

$$\begin{aligned} q &= \sum_{K=-\infty}^{\infty} \sum_{J=|K|}^{\infty} (2J+1)e^{-E_{JKM_J}/kT} \\ &= \sum_{K=-\infty}^{\infty} \sum_{J=|K|}^{\infty} (2J+1)e^{-hc\{BJ(J+1)+(A-B)K^2\}/kT} \\ &= \sum_{K=-\infty}^{\infty} e^{-\{hc(A-B)/kT\}K^2} \sum_{J=|K|}^{\infty} (2J+1)e^{-hcBJ(J+1)/kT} \end{aligned}$$

Now we assume that the temperature is so high that numerous states are occupied and that the sums may be approximated by integrals. Then

$$q = \int_{-\infty}^{\infty} e^{-\{hc(A-B)/kT\}K^2} \int_{|K|}^{\infty} (2J+1)e^{-hcBJ(J+1)/kT} \mathrm{d}J\mathrm{d}K$$

As before, the integral over J can be recognized as the integral of the derivative of a function, which is the function itself, so

$$\begin{aligned} \int_{|K|}^{\infty} (2J+1)e^{-hcBJ(J+1)/kT}\mathrm{d}J &= \int_{|K|}^{\infty} \left(-\frac{kT}{hcB}\right)\frac{\mathrm{d}}{\mathrm{d}J} e^{-hcBJ(J+1)/kT}\mathrm{d}J \\ &= \left(-\frac{kT}{hcB}\right) e^{-hcBJ(J+1)/kT}\Big|_{|K|}^{\infty} = \left(\frac{kT}{hcB}\right) e^{-hcB|K|(|K|+1)/kT} \\ &\approx \left(\frac{kT}{hcB}\right) e^{-hcBK^2/kT} \end{aligned}$$

In the last line we have supposed that $|K| \gg 1$ for most contributions. Now we can write

$$\begin{aligned} q &= \frac{kT}{hcB}\int_{-\infty}^{\infty} e^{-\{hc(A-B)/kT\}K^2} e^{-hcBK^2/kT}\mathrm{d}K \\ &= \frac{kT}{hcB}\int_{-\infty}^{\infty} e^{-\{hcA/kT\}K^2}\mathrm{d}K = \left(\frac{kT}{hcB}\right)\left(\frac{kT}{hcA}\right)^{1/2} \overbrace{\int_{-\infty}^{\infty} e^{-x^2}\mathrm{d}x}^{\pi^{1/2}} \\ &= \left(\frac{kT}{hc}\right)^{3/2}\left(\frac{\pi}{AB^2}\right)^{1/2} \end{aligned}$$

For an asymmetric rotor, one of the Bs is replaced by C, to give eqn 10.14b.

Synoptic table 10.1* Rotational and vibrational temperatures

Molecule	Mode	θ_V/K	θ_R/K
H_2		6330	88
HCl		4300	9.4
I_2		309	0.053
CO_2	ν_1	1997	0.561
	ν_2	3380	
	ν_3	960	

* For more values, see Table 6.2 in the *Data section* and use $hc/k = 1.439$ K cm.

A useful way of expressing the temperature above which the rotational approximation is valid is to introduce the **characteristic rotational temperature**, $\theta_R = hcB/k$. Then 'high temperature' means $T \gg \theta_R$ and under these conditions the rotational partition function of a linear molecule is simply T/θ_R. Some typical values of θ_R are shown in Table 10.1. The value for H_2 is abnormally high and we must be careful with the approximation for this molecule.

The general conclusion at this stage is that molecules with large moments of inertia (and hence small rotational constants and low characteristic rotational temperatures) have large rotational partition functions. The large value of q^R reflects the closeness in energy (compared with kT) of the rotational levels in large, heavy molecules, and the large number of them that are accessible at normal temperatures.

We must take care, however, not to include too many rotational states in the sum. For a homonuclear diatomic molecule or a symmetrical linear molecule (such as CO_2 or HC≡CH), a rotation through 180° results in an indistinguishable state of the molecule. Hence, the number of thermally accessible states is only half the number that can be occupied by a heteronuclear diatomic molecule, where rotation through 180° does result in a distinguishable state. Therefore, for a symmetrical linear molecule,

$$q^R = \frac{kT}{2hcB} = \frac{T}{2\theta_R} \tag{10.15a}$$

The equations for symmetrical and nonsymmetrical molecules can be combined into a single expression by introducing the **symmetry number**, σ, which is the number of indistinguishable orientations of the molecule. Then

$$q^R = \frac{kT}{\sigma hcB} = \frac{T}{\sigma\theta_R} \tag{10.15b}$$

For a heteronuclear diatomic molecule $\sigma = 1$; for a homonuclear diatomic molecule or a symmetrical linear molecule, $\sigma = 2$.

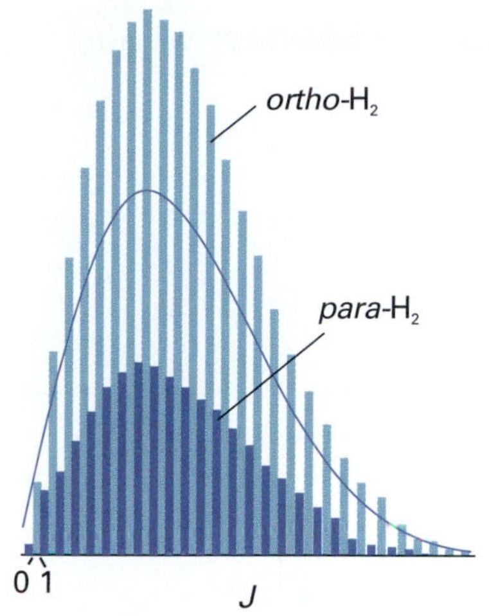

Fig. 10.3 The values of the individual terms $(2J+1)e^{-\beta hcBJ(J+1)}$ contributing to the mean partition function of a 3:1 mixture of *ortho-* and *para-*H_2. The partition function is the sum of all these terms. At high temperatures, the sum is approximately equal to the sum of the terms over all values of J, each with a weight of $\frac{1}{2}$. This is the sum of the contributions indicated by the curve.

Justification 10.2 *The origin of the symmetry number*

The quantum mechanical origin of the symmetry number is the Pauli principle, which forbids the occupation of certain states. We saw in Section 6.8, for example, that H_2 may occupy rotational states with even J only if its nuclear spins are paired (*para*-hydrogen), and odd J states only if its nuclear spins are parallel (*ortho*-hydrogen). There are three states of *ortho*-H_2 to each value of J (because there are three parallel spin states of the two nuclei).

To set up the rotational partition function we note that 'ordinary' molecular hydrogen is a mixture of one part *para*-H_2 (with only its even-J rotational states occupied) and three parts *ortho*-H_2 (with only its odd-J rotational states occupied). Therefore, the average partition function per molecule is

$$q^R = \tfrac{1}{4}\sum_{\text{even } J}(2J+1)e^{-\beta hcBJ(J+1)} + \tfrac{3}{4}\sum_{\text{odd } J}(2J+1)e^{-\beta hcBJ(J+1)}$$

The odd-J states are more heavily weighted than the even-J states (Fig. 10.3). From the illustration we see that we would obtain approximately the same answer for the partition function (the sum of all the populations) if each J term contributed half its normal value to the sum. That is, the last equation can be approximated as

$$q^R = \tfrac{1}{2}\sum_{J}(2J+1)e^{-\beta hcBJ(J+1)}$$

and this approximation is very good when many terms contribute (at high temperatures).

The same type of argument may be used for linear symmetrical molecules in which identical bosons are interchanged by rotation (such as CO_2). As pointed out in Section 6.8, if the nuclear spin of the bosons is 0, then only even-J states are admissible. Because only half the rotational states are occupied, the rotational partition function is only half the value of the sum obtained by allowing all values of J to contribute (Fig. 10.4).

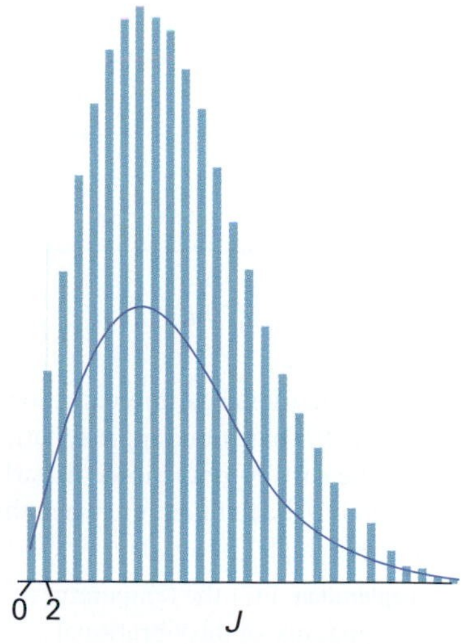

Fig. 10.4 The relative populations of the rotational energy levels of CO_2. Only states with even J values are occupied. The full line shows the smoothed, averaged population of levels.

Synoptic table 10.2* Symmetry numbers

Molecule	σ
H_2O	2
NH_3	3
CH_4	12
C_6H_6	12

* For more values, see Table 6.2 in the *Data section*.

The same care must be exercised for other types of symmetrical molecule, and for a nonlinear molecule we write

$$q^{R} = \frac{1}{\sigma}\left(\frac{kT}{hc}\right)^{3/2}\left(\frac{\pi}{ABC}\right)^{1/2} \qquad (10.16)$$

Some typical values of the symmetry numbers required are given in Table 10.2. The value $\sigma(H_2O) = 2$ reflects the fact that a 180° rotation about the bisector of the H—O—H angle interchanges two indistinguishable atoms. In NH_3, there are three indistinguishable orientations around the axis shown in (**1**). For CH_4, any of three 120° rotations about any of its four C—H bonds leaves the molecule in an indistinguishable state, so the symmetry number is $3 \times 4 = 12$. For benzene, any of six orientations around the axis perpendicular to the plane of the molecule leaves it apparently unchanged, as does a rotation of 180° around any of six axes in the plane of the molecule (three of which pass along each C—H bond and the remaining three pass through each C—C bond in the plane of the molecule). For the way that group theory is used to identify the value of the symmetry number, see Problem 10.17.

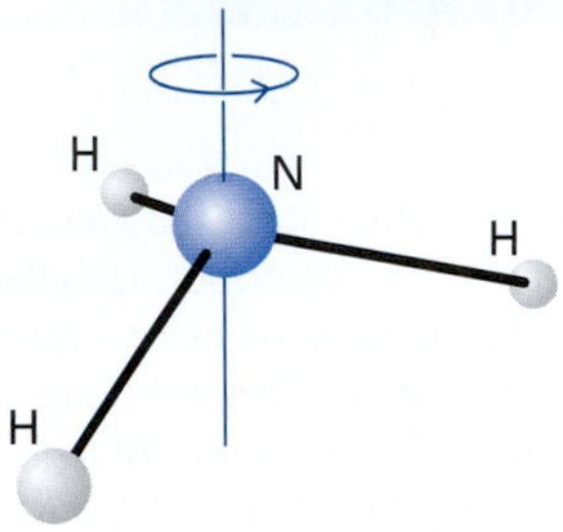

1

(c) The vibrational contribution

The vibrational partition function of a molecule is calculated by substituting the measured vibrational energy levels into the exponentials appearing in the definition of q^V, and summing them numerically. In a polyatomic molecule each normal mode (Section 6.14) has its own partition function (provided the anharmonicities are so small that the modes are independent). The overall vibrational partition function is the product of the individual partition functions, and we can write $q^V = q^V(1)q^V(2)\ldots$, where $q^V(K)$ is the partition function for the Kth normal mode and is calculated by direct summation of the observed spectroscopic levels.

If the vibrational excitation is not too great, the harmonic approximation may be made, and the vibrational energy levels written as

$$E_v = (v + \tfrac{1}{2})hc\tilde{\nu} \qquad v = 0, 1, 2, \ldots \qquad (10.17)$$

If, as usual, we measure energies from the zero-point level, then the permitted values are $\varepsilon_v = vhc\tilde{\nu}$ and the partition function is

$$q^V = \sum_v e^{-\beta vhc\tilde{\nu}} = \sum_v (e^{-\beta hc\tilde{\nu}})^v \qquad (10.18)$$

(because $e^{ax} = (e^x)^a$). We met this sum in Example 9.2 (which is no accident: the ladder-like array of levels in Fig. 9.3 is exactly the same as that of a harmonic oscillator). The series can be summed in the same way, and gives

$$q^V = \frac{1}{1 - e^{-\beta hc\tilde{\nu}}} \qquad (10.19)$$

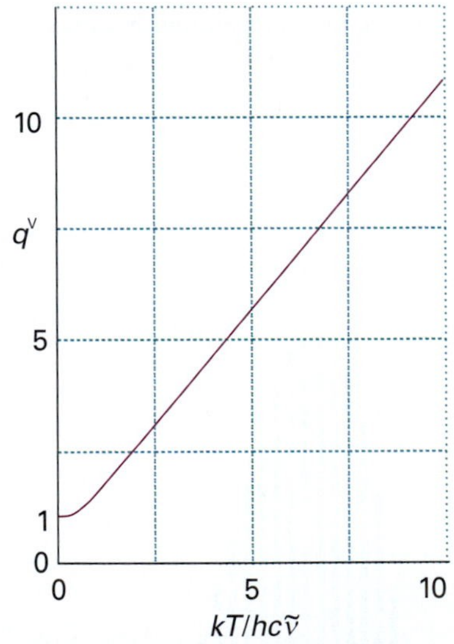

Fig. 10.5 The vibrational partition function of a molecule in the harmonic approximation. Note that the partition function is linearly proportional to the temperature when the temperature is high ($T \gg \theta_V$).

Exploration Plot the temperature dependence of the vibrational contribution to the molecular partition function for several values of the vibrational wavennumber. Estimate from your plots the temperature above which the harmonic oscillator is in the 'high temperature' limit.

This function is plotted in Fig. 10.5. In a polyatomic molecule, each normal mode gives rise to a partition function of this form.

Example 10.3 *Calculating a vibrational partition function*

The wavenumbers of the three normal modes of H_2O are 3656.7 cm^{-1}, 1594.8 cm^{-1}, and 3755.8 cm^{-1}. Evaluate the vibrational partition function at 1500 K.

Method Use eqn 10.19 for each mode, and then form the product of the three contributions. At 1500 K, $kT/hc = 1042.6$ cm^{-1}.

Answer We draw up the following table displaying the contributions of each mode:

Mode:	1	2	3
$\tilde{\nu}/\text{cm}^{-1}$	3656.7	1594.8	3755.8
$hc\tilde{\nu}/kT$	3.507	1.530	3.602
q^{V}	1.031	1.276	1.028

The overall vibrational partition function is therefore

$$q^{V} = 1.031 \times 1.276 \times 1.028 = 1.353$$

The three normal modes of H_2O are at such high wavenumbers that even at 1500 K most of the molecules are in their vibrational ground state. However, there may be so many normal modes in a large molecule that their excitation may be significant even though each mode is not appreciably excited. For example, a nonlinear molecule containing 10 atoms has $3N - 6 = 24$ normal modes (Section 6.14). If we assume a value of about 1.1 for the vibrational partition function of one normal mode, the overall vibrational partition function is about $q^{V} \approx (1.1)^{24} = 9.8$, which indicates significant vibrational excitation relative to a smaller molecule, such as H_2O.

Self-test 10.3 Repeat the calculation for CO_2, where the vibrational wavenumbers are 1388 cm^{-1}, 667.4 cm^{-1}, and 2349 cm^{-1}, the second being the doubly degenerate bending mode. [6.79]

In many molecules the vibrational wavenumbers are so great that $\beta hc\tilde{\nu} > 1$. For example, the lowest vibrational wavenumber of CH_4 is 1306 cm^{-1}, so $\beta hc\tilde{\nu} = 6.3$ at room temperature. C—H stretches normally lie in the range 2850 to 2960 cm^{-1}, so for them $\beta hc\tilde{\nu} \approx 14$. In these cases, $e^{-\beta hc\tilde{\nu}}$ in the denominator of q^{V} is very close to zero (for example, $e^{-6.3} = 0.002$), and the vibrational partition function for a single mode is very close to 1 ($q^{V} = 1.002$ when $\beta hc\tilde{\nu} = 6.3$), implying that only the zero-point level is significantly occupied.

Now consider the case of bonds so weak that $\beta hc\tilde{\nu} \ll kT$. When this condition is satisfied, the partition function may be approximated by expanding the exponential ($e^{x} = 1 + x + \cdots$):

$$q^{V} = \frac{1}{1 - (1 - \beta hc\tilde{\nu} + \cdots)} \tag{10.20}$$

That is, for weak bonds at high temperatures,

$$q^{V} = \frac{1}{\beta hc\tilde{\nu}} = \frac{kT}{hc\tilde{\nu}} \tag{10.21}$$

The temperatures for which eqn 10.21 is valid can be expressed in terms of the **characteristic vibrational temperature**, $\theta_V = hc\tilde{\nu}/k$ (Table 10.1). The value for H_2 is abnormally high because the atoms are so light and the vibrational frequency is correspondingly high. In terms of the vibrational temperature, 'high temperature' means $T \gg \theta_V$ and, when this condition is satisfied, $q^{V} = T/\theta_V$ (the analogue of the rotational expression).

(d) The electronic contribution

Electronic energy separations from the ground state are usually very large, so for most cases $q^{E} = 1$. An important exception arises in the case of atoms and molecules having

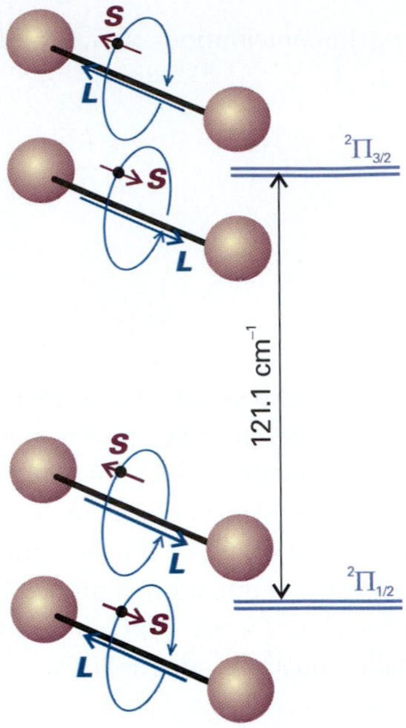

Fig. 10.6 The doubly degenerate ground electronic level of NO (with the spin and orbital angular momentum around the axis in opposite directions) and the doubly degenerate first excited level (with the spin and orbital momenta parallel). The upper level is thermally accessible at room temperature.

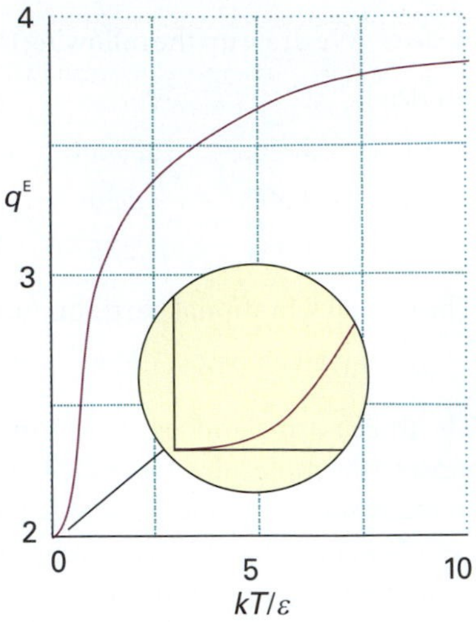

Fig. 10.7 The variation with temperature of the electronic partition function of an NO molecule. Note that the curve resembles that for a two-level system (Fig. 9.5), but rises from 2 (the degeneracy of the lower level) and approaches 4 (the total number of states) at high temperatures.

Exploration Plot the temperature dependence of the electronic partition function for several values of the energy separation ε between two doubly degenerate levels. From your plots, estimate the temperature at which the population of the excited level begins to increase sharply.

electronically degenerate ground states, in which case $q^E = g^E$, where g^E is the degeneracy of the electronic ground state. Alkali metal atoms, for example, have doubly degenerate ground states (corresponding to the two orientations of their electron spin), so $q^E = 2$.

Some atoms and molecules have low-lying electronically excited states. (At high enough temperatures, all atoms and molecules have thermally accessible excited states.) An example is NO, which has a configuration of the form . . . π^1 (see *Impact* I4.1). The orbital angular momentum may take two orientations with respect to the molecular axis (corresponding to circulation clockwise or counter-clockwise around the axis), and the spin angular momentum may also take two orientations, giving four states in all (Fig. 10.6). The energy of the two states in which the orbital and spin momenta are parallel (giving the $^2\Pi_{3/2}$ term) is slightly greater than that of the two other states in which they are antiparallel (giving the $^2\Pi_{1/2}$ term). The separation, which arises from spin–orbit coupling (Section 3.8), is only 121 cm^{-1}. Hence, at normal temperatures, all four states are thermally accessible. If we denote the energies of the two levels as $E_{1/2} = 0$ and $E_{3/2} = \varepsilon$, the partition function is

$$q^E = \sum_{\text{energy levels}} g_j e^{-\beta\varepsilon_j} = 2 + 2e^{-\beta\varepsilon} \qquad (10.22)$$

Figure 10.7 shows the variation of this function with temperature. At $T = 0$, $q^E = 2$, because only the doubly degenerate ground state is accessible. At high temperatures, q^E approaches 4 because all four states are accessible. At 25°C, $q^E = 3.1$.

(e) The overall partition function

The partition functions for each mode of motion of a molecule are collected in Table 10.3 at the end of the chapter. The overall partition function is the product of each contribution. For a diatomic molecule with no low-lying electronically excited states and $T \gg \theta_R$,

$$q = g^E\left(\frac{V}{\Lambda^3}\right)\left(\frac{T}{\sigma\theta_R}\right)\left(\frac{1}{1-e^{-T/\theta_v}}\right) \qquad (10.23)$$

Example 10.4 *Calculating a thermodynamic function from spectroscopic data*

Calculate the value of $G_m^{\ominus} - G_m^{\ominus}(0)$ for $H_2O(g)$ at 1500 K given that $A = 27.8778\ \text{cm}^{-1}$, $B = 14.5092\ \text{cm}^{-1}$, and $C = 9.2869\ \text{cm}^{-1}$ and the information in Example 10.3.

Method The starting point is eqn 10.9. For the standard value, we evaluate the translational partition function at $p^{\ominus}$ (that is, at 10^5 Pa exactly). The vibrational partition function was calculated in Example 10.3. Use the expressions in Table 17.3 for the other contributions.

Answer Because $m = 18.015$ u, it follows that $q_m^{T\ominus}/N_A = 1.706 \times 10^8$. For the vibrational contribution we have already found that $q^V = 1.352$. From Table 10.2 we see that $\sigma = 2$, so the rotational contribution is $q^R = 486.7$. Therefore,

$$\begin{aligned} G_m^{\ominus} - G_m^{\ominus}(0) &= -(8.3145\ \text{J K}^{-1}\ \text{mol}^{-1}) \times (1500\ \text{K}) \\ &\quad \times \ln\{(1.706 \times 10^8) \times 486.7 \times 1.352\} \\ &= -317.3\ \text{kJ mol}^{-1} \end{aligned}$$

Self-test 10.4 Repeat the calculation for CO_2. The vibrational data are given in Self-test 10.3; $B = 0.3902\ \text{cm}^{-1}$. [$-366.6\ \text{kJ mol}^{-1}$]

Comment 10.1
The text's web site contains links to on-line databases of atomic and molecular spectra.

Overall partition functions obtained from eqn 10.23 are approximate because they assume that the rotational levels are very close together and that the vibrational levels are harmonic. These approximations are avoided by using the energy levels identified spectroscopically and evaluating the sums explicitly.

Using statistical thermodynamics

We can now calculate any thermodynamic quantity from a knowledge of the energy levels of molecules: we have merged thermodynamics and spectroscopy. In this section, we indicate how to do the calculations for four important properties.

10.3 Mean energies

It is often useful to know the mean energy, $\langle\varepsilon\rangle$, of various modes of motion. When the molecular partition function can be factorized into contributions from each mode, the mean energy of each mode M (from eqn 9.29) is

$$\langle\varepsilon^M\rangle = -\frac{1}{q^M}\left(\frac{\partial q^M}{\partial\beta}\right)_V \qquad M = \text{T, R, V, or E} \qquad (10.24)$$

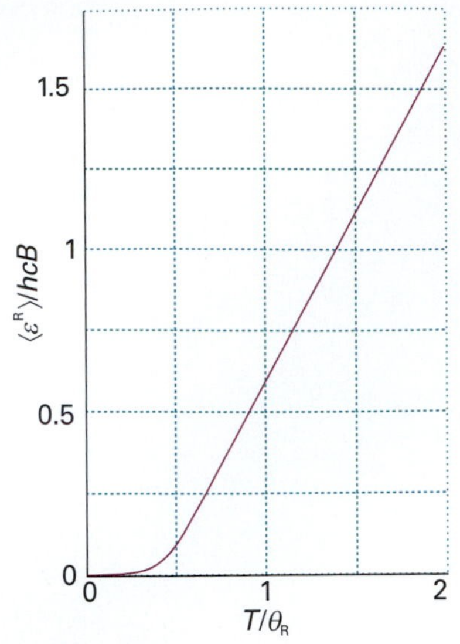

Fig. 10.8 The mean rotational energy of a nonsymmetrical linear rotor as a function of temperature. At high temperatures ($T \gg \theta_R$), the energy is linearly proportional to the temperature, in accord with the equipartition theorem.

Exploration Plot the temperature dependence of the mean rotational energy for several values of the rotational constant (for reasonable values of the rotational constant, see the *Data section*). From your plots, estimate the temperature at which the mean rotational energy begins to increase sharply.

(a) The mean translational energy

To see a pattern emerging, we consider first a one-dimensional system of length X, for which $q^T = X/\Lambda$, with $\Lambda = h(\beta/2\pi m)^{1/2}$. Then, if we note that Λ is a constant times $\beta^{1/2}$,

$$\langle \varepsilon^T \rangle = -\frac{\Lambda}{X}\left(\frac{\partial}{\partial \beta}\frac{X}{\Lambda}\right)_V = -\beta^{1/2}\frac{d}{d\beta}\left(\frac{1}{\beta^{1/2}}\right) = \frac{1}{2\beta} = \tfrac{1}{2}kT \qquad (10.25a)$$

For a molecule free to move in three dimensions, the analogous calculation leads to

$$\langle \varepsilon^T \rangle = \tfrac{3}{2}kT \qquad (10.25b)$$

Both conclusions are in agreement with the classical equipartition theorem that the mean energy of each quadratic contribution to the energy is $\tfrac{1}{2}kT$. Furthermore, the fact that the mean energy is independent of the size of the container is consistent with the thermodynamic result that the internal energy of a perfect gas is independent of its volume.

(b) The mean rotational energy

The mean rotational energy of a linear molecule is obtained from the partition function given in eqn 10.13. When the temperature is low ($T < \theta_R$), the series must be summed term by term, which gives

$$q^R = 1 + 3e^{-2\beta hcB} + 5e^{-6\beta hcB} + \cdots$$

Hence

$$\langle \varepsilon^R \rangle = \frac{hcB(6e^{-2\beta hcB} + 30e^{-6\beta hcB} + \cdots)}{1 + 3e^{-2\beta hcB} + 5e^{-6\beta hcB} + \cdots} \qquad (10.26a)$$

This function is plotted in Fig. 10.8. At high temperatures ($T \gg \theta_R$), q^R is given by eqn 10.15, and

$$\langle \varepsilon^R \rangle = -\frac{1}{q^R}\frac{dq^R}{d\beta} = -\sigma hc\beta B\frac{d}{d\beta}\frac{1}{\sigma hc\beta B} = \frac{1}{\beta} = kT \qquad (10.26b)$$

(q^R is independent of V, so the partial derivatives have been replaced by complete derivatives.) The high-temperature result is also in agreement with the equipartition theorem, for the classical expression for the energy of a linear rotor is $E_K = \tfrac{1}{2}I_\perp\omega_a^2 + \tfrac{1}{2}I_\perp\omega_b^2$. (There is no rotation around the line of atoms.) It follows from the equipartition theorem that the mean rotational energy is $2 \times \tfrac{1}{2}kT = kT$.

(c) The mean vibrational energy

The vibrational partition function in the harmonic approximation is given in eqn 10.19. Because q^V is independent of the volume, it follows that

$$\frac{dq^V}{d\beta} = \frac{d}{d\beta}\left(\frac{1}{1 - e^{-\beta hc\tilde{\nu}}}\right) = -\frac{hc\tilde{\nu}e^{-\beta hc\tilde{\nu}}}{(1 - e^{-\beta hc\tilde{\nu}})^2} \qquad (10.27)$$

and hence from

$$\langle \varepsilon^V \rangle = -\frac{1}{q^V}\frac{dq^V}{d\beta} = -(1 - e^{-\beta hc\tilde{\nu}})\left\{-\frac{hc\tilde{\nu}e^{-\beta hc\tilde{\nu}}}{(1 - e^{-\beta hc\tilde{\nu}})^2}\right\} = \frac{hc\tilde{\nu}e^{-\beta hc\tilde{\nu}}}{1 - e^{-\beta hc\tilde{\nu}}}$$

that

$$\langle \varepsilon^V \rangle = \frac{hc\tilde{\nu}}{e^{\beta hc\tilde{\nu}} - 1} \qquad (10.28)$$

The zero-point energy, $\frac{1}{2}hc\tilde{\nu}$, can be added to the right-hand side if the mean energy is to be measured from 0 rather than the lowest attainable level (the zero-point level). The variation of the mean energy with temperature is illustrated in Fig. 10.9. At high temperatures, when $T \gg \theta_V$, or $\beta hc\tilde{\nu} \ll 1$, the exponential functions can be expanded ($e^x = 1 + x + \cdots$) and all but the leading terms discarded. This approximation leads to

$$\langle \varepsilon^V \rangle = \frac{hc\tilde{\nu}}{(1 + \beta hc\tilde{\nu} + \cdots) - 1} \approx \frac{1}{\beta} = kT \qquad (10.29)$$

This result is in agreement with the value predicted by the classical equipartition theorem, because the energy of a one-dimensional oscillator is $E = \frac{1}{2}mv_x^2 + \frac{1}{2}kx^2$ and the mean energy of each quadratic term is $\frac{1}{2}kT$.

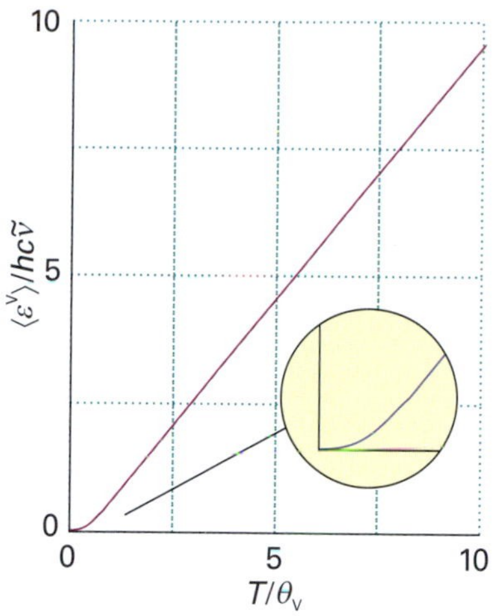

Fig. 10.9 The mean vibrational energy of a molecule in the harmonic approximation as a function of temperature. At high temperatures ($T \gg \theta_V$), the energy is linearly proportional to the temperature, in accord with the equipartition theorem.

Exploration Plot the temperature dependence of the mean vibrational energy for several values of the vibrational wavenumber (for reasonable values of the vibrational wavenumber, see the *Data section*). From your plots, estimate the temperature at which the mean vibrational energy begins to increase sharply.

10.4 Heat capacities

The constant-volume heat capacity is defined as $C_V = (\partial U/\partial T)_V$. The derivative with respect to T is converted into a derivative with respect to β by using

$$\frac{d}{dT} = \frac{d\beta}{dT}\frac{d}{d\beta} = -\frac{1}{kT^2}\frac{d}{d\beta} = -k\beta^2\frac{d}{d\beta} \qquad (10.30)$$

It follows that

$$C_V = -k\beta^2\left(\frac{\partial U}{\partial \beta}\right)_V \qquad (10.31a)$$

Because the internal energy of a perfect gas is a sum of contributions, the heat capacity is also a sum of contributions from each mode. The contribution of mode M is

$$C_V^M = N\left(\frac{\partial\langle\varepsilon^M\rangle}{\partial T}\right)_V = -Nk\beta^2\left(\frac{\partial\langle\varepsilon^M\rangle}{\partial\beta}\right)_V \qquad (10.31b)$$

(a) The individual contributions

The temperature is always high enough (provided the gas is above its condensation temperature) for the mean translational energy to be $\frac{3}{2}kT$, the equipartition value. Therefore, the molar constant-volume heat capacity is

$$C_{V,m}^T = N_A\frac{d(\frac{3}{2}kT)}{dT} = \frac{3}{2}R \qquad (10.32)$$

Translation is the only mode of motion for a monatomic gas, so for such a gas $C_{V,m} = \frac{3}{2}R = 12.47\ \mathrm{J\,K^{-1}\,mol^{-1}}$. This result is very reliable: helium, for example, has this value over a range of 2000 K. It follows from thermodynamic arguments (Volume 1, Section 2.5) that $C_{p,m} - C_{V,m} = R$, so for a monatomic perfect gas $C_{p,m} = \frac{5}{2}R$, and therefore

$$\gamma = \frac{C_p}{C_V} = \frac{5}{3} \qquad (10.33)°$$

When the temperature is high enough for the rotations of the molecules to be highly excited (when $T \gg \theta_R$), we can use the equipartition value kT for the mean rotational energy (for a linear rotor) to obtain $C_{V,m} = R$. For nonlinear molecules, the mean rotational energy rises to $\frac{3}{2}kT$, so the molar rotational heat capacity rises to $\frac{3}{2}R$ when $T \gg \theta_R$. Only the lowest rotational state is occupied when the temperature is very low, and then rotation does not contribute to the heat capacity. We can

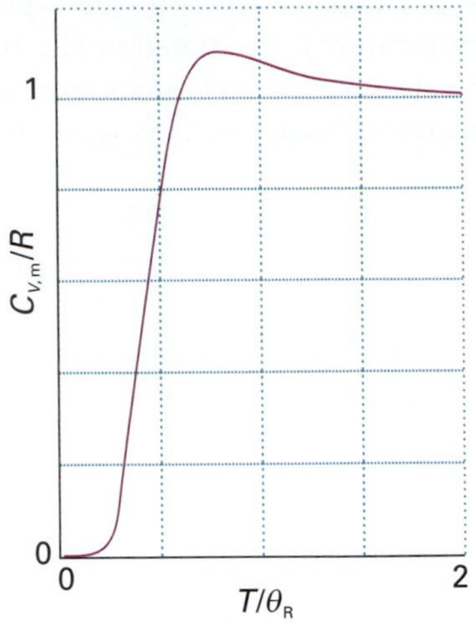

Fig. 10.10 The temperature dependence of the rotational contribution to the heat capacity of a linear molecule.

Exploration The *Living graphs* section of the text's web site has applets for the calculation of the temperature dependence of the rotational contribution to the heat capacity. Explore the effect of the rotational constant on the plot of $C^R_{V,m}$ against T.

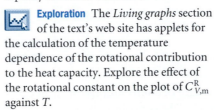

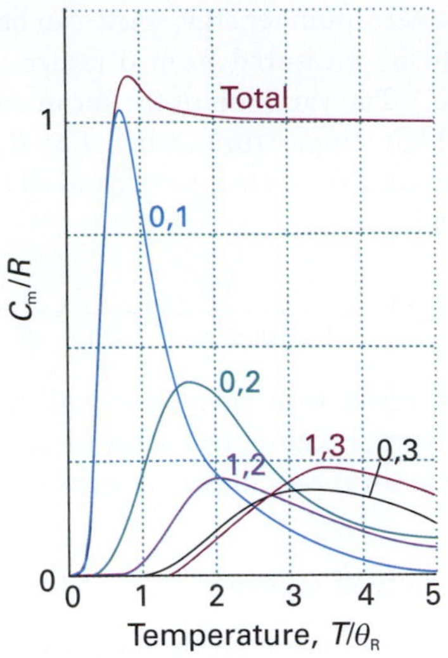

Fig. 10.11 The rotational heat capacity of a linear molecule can be regarded as the sum of contributions from a collection of two-level systems, in which the rise in temperature stimulates transitions between J levels, some of which are shown here. The calculation on which this illustration is based is sketched in Problem 10.19.

calculate the rotational heat capacity at intermediate temperatures by differentiating the equation for the mean rotational energy (eqn 10.26). The resulting (untidy) expression, which is plotted in Fig. 10.10, shows that the contribution rises from zero (when $T = 0$) to the equipartition value (when $T \gg \theta_R$). Because the translational contribution is always present, we can expect the molar heat capacity of a gas of diatomic molecules ($C^T_{V,m} + C^R_{V,m}$) to rise from $\frac{3}{2}R$ to $\frac{5}{2}R$ as the temperature is increased above θ_R. Problem 10.19 explores how the overall shape of the curve can be traced to the sum of thermal excitations between all the available rotational energy levels (Fig. 10.11).

Molecular vibrations contribute to the heat capacity, but only when the temperature is high enough for them to be significantly excited. The equipartition mean energy is kT for each mode, so the maximum contribution to the molar heat capacity is R. However, it is very unusual for the vibrations to be so highly excited that equipartition is valid, and it is more appropriate to use the full expression for the vibrational heat capacity, which is obtained by differentiating eqn 10.28:

Comment 10.2
Equation 10.34 is essentially the same as the Einstein formula for the heat capacity of a solid (eqn 1.7) with θ_V the Einstein temperature, θ_E. The only difference is that vibrations can take place in three dimensions in a solid.

$$C^V_{V,m} = Rf \qquad f = \left(\frac{\theta_V}{T}\right)^2 \left(\frac{e^{-\theta_V/2T}}{1 - e^{-\theta_V/T}}\right)^2 \tag{10.34}$$

where $\theta_V = hc\tilde{\nu}/k$ is the characteristic vibrational temperature. The curve in Fig. 10.12 shows how the vibrational heat capacity depends on temperature. Note that even when the temperature is only slightly above θ_V the heat capacity is close to its equipartition value.

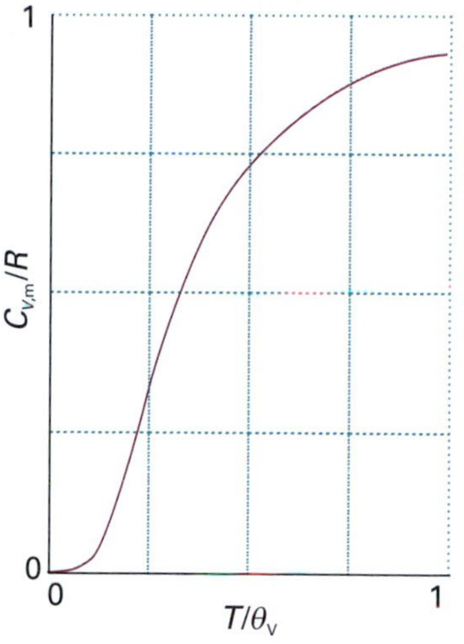

Fig. 10.12 The temperature dependence of the vibrational heat capacity of a molecule in the harmonic approximation calculated by using eqn 10.34. Note that the heat capacity is within 10 per cent of its classical value for temperatures greater than θ_V.

Exploration The *Living graphs* section of the text's web site has applets for the calculation of the temperature dependence of the vibrational contribution to the heat capacity. Explore the effect of the vibrational wavenumber on the plot of $C^V_{V,m}$ against T.

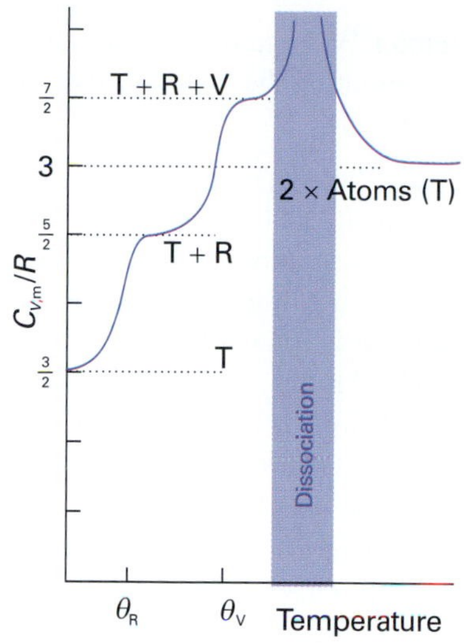

Fig. 10.13 The general features of the temperature dependence of the heat capacity of diatomic molecules are as shown here. Each mode becomes active when its characteristic temperature is exceeded. The heat capacity becomes very large when the molecule dissociates because the energy is used to cause dissociation and not to raise the temperature. Then it falls back to the translation-only value of the atoms.

(b) The overall heat capacity

The total heat capacity of a molecular substance is the sum of each contribution (Fig. 10.13). When equipartition is valid (when the temperature is well above the characteristic temperature of the mode, $T \gg \theta_M$) we can estimate the heat capacity by counting the numbers of modes that are active. In gases, all three translational modes are always active and contribute $\frac{3}{2}R$ to the molar heat capacity. If we denote the number of active rotational modes by ν_R^* (so for most molecules at normal temperatures $\nu_R^* = 2$ for linear molecules, and 3 for nonlinear molecules), then the rotational contribution is $\frac{1}{2}\nu_R^* R$. If the temperature is high enough for ν_V^* vibrational modes to be active, the vibrational contribution to the molar heat capacity is $\nu_V^* R$. In most cases $\nu_V^* \approx 0$. It follows that the total molar heat capacity is

$$C_{V,m} = \tfrac{1}{2}(3 + \nu_R^* + 2\nu_V^*)R \qquad (10.35)$$

Example 10.5 *Estimating the molar heat capacity of a gas*

Estimate the molar constant-volume heat capacity of water vapour at 100°C. Vibrational wavenumbers are given in Example 10.3; the rotational constants of an H_2O molecule are 27.9, 14.5, and 9.3 cm^{-1}.

Method We need to assess whether the rotational and vibrational modes are active by computing their characteristic temperatures from the data (to do so, use $hc/k = 1.439$ cm K).

Answer The characteristic temperatures (in round numbers) of the vibrations are 5300 K, 2300 K, and 5400 K; the vibrations are therefore not excited at 373 K. The three rotational modes have characteristic temperatures 40 K, 21 K, and 13 K, so they are fully excited, like the three translational modes. The translational contribution is $\frac{3}{2}R = 12.5$ J K^{-1} mol^{-1}. Fully excited rotations contribute a further 12.5 J K^{-1} mol^{-1}. Therefore, a value close to 25 J K^{-1} mol^{-1} is predicted. The experimental value is 26.1 J K^{-1} mol^{-1}. The discrepancy is probably due to deviations from perfect gas behaviour.

Self-test 10.5 Estimate the molar constant-volume heat capacity of gaseous I_2 at 25°C ($B = 0.037$ cm^{-1}; see Table 6.2 for more data). [29 J K^{-1} mol^{-1}]

10.5 Equations of state

The relation between p and Q in eqn 10.3 is a very important route to the equations of state of real gases in terms of intermolecular forces, for the latter can be built into Q. We have already seen (Example 10.1) that the partition function for a gas of independent particles leads to the perfect gas equation of state, $pV = nRT$. The equation of state of a real gas may be written

$$\frac{pV_m}{RT} = 1 + \frac{B}{V_m} + \frac{C}{V_m^2} + \cdots \tag{10.36}$$

where B is the second virial coefficient and C is the third virial coefficient.

The total kinetic energy of a gas is the sum of the kinetic energies of the individual molecules. Therefore, even in a real gas the canonical partition function factorizes into a part arising from the kinetic energy, which is the same as for the perfect gas, and a factor called the **configuration integral**, Z, which depends on the intermolecular potentials. We therefore write

$$Q = \frac{Z}{\Lambda^{3N}} \tag{10.37}$$

By comparing this equation with eqn 9.45 ($Q = q^N/N!$, with $q = V/\Lambda^3$), we see that for a perfect gas of atoms (with no contributions from rotational or vibrational modes)

$$Z = \frac{V^N}{N!} \tag{10.38}$$

For a real gas of atoms (for which the intermolecular interactions are isotropic), Z is related to the total potential energy E_P of interaction of all the particles by

$$Z = \frac{1}{N!}\int e^{-\beta E_P} d\tau_1 d\tau_2 \cdots d\tau_N \tag{10.39}$$

where $d\tau_i$ is the volume element for atom i. The physical origin of this term is that the probability of occurrence of each arrangement of molecules possible in the sample is given by a Boltzmann distribution in which the exponent is given by the potential energy corresponding to that arrangement.

Illustration 10.1 *Calculating a configuration integral*

When the molecules do not interact with one another, $E_P = 0$ and hence $e^{-\beta E_P} = 1$. Then

$$Z = \frac{1}{N!}\int d\tau_1 d\tau_2 \cdots d\tau_N = \frac{V^N}{N!}$$

because $\int d\tau = V$, where V is the volume of the container. This result coincides with eqn 10.39.

When we consider only interactions between pairs of particles the configuration integral simplifies to

$$Z = \tfrac{1}{2}\int e^{-\beta E_P} d\tau_1 d\tau_2 \tag{10.40}$$

The second virial coefficient then turns out to be

$$B = -\frac{N_A}{2V}\int f d\tau_1 d\tau_2 \tag{10.41}$$

The quantity f is the **Mayer *f*-function**: it goes to zero when the two particles are so far apart that $E_P = 0$. When the intermolecular interaction depends only on the separation r of the particles and not on their relative orientation or their absolute position in space, as in the interaction of closed-shell atoms in a uniform sample, the volume element simplifies to $4\pi r^2 dr$ (because the integrals over the angular variables in $d\tau = r^2 dr \sin\theta\, d\theta d\phi$ give a factor of 4π) and eqn 10.41 becomes

$$B = -2\pi N_A \int_0^\infty f r^2 dr \qquad f = e^{-\beta E_P} - 1 \tag{10.42}$$

The integral can be evaluated (usually numerically) by substituting an expression for the intermolecular potential energy.

We can illustrate how eqn 10.42 is used by considering the **hard-sphere potential**, which is infinite when the separation of the two molecules, r, is less than or equal to a certain value σ, and is zero for greater separations. Then

$$e^{-\beta E_P} = 0 \qquad f = -1 \quad \text{when} \quad r \le \sigma \quad (\text{and } E_P = \infty) \tag{10.43a}$$

$$e^{-\beta E_P} = 1 \qquad f = 0 \quad \text{when} \quad r > \sigma \quad (\text{and } E_P = 0) \tag{10.43b}$$

It follows from eqn 10.42 that the second virial coefficient is

$$B = 2\pi N_A \int_0^\sigma r^2 dr = \tfrac{2}{3}\pi N_A \sigma^3 \tag{10.44}$$

This calculation of B raises the question as to whether a potential can be found that, when the virial coefficients are evaluated, gives the van der Waals equation of state. Such a potential can be found for weak attractive interactions ($a \ll RT$): it consists of a hard-sphere repulsive core and a long-range, shallow attractive region (see Problem 10.15). A further point is that, once a second virial coefficient has been calculated for a given intermolecular potential, it is possible to calculate other thermodynamic properties that depend on the form of the potential. For example, it is possible to calculate the isothermal Joule–Thomson coefficient, μ_T (Volume 1, Section 3.8), from the thermodynamic relation

$$\lim_{p\to 0} \mu_T = B - T\frac{dB}{dT} \tag{10.45}$$

and from the result calculate the Joule–Thomson coefficient itself by using eqn 3.48 of Volume 1.

10.6 Molecular interactions in liquids

The starting point for the discussion of solids is the well ordered structure of a perfect crystal. The starting point for the discussion of gases is the completely disordered distribution of the molecules of a perfect gas. Liquids lie between these two extremes. We shall see that the structural and thermodynamic properties of liquids depend on the nature of intermolecular interactions and that an equation of state can be built in a similar way to that just demonstrated for real gases.

(a) The radial distribution function

The average relative locations of the particles of a liquid are expressed in terms of the **radial distribution function**, $g(r)$. This function is defined so that $g(r)r^2dr$ is the probability that a molecule will be found in the range dr at a distance r from another molecule. In a perfect crystal, $g(r)$ is a periodic array of sharp spikes, representing the certainty (in the absence of defects and thermal motion) that molecules (or ions) lie at definite locations. This regularity continues out to the edges of the crystal, so we say that crystals have **long-range order**. When the crystal melts, the long-range order is lost and, wherever we look at long distances from a given molecule, there is equal probability of finding a second molecule. Close to the first molecule, though, the nearest neighbours might still adopt approximately their original relative positions and, even if they are displaced by newcomers, the new particles might adopt their vacated positions. It is still possible to detect a sphere of nearest neighbours at a distance r_1, and perhaps beyond them a sphere of next-nearest neighbours at r_2. The existence of this **short-range order** means that the radial distribution function can be expected to oscillate at short distances, with a peak at r_1, a smaller peak at r_2, and perhaps some more structure beyond that.

The radial distribution function of the oxygen atoms in liquid water is shown in Fig. 10.14. Closer analysis shows that any given H_2O molecule is surrounded by other molecules at the corners of a tetrahedron. The form of $g(r)$ at 100°C shows that the intermolecular interactions (in this case, principally by hydrogen bonds) are strong enough to affect the local structure right up to the boiling point. Raman spectra indicate that in liquid water most molecules participate in either three or four hydrogen bonds. Infrared spectra show that about 90 per cent of hydrogen bonds are intact at the melting point of ice, falling to about 20 per cent at the boiling point.

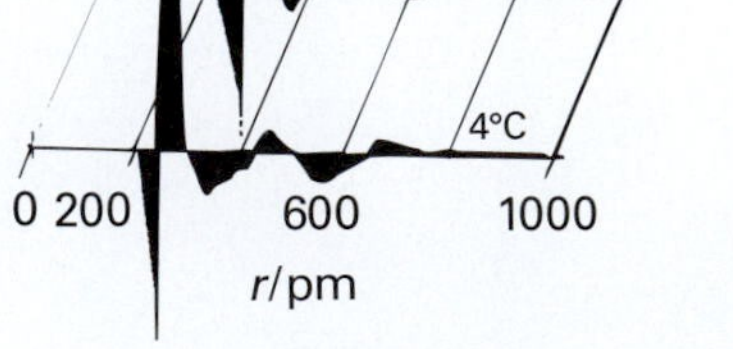

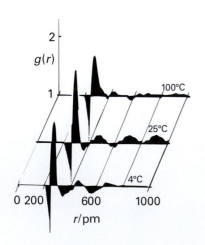

Fig. 10.14 The radial distribution function of the oxygen atoms in liquid water at three temperatures. Note the expansion as the temperature is raised. (A.H. Narten, M.D. Danford, and H.A. Levy, *Discuss. Faraday. Soc.* **43**, 97 (1967).)

The formal expression for the radial distribution function for molecules 1 and 2 in a fluid consisting of N particles is the somewhat fearsome equation

$$g(r_{12}) = \frac{\int\int \cdots \int e^{-\beta V_N} d\tau_3 d\tau_4 \ldots d\tau_N}{N^2 \int\int \cdots \int e^{-\beta V_N} d\tau_1 d\tau_2 \ldots d\tau_N} \tag{10.46}$$

where $\beta = 1/kT$ and V_N is the N-particle potential energy. Although fearsome, this expression is nothing more than the Boltzmann distribution for the relative locations of two molecules in a field provided by all the other molecules in the system.

(b) The calculation of $g(r)$

Because the radial distribution function can be calculated by making assumptions about the intermolecular interactions, it can be used to test theories of liquid structure. However, even a fluid of hard spheres without attractive interactions (a collection of ball-bearings in a container) gives a function that oscillates near the origin (Fig. 10.15), and one of the factors influencing, and sometimes dominating, the structure of a liquid is the geometrical problem of stacking together reasonably hard spheres. Indeed, the radial distribution function of a liquid of hard spheres shows more pronounced oscillations at a given temperature than that of any other type of liquid. The attractive part of the potential modifies this basic structure, but sometimes only quite weakly. One of the reasons behind the difficulty of describing liquids theoretically is the similar importance of both the attractive and repulsive (hard core) components of the potential.

There are several ways of building the intermolecular potential into the calculation of $g(r)$. Numerical methods take a box of about 10^3 particles (the number increases as computers grow more powerful), and the rest of the liquid is simulated by surrounding the box with replications of the original box (Fig. 10.16). Then, whenever a particle leaves the box through one of its faces, its image arrives through the opposite face. When calculating the interactions of a molecule in a box, it interacts with all the molecules in the box and all the periodic replications of those molecules and itself in the other boxes.

In the **Monte Carlo method**, the particles in the box are moved through small but otherwise random distances, and the change in total potential energy of the N particles in the box, ΔV_N, is calculated using the appropriate intermolecular potential. Whether or not this new configuration is accepted is then judged from the following rules:

1 If the potential energy is not greater than before the change, then the configuration is accepted.

If the potential energy is greater than before the change, then it is necessary to check if the new configuration is reasonable and can exist in equilibrium with configurations of lower potential energy at a given temperature. To make progress, we use the result that, at equilibrium, the ratio of populations of two states with energy separation ΔV_N is $e^{-\Delta V_N/kT}$. Because we are testing the viability of a configuration with a higher potential energy than the previous configuration in the calculation, $\Delta V_N > 0$ and the exponential factor varies between 0 and 1. In the Monte Carlo method, the second rule, therefore, is:

2 The exponential factor is compared with a random number between 0 and 1; if the factor is larger than the random number, then the configuration is accepted; if the factor is not larger, the configuration is rejected.

The configurations generated with Monte Carlo calculations can be used to construct $g(r)$ simply by counting the number of pairs of particles with a separation r and averaging the result over the whole collection of configurations.

In the **molecular dynamics** approach, the history of an initial arrangement is followed by calculating the trajectories of all the particles under the influence of the intermolecular potentials. To appreciate what is involved, we consider the motion of a particle in one dimension. We show in the following *Justification* that, after a time interval Δt, the position of a particle changes from x_{i-1} to a new value x_i given by

$$x_i = x_{i-1} + v_{i-1}\Delta t \qquad (10.47)$$

where v_{i-1} is the velocity of the atom when it was at x_{i-1}, its location at the start of the interval. The velocity at x_i is related to v_{i-1}, the velocity at the start of the interval, by

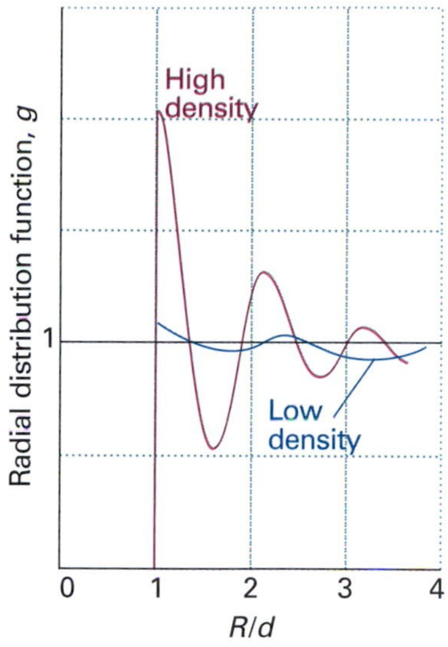

Fig. 10.15 The radial distribution function for a simulation of a liquid using impenetrable hard spheres (ball bearings).

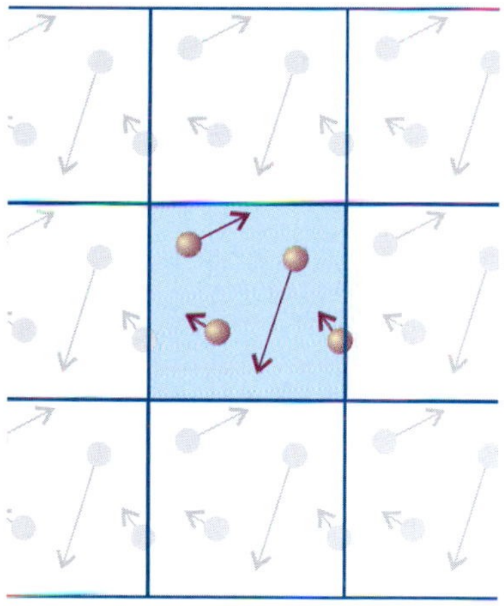

Fig. 10.16 In a two-dimensional simulation of a liquid that uses periodic boundary conditions, when one particle leaves the cell its mirror image enters through the opposite face.

$$v_i = v_{i-1} - m^{-1} \left.\frac{\mathrm{d}V_N(x)}{\mathrm{d}x}\right|_{x_{i-1}} \Delta t \qquad (10.48)$$

where the derivative of the potential energy $V_N(x)$ is evaluated at x_{i-1}. The time interval Δt is approximately 1 fs (10^{-15} s), which is shorter than the average time between collisions. The calculation of x_i and v_i is then repeated for tens of thousands of such steps. The time-consuming part of the calculation is the evaluation of the net force on the molecule arising from all the other molecules present in the system.

Justification 10.3 *Particle trajectories according to molecular dynamics*

Consider a particle of mass m moving along the x direction with an initial velocity v_1 given by

$$v_1 = \frac{\Delta x}{\Delta t}$$

If the initial and new positions of the atom are x_1 and x_2, then $\Delta x = x_2 - x_1$ and

$$x_2 = x_1 + v_1 \Delta t$$

The particle moves under the influence of a force arising from interactions with other atoms in the molecule. From Newton's second law of motion, we write the force F_1 at x_1 as

$$F_1 = ma_1$$

where the acceleration a_1 at x_1 is given by $a_1 = \Delta v/\Delta t$. If the initial and new velocities are v_1 and v_2, then $\Delta v = v_2 - v_1$ and

$$v_2 = v_1 + a_1\Delta t = v_1 + \frac{F_1}{m}\Delta t$$

Because $F = -\mathrm{d}V/\mathrm{d}x$, the force acting on the atom is related to the potential energy of interaction with other nearby atoms, the potential energy $V_N(x)$, by

$$F_1 = -\left.\frac{\mathrm{d}V_N(x)}{\mathrm{d}x}\right|_{x_1}$$

where the derivative is evaluated at x_1. It follows that

$$v_2 = v_1 - m^{-1}\left.\frac{\mathrm{d}V_N(x)}{\mathrm{d}x}\right|_{x_1}\Delta t$$

This expression generalizes to eqn 10.48 for the calculation of a velocity v_i from a previous velocity v_{i-1}.

Self-test 10.6 Consider a particle of mass m connected to a stationary wall with a spring of force constant k. Write an expression for the velocity of this particle once it is set into motion in the x direction from an equilibrium position x_0.

[$v_i = v_{i-1} + (k/m)(x_{i-1} - x_0)$]

A molecular dynamics calculation gives a series of snapshots of the liquid, and $g(r)$ can be calculated as before. The temperature of the system is inferred by computing the mean kinetic energy of the particles and using the equipartition result that

$$\langle \tfrac{1}{2}mv_q^2 \rangle = \tfrac{1}{2}kT \qquad (10.49)$$

for each coordinate q.

(c) The thermodynamic properties of liquids

Once $g(r)$ is known it can be used to calculate the thermodynamic properties of liquids. For example, the contribution of the pairwise additive intermolecular potential, V_2, to the internal energy is given by the integral

$$U = \frac{2\pi N^2}{V}\int_0^\infty g(r)V_2 r^2 \mathrm{d}r \qquad (10.50)$$

That is, U is essentially the average two-particle potential energy weighted by $g(r)r^2\mathrm{d}r$, which is the probability that the pair of particles have a separation between r and $r+\mathrm{d}r$. Likewise, the contribution that pairwise interactions make to the pressure is

$$\frac{pV}{nRT} = 1 - \frac{2\pi N}{kTV}\int_0^\infty g(r)v_2 r^2 \mathrm{d}r \qquad v_2 = r\frac{\mathrm{d}V_2}{\mathrm{d}r} \qquad (10.51a)$$

The quantity v_2 is called the **virial** (hence the term 'virial equation of state'). To understand the physical content of this expression, we rewrite it as

$$p = \frac{nRT}{V} - 2\pi\left(\frac{N}{V}\right)^2\int_0^\infty g(r)v_2 r^2 \mathrm{d}r \qquad (10.51b)$$

The first term on the right is the **kinetic pressure**, the contribution to the pressure from the impact of the molecules in free flight. The second term is essentially the internal pressure, $\pi_T = (\partial U/\partial V)_T$, introduced in Volume 1, Section 2.11, representing the contribution to the pressure from the intermolecular forces. To see the connection, we should recognize $-\mathrm{d}V_2/\mathrm{d}r$ (in v_2) as the force required to move two molecules apart, and therefore $-r(\mathrm{d}V_2/\mathrm{d}r)$ as the work required to separate the molecules through a distance r. The second term is therefore the average of this work over the range of pairwise separations in the liquid as represented by the probability of finding two molecules at separations between r and $r+\mathrm{d}r$, which is $g(r)r^2\mathrm{d}r$. In brief, the integral, when multiplied by the square of the number density, is the change in internal energy of the system as it expands, and therefore is equal to the internal pressure.

10.7 Residual entropies

Entropies may be calculated from spectroscopic data; they may also be measured experimentally. In many cases there is good agreement, but in some the experimental entropy is less than the calculated value. One possibility is that the experimental determination failed to take a phase transition into account (and a contribution of the form $\Delta_{trs}H/T_{trs}$ incorrectly omitted from the sum). Another possibility is that some disorder is present in the solid even at $T=0$. The entropy at $T=0$ is then greater than zero and is called the **residual entropy**.

The origin and magnitude of the residual entropy can be explained by considering a crystal composed of AB molecules, where A and B are similar atoms (such as CO, with its very small electric dipole moment). There may be so little energy difference between . . .AB AB AB AB. . . , . . .AB BA BA AB. . . , and other arrangements that the molecules adopt the orientations AB and BA at random in the solid. We can readily calculate the entropy arising from residual disorder by using the Boltzmann formula $S = k \ln W$. To do so, we suppose that two orientations are equally probable, and that the sample consists of N molecules. Because the same energy can be achieved in 2^N different ways (because each molecule can take either of two orientations), the total number of ways of achieving the same energy is $W = 2^N$. It follows that

$$S = k \ln 2^N = Nk \ln 2 = nR \ln 2 \qquad (10.52a)$$

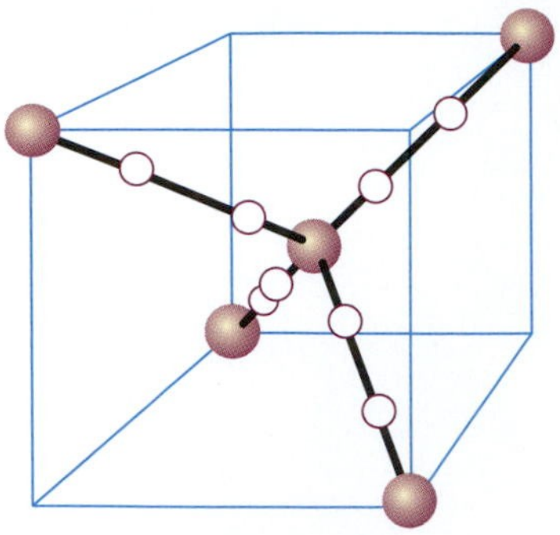

Fig. 10.17 The possible locations of H atoms around a central O atom in an ice crystal are shown by the white spheres. Only one of the locations on each bond may be occupied by an atom, and two H atoms must be close to the O atom and two H atoms must be distant from it.

We can therefore expect a residual molar entropy of $R \ln 2 = 5.8\ \mathrm{J\,K^{-1}\,mol^{-1}}$ for solids composed of molecules that can adopt either of two orientations at $T = 0$. If s orientations are possible, the residual molar entropy will be

$$S_m = R \ln s \qquad (10.52b)$$

An $FClO_3$ molecule, for example, can adopt four orientations with about the same energy (with the F atom at any of the four corners of a tetrahedron), and the calculated residual molar entropy of $R \ln 4 = 11.5\ \mathrm{J\,K^{-1}\,mol^{-1}}$ is in good agreement with the experimental value ($10.1\ \mathrm{J\,K^{-1}\,mol^{-1}}$). For CO, the measured residual entropy is $5\ \mathrm{J\,K^{-1}\,mol^{-1}}$, which is close to $R \ln 2$, the value expected for a random structure of the form . . .CO CO OC CO OC OC. . . .

Illustration 10.2 *Calculating a residual entropy*

Consider a sample of ice with N H_2O molecules. Each O atom is surrounded tetrahedrally by four H atoms, two of which are attached by short σ bonds, the other two being attached by long hydrogen bonds (Fig. 10.17). It follows that each of the $2N$ H atoms can be in one of two positions (either close to or far from an O atom as shown in Fig. 10.18), resulting in 2^{2N} possible arrangements. However, not all these arrangements are acceptable. Indeed, of the $2^4 = 16$ ways of arranging four H atoms around one O atom, only 6 have two short and two long OH distances and hence are acceptable. Therefore, the number of permitted arrangements is

$$W = 2^{2N}\left(\tfrac{6}{16}\right)^N = \left(\tfrac{3}{2}\right)^N$$

It then follows that the residual molar entropy is

$$S_m(0) \approx k \ln\left(\tfrac{3}{2}\right)^{N_A} = N_A k \ln\left(\tfrac{3}{2}\right) = R \ln\left(\tfrac{3}{2}\right) = 3.4\ \mathrm{J\,K^{-1}\,mol^{-1}}$$

which is in good agreement with the experimental value of $3.4\ \mathrm{J\,K^{-1}\,mol^{-1}}$. The model, however, is not exact because it ignores the possibility that next-nearest neighbours and those beyond can influence the local arrangement of bonds.

10.8 Equilibrium constants

The Gibbs energy of a gas of independent molecules is given by eqn 10.9 in terms of the molar partition function, $q_m = q/n$. The equilibrium constant K of a reaction is related to the standard Gibbs energy of reaction by $\Delta_r G^{\ominus} = -RT \ln K$. To calculate the equilibrium constant, we need to combine these two equations. We shall consider gas phase reactions in which the equilibrium constant is expressed in terms of the partial pressures of the reactants and products.

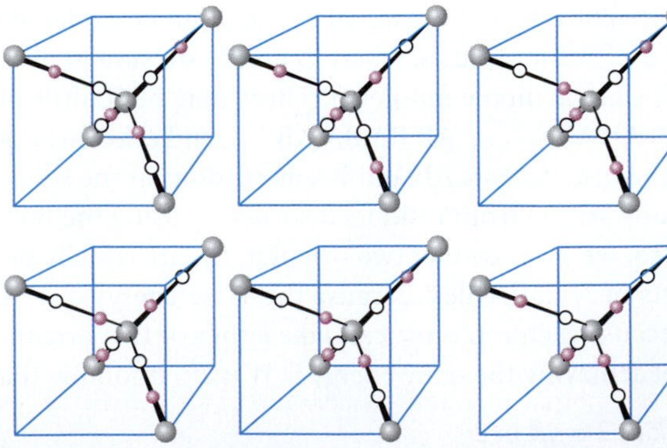

Fig. 10.18 The six possible arrangements of H atoms in the locations identified in Fig. 10.17. Occupied locations are denoted by red spheres and unoccupied locations by white spheres.

(a) The relation between K and the partition function

To find an expression for the standard reaction Gibbs energy we need expressions for the standard molar Gibbs energies, $G^{\ominus}/n$, of each species. For these expressions, we need the value of the molar partition function when $p = p^{\ominus}$ (where $p^{\ominus} = 1$ bar): we denote this **standard molar partition function** $q^{\ominus}_{m}$. Because only the translational component depends on the pressure, we can find $q^{\ominus}_{m}$ by evaluating the partition function with V replaced by $V^{\ominus}_{m}$, where $V^{\ominus}_{m} = RT/p^{\ominus}$. For a species J it follows that

$$G^{\ominus}_{m}(J) = G^{\ominus}_{m}(J,0) - RT \ln \frac{q^{\ominus}_{J,m}}{N_A} \qquad (10.53)^{\circ}$$

where $q^{\ominus}_{J,m}$ is the standard molar partition function of J. By combining expressions like this one (as shown in the *Justification* below), the equilibrium constant for the reaction

$$a\,A + b\,B \rightarrow c\,C + d\,D$$

is given by the expression

$$K = \frac{(q^{\ominus}_{C,m}/N_A)^c (q^{\ominus}_{D,m}/N_A)^d}{(q^{\ominus}_{A,m}/N_A)^a (q^{\ominus}_{B,m}/N_A)^b} e^{-\Delta_r E_0/RT} \qquad (10.54a)$$

where $\Delta_r E_0$ is the difference in molar energies of the ground states of the products and reactants (this term is defined more precisely in the *Justification*), and is calculated from the bond dissociation energies of the species (Fig. 10.19). In terms of the stoichiometric numbers introduced in Volume 1, Section 7.2, we would write

$$K = \left\{ \prod_J \left(\frac{q^{\ominus}_{J,m}}{N_A} \right)^{\nu_J} \right\} e^{-\Delta_r E_0/RT} \qquad (10.54b)$$

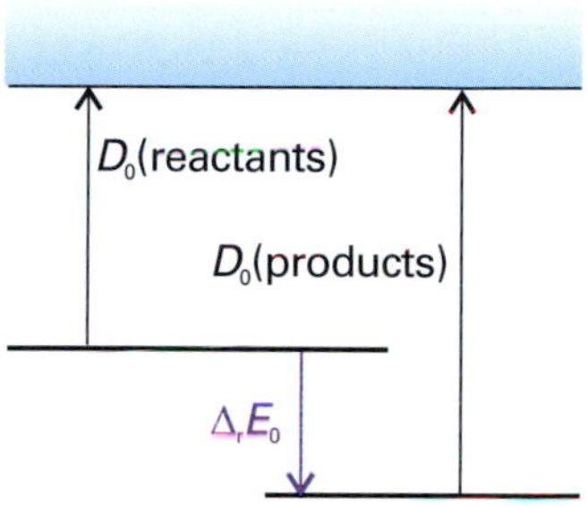

Fig. 10.19 The definition of $\Delta_r E_0$ for the calculation of equilibrium constants.

Justification 10.4 *The equilibrium constant in terms of the partition function 1*

The standard molar reaction Gibbs energy for the reaction is

$$\begin{aligned}\Delta_r G^{\ominus} &= cG^{\ominus}_{m}(C) + dG^{\ominus}_{m}(D) - aG^{\ominus}_{m}(A) - bG^{\ominus}_{m}(B) \\ &= cG^{\ominus}_{m}(C,0) + dG^{\ominus}_{m}(D,0) - aG^{\ominus}_{m}(A,0) - bG^{\ominus}_{m}(B,0) \\ &\quad - RT\left\{ c \ln \frac{q^{\ominus}_{C,m}}{N_A} + d \ln \frac{q^{\ominus}_{D,m}}{N_A} - a \ln \frac{q^{\ominus}_{A,m}}{N_A} - b \ln \frac{q^{\ominus}_{B,m}}{N_A} \right\}\end{aligned}$$

Because $G(0) = U(0)$, the first term on the right is

$$\Delta_r E_0 = cU^{\ominus}_{m}(C,0) + dU^{\ominus}_{m}(D,0) - aU^{\ominus}_{m}(A,0) - bU^{\ominus}_{m}(B,0) \qquad (10.55)$$

the reaction internal energy at $T = 0$ (a molar quantity).

Now we can write

$$\begin{aligned}\Delta_r G^{\ominus} &= \Delta_r E_0 - RT\left\{ \ln\left(\frac{q^{\ominus}_{C,m}}{N_A}\right)^c + \ln\left(\frac{q^{\ominus}_{D,m}}{N_A}\right)^d - \ln\left(\frac{q^{\ominus}_{A,m}}{N_A}\right)^a - \ln\left(\frac{q^{\ominus}_{B,m}}{N_A}\right)^b \right\} \\ &= \Delta_r E_0 - RT \ln \frac{(q^{\ominus}_{C,m}/N_A)^c (q^{\ominus}_{D,m}/N_A)^d}{(q^{\ominus}_{A,m}/N_A)^a (q^{\ominus}_{B,m}/N_A)^b} \\ &= -RT\left\{ \frac{\Delta_r E_0}{RT} + \ln \frac{(q^{\ominus}_{C,m}/N_A)^c (q^{\ominus}_{D,m}/N_A)^d}{(q^{\ominus}_{A,m}/N_A)^a (q^{\ominus}_{B,m}/N_A)^b} \right\}\end{aligned}$$

At this stage we can pick out an expression for K by comparing this equation with $\Delta_r G^{\ominus} = -RT \ln K$, which gives

$$\ln K = -\frac{\Delta_r E_0}{RT} + \ln \frac{(q^{\ominus}_{C,m}/N_A)^c (q^{\ominus}_{D,m}/N_A)^d}{(q^{\ominus}_{A,m}/N_A)^a (q^{\ominus}_{B,m}/N_A)^b}$$

This expression is easily rearranged into eqn 10.54a by forming the exponential of both sides.

(b) A dissociation equilibrium

We shall illustrate the application of eqn 10.54 to an equilibrium in which a diatomic molecule X_2 dissociates into its atoms:

$$X_2(g) \rightleftharpoons 2\,X(g) \qquad K = \frac{p_X^2}{p_{X_2} p^{\ominus}}$$

According to eqn 10.54 (with $a = 1$, $b = 0$, $c = 2$, and $d = 0$):

$$K = \frac{(q^{\ominus}_{X,m}/N_A)^2}{q^{\ominus}_{X_2,m}/N_A} e^{-\Delta_r E_0/RT} = \frac{(q^{\ominus}_{X,m})^2}{q^{\ominus}_{X_2,m} N_A} e^{-\Delta_r E_0/RT} \tag{10.56a}$$

with

$$\Delta_r E_0 = 2U^{\ominus}_m(X,0) - U^{\ominus}_m(X_2,0) = D_0(X{-}X) \tag{10.56b}$$

where $D_0(X{-}X)$ is the dissociation energy of the X—X bond. The standard molar partition functions of the atoms X are

$$q^{\ominus}_{X,m} = g_X \left(\frac{V^{\ominus}_m}{\Lambda_X^3}\right) = \frac{RTg_X}{p^{\ominus}\Lambda_X^3}$$

where g_X is the degeneracy of the electronic ground state of X and we have used $V^{\ominus}_m = RT/p^{\ominus}$. The diatomic molecule X_2 also has rotational and vibrational degrees of freedom, so its standard molar partition function is

$$q^{\ominus}_{X_2,m} = g_{X_2} \left(\frac{V^{\ominus}_m}{\Lambda_{X_2}^3}\right) q^R_{X_2} q^V_{X_2} = \frac{RTg_{X_2} q^R_{X_2} q^V_{X_2}}{p^{\ominus}\Lambda_{X_2}^3}$$

where g_{X_2} is the degeneracy of the electronic ground state of X_2. It follows from eqn 10.54 that the equilibrium constant is

$$K = \frac{kTg_X^2 \Lambda_{X_2}^3}{p^{\ominus} g_{X_2} q^R_{X_2} q^V_{X_2} \Lambda_X^6} e^{-D_0/RT} \tag{10.57}$$

where we have used $R/N_A = k$. All the quantities in this expression can be calculated from spectroscopic data. The Λs are defined in Table 10.3 and depend on the masses of the species and the temperature; the expressions for the rotational and vibrational partition functions are also available in Table 10.3 and depend on the rotational constant and vibrational wavenumber of the molecule.

Example 10.6 *Evaluating an equilibrium constant*

Evaluate the equilibrium constant for the dissociation $Na_2(g) \rightleftharpoons 2\,Na(g)$ at 1000 K from the following data: $B = 0.1547$ cm^{-1}, $\tilde{\nu} = 159.2$ cm^{-1}, $D_0 = 70.4$ kJ mol^{-1}. The Na atoms have doublet ground terms.

Method The partition functions required are specified in eqn 10.54. They are evaluated by using the expressions in Table 10.3. For a homonuclear diatomic molecule, $\sigma = 2$. In the evaluation of $kT/p^{\ominus}$ use $p^{\ominus} = 10^5$ Pa and 1 Pa m^3 = 1 J.

Answer The partition functions and other quantities required are as follows:

$\Lambda(\mathrm{Na_2}) = 8.14\ \mathrm{pm}$ $\Lambda(\mathrm{Na}) = 11.5\ \mathrm{pm}$
$q^{\mathrm{R}}(\mathrm{Na_2}) = 2246$ $q^{\mathrm{V}}(\mathrm{Na_2}) = 4.885$
$g(\mathrm{Na}) = 2$ $g(\mathrm{Na_2}) = 1$

Then, from eqn 10.54,

$$K = \frac{(1.38\times10^{-23}\ \mathrm{J\,K^{-1}})\times(1000\ \mathrm{K})\times4\times(8.14\times10^{-12}\ \mathrm{m})^3}{(10^5\ \mathrm{Pa})\times2246\times4.885\times(1.15\times10^{-11}\ \mathrm{m})^6}\times \mathrm{e}^{-8.47}$$
$$= 2.42$$

where we have used $1\ \mathrm{J} = 1\ \mathrm{kg\,m^2\,s^{-2}}$ and $1\ \mathrm{Pa} = 1\ \mathrm{kg\,m^{-1}\,s^{-1}}$.

Self-test 10.7 Evaluate K at 1500 K. [52]

(c) Contributions to the equilibrium constant

We are now in a position to appreciate the physical basis of equilibrium constants. To see what is involved, consider a simple R $\rightleftharpoons$ P gas-phase equilibrium (R for reactants, P for products).

Figure 10.20 shows two sets of energy levels; one set of states belongs to R, and the other belongs to P. The populations of the states are given by the Boltzmann distribution, and are independent of whether any given state happens to belong to R or to P. We can therefore imagine a single Boltzmann distribution spreading, without distinction, over the two sets of states. If the spacings of R and P are similar (as in Fig. 10.20), and P lies above R, the diagram indicates that R will dominate in the equilibrium mixture. However, if P has a high density of states (a large number of states in a given energy range, as in Fig. 10.21), then, even though its zero-point energy lies above that of R, the species P might still dominate at equilibrium.

It is quite easy to show (see the *Justification* below) that the ratio of numbers of R and P molecules at equilibrium is given by

$$\frac{N_{\mathrm{P}}}{N_{\mathrm{R}}} = \frac{q_{\mathrm{P}}}{q_{\mathrm{R}}}\mathrm{e}^{-\Delta_{\mathrm{r}}E_0/RT} \qquad (10.58\mathrm{a})$$

and therefore that the equilibrium constant for the reaction is

$$K = \frac{q_{\mathrm{P}}}{q_{\mathrm{R}}}\mathrm{e}^{-\Delta_{\mathrm{r}}E_0/RT} \qquad (10.58\mathrm{b})$$

just as would be obtained from eqn 10.54.

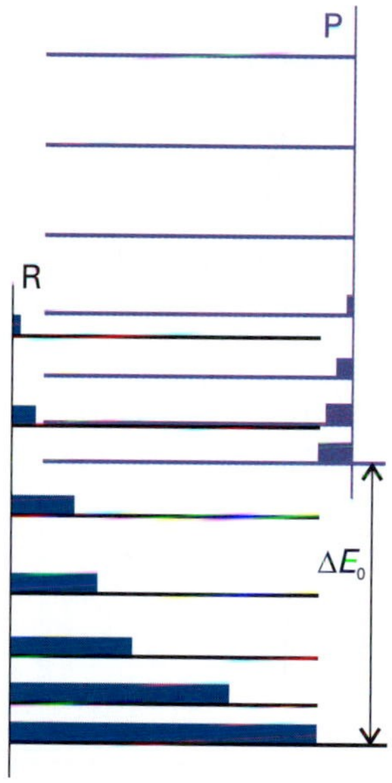

Fig. 10.20 The array of R(eactants) and P(roducts) energy levels. At equilibrium all are accessible (to differing extents, depending on the temperature), and the equilibrium composition of the system reflects the overall Bolzmann distribution of populations. As ΔE_0 increases, R becomes dominant.

Justification 10.5 *The equilibrium constant in terms of the partition function 2*

The population in a state i of the composite (R,P) system is

$$n_i = \frac{N\mathrm{e}^{-\beta\varepsilon_i}}{q}$$

where N is the total number of molecules. The total number of R molecules is the sum of these populations taken over the states belonging to R; these states we label r with energies ε_{r}. The total number of P molecules is the sum over the states

Comment 10.3

For an R $\rightleftharpoons$ P equilibrium, the V factors in the partition functions cancel, so the appearance of q in place of $q^{\ominus}$ has no effect. In the case of a more general reaction, the conversion from q to $q^{\ominus}$ comes about at the stage of converting the pressures that occur in K to numbers of molecules.

belonging to P; these states we label p with energies ε'_p (the prime is explained in a moment):

$$N_R = \sum_r n_r = \frac{N}{q}\sum_r e^{-\beta\varepsilon_r} \qquad N_P = \sum_p n_p = \frac{N}{q}\sum_p e^{-\beta\varepsilon'_p}$$

The sum over the states of R is its partition function, q_R, so

$$N_R = \frac{Nq_R}{q}$$

The sum over the states of P is also a partition function, but the energies are measured from the ground state of the combined system, which is the ground state of R. However, because $\varepsilon'_p = \varepsilon_p + \Delta\varepsilon_0$, where $\Delta\varepsilon_0$ is the separation of zero-point energies (as in Fig. 10.21),

$$N_P = \frac{N}{q}\sum_p e^{-\beta(\varepsilon_p+\Delta\varepsilon_0)} = \frac{N}{q}\left(\sum_p e^{-\beta\varepsilon_p}\right)e^{-\beta\Delta\varepsilon_0} = \frac{Nq_P}{q}e^{-\Delta_r E_0/RT}$$

The switch from $\Delta\varepsilon_0/k$ to $\Delta_r E_0/R$ in the last step is the conversion of molecular energies to molar energies.

The equilibrium constant of the R $\rightleftharpoons$ P reaction is proportional to the ratio of the numbers of the two types of molecule. Therefore,

$$K = \frac{N_P}{N_R} = \frac{q_P}{q_R}e^{-\Delta_r E_0/RT}$$

as in eqn 10.58b.

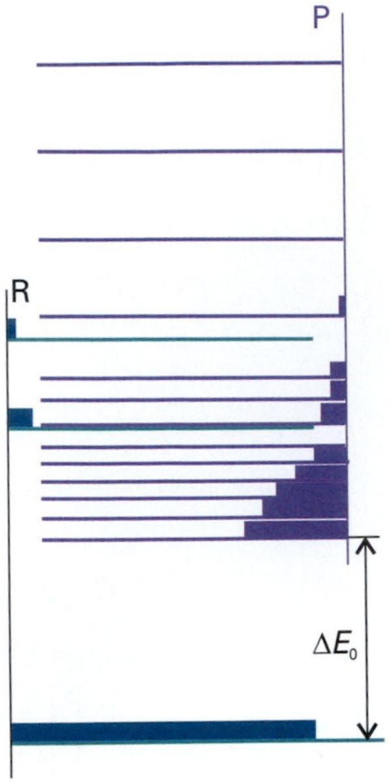

Fig. 10.21 It is important to take into account the densities of states of the molecules. Even though P might lie well above R in energy (that is, ΔE_0 is large and positive), P might have so many states that its total population dominates in the mixture. In classical thermodynamic terms, we have to take entropies into account as well as enthalpies when considering equilibria.

The content of eqn 10.58 can be seen most clearly by exaggerating the molecular features that contribute to it. We shall suppose that R has only a single accessible level, which implies that $q_R = 1$. We also suppose that P has a large number of evenly, closely spaced levels (Fig. 10.22). The partition function of P is then $q_P = kT/\varepsilon$. In this model system, the equilibrium constant is

$$K = \frac{kT}{\varepsilon}e^{-\Delta_r E_0/RT} \tag{10.59}$$

When $\Delta_r E_0$ is very large, the exponential term dominates and $K \ll 1$, which implies that very little P is present at equilibrium. When $\Delta_r E_0$ is small but still positive, K can exceed 1 because the factor kT/ε may be large enough to overcome the small size of the exponential term. The size of K then reflects the predominance of P at equilibrium on account of its high density of states. At low temperatures $K \ll 1$ and the system consists entirely of R. At high temperatures the exponential function approaches 1 and the pre-exponential factor is large. Hence P becomes dominant. We see that, in this endothermic reaction (endothermic because P lies above R), a rise in temperature favours P, because its states become accessible.

The model also shows why the Gibbs energy, G, and not just the enthalpy, determines the position of equilibrium. It shows that the density of states (and hence the entropy) of each species as well as their relative energies controls the distribution of populations and hence the value of the equilibrium constant.

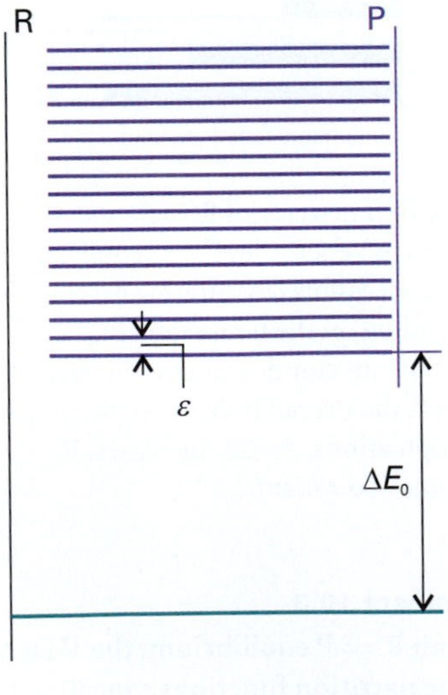

Fig. 10.22 The model used in the text for exploring the effects of energy separations and densities of states on equilibria. The products P can dominate provided ΔE_0 is not too large and P has an appreciable density of states.

Checklist of key ideas

- ☐ 1. The molecular partition function can be written as $q = q^{\mathrm{T}}q^{\mathrm{R}}q^{\mathrm{V}}q^{\mathrm{E}}$, with the contributions summarized in Table 10.3.
- ☐ 2. Thermodynamic functions can be expressed in terms of the partition function as summarized in Table 10.4.
- ☐ 3. The mean energy of a mode is $\langle\varepsilon^{\mathrm{M}}\rangle = -(1/q^{\mathrm{M}})(\partial q^{\mathrm{M}}/\partial\beta)_V$, with the contributions from each mode summarized in Table 10.5.
- ☐ 4. The contribution of a mode M to the constant-volume heat capacity is $C_V^{\mathrm{M}} = -Nk\beta(\partial\langle\varepsilon^{\mathrm{M}}\rangle/\partial\beta)_V$, with the contributions from each mode summarized in Table 10.5.
- ☐ 5. The overall heat capacity is written as $C_{V,\mathrm{m}} = \frac{1}{2}(3 + \nu_{\mathrm{R}}^* + 2\nu_{\mathrm{V}}^*)R$
- ☐ 6. The canonical partition function of a gas is $Q = Z/\Lambda^{3N}$, where Z is the configuration integral: $Z = V^N/N!$ for a perfect gas, and $Z = (1/N!)\int \mathrm{e}^{-\beta E_{\mathrm{P}}}\,\mathrm{d}\tau_1\mathrm{d}\tau_2\ldots\mathrm{d}\tau_N$ for a real gas.
- ☐ 7. In the virial equation of state, the second virial coefficient can be written as $B = -(N_{\mathrm{A}}/2V)\int f\mathrm{d}\tau_1\mathrm{d}\tau_2$, where the Mayer f-function is $f = \mathrm{e}^{-\beta E_{\mathrm{P}}} - 1$.
- ☐ 8. The radial distribution function, $g(r)$, where $g(r)r^2\mathrm{d}r$, is the probability that a molecule will be found in the range $\mathrm{d}r$ at a distance r from another molecule. The internal energy and pressure of a fluid may be expressed in terms of the radial distribution function (eqns 10.50 and 10.51, respectively).
- ☐ 9. The residual entropy is a non-zero entropy at $T = 0$ arising from molecular disorder.
- ☐ 10. The equilibrium constant can be written in terms of the partition function (eqn 10.54).

Further reading

Articles and texts

D. Chandler, *Introduction to modern statistical mechanics*. Oxford University Press (1987).

K.A. Dill and S. Bromberg, *Molecular driving forces: statistical thermodynamics in chemistry and biology*. Garland Publishing (2002).

T.L. Hill, *An introduction to statistical thermodynamics*. Dover, New York (1986).

D.A. McQuarrie and J.D. Simon, *Molecular thermodynamics*. University Science Books, Sausalito (1999).

B. Widom, *Statistical mechanics: a concise introduction for chemists*. Cambridge University Press (2002).

Table 10.3 Contributions to the molecular partition function

Mode	Expression	Value
Translation	$q^{\mathrm{T}} = \dfrac{V}{\Lambda^3}$ $\Lambda = \dfrac{h}{(2\pi mkT)^{1/2}}$	$\Lambda/\mathrm{pm} = \dfrac{1749}{(T/\mathrm{K})^{1/2}(M/\mathrm{g\,mol^{-1}})^{1/2}}$
	$\dfrac{q_{\mathrm{m}}^{\mathrm{T}\ominus}}{N_{\mathrm{A}}} = \dfrac{kT}{p^{\ominus}\Lambda^3}$	$\dfrac{q_{\mathrm{m}}^{\mathrm{T}\ominus}}{N_{\mathrm{A}}} = 2.561 \times 10^{-2}(T/\mathrm{K})^{5/2}(M/\mathrm{g\,mol^{-1}})^{3/2}$
Rotation		
Linear molecules	$q^{\mathrm{R}} = \dfrac{kT}{\sigma hcB} = \dfrac{T}{\theta_{\mathrm{R}}}$ $\theta_{\mathrm{R}} = \dfrac{hcB}{k}$	$q^{\mathrm{R}} = \dfrac{0.6950}{\sigma} \times \dfrac{T/\mathrm{K}}{(B/\mathrm{cm^{-1}})}$
Nonlinear molecules	$q^{\mathrm{R}} = \dfrac{1}{\sigma}\left(\dfrac{kT}{hc}\right)^{3/2}\left(\dfrac{\pi}{ABC}\right)^{1/2}$	$q^{\mathrm{R}} = \dfrac{1.0270}{\sigma} \times \dfrac{(T/\mathrm{K})^{3/2}}{(ABC/\mathrm{cm^{-3}})^{1/2}}$
Vibration	$q^{\mathrm{V}} = \dfrac{1}{1-\mathrm{e}^{-hc\tilde{\nu}/kT}} = \dfrac{1}{1-\mathrm{e}^{-\theta_{\mathrm{V}}/T}}$	
	$\theta_{\mathrm{V}} = \dfrac{hc\tilde{\nu}}{k} = \dfrac{h\nu}{k}$	
	For $T \gg \theta_{\mathrm{V}}$, $q^{\mathrm{V}} = \dfrac{kT}{hc\tilde{\nu}} = \dfrac{T}{\theta_{\mathrm{V}}}$	$q^{\mathrm{V}} = 0.695 \times \dfrac{T/\mathrm{K}}{\tilde{\nu}/\mathrm{cm^{-1}}}$
Electronic	$q^{\mathrm{E}} = g_0$ [+ higher terms] where g_0 is the degeneracy of the electronic ground state	

Note that $\beta = 1/kT$.

Table 10.4 Thermodynamic functions in terms of the partition function

Function	Expression: General case	Expression: Independent molecules*	
Internal energy	$U(T)-U(0)=-\left(\dfrac{\partial \ln Q}{\partial \beta}\right)_V$	$U(T)-U(0)=-N\left(\dfrac{\partial \ln q}{\partial \beta}\right)_V$	
Entropy	$S=\dfrac{U(T)-U(0)}{T}+k\ln Q$	$S=\dfrac{U(T)-U(0)}{T}+Nk\ln q$	(a)
		$S=\dfrac{U(T)-U(0)}{T}+Nk\ln\dfrac{eq}{N}$	(b)
Helmholtz energy	$A(T)-A(0)=-kT\ln Q$	$A(T)-A(0)=-NkT\ln q$	(a)
		$A(T)-A(0)=-NkT\ln\dfrac{eq}{N}$	(b)
Pressure	$p=kT\left(\dfrac{\partial \ln Q}{\partial V}\right)_T$	$p=NkT\left(\dfrac{\partial \ln q}{\partial V}\right)_T$	(b)
Enthalpy	$H(T)-H(0)=-\left(\dfrac{\partial \ln Q}{\partial \beta}\right)_V+kTV\left(\dfrac{\partial \ln Q}{\partial V}\right)_T$	$H(T)-H(0)=-N\left(\dfrac{\partial \ln q}{\partial \beta}\right)_V+NkTV\left(\dfrac{\partial \ln q}{\partial V}\right)_T$	
Gibbs energy	$G(T)-G(0)=-kT\ln Q+kTV\left(\dfrac{\partial \ln Q}{\partial V}\right)_T$	$G(T)-G(0)=-NkT\ln q+NkTV\left(\dfrac{\partial \ln q}{\partial V}\right)_T$	(a)
		$G(T)-G(0)=-NkT\ln\dfrac{eq}{N}+NkTV\left(\dfrac{\partial \ln q}{\partial V}\right)_T$	(b)

* (a) is for distinguishable particles (from $Q=q^N$), (b) for indistinquishable particles (from $Q=q^N/N!$).

Table 10.5 Contributions to mean energies and heat capacities

Mode	Expression: Mean energy	Expression: Heat capacity*
General mode, M	$\langle\varepsilon^{\mathrm{M}}\rangle=-\left(\dfrac{\partial \ln q^{\mathrm{M}}}{\partial \beta}\right)_V=-\dfrac{1}{q^{\mathrm{M}}}\left(\dfrac{\partial q^{\mathrm{M}}}{\partial \beta}\right)_V$	$C_V^{\mathrm{M}}=-Nk\beta^2\left(\dfrac{\partial\langle\varepsilon^{\mathrm{M}}\rangle}{\partial \beta}\right)_V$
Translation	$\langle\varepsilon^{\mathrm{T}}\rangle=\frac{3}{2}kT$	$C_V^{\mathrm{T}}=\frac{3}{2}nR$
Rotation ($T\gg\theta_{\mathrm{R}}$)	$\langle\varepsilon^{\mathrm{R}}\rangle=kT$, *linear molecules*	$C^{\mathrm{R}}=nR$, *linear molecules*
	$\langle\varepsilon^{\mathrm{R}}\rangle=\frac{3}{2}kT$, *nonlinear molecules*	$C^{\mathrm{R}}=\frac{3}{2}nR$, *nonlinear molecules*
Vibration	$\langle\varepsilon^{\mathrm{V}}\rangle=\dfrac{hc\nu}{e^{-\theta_{\mathrm{V}}/T}-1}=\dfrac{h\nu}{e^{-\theta_{\mathrm{V}}/T}-1}$	$C^{\mathrm{V}}=nRf$, $f=\left(\dfrac{\theta_{\mathrm{V}}}{T}\right)^2\dfrac{e^{-\theta_{\mathrm{V}}/T}}{(1-e^{-\theta_{\mathrm{V}}/T})^2}$
Vibration ($T\gg\theta_{\mathrm{V}}$)	$\langle\varepsilon^{\mathrm{V}}\rangle=kT$	$C^{\mathrm{V}}=nR$

* No distinction need be made between C_V and C_p for internal modes.

Each end-of-chapter discussion question, exercise, and problem for which a solution is available in the Student Solution Manual has a corresponding bracketed number referring to the Student Solutions Manual number for this problem, for example [SSM 17.1]. Each end-of-chapter exercise and problem for which a solution is available in the Instructors' Solution Manual has a corresponding bracketed number referring to the Instructor Solutions Manual number for this problem, for example [ISM 17.2].

Discussion questions

10.1 Discuss the limitations of the expressions $q^R = kT/hcB$, $q^V = kT/hc\tilde{\nu}$, and $q^E = g^E$. [SSM **17.1**]

10.2 Explain the origin of the symmetry number. [ISM **17.2**]

10.3 Explain the origin of residual entropy. [SSM **17.3**]

10.4 Use concepts of statistical thermodynamics to describe the molecular features that determine the magnitudes of the constant-volume molar heat capacity of a molecular substance. [ISM **17.4**]

10.5 Use concepts of statistical thermodynamics to describe the molecular features that lead to the equations of state of perfect and real gases. [SSM **17.5**]

10.6 Describe how liquids are investigated by using concepts of statistical thermodynamics. [ISM **17.6**]

10.7 Use concepts of statistical thermodynamics to describe the molecular features that determine the magnitudes of equilibrium constants and their variation with temperature. [SSM **17.7**]

Exercises

10.1a Use the equipartition theorem to estimate the constant-volume molar heat capacity of (a) I_2, (b) CH_4, (c) C_6H_6 in the gas phase at 25°C. [ISM **17.1a**]

10.1b Use the equipartition theorem to estimate the constant-volume molar heat capacity of (a) O_3, (b) C_2H_6, (c) CO_2 in the gas phase at 25°C. [SSM **17.1b**]

10.2a Estimate the values of $\gamma = C_p/C_V$ for gaseous ammonia and methane. Do this calculation with and without the vibrational contribution to the energy. Which is closer to the expected experimental value at 25°C? [ISM **17.2a**]

10.2b Estimate the value of $\gamma = C_p/C_V$ for carbon dioxide. Do this calculation with and without the vibrational contribution to the energy. Which is closer to the expected experimental value at 25°C? [SSM **17.2b**]

10.3a Estimate the rotational partition function of HCl at (a) 25°C and (b) 250°C. [ISM **17.3a**]

10.3b Estimate the rotational partition function of O_2 at (a) 25°C and (b) 250°C. [SSM **17.3b**]

10.4a Give the symmetry number for each of the following molecules: (a) CO, (b) O_2, (c) H_2S, and (d) SiH_4, (e) $CHCl_3$. [ISM **17.4a**]

10.4b Give the symmetry number for each of the following molecules: (a) CO_2, (b) O_3, (c) SO_3, (d) SF_6, and (e) Al_2Cl_6. [SSM **17.4b**]

10.5a Calculate the rotational partition function of H_2O at 298 K from its rotational constants 27.878 cm^{-1}, 14.509 cm^{-1}, and 9.287 cm^{-1}. Above what temperature is the high-temperature approximation valid to within 10 per cent of the true value? [ISM **17.5a**]

10.5b Calculate the rotational partition function of SO_2 at 298 K from its rotational constants 2.027 36 cm^{-1}, 0.344 17 cm^{-1}, and 0.293 535 cm^{-1}. Above what temperature is the high-temperature approximation valid to within 10 per cent of the true value? [SSM **17.5b**]

10.6a From the results of Exercise 10.5a, calculate the rotational contribution to the molar entropy of gaseous water at 25°C. [ISM **17.6a**]

10.6b From the results of Exercise 10.5b, calculate the rotational contribution to the molar entropy of sulfur dioxide at 25°C. [SSM **17.6b**]

10.7a Calculate the rotational partition function of CH_4 (a) by direct summation of the energy levels at 298 K and 500 K, and (b) by the high-temperature approximation. Take $B = 5.2412$ cm^{-1}. [ISM **17.7a**]

10.7b Calculate the rotational partition function of CH_3CN (a) by direct summation of the energy levels at 298 K and 500 K, and (b) by the high-temperature approximation. Take $A = 5.28$ cm^{-1} and $B = 0.307$ cm^{-1}. [SSM **17.7b**]

10.8a The bond length of O_2 is 120.75 pm. Use the high-temperature approximation to calculate the rotational partition function of the molecule at 300 K. [ISM **17.8a**]

10.8b The NOF molecule is an asymmetric rotor with rotational constants 3.1752 cm^{-1}, 0.3951 cm^{-1}, and 0.3505 cm^{-1}. Calculate the rotational partition function of the molecule at (a) 25°C, (b) 100°C. [SSM **17.8b**]

10.9a Plot the molar heat capacity of a collection of harmonic oscillators as a function of T/θ_V, and predict the vibrational heat capacity of ethyne at (a) 298 K, (b) 500 K. The normal modes (and their degeneracies in parentheses) occur at wavenumbers 612(2), 729(2), 1974, 3287, and 3374 cm^{-1}. [ISM **17.9a**]

10.9b Plot the molar entropy of a collection of harmonic oscillators as a function of T/θ_V, and predict the standard molar entropy of ethyne at (a) 298 K, (b) 500 K. For data, see the preceding exercise. [SSM **17.9b**]

10.10a A CO_2 molecule is linear, and its vibrational wavenumbers are 1388.2 cm^{-1}, 667.4 cm^{-1}, and 2349.2 cm^{-1}, the last being doubly degenerate and the others non-degenerate. The rotational constant of the molecule is 0.3902 cm^{-1}. Calculate the rotational and vibrational contributions to the molar Gibbs energy at 298 K. [ISM **17.10a**]

10.10b An O_3 molecule is angular, and its vibrational wavenumbers are 1110 cm^{-1}, 705 cm^{-1}, and 1042 cm^{-1}. The rotational constants of the molecule are 3.553 cm^{-1}, 0.4452 cm^{-1}, and 0.3948 cm^{-1}. Calculate the rotational and vibrational contributions to the molar Gibbs energy at 298 K. [SSM **17.10b**]

10.11a The ground level of Cl is ${}^2P_{3/2}$ and a ${}^2P_{1/2}$ level lies 881 cm^{-1} above it. Calculate the electronic contribution to the heat capacity of Cl atoms at (a) 500 K and (b) 900 K. [ISM **17.11a**]

10.11b The first electronically excited state of O_2 is ${}^1\Delta_g$ and lies 7918.1 cm^{-1} above the ground state, which is ${}^3\Sigma_g^-$. Calculate the electronic contribution to the molar Gibbs energy of O_2 at 400 K. [SSM **17.11b**]

10.12a The ground state of the Co^{2+} ion in $CoSO_4\cdot 7H_2O$ may be regarded as ${}^4T_{9/2}$. The entropy of the solid at temperatures below 1 K is derived almost entirely from the electron spin. Estimate the molar entropy of the solid at these temperatures. [ISM **17.12a**]

10.12b Estimate the contribution of the spin to the molar entropy of a solid sample of a *d*-metal complex with $S = \frac{5}{2}$. [SSM **17.12b**]

10.13a Calculate the residual molar entropy of a solid in which the molecules can adopt (a) three, (b) five, (c) six orientations of equal energy at $T = 0$. [ISM **17.13a**]

10.13b Suppose that the hexagonal molecule $C_6H_nF_{6-n}$ has a residual entropy on account of the similarity of the H and F atoms. Calculate the residual for each value of *n*. [SSM **17.13b**]

10.14a Calculate the equilibrium constant of the reaction $I_2(g) \rightleftharpoons 2\,I(g)$ at 1000 K from the following data for I_2: $\tilde{\nu} = 214.36$ cm^{-1}, $B = 0.0373$ cm^{-1}, $D_e = 1.5422$ eV. The ground state of the I atoms is ${}^2P_{3/2}$, implying fourfold degeneracy. [ISM **17.14a**]

10.14b Calculate the value of K at 298 K for the gas-phase isotopic exchange reaction $2\,{}^{79}Br^{81}Br\ {}^{79}Br^{79} \rightleftharpoons Br + {}^{81}Br^{81}Br$. The Br_2 molecule has a non-degenerate ground state, with no other electronic states nearby. Base the calculation on the wavenumber of the vibration of ${}^{79}Br^{81}Br$, which is 323.33 cm^{-1}. [SSM **17.14b**]

Problems*

Numerical problems

10.1 The NO molecule has a doubly degenerate electronic ground state and a doubly degenerate excited state at $121.1\ \text{cm}^{-1}$. Calculate the electronic contribution to the molar heat capacity of the molecule at (a) 50 K, (b) 298 K, and (c) 500 K. [SSM **17.1**]

10.2 Explore whether a magnetic field can influence the heat capacity of a paramagnetic molecule by calculating the electronic contribution to the heat capacity of an NO_2 molecule in a magnetic field. Estimate the total constant-volume heat capacity using equipartition, and calculate the percentage change in heat capacity brought about by a 5.0 T magnetic field at (a) 50 K, (b) 298 K. [ISM **17.2**]

10.3 The energy levels of a CH_3 group attached to a larger fragment are given by the expression for a particle on a ring, provided the group is rotating freely. What is the high-temperature contribution to the heat capacity and entropy of such a freely rotating group at 25°C? The moment of inertia of CH_3 about its three-fold rotation axis (the axis that passes through the C atom and the centre of the equilateral triangle formed by the H atoms) is $5.341 \times 10^{-47}\ \text{kg m}^2$). [SSM **17.3**]

10.4 Calculate the temperature dependence of the heat capacity of p-H_2 (in which only rotational states with even values of J are populated) at low temperatures on the basis that its rotational levels $J = 0$ and $J = 2$ constitute a system that resembles a two-level system except for the degeneracy of the upper level. Use $B = 60.864\ \text{cm}^{-1}$ and sketch the heat capacity curve. The experimental heat capacity of p-H_2 does in fact show a peak at low temperatures. [ISM **17.4**]

10.5 The pure rotational microwave spectrum of HCl has absorption lines at the following wavenumbers (in cm^{-1}): 21.19, 42.37, 63.56, 84.75, 105.93, 127.12 148.31 169.49, 190.68, 211.87, 233.06, 254.24, 275.43, 296.62, 317.80, 338.99, 360.18, 381.36, 402.55, 423.74, 444.92, 466.11, 487.30, 508.48. Calculate the rotational partition function at 25°C by direct summation. [SSM **17.5**]

10.6 Calculate the standard molar entropy of N_2(g) at 298 K from its rotational constant $B = 1.9987\ \text{cm}^{-1}$ and its vibrational wavenumber $\tilde{\nu} = 2358\ \text{cm}^{-1}$. The thermochemical value is $192.1\ \text{J K}^{-1}\ \text{mol}^{-1}$. What does this suggest about the solid at $T = 0$? [ISM **17.6**]

10.7‡ J.G. Dojahn, E.C.M. Chen, and W.E. Wentworth (*J. Phys. Chem.* **100**, 9649 (1996)) characterized the potential energy curves of the ground and electronic states of homonuclear diatomic halogen anions. The ground state of F_2^- is $^2\Sigma_u^+$ with a fundamental vibrational wavenumber of $450.0\ \text{cm}^{-1}$ and equilibrium internuclear distance of 190.0 pm. The first two excited states are at 1.609 and 1.702 eV above the ground state. Compute the standard molar entropy of F_2^- at 298 K. [SSM **17.7**]

10.8‡ In a spectroscopic study of buckminsterfullerene C_{60}, F. Negri, G. Orlandi, and F. Zerbetto (*J. Phys. Chem.* **100**, 10849 (1996)) reviewed the wavenumbers of all the vibrational modes of the molecule. The wavenumber for the single A_u mode is $976\ \text{cm}^{-1}$; wavenumbers for the four threefold degenerate T_{1u} modes are 525, 578, 1180, and $1430\ \text{cm}^{-1}$; wavenumbers for the five threefold degenerate T_{2u} modes are 354, 715, 1037, 1190, and $1540\ \text{cm}^{-1}$; wavenumbers for the six fourfold degenerate G_u modes are 345, 757, 776, 963, 1315, and $1410\ \text{cm}^{-1}$; and wavenumbers for the seven fivefold degenerate H_u modes are 403, 525, 667, 738, 1215, 1342, and $1566\ \text{cm}^{-1}$. How many modes have a vibrational temperature θ_V below 1000 K? Estimate the molar constant-volume heat capacity of C_{60} at 1000 K, counting as active all modes with θ_V below this temperature. [ISM **17.8**]

10.9‡ Treat carbon monoxide as a perfect gas and apply equilibrium statistical thermodynamics to the study of its properties, as specified below, in the temperature range 100–1000 K at 1 bar. $\tilde{\nu} = 2169.8\ \text{cm}^{-1}$, $B = 1.931\ \text{cm}^{-1}$, and $D_0 = 11.09$ eV; neglect anharmonicity and centrifugal distortion. (a) Examine the probability distribution of molecules over available rotational and vibrational states. (b) Explore numerically the differences, if any, between the rotational molecular partition function as calculated with the discrete energy distribution and that calculated with the classical, continuous energy distribution. (c) Calculate the individual contributions to $U_m(T) - U_m(100\ \text{K})$, $C_{V,m}(T)$, and $S_m(T) - S_m(100\ \text{K})$ made by the translational, rotational, and vibrational degrees of freedom. [SSM **17.9**]

10.10 Calculate and plot as a function of temperature, in the range 300 K to 1000 K, the equilibrium constant for the reaction $CD_4(g) + HCl(g) \rightleftharpoons CHD_3(g) + DCl(g)$ using the following data (numbers in parentheses are degeneracies): $\tilde{\nu}(CHD_3)/\text{cm}^{-1} = 2993(1), 2142(1), 1003(3), 1291(2), 1036(2)$; $\tilde{\nu}(CD_4)/\text{cm}^{-1} = 2109(1), 1092(2), 2259(3), 996(3)$; $\tilde{\nu}(HCl)/\text{cm}^{-1} = 2991$; $\tilde{\nu}(DCl)/\text{cm}^{-1} = 2145$; $B(HCl)/\text{cm}^{-1} = 10.59$; $B(DCl)/\text{cm}^{-1} = 5.445$; $B(CHD_3)/\text{cm}^{-1} = 3.28$; $A(CHD_3)/\text{cm}^{-1} = 2.63$, $B(CD_4)/\text{cm}^{-1} = 2.63$. [ISM **17.10**]

10.11 The exchange of deuterium between acid and water is an important type of equilibrium, and we can examine it using spectroscopic data on the molecules. Calculate the equilibrium constant at (a) 298 K and (b) 800 K for the gas-phase exchange reaction $H_2O + DCl \rightleftharpoons HDO + HCl$ from the following data: $\tilde{\nu}(H_2O)/\text{cm}^{-1} = 3656.7, 1594.8, 3755.8$; $\tilde{\nu}(HDO)/\text{cm}^{-1} = 2726.7, 1402.2, 3707.5$; $A(H_2O)/\text{cm}^{-1} = 27.88$; $B(H_2O)/\text{cm}^{-1} = 14.51$; $C(H_2O)/\text{cm}^{-1} = 9.29$; $A(HDO)/\text{cm}^{-1} = 23.38$; $B(HDO)/\text{cm}^{-1} = 9.102$; $C(HDO)/\text{cm}^{-1} = 6.417$; $B(HCl)/\text{cm}^{-1} = 10.59$; $B(DCl)/\text{cm}^{-1} = 5.449$; $\tilde{\nu}(HCl)/\text{cm}^{-1} = 2991$; $\tilde{\nu}(DCl)/\text{cm}^{-1} = 2145$. [SSM **17.11**]

Theoretical problems

10.12 Derive the Sackur–Tetrode equation for a monatomic gas confined to a two-dimensional surface, and hence derive an expression for the standard molar entropy of condensation to form a mobile surface film. [ISM **17.12**]

10.13‡ For H_2 at very low temperatures, only translational motion contributes to the heat capacity. At temperatures above $\theta_R = hcB/k$, the rotational contribution to the heat capacity becomes significant. At still higher temperatures, above $\theta_V = h\nu/k$, the vibrations contribute. But at this latter temperature, dissociation of the molecule into the atoms must be considered. (a) Explain the origin of the expressions for θ_R and θ_V, and calculate their values for hydrogen. (b) Obtain an expression for the molar constant-pressure heat capacity of hydrogen at all temperatures taking into account the dissociation of hydrogen. (c) Make a plot of the molar constant-pressure heat capacity as a function of temperature in the high-temperature region where dissociation of the molecule is significant. [SSM **17.13**]

10.14 Derive expressions for the internal energy, heat capacity, entropy, Helmholtz energy, and Gibbs energy of a harmonic oscillator. Express the results in terms of the vibrational temperature, θ_V and plot graphs of each property against T/θ_V. [ISM **17.14**]

10.15 Suppose that an intermolecular potential has a hard-sphere core of radius r_1 and a shallow attractive well of uniform depth ε out to a distance r_2. Show, by using eqn 10.42 and the condition $\varepsilon \ll kT$, that such a model is

* Problems denoted with the symbol ‡ were supplied by Charles Trapp, Carmen Giunta, and Marshall Cady.

approximately consistent with a van der Waals equation of state when $b \ll V_m$, and relate the van der Waals parameters and the Joule–Thomson coefficient to the parameters in this model. [SSM **17.15**]

10.16‡ (a) Show that the number of molecules in any given rotational state of a linear molecule is given by $N_J = C(2J+1)e^{-hcBJ(J+1)/kT}$, where C is independent of J. (b) Use this result to derive eqn 6.39 for the value of J of the most highly populated rotational level. (c) Estimate the temperature at which the spectrum of HCl shown in Fig. 6.44 was obtained. (d) What is the most highly populated level of a spherical rotor at a temperature T? [ISM **17.16**]

10.17 A more formal way of arriving at the value of the symmetry number is to note that σ is the order (the number of elements) of the *rotational subgroup* of the molecule, the point group of the molecule with all but the identity and the rotations removed. The rotational subgroup of H_2O is $\{E, C_2\}$, so $\sigma = 2$. The rotational subgroup of NH_3 is $\{E, 2C_3\}$, so $\sigma = 3$. This recipe makes it easy to find the symmetry numbers for more complicated molecules. The rotational subgroup of CH_4 is obtained from the T character table as $\{E, 8C_3, 3C_2\}$, so $\sigma = 12$. For benzene, the rotational subgroup of D_{6h} is $\{E, 2C_6, 2C_3, C_2, 3C_2', 3C_2''\}$, so $\sigma = 12$. (a) Estimate the rotational partition function of ethene at 25°C given that $A = 4.828\ \text{cm}^{-1}$, $B = 1.0012\ \text{cm}^{-1}$, and $C = 0.8282\ \text{cm}^{-1}$. (b) Evaluate the rotational partition function of pyridine, C_5H_5N, at room temperature ($A = 0.2014\ \text{cm}^{-1}$, $B = 0.1936\ \text{cm}^{-1}$, $C = 0.0987\ \text{cm}^{-1}$). [SSM **17.17**]

10.18 Although expressions like $\langle\varepsilon\rangle = -\text{d}\ln q/\text{d}\beta$ are useful for formal manipulations in statistical thermodynamics, and for expressing thermodynamic functions in neat formulas, they are sometimes more trouble than they are worth in practical applications. When presented with a table of energy levels, it is often much more convenient to evaluate the following sums directly:

$$q = \sum_j e^{-\beta\varepsilon_j} \qquad \dot{q} = \sum_j \beta\varepsilon_j e^{-\beta\varepsilon_j} \qquad \ddot{q} = \sum_j (\beta\varepsilon_j)^2 e^{-\beta\varepsilon_j}$$

(a) Derive expressions for the internal energy, heat capacity, and entropy in terms of these three functions. (b) Apply the technique to the calculation of the electronic contribution to the constant-volume molar heat capacity of magnesium vapour at 5000 K using the following data:

Term	^1S	$^3P_0$	$^3P_1$	$^3P_2$	$^1P_1$	$^3S_1$
Degeneracy	1	1	3	5	3	3
$\tilde{\nu}/\text{cm}^{-1}$	0	21 850	21 870	21 911	35 051	41 197

[ISM **17.18**]

10.19 Show how the heat capacity of a linear rotor is related to the following sum:

$$\zeta(\beta) = \frac{1}{q^2}\sum_{J,J'}\{\varepsilon(J) - \varepsilon(J')\}^2 g(J)g(J')e^{-\beta\{\varepsilon(J)+\varepsilon(J')\}}$$

by

$$C = \tfrac{1}{2}Nk\beta^2\zeta(\beta)$$

where the $\varepsilon(J)$ are the rotational energy levels and $g(J)$ their degeneracies. Then go on to show graphically that the total contribution to the heat capacity of a linear rotor can be regarded as a sum of contributions due to transitions $0\to1$, $0\to2$, $1\to2$, $1\to3$, etc. In this way, construct Fig. 10.11 for the rotational heat capacities of a linear molecule. [SSM **17.19**]

10.20 Set up a calculation like that in Problem 10.19 to analyse the vibrational contribution to the heat capacity in terms of excitations between levels and illustrate your results graphically in terms of a diagram like that in Fig. 10.11. [ISM **17.20**]

10.21 Determine whether a magnetic field can influence the value of an equilibrium constant. Consider the equilibrium $I_2(g) \rightleftharpoons 2\,I(g)$ at 1000 K, and calculate the ratio of equilibrium constants $K(\mathcal{B})/K$, where $K(\mathcal{B})$ is the equilibrium constant when a magnetic field $\mathcal{B}$ is present and removes the degeneracy of the four states of the $^2P_{3/2}$ level. Data on the species are given in Exercise 10.14a. The electronic g value of the atoms is $\frac{4}{3}$. Calculate the field required to change the equilibrium constant by 1 per cent. [SSM **17.21**]

10.22 The heat capacity ratio of a gas determines the speed of sound in it through the formula $c_s = (\gamma RT/M)^{1/2}$, where $\gamma = C_p/C_V$ and M is the molar mass of the gas. Deduce an expression for the speed of sound in a perfect gas of (a) diatomic, (b) linear triatomic, (c) nonlinear triatomic molecules at high temperatures (with translation and rotation active). Estimate the speed of sound in air at 25°C. [ISM **17.22**]

Applications: to biology, materials science, environmental science, and astrophysics

10.23 An average human DNA molecule has 5×10^8 binucleotides (rungs on the DNA ladder) of four different kinds. If each rung were a random choice of one of these four possibilities, what would be the residual entropy associated with this typical DNA molecule? [SSM **17.23**]

10.24 It is possible to write an approximate expression for the partition function of a protein molecule by including contributions from only two states: the native and denatured forms of the polymer. Proceeding with this crude model gives us insight into the contribution of denaturation to the heat capacity of a protein. It follows from *Illustration* 9.4 that the total energy of a system of N protein molecules is

$$E = \frac{N\varepsilon e^{-\varepsilon/kT}}{1 + e^{-\varepsilon/kT}}$$

where ε is the energy separation between the denatured and native forms. (a) Show that the constant-volume molar heat capacity is

$$C_{V,m} = \frac{R(\varepsilon_m/RT)^2 e^{-\varepsilon_m/RT}}{(1 + e^{-\varepsilon_m/RT})^2}$$

Hint. For two functions f and g, the quotient rule of differentiation states that $\text{d}(f/g)/\text{d}x = (1/g)\text{d}f/\text{d}x - (f/g^2)\text{d}g/\text{d}x$. (b) Plot the variation of $C_{V,m}$ with temperature. (c) If the function $C_{V,m}(T)$ has a maximum or minimum, derive an expression for the temperature at which it occurs. [ISM **17.24**]

10.25‡ R. Viswanathan, R.W. Schmude, Jr., and K.A. Gingerich (*J. Phys. Chem.* **100**, 10784 (1996)) studied thermodynamic properties of several boron–silicon gas-phase species experimentally and theoretically. These species can occur in the high-temperature chemical vapour deposition (CVD) of silicon-based semiconductors. Among the computations they reported was computation of the Gibbs energy of BSi(g) at several temperatures based on a $^4\Sigma^-$ ground state with equilibrium internuclear distance of 190.5 pm and fundamental vibrational wavenumber of 772 cm^{-1} and a $^2P_0$ first excited level 8000 cm^{-1} above the ground level. Compute the standard molar Gibbs energy $G_m^\ominus(2000\ \text{K}) - G_m^\ominus(0)$. [SSM **17.25**]

10.26‡ The molecule Cl_2O_2, which is believed to participate in the seasonal depletion of ozone over Antarctica, has been studied by several means. M. Birk, R.R. Friedl, E.A. Cohen, H.M. Pickett, and S.P. Sander (*J. Chem. Phys.* **91**, 6588 (1989)) report its rotational constants (actually cB) as 13 109.4, 2409.8, and 2139.7 MHz. They also report that its rotational spectrum indicates a molecule with a symmetry number of 2. J. Jacobs, M. Kronberg, H.S.P. Möller, and H. Willner (*J. Amer. Chem. Soc.* **116**, 1106 (1994)) report its vibrational wavenumbers as 753, 542, 310, 127, 646, and 419 cm^{-1}. Compute $G_m^\ominus(200\ \text{K}) - G_m^\ominus(0)$ of Cl_2O_2. [ISM **17.26**]

10.27‡ J. Hutter, H.P. Lüthi, and F. Diederich (*J. Amer. Chem. Soc.* **116**, 750 (1994)) examined the geometric and vibrational structure of several carbon molecules of formula C_n. Given that the ground state of C_3, a molecule found in interstellar space and in flames, is an angular singlet with moments of inertia 39.340, 39.032, and 0.3082 u Å^2 (where 1 Å = 10^{-10} m) and with vibrational wavenumbers of 63.4, 1224.5, and 2040 cm^{-1}, compute $G_m^\ominus(10.00\ \text{K}) - G_m^\ominus(0)$ and $G_m^\ominus(1000\ \text{K}) - G_m^\ominus(0)$ for C_3. [SSM **17.27**]

Appendix 1 Quantities, units, and notational conventions

A1

The result of a measurement is a **physical quantity** (such as mass or density) that is reported as a numerical multiple of an agreed **unit**:

physical quantity = numerical value × unit

For example, the mass of an object may be reported as $m = 2.5$ kg and its density as $d = 1.010$ kg dm^{-3} where the units are, respectively, 1 kilogram (1 kg) and 1 kilogram per decimetre cubed (1 kg dm^{-3}). Units are treated like algebraic quantities, and may be multiplied, divided, and cancelled. Thus, the expression (physical quantity)/unit is simply the numerical value of the measurement in the specified units, and hence is a dimensionless quantity. For instance, the mass reported above could be denoted m/kg = 2.5 and the density as d/(kg dm^{-3}) = 1.01.

Physical quantities are denoted by italic or (sloping) Greek letters (as in m for mass and Π for osmotic pressure). Units are denoted by Roman letters (as in m for metre).

Names of quantities

A **substance** is a distinct, pure form of matter. The **amount of substance**, n (more colloquially 'number of moles' or 'chemical amount'), in a sample is reported in terms of the **mole** (mol): 1 mol is the amount of substance that contains as many objects (atoms, molecules, ions, or other specified entities) as there are atoms in exactly 12 g of carbon-12. This number is found experimentally to be approximately 6.02×10^{23} (see the endpapers for more precise values). If a sample contains N entities, the amount of substance it contains is $n = N/N_A$, where N_A is the Avogadro constant: $N_A = 6.02 \times 10^{23}$ mol^{-1}. Note that N_A is a quantity with units, not a pure number.

An **extensive property** is a property that depends on the amount of substance in the sample. Two examples are mass and volume. An **intensive property** is a property that is independent of the amount of substance in the sample. Examples are temperature, mass density (mass divided by volume), and pressure.

A **molar property**, X_m, is the value of an extensive property, X, of the sample divided by the amount of substance present in the sample: $X_m = X/n$. A molar property is intensive. An example is the **molar volume**, V_m, the volume of a sample divided by the amount of substance in the sample (the volume per mole). The one exception to the notation X_m is the **molar mass**, which is denoted M. The molar mass of an element is the mass per mole of its atoms. The molar mass of a molecular compound is the mass per mole of molecules, and the molar mass of an ionic compound is the mass per mole of formula units. A **formula unit** of an ionic compound is an assembly of ions corresponding to the chemical formula of the compound; so the formula unit NaCl consists of one Na^+ ion and one Cl^- ion. The names *atomic weight* and *molecular weight* are still widely used in place of molar mass (often with the units omitted), but we shall not use them in this text.

The **molar concentration** ('molarity') of a solute in a solution is the amount ofsubstance of the solute divided by the volume of the solution. Molar concentration is usually expressed in moles per decimetre cubed (mol dm^{-3} or mol L^{-1}; 1 dm^3 is identical to 1 L). A solution in which the molar concentration of the solute is 1 mol dm^{-3} is prepared by dissolving 1 mol of the solute in sufficient solvent to prepare 1 dm^3 of solution. Such a solution is widely called a '1 molar' solution and denoted 1 M. The term **molality** refers to the amount of substance of solute divided by the mass of solvent used to prepare the solution. Its units are typically moles of solute per kilogram of solvent (mol kg^{-1}).

Units

In the **International System** of units (SI, from the French *Système International d'Unités*), the units are formed from seven **base units** listed in Table A1.1. All other physical quantities may be expressed as combinations of these physical quantities and reported in terms of **derived units**. Thus, volume is $(\text{length})^3$ and may be reported as a multiple of 1 metre cubed (1 m^3), and density, which is mass/volume, may be reported as a multiple of 1 kilogram per metre cubed (1 kg m^{-3}).

A number of derived units have special names and symbols. The names of units derived from names of people are lower case (as in torr, joule, pascal, and kelvin), but their symbols are upper case (as in Torr, J, Pa, and K). Among the most important for our purposes are those listed in Table A1.2. In all cases (both for base and derived quantities), the units may be modified by a prefix that denotes a factor of a power of 10. In a perfect world, Greek prefixes of units are upright (as in μm) and sloping for physical properties (as in μ for chemical potential), but available typefaces are not always so obliging. Among the most common prefixes are those listed in Table A1.3. Examples of the use of these prefixes are

$$1\ \text{nm} = 10^{-9}\ \text{m} \qquad 1\ \text{ps} = 10^{-12}\ \text{s} \qquad 1\ \mu\text{mol} = 10^{-6}\ \text{mol}$$

Table A1.1 The SI base units

Physical quantity	Symbol for quantity	Base unit
Length	l	metre, m
Mass	M	kilogram, kg
Time	t	second, s
Electric current	I	ampere, A
Thermodynamic temperature	T	kelvin, K
Amount of substance	n	mole, mol
Luminous intensity	I	candela, cd

Table A1.2 A selection of derived units

Physical quantity	Derived unit*	Name of derived unt
Force	1 kg m s^{-2}	newton, N
Pressure	1 kg m^{-1} s^{-2}	pascal, Pa
	1 N m^{-2}	
Energy	1 kg m^2 s^{-2}	joule, J
	1 N m	
	1 Pa m^3	
Power	kg m^2 s^{-3}	watt, W
	1 J s^{-1}	

* Equivalent definitions in terms of derived units are given following the definition in terms of base units.

Table A1.3 Common SI prefixes

Prefix	z	a	f	p	n	μ	m	c	d
Name	zepto	atto	femto	pico	nano	micro	milli	centi	deci
Factor	10^{-21}	10^{-18}	10^{-15}	10^{-12}	10^{-9}	10^{-6}	10^{-3}	10^{-2}	10^{-1}
Prefix	k	M	G	T					
Name	kilo	mega	giga	tera					
Factor	10^3	10^6	10^9	10^{12}					

The kilogram (kg) is anomalous: although it is a base unit, it is interpreted as 10^3 g, and prefixes are attached to the gram (as in 1 mg = 10^{-3} g). Powers of units apply to the prefix as well as the unit they modify:

$$1\ \text{cm}^3 = 1\ (\text{cm})^3 = 1\ (10^{-2}\ \text{m})^3 = 10^{-6}\ \text{m}^3$$

Note that 1 cm^3 does not mean 1 c(m^3). When carrying out numerical calculations, it is usually safest to write out the numerical value of an observable as powers of 10.

There are a number of units that are in wide use but are not a part of the International System. Some are exactly equal to multiples of SI units. These include the *litre* (L), which is exactly 10^3 cm^3 (or 1 dm^3) and the *atmosphere* (atm), which is exactly 101.325 kPa. Others rely on the values of fundamental constants, and hence are liable to change when the values of the fundamental constants are modified by more accurate or more precise measurements. Thus, the size of the energy unit *electronvolt* (eV), the energy acquired by an electron that is accelerated through a potential difference of exactly 1 V, depends on the value of the charge of the electron, and the present (2005) conversion factor is 1 eV = 1.602 177 33 × 10^{-19} J. Table A1.4 gives the conversion factors for a number of these convenient units.

Notational conventions

We use SI units and IUPAC conventions throughout (see *Further reading*), except in a small number of cases. The default numbering of equations is (*C.n*), where *C* is the chapter; however, [*C.n*] is used to denote a definition and {*C.n*} is used to indicate that a variable x should be interpreted as $x/x^{\ominus}$, where $x^{\ominus}$ is a standard value. A subscript r

Table A1.4 Some common units

Physical quantity	Name of unit	Symbol for unit	Value*
Time	minute	min	60 s
	hour	h	3600 s
	day	d	86 400 s
Length	ångström	Å	10^{-10} m
Volume	litre	L, l	1 dm^3
Mass	tonne	t	10^3 kg
Pressure	bar	bar	10^5 Pa
	atmosphere	atm	101.325 kPa
Energy	electronvolt	eV	$1.602\,177\,33 \times 10^{-19}$ J
			96.485 31 kJ mol^{-1}

* All values in the final column are exact, except for the definition of 1 eV, which depends on the measured value of *e*.

attached to an equation number indicates that the equation is valid only for a reversible change. A superscript ° indicates that the equation is valid for an ideal system, such as a perfect gas or an ideal solution.

We use

$$p^{\ominus} = 1 \text{ bar} \qquad b^{\ominus} = 1 \text{ mol kg}^{-1} \qquad 1\, c^{\ominus} = 1 \text{ mol dm}^{-3}$$

When referring to temperature, T denotes a thermodynamic temperature (for instance, on the Kelvin scale) and θ a temperature on the Celsius scale.

For numerical calculations, we take special care to use the proper number of significant figures. Unless otherwise specified, assume that zeros in data like 10, 100, 1000, etc are significant (that is, interpret such data as 10., 100., 100., etc).

Further reading

I.M. Mills (ed.), *Quantities, units, and symbols in physical chemistry.* Blackwell Scientific, Oxford (1993).

Appendix 2 Mathematical techniques

A2

Basic procedures

A2.1 Logarithms and exponentials

The **natural logarithm** of a number x is denoted $\ln x$, and is defined as the power to which e = 2.718 . . . must be raised for the result to be equal to x. It follows from the definition of logarithms that

$$\ln x + \ln y + \cdots = \ln xy \cdots \quad \text{(A2.1)}$$

$$\ln x - \ln y = \ln(x/y) \quad \text{(A2.2)}$$

$$a \ln x = \ln x^a \quad \text{(A2.3)}$$

We also encounter the **common logarithm** of a number, $\log x$, the logarithm compiled with 10 in place of e. Common logarithms follow the same rules of addition and subtraction as natural logarithms. Common and natural logarithms are related by

$$\ln x = \ln 10 \log x \approx 2.303 \log x \quad \text{(A2.4)}$$

The **exponential function**, e^x, plays a very special role in the mathematics of chemistry. The following properties are important:

$$e^x e^y e^z \ldots = e^{x+y+z+\cdots} \quad \text{(A2.5)}$$

$$e^x/e^y = e^{x-y} \quad \text{(A2.6)}$$

$$(e^x)^a = e^{ax} \quad \text{(A2.7)}$$

A2.2 Complex numbers and complex functions

Complex numbers have the form

$$z = x + iy \quad \text{(A2.8)}$$

where $i = (-1)^{1/2}$. The real numbers x and y are, respectively, the real and imaginary parts of z, denoted $\text{Re}(z)$ and $\text{Im}(z)$. We write the **complex conjugate** of z, denoted z^*, by replacing i by −i:

$$z^* = x - iy \quad \text{(A2.9)}$$

The **absolute value** or **modulus** of the complex number z is denoted $|z|$ and is given by:

$$|z| = (z^* z)^{1/2} = (x^2 + y^2)^{1/2} \quad \text{(A2.10)}$$

The following rules apply for arithmetic operations involving complex numbers:

1 Addition. If $z = x + iy$ and $z' = x' + iy'$, then

$$z \pm z' = (x \pm x') + i(y \pm y') \quad \text{(A2.11)}$$

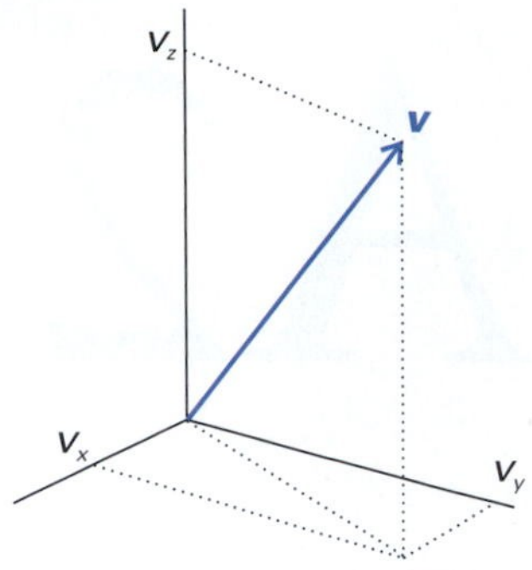

Fig. A2.1 The vector v has components v_x, v_y, and v_z on the x, y, and z axes with magnitudes v_x, v_y, and v_z, respectively.

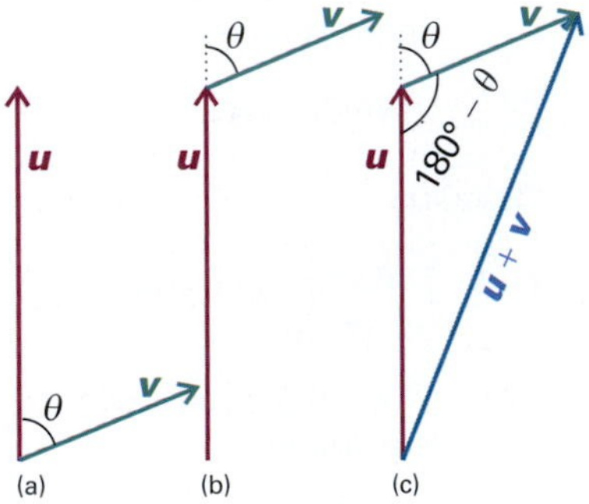

Fig. A2.2 (a) The vectors v and u make an angle θ. (b) To add u to v, we first join the tail of u to the head of v, making sure that the angle θ between the vectors remains unchanged. (c) To finish the process, we draw the resultant vector by joining the tail of u to the head of v.

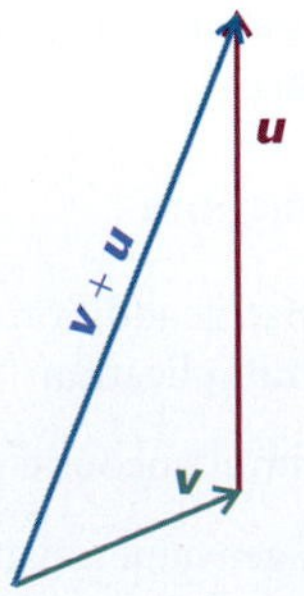

Fig. A2.3 The result of adding the vector v to the vector u, with both vectors defined in Fig. A2.2a. Comparison with the result shown in Fig. A2.2c for the addition of u to v shows that reversing the order of vector addition does not affect the result.

2 Multiplication. For z and z' defined above,

$$z \times z' = (x + iy)(x' + iy') = (xx' - yy') + i(xy' + yx') \tag{A2.12}$$

3 Division. For z and z' defined above,

$$\frac{z}{z'} = \frac{z(z')^*}{|z'|^2} \tag{A2.13}$$

Functions of complex arguments are useful in the discussion of wave equations (Chapter 1). We write the complex conjugate, f^*, of a complex function, f, by replacing i wherever it occurs by −i. For instance, the complex conjugate of e^{ix} is e^{-ix}.

Complex exponential functions may be written in terms of trigonometric functions. For example,

$$e^{\pm ix} = \cos x \pm i \sin x \tag{A2.14}$$

which implies that

$$\cos x = \tfrac{1}{2}(e^{ix} + e^{-ix}) \tag{A2.15}$$

$$\sin x = -\tfrac{1}{2}i(e^{ix} - e^{-ix}) \tag{A2.16}$$

A2.3 Vectors

A vector quantity has both magnitude and direction. The vector shown in Fig. A2.1 has components on the x, y, and z axes with magnitudes v_x, v_y, and v_z, respectively. The vector may be represented as

$$\boldsymbol{v} = v_x\boldsymbol{i} + v_y\boldsymbol{j} + v_z\boldsymbol{k} \tag{A2.17}$$

where $\boldsymbol{i}$, $\boldsymbol{j}$, and $\boldsymbol{k}$ are **unit vectors**, vectors of magnitude 1, pointing along the positive directions on the x, y, and z axes. The magnitude of the vector is denoted v or $|\boldsymbol{v}|$ and is given by

$$v = (v_x^2 + v_y^2 + v_z^2)^{1/2} \tag{A2.18}$$

Using this representation, we can define the following vector operations:

1 Addition and subtraction. If $\boldsymbol{v} = v_x\boldsymbol{i} + v_y\boldsymbol{j} + v_z\boldsymbol{k}$ and $\boldsymbol{u} = u_x\boldsymbol{i} + u_y\boldsymbol{j} + u_z\boldsymbol{k}$, then

$$\boldsymbol{v} \pm \boldsymbol{u} = (v_x \pm u_x)\boldsymbol{i} + (v_y \pm u_y)\boldsymbol{j} + (v_z \pm u_z)\boldsymbol{k} \tag{A2.19}$$

A graphical method for adding and subtracting vectors is sometimes desirable, as we saw in Chapter 3. Consider two vectors $\boldsymbol{v}$ and $\boldsymbol{u}$ making an angle θ (Fig. A2.2a). The first step in the addition of $\boldsymbol{u}$ to $\boldsymbol{v}$ consists of joining the tail of $\boldsymbol{u}$ to the head of $\boldsymbol{v}$, as shown in Fig. A2.2b. In the second step, we draw a vector $\boldsymbol{v}_{res}$, the **resultant vector**, originating from the tail of $\boldsymbol{v}$ to the head of $\boldsymbol{u}$, as shown in Fig. A2.2c. Reversing the order of addition leads to the same result. That is, we obtain the same $\boldsymbol{v}_{res}$ whether we add $\boldsymbol{u}$ to $\boldsymbol{v}$ (Fig. A2.2c) or $\boldsymbol{v}$ to $\boldsymbol{u}$ (Fig. A2.3).

To calculate the magnitude of $\boldsymbol{v}_{res}$, we note that $\boldsymbol{v}$, $\boldsymbol{u}$, and $\boldsymbol{v}_{res}$ form a triangle and that we know the magnitudes of two of its sides (v and u) and of the angle between them (180° − θ; see Fig. A2.2c). To calculate the magnitude of the third side, v_{res}, we make use of the *law of cosines*, which states that:

For a triangle with sides a, b, and c, and angle C facing side c:

$$c^2 = a^2 + b^2 - 2ab \cos C$$

This law is summarized graphically in Fig. A2.4 and its application to the case shown in Fig. A2.2c leads to the expression

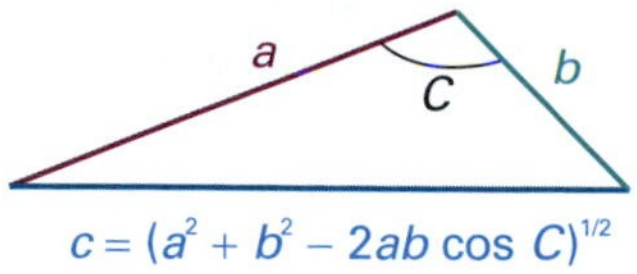

Fig. A2.4 The graphical representation of the law of cosines.

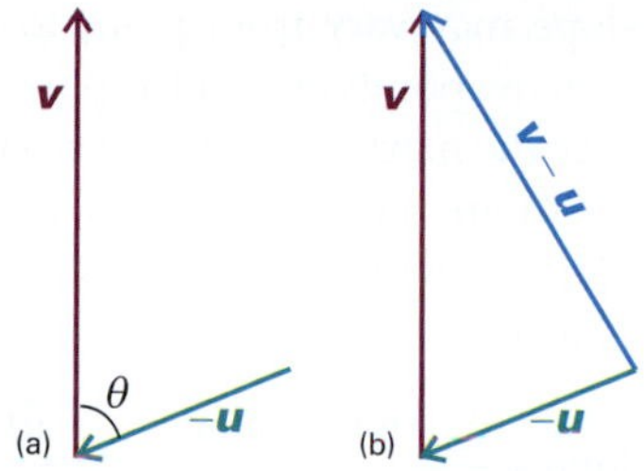

Fig. A2.5 The graphical method for subtraction of the vector $\boldsymbol{u}$ from the vector $\boldsymbol{v}$ (as shown in Fig. A2.2a) consists of two steps: (a) reversing the direction of $\boldsymbol{u}$ to form $-\boldsymbol{u}$, and (b) adding $-\boldsymbol{u}$ to $\boldsymbol{v}$.

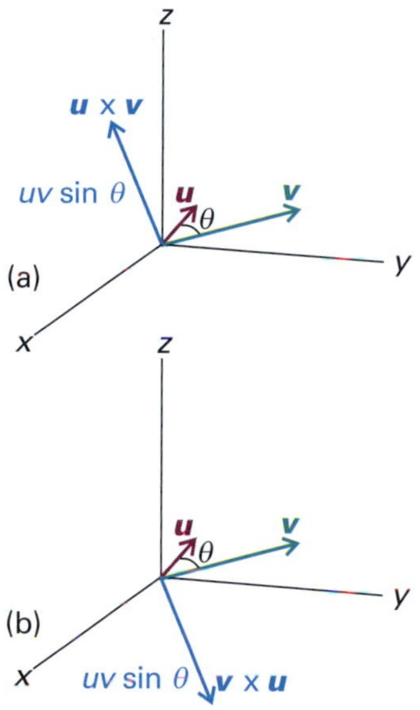

Fig. A2.6 The direction of the cross-products of two vectors $\boldsymbol{u}$ and $\boldsymbol{v}$ with an angle θ between them: (a) $\boldsymbol{u}\times\boldsymbol{v}$ and (b) $\boldsymbol{v}\times\boldsymbol{u}$. Note that the cross-product, and the unit vector $\boldsymbol{l}$ of eqn A2.21, are perpendicular to both $\boldsymbol{u}$ and $\boldsymbol{v}$ but the direction depends on the order in which the product is taken.

$$v_{\rm res}^2 = v^2 + u^2 - 2vu\cos(180° - \theta)$$

Because $\cos(180° - \theta) = -\cos\theta$, it follows after taking the square-root of both sides of the preceding expression that

$$v_{\rm res} = (v^2 + u^2 + 2vu\cos\theta)^{1/2} \qquad \text{(A2.20)}$$

The subtraction of vectors follows the same principles outlined above for addition. Consider again the vectors shown in Fig. A2.2a. We note that subtraction of $\boldsymbol{u}$ from $\boldsymbol{v}$ amounts to addition of $-\boldsymbol{u}$ to $\boldsymbol{v}$. It follows that in the first step of subtraction we draw $-\boldsymbol{u}$ by reversing the direction of $\boldsymbol{u}$ (Fig. A2.5a) Then, the second step consists of adding the $-\boldsymbol{u}$ to $\boldsymbol{v}$ by using the strategy shown in Fig. A2.2c: we draw a resultant vector $\boldsymbol{v}_{\rm res}$ by joining the tail of $-\boldsymbol{u}$ to the head of $\boldsymbol{v}$.

2 Multiplication. There are two ways to multiply vectors. In one procedure, the **cross-product** of two vectors $\boldsymbol{u}$ and $\boldsymbol{v}$ is a vector defined as

$$\boldsymbol{u}\times\boldsymbol{v} = (uv\sin\theta)\boldsymbol{l} \qquad \text{(A2.21a)}$$

where θ is the angle between the two vectors and $\boldsymbol{l}$ is a unit vector perpendicular to both $\boldsymbol{u}$ and $\boldsymbol{v}$, with a direction determined as in Fig. A2.6. An equivalent definition is

$$\boldsymbol{u}\times\boldsymbol{v} = \begin{vmatrix} \boldsymbol{i} & \boldsymbol{j} & \boldsymbol{k} \\ u_x & u_y & u_z \\ v_x & v_y & v_z \end{vmatrix} = (u_y v_z - u_z v_y)\boldsymbol{i} - (u_x v_z - u_z v_x)\boldsymbol{j} + (u_x v_y - u_y v_x)\boldsymbol{k} \qquad \text{(A2.21b)}$$

where the structure in the middle is a determinant (see below). The second type of vector multiplication is the **scalar product** (or **dot product**) of two vectors $\boldsymbol{u}$ and $\boldsymbol{v}$:

$$\boldsymbol{u}\cdot\boldsymbol{v} = uv\cos\theta \qquad \text{(A2.22)}$$

As its name suggests, the scalar product of two vectors is a scalar.

Calculus

A2.4 Differentiation and integration

Rates of change of functions—slopes of their graphs—are best discussed in terms of the infinitesimal calculus. The slope of a function, like the slope of a hill, is obtained by dividing the rise of the hill by the horizontal distance (Fig. A2.7). However, because

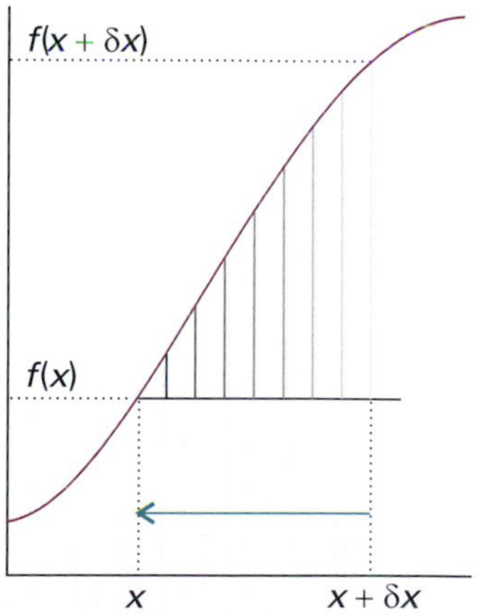

Fig. A2.7 The slope of $f(x)$ at x, df/dx, is obtained by making a series of approximations to the value of $f(x + \delta x) - f(x)$ divided by the change in x, denoted δx, and allowing δx to approach 0 (as indicated by the vertical lines getting closer to x.

the slope may vary from point to point, we should make the horizontal distance between the points as small as possible. In fact, we let it become infinitesimally small —hence the name *infinitesimal* calculus. The values of a function f at two locations x and $x+\delta x$ are $f(x)$ and $f(x+\delta x)$, respectively. Therefore, the slope of the function f at x is the vertical distance, which we write δf, divided by the horizontal distance, which we write δx:

$$\text{Slope} = \frac{\text{rise in value}}{\text{horizontal distance}} = \frac{\delta f}{\delta x} = \frac{f(x+\delta x) - f(x)}{\delta x} \qquad \text{(A2.23)}$$

The slope at x itself is obtained by letting the horizontal distance become zero, which we write $\lim \delta x \to 0$. In this limit, the δ is replaced by a d, and we write

$$\text{Slope at } x = \frac{\mathrm{d}f}{\mathrm{d}x} = \lim_{\delta x \to 0} \frac{f(x+\delta x) - f(x)}{\delta x} \qquad \text{(A2.24)}$$

To work out the slope of any function, we work out the expression on the right: this process is called **differentiation** and the expression for $\mathrm{d}f/\mathrm{d}x$ is the **derivative** of the function f with respect to the variable x. Some important derivatives are given inside the front cover of the text. Most of the functions encountered in chemistry can be differentiated by using the following rules (noting that in these expressions, derivatives $\mathrm{d}f/\mathrm{d}x$ are written as $\mathrm{d}f$):

Rule 1 For two functions f and g:

$$\mathrm{d}(f+g) = \mathrm{d}f + \mathrm{d}g \qquad \text{(A2.25)}$$

Rule 2 (the product rule) For two functions f and g:

$$\mathrm{d}(fg) = f\,\mathrm{d}g + g\,\mathrm{d}f \qquad \text{(A2.26)}$$

Rule 3 (the quotient rule) For two functions f and g:

$$\mathrm{d}\frac{f}{g} = \frac{1}{g}\mathrm{d}f - \frac{f}{g^2}\mathrm{d}g \qquad \text{(A2.27)}$$

Rule 4 (the chain rule) For a function $f = f(g)$, where $g = g(t)$,

$$\frac{\mathrm{d}f}{\mathrm{d}t} = \frac{\mathrm{d}f}{\mathrm{d}g}\frac{\mathrm{d}g}{\mathrm{d}t} \qquad \text{(A2.28)}$$

The area under a graph of any function f is found by the techniques of **integration**. For instance, the area under the graph of the function f drawn in Fig. A2.8 can be written as the value of f evaluated at a point multiplied by the width of the region, δx, and then all those products $f(x)\delta x$ summed over all the regions:

$$\text{Area between } a \text{ and } b = \sum f(x)\delta x$$

When we allow δx to become infinitesimally small, written $\mathrm{d}x$, and sum an infinite number of strips, we write:

$$\text{Area between } a \text{ and } b = \int_a^b f(x)\mathrm{d}x \qquad \text{(A2.29)}$$

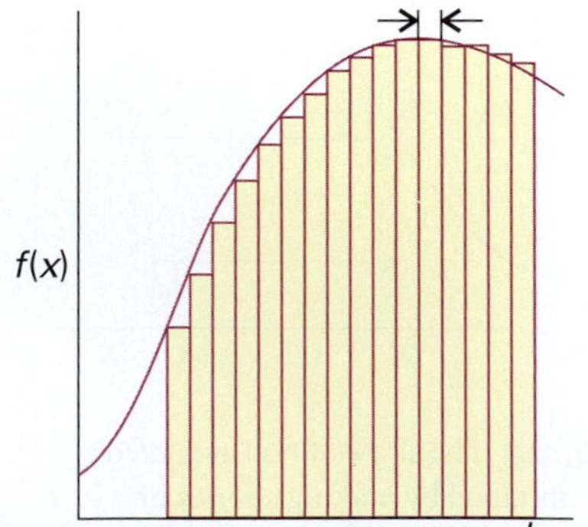

Fig. A2.8 The shaded area is equal to the definite integral of $f(x)$ between the limits a and b.

The elongated S symbol on the right is called the **integral** of the function f. When written as $\int$ alone, it is the **indefinite integral** of the function. When written with limits (as in eqn A2.29), it is the **definite integral** of the function. The definite integral is the indefinite integral evaluated at the upper limit (b) minus the indefinite integral evaluated at the lower limit (a). The **average value** of a function $f(x)$ in the range $x = a$ to $x = b$ is

$$\text{Average value of } f(x) \text{ from } a \text{ to } b = \frac{1}{b-a}\int_a^b f(x)\mathrm{d}x \tag{A2.30}$$

The **mean value theorem** states that a continuous function has its mean value at least once in the range.

Integration is the inverse of differentiation. That is, if we integrate a function and then differentiate the result, we get back the original function. Some important integrals are given on the front back cover of the text. Many other standard forms are found in tables (see *Further reading*) and it is also possible to calculate definite and indefinite integrals with mathematical software. Two integration techniques are useful:

Technique 1 (integration by parts) For two functions f and g:

$$\int f\frac{\mathrm{d}g}{\mathrm{d}x}\mathrm{d}x = fg - \int g\frac{\mathrm{d}f}{\mathrm{d}x}\mathrm{d}x \tag{A2.31}$$

Technique 2 (method of partial fractions) To solve an integral of the form $\int \frac{1}{(a-x)(b-x)}\mathrm{d}x$, where a and b are constants, we write

$$\frac{1}{(a-x)(b-x)} = \frac{1}{b-a}\left(\frac{1}{a-x} - \frac{1}{b-x}\right)$$

and integrate the expression on the right. It follows that

$$\int \frac{\mathrm{d}x}{(a-x)(b-x)} = \frac{1}{b-a}\left[\int \frac{\mathrm{d}x}{a-x} - \int \frac{\mathrm{d}x}{b-x}\right]$$

$$= \frac{1}{b-a}\left(\ln\frac{1}{a-x} - \ln\frac{1}{b-x}\right) + \text{constant} \tag{A2.32}$$

A2.5 Power series and Taylor expansions

A **power series** has the form

$$c_0 + c_1(x-a) + c_2(x-a)^2 + \cdots + c_n(x-a)^n + \cdots = \sum_{n=0}^{\infty} c_n(x-a)^n \tag{A2.33}$$

where c_n and a are constants. It is often useful to express a function $f(x)$ in the vicinity of $x = a$ as a special power series called the **Taylor series**, or **Taylor expansion**, which has the form:

$$f(x) = f(a) + \left(\frac{\mathrm{d}f}{\mathrm{d}x}\right)_a (x-a) + \frac{1}{2!}\left(\frac{\mathrm{d}^2f}{\mathrm{d}x^2}\right)_a (x-a)^2 + \cdots + \frac{1}{n!}\left(\frac{\mathrm{d}^nf}{\mathrm{d}x^n}\right)_a (x-a)^n$$

$$= \sum_{n=0}^{\infty} \frac{1}{n!}\left(\frac{\mathrm{d}^nf}{\mathrm{d}x^n}\right)_a (x-a)^n \tag{A2.34}$$

where $n!$ denotes a **factorial**, given by

$$n! = n(n-1)(n-2)\ldots 1 \tag{A2.35}$$

By definition $0! = 1$. The following Taylor expansions are often useful:

$$\frac{1}{1+x} = 1 - x + x^2 + \cdots$$

$$\mathrm{e}^x = 1 + x + \tfrac{1}{2}x^2 + \cdots$$

$$\ln x = (x-1) - \tfrac{1}{2}(x-1)^2 + \tfrac{1}{3}(x-1)^3 - \tfrac{1}{4}(x-1)^4 + \cdots$$

$$\ln(1+x) = x - \tfrac{1}{2}x^2 + \tfrac{1}{3}x^3 - \cdots$$

If $x \ll 1$, then $(1+x)^{-1} \approx 1 - x$, $\mathrm{e}^x \approx 1 + x$, and $\ln(1+x) \approx x$.

A2.6 Partial derivatives

A **partial derivative** of a function of more than one variable, such as $f(x,y)$, is the slope of the function with respect to one of the variables, all the other variables being held constant. Although a partial derivative shows how a function changes when one variable changes, it may be used to determine how the function changes when more than one variable changes by an infinitesimal amount. Thus, if f is a function of x and y, then when x and y change by $\mathrm{d}x$ and $\mathrm{d}y$, respectively, f changes by

$$\mathrm{d}f = \left(\frac{\partial f}{\partial x}\right)_y \mathrm{d}x + \left(\frac{\partial f}{\partial y}\right)_x \mathrm{d}y \tag{A2.36}$$

where the symbol ∂ is used (instead of d) to denote a partial derivative. The quantity $\mathrm{d}f$ is also called the **differential** of f. For example, if $f = ax^3y + by^2$, then

$$\left(\frac{\partial f}{\partial x}\right)_y = 3ax^2y \qquad \left(\frac{\partial f}{\partial y}\right)_x = ax^3 + 2by$$

Then, when x and y undergo infinitesimal changes, f changes by

$$\mathrm{d}f = 3ax^2y\mathrm{d}x + (ax^3 + 2by)\mathrm{d}y$$

Partial derivatives may be taken in any order:

$$\frac{\partial^2 f}{\partial x \partial y} = \frac{\partial^2 f}{\partial y \partial x} \tag{A2.37}$$

For the function f given above, it is easy to verify that

$$\left(\frac{\partial}{\partial y}\left(\frac{\partial f}{\partial x}\right)_y\right)_x = 3ax^2 \qquad \left(\frac{\partial}{\partial x}\left(\frac{\partial f}{\partial y}\right)_x\right)_y = 3ax^2$$

In the following, z is a variable on which x and y depend (for example, x, y, and z might correspond to p, V, and T).

Relation 1 When x is changed at constant z:

$$\left(\frac{\partial f}{\partial x}\right)_z = \left(\frac{\partial f}{\partial x}\right)_y + \left(\frac{\partial f}{\partial y}\right)_x \left(\frac{\partial y}{\partial x}\right)_z \tag{A2.38}$$

Relation 2

$$\left(\frac{\partial y}{\partial x}\right)_z = \frac{1}{(\partial x/\partial y)_z} \tag{A2.39}$$

Relation 3

$$\left(\frac{\partial x}{\partial y}\right)_z = -\left(\frac{\partial x}{\partial z}\right)_y \left(\frac{\partial z}{\partial y}\right)_x \tag{A2.40}$$

By combining this relation and Relation 2 we obtain the **Euler chain relation**:

$$\left(\frac{\partial y}{\partial x}\right)_z \left(\frac{\partial x}{\partial z}\right)_y \left(\frac{\partial z}{\partial y}\right)_x = -1 \tag{A2.41}$$

Relation 4 This relation establishes whether or not $\mathrm{d}f$ is an **exact differential**.

$$\mathrm{d}f = g(x,y)\mathrm{d}x + h(x,y)\mathrm{d}y \qquad \text{is exact if} \quad \left(\frac{\partial g}{\partial y}\right)_x = \left(\frac{\partial h}{\partial x}\right)_y \tag{A2.42}$$

If $\mathrm{d}f$ is exact, its integral between specified limits is independent of the path.

A2.7 Functionals and functional derivatives

Just as a function f can be regarded as a set of mathematical procedures that associates a number $f(x)$ to a specified value of a variable x, so a **functional** G gives a prescription for associating a number $G[f]$ to a function $f(x)$ over a specified range of the variable x. That is, the functional is a function of a function. Functionals are important in quantum chemistry. We saw in Chapter 4 that the energy of a molecule is a functional of the electron density, which in turn is a function of the position.

To make the following discussion more concrete, consider the functional

$$G[f] = \int_0^1 f(x)^2\,dx \tag{A2.43}$$

If we let $f(x) = x$, then $G[f] = \frac{1}{3}$ over the range $0 \le x \le 1$. However, if $f(x) = \sin \pi x$, then $G[f] = \frac{1}{2}$ over the range $0 \le x \le 1$.

Just as the derivative of a function $f(x)$ tells us how the function changes with small changes δx in the variable x, so a **functional derivative** tells us about the variation δG of a functional $G[f]$ with small changes δf in the function $f(x)$. By analogy with eqn A2.24, we can write the following definition of the functional derivative as

$$\frac{\delta G}{\delta f} = \lim_{\delta f \to 0} \frac{G[f + \delta f] - G[f]}{\delta f} \tag{A2.44}$$

However, this equation does not give us a simple method for calculating the functional derivative. It can be shown that an alternative definition of $\delta G/\delta f$ is (see *Further reading*):

$$G[f + \delta f] - G[f] = \int_a^b \left(\frac{\delta G}{\delta f}\delta f(x)\right)dx \tag{A2.45}$$

where the integral is evaluated in the range over which x varies.

To see how eqn A2.45 is used to calculate a functional derivative, consider the functional given by eqn A2.43. We begin by writing

$$G[f + \delta f] = \int_0^1 \{f(x) + \delta f(x)\}^2 dx = \int_0^1 \{f(x)^2 + 2f(x)\delta f(x) + \delta f(x)^2\}dx$$

$$= \int_0^1 \{f(x)^2 + 2f(x)\delta f(x)\}dx = G[f] + \int_0^1 2f(x)\delta f(x)dx$$

where we have ignored the minute contribution from δf^2 to arrive at the penultimate expression and then used eqn A2.43 to write the final expression. It follows that

$$G[f + \delta f] - G[f] = \int_0^1 2f(x)\delta f(x)dx$$

By comparing this expression with eqn A2.45, we see that the functional derivative is

$$\frac{\delta G}{\delta f} = 2f(x)$$

A2.8 Undetermined multipliers

Suppose we need to find the maximum (or minimum) value of some function f that depends on several variables $x_1, x_2, \ldots, x_n$. When the variables undergo a small change from x_i to $x_i + \delta x_i$ the function changes from f to $f + \delta f$, where

$$\delta f = \sum_i^n \left(\frac{\partial f}{\partial x_i}\right) \delta x_i \tag{A2.46}$$

At a minimum or maximum, $\delta f = 0$, so then

$$\sum_i^n \left(\frac{\partial f}{\partial x_i}\right) \delta x_i = 0 \tag{A2.47}$$

If the x_i were all independent, all the δx_i would be arbitrary, and this equation could be solved by setting each $(\partial f / \partial x_i) = 0$ individually. When the x_i are not all independent, the δx_i are not all independent, and the simple solution is no longer valid. We proceed as follows.

Let the constraint connecting the variables be an equation of the form $g = 0$. For example, in Chapter 9, one constraint was $n_0 + n_1 + \cdots = N$, which can be written

$$g = 0, \qquad \text{with} \quad g = (n_0 + n_1 + \cdots) - N$$

The constraint $g = 0$ is always valid, so g remains unchanged when the x_i are varied:

$$\delta g = \sum_i \left(\frac{\partial g}{\partial x_i}\right) \delta x_i = 0 \tag{A2.48}$$

Because δg is zero, we can multiply it by a parameter, λ, and add it to eqn A2.47:

$$\sum_i^n \left\{ \left(\frac{\partial f}{\partial x_i}\right) + \lambda \left(\frac{\partial g}{\partial x_i}\right) \right\} \delta x_i = 0 \tag{A2.49}$$

This equation can be solved for one of the δx, δx_n for instance, in terms of all the other δx_i. All those other δx_i $(i = 1, 2, \ldots n - 1)$ are independent, because there is only one constraint on the system. But here is the trick: λ is arbitrary; therefore we can choose it so that the coefficient of δx_n in eqn A2.49 is zero. That is, we choose λ so that

$$\left(\frac{\partial f}{\partial x_n}\right) + \lambda \left(\frac{\partial g}{\partial x_n}\right) = 0 \tag{A2.50}$$

Then eqn A2.49 becomes

$$\sum_i^{n-1} \left\{ \left(\frac{\partial f}{\partial x_i}\right) + \lambda \left(\frac{\partial g}{\partial x_i}\right) \right\} \delta x_i = 0 \tag{A2.51}$$

Now the $n - 1$ variations δx_i *are* independent, so the solution of this equation is

$$\left(\frac{\partial f}{\partial x_i}\right) + \lambda \left(\frac{\partial g}{\partial x_i}\right) = 0 \qquad i = 1, 2, \ldots, n-1 \tag{A2.52}$$

However, eqn A2.50 has exactly the same form as this equation, so the maximum or minimum of f can be found by solving

$$\left(\frac{\partial f}{\partial x_i}\right) + \lambda \left(\frac{\partial g}{\partial x_i}\right) = 0 \qquad i = 1, 2, \ldots, n \tag{A2.53}$$

The use of this approach was illustrated in the text for two constraints and therefore two undetermined multipliers λ_1 and λ_2 (α and $-\beta$).

The multipliers λ cannot always remain undetermined. One approach is to solve eqn A2.50 instead of incorporating it into the minimization scheme. In Chapter 9 we used the alternative procedure of keeping λ undetermined until a property was calculated for which the value was already known. Thus, we found that $\beta = 1/kT$ by calculating the internal energy of a perfect gas.

A2.9 Differential equations

(a) Ordinary differential equations

An **ordinary differential equation** is a relation between derivatives of a function of one variable and the function itself, as in

$$a\frac{d^2y}{dx^2}+b\frac{dy}{dx}+cy=0 \qquad \text{(A2.54)}$$

The coefficients a, b, etc. may be functions of x. The **order** of the equation is the order of the highest derivative that occurs in it, so eqn A2.54 is a second-order equation. Only rarely in science is a differential equation of order higher than 2 encountered. A solution of a differential equation is an expression for y as a function of x. The process of solving a differential equation is commonly termed 'integration', and in simple cases simple integration can be employed to find $y(x)$. A **general solution** of a differential equation is the most general solution of the equation and is expressed in terms of a number of constants. When the constants are chosen to accord with certain specified **initial conditions** (if one variable is the time) or certain **boundary conditions** (to fulfil certain spatial restrictions on the solutions), we obtain the **particular solution** of the equation. A first–order differential equation requires the specification of *one* boundary (or initial) condition; a second–order differential equation requires the specification of *two* such conditions, and so on.

First-order differential equations may often be solved by direct integration. For example, the equation

$$\frac{dy}{dx}=axy$$

with a constant may be rearranged into

$$\frac{dy}{y}=ax\mathrm{d}x$$

and then integrated to

$$\ln y=\tfrac{1}{2}ax^2+A$$

where A is a constant. If we know that $y=y_0$ when $x=0$ (for instance), then it follows that $A=\ln y_0$, and hence the particular solution of the equation is

$$\ln y=\tfrac{1}{2}ax^2+\ln y_0$$

This expression rearranges to

$$y=y_0e^{ax^2/2}$$

First-order equations of a more complex form can often be solved by the appropriate substitution. For example, it is sensible to try the substitution $y=sx$, and to change the variables from x and y to x and s. An alternative useful transformation is to write $x=u+a$ and $y=v+b$, and then to select a and b to simplify the form of the resulting expression.

Solutions to complicated differential equations may also be found by referring to tables (see *Further reading*). For example, first-order equations of the form

$$\frac{dy}{dx}+yf(x)=g(x) \qquad \text{(A2.55)}$$

appear in the study of chemical kinetics. The solution is given by

$$ye^{\int f(x)\mathrm{d}x} = \int e^{\int f(x)\mathrm{d}x} g(x)\mathrm{d}x + \text{constant} \quad \text{(A2.56)}$$

Mathematical software is now capable of finding analytical solutions of a wide variety of differential equations.

Second-order differential equations are in general much more difficult to solve than first-order equations. One powerful approach commonly used to lay siege to second-order differential equations is to express the solution as a power series:

$$y = \sum_{n=0}^{\infty} c_n x^n \quad \text{(A2.57)}$$

and then to use the differential equation to find a relation between the coefficients. This approach results, for instance, in the Hermite polynomials that form part of the solution of the Schrödinger equation for the harmonic oscillator (Section 2.4). All the second-order differential equations that occur in this text are tabulated in compilations of solutions or can be solved with mathematical software, and the specialized techniques that are needed to establish the form of the solutions may be found in mathematical texts.

(b) Numerical integration of differential equations

Many of the differential equations that describe physical phenomena are so complicated that their solutions cannot be cast as functions. In such cases, we resort to numerical methods, in which approximations are made in order to integrate the differential equation. Software packages are now readily available that can be used to solve almost any equation numerically. The general form of such programs to solve $\mathrm{d}f/\mathrm{d}x = g(x)$, for instance, replaces the infinitesimal quantity $\mathrm{d}f = g(x)\mathrm{d}x$ by the small quantity $\Delta f = g(x)\Delta x$, so that

$$f(x+\Delta x) \approx f(x) + g(x)\Delta x$$

and then proceeds numerically to step along the x-axis, generating $f(x)$ as it goes. The actual algorithms adopted are much more sophisticated than this primitive scheme, but stem from it. Among the simple numerical methods, the fourth-order **Runge–Kutta method** is one of the most accurate.

The *Further reading* section lists monographs that discuss the derivation of the fourth-order Runge–Kutta method. Here we illustrate the procedure with a first-order differential equation of the form:

$$\frac{\mathrm{d}y}{\mathrm{d}x} = f(x,y) \quad \text{(A2.58)}$$

Differential equations of this kind occur widely in chemical kinetics (Chapters 9 and 10 of Volume 1).

To obtain an approximate value of the integral of eqn A2.58, we proceed by rewriting it in terms of finite differences instead of differentials:

$$\frac{\Delta y}{\Delta x} = f(x,y)$$

where Δy may also be written as $y(x + \Delta x) - y(x)$. The fourth-order Runge–Kutta method is based on the following approximation:

$$y(x+\Delta x) = y(x) + \tfrac{1}{6}(k_1 + 2k_2 + 2k_3 + k_4) \quad \text{(A2.59)}$$

where

$$k_1 = f(x,y)\Delta x \quad \text{(A2.60a)}$$

$$k_2 = f(x+\tfrac{1}{2}\Delta x, y+\tfrac{1}{2}k_1)\Delta x \quad \text{(A2.60b)}$$

$$k_3 = f(x+\tfrac{1}{2}\Delta x, y+\tfrac{1}{2}k_2)\Delta x \quad \text{(A2.60c)}$$

$$k_4 = f(x+\Delta x, y+k_3)\Delta x \quad \text{(A2.60d)}$$

Therefore, if we know the functional form of $f(x,y)$ and $y(0)$, we can use eqns A2.60(a–d) to calculate values of y for a range of x values. The process can be automated easily with an electronic spreadsheet or with mathematical software. The accuracy of the calculation increases with decreasing values of the increment Δx.

(c) Partial differential equations

A **partial differential equation** is a differential in more than one variable. An example is

$$\frac{\partial^2 y}{\partial t^2} = a\frac{\partial^2 y}{\partial x^2} \quad \text{(A2.61)}$$

with y a function of the two variables x and t. In certain cases, partial differential equations may be separated into ordinary differential equations. Thus, the Schrödinger equation for a particle in a two-dimensional square well (Section 2.2) may be separated by writing the wavefunction, $\psi(x,y)$, as the product $X(x)Y(y)$, which results in the separation of the second-order partial differential equation into two second-order differential equations in the variables x and y. A good guide to the likely success of such a **separation of variables** procedure is the symmetry of the system.

Statistics and probability

Throughout the text, we use several elementary results from two branches of mathematics: **probability theory**, which deals with quantities and events that are distributed randomly, and **statistics**, which provides tools for the analysis of large collections of data. Here we introduce some of the fundamental ideas from these two fields.

A2.10 Random selections

Combinatorial functions allow us to express the number of ways in which a system of particles may be configured; they are especially useful in statistical thermodynamics (Chapters 9 and 10). Consider a simple coin-toss problem. If n coins are tossed, the number $N(n,i)$ of outcomes that have i heads and $(n-i)$ tails, regardless of the order of the results, is given by the coefficients of the **binomial expansion** of $(1+x)^n$:

$$(1+x)^n = 1+\sum_{i=1}^{n} N(n,i)x^i, \qquad N(n,i) = \frac{n!}{(n-i)!i!} \quad \text{(A2.62)}$$

The numbers $N(n,i)$, which are sometimes denoted $\binom{n}{i}$, are also called **binomial coefficients.**

Suppose that, unlike the coin-toss problem, there are more than two possible results for each event. For example, there are six possible results for the roll of a die. For n rolls of the die, the number of ways, W, that correspond to n_1 occurrences of the number 1, n_2 occurrences of the number 2, and so on, is given by

$$W = \frac{n!}{n_1!n_2!n_3!n_4!n_5!n_6!}, \qquad n = \sum_{i=1}^{6} n_i$$

This is an example of a **multinomial coefficient**, which has the form

$$W = \frac{n!}{n_1!n_2!\ldots n_m!}, \qquad n = \sum_{i=1}^{m} n_i \tag{A2.63}$$

where W is the number of ways of achieving an outcome, n is the number of events, and m is the number of possible results. In Chapter 9 we use the multinomial coefficient to determine the number of ways to configure a system of identical particles given a specific distribution of particles into discrete energy levels.

In chemistry it is common to deal with a very large number of particles and outcomes and it is useful to express factorials in different ways. We can simplify factorials of large numbers by using **Stirling's approximation**:

$$n! \approx (2\pi)^{1/2} n^{n+\frac{1}{2}} e^{-n} \tag{A2.64}$$

The approximation is in error by less than 1 per cent when n is greater than about 10. For very large values of n, it is possible to use another form of the approximation:

$$\ln n! \approx n \ln n - n \tag{A2.65}$$

A2.11 Some results of probability theory

Here we develop two general results of probability theory: the mean value of a variable and the mean value of a function. The calculation of mean values is useful in the description of random coils and molecular diffusion.

The mean value (also called the *expectation value*) $\langle X \rangle$ of a variable X is calculated by first multiplying each discrete value x_i that X can have by the probability p_i that x_i occurs and then summing these products over all possible N values of X:

$$\langle X \rangle = \sum_{i=1}^{N} x_i p_i$$

When N is very large and the x_i values are so closely spaced that X can be regarded as varying continuously, it is useful to express the probability that X can have a value between x and $x + dx$ as

Probability of finding a value of X between x and $x + dx = f(x)dx$

where the function $f(x)$ is the *probability density*, a measure of the distribution of the probability values over x, and dx is an infinitesimally small interval of x values. It follows that the probability that X has a value between $x = a$ and $x = b$ is the integral of the expression above evaluated between a and b:

Probability of finding a value of X between a and $b = \int_a^b f(x)dx$

The mean value of the continuously varying X is given by

$$\langle X \rangle = \int_{-\infty}^{+\infty} x f(x) dx \tag{A2.66}$$

This expression is similar to that written for the case of discrete values of X, with $f(x)dx$ as the probability term and integration over the closely spaced x values replacing summation over widely spaced x_i.

The mean value of a function $g(X)$ can be calculated with a formula similar to that for $\langle X \rangle$:

$$\langle g(X) \rangle = \int_{-\infty}^{+\infty} g(x)f(x)\mathrm{d}x \tag{A2.67}$$

Matrix algebra

A **matrix** is an array of numbers. Matrices may be combined together by addition or multiplication according to generalizations of the rules for ordinary numbers. Most numerical matrix manipulations are now carried out with mathematical software.

Consider a square matrix $\boldsymbol{M}$ of n^2 numbers arranged in n columns and n rows. These n^2 numbers are the **elements** of the matrix, and may be specified by stating the row, r, and column, c, at which they occur. Each element is therefore denoted M_{rc}. For example, in the matrix

$$\boldsymbol{M} = \begin{pmatrix} 1 & 2 \\ 3 & 4 \end{pmatrix}$$

the elements are $M_{11} = 1$, $M_{12} = 2$, $M_{21} = 3$, and $M_{22} = 4$. This is an example of a 2×2 matrix. The **determinant**, $|\boldsymbol{M}|$, of this matrix is

$$|\boldsymbol{M}| = \begin{vmatrix} 1 & 2 \\ 3 & 4 \end{vmatrix} = 1 \times 4 - 2 \times 3 = -2$$

A **diagonal matrix** is a matrix in which the only nonzero elements lie on the major diagonal (the diagonal from M_{11} to M_{nn}). Thus, the matrix

$$\boldsymbol{D} = \begin{pmatrix} 1 & 0 & 0 \\ 0 & 2 & 0 \\ 0 & 0 & 1 \end{pmatrix}$$

is diagonal. The condition may be written

$$M_{rc} = m_r \delta_{rc} \tag{A2.68}$$

where δ_{rc} is the **Kronecker delta**, which is equal to 1 for $r = c$ and to 0 for $r \neq c$. In the above example, $m_1 = 1$, $m_2 = 2$, and $m_3 = 1$. The **unit matrix**, **1** (and occasionally **I**), is a special case of a diagonal matrix in which all nonzero elements are 1.

The **transpose** of a matrix $\boldsymbol{M}$ is denoted $\boldsymbol{M}^{\mathrm{T}}$ and is defined by

$$M^{\mathrm{T}}_{mn} = M_{nm} \tag{A2.69}$$

Thus, for the matrix $\boldsymbol{M}$ we have been using,

$$\boldsymbol{M}^{\mathrm{T}} = \begin{pmatrix} 1 & 3 \\ 2 & 4 \end{pmatrix}$$

Matrices are very useful in chemistry. They simplify some mathematical tasks, such as solving systems of simultaneous equations, the treatment of molecular symmetry (Chapter 5), and quantum mechanical calculations (Chapter 4).

A2.12 Matrix addition and multiplication

Two matrices $\boldsymbol{M}$ and $\boldsymbol{N}$ may be added to give the sum $\boldsymbol{S} = \boldsymbol{M} + \boldsymbol{N}$, according to the rule

$$S_{rc} = M_{rc} + N_{rc} \tag{A2.70}$$

(that is, corresponding elements are added). Thus, with $\boldsymbol{M}$ given above and

$$\boldsymbol{N}=\begin{pmatrix}5 & 6\\ 7 & 8\end{pmatrix}$$

the sum is

$$\boldsymbol{S}=\begin{pmatrix}1 & 2\\ 3 & 4\end{pmatrix}+\begin{pmatrix}5 & 6\\ 7 & 8\end{pmatrix}=\begin{pmatrix}6 & 8\\ 10 & 12\end{pmatrix}$$

Two matrices may also be multiplied to give the product $\boldsymbol{P}=\boldsymbol{MN}$ according to the rule

$$P_{rc}=\sum_n M_{rn}N_{nc} \tag{A2.71}$$

For example, with the matrices given above,

$$\boldsymbol{P}=\begin{pmatrix}1 & 2\\ 3 & 4\end{pmatrix}\begin{pmatrix}5 & 6\\ 7 & 8\end{pmatrix}=\begin{pmatrix}1\times 5+2\times 7 & 1\times 6+2\times 8\\ 3\times 5+4\times 7 & 3\times 6+4\times 8\end{pmatrix}=\begin{pmatrix}19 & 22\\ 43 & 50\end{pmatrix}$$

It should be noticed that in general $\boldsymbol{MN}\neq\boldsymbol{NM}$, and matrix multiplication is in general non-commutative.

The **inverse** of a matrix $\boldsymbol{M}$ is denoted $\boldsymbol{M}^{-1}$, and is defined so that

$$\boldsymbol{MM}^{-1}=\boldsymbol{M}^{-1}\boldsymbol{M}=\boldsymbol{1} \tag{A2.72}$$

The inverse of a matrix can be constructed by using mathematical software, but in simple cases the following procedure can be carried through without much effort:

1 Form the determinant of the matrix. For example, for our matrix $\boldsymbol{M}$, $|\boldsymbol{M}|=-2$.

2 Form the transpose of the matrix. For example, $\boldsymbol{M}^{\mathrm{T}}=\begin{pmatrix}1 & 3\\ 2 & 4\end{pmatrix}$.

3 Form $\tilde{\boldsymbol{M}}'$, where $\tilde{\boldsymbol{M}}'_{rc}$ is the **cofactor** of the element M_{rc}, that is, it is the determinant formed from $\boldsymbol{M}$ with the row r and column c struck out. For example,

$$\tilde{\boldsymbol{M}}'=\begin{pmatrix}4 & -2\\ -3 & 1\end{pmatrix}$$

4 Construct the inverse as $\boldsymbol{M}^{-1}=\tilde{\boldsymbol{M}}'/|\boldsymbol{M}|$. For example,

$$\boldsymbol{M}^{-1}=\frac{1}{-2}\begin{pmatrix}4 & -2\\ -3 & 1\end{pmatrix}=\begin{pmatrix}-2 & 1\\ \frac{3}{2} & -\frac{1}{2}\end{pmatrix}$$

A2.13 Simultaneous equations

A set of n **simultaneous equations**

$$\begin{aligned}
&a_{11}x_1+a_{12}x_2+\cdots+a_{1n}x_n=b_1\\
&a_{21}x_1+a_{22}x_2+\cdots+a_{2n}x_n=b_2\\
&\vdots\\
&a_{n1}x_1+a_{n2}x_2+\cdots+a_{nn}x_n=b_n
\end{aligned} \tag{A2.73}$$

can be written in matrix notation if we introduce the **column vectors** $\boldsymbol{x}$ and $\boldsymbol{b}$:

$$\boldsymbol{x}=\begin{pmatrix}x_1\\ x_2\\ \vdots\\ x_n\end{pmatrix}\qquad \boldsymbol{b}=\begin{pmatrix}b_1\\ b_2\\ \vdots\\ b_n\end{pmatrix}$$

Then, with the $\boldsymbol{a}$ matrix of coefficients a_{rc}, the n equations may be written as

$$\boldsymbol{ax} = \boldsymbol{b} \tag{A2.74}$$

The formal solution is obtained by multiplying both sides of this matrix equation by $\boldsymbol{a}^{-1}$, for then

$$\boldsymbol{x} = \boldsymbol{a}^{-1}\boldsymbol{b} \tag{A2.75}$$

A2.14 Eigenvalue equations

An **eigenvalue equation** is a special case of eqn A2.74 in which

$$\boldsymbol{ax} = \lambda\boldsymbol{x} \tag{A2.76}$$

where λ is a constant, the **eigenvalue**, and $\boldsymbol{x}$ is the **eigenvector**. In general, there are n eigenvalues $\lambda^{(i)}$, and they satisfy the n simultaneous equations

$$(\boldsymbol{a} - \lambda\mathbf{1})\boldsymbol{x} = 0 \tag{A2.77}$$

There are n corresponding eigenvectors $\boldsymbol{x}^{(i)}$. Equation A2.77 has a solution only if the determinant of the coefficients is zero. However, this determinant is just $|\boldsymbol{a} - \lambda\mathbf{1}|$, so the n eigenvalues may be found from the solution of the **secular equation**:

$$|\boldsymbol{a} - \lambda\mathbf{1}| = 0 \tag{A2.78}$$

The n eigenvalues the secular equation yields may be used to find the n eigenvectors. These eigenvectors (which are $n \times 1$ matrices), may be used to form an $n \times n$ matrix $\boldsymbol{X}$. Thus, because

$$\boldsymbol{x}^{(1)} = \begin{pmatrix} x_1^{(1)} \\ x_2^{(1)} \\ \vdots \\ x_n^{(1)} \end{pmatrix} \qquad \boldsymbol{x}^{(2)} = \begin{pmatrix} x_1^{(2)} \\ x_2^{(2)} \\ \vdots \\ x_n^{(2)} \end{pmatrix} \qquad \text{etc.}$$

we may form the matrix

$$\boldsymbol{X} = (\boldsymbol{x}^{(1)}, \boldsymbol{x}^{(2)}, \ldots, \boldsymbol{x}^{(n)}) = \begin{pmatrix} x_1^{(1)} & x_1^{(2)} & \cdots & x_1^{(n)} \\ x_2^{(1)} & x_2^{(2)} & \cdots & x_2^{(n)} \\ \vdots & \vdots & & \vdots \\ x_n^{(1)} & x_n^{(2)} & \cdots & x_n^{(n)} \end{pmatrix}$$

so that $X_{rc} = x_r^{(c)}$. If further we write $\Lambda_{rc} = \lambda_r\delta_{rc}$, so that $\boldsymbol{\Lambda}$ is a diagonal matrix with the elements $\lambda_1, \lambda_2, \ldots, \lambda_n$ along the diagonal, then all the eigenvalue equations $\boldsymbol{ax}^{(i)} = \lambda_i\boldsymbol{x}^{(i)}$ may be confined into the single equation

$$\boldsymbol{aX} = \boldsymbol{X\Lambda} \tag{A2.79}$$

because this expression is equal to

$$\sum_n a_{rn}X_{nc} = \sum_n X_{rn}\Lambda_{nc}$$

or

$$\sum_n a_{rn}x_n^{(c)} = \sum_n x_r^{(n)}\lambda_n\delta_{nc} = \lambda_c x_r^{(c)}$$

as required. Therefore, if we form $\boldsymbol{X}^{-1}$ from $\boldsymbol{X}$, we construct a **similarity transformation**

$$\boldsymbol{\Lambda} = \boldsymbol{X}^{-1}\boldsymbol{aX} \tag{A2.80}$$

that makes $\boldsymbol{a}$ diagonal (because $\boldsymbol{\Lambda}$ is diagonal). It follows that if the matrix $\boldsymbol{X}$ that causes $\boldsymbol{X}^{-1}\boldsymbol{aX}$ to be diagonal is known, then the problem is solved: the diagonal matrix so produced has the eigenvalues as its only nonzero elements, and the matrix $\boldsymbol{X}$ used to bring about the transformation has the corresponding eigenvectors as its columns. The solutions of eigenvalue equations are best found by using mathematical software.

Further reading

P.W. Atkins, J.C. de Paula, and V.A. Walters, *Explorations in physical chemistry.* W.H. Freeman and Company, New York (2005).

J.R. Barrante, *Applied mathematics for physical chemistry.* Prentice Hall, Upper Saddle River (2004).

M.P. Cady and C.A. Trapp, *A Mathcad primer for physical chemistry.* W.H. Freeman and Company, New York (2000).

R.G. Mortimer, *Mathematics for physical chemistry.* Academic Press, San Diego (2005).

D.A. McQuarrie, *Mathematical methods for scientists and engineers.* University Science Books, Sausalito (2003).

D. Zwillinger (ed.), *CRC standard mathematical tables and formulae.* CRC Press, Boca Raton (1996).

Appendix 3 Essential concepts of physics

A3

Energy

The central concept of all explanations in physical chemistry, as in so many other branches of physical science, is that of **energy**, the capacity to do work. We make use of the apparently universal law of nature that *energy is conserved*, that is, energy can neither be created nor destroyed. Although energy can be transferred from one location to another, the total energy is constant.

A3.1 Kinetic and potential energy

The **kinetic energy**, E_K, of a body is the energy the body possesses as a result of its motion. For a body of mass m travelling at a speed v,

$$E_K = \tfrac{1}{2}mv^2 \qquad \text{(A3.1)}$$

The **potential energy**, E_P or V, of a body is the energy it possesses as a result of its position. The zero of potential energy is arbitrary. For example, the gravitational potential energy of a body is often set to zero at the surface of the Earth; the electrical potential energy of two charged particles is set to zero when their separation is infinite. No universal expression for the potential energy can be given because it depends on the type of interaction the body experiences. One example that gives rise to a simple expression is the potential energy of a body of mass m in the gravitational field close to the surface of the Earth (a gravitational field acts on the mass of a body). If the body is at a height h above the surface of the Earth, then its potential energy is mgh, where g is a constant called the **acceleration of free fall**, $g = 9.81\ \text{m s}^{-2}$, and $V = 0$ at $h = 0$ (the arbitrary zero mentioned previously).

The total energy is the sum of the kinetic and potential energies of a particle:

$$E = E_K + E_P \qquad \text{(A3.2)}$$

A3.2 Energy units

The SI unit of energy is the *joule* (J), which is defined as

$$1\ \text{J} = 1\ \text{kg m}^2\ \text{s}^{-2} \qquad \text{(A3.3)}$$

Calories (cal) and kilocalories (kcal) are still encountered in the chemical literature: by definition, 1 cal = 4.184 J. An energy of 1 cal is enough to raise the temperature of 1 g of water by 1°C.

The rate of change of energy is called the **power**, P, expressed as joules per second, or *watt*, W:

$$1\ \text{W} = 1\ \text{J s}^{-1} \qquad \text{(A3.4)}$$

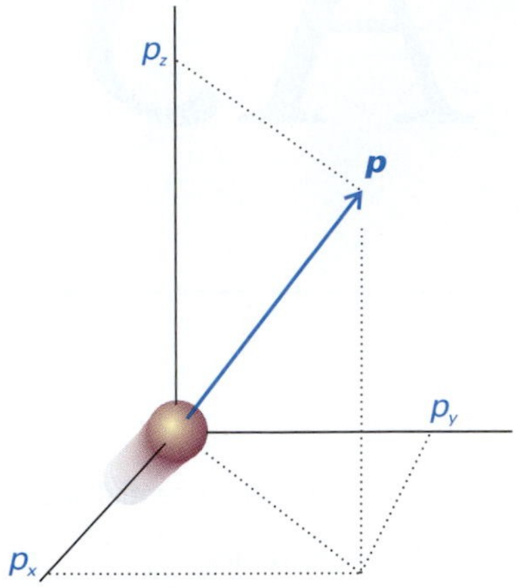

Fig. A3.1 The linear momentum of a particle is a vector property and points in the direction of motion.

Classical mechanics

Classical mechanics describes the behaviour of objects in terms of two equations. One expresses the fact that the total energy is constant in the absence of external forces; the other expresses the response of particles to the forces acting on them.

A3.3 The trajectory in terms of the energy

The **velocity**, $\boldsymbol{v}$, of a particle is the rate of change of its position:

$$\boldsymbol{v} = \frac{d\boldsymbol{r}}{dt} \quad \text{(A3.5)}$$

The velocity is a vector, with both direction and magnitude. The magnitude of the velocity is the **speed**, v. The **linear momentum**, $\boldsymbol{p}$, of a particle of mass m is related to its velocity, $\boldsymbol{v}$, by

$$\boldsymbol{p} = m\boldsymbol{v} \quad \text{(A3.6)}$$

Like the velocity vector, the linear momentum vector points in the direction of travel of the particle (Fig. A3.1). In terms of the linear momentum, the total energy of a particle is

$$E = \frac{p^2}{2m} + V(x) \quad \text{(A3.7)}$$

This equation can be used to show that a particle will have a definite **trajectory**, or definite position and momentum at each instant. For example, consider a particle free to move in one direction (along the x-axis) in a region where $V = 0$ (so the energy is independent of position). Because $v = dx/dt$, it follows from eqns A3.6 and A3.7 that

$$\frac{dx}{dt} = \left(\frac{2E_K}{m}\right)^{1/2} \quad \text{(A3.8)}$$

A solution of this differential equation is

$$x(t) = x(0) + m\left(\frac{2E_K}{m}\right)^{1/2} t \quad \text{(A3.9)}$$

The linear momentum is a constant:

$$p(t) = mv(t) = m\frac{dx}{dt} = (2mE_K)^{1/2} \quad \text{(A3.10)}$$

Hence, if we know the initial position and momentum, we can predict all later positions and momenta exactly.

A3.4 Newton's second law

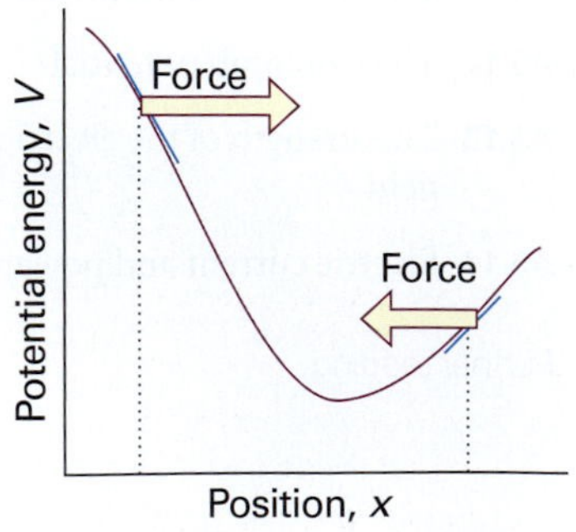

Fig. A3.2 The force acting on a particle is determined by the slope of the potential energy at each point. The force points in the direction of lower potential energy.

The **force**, F, experienced by a particle free to move in one dimension is related to its potential energy, V, by

$$F = -\frac{dV}{dx} \quad \text{(A3.11a)}$$

This relation implies that the direction of the force is towards decreasing potential energy (Fig. A3.2). In three dimensions,

$$\boldsymbol{F} = -\nabla V \qquad \nabla = \boldsymbol{i}\frac{\partial}{\partial x} + \boldsymbol{j}\frac{\partial}{\partial y} + \boldsymbol{k}\frac{\partial}{\partial z} \tag{A3.11b}$$

Newton's second law of motion states that *the rate of change of momentum is equal to the force acting on the particle.* In one dimension:

$$\frac{\mathrm{d}p}{\mathrm{d}t} = F \tag{A3.12a}$$

Because $p = m(\mathrm{d}x/\mathrm{d}t)$ in one dimension, it is sometimes more convenient to write this equation as

$$m\frac{\mathrm{d}^2x}{\mathrm{d}t^2} = F \tag{A3.12b}$$

The second derivative, $\mathrm{d}^2x/\mathrm{d}t^2$, is the **acceleration** of the particle, its rate of change of velocity (in this instance, along the x-axis). It follows that, if we know the force acting everywhere and at all times, then solving eqn A3.12 will also give the trajectory. This calculation is equivalent to the one based on E, but is more suitable in some applications. For example, it can be used to show that, if a particle of mass m is initially stationary and is subjected to a constant force F for a time τ, then its kinetic energy increases from zero to

$$E_{\mathrm{K}} = \frac{F^2\tau^2}{2m} \tag{A3.13}$$

and then remains at that energy after the force ceases to act. Because the applied force, F, and the time, τ, for which it acts may be varied at will, the solution implies that the energy of the particle may be increased to any value.

A3.5 Rotational motion

The rotational motion of a particle about a central point is described by its **angular momentum**, $\boldsymbol{J}$. The angular momentum is a vector: its magnitude gives the rate at which a particle circulates and its direction indicates the axis of rotation (Fig. A3.3). The magnitude of the angular momentum, J, is given by the expression

$$J = I\omega \tag{A3.14}$$

where ω is the **angular velocity** of the body, its rate of change of angular position (in radians per second), and I is the **moment of inertia**. The analogous roles of m and I, of v and ω, and of p and J in the translational and rotational cases, respectively, should be remembered, because they provide a ready way of constructing and recalling equations. For a point particle of mass m moving in a circle of radius r, the moment of inertia about the axis of rotation is given by the expression

$$I = mr^2 \tag{A3.15}$$

To accelerate a rotation it is necessary to apply a **torque**, $\boldsymbol{T}$, a twisting force. Newton's equation is then

$$\frac{\mathrm{d}\boldsymbol{J}}{\mathrm{d}t} = \boldsymbol{T} \tag{A3.16}$$

If a constant torque is applied for a time τ, the rotational energy of an initially stationary body is increased to

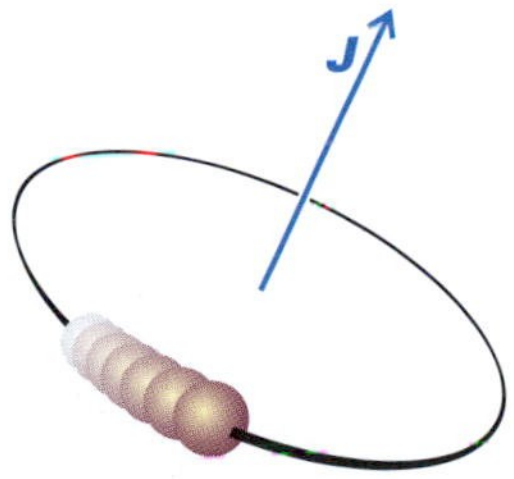

Fig. A3.3 The angular momentum of a particle is represented by a vector along the axis of rotation and perpendicular to the plane of rotation. The length of the vector denotes the magnitude of the angular momentum. The direction of motion is clockwise to an observer looking in the direction of the vector.

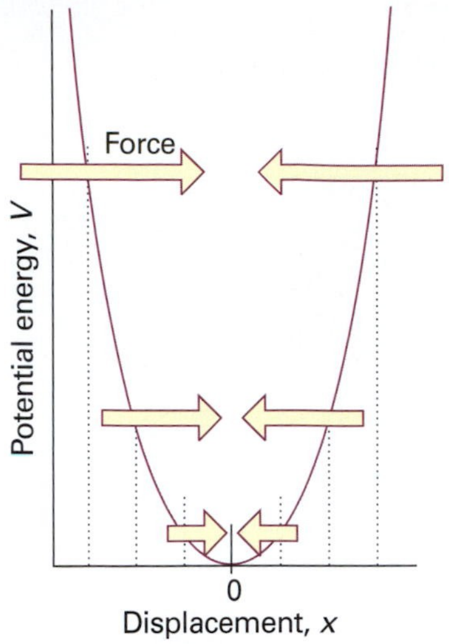

Fig. A3.4 The force acting on a particle that undergoes harmonic motion. The force is directed towards zero displacement and is proportional to the displacement. The corresponding potential energy is parabolic (proportional to x^2).

$$E_K = \frac{T^2\tau^2}{2I} \tag{A3.17}$$

The implication of this equation is that an appropriate torque and period for which it is applied can excite the rotation to an arbitrary energy.

A3.6 The harmonic oscillator

A **harmonic oscillator** consists of a particle that experiences a restoring force proportional to its displacement from its equilibrium position:

$$F = -kx \tag{A3.18}$$

An example is a particle joined to a rigid support by a spring. The constant of proportionality k is called the **force constant**, and the stiffer the spring the greater the force constant. The negative sign in F signifies that the direction of the force is opposite to that of the displacement (Fig. A3.4).

The motion of a particle that undergoes harmonic motion is found by substituting the expression for the force, eqn A3.18, into Newton's equation, eqn A3.12b. The resulting equation is

$$m\frac{d^2x}{dt^2} = -kx$$

A solution is

$$x(t) = A \sin \omega t \qquad p(t) = m\omega A \cos \omega t \qquad \omega = (k/m)^{1/2} \tag{A3.19}$$

These solutions show that the position of the particle varies **harmonically** (that is, as $\sin \omega t$) with a frequency $\nu = \omega/2\pi$. They also show that the particle is stationary ($p = 0$) when the displacement, x, has its maximum value, A, which is called the **amplitude** of the motion.

The total energy of a classical harmonic oscillator is proportional to the square of the amplitude of its motion. To confirm this remark we note that the kinetic energy is

$$E_K = \frac{p^2}{2m} = \frac{(m\omega A \cos \omega t)^2}{2m} = \tfrac{1}{2}m\omega^2 A^2 \cos 2\omega t \tag{A3.20}$$

Then, because $\omega = (k/m)^{1/2}$, this expression may be written

$$E_K = \tfrac{1}{2}kA^2 \cos^2\omega t \tag{A3.21}$$

The force on the oscillator is $F = -kx$, so it follows from the relation $F = -dV/dx$ that the potential energy of a harmonic oscillator is

$$V = \tfrac{1}{2}kx^2 = \tfrac{1}{2}kA^2 \sin^2\omega t \tag{A3.22}$$

The total energy is therefore

$$E = \tfrac{1}{2}kA^2 \cos^2\omega t + \tfrac{1}{2}kA^2 \sin^2\omega t = \tfrac{1}{2}kA^2 \tag{A3.23}$$

(We have used $\cos^2\omega t + \sin^2\omega t = 1$.) That is, the energy of the oscillator is constant and, for a given force constant, is determined by its maximum displacement. It follows that the energy of an oscillating particle can be raised to any value by stretching the spring to any desired amplitude A. Note that the frequency of the motion depends only on the inherent properties of the oscillator (as represented by k and m) and is independent of the energy; the amplitude governs the energy, through $E = \tfrac{1}{2}kA^2$, and is independent of the frequency. In other words, the particle will oscillate at the same frequency regardless of the amplitude of its motion.

Waves

Waves are disturbances that travel through space with a finite velocity. Examples of disturbances include the collective motion of water molecules in ocean waves and of gas particles in sound waves. Waves can be characterized by a **wave equation**, a differential equation that describes the motion of the wave in space and time. **Harmonic waves** are waves with displacements that can be expressed as sine or cosine functions. These concepts are used in classical physics to describe the wave character of electromagnetic radiation, which is the focus of the following discussion.

A3.7 The electromagnetic field

In classical physics, electromagnetic radiation is understood in terms of the **electromagnetic field**, an oscillating electric and magnetic disturbance that spreads as a harmonic wave through empty space, the vacuum. The wave travels at a constant speed called the *speed of light*, c, which is about 3×10^8 m s^{-1}. As its name suggests, an electromagnetic field has two components, an **electric field** that acts on charged particles (whether stationary of moving) and a **magnetic field** that acts only on moving charged particles. The electromagnetic field is characterized by a **wavelength**, λ (lambda), the distance between the neighbouring peaks of the wave, and its **frequency**, ν (nu), the number of times per second at which its displacement at a fixed point returns to its original value (Fig. A3.5). The frequency is measured in *hertz*, where 1 Hz = 1 s^{-1}. The wavelength and frequency of an electromagnetic wave are related by

$$\lambda\nu = c \qquad \text{(A3.24)}$$

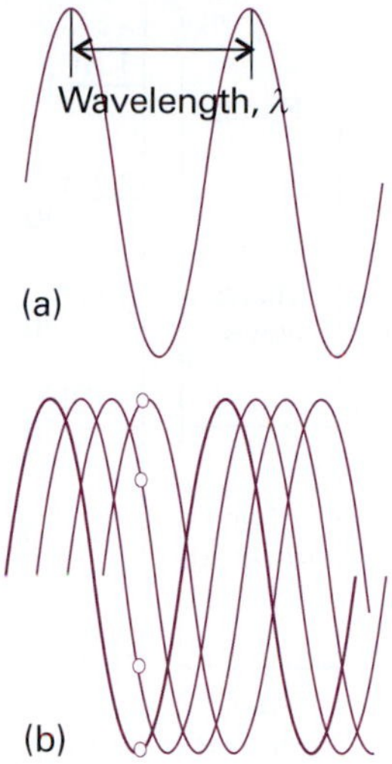

Fig. A3.5 (a) The wavelength, λ, of a wave is the peak-to-peak distance. (b) The wave is shown travelling to the right at a speed c. At a given location, the instantaneous amplitude of the wave changes through a complete cycle (the four dots show half a cycle) as it passes a given point. The frequency, ν, is the number of cycles per second that occur at a given point. Wavelength and frequency are related by $\lambda\nu = c$.

Therefore, the shorter the wavelength, the higher the frequency. The characteristics of a wave are also reported by giving the **wavenumber**, $\tilde{\nu}$ (nu tilde), of the radiation, where

$$\tilde{\nu} = \frac{\nu}{c} = \frac{1}{\lambda} \qquad \text{(A3.25)}$$

A wavenumber can be interpreted as the number of complete wavelengths in a given length. Wavenumbers are normally reported in reciprocal centimetres (cm^{-1}), so a wavenumber of 5 cm^{-1} indicates that there are 5 complete wavelengths in 1 cm. The classification of the electromagnetic field according to its frequency and wavelength is summarized in Fig. A3.6.

A3.8 Features of electromagnetic radiation

Consider an electromagnetic disturbance travelling along the x direction with wavelength λ and frequency ν. The functions that describe the oscillating electric field, $\mathcal{E}(x,t)$, and magnetic field, $\mathcal{B}(x,t)$, may be written as

$$\mathcal{E}(x,t) = \mathcal{E}_0\cos\{2\pi\nu t - (2\pi/\lambda)x + \phi\} \qquad \text{(A3.26a)}$$

$$\mathcal{B}(x,t) = \mathcal{B}_0\cos\{2\pi\nu t - (2\pi/\lambda)x + \phi\} \qquad \text{(A3.26b)}$$

where $\mathcal{E}_0$ and $\mathcal{B}_0$ are the amplitudes of the electric and magnetic fields, respectively, and the parameter ϕ is the **phase** of the wave, which varies from $-\pi$ to π and gives the relative location of the peaks of two waves. If two waves, in the same region of space, with the same wavelength are shifted by $\phi = \pi$ or $-\pi$ (so the peaks of one wave coincide with the troughs of the other), then the resultant wave will have diminished

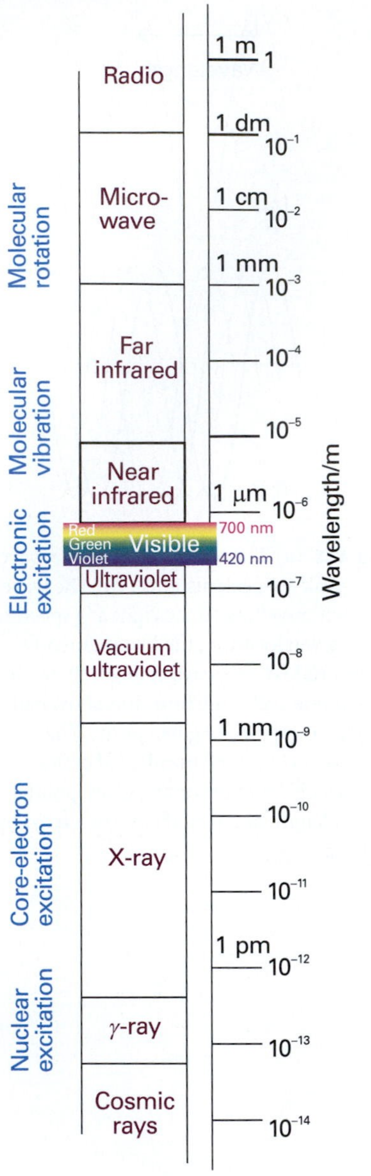

Fig. A3.6 The regions of the electromagnetic spectrum and the types of excitation that give rise to each region.

amplitudes. The waves are said to interfere destructively. A value of $\phi = 0$ (coincident peaks) corresponds to constructive interference, or the enhancement of the amplitudes.

Equations A3.26a and A3.26b represent electromagnetic radiation that is **plane-polarized**; it is so called because the electric and magnetic fields each oscillate in a single plane (in this case the *xy*-plane, Fig. A3.7). The plane of polarization may be orientated in any direction around the direction of propagation (the *x*-direction in Fig. A3.7), with the electric and magnetic fields perpendicular to that direction (and perpendicular to each other). An alternative mode of polarization is **circular polarization**, in which the electric and magnetic fields rotate around the direction of propagation in either a clockwise or a counterclockwise sense but remain perpendicular to it and each other.

It is easy to show by differentiation that eqns A3.26a and A3.26b satisfy the following equations:

$$\frac{\partial^2}{\partial x^2}\psi(x,t) = -\frac{4\pi^2}{\lambda^2}\psi(x,t) \qquad \frac{\partial^2}{\partial t^2}\psi(x,t) = -4\pi^2\nu^2\psi(x,t) \tag{A3.27}$$

where $\psi(x,t)$ is either $\mathcal{E}(x,t)$ or $\mathcal{B}(x,t)$.

According to classical electromagnetic theory, the intensity of electromagnetic radiation is proportional to the square of the amplitude of the wave. For example, the light detectors discussed in *Further information* 9.1 are based on the interaction between the electric field of the incident radiation and the detecting element, so light intensities are proportional to $\mathcal{E}_0^2$.

A3.9 Refraction

A beam of light changes direction ('bends') when it passes from one transparent medium to another. This effect, called **refraction**, depends on the **refractive index**, n_r, of the medium, the ratio of the speed of light in a vacuum, c, to its speed c' in the medium:

$$n_r = \frac{c}{c'} \qquad [A3.28]$$

It follows from the Maxwell equations (see *Further reading*), that the refractive index at a (visible or ultraviolet) specified frequency is related to the relative permittivity ε_r at that frequency by

$$n_r = \varepsilon_r^{1/2} \tag{A3.29}$$

Table A3.1 lists refractive indices of some materials.

Because the relative permittivity of a medium is related to its polarizability eqn 20.10, the refractive index is related to the polarizability. To see why this is so, we need to realize that propagation of light through a medium induces an oscillating dipole moment, which then radiates light of the same frequency. The newly generated radiation is delayed slightly by this process, so it propagates more slowly through the medium than through a vacuum. Because photons of high-frequency light carry more energy than those of low-frequency light, they can distort the electronic distributions of the molecules in their path more effectively. Therefore, after allowing for the loss of contributions from low-frequency modes of motion, we can expect the electronic polarizabilities of molecules, and hence the refractive index, to increase as the incident frequency rises towards an absorption frequency. This dependence on frequency is the origin of the dispersion of white light by a prism: the refractive index is greater for blue light than for red, and therefore the blue rays are bent more than the red. The term **dispersion** is a term carried over from this phenomenon to mean the variation of the refractive index, or of any property, with frequency.

A3.10 Optical activity

The concept of refractive index is closely related to the property of optical activity. An **optically active** substance is a substance that rotates the plane of polarization of plane-polarized light. To understand this effect, it is useful to regard the incident plane-polarized beam as a superposition of two oppositely rotating circularly polarized components. By convention, in right-handed circularly polarized light the electric vector rotates clockwise as seen by an observer facing the oncoming beam (Fig. A3.8). On entering the medium, one component propagates faster than the other if their refractive indices are different. If the sample is of length l, the difference in the times of passage is

$$\Delta t = \frac{l}{c_L} - \frac{l}{c_R}$$

where c_R and c_L are the speeds of the two components in the medium. In terms of the refractive indices, the difference is

$$\Delta t = (n_R - n_L)\frac{l}{c}$$

The phase difference between the two components when they emerge from the sample is therefore

$$\Delta\theta = 2\pi\nu\Delta t = \frac{2\pi c\Delta t}{\lambda} = (n_R - n_L)\frac{2\pi l}{\lambda}$$

where λ is the wavelength of the light. The two rotating electric vectors have a different phase when they leave the sample from the value they had initially, so their superposition gives rise to a plane-polarized beam rotated through an angle $\Delta\theta$ relative to the plane of the incoming beam. It follows that the angle of optical rotation is proportional to the difference in refractive index, $n_R - n_L$. A sample in which these two refractive indices are different is said to be **circularly birefringent**.

To explain why the refractive indices depend on the handedness of the light, we must examine why the polarizabilities depend on the handedness. One interpretation is that, if a molecule is helical (such as a polypeptide α-helix described in Section 19.7) or a crystal has molecules in a helical arrangement (as in a cholesteric liquid crystal, as described in *Impact* I6.1), its polarizability depends on whether or not the electric field of the incident radiation rotates in the same sense as the helix.

Associated with the circular birefringence of the medium is a difference in absorption intensities for right- and left-circularly polarized radiation. This difference is known as *circular dichroism*, which is explored in Chapter 7.

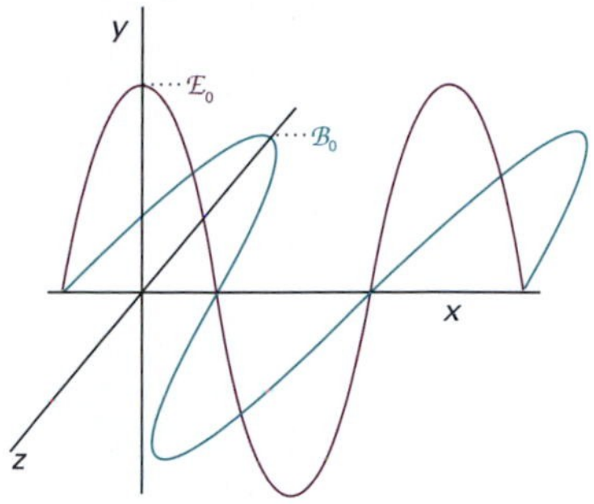

Fig. A3.7 Electromagnetic radiation consists of a wave of electric and magnetic fields perpendicular to the direction of propagation (in this case the x-direction), and mutually perpendicular to each other. This illustration shows a plane-polarized wave, with the electric and magnetic fields oscillating in the xy- and xz-planes, respectively.

Synoptic table A3.1* Refractive indices relative to air at 20°C

	434 nm	589 nm	656 nm
C_6H_6(l)	1.524	1.501	1.497
CS_2(l)	1.675	1.628	1.618
H_2O(l)	1.340	1.333	1.331
KI(s)	1.704	1.666	1.658

* More values are given in the *Data section*.

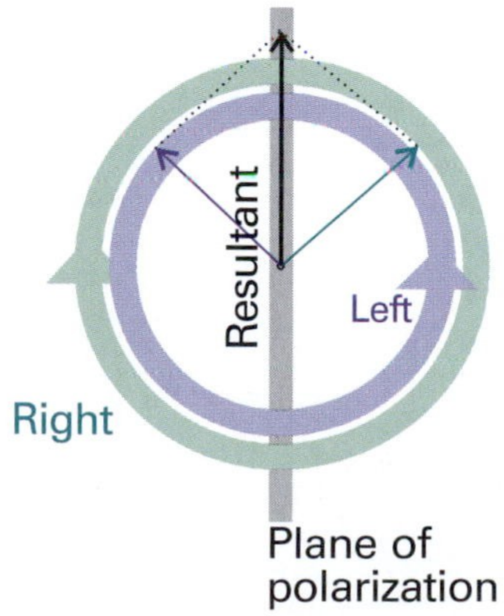

Fig. A3.8 The superposition of left and right circularly polarized light as viewed from an observer facing the oncoming beam.

Electrostatics

Electrostatics is the study of the interactions of stationary electric charges. The elementary charge, the magnitude of charge carried by a single electron or proton, is $e \approx 1.60 \times 10^{-19}$ C. The magnitude of the charge per mole is Faraday's constant: $F = N_A e = 9.65 \times 10^4$ C mol^{-1}.

A3.11 The Coulomb interaction

If a point charge q_1 is at a distance r in a vacuum from another point charge q_2, then their potential energy is

$$V = \frac{q_1 q_2}{4\pi\varepsilon_0 r} \tag{A3.30}$$

The constant ε_0 is the **vacuum permittivity**, a fundamental constant with the value $8.85 \times 10^{-12}\ \mathrm{C^2\ J^{-1}\ m^{-1}}$. This very important relation is called the **Coulomb potential energy** and the interaction it describes is called the **Coulomb interaction** of two charges. The Coulomb potential energy is equal to the work that must be done to bring up a charge q_1 from infinity to a distance r from a charge q_2.

It follows from eqns A3.5 and A3.30 that the electrical force, $\boldsymbol{F}$, exerted by a charge q_1 on a second charge q_2 has magnitude

$$F = \frac{q_1 q_2}{4\pi\varepsilon_0 r^2} \tag{A3.31}$$

The force itself is a vector directed along the line joining the two charges. With charge in coulombs and distance in metres, the force is obtained in newtons.

In a medium other than a vacuum, the potential energy of interaction between two charges is reduced, and the vacuum permittivity is replaced by the **permittivity**, ε, of the medium.

A3.12 The Coulomb potential

The potential energy of a charge q_1 in the presence of another charge q_2 can be expressed in terms of the **Coulomb potential**, ϕ:

$$V = q_1\phi \qquad \phi = \frac{q_2}{4\pi\varepsilon_0 r} \tag{A3.32}$$

The units of potential are joules per coulomb, $\mathrm{J\ C^{-1}}$, so, when φ is multiplied by a charge in coulombs, the result is in joules. The combination joules per coulomb occurs widely in electrostatics, and is called a *volt*, V:

$$1\ \mathrm{V} = 1\ \mathrm{J\ C^{-1}} \tag{A3.33}$$

If there are several charges $q_2, q_3, \ldots$ present in the system, the total potential experienced by the charge q_1 is the sum of the potential generated by each charge:

$$\phi = \phi_2 + \phi_3 + \cdots \tag{A3.34}$$

When the charge distribution is more complex than a single point–like object, the Coulomb potential is described in terms of a charge density, ρ. With charge in coulomb and length in metres, the charge density is expressed in coulombs per metre-cubed ($\mathrm{C\ m^{-3}}$). The electric potential arising from a charge distribution with density ρ is the solution to **Poisson's equation:**

$$\nabla^2\phi = -\rho/\varepsilon_0 \tag{A3.35}$$

where $\nabla^2 = (\partial^2/\partial x^2 + \partial^2/\partial y^2 + \partial^2/\partial z^2)$. If the distribution is spherically symmetrical, then so too is ϕ and eqn A3.35 reduces to the form used in *Further information* 5.1.

A3.13 The strength of the electric field

Just as the potential energy of a charge q_1 can be written $V = q_1\phi$, so the magnitude of the force on q_1 can be written $F = q_1\mathcal{E}$, where $\mathcal{E}$ is the magnitude of the electric field strength arising from q_2 or from some more general charge distribution. The electric field strength (which, like the force, is actually a vector quantity) is the negative gradient of the electric potential:

$$\mathcal{E} = -\nabla\phi \tag{A3.36}$$

A3.14 Electric current and power

The motion of charge gives rise to an electric **current**, I. Electric current is measured in *ampere*, A, where

$$1\ \mathrm{A} = 1\ \mathrm{C\ s^{-1}} \tag{A3.37}$$

If the current flows from a region of potential ϕ_i to ϕ_f, through a **potential difference** $\Delta\phi = \phi_f - \phi_i$, the rate of doing work is the current (the rate of transfer of charge) multiplied by the potential difference, $I\Delta\phi$. The rate of doing electrical work is the **electrical power**, P, so

$$P = I\Delta\phi \tag{A3.38}$$

With current in amperes and the potential difference in volts, the power works out in watts. The total energy, E, supplied in an interval Δt is the power (the rate of energy supply) multiplied by the duration of the interval:

$$E = P\Delta t = I\Delta\phi\Delta t \tag{A3.39}$$

The energy is obtained in joules with the current in amperes, the potential difference in volts, and the time in seconds.

Further reading

R.P. Feynman, R.B. Leighton, and M. Sands, *The Feynman lectures on physics*. Vols 1–3. Addison–Wesley, Reading (1966).

G.A.D. Ritchie and D.S. Sivia, *Foundations of physics for chemists*. Oxford Chemistry Primers, Oxford University Press (2000).

W.S. Warren, *The physical basis of chemistry*. Academic Press, San Diego (2000).

R. Wolfson and J.M. Pasachoff, *Physics for scientists and engineers*. Benjamin Cummings, San Francisco (1999).

Data section

Contents

The following is a directory of all tables in the text; those included in this *Data section* are marked with an asterisk. The remainder will be found on the pages indicated.

The following tables reproduce and expand the data given in the short tables in the text, and follow their numbering. Standard states refer to a pressure of $p^\ominus = 1$ bar. The general references are as follows:

AIP: D.E. Gray (ed.), *American Institute of Physics handbook*. McGraw Hill, New York (1972).

AS: M. Abramowitz and I.A. Stegun (ed.), *Handbook of mathematical functions*. Dover, New York (1963).

E: J. Emsley, *The elements*. Oxford University Press (1991).

HCP: D.R. Lide (ed.), *Handbook of chemistry and physics*. CRC Press, Boca Raton (2000).

JL: A.M. James and M.P. Lord, *Macmillan's chemical and physical data*. Macmillan, London (1992).

KL: G.W.C. Kaye and T.H. Laby (ed.), *Tables of physical and chemical constants*. Longman, London (1973).

LR: G.N. Lewis and M. Randall, resived by K.S. Pitzer and L. Brewer, *Thermodynamics*. McGraw-Hill, New York (1961).

NBS: *NBS tables of chemical thermodynamic properties*, published as *J. Phys. and Chem. Reference Data*, 11, Supplement 2 (1982).

RS: R.A. Robinson and R.H. Stokes, *Electrolyte solutions*. Butterworth, London (1959).

TDOC: J.B. Pedley, J.D. Naylor, and S.P. Kirby, *Thermochemical data of organic compounds*. Chapman & Hall, London (1986).

Physical properties of selected materials

	$\rho/(\mathrm{g\,cm^{-3}})$ at 293 K†	T_f/K	T_b/K
Elements			
Aluminium(s)	2.698	933.5	2740
Argon(g)	1.381	83.8	87.3
Boron(s)	2.340	2573	3931
Bromine(l)	3.123	265.9	331.9
Carbon(s, gr)	2.260	3700s	
Carbon(s, d)	3.513		
Chlorine(g)	1.507	172.2	239.2
Copper(s)	8.960	1357	2840
Fluorine(g)	1.108	53.5	85.0
Gold(s)	19.320	1338	3080
Helium(g)	0.125		4.22
Hydrogen(g)	0.071	14.0	20.3
Iodine(s)	4.930	386.7	457.5
Iron(s)	7.874	1808	3023
Krypton(g)	2.413	116.6	120.8
Lead(s)	11.350	600.6	2013
Lithium(s)	0.534	453.7	1620
Magnesium(s)	1.738	922.0	1363
Mercury(l)	13.546	234.3	629.7
Neon(g)	1.207	24.5	27.1
Nitrogen(g)	0.880	63.3	77.4
Oxygen(g)	1.140	54.8	90.2
Phosphorus(s, wh)	1.820	317.3	553
Potassium(s)	0.862	336.8	1047
Silver(s)	10.500	1235	2485
Sodium(s)	0.971	371.0	1156
Sulfur(s, α)	2.070	386.0	717.8
Uranium(s)	18.950	1406	4018
Xenon(g)	2.939	161.3	166.1
Zinc(s)	7.133	692.7	1180

	$\rho/(\mathrm{g\,cm^{-3}})$ at 293 K†	T_f/K	T_b/K
Inorganic compounds			
$CaCO_3$(s, calcite)	2.71	1612	1171d
$CuSO_4\cdot5H_2O$(s)	2.284	383($-H_2O$)	423($-5H_2O$)
HBr(g)	2.77	184.3	206.4
HCl(g)	1.187	159.0	191.1
HI(g)	2.85	222.4	237.8
H_2O(l)	0.997	273.2	373.2
D_2O(l)	1.104	277.0	374.6
NH_3(g)	0.817	195.4	238.8
KBr(s)	2.750	1003	1708
KCl(s)	1.984	1049	1773s
NaCl(s)	2.165	1074	1686
H_2SO_4(l)	1.841	283.5	611.2
Organic compounds			
Acetaldehyde, CH_3CHO(l, g)	0.788	152	293
Acetic acid, CH_3COOH(l)	1.049	289.8	391
Acetone, $(CH_3)_2CO$(l)	0.787	178	329
Aniline, $C_6H_5NH_2$(l)	1.026	267	457
Anthracene, $C_{14}H_{10}$(s)	1.243	490	615
Benzene, C_6H_6(l)	0.879	278.6	353.2
Carbon tetrachloride, CCl_4(l)	1.63	250	349.9
Chloroform, $CHCl_3$(l)	1.499	209.6	334
Ethanol, C_2H_5OH(l)	0.789	156	351.4
Formaldehyde, HCHO(g)		181	254.0
Glucose, $C_6H_{12}O_6$(s)	1.544	415	
Methane, CH_4(g)		90.6	111.6
Methanol, CH_3OH(l)	0.791	179.2	337.6
Naphthalene, $C_{10}H_8$(s)	1.145	353.4	491
Octane, C_8H_{18}(l)	0.703	216.4	398.8
Phenol, C_6H_5OH(s)	1.073	314.1	455.0
Sucrose, $C_{12}H_{22}O_{11}$(s)	1.588	457d	

d: decomposes; s: sublimes; Data: AIP, E, HCP, KL. † For gases, at their boiling points.

Masses and natural abundances of selected nuclides

Nuclide		m/u	Abundance/%
H	^{1}H	1.0078	99.985
	^{2}H	2.0140	0.015
He	^{3}He	3.0160	0.000 13
	^{4}He	4.0026	100
Li	^{6}Li	6.0151	7.42
	^{7}Li	7.0160	92.58
B	^{10}B	10.0129	19.78
	^{11}B	11.0093	80.22
C	^{12}C	12*	98.89
	^{13}C	13.0034	1.11
N	^{14}N	14.0031	99.63
	^{15}N	15.0001	0.37
O	^{16}O	15.9949	99.76
	^{17}O	16.9991	0.037
	^{18}O	17.9992	0.204
F	^{19}F	18.9984	100
P	^{31}P	30.9738	100
S	^{32}S	31.9721	95.0
	^{33}S	32.9715	0.76
	^{34}S	33.9679	4.22
Cl	^{35}Cl	34.9688	75.53
	^{37}Cl	36.9651	24.4
Br	^{79}Br	78.9183	50.54
	^{81}Br	80.9163	49.46
I	^{127}I	126.9045	100

* Exact value.

Table 2.2 The error function

z	erf z	z	erf z
0	0	0.45	0.475 48
0.01	0.011 28	0.50	0.520 50
0.02	0.022 56	0.55	0.563 32
0.03	0.033 84	0.60	0.603 86
0.04	0.045 11	0.65	0.642 03
0.05	0.056 37	0.70	0.677 80
0.06	0.067 62	0.75	0.711 16
0.07	0.078 86	0.80	0.742 10
0.08	0.090 08	0.85	0.770 67
0.09	0.101 28	0.90	0.796 91
0.10	0.112 46	0.95	0.820 89
0.15	0.168 00	1.00	0.842 70
0.20	0.222 70	1.20	0.910 31
0.25	0.276 32	1.40	0.952 28
0.30	0.328 63	1.60	0.976 35
0.35	0.379 38	1.80	0.989 09
0.40	0.428 39	2.00	0.995 32

Data: AS.

Table 3.2 Screening constants for atoms; values of $Z_{\text{eff}} = Z - \sigma$ for neutral ground-state atoms

	H							He
1s	1							1.6875
	Li	**Be**	**B**	**C**	**N**	**O**	**F**	**Ne**
1s	2.6906	3.6848	4.6795	5.6727	6.6651	7.6579	8.6501	9.6421
2s	1.2792	1.9120	2.5762	3.2166	3.8474	4.4916	5.1276	5.7584
2p			2.4214	3.1358	3.8340	4.4532	5.1000	5.7584
	Na	**Mg**	**Al**	**Si**	**P**	**S**	**Cl**	**Ar**
1s	10.6259	11.6089	12.5910	13.5745	14.5578	15.5409	16.5239	17.5075
2s	6.5714	7.3920	8.3736	9.0200	9.8250	10.6288	11.4304	12.2304
2p	6.8018	7.8258	8.9634	9.9450	10.9612	11.9770	12.9932	14.0082
3s	2.5074	3.3075	4.1172	4.9032	5.6418	6.3669	7.0683	7.7568
3p			4.0656	4.2852	4.8864	5.4819	6.1161	6.7641

Data: E. Clementi and D.L. Raimondi, *Atomic screening constants from SCF functions.* IBM Res. Note NJ-27 (1963). *J. chem. Phys.* **38**, 2686 (1963).

Table 3.3 Ionization energies, $I/(\text{kJ mol}^{-1})$

H							He
1312.0							2372.3
							5250.4
Li	Be	B	C	N	O	F	Ne
513.3	899.4	800.6	1086.2	1402.3	1313.9	1681	2080.6
7298.0	1757.1	2427	2352	2856.1	3388.2	3374	3952.2
Na	Mg	Al	Si	P	S	Cl	Ar
495.8	737.7	577.4	786.5	1011.7	999.6	1251.1	1520.4
4562.4	1450.7	1816.6	1577.1	1903.2	2251	2297	2665.2
		2744.6		2912			
K	Ca	Ga	Ge	As	Se	Br	Kr
418.8	589.7	578.8	762.1	947.0	940.9	1139.9	1350.7
3051.4	1145	1979	1537	1798	2044	2104	2350
		2963	2735				
Rb	Sr	In	Sn	Sb	Te	I	Xe
403.0	549.5	558.3	708.6	833.7	869.2	1008.4	1170.4
2632	1064.2	1820.6	1411.8	1794	1795	1845.9	2046
		2704	2943.0	2443			
Cs	Ba	Tl	Pb	Bi	Po	At	Rn
375.5	502.8	589.3	715.5	703.2	812	930	1037
2420	965.1	1971.0	1450.4	1610			
		2878	3081.5	2466			

Data: E.

Table 3.4 Electron affinities, $E_{ea}/(\text{kJ mol}^{-1})$

H							He
72.8							−21
Li	Be	B	C	N	O	F	Ne
59.8	≤0	23	122.5	−7	141	322	−29
					−844		
Na	Mg	Al	Si	P	S	Cl	Ar
52.9	≤0	44	133.6	71.7	200.4	348.7	−35
					−532		
K	Ca	Ga	Ge	As	Se	Br	Kr
48.3	2.37	36	116	77	195.0	324.5	−39
Rb	Sr	In	Sn	Sb	Te	I	Xe
46.9	5.03	34	121	101	190.2	295.3	−41
Cs	Ba	Tl	Pb	Bi	Po	At	Rn
45.5	13.95	30	35.2	101	186	270	−41

Data: E.

Table 4.2 Bond lengths, R_e/pm

(a) Bond lengths in specific molecules

Br_2	228.3
Cl_2	198.75
CO	112.81
F_2	141.78
H_2^+	106
H_2	74.138
HBr	141.44
HCl	127.45
HF	91.680
HI	160.92
N_2	109.76
O_2	120.75

(b) Mean bond lengths from covalent radii*

H	37						
C	77(1)	N	74(1)	O	66(1)	F	64
	67(2)		65(2)		57(2)		
	60(3)						
Si	118	P	110	S	104(1)	Cl	99
					95(2)		
Ge	122	As	121	Se	104	Br	114
		Sb	141	Te	137	I	133

* Values are for single bonds except where indicated otherwise (values in parentheses). The length of an A—B covalent bond (of given order) is the sum of the corresponding covalent radii.

Table 4.3a Bond dissociation enthalpies, $\Delta H^{\ominus}$(A—B)/(kJ mol^{-1}) at 298 K

Diatomic molecules

H—H	436	F—F	155	Cl—Cl	242	Br—Br	193	I—I	151
O=O	497	C≡O	1076	N≡N	945				
H—O	428	H—F	565	H—Cl	431	H—Br	366	H—I	299

Polyatomic molecules

H—CH_3	435	H—NH_2	460	H—OH	492	H—C_6H_5	469
H_3C—CH_3	368	H_2C=CH_2	720	HC≡CH	962		
HO—CH_3	377	Cl—CH_3	352	Br—CH_3	293	I—CH_3	237
O=CO	531	HO—OH	213	O_2N—NO_2	54		

Data: HCP, KL.

Table 4.3b Mean bond enthalpies, $\Delta H^{\ominus}(\text{A—B})/(\text{kJ mol}^{-1})$

	H	C	N	O	F	Cl	Br	I	S	P	Si
H	436										
C	412	348(i)									
		612(ii)									
		838(iii)									
		518(a)									
N	388	305(i)	163(i)								
		613(ii)	409(ii)								
		890(iii)	946(iii)								
O	463	360(i)	157	146(i)							
		743(ii)		497(ii)							
F	565	484	270	185	155						
Cl	431	338	200	203	254	242					
Br	366	276				219	193				
I	299	238				210	178	151			
S	338	259			496	250	212		264		
P	322									201	
Si	318		374	466							226

(i) Single bond, (ii) double bond, (iii) triple bond, (a) aromatic.
Data: HCP and L. Pauling, *The nature of the chemical bond*. Cornell University Press (1960).

Table 4.4 Pauling (*italics*) and Mulliken electronegativities

H							He
2.20							
3.06							
Li	Be	B	C	N	O	F	Ne
0.98	*1.57*	*2.04*	*2.55*	*3.04*	*3.44*	*3.98*	
1.28	1.99	1.83	2.67	3.08	3.22	4.43	4.60
Na	Mg	Al	Si	P	S	Cl	Ar
0.93	*1.31*	*1.61*	*1.90*	*2.19*	*2.58*	*3.16*	
1.21	1.63	1.37	2.03	2.39	2.65	3.54	3.36
K	Ca	Ga	Ge	As	Se	Br	Kr
0.82	*1.00*	*1.81*	*2.01*	*2.18*	*2.55*	*2.96*	*3.0*
1.03	1.30	1.34	1.95	2.26	2.51	3.24	2.98
Rb	Sr	In	Sn	Sb	Te	I	Xe
0.82	*0.95*	*1.78*	*1.96*	*2.05*	*2.10*	*2.66*	*2.6*
0.99	1.21	1.30	1.83	2.06	2.34	2.88	2.59
Cs	Ba	Tl	Pb	Bi			
0.79	*0.89*	*2.04*	*2.33*	*2.02*			

Data: Pauling values: A.L. Allred, *J. Inorg. Nucl. Chem.* **17**, 215 (1961); L.C. Allen and J.E. Huheey, *ibid.*, **42**, 1523 (1980). Mulliken values: L.C. Allen, *J. Am. Chem. Soc.* **111**, 9003 (1989). The Mulliken values have been scaled to the range of the Pauling values.

Table 6.2 Properties of diatomic molecules

	$\tilde{\nu}_0$/cm^{-1}	θ_V/K	B/cm^{-1}	θ_R/K	r/pm	k/(N m^{-1})	D/(kJ mol^{-1})	σ
$^1H_2^+$	2321.8	3341	29.8	42.9	106	160	255.8	2
$^1H_2$	4400.39	6332	60.864	87.6	74.138	574.9	432.1	2
$^2H_2$	3118.46	4487	30.442	43.8	74.154	577.0	439.6	2
$^1H^{19}F$	4138.32	5955	20.956	30.2	91.680	965.7	564.4	1
$^1H^{35}Cl$	2990.95	4304	10.593	15.2	127.45	516.3	427.7	1
$^1H^{81}Br$	2648.98	3812	8.465	12.2	141.44	411.5	362.7	1
$^1H^{127}I$	2308.09	3321	6.511	9.37	160.92	313.8	294.9	1
$^{14}N_2$	2358.07	3393	1.9987	2.88	109.76	2293.8	941.7	2
$^{16}O_2$	1580.36	2274	1.4457	2.08	120.75	1176.8	493.5	2
$^{19}F_2$	891.8	1283	0.8828	1.27	141.78	445.1	154.4	2
$^{35}Cl_2$	559.71	805	0.2441	0.351	198.75	322.7	239.3	2
$^{12}C^{16}O$	2170.21	3122	1.9313	2.78	112.81	1903.17	1071.8	1
$^{79}Br^{81}Br$	323.2	465	0.0809	10.116	283.3	245.9	190.2	1

Data: AIP.

Table 6.3 Typical vibrational wavenumbers, $\tilde{\nu}$/cm^{-1}

C—H stretch	2850–2960
C—H bend	1340–1465
C—C stretch, bend	700–1250
C=C stretch	1620–1680
C≡C stretch	2100–2260
O—H stretch	3590–3650
H-bonds	3200–3570
C=O stretch	1640–1780
C≡N stretch	2215–2275
N—H stretch	3200–3500
C—F stretch	1000–1400
C—Cl stretch	600–800
C—Br stretch	500–600
C—I stretch	500
CO_3^{2-}	1410–1450
NO_3^-	1350–1420
NO_2^-	1230–1250
SO_4^{2-}	1080–1130
Silicates	900–1100

Data: L.J. Bellamy, *The infrared spectra of complex molecules* and *Advances in infrared group frequencies*. Chapman and Hall.

Table 7.1 Colour, frequency, and energy of light

Colour	λ/nm	ν/(10^{14} Hz)	$\tilde{\nu}$/(10^4 cm^{-1})	E/eV	E/(kJ mol^{-1})
Infrared	>1000	<3.00	<1.00	<1.24	<120
Red	700	4.28	1.43	1.77	171
Orange	620	4.84	1.61	2.00	193
Yellow	580	5.17	1.72	2.14	206
Green	530	5.66	1.89	2.34	226
Blue	470	6.38	2.13	2.64	254
Violet	420	7.14	2.38	2.95	285
Near ultraviolet	300	10.0	3.33	4.15	400
Far ultraviolet	<200	>15.0	>5.00	>6.20	>598

Data: J.G. Calvert and J.N. Pitts, *Photochemistry*. Wiley, New York (1966).

Table 7.3 Absorption characteristics of some groups and molecules

Group	$\tilde{\nu}_{max}/(10^4\ cm^{-1})$	λ_{max}/nm	$\varepsilon_{max}/(dm^3\ mol^{-1}\ cm^{-1})$
C=C ($\pi^* \leftarrow \pi$)	6.10	163	1.5×10^4
	5.73	174	5.5×10^3
C=O ($\pi^* \leftarrow n$)	3.7–3.5	270–290	10–20
—N=N—	2.9	350	15
	>3.9	<260	Strong
$—NO_2$	3.6	280	10
	4.8	210	1.0×10^4
$C_6H_5—$	3.9	255	200
	5.0	200	6.3×10^3
	5.5	180	1.0×10^5
$[Cu(OH_2)_6]^{2+}(aq)$	1.2	810	10
$[Cu(NH_3)_4]^{2+}(aq)$	1.7	600	50
H_2O ($\pi^* \leftarrow n$)	6.0	167	7.0×10^3

Table 8.2 Nuclear spin properties

Nuclide	Natural abundance %	Spin I	Magnetic moment μ/μ_N	g-value	$\gamma/(10^7\ T^{-1}\ s^{-1})$	NMR frequency at 1 T, ν/MHz
^1n*		$\frac{1}{2}$	−1.9130	−3.8260	−18.324	29.164
^1H	99.9844	$\frac{1}{2}$	2.792 85	5.5857	26.752	42.576
^2H	0.0156	1	0.857 44	0.857 45	4.1067	6.536
^3H*		$\frac{1}{2}$	2.978 96	−4.2553	−20.380	45.414
^{10}B	19.6	3	1.8006	0.6002	2.875	4.575
^{11}B	80.4	$\frac{3}{2}$	2.6886	1.7923	8.5841	13.663
^{13}C	1.108	$\frac{1}{2}$	0.7024	1.4046	6.7272	10.708
^{14}N	99.635	1	0.403 76	0.403 56	1.9328	3.078
^{17}O	0.037	$\frac{5}{2}$	−1.893 79	−0.7572	−3.627	5.774
^{19}F	100	$\frac{1}{2}$	2.628 87	5.2567	25.177	40.077
^{31}P	100	$\frac{1}{2}$	1.1316	2.2634	10.840	17.251
^{33}S	0.74	$\frac{3}{2}$	0.6438	0.4289	2.054	3.272
^{35}Cl	75.4	$\frac{3}{2}$	0.8219	0.5479	2.624	4.176
^{37}Cl	24.6	$\frac{3}{2}$	0.6841	0.4561	2.184	3.476

* Radioactive.

μ is the magnetic moment of the spin state with the largest value of m_I: $\mu = g_I \mu_N I$ and μ_N is the nuclear magneton (see inside front cover).

Data: KL and HCP.

Table 8.3 Hyperfine coupling constants for atoms, a/mT

Nuclide	Spin	Isotropic coupling	Anisotropic coupling
^{1}H	$\frac{1}{2}$	50.8(1*s*)	
^{2}H	1	7.8(1*s*)	
^{13}C	$\frac{1}{2}$	113.0(2*s*)	6.6(2*p*)
^{14}N	1	55.2(2*s*)	4.8(2*p*)
^{19}F	$\frac{1}{2}$	1720(2*s*)	108.4(2*p*)
^{31}P	$\frac{1}{2}$	364(3*s*)	20.6(3*p*)
^{35}Cl	$\frac{3}{2}$	168(3*s*)	10.0(3*p*)
^{37}Cl	$\frac{3}{2}$	140(3*s*)	8.4(3*p*)

Data: P.W. Atkins and M.C.R. Symons, *The structure of inorganic radicals*. Elsevier, Amsterdam (1967).

Table A3.1 Refractive indices relative to air at 20°C

	434 nm	589 nm	656 nm
Benzene	1.5236	1.5012	1.4965
Carbon tetrachloride	1.4729	1.4676	1.4579
Carbon disulfide	1.6748	1.6276	1.6182
Ethanol	1.3700	1.3618	1.3605
KCl(s)	1.5050	1.4904	1.4973
KI(s)	1.7035	1.6664	1.6581
Methanol	1.3362	1.3290	1.3277
Methylbenzene	1.5170	1.4955	1.4911
Water	1.3404	1.3330	1.3312

Data: AIP.

Character tables

The groups C_1, C_s, C_i

C_1 (1)	E	$h=1$
A	1	

$C_s=C_h$ (m)	E	σ_h	$h=2$	
A′	1	1	x, y, R_z	x^2, y^2, z^2, xy
A″	1	−1	z, R_x, R_y	yz, xz

$C_i=S_2$ ($\bar{1}$)	E	i	$h=2$	
A_g	1	1	R_x, R_y, R_z	$x^2, y^2, z^2, xy, xz, yz$
A_u	1	−1	x, y, z	

The groups C_{nv}

C_{2v}, $2mm$	E	C_2	σ_v	σ'_v	$h=4$	
A_1	1	1	1	1	z, z^2, x^2, y^2	
A_2	1	1	−1	−1	xy	R_z
B_1	1	−1	1	−1	x, xz	R_y
B_2	1	−1	−1	1	y, yz	R_x

C_{3v}, $3m$	E	$2C_3$	$3\sigma_v$	$h=6$	
A_1	1	1	1	z, z^2, x^2+y^2	
A_2	1	1	−1		R_z
E	2	−1	0	$(x, y), (xy, x^2-y^2)$ (xz, yz)	(R_x, R_y)

C_{4v}, $4mm$	E	C_2	$2C_4$	$2\sigma_v$	$2\sigma_d$	$h=8$	
A_1	1	1	1	1	1	z, z^2, x^2+y^2	
A_2	1	1	1	−1	1		R_z
B_1	1	1	−1	1	−1	x^2-y^2	
B_2	1	1	−1	−1	1	xy	
E	2	−2	0	0	0	$(x, y), (xz, yz)$	(R_x, R_y)

C_{5v}	E	$2C_5$	$2C_5^2$	$5\sigma_v$	$h=10, \alpha=72°$	
A_1	1	1	1	1	z, z^2, x^2+y^2	
A_2	1	1	1	−1		R_z
E_1	2	$2\cos\alpha$	$2\cos 2\alpha$	0	$(x, y), (xz, yz)$	(R_x, R_y)
E_2	2	$2\cos 2\alpha$	$2\cos\alpha$	0	(xy, x^2-y^2)	

C_{6v}, $6mm$	E	C_2	$2C_3$	$2C_6$	$3\sigma_d$	$3\sigma_v$	$h=12$	
A_1	1	1	1	1	1	1	z, z^2, x^2+y^2	
A_2	1	1	1	1	−1	1		R_z
B_1	1	−1	1	−1	−1	1		
B_2	1	−1	1	−1	1	−1		
E_1	2	−2	−1	1	0	0	$(x, y), (xz, yz)$	(R_x, R_y)
E_2	2	2	−1	−1	0	0	(xy, x^2-y^2)	

$C_{\infty v}$	E	$2C_\phi$†	$\infty\sigma_v$	$h=\infty$	
$A_1(\Sigma^+)$	1	1	1	z, z^2, x^2+y^2	
$A_2(\Sigma^-)$	1	1	−1		R_z
$E_1(\Pi)$	2	$2\cos\phi$	0	$(x, y), (xz, yz)$	(R_x, R_y)
$E_2(\Delta)$	2	$2\cos 2\phi$	0	(xy, x^2-y^2)	

† There is only one member of this class if $\phi=\pi$.

The groups D_n

D_2, 222	E	C_2^z	C_2^y	C_2^x	$h=4$	
A_1	1	1	1	1	x^2, y^2, z^2	
B_1	1	1	−1	−1	z, xy	R_z
B_2	1	−1	1	−1	y, xz	R_y
B_3	1	−1	−1	1	x, yz	R_x

D_3, 32	E	$2C_3$	$3C_2'$	$h=6$	
A_1	1	1	1	z^2, x^2+y^2	
A_2	1	1	−1	z	R_z
E	2	−1	0	$(x, y), (xz, yz), (xy, x^2-y^2)$	(R_x, R_y)

D_4, 422	E	C_2	$2C_4$	$2C_2'$	$2C_2''$	$h=8$	
A_1	1	1	1	1	1	z^2, x^2+y^2	
A_2	1	1	1	−1	−1	z	R_z
B_1	1	1	−1	1	−1	x^2-y^2	
B_2	1	1	−1	−1	1	xy	
E	2	−2	0	0	0	$(x, y), (xz, yz)$	(R_x, R_y)

The groups D_{nh}

D_{3h}, $\bar{6}2m$	E	σ_h	$2C_3$	$2S_3$	$3C_2'$	$3\sigma_v$	$h=12$	
A_1'	1	1	1	1	1	1	z^2, x^2+y^2	
A_2'	1	1	1	1	−1	−1		R_z
A_1''	1	−1	1	−1	1	−1		
A_2''	1	−1	1	−1	−1	1	z	
E'	2	2	−1	−1	0	0	$(x, y), (xy, x^2-y^2)$	
E''	2	−2	−1	1	0	0	(xz, yz)	(R_x, R_y)

D_{4h}, 4/*mmm*	E	$2C_4$	C_2	$2C_2'$	$2C_2''$	i	$2S_4$	σ_h	$2\sigma_v$	$2\sigma_d$	$h=16$	
A_{1g}	1	1	1	1	1	1	1	1	1	1	x^2+y^2, z^2	
A_{2g}	1	1	1	−1	−1	1	1	1	−1	−1		R_z
B_{1g}	1	−1	1	1	−1	1	−1	1	1	−1	x^2-y^2	
B_{2g}	1	−1	1	−1	1	1	−1	1	−1	1	xy	
E_g	2	0	−2	0	0	2	0	−2	0	0	(xz, yz)	(R_x, R_y)
A_{1u}	1	1	1	1	1	−1	−1	−1	−1	−1		
A_{2u}	1	1	1	−1	−1	−1	−1	−1	1	1	z	
B_{1u}	1	−1	1	1	−1	−1	1	−1	−1	1		
B_{2u}	1	−1	1	−1	1	−1	1	−1	1	−1		
E_u	2	0	−2	0	0	−2	0	2	0	0	(x, y)	

D_{5h}	E	$2C_5$	$2C_5^2$	$5C_2$	σ_h	$2S_5$	$2S_5^3$	$5\sigma_v$	$h=20$	$\alpha=72°$
A_1'	1	1	1	1	1	1	1	1	x^2+y^2, z^2	
A_2'	1	1	1	−1	1	1	1	−1		R_z
E_1'	2	$2\cos\alpha$	$2\cos 2\alpha$	0	2	$2\cos\alpha$	$2\cos 2\alpha$	0	(x, y)	
E_2'	2	$2\cos 2\alpha$	$2\cos\alpha$	0	2	$2\cos 2\alpha$	$2\cos\alpha$	0	(x^2-y^2, xy)	
A_1''	1	1	1	1	−1	−1	−1	−1		
A_2''	1	1	1	−1	−1	−1	−1	1	z	
E_1''	2	$2\cos\alpha$	$2\cos 2\alpha$	0	−2	$-2\cos\alpha$	$-2\cos 2\alpha$	0	(xz, yz)	(R_x, R_y)
E_2''	2	$2\cos 2\alpha$	$2\cos\alpha$	0	−2	$-2\cos 2\alpha$	$-2\cos\alpha$	0		

$D_{\infty h}$	E	$2C_\phi$	$\infty C_2'$	i	$2iC_\infty$	iC_2'	$h=\infty$	
$A_{1g}(\Sigma_g^+)$	1	1	1	1	1	1	z^2, x^2+y^2	
$A_{1u}(\Sigma_u^+)$	1	1	1	−1	−1	−1	z	
$A_{2g}(\Sigma_g^-)$	1	1	−1	1	1	−1		R_z
$A_{2u}(\Sigma_u^-)$	1	1	−1	−1	1	1		
$E_{1g}(\Pi_g)$	2	$2\cos\phi$	0	2	$-2\cos\phi$	0	(xz, yz)	(R_x, R_y)
$E_{1u}(\Pi_u)$	2	$2\cos\phi$	0	−2	$2\cos\phi$	0	(x, y)	
$E_{2g}(\Delta_g)$	2	$2\cos 2\phi$	0	2	$2\cos 2\phi$	0	(xy, x^2-y^2)	
$E_{2u}(\Delta_u)$	2	$2\cos 2\phi$	0	−2	$-2\cos 2\phi$	0		
⋮								

The cubic groups

T_d, $\bar{4}3m$	E	$8C_3$	$3C_2$	$6\sigma_d$	$6S_4$	$h=24$	
A_1	1	1	1	1	1	$x^2+y^2+z^2$	
A_2	1	1	1	−1	−1		
E	2	−1	2	0	0	$(3z^2-r^2, x^2-y^2)$	
T_1	3	0	−1	−1	1		(R_x, R_y, R_z)
T_2	3	0	−1	1	−1	(x, y, z), (xy, xz, yz)	

O_h $(m3m)$	E	$8C_3$	$6C_2$	$6C_2$	$3C_2\,(=C_4^2)$	i	$6S_4$	$8S_6$	$3\sigma_h$	$6\sigma_d$	$h=48$	
A_{1g}	1	1	1	1	1	1	1	1	1	1		$x^2+y^2+z^2$
A_{2g}	1	1	−1	−1	1	1	−1	1	1	−1		
E_g	2	−1	0	0	2	2	0	−1	2	0		$(2z^2-x^2-y^2, x^2-y^2)$
T_{1g}	3	0	−1	1	−1	3	1	0	−1	−1	(R_x, R_y, R_z)	
T_{2g}	3	0	1	−1	−1	3	−1	0	−1	1		(xy, yz, xy)
A_{1u}	1	1	1	1	1	−1	−1	−1	−1	−1		
A_{2u}	1	1	−1	−1	1	−1	1	−1	−1	1		
E_u	2	−1	0	0	2	−2	0	1	−2	0		
T_{1u}	3	0	−1	1	−1	−3	−1	0	1	1	(x, y, z)	
T_{2u}	3	0	1	−1	−1	−3	1	0	1	−1		

The icosahedral group

I	E	$12C_5$	$12C_5^2$	$20C_3$	$15C_2$	$h=60$	
A	1	1	1	1	1		$z^2+y^2+z^2$
T_1	3	$\frac{1}{2}(1+\sqrt{5})$	$\frac{1}{2}(1-\sqrt{5})$	0	−1	(x, y, z)	
T_2	3	$\frac{1}{2}(1-\sqrt{5})$	$\frac{1}{2}(1+\sqrt{5})$	0	−1	(R_x, R_y, R_z)	
G	4	−1	−1	1	0		
G	5	0	0	−1	1		$(2z^2-x^2-y^2, x^2-y^2, xy, yz, zx)$

Further information: P.W. Atkins, M.S. Child, and C.S.G. Phillips, *Tables for group theory*. Oxford University Press (1970).

Answers to b) exercises

Chapter 1

1.1 $v = 1.3 \times 10^{-5}$ m s^{-1}.

1.2 $p = 1.89 \times 10^{-27}$ kg m s^{-1}, $v = 0.565$ m s^{-1}.

1.3 $\Delta x = 5.8 \times 10^{-6}$ m.

1.4 (a) $E = 0.93 \times 10^{-19}$ J, E (per mole) = 598 kJ mol^{-1};
(b) $E = 1.32 \times 10^{-15}$ J, E (per mole) = 7.98×10^{5} kJ mol^{-1};
(c) $E = 1.99 \times 10^{-23}$ J, E (per mole) = 0.012 kJ mol^{-1}.

1.5 (a) $v = 0.499$ m s^{-1}; (b) $v = 665$ m s^{-1}; (c) $v = 9.98 \times 10^{-6}$ m s^{-1}.

1.6 $v = 158$ m s^{-1}.

1.7 (a) 3.52×10^{17} s^{-1}; (b) 3.52×10^{18} s^{-1}.

1.8 (a) 0; (b) $E_K = 6.84 \times 10^{-19}$ J; $v = 1.23 \times 10^{6}$ m s^{-1}.

1.9 (a) $E = 2.65 \times 10^{-19}$ J, or 160 kJ mol^{-1}; (b) $E = 3.00 \times 10^{-19}$ J, or 181 kJ mol^{-1}; (c) $E = 6.62 \times 10^{-31}$ J, or 4.0×10^{-10} kJ mol^{-1}.

1.10 (a) $\lambda = 1.23 \times 10^{-10}$ m; (b) $\lambda = 3.9 \times 10^{-11}$ m; (c) $\lambda = 3.88 \times 10^{-1}$ m.

1.12 $\Delta x = 100$ pm, speed (Δv): 5.8×10^{5} m s^{-1}.

1.13 1.67×10^{-16} J.

1.14 $\hbar$.

Chapter 2

2.1 (a) 2.14×10^{-19} J, 1.34 eV, 1.08×10^{4} cm^{-1}, 129 kJ mol^{-1};
(b) 3.48×10^{-19} J, 2.17 eV, 1.75×10^{4} cm^{-1}, 210 kJ mol^{-1}.

2.2 (a) $P = 0.031$; (b) $P = 0.029$.

2.3 $p = 0, p^2 = \dfrac{h^2}{L^2}$.

2.4 $L = \left(\dfrac{3}{8}\right)^{1/2} \dfrac{h}{mc} = \left(\dfrac{3}{8}\right)^{1/2} \lambda_c$.

2.5 $x = \dfrac{L}{10}, \dfrac{3L}{10}, \dfrac{L}{2}, \dfrac{7L}{10}, \dfrac{9L}{10}$.

2.6 6.

2.7 $n = 7.26 \times 10^{10}$, $\Delta E = 1.71 \times 10^{-31}$ J, m = 27.5 pm, the particle behaves classically.

2.8 $E_0 = 3.92 \times 10^{-21}$ J.

2.9 $k = 260$ N m^{-1}.

2.10 $\lambda = 13.2$ μm.

2.11 $\lambda = 18.7$ μm.

2.12 (a) $\Delta E = 2.2 \times 10^{-29}$ J; (b) $\Delta E = 3.14 \times 10^{-20}$ J.

2.14 0, $\pm 0.96\alpha$, or $\pm 2.02\alpha$.

2.15 $E_0 = 2.3421 \times 10^{-20}$ J.

2.17 Magnitude = 2.58×10^{-34} J s; possible projections = 0, $\pm 1.0546 \times 10^{-34}$ J s and $\pm 2.1109 \times 10^{-34}$ J s.

Chapter 3

3.1 $I = 12.1$ eV.

3.2 $r = 11.5a_0/Z$, $r = 3.53a_0/Z$, $r = 0$.

3.3 $r = 0$, $r = 1.382a_0$, $r = 3.618a_0$.

3.4 $N = \dfrac{1}{4\sqrt{2\pi a_0^3}}$.

3.5 $\langle V \rangle = -\dfrac{Z^2 e^2}{16\pi\varepsilon_0 a_0}$, $\langle E_K \rangle = \dfrac{\hbar^2 Z^2}{8 m a_0^2}$.

3.6 $P_{3s} = 4\pi r^2 \left(\dfrac{1}{4\pi}\right) \times \left(\dfrac{1}{243}\right) \times \left(\dfrac{Z}{\alpha_0}\right)^3 \times (6 - 6\rho + \rho^2)^2 e^{-\rho}$,
$r = 0.74\, a_0/Z$, $4.19\, a_0/Z$ and $13.08\, a_0/Z$.

3.7 $r = 1.76\, a_0/Z$.

3.8 (a) angular momentum = $6\frac{1}{2}h = 2.45 \times 10^{-34}$ J s, angular nodes = 2, radial nodes = 1;
(b) angular momentum = $2\frac{1}{2}h = 1.49 \times 10^{-34}$ J s, angular nodes = 1, radial nodes = 0;
(c) angular momentum = $2\frac{1}{2}h = 1.49 \times 10^{-34}$ J s, angular nodes = 1, radial nodes = 1.

3.9 (a) $j = \frac{1}{2}, \frac{3}{2}$; (b) $j = \frac{9}{2}, \frac{11}{2}$.

3.10 $J = 8,7,6,5,4,3,2$.

3.11 (a) $g = 1$; (b) $g = 64$; (c) $g = 25$.

3.12 The letter F indicates that the total orbital angular momentum quantum number L is 3; the superscript 3 is the multiplicity of the term, $2S + 1$, related to the spin quantum number $S = 1$; and the subscript 4 indicates the total angular momentum quantum number J.

3.13 (a) $r = 110$ pm, $r = 20.1$ ppm; (b) $r = 86$ pm, $r = 29.4$ pm.

3.14 (a) forbidden; (b) allowed; (c) forbidden.

3.15 (a) [Ar]$3d^3$; (b) For $S = \frac{3}{2}$, $M_s = \pm\frac{1}{2}$ and $\pm\frac{3}{2}$, for $S = \frac{1}{2}$, $M_s = \pm\frac{1}{2}$.

3.16 (a) $S = 2,1,0$; multiplicities = 5,3,1, respectively. (b) $S' = \frac{5}{2}, \frac{3}{2}, \frac{1}{2}$; multiplicities = 6,4,2 respectively.

3.17 $^1F_3$; $^3F_4$; $^3F_3$; $^3F_2$; $^1D_2$; $^3D_3$; $^3D_2$; $^3D_1$; $^1P_1$; $^3P_2$; $^3P_1$; $^3P_0$, the $^3F_2$ set of terms are the lower in energy.

3.18 (a) $J = 3$, 2 and 1, with 7, 5 and 3 states respectively;
(b) $J = \frac{7}{2}, \frac{5}{2}, \frac{3}{2}, \frac{1}{2}$, with 8, 6, 4 and 2 states respectively;
(c) $J = \frac{9}{2}, \frac{7}{2}$, with 10 and 8 states respectively.

3.19 (a) $^2D_{5/2}$ and $^2D_{3/2}$ (b) $^2P_{3/2}$ and $^2P_{1/2}$.

Chapter 4

4.1 (a) $1\sigma^2\, 2\sigma^{*1}$; (b) $1\sigma^2\, 2\sigma^{*2}\, 1\pi^4\, 3\sigma^2$; (c) $1\sigma^2\, 2\sigma^{*2}\, 3\sigma^2 1\pi^4\, 2\pi^{*2}$.

4.2 (a) $1\sigma^2\, 2\sigma^{*2}\, 3\sigma^2\, 1\pi^4\, 2\pi^{*4}$; (b) $1\sigma^2\, 2\sigma^{*2}\, 1\pi^4\, 3\sigma^2$;
(c) $1\sigma^2\, 2\sigma^{*2}\, 3\sigma^2\, 1\pi^4\, 2\pi^{*3}$.

4.3 (a) C_2 and CN; (b) NO, O_2 and F_2.

4.4 BrCl is likely to have a shorter bond length than $BrCl^-$; it has a bond order of 1, while $BrCl^-$ has a bond order of 1/2.

4.5 The sequence O_2^+, O_2, O_2^-, O_2^{2-} has progressively longer bonds.

4.6 $N = \left(\dfrac{1}{1 + 2\lambda S + \lambda^2}\right)^{1/2}$.

4.7 $a = -\dfrac{0.145S + 0.844}{0.145 + 0.844S} b$, $N(0.844A - 0.145B)$.

4.8 Not appropriate.

4.9 $E_{\text{trial}} = \dfrac{-\mu e^4}{12\pi^3 \varepsilon_0^2 \hbar^2}$.

4.10 3.39×10^{-16} J.

4.12 (a) $a_{2u}^2 e_{1g}^4 e_{2u}^1$, $E = 7\alpha + 7\beta$; (b) $a_{2u}^2 e_{1g}^3$, $E = 5\alpha + 7\beta$.

4.13 (a) 19.31368β; (b) 19.44824β.

Chapter 5

5.1 CCl_4 has 4 C_3 axes (each C—Cl axis), 3 C_2 axes (bisecting Cl—C—Cl angles), 3 S_4 axes (the same as the C_2 axes), and 6 dihedral mirror planes (each Cl—C—Cl plane).

5.2 (a) CH_3Cl.

5.3 Yes, it is zero.

5.4 Forbidden.

5.6 T_d has S_4 axes and mirror planes ($= S_1$), T_h has a centre of inversion ($= S_2$).

5.8 (a) $C_{\infty v}$; (b) D_3; (c) C_{4v}, C_{2v}; (d) C_s.

5.9 (a) D_{2h}; (b) D_{2h}; (c) (i) C_{2v}; (ii) C_{2v}; (iii) D_{2h}.

5.10 (a) $C_{\infty v}$; (b) D_{5h}; (c) C_{2v}; (d) D_{3h}; (e) O_h; (f) T_d.

5.11 (a) *ortho*-dichlorobenzene, *meta*-dichlorobenzene, HF and XeO_2F_2; (b) none are chiral.

5.12 NO_3^-: p_x and p_y, SO_3: all d orbitals except d_z^2.

5.13 A_2.

5.14 (a) B_{3u}(x-polarized), B_{2u}(y-polarized), B_{1u}(z-polarized); (b) A_{1u} or E_{1u}.

5.15 Yes, it is zero.

Chapter 6

6.1 (a) 7.73×10^{-32} J m^{-3} s; (b) $\nu = 6.2 \times 10^{-28}$ J m^{-3} s.

6.2 $s = 6.36 \times 10^7$ m s^{-1}.

6.3 (a) 1.59 ns; (b) 2.48 ps.

6.4 (a) 160 MHz; (b) 16 MHz.

6.5 $\nu = 3.4754 \times 10^{11}$ s^{-1}.

6.6 (a) $I = 3.307 \times 10^{-47}$ kg m^2; (b) $R = 141$ pm.

6.7 (a) $I = 5.420 \times 10^{-46}$ kg m^2; (b) $R = 162.8$ pm.

6.8 $R = 116.21$ pm.

6.9 $R_{CO} = 116.1$ pm; $R_{CS} = 155.9$ pm.

6.10 $\tilde{\nu}_{Stokes} = 20\,603$ cm^{-1}.

6.11 $R = 141.78$ pm.

6.12 (a) H_2O, C_{2v}; (b) H_2O_2, C_2; (c) NH_3, C_{3V}; (d) N_2O, $C_{\infty V}$.

6.13 (a) CH_2Cl_2; (b) CH_3CH_3; (d) N_2O.

6.14 $k = 0.71$ N m^{-1}.

6.15 28.4 per cent.

6.16 $k = 245.9$ N m^{-1}.

6.17 (a) 0.212; (b) 0.561.

6.18 DF: $\tilde{\nu} = 3002.3$ cm^{-1}, DCl: $\tilde{\nu} = 2143.7$ cm^{-1}, DBr: $\tilde{\nu} = 1885.8$ cm^{-1}, DI: $\tilde{\nu} = 1640.1$ cm^{-1}.

6.19 $\tilde{\nu} = 2374.05$ cm^{-1}, $x_e = 6.087 \times 10^{-3}$.

6.20 $D_0 = 3.235 \times 10^4$ cm^{-1} = 4.01 eV.

6.21 $\tilde{\nu}_R = 2347.16$ cm^{-1}.

6.22 (a) CH_3CH_3; (b) CH_4; (c) CH_3Cl.

6.23 (a) 30; (b) 42; (c) 13.

6.24 (a) IR active = $A_2'' + E'$, Raman active = $A_1 + E'$; (b) IR active = $A_1 + E$, Raman active = $A_1 + E$.

6.25 (a) IR active; (b) Raman active.

6.26 $A_{1g} + A_{2g} + E_{1u}$.

Chapter 7

7.1 multiplicity = 3, parity = u.

7.2 22.2 per cent.

7.3 $\varepsilon = 7.9 \times 10^5$ cm^2 mol^{-1}.

7.4 1.33×10^{-3} mol dm^{-3}.

7.5 $A = 1.56 \times 10^8$ dm^3 mol^{-1} cm^{-2}.

7.6 Rise.

7.7 $\varepsilon = 522$ dm^3 mol^{-1} cm^{-1}.

7.8 $\varepsilon = 128$ dm^3 mol^{-1} cm^{-1}, $T = 0.13$.

7.9 (a) 0.010 cm; (b) 0.033 cm.

7.10 (a) 1.39×10^8 dm^3 mol^{-1} cm^{-2}; (b) 1.39×10^9 m mol^{-1}.

7.11 Stronger.

Chapter 8

8.1 $\nu = 649$ MHz.

8.2 $E_{m_I} = -2.35 \times 10^{-26}$ J, 0, $+2.35 \times 10^{-26}$ J.

8.3 47.3 MHz.

8.4 (a) $\Delta E = 2.88 \times 10^{-26}$ J; (b) $\Delta E = 5.77 \times 10^{-24}$ J.

8.5 3.523 T.

8.6

	$\mathcal{B}$/T	^{14}N	^{19}F	^{31}P
	g_I	0.40356	5.2567	2.2634
(a)	300 MHz	97.5	7.49	17.4
(b)	750 MHz	244	18.7	43.5

8.7 (a) 4.3×10^{-7}; (b) 2.2×10^{-6}; (c) 1.34×10^{-5}.

8.8 (a) δ is independent of both $\mathcal{B}$ and υ. (b) $\dfrac{\nu - \nu°(800\text{ MHz})}{\nu - \nu°(60\text{ MHz})} = 13$.

8.9 (a) 4.2×10^{-6} T; (b) 3.63×10^{-5} T. Spectrum appears narrower at 650 MHz.

8.11 2.9×10^3 s^{-1}.

8.14 (a) The H and F nuclei are both chemically and magnetically equivalent. (b) The P and H nuclei chemically and magnetically equivalent in both the *cis*- and *trans*-forms.

8.15 $\mathcal{B}_1 = 9.40 \times 10^{-4}$ T, 6.25 μs.

8.16 1.3 T.

8.17 $g = 2.0022$.

8.18 2.2 mT, $g = 1.992$.

8.19 Eight equal parts at $\pm 1.445 \pm 1.435 \pm 1.055$ mT from the centre, namely: 328.865, 330.975, 331.735, 331.755, 333.845, 333.865, 334.625 and 336.735 mT.

8.21 (a) 332.3 mT; (b)1209 mT.

8.22 $I = 1$.

Chapter 9

9.1 $T = 623$ K.

9.2 (a) 15.9 pm, 5.04 pm; (b) 2.47×10^{26}, 7.82×10^{27}.

9.3 $\dfrac{q_{Xe}}{q_{He}} = 187.9$.

9.4 $q = 4.006$.

9.5 $E = 7.605$ kJ mol^{-1}.

9.6 213 K.

9.7 (a) 0.997, 0.994; (b) 0.999 99, 0.999 98.

9.8 (a) (i) $\dfrac{n_2}{n_1} = 1.39 \times 10^{-11}$, $\dfrac{n_3}{n_1} = 1.93 \times 10^{-22}$;

(ii) $\dfrac{n_2}{n_1} = 0.368$, $\dfrac{n_3}{n_1} = 0.135$;

(iii) $\dfrac{n_2}{n_1} = 0.779$, $\dfrac{n_3}{n_1} = 0.607$;

(b) $q = 1.503$; (c) $U_m = 88.3$ J mol^{-1};
(d) $C_V = 3.53$ J K^{-1} mol^{-1}; (e) $S_m = 6.92$ J K^{-1} mol^{-1}.

9.9 7.26 K.

9.10 (a) 147 J K^{-1} mol^{-1}; (b) 169.6 J K^{-1} mol^{-1}.

9.11 10.7 J K^{-1} mol^{-1}.

9.12 (a)

Chapter 10

10.1 (a) $O_3 : 3R$ [experimental = $3.7R$]
(b) $C_2H_6 : 4R$ [experimental = $6.3R$]
(c) $CO_2 : \frac{5}{2}R$ [experimental = $4.5R$]

10.2 With vibrations: 115, Without vibrations: 140, Experimental: 1.29.

10.3 (a) 143; (b) 251.

10.4 (a) 2; (b) 2; (c) 6; (d) 24; (e) 4.

10.5 $q^R = 5837$, $\theta_R = 0.8479$ K, $T = 0.3335$ K.

10.6 $S_m^R = 84.57$ J K^{-1} mol^{-1}.

10.7 (a) At 298 K, $q^R = 2.50 \times 10^3$. At 500 K, $q^R = 5.43 \times 10^3$;
(b) At 298 K, $q = 2.50 \times 10^3$. At 500 K, $q = 5.43 \times 10^3$.

10.8 (a) At 25°C, $q^R = 7.97 \times 10^3$; (b) At 100°C, $q^R = 1.12 \times 10^4$.

10.9 (a) At 298 K, $S_m = 5.88$ J mol^{-1} K^{-1}.
(b) At 500 K, $S_m = 16.48$ J mol^{-1} K^{-1}.

10.10 $G_m^R - G_m^R(0) = -20.1$ kJ mol^{-1}, $G_m^V - G_m^V(0) = -0.110$ kJ mol^{-1}.

10.11 -3.65 kJ mol^{-1}.

10.12 $S_m = 14.9$ J mol^{-1} K^{-1}.

10.14 $K \approx 0.25$.

Answers to selected odd problems

Chapter 1

1.1 (a) $\Delta E = 1.6\times10^{33}$ J m^{-3}; (b) $\Delta E = 2.5\times10^{-4}$ J m^{-3}.

1.3 (a) $\theta_E = 2231$ K, $\frac{C_V}{3R} = 0.031$; (b) $\theta_E = 343$ K, $\frac{C_V}{3R} = 0.897$.

1.5 (a) 9.0×10^{-6}; (b) 1.2×10^{-6}.

1.13 (a) $N=\left(\frac{2}{L}\right)^{1/2}$; (b) $N=\frac{1}{c(2L)^{1/2}}$; (c) $N=\frac{1}{(\pi a^3)^{1/2}}$; (d) $N=\frac{1}{(32\pi a^5)^{1/2}}$.

1.15 (a) Yes, eigenvalue = ik; (b) No; (c) Yes, eigenvalue = 0; (d) No; (e) No.

1.17 (a) Yes, eigenvalue $= -k^2$;
(b) Yes, eigenvalue $= -k^2$;
(c) Yes, eigenvalue = 0;
(d) Yes, eigenvalue = 0;
(e) No.

Hence, (a,b,c,d) are eigenfunctions of $\frac{d^2}{dx^2}$; (b,d) are eigenfunctions of $\frac{d^2}{dx^2}$, but not of $\frac{d}{dx}$.

1.19 $\frac{\hbar^2k^2}{2m}$.

1.21 (a) $r=6a_0$, $r^2=42a_0^2$; (b) $r=5a_0$, $r^2=30a_0^2$.

1.27 (a) $\lambda_{\text{relativistic}} = 5.35$ pm.

1.29 (a) Methane is unstable above 825 K; (b) λ_{max} (1000 K) = 2880 nm; (c) Excitance ratio $= 7.7\times10^{-4}$; Energy density ratio $= 8.8\times10^{-3}$; (d) 2.31×10^{-7}, it hardly shines.

Chapter 2

2.1 $E_2 - E_1 = 1.24\times10^{-39}$ J, $n = 2.2\times10^{9}$ J, $E_n - E_{n-1} = 1.8\times10^{-30}$ J.

2.3 $E_1 = 1.30\times10^{-22}$ J, minimum angular momentum $= \pm\eta$.

2.5 (a) $E_1^{(1)} = \frac{\varepsilon a}{L} + \frac{\varepsilon}{\pi}\sin\left(\frac{\pi a}{L}\right)$; (b) $E_1^{(1)} = \frac{\varepsilon}{10} + \frac{\varepsilon}{\pi}\sin\left(\frac{\pi}{10}\right) = 0.1984\varepsilon$.

2.11 (a) $P = \frac{N^2}{2\kappa}$; (b) $\langle x\rangle = \frac{N^2}{4\kappa^2}$.

2.13 $\langle T\rangle = \frac{1}{2}\left(v+\frac{1}{2}\right)\hbar\omega$.

2.15 (a) $\delta x = L\left(\frac{1}{12} - \frac{1}{2\pi^2n^2}\right)^{1/2}$, $\delta p = \frac{nh}{2L}$;

(b) $\delta x = \left[\left(v+\frac{1}{2}\right)\frac{\hbar}{\omega m}\right]^{1/2}$, $\delta p = \left[\left(v+\frac{1}{2}\right)\hbar\omega m\right]^{1/2}$.

2.19 $\langle T\rangle = -\tfrac{1}{2}\langle V\rangle$.

2.23 (a) $E=0$, angular momentum = 0; (b) $E=\frac{3\hbar^2}{I}$, angular momentum $= 6^{1/2}\hbar$; (c) $E=\frac{6\hbar^2}{I}$, angular momentum $= 2\sqrt{3}\hbar$.

2.25 $\theta = \arccos\frac{m_l}{\{l(l+1)\}^{1/2}}$, 54°44′.

2.31 (a) $\Delta E = 3.3\times10^{-19}$ J; (b) $v = 4.95\times10^{-14}$ J s^{-1}; (c) lower, increases.

2.33 $\omega = 2.68\times10^{-14}$ J s^{-1}.

2.35 (a) $l_z = 5.275\times10^{-34}$ J s, $E_{\pm5} = 1.39\times10^{-24}$ J; (b) $\nu = 9.2\times10^{8}$ Hz.

2.37 $F = 5.8\times10^{-11}$ N.

Chapter 3

3.1 $n_2 \to 6$, transitions occur at 12 372 nm, 7503 nm, 5908 nm, 5129 nm, . . . , 3908 nm (at $n_2 = 15$), converging to 3282 nm as $n_2\to\infty$.

3.3 $R_{Li^{2+}} = 987\ 663$ cm^{-1}, the Balmer transitions lie at $\tilde{\nu} = 137\ 175$ cm^{-1}, 85 187 cm^{-1}, 122.5 eV.

3.5 $^2P_{1/2}$ and $^2P_{3/2}$, of which the former has the lower energy, $^2P_{3/2}$ and $^2P_{5/2}$ of which the former has the lower energy, the ground state will be $^2P_{3/2}$.

3.7 3.3429×10^{-24} kg, $\frac{I_D}{I_H} = 1.000\ 272$.

3.9 (a) $\Delta\tilde{\nu} = 0.9$ cm^{-1}, (b) Normal Zeeman splitting is small compared to the difference in energy of the states involved in the transition.

3.11 ±106 pm.

3.13 (b) For 3s, $\rho_{node} = 3+\sqrt{3}$ and $\rho_{node} = 3-\sqrt{3}$, no nodal plane; for $3p_x$, $\rho_{node} = 0$ and $\rho_{node} = 4$, yz nodal plane ($\phi = 90°$); for $3d_{xy}$, $\rho_{node} = 0$, xz nodal plane ($\phi = 0$) and yz nodal plane ($\phi = 90°$); (c) $\langle r\rangle_{3s} = \frac{27a_0}{2}$.

3.17 $\langle r^{-1}\rangle_{1s} = \frac{Z}{a_0}$; (b) $\langle r^{-1}\rangle_{2s} = \frac{Z}{4a_0}$; (c) $\langle r^{-1}\rangle_{2p} = \frac{Z}{4a_0}$.

3.25 The wavenumbers for $n=3\to n=2$: ^{4}He = 60 957.4 cm^{-1}, ^{3}He = 60 954.7 cm^{-1}. The wavenumbers for $n=2\to n=1$: ^{4}He = 329 170 cm^{-1}, ^{3}He = 329 155 cm^{-1}.

3.27 (a) receeding; $s = 3.381\times10^{5}$ ms^{-1}.

Chapter 4

4.3 $R = 2.1a_0$.

4.7 (a) $P = 8.6\times10^{-7}$ / $P = 2.0\times10^{-6}$;
(b) $P = 8.6\times10^{-7}$ / $P = 2.0\times10^{-6}$;
(c) $P = 3.7\times10^{-7}$ / $P = 0$;
(d) $P = 4.9\times10^{-7}$/ $P = 5.5\times10^{-7}$.

4.13 Delocalization energy $= 2\{E_{\text{with resonance}} - E_{\text{without resonance}}\}$ $= \{(\alpha_O-\alpha_N)^2+12\beta^2\}^{1/2} - \{(\alpha_O-\alpha_N)^2+4\beta^2\}^{1/2}$

4.15 (a) C_2H_4: −3.813, C_4H_6: −4.623, C_6H_8: −5.538, C_8H_{10}: −5.873; (b) 8.913 eV.

4.29 (a) linear relationship; (b) $E^{\ominus} = -0.122$ V; (c) $E^{\ominus} = -0.174$ V, ubiquinone a better oxidizing agent than plastiquinone.

Chapter 5

5.1 (a) D_{3d}; (b) chair: D_{3d}, boat: C_{2V}; (c) D_{2h}; (d) D_3; (d) D_{4d}; (i) Polar: Boat C_6H_{12}; (ii) Chiral: $[Co(en)_3]^{3+}$.

5.3 $C_2\sigma_h = i$.

5.7

	1	σ_x	σ_y	σ_z
1	1	σ_x	σ_y	σ_z
σ_x	σ_x	1	$i\sigma_x$	$-i\sigma_y$
σ_y	σ_y	$i\sigma_z$ $i\sigma_z$	1	$i\sigma_x$
σ_z	σ_z	$i\sigma_y$	$i\sigma_z$ $i\sigma_z$	1

The matrices do not form a group since the products $i\sigma_z$, $i\sigma_y$, $i\sigma_x$ and their negatives are not among the four given matrices.

5.9 All five *d* orbitals may contribute to bonding. (b) All except $A_2(d_{xy})$ may participate in bonding.

5.11 (a) D_{2h} ; (b) (i) Staggered: C_{2h}; (ii) Eclipsed: C_{2v}.

5.13 (a) $C_{2v}, f \rightarrow 2A_1 + A_2 + 2B_1 + 2B_2$;
(b) $C_{3v}, f \rightarrow + A_2 + 3E$;
(c) $T_d, f \rightarrow A_1 + T_1 + T_2$;
(d) $O_h, f \rightarrow A_{2u} + T_{1u} + T_{2u}$.
Lanthanide ion (a) tetrahedral complex: $f \rightarrow A_1 + T_1 + T_2$ in T_d symmetry, and there is one nondegenerate orbital and two sets of triply degenerate orbitals. (b) octahedral complex: $f \rightarrow A_{2u} + T_{1u} + T_{2u}$, and the pattern of splitting is the same.

5.15 irreducible representations: $4A_1 + 2B_1 + 3B_2 + A_2$

Chapter 6

6.1

T/K	E/J m^{-3}	E_{class}/J m^{-3}
(a) 1500	2.136×10^{-6}	2.206
(b) 2500	9.884×10^{-4}	3.676
(c) 5800	3.151×10^{-1}	8.528

6.3 $\tau = \dfrac{1}{z} = \dfrac{kT}{4\sigma p}\left(\dfrac{\pi m}{kT}\right)^{1/2}$, $\delta\nu \approx 700$ MHz, below 1 Torr.

6.5 $R_0 = 112.83$ pm, $R_1 = 123.52$ pm.

6.7 $I = 2.728 \times 10^{-47}$ kg m^2, $R = 129.5$ pm, hence we expect lines at 10.56, 21.11, 31.67, . . . cm^{-1}.

6.9 218 pm.

6.11 $B = 14.35$ m^{-1}, $J_{max} = 26$ at 298 K, $J_{max} = 15$ at 100 K.

6.13 linear.

6.15 (a) 5.15 eV; (b) 5.20 eV.

6.17 (a) $\tilde{\nu} = 152$ m^{-1}, $k = 2.72 \times 10^{-4}$ kg s^{-2}, $I = 2.93 \times 10^{-46}$ kg m^2, $B = 95.5$ m^{-1}.
(b) $x_e = 0.96$.

6.19 (a) C_{3v}; (b) 9; (c) $2A_1 + A_2 + 3E$. (d) All but the A_2 mode are infrared active. (e) All but the A_2 mode may be Raman active.

6.23 (a) spherical rotor; (b) symmetric rotor; (c) linear rotor; (d) asymmetric rotor; (e) symmetric rotor; (f) asymmetric rotor.

6.25 $HgCl_2$: 230, $HgBr_2$: 240, HgI_2: 250 pm.

6.27 (a) infrared active; (b) 796 cm^{-1}; (c) O_2: 2, O_2^-: 1.5, O_2^{2-}: 1; (d) $Fe_2^{3+}O_2^{2-}$; (e) Structures 6 and 7 are consistent with this observation, but structures 4 and 5 are not.

6.29 $s = 0.0768$ c, $T = 8.34 \times 10^5$ K.

6.31 $B = 2.031$ cm^{-1}; $T = 2.35$ K.

Chapter 7

7.1 49 364 cm^{-1}.

7.3 14 874 cm^{-1}.

7.5 $\mathcal{A} = 1.1 \times 10^6$ dm^3 mol^{-1} cm^{-2}, Excitations from A_1 to A_1, B_1, and B_2 terms are allowed.

7.7 5.06 eV.

7.9

Hydrocarbon	E_{HOMO}/eV*
Benzene	−9.7506
Biphenyl	−8.9169
Naphthalene	−8.8352
Phenanthrene	−8.7397
Pyrene	−8.2489
Anthracene	−8.2477

7.11 (a) $\dfrac{n}{V} = 1.7 \times 10^{-9}$ mol dm^{-3}, (b) $N = 6.0 \times 10^2$.

7.15 The transition moves toward the red as the chain lengthens and the apparent color of the dye shifts towards blue.

7.21 (a) 3 + 1, 3 + 3; (b) 4 + 4, 2 + 2.

7.23 4.4×10^3.

7.25 $\mathcal{A} = 1.24 \times 10^5$ dm^3 mol^{-1} dm^{-2}.

7.27 (a) $\mathcal{A} = 2.24 \times 10^5$ dm^3 mol^{-1} cm^{-2}; (b) $A = 0.185$;
(c) $\varepsilon = 135$ dm^3 mol^{-1} cm^{-1}.

7.29 $V_1 - V_0 = 3.1938$ eV, $\tilde{\nu}_1 - \tilde{\nu}_0 = 79.538$ cm^{-1}, $\tilde{\nu}_0 = 2034.3$ cm^{-1}, $T_{eff} = 1321$ K.

Chapter 8

8.1 $B_0 = 10.3$ T, $\dfrac{\delta N}{N} \approx 2.42 \times 10^{-5}$, β state lies lower.

8.3 300×10^{-6} Hz ± 10 Hz, 0.29 s.

8.5 Both fit the data equally well.

8.9 Width of the CH_3 spectrum is $3a_H = 6.9$ mT. The width of the CD_3 spectrum is $6a_D$. The overall width is $6a_D = 2.1$ mT.

8.11 P(N2s) = 0.10, P(N2p_z) = 0.38, total probability: P(N) = 0.48, P(O) = 0.52, hybridization ratio = 3.8, $\Phi = 131°$.

8.13 $\sigma_d = \dfrac{e^2\mu_0 Z}{12\pi m_e a_0} = 1.78 \times 10^{-5}\, Z$.

8.15 $R = 158$ pm.

8.17 $I(\omega) \approx \dfrac{1}{2}\dfrac{A\tau}{1 + (\omega_0 - \omega)^2\tau^2}$.

8.21 $\dfrac{-g_I\mu_N\mu_0 m_I}{4\pi R^3}(\cos^2\theta_{max} + \cos\theta_{max})$, $\langle\mathcal{B}_{nucl}\rangle = 0.58$ mT.

Chapter 9

9.1 $W = 2 \times 10^{40}$, $S = 1.282 \times 10^{-21}$ J K^{-1}, $S_1 = 0.637 \times 10^{-21}$ J K^{-1}, $S_2 = 0.645 \times 10^{-21}$ J K^{-1}.

9.3 $\dfrac{\Delta W}{W} \approx 2.4 \times 10^{25}$.

9.5 $T = 3.5 \times 10^{-15}$ K, $q = 7.41$.

9.7 (a) (i) $q = 5.00$; (ii) $q = 6.26$;
(b) $p_0 = 1.00$ at 298 K, $p_0 = 0.80$ at 5000 K; $p_2 = 6.5 \times 10^{-11}$ at 298 K, $p_2 = 0.12$ at 5000 K.
(c) (i) $S_m = 13.38\ \mathrm{J\ K^{-1}\ mol^{-1}}$, (ii) $S_m = 18.07\ \mathrm{J\ K^{-1}\ mol^{-1}}$.

9.9 (a) $p_0 = 0.64$, $p_1 = 0.36$; (b) 0.52 kJ mol^{-1}. At 300 K, $S_m = 11.2\ \mathrm{J\ K^{-1}\ mol^{-1}}$, At 500 K, $S_m = 11.4\ \mathrm{J\ K^{-1}\ mol^{-1}}$.

9.11 (a) At 100 K: $q = 1.049$, $p_0 = 0.953$, $p_1 = 0.044$, $p_2 = 0.002$; $U_m - U_m(0) = 123\ \mathrm{J\ mol^{-1}}$, $S_m = 1.63\ \mathrm{J\ K^{-1}\ mol^{-1}}$.
(b) At 298 K: $q = 1.55$, $p_0 = 0.645$, $p_1 = 0.230$, $p_2 = 0.083$, $U_m - U_m(0) = 1348\ \mathrm{J\ mol^{-1}}$, $S_m = 8.17\ \mathrm{J\ K^{-1}\ mol^{-1}}$.

9.13 Most probable configurations are {2, 2, 0, 1, 0, 0} and {2, 1, 2, 0, 0, 0} jointly.

9.15 (a) $T = 160$ K.

Chapter 10

10.1 (a) 0.351; (b) 0.079; (c) 0.029.

10.3 $C_{V,m} = 4.2\ \mathrm{J\ K^{-1}\ mol^{-1}}$, $S_m = 15\ \mathrm{J\ K^{-1}\ mol^{-1}}$.

10.5 $q = 19.90$.

10.7 $S_m^{\ominus} = 199.4\ \mathrm{J\ mol^{-1}\ K^{-1}}$.

10.11 At 298 K: $K = 3.89$. At 800 K: $K = 2.41$.

10.13 (a) $\theta_R = 87.55$ K, $\theta_V = 6330$ K.

10.16 (b) $J_{max} = \left(\dfrac{kT}{2hcB}\right)^{1/2} - \dfrac{1}{2}$; (c) $T \approx 374$ K.

10.17 (a) $q^R = 660.6$; (b) $q^R = 4.26 \times 10^4$.

10.23 $S = 9.57 \times 10^{-15}\ \mathrm{J\ K^{-1}}$.

10.25 $G_m^{\ominus} - G_m^{\ominus}(0) = 513.5\ \mathrm{kJ\ mol^{-1}}$.

10.27 At 10 K, $G_m^{\ominus} - G_m^{\ominus}(0) = 660.8\ \mathrm{J\ mol^{-1}}$.
At 1000 K, $G_m^{\ominus} - G_m^{\ominus}(0) = 241.5\ \mathrm{kJ\ mol^{-1}}$.

Index

(T) denotes a table in the Data Section.

Q

R

S

T

U

V

W

X

Z